Approximate Computing

Weiqiang Liu • Fabrizio Lombardi
Editors

Approximate Computing

Editors
Weiqiang Liu
Nanjing University of Aeronautics and Astronautics
Nanjing, China

Fabrizio Lombardi
Northeastern University
Boston, MA, USA

ISBN 978-3-030-98349-9 ISBN 978-3-030-98347-5 (eBook)
https://doi.org/10.1007/978-3-030-98347-5

This Springer imprint is published by the registered company Springer Nature Switzerland AG
The registered company address is: Gewerbestrasse 11, 6330 Cham, Switzerland

Preface

Computing systems at all scales (from mobile handheld devices to supercomputers, servers, and large cloud-based data centers) have seen significant performance gains, mostly through the continuous shrinking of the complementary metal-oxide semiconductor (CMOS) feature size that has doubled the number of transistors on a chip with every technology generation. However, power dissipation has become the fundamental barrier to scale computing performance across all platforms. As the classical Dennard scaling is coming to an end, reduction in on-chip power consumption as well as a throughput increase (as per Moore's Law) have become serious challenges. Computation at the nanoscales necessitates fundamentally different approaches. These approaches rely on different computational paradigms that exploit features in the targeted set of applications as well as exploiting unique interactions between hardware, software, and the processing algorithms of a computing system.

Approximate computing has been proposed as a novel paradigm for efficient and low-power design at nanoscales [1]. Efficiency is related to the computation of approximate results with at least comparable performance and lower power consumption compared to the fully accurate counterpart. Therefore, approximate computing generates results that are good enough rather than always fully accurate. Although computational errors generally are not desirable, applications such as multimedia, signal processing, machine learning, pattern recognition, and data mining are tolerant to the occurrence of some errors; this application-driven microcosm is shown in Fig. 1. Therefore, approximate computing is mostly applicable to computing systems that are related to human perception/cognition and have inherent error resilience. Many of these applications are based on statistical or probabilistic computation, where different approximations can be made to better suit the desired objectives. Therefore, it is possible to achieve not only energy efficiency but also a simpler design and lower latency, while relaxing the accuracy requirement for these applications. The basic principles of approximate computing are found across the entire technology design stack: devices exhibit approximate behavior at reduced dimensions, circuits can be redesigned to better fit specific operational features, and

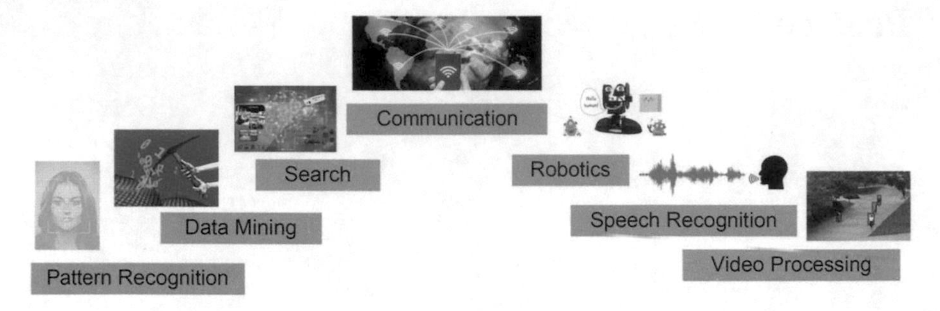

Fig. 1 Error-tolerant applications amenable to approximate computing

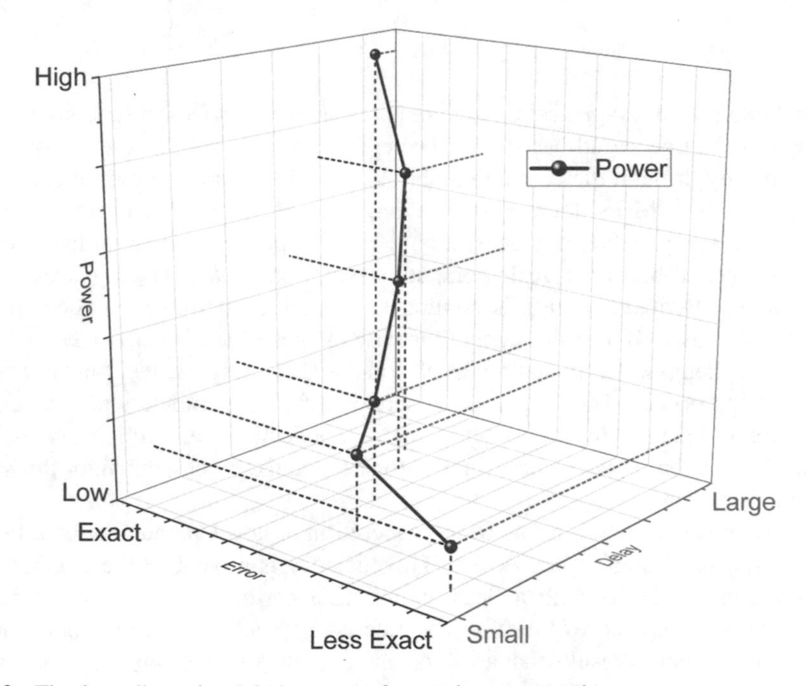

Fig. 2 The three-dimensional design space of approximate computing

modules and systems can be operated at a level to guarantee an acceptable accuracy (Fig. 2).

Approximate techniques have been studied at several levels including nanoscale hardware (devices, circuits, and architectures), software/algorithms, programming languages, logic synthesis, and automated design processes. Various computing and memory architectures have been proposed for several applications. At hardware level, the design of approximate arithmetic units has received significant research interest. Design metrics and analytical approaches have been proposed for the evaluation of approximate circuits such as adders and multipliers [2, 3]. Approximate algorithms and systems have been studied for emerging computing applications, such as deep neural networks [4]. Design automation, test, and security issues of approximate computing systems have also been investigated [5–7]. It is generally accepted that the full potential of approximate computing cannot be fully exploited without considering the interactions between these different levels under often conflicting constraints, such as error analysis/control. Approximate computing accomplishes a reduction in power consumption and higher performance by carefully controlling the introduced errors based on an error analysis to ensure that the measurable quality features of an error-tolerant application are still fully met. Generally, such error analysis involves the least significant parts of the computational media, either hardware and/or software; then, errors are selectively introduced in the design at the appropriate level(s) to achieve the desired benefits and meet the trade-offs between figures of merits such as power, accuracy, and delay.

Approximate computing has received significant attention from both research and industrial communities in the past few years due to the challenge of designing power-efficient systems for emerging applications. The research community from electronic engineering, computer science and computer engineering, and many other related areas has widely expanded on a worldwide basis to study approximate techniques at different levels.

It should also be noted that leading companies (such as IBM, Google, and Samsung) are involved in experimental research and implementing commercial products and services based on approximate computing. For example, Google's Tensor Processing Units (TPUs) use approximate computing to reduce their power consumption [8]. IBM has fabricated an AI accelerator chip to achieve multi-TOPs performance by applying multiple approximate techniques (including hardware, architecture, algorithms, and programs) [9]. Samsung has adopted approximate multipliers in the DSP block of a high-speed transceiver [10]. An increased number of funding agencies (such as NSF, DARPA, NSFC, and the EU Framework Horizon 2020) are sponsoring research in this field.

Although approximate computing has gained significant attention from both academic and industrial communities in the past decade, it still requires significant efforts for implementation as mainstream computing paradigm for energy-efficient and high-performance systems This book focuses initially on approximate arithmetic and then on new applications of approximate computing such as machine learning, security, and communication. This book extends the coverage of existing books and articles found in the technical literature such as the special issue published by *Proceedings of the IEEE* and other leading journals. Therefore, this book explores the technological contributions and developments at various levels of this new emerging paradigm. With this book, it is expected to collect state-of-the-art progress in this area. This has been achieved by chapters covering the entire spectrum of research activities in approximate computing, so bridging device, circuit, architecture, and system levels.

The chapters in this book are organized into five parts. The first part presents approximate arithmetic circuits that are the fundamental components for approximate computing systems, and approximate arithmetic circuits based on both conventional and emerging devices are included. The second part introduces the design automation and testing techniques to facilitate approximate design. The next three parts cover a wide range of applications that utilize approximate computing, including security, neural networks, machine learning, digital signal processing, and communication. The security and testing issues of approximate computing itself are also studied in Part II and Part III.

Although approximate computing is not yet at a fully mature stage, it is very promising for future highly energy-efficient and high-performance error-tolerant applications; current approximate computing techniques have been used in a number of AI-based applications and services by leading companies. There is also an increasing trend in EDA and the software engineering community to support approximate computing designs.

In the coming decade, it is anticipated that approximate computing techniques will likely be employed in energy-efficient systems for applications such as AI and signal processing [8–10]. Furthermore, new approximate computing applications will be exploited; for instance, approximate computing can also provide new approaches for hardware security and post-quantum cryptography based on lattice [7]. It is believed more applications will utilize approximate computing due to power limitations. Therefore, it is likely that there will be a growing demand for approximate computing with diverse contributions, hardware designer, system developer, test engineer, and research and teaching community to enable such computing as a mainstream paradigm.Bibliography

References

1. Liu W, Lombardi F, Schulte M. A retrospective and prospective view of approximate computing. Proc IEEE. 2020;108(3):394–9.
2. Jiang H, Santiago FJH, Mo H, Liu L, Han J. Approximate arithmetic circuits: a survey, characterization and recent applications. Proc IEEE. 2020;108(12):2108–35.
3. Liu W, Qian L, Wang C, Jiang H, Han J, Lombardi F. Design of approximate radix-4 booth multipliers for error-tolerant computing. IEEE Trans Comput. 2017;66(8):1435–41.
4. Venkataramani S, Xiao S, Wang N, Chen C-Y, Choi J, Kang M, Agarwal A, et al. Efficient AI system design with cross-layer approximate computing. Proc IEEE. 2020;108(12):2232–50.
5. Scarabottolo I, Ansaloni G, Constantinides GA, Pozzi L, Reda S. Approximate logic synthesis: a survey. Proc IEEE. 2020;108(12):2195–213.
6. Traiola M, Virazel A, Girard P, Barbareschi M, Bosio A. A survey of testing techniques for approximate integrated circuits. Proc IEEE. 2020;108(12):2178–94.
7. Liu W, Gu C, Qu G, O'Neill M, Montuschi P, Lombardi F. Security in approximate computing and approximate computing for security: challenges and opportunities. Proc IEEE. 2020;108(12):2214–31.
8. Jouppi NP, Young C, Patil N, Patterson D, Agrawal G, Bajwa R, Bates S, et al. In-datacenter performance analysis of a tensor processing unit. In: Proceedings of the 44th Annual International Symposium on Computer Architecture (ISCA). ACM; 2017. p. 1–12.
9. Fleischer B, Shukla S, Ziegler M, Silberman J, Oh J, Srinivasan V, Choi J, et al. A scalable multi-TeraOPS deep learning processor core for AI trainina and inference. In: Proceedings of the IEEE Symposium on VLSI Circuits. IEEE; 2018. p. 35–6.
10. Yoo B-J, Lim D-H, Pang H, Lee J-H, Baek S-Y, Kim N, Choi D-H, et al. 6.4 A 56Gb/s 7.7 mW/Gb/s PAM-4 wireline transceiver in 10nm FinFET using MM-CDR-based ADC timing skew control and low-power DSP with approximate multiplier. In: Proceedings of the IEEE International Solid-State Circuits Conference (ISSCC). IEEE; 2020. p. 122–4.

Acknowledgment

This book is dedicated to our families whose supports have made it possible. The research of Weiqiang Liu is supported by NSFC (62022041 and 61871216), while the research of Fabrizio Lombardi is supported by NSF under CCF-1953961 and 1812467 grants.

Contents

About the Editors

Fabrizio Lombardi graduated in 1977 from the University of Essex (UK) with a BSc (Hons) in electronic engineering. In 1977, he joined the Microwave Research Unit at University College London, where he received his master's in microwaves and modern optics (1978), diploma in microwave engineering (1978), and PhD from the University of London (1982). He is currently holder of the International Test Conference (ITC) Endowed Chair at Northeastern University, Boston. Currently, Dr. Lombardi is the president of the IEEE Nanotechnology Council (NTC); in 2021, he was second vice president of the IEEE Computer Society (CS). He is also a member of the IEEE Publication Services and Products Board (PSPB) (2019–2023). Dr. Lombardi was the vice president of publications of the IEEE CS (2019–2020) and the NTC (2020). He has been appointed on executive boards of many non-profit organizations (such as Computing-in-the-Core, now code.org, and the non-partisan advocacy coalition for K-12 Computer Science education) as well as the Computer Society (as an elected two-term member of its board of governors (2012–2017)) and the IEEE (as an appointed member of the Future Directions Committee (2014–2017)).

In the past, Dr. Lombardi has been a two-term editor-in-chief (2007–2010), associate editor-in-chief (2000–2006), and associate editor (1996–2000) of *IEEE Transactions on Computers*, the inaugural two-term editor-in-chief of *IEEE Transactions on Emerging Topics in Computing* (2013–2017), and editor-in-chief of

IEEE Transactions on Nanotechnology (2014–2019) as well as member of the editorial boards of the *ACM Journal of Emerging Technologies in Computing Systems*, *IEEE Design & Test*, and *IEEE Transactions on CAD of ICAS*. Dr. Lombardi has twice been a distinguished visitor of the IEEE Computer Society (1990–1993 and 2001–2004).

Dr. Lombardi is a Fellow of the IEEE for "*contributions to testing and fault tolerance of digital systems*." He was the recipient of the 2011 Meritorious Service Award and elevated to Golden Core membership in the same year by the IEEE CS; he was the chair of the 2016 and 2017 IEEE CS Fellow Evaluation Committee. He has been awarded the 2019 NTC Distinguished Service Award, the 2019 "Spirit of the CS" Award, and the 2021 T Michael Elliott Distinguished Service Certificate from the CS. Dr. Lombardi has received many professional awards: the Visiting Fellowship at the British Columbia Advanced System Institute, University of Victoria, Canada (1988); the Texas Experimental Engineering Station Research Fellowship (1991–1992, 1997–1998); the Halliburton Professorship (1995); the Outstanding Engineering Research Award at Northeastern University (2004); and an International Research Award from the Ministry of Science and Education of Japan (1993–1999). Dr. Lombardi was the recipient of the 1985/86 Research Initiation Award from the IEEE/Engineering Foundation and a Silver Quill Award from Motorola-Austin (1996). Together with his students, his manuscripts have been selected for best paper awards at technical events/meeting such as IEEE DFT and IEEE/ACM Nanoarch.

Dr. Lombardi has been involved in organizing many international symposia, conferences, and workshops sponsored by professional organizations and has been guest editor of special issues in archival journals and magazines. His research interests are emerging computing paradigms and technologies (mostly nanoscale circuits and magnetic devices), memory systems, VLSI design, and fault/defect tolerance of digital systems. He has extensively published in these areas and coauthored/edited ten books.

Weiqiang Liu is currently a professor and the vice dean of the College of Electronic and Information Engineering and the College of Integrated Circuits at Nanjing University of Aeronautics and Astronautics (NUAA), Nanjing, China. He received his BSc degree in information engineering from NUAA and PhD degree in electronic engineering from Queen's University Belfast (QUB), Belfast, the UK, in 2006 and 2012, respectively. In December 2013, he joined the College of Electronic and Information Engineering, NUAA, where he is now a full professor.

His research interest is energy-efficient and secure computing integrated circuits and systems that include approximate computing, computer arithmetic, hardware security, VLSI design for DSP and cryptography, and mixed-signal integrated circuits. His research has been funded by Natural Science Foundation China (NSFC), State Grid Corporation of China, and Natural Science Foundation of Jiangsu Province, among others. He has published one research book by Artech House and over 190 leading journal and conference papers (over 70 IEEE and ACM journals including 8 invited papers). His papers were selected as the Highlight Paper of IEEE Transactions on Circuits and Systems I: Regular Papers in the 2021 January Issue, the two Feature Papers of IEEE Circuits and Systems Magazine in the 2021 4th issue and IEEE Transactions on Computers in the 2017 December issue, and IET Computer & Design Techniques Editor's Choice Award in 2021, the Best Paper Candidates of ISCAS 2011, GLSVLSI 2015 and GLSVLSI 2022. He received the prestigious Excellent Young Scientists Fund from NSFC in 2020 and was listed in Stanford University's 2020 list of the top 2% scientists in the world (computer hardware and architecture).

Dr. Liu is currently the Vice President-Elect for Technical Activities (2022–2023) of the IEEE Nanotechnology Council (NTC). He has served as a steering committee member of *IEEE Transactions on VLSI Systems* and *IEEE Transactions on Multi-Scale Computing Systems*; associate editor of *IEEE Transactions on Circuits and Systems I: Regular Paper*, *IEEE Transactions on Emerging Topic in Computing and Computers* (TETC), *IEEE Transactions on Computers*, *IEEE Open*

Journal of Computer Society, and *IET Computers & Digital Techniques*; guest editor of *Proceedings of the IEEE and IEEE Nanotechnology*; and member of 2020 TETC and AsianHOST 2021 Best Paper Award Committee. He is the program co-chair of IEEE Symposium on Computer Arithmetic (ARITH 2020) and ACM/IEEE Symposium on Nano Architecture (NANOARCH 2022) program member for a number of international conferences including DAC, ARITH, DATE, ASP-DAC, ISCAS, ASAP, ISVLSI, ICCD, GLSVLSI, AsianHOST, NANORACH, AICAS, SiPS, NMDC, and ICONIP. He was a orgnizer and speaker of DAC 2022, DATE 2022, IEEE ISCAS 2021, and COINS Tutorial. Dr. Liu is a member of both Circuits & Systems for Communications (CASCOM) Technical Committee and VLSI Systems and Applications (VSA) Technical Committee, IEEE Circuits and Systems Society. He is a senior member of the IEEE, CIE, and CCF.

Part I
Introduction: Approximate Arithmetic Circuits

Weiqiang Liu and Fabrizio Lombardi

Part I deals with approximate circuits when computing many arithmetic functions. The first chapter, titled "Approximate Arithmetic Circuits: Design and Applications" (authored by Ke Chen, Weiqiang Liu, and Fabrizio Lombardi), introduces and evaluates various basic arithmetic units (such as adder, multiplier, and divider). Also, error compensation (and related hardware schemes) is discussed for improving performance of an approximate design with respect to error management. Furthermore, widely used applications utilizing approximate arithmetic circuits (such as neural networks, digital signal processing, digital image processing, and N modular redundancy) are presented to show that substantial improvements related to different metrics (such as power dissipation, accuracy, and delay) can be achieved using approximate arithmetic circuits.

These topics are further expanded in the next chapter, "An Automated Logic Level Framework for Approximate Modular Arithmetic Circuits," by Haroon Waris, Chenghua Wang, and Weiqiang Liu. In this chapter, the authors present an automated framework (denoted as approximate solution finder, ASF) that exploits logic-level approximations under a given error constraint to design approximate arithmetic circuits. This methodology improves on current manual design techniques and is evaluated in error-tolerant case studies, such as image sharpening and canny edge detection.

A different aspect of approximate circuit design is addressed in the third chapter, "Approximate Multiplier Design for Energy Efficiency: From Circuit to Algorithm" (authored by Ying Wu, Chuangtao Chen, Chenyi Wen, Weikang Qian, Xunzhao Yin, and Cheng Zhuo). It deals with a specific arithmetic circuit (namely a multiplier) and establishes that its energy efficiency largely depends on how and where the inaccuracy is introduced. This is roughly categorized at different design levels; such large design space inevitably incurs into complexity issues and challenges when selecting the appropriate multiplier for a particular application. This chapter therefore provides a comprehensive review of the state-of-the-art designs of approximate multipliers for future investigation.

The topic of number formats for arithmetic circuits is reviewed in "Low-Precision Floating-Point Formats: From General-Purpose to Application-Specific" by Amir Sabbagh Molahosseini, Leonel Sousa, Azadeh Alsadat Emrani Zarandi, and Hans Vandierendonck. Low-precision floating-point (FP) formats are used to enhance performance by providing the least precision to meet the requirements of an application. For small bit widths, a careful adjustment of precision and a dynamic range are required, hence such applications require a bespoke number format. This chapter provides a comprehensive review of low-precision FP formats identifying the numerical features (e.g., dynamic range and precision) of each format as well as their usage in the target applications, while assessing accuracy and performance. Finally, guidelines are provided to design high-performance and efficient applications via customized FP formats.

The next two chapters deal with the utilization of emerging technologies for approximate arithmetic circuit design. The first of these chapters is "Spintronic Solutions for Approximate Computing," authored by You Wang, Hao Cai, Kaili Zhang, Bo Wu, Bo Liu, Deming Zhang, and Weisheng Zhao. With features such as non-volatility, fast access speed, high scalability, and current-induced thresholding operation, spintronic devices are promising candidates for approximate computing. This chapter exploits the application of approximate techniques in spintronic device–based circuit designs for energy-efficient in-memory processing. Approximate techniques based on spintronic devices are explored for both traditional full adders and write-only bitwise full adders at a substantially reduced complexity. Simulation results are provided to show that energy can be significantly reduced with a negligible loss in output quality.

The last chapter is "Majority Logic Based Approximate Multipliers for Error-Tolerant Applications," authored by Tingting Zhang, Honglan Jiang, Weiqiang Liu, Fabrizio Lombardi, Leibo Liu, Seok-Bum Ko, and Jie Han. As many emerging nanotechnologies extensively assemble circuits based on voter-based majority logic (ML), the authors aim to provide a comprehensive investigation of approximate unsigned and signed multiplier designs based on ML. Approximate partial product generation, reduction, and compression are discussed, by specifically employing complementary strategies guided by an analysis of error effects for compensating the resulting accuracy loss. Approximate multiplier designs are then comparatively evaluated based on error and circuit characteristics. Image processing and neural networks (as case studies of error-tolerant applications) are presented to show the validity and advantages of the proposed designs.

Chapter 1
Approximate Arithmetic Circuits: Design and Applications

Ke Chen, Weiqiang Liu, and Fabrizio Lombardi

1 Introduction

Despite the advances in semiconductor technologies and the development of energy-efficient design techniques, the overall energy consumption of computer systems is still rapidly growing at an alarming rate to process an ever-increasing amount of information. In particular, as computer systems become pervasive, they are increasingly used to interacting with the physical world and processing large amounts of data from various sources. Energy efficiency has become the paramount concern in the design of computing systems. At the same time, as the computing systems become increasingly embedded and mobile, computational tasks include a growing set of applications that involve media processing (audio, video, graphics, and image), recognition, and data mining. A common characteristic of the above class of applications is that a perfect result is often unnecessary, and an approximate or less-than-optimal result is sufficient. As one of the most promising energy-efficient computing paradigms, approximate computing has gained much research attention in the past few years.

Approximate computation can be used in many fault-tolerant computing systems where the computation does not require a completely accurate result, and an approximation result with errors is still acceptable. Approximate computing techniques can be found in digital signal processing [1], memory [2], logic synthesis [3], in-memory computing [4], security [5], and other fields. However, as the fundamental level, circuit-level approximate arithmetic circuit is the essential building blocks for the

K. Chen (✉) · W. Liu
Nanjing University of Aeronautics and Astronautics, Nanjing, China
e-mail: chen.ke@nuaa.edu.cn

F. Lombardi
Northeastern University, Boston, MA, USA

W. Liu, F. Lombardi (eds.), *Approximate Computing*,
https://doi.org/10.1007/978-3-030-98347-5_1

high-level system and applications. Thus, this chapter will introduce approximation techniques for the basic arithmetic unit like adder, multiplier, and divider. Besides, the applications based on circuit-level arithmetic will also be presented.

2 Approximate Arithmetic Unit

2.1 Approximate Adder

For approximate adders, Lu proposed an approximate adder called LUA [6] to improve microprocessors' performance in 2004. After that, approximate adders have been studied extensively, and various approximate adders have been proposed. Gupta et al. [7] proposed an approximate mirror adder (AMA) to obtain different approximate full adders by deleting some transistors inside the full adder. The low-part-OR adder (LOA) proposed by Mahdiani et al. [8] computes the exact full adder for the most significant bits (MSBs) of the operands, while the least significant bits (LSBs) are calculated with OR gates. The error-tolerant adder (ETA) proposed by Zhu et al. [9] also computes the MSBs exactly, and the LSBs are processed with a modified XOR gate (Fig. 1.1).

The two approximate adders mentioned above can reduce the complexity of the circuit structure. However, their error rates are extremely high. Thus, Kahng and Kang [10] proposed an accuracy-configurable adder (ACA), allowing it to choose between the exact and approximate modes of operation to adjust the accuracy. Nevertheless, ACA consumes more power and area since it employs additional modules for mode control.

The carry-skip adder (CSA) [11] and the speculative carry select addition (SCSA) [12] split the input operand into submodules. Each submodule is computed

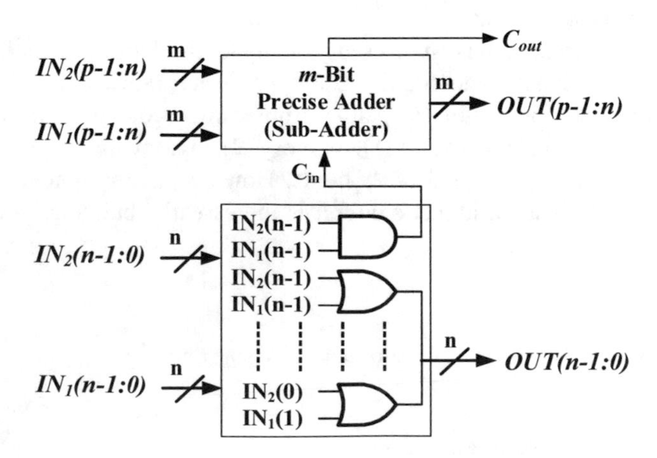

Fig. 1.1 LOA structure proposed by Mahdiani et al. [8]

using the carry signal from the previous submodule. This approach can effectively reduce the critical path delay by cutting the carry chain, but at the same time, they introduce a significant relative error compared with the exact results when the inexact carry signal occurs at the MSB part. Therefore, these two approximate adders are not good candidates in applications with relatively high accuracy requirements.

2.2 *Approximate Multiplier*

A typical binary multiplier usually consists of partial product generation modules, partial product compression part, and a carry propagation adder. The approximation methods of multipliers' design are more flexible. In this section, three approximation techniques are introduced: (1) input operands approximation, (2) partial product generation approximation, and (3) using approximate compressors.

The approximation to the operands originates from the logarithmic multiplier (LM) proposed by Mitchell in the 1960s. The idea is to convert multiplication to addition in the logarithmic domain, which has low power consumption and low accuracy in the meantime. Ansari et al. [13] proposed an improved logarithmic multiplier (ILM) for neural networks. ILM can reduce the power consumption by 24% compared to the conventional Mitchell logarithmic multiplier, but the mean relative error distance (MRED) is also doubled. Yin et al. [14] proposed an approximate logarithm multiply-accumulate (MAC) for CNN. First, the input operand is dynamically intercepted using the leading one detection unit (LOD), and the interception method is considered to use the approximation method of truncation and end-setting "1." The truncated approximate data is obtained and then converted into a logarithmic format, and the logarithmic multiplier converts the multiplication into addition and then compensates and shifts the logarithmic multiplication result to obtain the final output result. The power-delay product (PDP) of the proposed MACs is decreased by up to 54.07%.

Kulkarni et al. [15] proposed a 2 × 2 approximate multiplier. It is implemented based on the approximate truth table. The approximate 2 × 2 multiplier is the essential building block for long bit-width multipliers to generate the approximate partial product. Figure 1.2 shows a schematic diagram of this 2-bit multiplier constituting a 4-bit multiplier, where $A_L \times X_L$ denotes a 2-bit multiplication operation. However, the method proposed by Kulkarni et al. [15] can only work with the unsigned operands.

Based on this, the 4 × 4 and 8 × 8 multiplication arrays can also be built by the 2 × 2 multiplier submodule [16]. The 4 × 4 array is composed of four 2 × 2 multiplication modules. Approximate multipliers with different error features are selected for these four modules. The overall 4 × 4 multiplier can be achieved in which the final error is close to 0. The design idea of 8 × 8 is similar. An exploration algorithm platform for selecting the approximate multipliers is also presented in this paper. This approximate MAC reduces power consumption by about 13% in

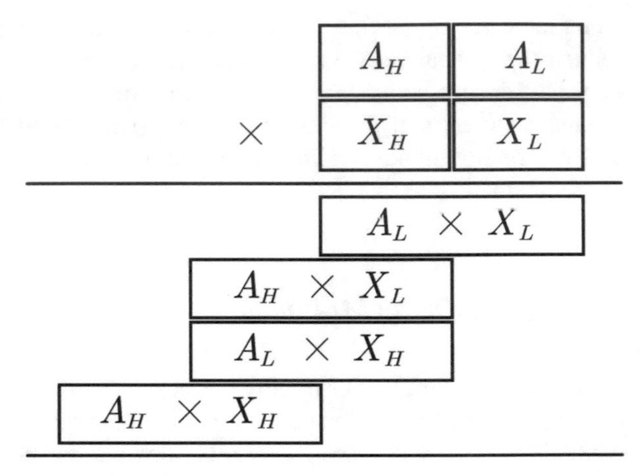

Fig. 1.2 4-bit multiplier built by 2 × 2 multiplier by Kulkarni et al. [15]

wireless astronomical calibration processing applications. This design works well for low bit-width multiplication. With the multiplier bit-width increasing, the power reduction becomes less and less significant.

As the most commonly used algorithm in signed number multiplication, the Booth algorithm can significantly reduce partial products. Various approximate multipliers are proposed based on the approximate Booth coding algorithm. Jiang et al. [17] proposed an approximate radix-8 Booth multiplier, which solves the problem that the ± 3 coding result is difficult to achieve by approximation, and the corresponding error compensation and error correction circuits are proposed. Qian et al. [18] proposed a novel radix-4 approximate Booth coding circuit by truth table analysis; Liu et al. [19] proposed a radix-4 approximate Booth coding circuit with high performance based on the structure by Qian et al. [18], and a metric called error factor is introduced to analyze the compromise effect between accuracy and performance. Liu et al. [20] apply the approximate Booth coding and approximate 4-2 compressor ideas to a redundant number binary multiplier.

This approximation technique for Booth multipliers is similar. It is further explained using the structure by Liu et al. [19] as an example. Although the Booth algorithm can significantly reduce the number of multiplier's partial products, the Booth encoder and decoder have a considerable hardware cost. Thus, Liu et al. [19] present an approximate radix-4 Booth encoder design to simplify the structure of the partial product generation circuit. Figure 1.3 shows the approximate radix-4 Booth multiplier proposed in this paper, reducing the area by 53% (R4ABE1) and 88% (R4ABE2) compared to the encoder with the conventional Booth multiplier. When applied to a 16-bit multiplier, the power consumption can be reduced by 19% and 25%, respectively, when the approximated bits are in the lower 16 bits.

The approximate 4-2 compressor was first proposed as a basic unit and presented two approximate 4-2 compressor designs [21]. The approximation method of cutting the carry propagation chain was widely used in later designs. The conventional 4-2 compressor generally consists of two full adders cascaded to output two carry

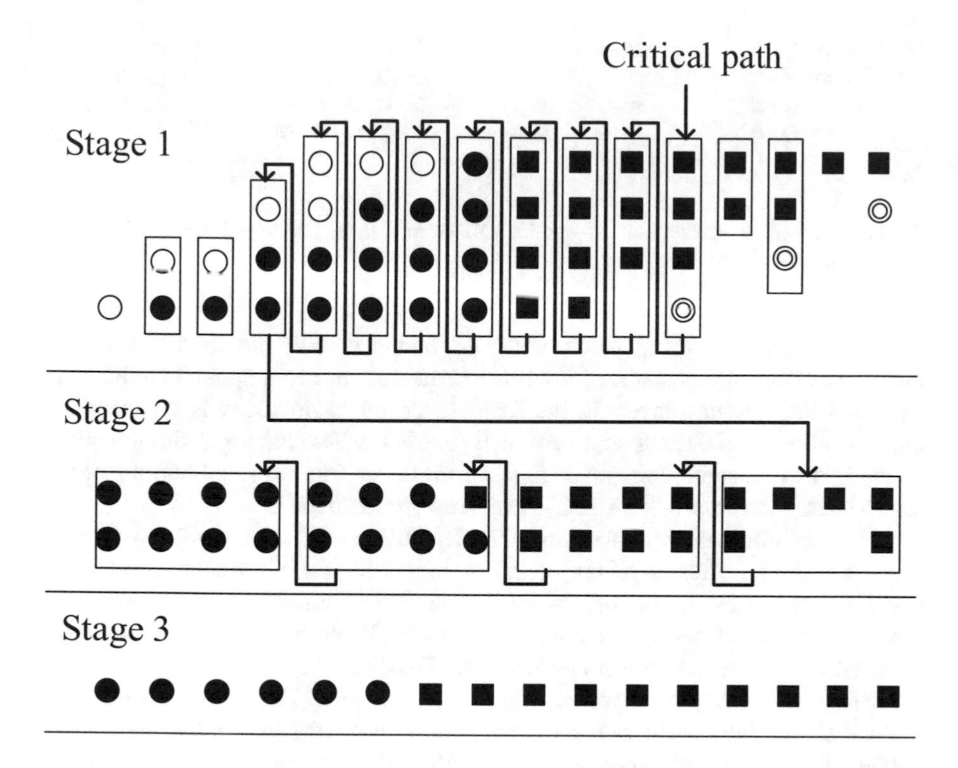

Fig. 1.3 Approximate radix-4 Booth multiplier proposed by Liu et al. [19]

signals. One of the carry signals is “1” only when the input is all 1’s, which has a very low probability of occurring, so an approximate 4-2 compressor with only one carry signal has been proposed by Momeni et al. [21].

There are also compressors with large input sizes. Marimuthu et al. [22] presented a 15-4 compressor, which consists of three approximate 5-3 compressors as submodules, suitable for the compression of multipliers with large bit widths.

2.3 Approximate Divider

The division is more complex than multiplication. Division operation requires quotient selection and overflow detection. Moreover, the division is sequential, while multiplication can be executed as a parallel addition of multiple operands. Conventional array dividers are implemented based on subtractor modules. Non-restoring array dividers and restoring array dividers are the two primary types of implementations [23].

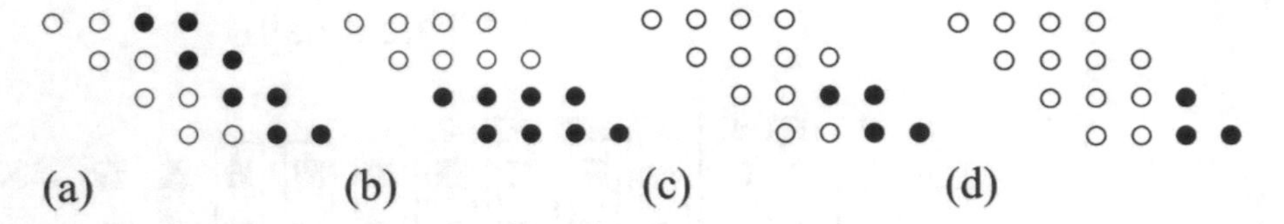

Fig. 1.4 Four replacement types for 8- to 4-bit AXDnr: (**a**) VR, (**b**) HR, (**c**) SR, and (**d**) TR. Black cells denote the replaced cells [25]

Based on the exact array divider structure, many approximate divider structures have been designed. The main approximation methods used are applied on the logic layer and the structure layer. In the logic layer, the main focus is on the logic approximation for the exact subtractor cell (EXSC), where the logic simplification is achieved by changing the values in the truth table. Three approximations of the approximate subtractor cell (AXSC) have been proposed [24].

4 (3) transistors can implement an XOR (XNOR) gate. Chen et al. [24] presented three AXSC structures: AXSC1 contains only 7-8; AXSC2 contains 6-8; and AXSC3 contains 5-6 transistors, while the number of transistors in EXSC is 10. However, the accuracy requirement should be considered while using the approximation method to reduce power consumption. For division, the quotient value has a much higher accuracy priority than the remainder value. Therefore, AXSC1 and AXSC2 follow the consideration of maintaining accuracy by approximating the Boolean function of the remainder without affecting the accuracy of the quotient. Of course, for specific applications requiring a higher remainder accuracy, AXSC2 can be chosen.

Chen et al. [24] proposed two approaches of approximation. The first one is the replacement. The approximated AXSCs replace the EXSCs to reduce the circuit complexity and power consumption. Four replacement schemes have been proposed by Chen et al. [25], which are shown in Fig. 1.4: vertical replacement (VR), horizontal replacement (HR), square replacement (SR), and triangular replacement (TR).

The second approximation method for array divider is truncation. Truncation completely deletes at least one cell (instead of replacement as described above). The input X of the removed cell is left unchanged and moved down the rest of the output direction, while the input Y is discarded. Similar to the replacement scheme, the truncation scheme has four configurations, namely, vertical truncation (VT), horizontal truncation (HT), square truncation (ST), and triangle truncation (TT). Compared with the replacement policy, truncation significantly reduces hardware resource consumption and considerably power consumption. However, the accuracy loss is significant.

Like the multiplier, a divider can be implemented with a logarithmic structure, which provides a better trade-off between accuracy and power consumption.

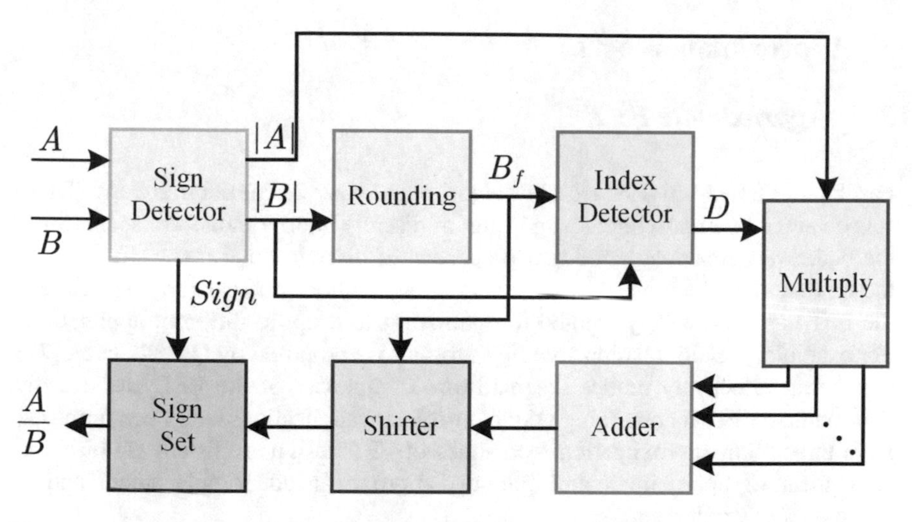

Fig. 1.5 SEERAD approximate logarithmic divider structure [26]

In the design of approximate dividers, a high-speed energy-efficient approximate divider (SEERAD) [26] can significantly reduce the delay and power consumption of the divider. In the SEERAD algorithm, the division operation (i.e., *A*/*B*) can be expressed in the form of $2^{K+L}/D$ by rounding the divisor to the nearest integer number (i.e., Br), where *K* denotes the position of the highest valid "1" of *B* in binary and *L* and *D* are predefined constant integers. Thus, in this approach, the division operation is approximated as follows:

$$\frac{A}{B} \approx \frac{A}{B_{\mathrm{r}}} = \frac{D \times A}{2^{K+L}}$$

The setting of parameters *D* and *L* plays a decisive role in the accuracy of the overall divider. For each input bit length, to determine the parameters *D* and *L*, an iterative search should be performed to determine the most appropriate combination of *D* and *L*. During the search, *A*'s value is taken iteratively from 0 to the maximum value, and then, the error of dividing *A* by B_{r} is compared to dividing by *B*. The block diagram of the final SEERAD hardware implementation of the approximate divider is shown in Fig. 1.5.

3 Approximate Blocks

3.1 Approximate FFT

Fast Fourier transform (FFT) is a fast algorithm of discrete Fourier transform (DFT). It is obtained by improving the algorithm of discrete Fourier transform according to the odd, even, imaginary, and real properties of discrete Fourier transform. Since the FFT can be widely used in various DSP applications, the hardware structure of the FFT has been widely studied to optimize it to adapt to different applications. For example, a multiplication-free FFT structure is proposed by Qureshi et al. [27], which can effectively reduce the hardware complexity of the FFT. Besides, the approximate FFTs for emerging DSP fault-tolerant applications have been proposed. Here three main approximation techniques of FFT design are listed: (1) bit-width adjustment, (2) approximate multiplier and approximate adder replacement, and (3) approximate twiddle factors.

For bit-width adjustment, Fig. 1.6 shows the bit-width selection algorithm for resource-intensive optimization proposed by Liao et al. [28]. First, the measurement scheme for the maximum error tolerance in the chosen design is specified, i.e., the metric for measuring the quality of result (QoR). In the first iteration, only the bit width of the first level is reduced within the error tolerance, and the bit width of the other levels remains unchanged; after that, the second iteration is carried out. The entire bit-width adjustment process for resource optimization is completed when the threshold constraint is approached. For the scenarios that require a high-performance design with sufficient resources, the performance-optimized bit-width selection algorithm needs to balance the performance of each level to avoid the bottleneck in certain stages. In this way, the bit width of each level should be similarly equal. After finding the properly balanced bit width, the previous resource-intensive optimization can be applied until the threshold is approached. These two bit-width selection algorithms can effectively save hardware resources and increase frequency. The resource-optimized bit-width selection algorithm can save hardware

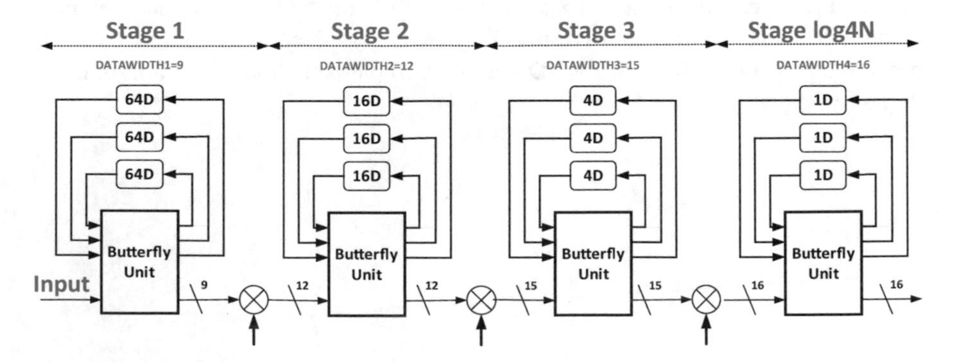

Fig. 1.6 256-point FFT with bit-width adjustment algorithm [28]

resources up to 25% or more, with a slight increase in frequency; the performance-optimized bit-width selection algorithm does not save as many resources as the former, but the frequency increase can be up to 24%.

FFT operations contain many multiplication operations, so the approximate design of multipliers is of interest for FFT. Xiao et al. [29] implement low-power FFT by designing approximate 4-2 compressors and approximate compressed truncated multipliers. The approximation-based 512-point FFT processor reduces the kernel area by 15.24%, 8.32%, and 35.41% compared to the conventional 512-point FFT by Huang and Chen [30] and Chen et al. [31], respectively.

The multiplier in the complex multiplication unit of the FFT is approximated to replace the exact multiplier [32]. Four radix-4 Booth multipliers with different approximation levels are proposed to reduce the hardware complexity (as shown in Fig. 1.7). The pipeline FFT and the parallel FFT based on the proposed approximate multipliers are implemented and extensively evaluated. Compared with the state-of-the-art FFT designs, the LUT's amount is reduced up to 20.3% and 29.1% for pipeline and parallel FFTs, respectively. The power is reduced up to 69.9% for pipeline FFT, and the delay is reduced up to 45.7%.

For the parallel FFT, the twiddle factors are fixed, which the canonical signed digit (CSD) encoding number can be employed as the twiddle factors. Discrete cosine transforms (DCTs), which also use CSD encoding, use an approximate CSD design [33]. This design can reduce the hardware consumption by up to 51.5%, so the CSD multiplier used in the FFT can also reduce the hardware consumption effectively by referring to this design.

3.2 Approximate MAC

Compared to approximate multipliers, there is not much research on approximate multiply-accumulate (MAC) modules. In recent years, approximate MAC has received much attention from researchers due to further research on deep neural networks, and Emer et al. [34] show that more than 90% of operations in deep neural networks are convolutional operations. As the dominated operation, the MAC consumes the most energy. The design of approximate MAC is mainly done by introducing approximate multiplier and approximate adder.

Figure 1.8 shows the approximate multiplier for MAC proposed by Esposito et al. [35]. Figure 1.8a shows the exact partial product array of 8-bit signed numbers multiplied by unsigned numbers, and Fig. 1.8b approximates every two partial products in each column of the array by taking OR gates and omitting the AND-gate part of the addition, which can eventually reduce the stages of the partial product array. For further approximation, the LSB part of the partial product is truncated, and the truncation length can be adjusted according to the accuracy requirement. Four truncation lengths are tested in this paper. To further reduce the error of the approximate multiplication when performing the accumulation, the average error is compensated into the accumulation. This approximate MAC is applied to image

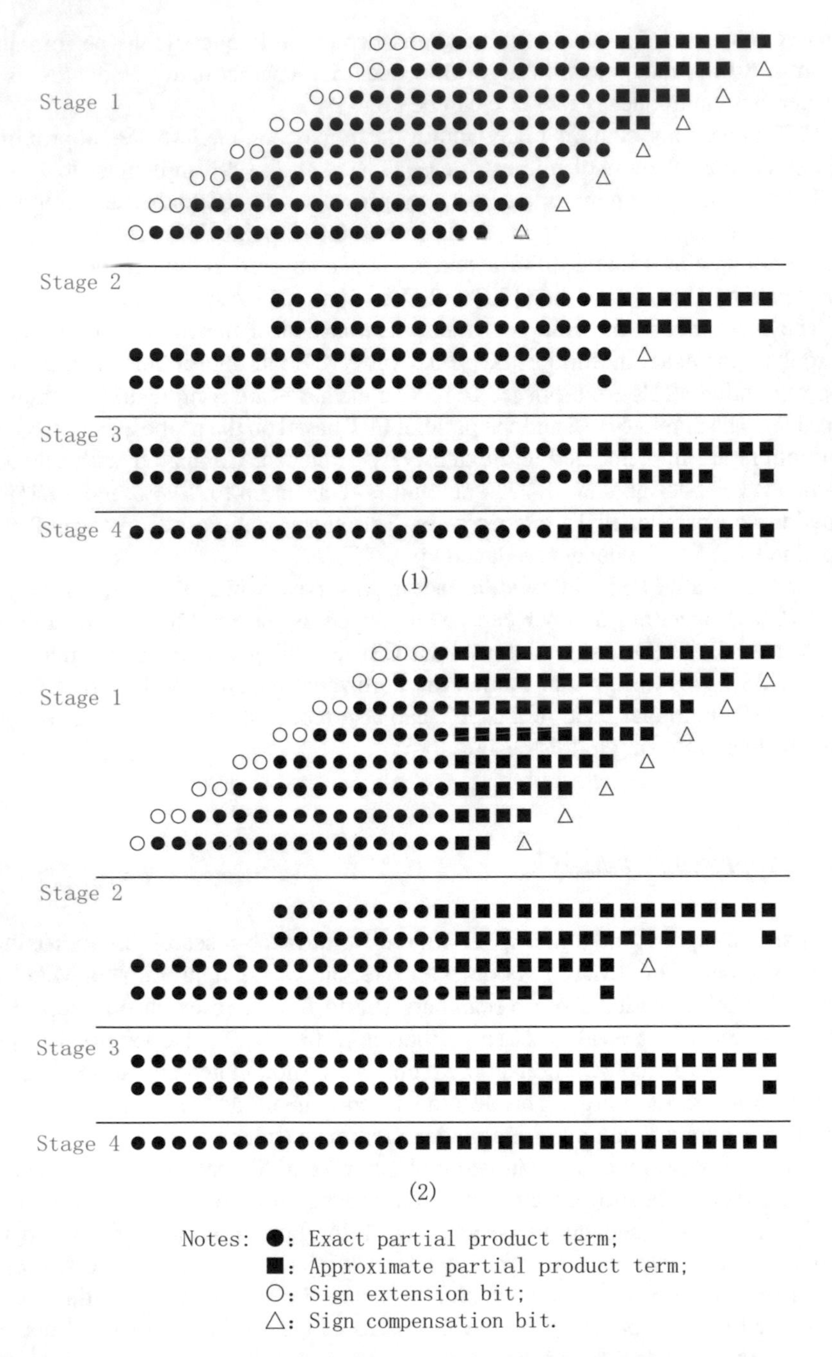

Fig. 1.7 16 × 16 Booth multiplier using an approximate regular partial product array: (1) 8 LSBs of the partial product are approximate. (2) 16 LSBs of the partial product are approximate

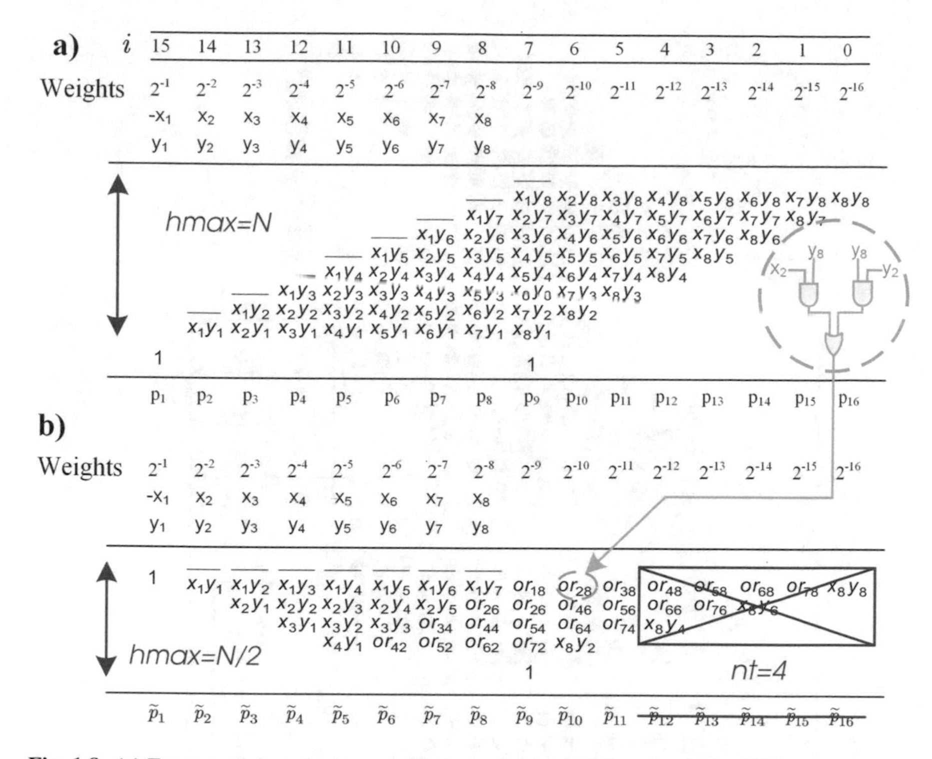

Fig. 1.8 (**a**) Exact partial product array; (**b**) approximate partial product array [35]

processing, and the power consumption can be reduced by 30–70% compared to conventional MACs while maintaining image quality.

Yang et al. [36] designed an $8 \times 8 + 16$-bit MAC cell by inserting the accumulation into the partial product array, as shown in Fig. 1.9. In the array compression, the two rows of partial products accumulation are replaced by OR operation instead of addition. The array is suitable for unsigned multiplication and accumulation. For signed calculation, the operands will be detected by the sign bit and converted to unsigned numbers. The conversion can be approximated by using 1's complementary number instead of 2's complementary number. This design can reduce power consumption by about 43% for unsigned MAC and 18% for signed MAC.

4 Applications Based on Approximate Arithmetic

4.1 Approximate Redundancy

A soft error may occur due to a strike by a high-energy particle and manifests itself as a transient bit reversal in the logic value of a circuit node. It has become

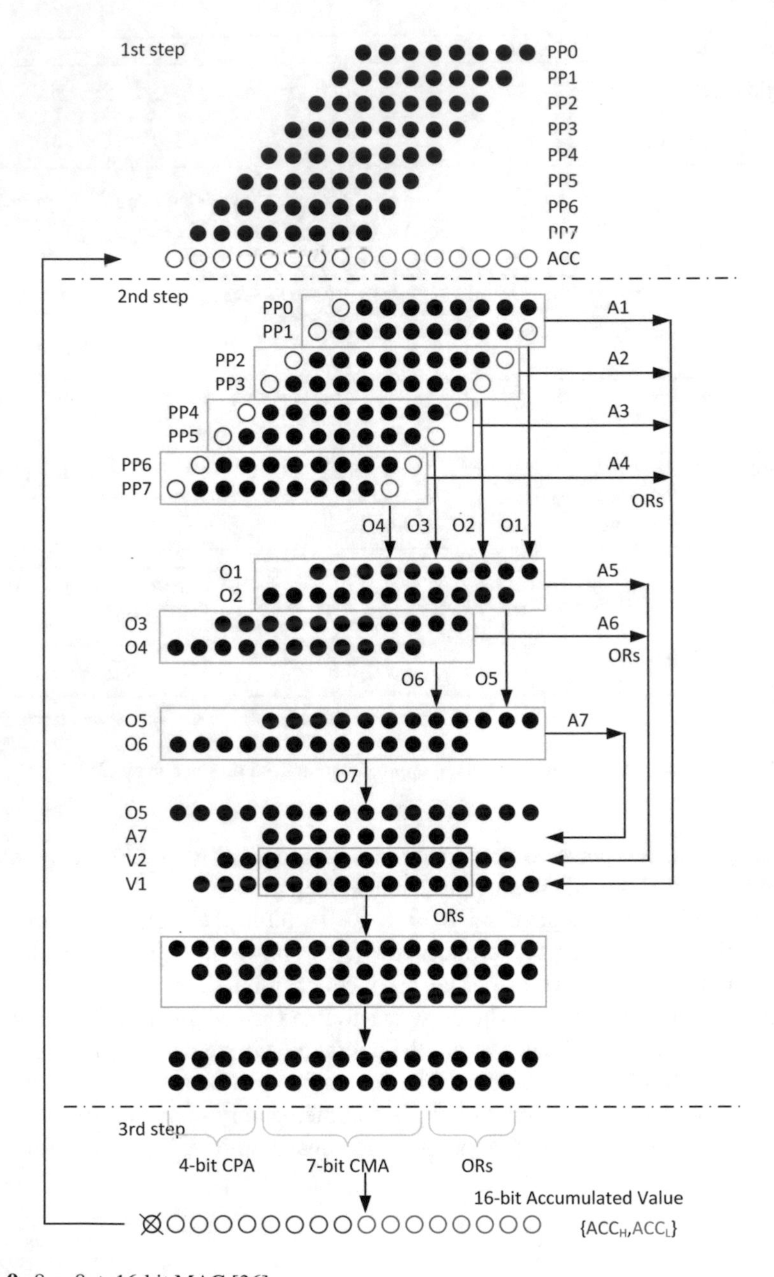

Fig. 1.9 $8 \times 8 + 16$-bit MAC [36]

a significant concern in the design of nanoscale digital integrated circuits [37]. Redundancy techniques are adequate to address soft errors; they are commonly used for designing dependable systems to ensure high reliability and availability [38, 39]. One of the most effective fault-tolerant design schemes is the so-called N-modular

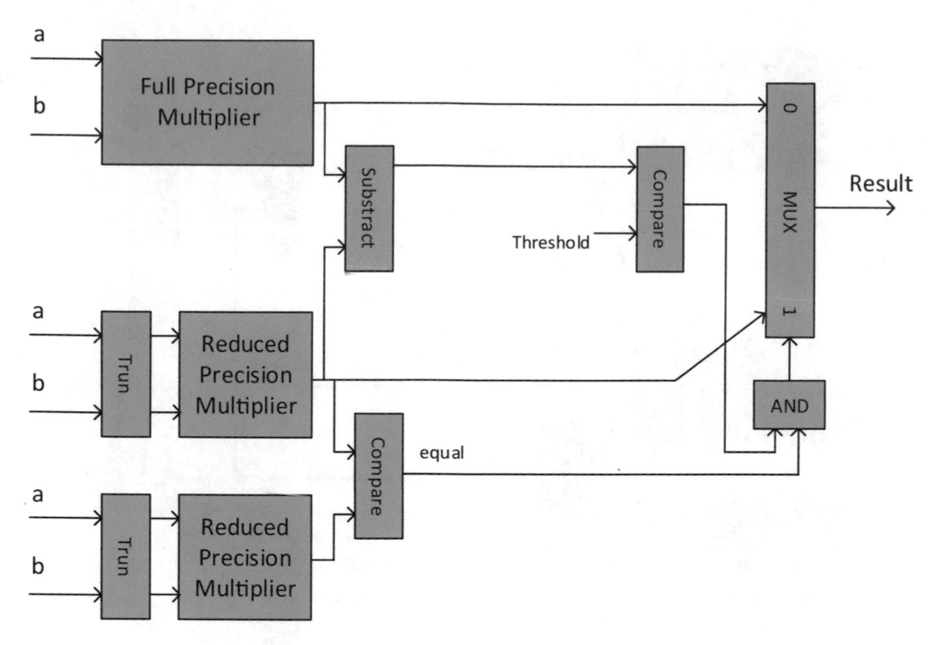

Fig. 1.10 Conventional RPR scheme

redundancy (NMR); in an NMR scheme, N copies of a module are utilized [40]. However, the hardware cost is the primary concern of the NMR technique (Fig. 1.10).

An alternative to complete redundancy such as triplication or duplication is to use reduced precision redundancy (RPR) [41, 42]. RPR uses a full precision version of the circuit and two reduced precision ones with a voting logic to correct errors. RPR significantly lowers the overhead because the reduced precision copies are much smaller than the full precision implementation.

Chen et al. [43] proposed a scheme based on the observation that for signed integers in 2's complement format, the truncation error in a multiplication depends on the sign of the operands. When the two operands (a and b) have the same sign, the errors introduced by truncation add. However, when the operands have different signs, errors will compensate. Therefore, the difference between the full precision and the reduced precision copies will be significant. This can be exploited by using different thresholds depending on the sign of the operands. Compared with the conventional RPR scheme, the results show that the proposed modified RPR can significantly reduce the mean square error (MSE) at the output when the circuit is affected by a soft error, and the implementation overhead of the proposed schemes is extremely low (Fig. 1.11).

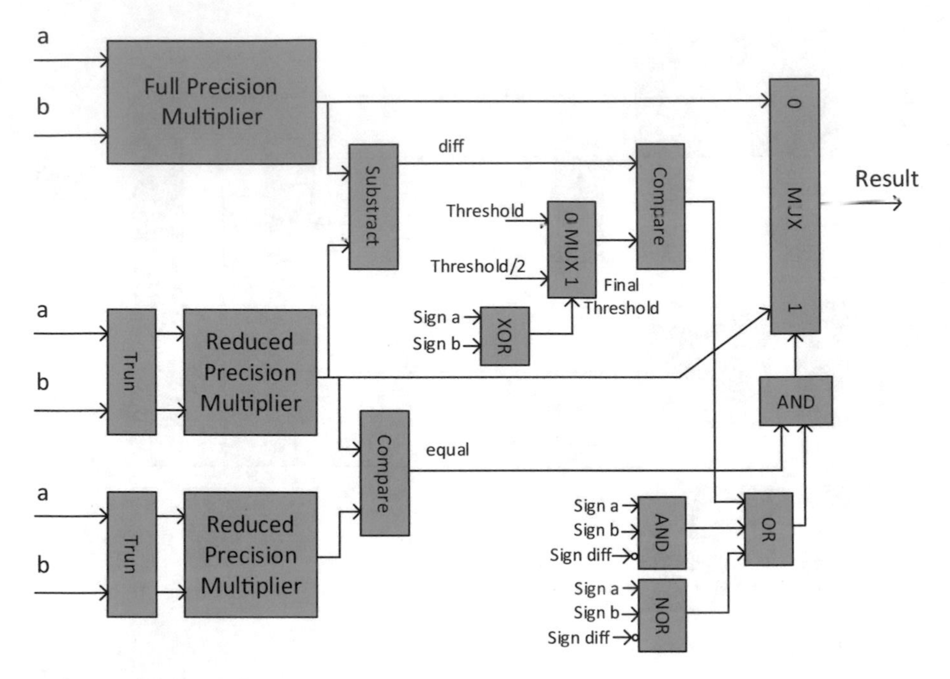

Fig. 1.11 Modified RPR scheme [43]

4.2 Approximate 2D Convolver for Image Processing

Image and video processing requires the computation of two-dimensional (2D) convolution for many applications, such as filtering, restoration, feature recognition, object tracking, and template matching [44]. Chen et al. [45] propose an approximate 2D convolver in which both circuit and algorithm approximate techniques are utilized in the design. These techniques allow substantial reductions in figures of merit (such as power dissipation and circuit complexity) and keep the loss of accuracy as nearly as desired. Truncation and voltage over scaling are used as circuit techniques, while bit-width reduction is utilized at the algorithm level. These different techniques are related to the configuration of the convolver by which its operation can be configured to meet different and often contrasting figures of merit (Fig. 1.12).

Chen et al. [46] utilize a bit-width reduction process as an algorithm-based technique for approximate computing. A greedy process is used in this paper to reduce the bit width. Starting from the low weight coefficients, the bit width is reduced by a single level for pixels corresponding to every coefficient until the estimated PSNR is just less than T. This process continues until a candidate (if it exists) satisfies the image quality constraint in a so-called configuration for computing the convolution. Besides, the simple error compensation method

Fig. 1.12 (**a**) Exact convolution (100% power consumption); (**b**) config 1 (69.3% power consumption, 49.03 dB); (**c**) config 2 (48.4% power consumption, 40.09 dB); (**d**) config 3 (35.6% power consumption, 30.87 dB)

is introduced, and the voltage scaling way of approximation is evaluated. The configuration achieved up to 69.3% reduction compared with an exact convolver.

5 Error Compensation for Approximate Arithmetic

In many digital signal processing algorithms, like multiplication, the bit width is generally fixed to prevent the increase in bit width of the result of each multiplication. For fixed-width multipliers, direct truncation is usually used to ignore the partial product of lower effective bits, reducing the area and power consumption of the multiplier but introducing a more significant truncation error at the same time. To achieve a balanced design between accuracy and hardware consumption, various error compensation circuits have been proposed to reduce the truncation error of the array multiplier and Booth multiplier:

1. Constant value compensation and adaptive compensation

Chen et al. [47] propose an input correlation method using probabilistic, statistical, and linear regression analysis to find approximate compensation values. The error compensation circuit consists of a partial product of the highest weights in the truncated section and a constant compensation value. The compensation value is obtained according to the input and truncated bit widths calculated in advance by simulation statistics. Schulte and Swartzlander [48] consider the truncation and rounding errors by superimposing the weights into a constant compensation for the partial product. For the constant compensation method, the error is generally larger. King and Swartzlander [49] changed the compensation constants into multiple correction terms (adaptive compensation) based on Schulte and Swartzlander [48] and generated multiple compensation terms superimposed into the partial product based on the truncated partial product. For constant and multiple correction terms, compensation is generally used when the truncation length is equal to the input width, in which case the hardware circuit is implemented only by adding a full adder. The absolute error of constant compensation is maximum when the truncated partial product is maximum, and the absolute error of multi-correction-term compensation is maximum when the truncated partial product is zero. Stine and Duverne [50] proposed a hybrid compensation method combining constant compensation and multi-correction-term compensation, and the ratio of the two compensation methods is selected by introducing a scaling parameter. Compared with the one proposed by Schulte and Swartzlander [48] and King and Swartzlander [49], the average and maximum absolute errors obtained by this compensation method are smaller. Different adaptive error compensation methods were proposed and applied to the Baugh–Wooley array multiplier design in Van and Yang [51] and Petra et al. [52], respectively. The above methods are mainly applied to array multipliers and cannot be directly applied to Booth coded multipliers.

2. Simplifying the low weight partial product logic

For this type of compensation scheme, the partial product array is typically divided into three parts: exact, approximate, and truncated parts. According to Petra et al. [52] and Wey and Wang [53], the second-highest truncated weight is considered to calculate the carry value instead of directly truncation. The study of Kumar and Sahoo [54] is further optimized based on the study of Wey and Wang [53] to have lower compensation error and lower hardware complexity, especially when the number of multiplier input bits increases. Cho et al. [55] and Wang et al. [56] divided the truncated part of Booth partial product into primary and secondary truncated parts and the primary truncated part adjacent to the higher bits was retained, while a simple gate circuit approximated the truncated error of the secondary truncated part.

6 Conclusion

This chapter provides an overview of circuit-level approximation techniques commonly used in several computing scenarios. It mainly includes approximation circuit design methods for operand approximation processing, approximation partial product generation, approximation compressor, approximation compression tree, and partial product truncation for basic arithmetic units such as multipliers; design methods for bit-width optimization selection algorithms, approximation arithmetic unit replacement, and rotation factor approximation for FFT operations; and approximation MAC unit design methods for AI applications. Besides, the applications like reduced precision redundancy and image processing implemented with approximate arithmetic are presented. At last, two error compensation methods are introduced. However, the approximate design is highly related to the application. The systematic design method should be investigated in the future.

References

1. Jiang H, Santiago FJH, Mo H, et al. Approximate arithmetic circuits: a survey, characterization, and recent applications. Proc IEEE. 2020;108:2108–35. https://doi.org/10.1109/JPROC.2020.3006451.
2. Amanollahi S, Kamal M, Afzali-Kusha A, Pedram M. Circuit-level techniques for logic and memory blocks in approximate computing systemsx. Proc IEEE. 2020;108:2150–77. https://doi.org/10.1109/JPROC.2020.3020792.
3. Scarabottolo I, Ansaloni G, Constantinides GA, et al. Approximate logic synthesis: a survey. Proc IEEE. 2020;108:2195–213. https://doi.org/10.1109/JPROC.2020.3014430.
4. Kang M, Gonugondla SK, Shanbhag NR. Deep in-memory architectures in SRAM: an analog approach to approximate computing. Proc IEEE. 2020;108:2251–75. https://doi.org/10.1109/JPROC.2020.3034117.
5. Liu W, Gu C, O'Neill M, et al. Security in approximate computing and approximate computing for security: challenges and opportunities. Proc IEEE. 2020;108:2214–31. https://doi.org/10.1109/JPROC.2020.3030121.
6. Lu S-L. Speeding up processing with approximation circuits. Computer. 2004;37:67–73.
7. Gupta V, Mohapatra D, Park SP, et al. IMPACT: IMPrecise adders for low-power approximate computing. In: IEEE/ACM international symposium on low power electronics and design. IEEE; 2011. p. 409–14.
8. Mahdiani HR, Ahmadi A, Fakhraie SM, Lucas C. Bio-inspired imprecise computational blocks for efficient VLSI implementation of soft-computing applications. IEEE Trans Circuits Syst I: Regular Papers. 2009;57:850–62.
9. Zhu N, Goh WL, Zhang W, et al. Design of low-power high-speed truncation-error-tolerant adder and its application in digital signal processing. IEEE Trans Very Large Scale Integr Syst. 2009;18:1225–9.
10. Kahng AB, Kang S. Accuracy-configurable adder for approximate arithmetic designs. In: Proceedings of the 49th annual design automation conference; 2012. p. 820–5.
11. Kim Y, Zhang Y, Li P. An energy efficient approximate adder with carry skip for error resilient neuromorphic VLSI systems. In: 2013 IEEE/ACM international conference on computer-aided design (ICCAD). IEEE; 2013. p. 130–7.
12. Du K, Varman P, Mohanram K. High performance reliable variable latency carry select addition. In: 2012 design, automation & test in Europe conference & exhibition (DATE). IEEE; 2012. p. 1257–62.
13. Ansari MS, Cockburn BF, Han J. An improved logarithmic multiplier for energy-efficient neural computing. IEEE Trans Comput. 2020;70:614–25.
14. Yin P, Wang C, Waris H, et al. Design and analysis of energy-efficient dynamic range approximate logarithmic multipliers for machine learning. IEEE Trans Sustain Comput. 2020;6:612.

15. Kulkarni P, Gupta P, Ercegovac M. Trading accuracy for power with an underdesigned multiplier architecture. In: 2011 24th international conference on VLSI design. IEEE; 2011. p. 346–51.
16. Gillani GA, Hanif MA, Verstoep B, et al. MACISH: designing approximate MAC accelerators with internal-self-healing. IEEE Access. 2019;7:77142–60.
17. Jiang H, Han J, Qiao F, Lombardi F. Approximate radix-8 booth multipliers for low-power and high-performance operation. IEEE Trans Comput. 2015;65:2638–44.
18. Qian L, Wang C, Liu W, et al. Design and evaluation of an approximate Wallace-Booth multiplier. In: 2016 IEEE international symposium on circuits and systems (ISCAS). IEEE; 2016. p. 1974–7.
19. Liu W, Qian L, Wang C, et al. Design of approximate radix-4 booth multipliers for error-tolerant computing. IEEE Trans Comput. 2017;66:1435–41.
20. Liu W, Cao T, Yin P, et al. Design and analysis of approximate redundant binary multipliers. IEEE Trans Comput. 2018;68:804–19.
21. Momeni A, Han J, Montuschi P, Lombardi F. Design and analysis of approximate compressors for multiplication. IEEE Trans Comput. 2014;64:984–94.
22. Marimuthu R, Rezinold YE, Mallick PS. Design and analysis of multiplier using approximate 15-4 compressor. IEEE Access. 2016;5:1027–36.
23. Parhami B. Computer arithmetic. Oxford University Press; 2010.
24. Chen L, Han J, Liu W, Lombardi F. Design of approximate unsigned integer non-restoring divider for inexact computing. In: Proceedings of the 25th edition on Great Lakes symposium on VLSI; 2015a. p. 51–6.
25. Chen L, Han J, Liu W, Lombardi F. On the design of approximate restoring dividers for error-tolerant applications. IEEE Trans Comput. 2015b;65:2522–33.
26. Zendegani R, Kamal M, Fayyazi A, et al. SEERAD: a high speed yet energy-efficient rounding-based approximate divider. In: 2016 design, automation & test in Europe conference & exhibition (DATE). IEEE; 2016. p. 1481–4.
27. Qureshi F, Ali M, Takala J. Multiplierless reconfigurable processing element for mixed radix-2/3/4/5 FFTs. In: 2017 IEEE international workshop on signal processing systems (SiPS). IEEE; 2017. p. 1–6.
28. Liao Q, Liu W, Qiao F, et al. Design of approximate FFT with bit-width selection algorithms. In: 2018 IEEE international symposium on circuits and systems (ISCAS). IEEE; 2018. p. 1–5.
29. Xiao H, Yin X, Wu N, et al. VLSI design of low-cost and high-precision fixed-point reconfigurable FFT processors. IET Comput Digital Tech. 2018;12:105–10.
30. Huang S-J, Chen S-G. A green FFT processor with 2.5-GS/s for IEEE 802.15. 3c (WPANs). In: The 2010 international conference on green circuits and systems. IEEE; 2010. p. 9–13.
31. Chen Y, Lin Y-W, Tsao Y-C, Lee C-Y. A 2.4-Gsample/s DVFS FFT processor for MIMO OFDM communication systems. IEEE J Solid State Circuits. 2008;43:1260–73.
32. Du J, Chen K, Yin P, et al. Design of an approximate FFT processor based on approximate complex multipliers. In: 2021 IEEE computer society annual symposium on VLSI (ISVLSI). IEEE; 2021. p. 308–13.
33. Cai L, Qian Y, He Y, Feng W. Design of approximate multiplierless DCT with CSD encoding for image processing. In: 2021 IEEE international symposium on circuits and systems (ISCAS). IEEE; 2021. p. 1–4.
34. Emer J, Sze V, Chen Y-H, Yang T-J. Hardware architectures for deep neural networks. CICS/MTL Tutorial, Mar. 2017;27:258.
35. Esposito D, Strollo AGM, Alioto M. Low-power approximate MAC unit. In: 2017 13th conference on Ph. D. research in microelectronics and electronics (PRIME). IEEE; 2017. p. 81–4.
36. Yang T, Sato T, Ukezono T. An approximate multiply-accumulate unit with low power and reduced area. In: 2019 IEEE computer society annual symposium on VLSI (ISVLSI). IEEE; 2019. p. 385–90.
37. Baumann R. Soft errors in advanced computer systems. IEEE Des Test Comput. 2005;22:258–66.

38. Vaidya NF, Pradhan DK. Fault-tolerant design strategies for high reliability and safety. IEEE Trans Comput. 1993;42:1195–206.
39. von Neumann J. Probabilistic logics and the synthesis of reliable organisms from unreliable components. In: Automata studies.(AM-34), vol. 34. Princeton University Press; 2016. p. 43–98.
40. Pierce WH. Failure-tolerant computer design. Academic Press; 2014.
41. Shim B, Shanbhag NR. Energy-efficient soft error-tolerant digital signal processing. IEEE Trans Very Large Scale Integr Syst. 2006;14:336–48.
42. Shim B, Sridhara SR, Shanbhag NR. Reliable low-power digital signal processing via reduced precision redundancy. IEEE Trans Very Large Scale Integr Syst. 2004;12:497–510.
43. Chen K, Chen L, Reviriego P, Lombardi F. Efficient implementations of reduced precision redundancy (RPR) multiply and accumulate (mac). IEEE Trans Comput. 2018a;68:784–90.
44. Bosi B, Bois G, Savaria Y. Reconfigurable pipelined 2-D convolvers for fast digital signal processing. IEEE Trans Very Large Scale Integr Syst. 1999;7:299–308.
45. Chen K, Lombardi F, Han J. Design and analysis of an approximate 2D convolver. In: 2016 IEEE international symposium on defect and fault tolerance in VLSI and nanotechnology systems (DFT). IEEE; 2016. p. 31–4.
46. Chen K, Han J, Montuschi P, et al. Design and application of an approximate 2-D Convolver with error compensation. In: 2018 IEEE international symposium on circuits and systems (ISCAS). IEEE; 2018b. p. 1–5.
47. Chen Y-H, Chang T-Y, Jou R-Y. A statistical error-compensated Booth multipliers and its DCT applications. In: TENCON 2010–2010 IEEE region 10 conference. IEEE; 2010. p. 1146–9.
48. Schulte MJ, Swartzlander EE. Truncated multiplication with correction constant [for DSP]. In: Proceedings of IEEE workshop on VLSI signal processing. IEEE; 1993. p. 388–96.
49. King EJ, Swartzlander EE. Data-dependent truncation scheme for parallel multipliers. In: Conference record of the thirty-first Asilomar conference on signals, systems and computers (Cat. No. 97CB36136). IEEE; 1997. p. 1178–82.
50. Stine JE, Duverne OM. Variations on truncated multiplication. In: Euromicro symposium on digital system design, 2003. Proceedings. IEEE; 2003. p. 112–9.
51. Van L-D, Yang C-C. Generalized low-error area-efficient fixed-width multipliers. IEEE Trans Circuits Syst I: Regular Papers. 2005;52:1608–19.
52. Petra N, de Caro D, Garofalo V, et al. Truncated binary multipliers with variable correction and minimum mean square error. IEEE Trans Circuits Syst I Regular Papers. 2009;57:1312–25.
53. Wey I-C, Wang C-C. Low-error and hardware-efficient fixed-width multiplier by using the dual-group minor input correction vector to lower input correction vector compensation error. IEEE Trans Very Large Scale Integr Syst. 2011;20:1923–8.
54. Kumar GG, Sahoo SK. Power-efficient compensation circuit for fixed-width multipliers. IET Circuits Dev Syst. 2020;14:505–9.
55. Cho K-J, Lee K-C, Chung J-G, Parhi KK. Design of low-error fixed-width modified booth multiplier. IEEE Tran Very Large Scale Integr Syst. 2004;12:522–31.
56. Wang J-P, Kuang S-R, Liang S-C. High-accuracy fixed-width modified Booth multipliers for lossy applications. IEEE Trans Very Large Scale Integr Syst. 2009;19:52–60.

Chapter 2
An Automated Logic-Level Framework for Approximate Modular Arithmetic Circuits

Haroon Waris, Chenghua Wang, and Weiqiang Liu

1 Introduction

Exact and fast computing systems have been the dominating objective of digital chip designers; however, with the reduction in technology node, it is becoming difficult to meet the hardware requirements within a specified power budget. Approximate computing [1] is one such alternative that has been investigated to mitigate the aforementioned challenge. In general, inexactness is not desired, but many applications can tolerate noisy inputs and do not require a precise result. For example, in classification problems, a precise output is often not a strict requirement. Similarly, in image processing, it is difficult for an end user to perceive minor losses in image quality [2]. Therefore, machine learning, multi-media, recognition, communication, and data mining are some of the target applications for approximate computing [3].

In recent years, workloads have dramatically changed allowing an approximation to be integrated into a computational task. Xu et al. [4] have presented a generalized framework of approximate computing for error-tolerant application. The identification of error-flexible parts is one of the crucial steps for this paradigm; software, architecture, and hardware layers are equally good candidates to be considered for approximation [5]. At the hardware layer, approximation refers mostly to approximate arithmetic units that require large power dissipation in embedded systems [6]. Arithmetic units are present in almost every central processing unit (CPU), application specific processors, and graphics processing unit (GPU); they are also the main source of power consumption in these systems. Therefore, most

H. Waris (✉) · C. Wang · W. Liu
Nanjing University of Aeronautics and Astronautics, Nanjing, China
e-mail: haroonwaris@nuaa.edu.cn; chwang@nuaa.edu.cn; liuweiqiang@nuaa.edu.cn

W. Liu, F. Lombardi (eds.), *Approximate Computing*,
https://doi.org/10.1007/978-3-030-98347-5_2

of the research in hardware approximation has been directed toward the design of approximate arithmetic units.

Significant research has been conducted on the design of approximate (8/16/32 bit) multipliers. Truncation and rounding based scalable approximate multiplier (TOSAM) has been presented in [7]. This approach truncates input operands based on the leading one bit position, thus, reducing the number of partial products. Jiang et al. [8] have proposed a novel approximate multiplier by accumulating the partial products using a simple tree of approximate adders. Moreover, for a better accuracy, error compensation is implemented using error signals. Mitchell's logarithmic multiplication has been used in [9] to propose customizable approximate multipliers. Design techniques (exact zero computation, leading one detector, and efficient shift amount calculation) are then used by performing inferences on convolutional neural networks. Leon et al. [10] have combined high-radix encoding, perforation, and truncation to propose power-efficient approximate multipliers. Moreover, these designs improve the resolution of the error-energy Pareto front. Approximate radix-4 Booth multipliers have been proposed in [11] for low-power and high-performance operation. The proposed multiplier provides ease in the generation of partial products; therefore, a reduced hardware complexity has been achieved compared to the existing state-of-the-art designs. Recently, signed multipliers have been proposed [12] for FPGA-based systems. As FPGA provides design granularity at the Look-up table (LUT-level), therefore, to achieve power-optimized hardware implementations, LUTs-based design approach has been presented.

A lot of technical literature is available for manual approximate design methodologies; however, very less research is dedicated to automate this design process. The use of embedded design automation tool in the design of approximate multipliers reduces the design efforts to tailor them for a particular accuracy requirement. Manual design approach leads to limited trade-off options as they have restricted number of design variants. Moreover, manually designing new approximate design versions increases the design time whenever the global quality constraint varies. Therefore, an intensive study is required to automate the design approach for approximate arithmetic circuits to achieve various approximate versions that satisfy a given quality constraint. Furthermore, multiple optimized approximate designs either for energy, area, or delay can be achieved with reduced design time.

The paper is organized as follows. Existing state-of-the-art works are discussed in Sect. 2. Section 3 formulates the problem and discusses the essence used to realize the solution. Section 4 presents the automated framework. The design exploration of approximate multipliers is explained in Sect. 5. Section 6 describes the comparison of ASF based multipliers with that of state-of-the-art designs. In Sect. 7, the performance of proposed approximate multipliers is evaluated using image sharpening and Canny edge applications. Section 8 concludes this paper.

2 Related Work

The very first automated approach has been presented by Shin et al., which approximates the input function. The heuristic approach to search for all possible approximate solutions make this process very computational expensive. [14] has presented a systematic logic synthesis of approximate circuits (SALSA), which defines a quality constraint to establish an approximate synthesis problem. Lingamneni et al. [15] have used probabilistic design techniques and a probabilistic logic minimization (PLM) methodology to design inexact circuits. Samadi et al. [16] have introduced automation at software layer; the so-called self-tuning approximation for graphics engines (SAGE) takes advantage of GPU microarchitecture and operates based on approximated kernels. In [17], the approximate logic synthesis presented in [13] is further enhanced by constraining the error, i.e., the greedy approximate logic synthesis (GALS). An automatic methodology for sequential logic approximation (ASLAN) [18] is the first design approach to achieve approximation in sequential circuits. Nepal et al. [19] have presented a technique for automated behavioral synthesis of approximate computing circuits (ABACUS), which uses a behavioral description to search optimal approximate designs.

In [20], register-transfer level (RTL) representations are used to propose approximate designs. Initially, RTL description of a circuit is used to construct an abstract syntax. Then, data type simplifications, arithmetic approximations, and loop transformations are applied to create approximate designs. [21] has used the truth table folding technique to perform circuit approximations. Matrix-based circuit factorization is used and includes both semi-ring (OR) and field (XOR) algebraic implementations. Symmetric approximation has been proposed in [22]; the original function f is approximated by the closest totally symmetric function. Binary decision diagrams and the Boolean function are used to generate the characteristic vector of a symmetric function. Venkataramni et al. [23] have used the concept of observability don't cares (ODCs) to minimize circuit complexity. The set of input values for which the outputs remain insensitive are then approximated to simplify the node in a logic circuit. [24] has presented an approximate simplification for a Boolean network. A Boolean network consists of multiple nodes; this methodology performs literal reduction from the original expression of a node while keeping error in bounds. Recently [25], a heuristic method based on optimal set of input combinations (SICC) is proposed to approximate two-level logic synthesis.

Most of the previous works on the design automation of approximate arithmetic circuits are focused around single-output designs. Recently in [25], a heuristic method based on optimal set of input combinations (SICC) is proposed to approximate two-level logic synthesis. However, it is not easy to automate the multiple-output designs under the specified ER constraint; thus, we believe an intensive study is required to address this problem. The algorithm presented in [25] can only take up to 20 inputs due to memory limitations; this implies 16×16 multipliers cannot be designed using the proposed approach. Moreover, the comparison of approximate logic synthesis results with that of state-of-the-art manual approximate circuit designs is not studied in detail. The well-known SALSA and PLM compared their approximate synthesis results with that of exact counterparts, whereas recently proposed [25] approach used IWLS93 benchmark suit to show the performance gain. The above-mentioned challenges are addressed in this brief; first problem is catered by using modular design approach to built large-size multipliers and later, comparison is performed with the state-of-the-art approximate multipliers designed using manual approach.

3 Problem Formulation and Approach

Arithmetic designs can be represented in their corresponding truth table (TT) representation. Single-output designs refer to the terminology in which one TT represents the output of the design whereas multiple-output designs use more than one TTs for their complete output representation. Generally, K-map representation is used to identify the adjacent group of 1's present in the TT. Area of the encircled groups in the K-map can only be power of two (i.e., 1,2,4,8...). Next, we first present the problem statement and then basic strategy used to accomplish the solution.

3.1 Problem Statement

Consider an original TT, we want to find an approximate TTx with the least number of transistor count (TC) under a given error bound. The allowable error E is governed by $\text{ER}\times 2^n$, where ER is the error rate and n are the TT inputs. The outputs that can differ from the original TT should be no more than E. Note, this error bound does not depend on the number of outputs of the function; therefore, both single- and multiple-output designs can be governed using this error constraint. Note that, the transistor counts for CMOS logic gates used in this work are as follows: Inverter (2), N-Input NAND (2N), N-Input AND and OR (2N+2) [26].

3.2 Basic Approach

Bit flips in a TT can be introduced in two ways: 1-to-0 complement and 1-to-0 complement. The study presented in [27] dictates that changing a 1 to a 0 and from 0 to a 1 in the same TT introduces both positive and negative errors; therefore, errors can complement each other while achieving the reduced hardware complexity. Both cases are considered to see the impact of bit flips on the TT under the specified error. The widely used quality measure for the sum-of-product (SOP) expression is the number of literals [28]. All the variables (with or without negation) of an SOP are referred to as literals. Therefore, the complexity of TTx is established in terms of number of literals as well as transistor count.

3.2.1 1-to-0 Complement

Assume that an upper bound on the ER is 1/16. This implies that only one output against a certain input combination can be approximated in a TT. Consider a four-input TT (A,B,C,D) as shown in Table 2.1. Minterms corresponding to 1 are encircled in groups; three groups are formed and Boolean expression is shown in (2.1).

$$F = A'B + BD + ACD \tag{2.1}$$

Approximating the minterm group (marked in blue circle in Table 2.1) results in a new approximate TTx (Table 2.2). Dimensional reduction is achieved for the F_{apx} as only 3-input variables (A,B,D) are required for its representation (2.2). The normalized gate complexity gives the TC for (2.1) as 34. However, the TC for the

Table 2.1 1-to-0 complement (TT)

AB \ CD	00	01	11	10
00	0	0	0	0
01	1	1	1	1
11	0	1	1	0
10	0	0	1	0

Table 2.2 1-to-0 complement (TTx)

AB \ CD	00	01	11	10
00	0	0	0	0
01	1	1	1	1
11	0	1	1	0
10	0	0	0	0

F_{apx} is 20; a 41% reduction in hardware complexity is achieved with an error rate of 6.25%.

$$F_{apx} = A'B + BD \tag{2.2}$$

3.2.2 0-to-1 Complement

Consider ER is bounded by 1/16. This means that approximate TTx could have only one wrong output. Table 2.3 shows a four-input TT (A,B,C,D) and relevant Boolean expression is shown in (2.3).

$$F = A'B'C'D + BC'D + ABD + BCD' \tag{2.3}$$

The output marked with a red circle in Table 2.3 is different with respect to its adjacent neighboring outputs. Approximating that output to 1 (Table 2.4) results in 4 minterm groups, and corresponding Boolean expression is shown in (2.4). The complexity of F in terms of number of literals is 13; however, for the F_{apx} the sum

Table 2.3 0-to-1 complement (TT)

AB \ CD	00	01	11	10
00	0	0	1	0
01	0	1	0	1
11	0	1	1	1
10	0	0	0	0

Table 2.4 0-to-1 complement (TTx)

AB \ CD	00	01	11	10
00	0	0	1	0
01	0	1	1	1
11	0	1	1	1
10	0	0	0	0

of literals is 7, a 46% reduction in hardware complexity is achieved with an error rate of 6.25%.

$$F_{apx} = BD + BC + A'CD \tag{2.4}$$

Remarks 0-to-1/1-to-0 complement of certain outputs in an original TT results in a reduction of minterm groups corresponding to 1. The resultant approximate TTx requires less number of minterm groups for its representation as compared to the original TT. Consequently, the reduction in hardware complexity is achieved.

4 The Proposed Automated Framework

The approximate solution finder (ASF) realizes multiple approximate solutions for a given logic function. The truth table of each output bit of an exact multiplier does not depend on the multiplier structure (i.e., Wallace or Booth). Therefore, it is expected that introducing errors to the truth tables of an exact multiplier results in better speed or power compared to the exact multiplier regardless of its multiplication algorithm. In addition, the number of partial products of 8-bit ASF is equal to four, which is similar to Booth multiplier rather than Wallace. Therefore, the power and energy consumption of the proposed multipliers will be better compared to Booth multiplier while the former has a delay overhead. Table 2.5 lists the acronyms used in the description of ASF and Fig. 2.1 describes the flow chart of ASF. *I* denotes the number of input variables of the exact Truth Table (TT) and *E* is the largest error (calculated using ER$\times 2^n$ constraint) that can be introduced in the TT. All three inputs *(I, E, and TT)* are used by the symmetry finder (SF) and mine sweeper (MS) algorithms; these algorithms search for all possible approximate solutions (Lines 6-8 in Algorithm 1). The approximation is subject to the error constraint of ER$\times 2^n$. Assume that ER of 1/4 and n = 4; this implies we can only approximate four outputs of a TT. The approximate TT (TT_x) is then simplified prior to establish the hardware resources; for simplicity and beyond the scope of

Table 2.5 Description of acronyms used in proposed automated framework

Acronym	Description
I	Number of input variables of a truth table
E	Error bound in a truth table
TC	Transistor count
SF	Symmetry finder algorithm
MS	Mine sweeper algorithm
EO	Erroneous output
TT	Exact truth table
TTx	Approximate truth table
ASF	Approximate solution finder

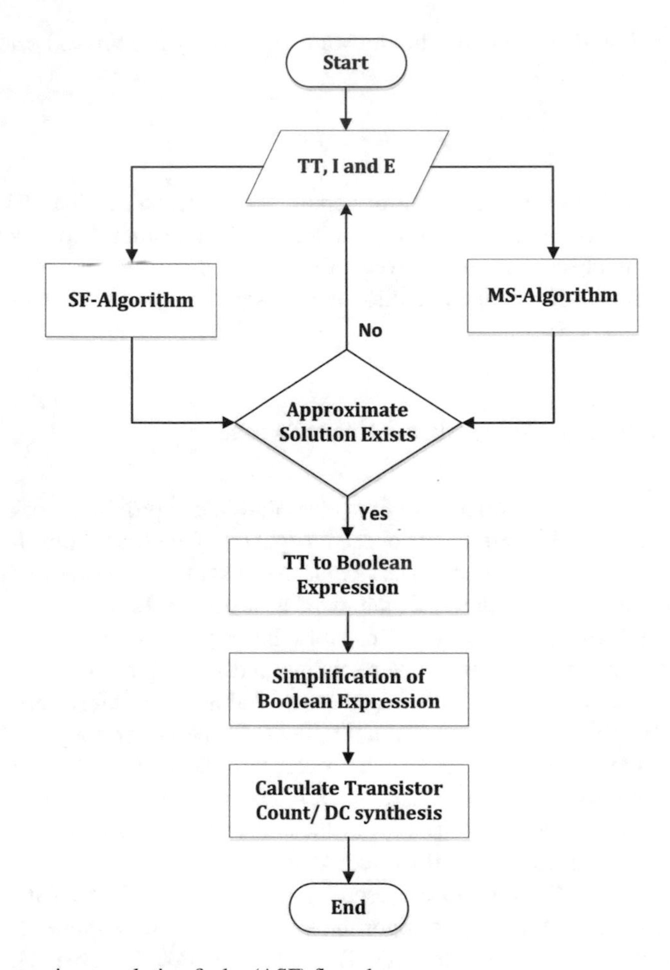

Fig. 2.1 Approximate solution finder (ASF) flow chart

this paper an off-the-shelf synthesis tool Espresso [29] is used to find the minimized Boolean expressions (Lines 10–11 in Algorithm 1). In the final step, the hardware complexity is calculated and all approximate solutions are ranked with respect to the least resource consumption (Lines 12–13 in Algorithm 1). The Pseudo-code of ASF is shown in Algorithm 1 while the SF and MS algorithms are discussed in detail (using an example) in the following subsections.

4.1 Symmetry Finder (SF) Algorithm

The SF finds the symmetric patterns of a TT. The algorithm is explained using an example. An input TT of length = 16 (Fig. 2.2) with four-input variables is

Algorithm 1 ASF pseudo-code

Require:
1: $TT(TruthTable) \in$ Representing particular design
2: $I \in$ No of input variables
3: $E \in$ Amount of error that can be introduced in TT
4: **procedure** ASF (I, E, TT)
5: $\quad SF_{tt} \leftarrow SF_{Algotihm}\ (I, E, TT)$
6: $\quad MS_{tt} \leftarrow MS_{Algotihm}\ (TT, E)$
7: $\quad Approx_{tt} \leftarrow Combine\ (SF_{tt}, MS_{tt})$
8: $\quad$ **if** $(Soution == Exists)$ **then**
9: $\quad\quad Bool_{Express} \leftarrow Convert\ (Approx_{tt})$
10: $\quad\quad Bool_{Smp} \leftarrow Simplify\ (Bool_{Express})$
11: $\quad\quad TC \leftarrow HW_{complexity}\ (Bool_{Smp})$
12: $\quad\quad Approx_{Sol} \leftarrow Rank\ (Bool_{Smp}, TC)$
13: $\quad$ **else**
14: $\quad\quad$ *Change E* {N}o Approx Solution
15: $\quad$ **end if**
16: **end procedure**

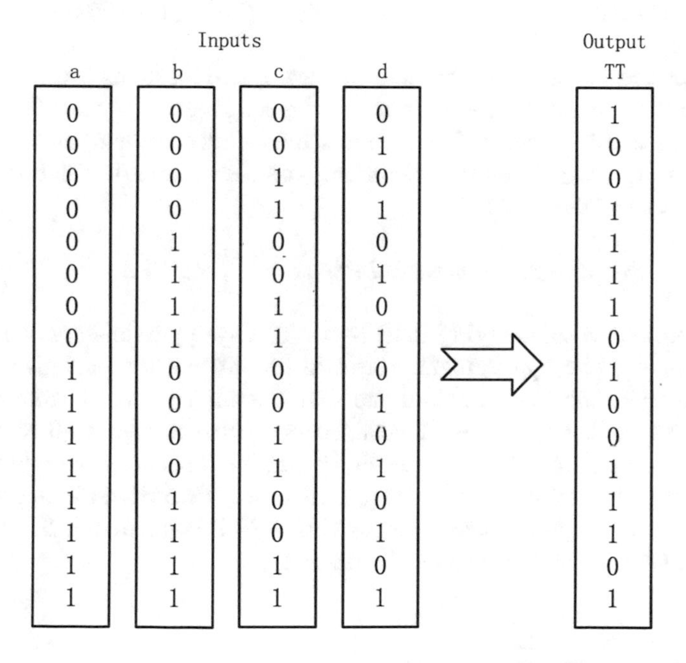

Inputs a	b	c	d	Output TT
0	0	0	0	1
0	0	0	1	0
0	0	1	0	0
0	0	1	1	1
0	1	0	0	1
0	1	0	1	1
0	1	1	0	1
0	1	1	1	0
1	0	0	0	1
1	0	0	1	0
1	0	1	0	0
1	0	1	1	1
1	1	0	0	1
1	1	0	1	1
1	1	1	0	0
1	1	1	1	1

$$y = acd + \bar{b}cd + \bar{c}\bar{d} + \bar{a}b\bar{d} + b\bar{c}$$

Fig. 2.2 Truth Table example for the Symmetry Finder algorithm (length = 16 with four-input variables)

considered and allowable error is kept to 3 (so an approximate TT can have maximum 3 wrong outputs).

Step-1: The original TT (Fig. 2.3a) is divided into two blocks with respect to each input variable (Lines 5-8 in Algorithm 2). As the considered example has four-input variables; therefore, four sub-blocks are generated (Fig. 2.3b–e). Figure 2.3b shows the results for the most significant variable *d*, the first half corresponds to elements with *d* variable having value 1'b0, while the elements in the second half refers to the *d* variable with value 1'b1. Therefore, these two halves have a one-to-one correspondence with respect to the specified input variable. Similarly, the first half in Fig. 2.3e corresponds to the least significant variable *a* with value 1'b0 whereas the second half consists of elements with *a* as 1'b1.

Step-2: Each sub-block is independently analyzed and symmetry is searched within the two halves (Lines 9 in Algorithm 2). For example, in Fig. 2.3b the last two locations have different elements. Similarly (Fig. 2.3c–e) have (4,6,6) differences, respectively. As we have the error bound of three; therefore, out of four sub-blocks only Fig. 2.3b is the best candidate (having two differences) for approximation.

Step-3: The selected sub-block has a difference of two locations between two halves; therefore, symmetry can be introduced among them in 4 possible ways (Lines 10–12 in Algorithm 2). Let d denote the differences between the two halves; then Eq. (2.5) shows the maximum possible symmetric solutions by taking into account the error E.

$$No\,of\ Approximate\ Solutions = 2^d,\ \forall\, d \leq E \qquad (2.5)$$

The first approximate solution (Fig. 2.3f) is obtained by approximating two bits of the second half, while approximating the two bits of the first half gives the second solution (Fig. 2.3g). For the third and fourth solutions, one bit from both halves is approximated. Figure 2.3h is obtained by approximating 1'b0 to 1'b1 whereas approximating 1'b1 to 1'b0 results in Fig. 2.3i. All resulting approximate TTs are symmetrical; therefore, complexity reduction of 30%, 54%, 61%, and 46% is achieved by the four approximate solutions (Fig. 2.3f–i) compared to the exact TT. The pseudo-Code of SF is shown in Algorithm 2.

4.2 Mine Sweeper (MS) Algorithm

Isolated bits either 1'b1 or 1'b0 in a TT increases the hardware complexity. To overcome this problem these bits need to be adjusted with respect to their neighbors. The MS algorithm considers the transition of bits from a binary value to inverse;

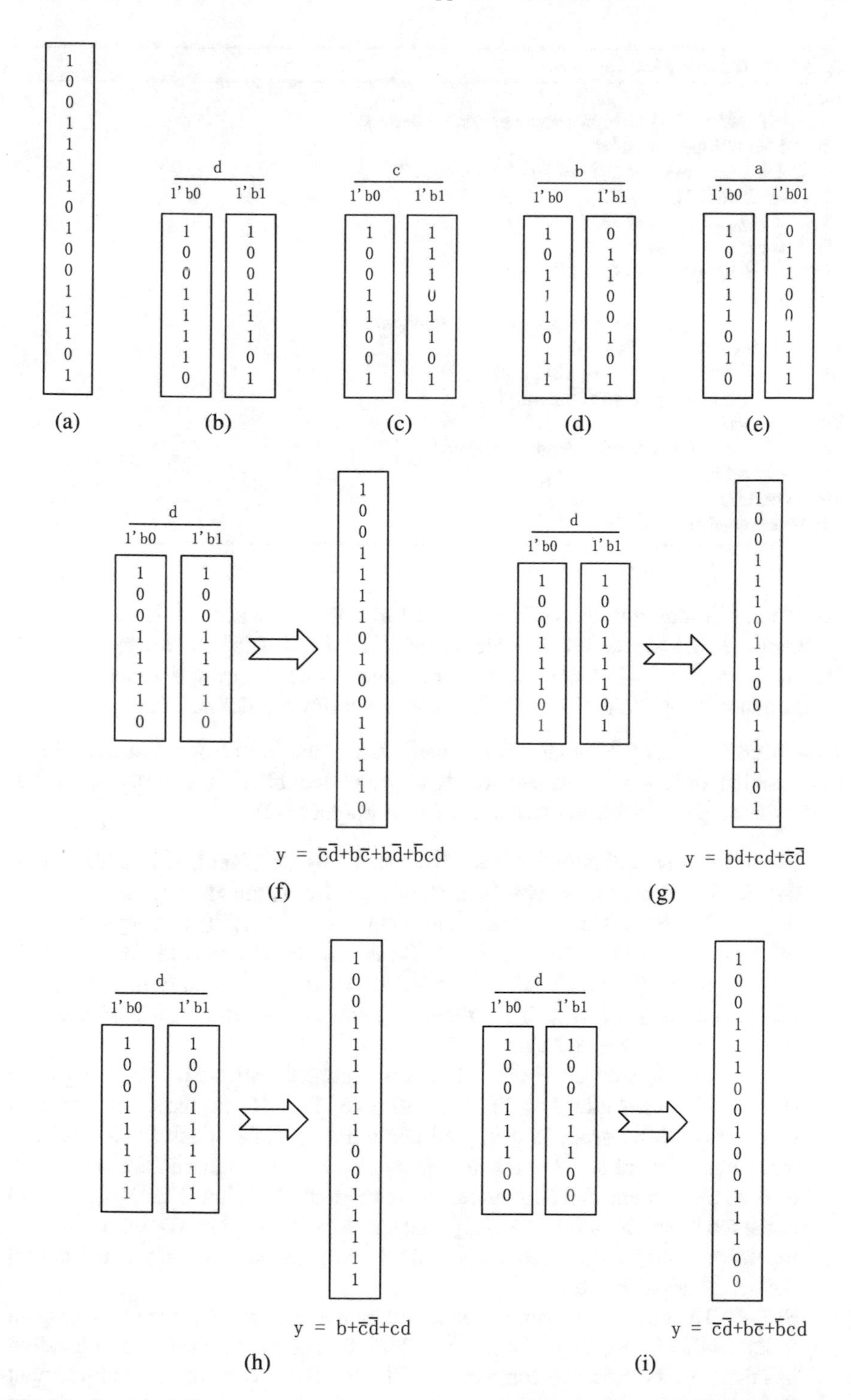

Fig. 2.3 Graphical illustration of the symmetry finder (SF) algorithm. (**a**) Original truth table. (**b–e**) Sub-blocks of original TT with respect to input variables d, b, c, and a. (**f–i**) Approximate solutions with complexity reduction of 30%, 54%, 61%, and 46% compared to the exact truth table

Algorithm 2 SF pseudo-code

Require:
1: $TT(TruthTable) \in$ Representing particular design
2: $I \in$ No of input variables
3: $E \in$ Amount of error that can be introduced in TT
4: **procedure** SF (I, E, TT)
5: $\quad TT_{Blocks} \leftarrow Find\ (I)$
6: $\quad$ **for** $i \leftarrow 1, I$ **do**
7: $\quad\quad Upperpart \leftarrow I_{1'b1}\ (TT_{Blocks})$
8: $\quad\quad Lowerpart \leftarrow I_{1'b0}\ (TT_{Blocks})$
9: $\quad\quad d \leftarrow Differences\ (Upperpart, Lowerpart)$
10: $\quad\quad$ **if** $(d <= E)$ **then**
11: $\quad\quad\quad Symmetric_{sol} \leftarrow Calculate_{all}$
12: $\quad\quad\quad SF_{TT} \leftarrow Map\ (Symmetric_{sol})$
13: $\quad\quad$ **else**
14: $\quad\quad\quad i = i + 1$ {I}f No Symmetry found
15: $\quad\quad$ **end if**
16: $\quad$ **end for**
17: **end procedure**

therefore, MS may not lead to large errors for most of the arithmetic circuits. The graphical illustration of the MS algorithm is depicted using an example. TT of length = 16 (Fig. 2.4) is taken as an input, and allowable error is limited to one. To introduce the MS algorithm, we first give the following definition.

Definition 1 For any N-input TT, if a particular output corresponding to an input combination differs with respect to its adjacent neighboring outputs, then that particular output is referred to as an erroneous output (EO).

Step-1: Erroneous outputs (EOs, 1'b1 and 1'b0) are identified in a TT (Lines 4–5 in Algorithm 3). Figure 2.5a shows the Karnaugh map (K-map) of the original function. The two solid-line rectangles (Fig. 2.5b) correspond to the SOP expression ($Y = ACD + ABC$). Based on Definition-1, the bit at the K-map locations (6 (Fig. 2.5c), 9 (Fig. 2.5d), 10 (Fig. 2.5e), and 15 (Fig. 2.5f)) differs with respect to their neighbors. Figure 2.5g shows all the prospective EOs present in the exact TT.

Step-2: In this step all identified EOs are graded with respect to neighbors (Lines 6-11 in Algorithm 3). For example, the EO at the sixth location (Fig. 2.5h) can be grouped using red circle and with the black (dotted) circle. Red circle is more suitable for its pairing as all three neighbors are with 1'b0. Similar phenomena can be seen for EO at the ninth location (Fig. 2.5i). For EO at the fourteenth location (Fig. 2.5j), red circle has two neighbors with 1'b0 and black circle has two neighbors with 1'b1. Figure 2.5k shows that this EO has all the neighbors with 1'b1.

Step-3: The graded EOs are sorted based on the largest difference with respect to its neighbors (Lines 12–14 in Algorithm 3). The EO bit at the tenth location has the largest difference compared to all other EOs. Therefore, bit adjustment

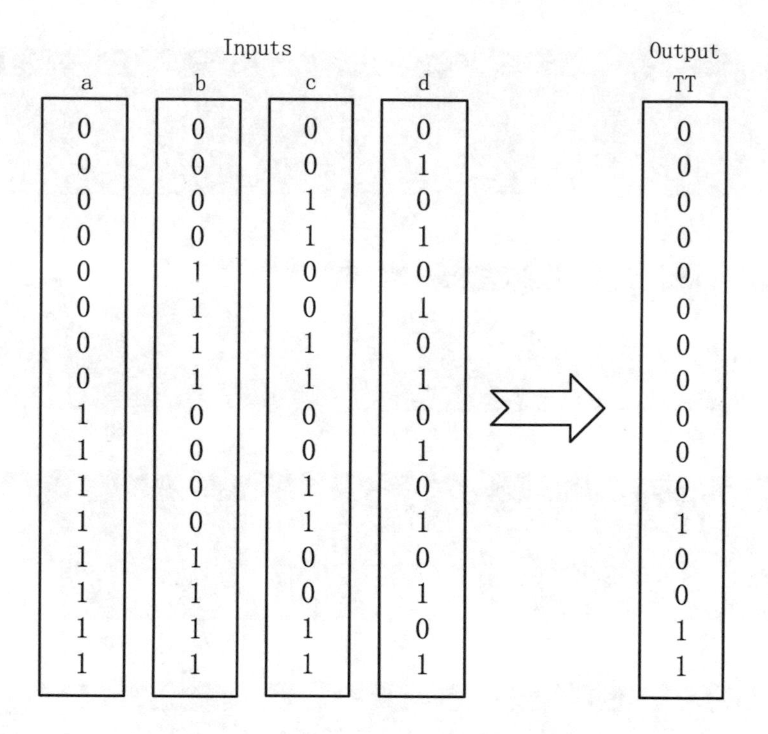

Fig. 2.4 Truth table example for the mine sweeper (MS) algorithm (length = 16 with four-input variables)

Algorithm 3 MS pseudo-code

Require:
1: $TT(TruthTable) \in$ Representing particular design
2: $E \in$ Amount of error that can be introduced in TT
3: **procedure** MS (TT, E)
4: $\quad Isolated_{1'b1} \leftarrow Find_{1'b1}\ (TT)$
5: $\quad Isolated_{1'b0} \leftarrow Find_{1'b0}\ (TT)$
6: $\quad L_{TT} \leftarrow No of\ Entries\ (TT)$
7: $\quad$ **for** $i \leftarrow 1, L_{TT}$ **do**
8: $\qquad Rank_{1'b1} \leftarrow Neighbor\ (Isolated_{1'b1})$
9: $\qquad Rank_{1'b0} \leftarrow Neighbor\ (Isolated_{1'b0})$
10: $\quad$ **end for**
11: $\quad Global_{Rank} \leftarrow Merge\ (Rank_{1'b1}, Rank_{1'b0})$
12: $\quad$ **while** $(E >= 0)$ **do**
13: $\qquad TT_{Modified} \leftarrow Error\ (TT, Global_{Rank})$
14: $\qquad MS_{tt} \leftarrow Map\ (TT_{Modified})$
15: $\qquad E = E - 1$
16: $\quad$ **end while**
17: **end procedure**

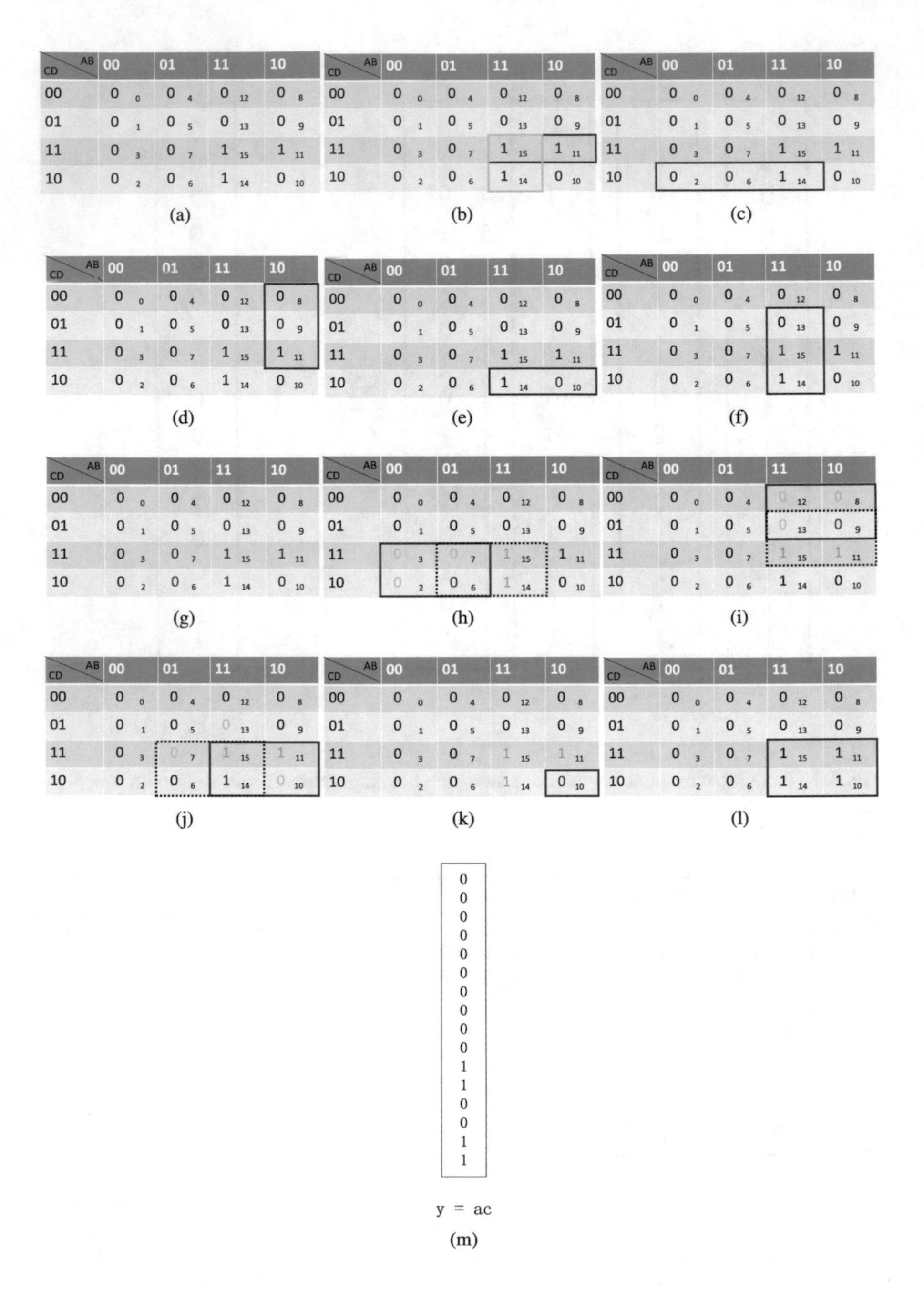

Fig. 2.5 Graphical illustration of the mine sweeper (MS) algorithm. (**a**) Original K-map. (**b**) SOP expression, Y = ACD+ ABC. (**c–g**) Erroneous outputs based on Definition-1. (**h–j**) Grading of erroneous outputs with respect to neighbors. (**k**) Erroneous output with the largest difference with respect to neighbors. (**l**) Complementing the output of the input minterm, AB'CD'. (**m**) Approximate truth table with complexity reduction of 60% compared to the exact truth table

is performed (from 1'b0 to 1'b1) for this EO (Fig. 2.5l), and this process is constrained by the value E. Figure 2.5m shows the approximate TT. The pseudo-Code of MS is shown in Algorithm 3.

5 Design Exploration of Approximate Arithmetic Units

In this section, initially the design exploration of approximate 2×2 and 4×4 multipliers is pursued using ASF tool. Later, large-size multipliers are designed using approximate multiplier blocks.

5.1 2×2 Multipliers

The result of a 2-bit multiplier can be represented using four TTs, where each TT represents the output of a product bit. These four TTs are passed to the ASF, Table 2.6 shows the output provided by the ASF. The outputs against input combinations $(0111)_2$ and $(1111)_2$ are approximated and corresponding reduced SOP expressions (for variable c1 and c2) are shown in (2.6) and (2.7), respectively. The number of literals in $c1_{approx}$ is 5 while $c2_{approx}$ has 2 literals. Therefore, achieving an improvement of 41% and 71% in terms of hardware complexity with an error rate of 12.5%.

Table 2.6 Approximate 2×2 multiplier

Inputs	Approx. truth table				Output (Dec.)	
abcd	c3	c2	c1	c0	Exact	Approx
0000	0✓	0✓	0✓	0✓	0	0
0001	0✓	0✓	0✓	0✓	0	0
0010	0✓	0✓	0✓	0✓	0	0
0011	0✓	0✓	0✓	0✓	0	0
0100	0✓	0✓	0✓	0✓	0	0
0101	0✓	0✓	0✓	1✓	1	1
0110	0✓	0✓	1✓	0✓	2	2
0111	0✓	0✓	0×	1✓	3	1
1000	0✓	0✓	0✓	0✓	0	0
1001	0✓	0✓	1✓	0✓	2	2
1010	0✓	1✓	0✓	0✓	2	2
1011	0✓	1✓	1✓	0✓	6	6
1100	0✓	0✓	0✓	0✓	0	0
1101	0✓	0✓	1✓	1✓	3	3
1110	0✓	1✓	1✓	0✓	6	6
1111	0×	1×	1×	1✓	9	7

$$\left.\begin{aligned} c1_{exact} &= bc\overline{d} + a\overline{c}d + a\overline{b}d + \overline{a}bc \\ c1_{approx.} &= ad + bc\overline{d} \end{aligned}\right\} \tag{2.6}$$

$$\left.\begin{aligned} c2_{exact} &= a\overline{b}cd + ab\overline{d} \\ c2_{approx.} &= ac \end{aligned}\right\} \tag{2.7}$$

5.2 4×4 Multipliers

In case of a 4×4 multiplier, the result of the product comes in 8-bits; therefore, 8 TTs are needed (each TT has a length of $2^8 = 256$) for the design exploration. These 8 TTs are provided as input to ASF. Table 2.7 shows the approximate TTs generated against the allowable error E. The allowable error for each bit position of the approximate 4x4 multiplier is the least value against which the proposed ASF provides the approximate solutions. Exhaustive simulations of the ASF are performed to establish the allowable error values for all truth tables. For example, in Table 2.7, for p1 and error value = 2, ASF finds seven different approximation solutions. However, by keeping this error value to one, ASF provides no approximate solution. The number of approximation solutions can be increased by relaxing the error bound.

TT (p0) gives no approximate solution because it is already the best solution. For the TTs (p1 and p2), the approximate TTs have the least error. Furthermore, for the TTs (p3, p4 and p5) a single approximate solution exists within a specified E. Similarly, for the TTs (p6 and p7) ASF finds a number of approximate solutions. The largest error is introduced for p4, $E = 18$, the corresponding error rate is only 18/256 = 7.03%. For a particular TT, against a defined E value several TTx are generated. The hardware resources for all the TTx are established, and the solution with the least hardware resource consumption is selected for further steps. The reduction achieved in hardware complexity is shown in Table 2.8. For (p1, p6, and p7) a

Table 2.7 Approximate solutions for an exact 4 × 4 multiplier with $E > 0$

Exact TT	Error values	No. of approximate TTx
p1	02	07
p2	02	05
p3	12	01
p4	18	01
p5	16	01
p6	12	11
p7	04	09

p1,p2...p7 are the truth table representation of an exact 4 × 4 multiplier

Table 2.8 Area improvement in a 4 × 4 multiplier

Area (μm^2)			
Truth table	Exact	Approx.	Improvement (%)
p0	0.93	0	0
p1	10.13	4.4	57
p2	8.95	6.26	30
p3	15.13	9.68	36
p4	20.03	13.42	33
p5	14.45	9.54	34
p6	22.38	7.83	65
p7	11.86	3.52	70

Exact p0 is already the most optimized SOP representation

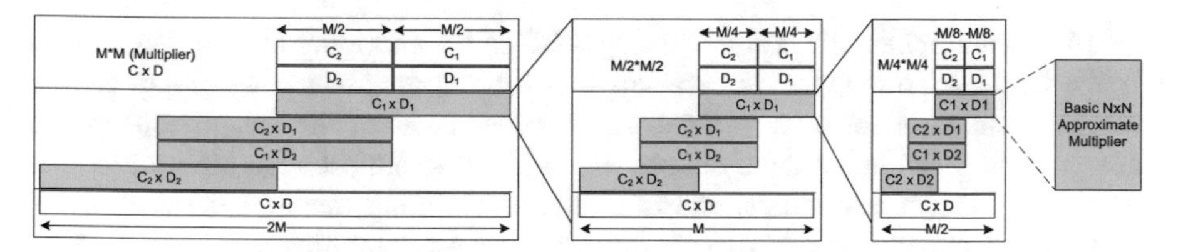

Fig. 2.6 An M×M multiplier designed using N×N approximate multiplier

reduction of over 50% is achieved in terms of circuit complexity while for (p2, p3, p4, and p5) this reduction is over 30%.

5.3 *Larger Multipliers*

Figure 2.6 shows the block diagram of building 16×16 multipliers using 2×2 multiplier as basic block. In this manuscript, the 4×4 multiplier is used as a basic unit to design 8×8 multipliers. Later, 16×16 multipliers are designed using 8×8 multipliers. For the 8×8 multiplier, the 8-bit operands A and B are divided into upper (a_H, b_H) and lower (a_L, b_L) 4-bits operands. The four partial products PP_0 to PP_3 are configured in different combinations to design 8×8 multipliers of varying accuracy as shown in Table 2.9.

The best approximated 4×4 multiplier is used in all ASF-M8-x multipliers. ASF-M8-1 is the least approximated multiplier using only one approximate 4×4 block, while ASF-M8-4 is designed using all approximate blocks of 4×4. The worst case error (WCE), if approximation is used in generation of all the partial products is shown in (2.8), where n = 4. Therefore, the ASF-M8-x designs have an accuracy-area trade-off, which can be utilized based on an application requirement.

$$WCE_{8\times 8} = 2^{2n} WCE_{PP3} + 2^n WCE_{PP2} + 2^n WCE_{PP1} + WCE_{PP0} \tag{2.8}$$

Table 2.9 8 × 8 multiplier variants using combinations of approximate 4 × 4 building block and 16 × 16 multiplier variants using combinations of approximate 8 × 8 multiplier

Size	Design name	PP0	PP1	PP2	PP3
8 × 8	ASF-M8-1	Approx	Exact	Exact	Exact
	ASF-M8-2	Approx	Approx	Exact	Exact
	ASF-M8-3	Approx	Approx	Approx	Exact
	ASF-M8-4	Approx	Approx	Approx	Approx
16 × 16	ASF-M16-1	ASF-M8-1	ASF-M8-1	ASF-M8-1	ASF-M8-1
	ASF-M16-2	ASF-M8-2	ASF-M8-2	ASF-M8-2	ASF-M8-2
	ASF-M16-3	ASF-M8-3	ASF-M8-3	ASF-M8-3	ASF-M8-3
	ASF-M16-4	ASF-M8-4	ASF-M8-4	ASF-M8-4	ASF-M8-4

16×16 multipliers are designed using ASF-M8-x multiplier blocks. The design space for 16-bit multipliers is very large as 8-bit approximate multipliers can be combined in different ways. Therefore, to cover most of the design aspects we are proposing following three classes of multipliers: almost accurate multipliers, hardware efficient multipliers, and multipliers exhibiting trade-off between hardware and accuracy. ASF-M16-1 is the most accurate multiplier designed using ASF-M8-1 multiplier while ASF-M16-2 and ASF-M16-3 multipliers provide a good hardware-accuracy trade-off using ASF-M8-2 and ASF-M8-3 multipliers, respectively. ASF-M16-4 is the resource efficient multiplier designed using ASF-M8-4 multipliers. Table 2.9 shows the 16×16 multipliers designed using 8×8 multipliers as basic building block.

6 Performance Evaluation

Compared multipliers include existing state-of-the-art approximate multipliers [30–35] and EvoApprox8b [36] (designed using multi-objective Cartesian genetic programming). Wallace multiplier is used as a baseline exact multiplier for comparison with the approximate multipliers. Note, all three considered EvoApprox (8-bit) designs have a large delay compared to exact multiplier [36]. However, as these designs have a comparable MRED with the proposed ASF multiplies; therefore, they are included for comparison. All designs are implemented in Verilog and synthesized at the 45nm technology node using Synopsys Design Compiler. The standard CMOS library is used with basic logic gates; optimization primitive is enabled (optimized for power). The supply voltage is 1.25V and the nominal room temperature is maintained. The power dissipation is found by passing uniformly distributed input combinations, and a back annotated switching file is generated. For the error analysis, the error rate (ER), mean relative error distance (MRED), and the normalized mean error distance (NMED) are considered [37]. ER is the percentage of generating an approximate result against all inputs and RED is defined as

$$\left.\begin{array}{l} ED = |O' - O| \\ RED = ED/O \end{array}\right\} \quad (2.9)$$

where ED is the error distance, and O and O' are the exact and approximate results. MATLAB is used to obtain both the MRED and NMED of approximate designs against all possible input combinations.

6.1 Error Analysis

The most accurate proposed design, i.e., ASF-M8-1 has a small MRED (Table 2.10) compared to other approximate ASF designs. ASF-M8-2/3 have a moderate MRED and ER while ASF-M8-4 has a higher MRED. The ASF approximated designs have a variable accuracy, therefore, can be used in wide-range of applications (requiring low error as well as tolerating high error). ASF-M8-x are designed using the recursive topology as used to build M8-x multipliers [31]. The better error metrics for ASF-M8-x multipliers are because the basic 4x4 multiplier has a lower ER. IWM has a least MRED among all other approximate designs as in the partial product reduction only one approximate 4:2 counter is used. The mul8u_14VP is from the pareto optimal subset of MRED vs power and has a comparable MRED with ASF-M8-1. While mul8u_PKY and mul8u_L40 are from the optimal subset of ER vs power and have a comparable MRED with ASF-M8-3 and ASF-M8-4, respectively. Note, ASF-M16-x follows the same trends as of ASF-M8-x (Table 2.10). Evoapprox (16bit) designs are from the pareto optimal subset of MRE vs power. LOAM (16-bit) has a better MRED than ASF-M16-1. As a matter of fact, ASF-M8-1 and ASF-M16-1 have the similar MRED metric value. Moreover, the ASF-M16-x designs still exhibit a lower MRED than M16-x designs [31]. 8/16 bit hybrid multipliers [34] have a moderate MRED while Approx_mul [35] exhibits a high MRED.

6.2 Hardware Analysis

Table 2.10 shows the results for the area, delay, power, and PDP of the proposed and other approximate multiplier designs. ASF-M8-4 exhibits least PDP among all other 8-bit approximate designs. Moreover, ASF-M8-4 is also the fastest design, it shows an improvement of 38% compared to the exact Wallace multiplier. The least number of approximate 4×4 blocks are used in both ASF-M8-1 and M8-5 [32]; however, ASF-M8-1 has a better PDP, i.e., an improvement of 15%. Even our moderate approximate designs, ASF-M8-2/3, have a relative small PDP values and are among the most energy-efficient designs. The hardware resources of 16-bit ASF multipliers are also shown in Table 2.10. The so-called mul6u_HGK is the fastest and power-efficient design when compared with all other approximate designs.

Table 2.10 Error and hardware complexity of the proposed and existing (8 × 8 and 16 × 16) approximate multipliers

Size	Designs	Area (μm^2)	Delay (ns)	Power (μW)	PDP (fJ)	ER (%)	NMED	MRED
8 × 8	Wallace$_{8x8}$	690.47	1.32	378.12	499.11	–	-	-
	ASF-M8-1	370.25	0.85	161.48	137.25	28.31	5.4×10^{-5}	0.0010
	ASF-M8-2	344.21	0.84	136.92	115.01	47.95	3.5×10^{-4}	0.0084
	ASF-M8-3	295.34	0.83	116.82	96.96	62.41	1.8×10^{-5}	0.0156
	ASF-M8-4	252.46	0.82	103.58	84.93	70.12	1.5×10^{-2}	0.0531
	UDM [30]	369.42	1.00	153.82	153.82	47.09	1.4×10^{-2}	0.0328
	IWM [31]	401.65	0.93	162.94	151.53	2.68	0.29×10^{-3}	0.0004
	M8-1 [32]	288.45	0.84	126.83	106.53	73.17	1.9×10^{-2}	0.0649
	M8-3 [32]	321.56	0.85	144.53	122.85	66.36	2.1×10^{-3}	0.0170
	M8-5 [32]	398.45	0.87	187.32	162.96	36.22	6.8×10^{-5}	0.0013
	LOAM [33]	256.61	0.93	112.84	104.94	75.91	2.0×10^{-3}	0.0181
	Hybrid [34]	180.93	0.83	143.58	119.17	78.00	4.0×10^{-3}	0.0225
	Approx_mul [35]	179.23	0.84	161.75	135.87	99.00	5.9×10^{-2}	0.0418
	mul8u_14VP [a]	632.51	1.38	341.78	471.65	39.26	7.2×10^{-5}	0.0014
	mul8u_PKY [a]	182.73	1.42	232.95	330.78	64.73	2.4×10^{-3}	0.0199
	mul8u_L40 [a]	415.85	1.48	175.28	259.41	74.91	2.7×10^{-2}	0.0746
16 × 16	Wallace$_{16x16}$	1364.80	2.70	721.34	1947.61	–	-	-
	ASF-M16-1	816.25	1.80	364.95	656.91	68.12	4.2×10^{-6}	0.0010
	ASF-M16-2	743.29	1.74	289.41	503.57	79.87	2.9×10^{-4}	0.0098
	ASF-M16-3	647.18	1.69	251.37	424.81	87.64	0.9×10^{-3}	0.0130
	ASF-M16-4	591.48	1.63	229.40	373.92	95.23	4.3×10^{-2}	0.0624
	UDM [30]	807.95	2.01	681.54	1369.8	80.99	1.4×10^{-2}	0.0333
	IWM [31]	915.32	1.87	364.85	682.26	5.45	0.29×10^{-3}	0.0006
	M16-1 [32]	603.81	1.65	279.31	460.86	96.71	5.7×10^{-2}	0.0644
	M16-3 [32]	680.97	1.66	316.73	525.77	94.74	1.2×10^{-3}	0.0168
	M16-5 [32]	831.57	1.82	385.91	702.35	72.49	5.1×10^{-6}	0.0013
	LOAM [33]	1008.2	1.70	423.68	720.25	96.48	0.1×10^{-3}	0.0008
	Hybrid [34]	607.23	1.43	546.23	781.10	99.00	3.3×10^{-3}	0.0096
	Approx_mul [35]	357.29	1.23	235.81	290.04	100.0	5.9×10^{-2}	0.1630
	mul16u_3BB[a]	1576.9	1.89	961.28	1816.8	99.99	5.6×10^{-6}	0.0017
	mul16u_6NY [a]	850.21	1.64	489.35	802.53	100.0	1.4×10^{-3}	0.0134
	mul16u_HGK [a]	362.78	1.01	154.83	156.37	100.0	3.9×10^{-2}	0.0915

[a] They represent the Evoapprox8b designs [36]

ASF-M16-x designs exhibit a similar trend as in ASF-M8-x when compared with the recent proposed modular designs [32]. The 16-bit hybrid multiplier [34] shows better delay compared to ASF designs; however, it has a higher power consumption. Approx_mul (16-bit) [35] has the least area and shows an improvement of 54% in delay compared to the exact 16-bit Wallace multiplier.

6.3 *MRED vs PDP Comparison*

Approximate designs are further compared in terms of MRED and PDP. Since all the designs have a close MRED values, therefore, for better clarity logarithmic scale is used to plot MRED. Note, use of logarithmic scale has shifted the designs with small MRED to right corner and high MRED based designs to the left corner. To make the analysis meaningful, the considered approximate multipliers are characterized into three groups: low, high, and moderate MRED based designs. ASF-M8-1, ASF-M8-3, and ASF-M8-4 exhibit least PDP among their respective MRED groups (Fig. 2.7a). They show an improvement of 16%, 18%, and 20% compared to previous best PDP-based designs (M8-5, Hybrid, and M8-1). Similarly for the 16bit approximate multipliers, ASF-M16-1 and ASF-M16-3 have a least PDP among the designs exhibiting lower and moderate MRED, respectively (Fig. 2.7b). An improvement up to 19% is achieved compared to previous best PDP-based designs (M16-5 and M16-3). Among the high MRED based 16-bit designs, ASF-M16-4 provides a good trade-off. Although it exhibits a high PDP, it has the least MRED compared to three other approximate designs (M16-1, Approx_mul, and mul6u_HGK) of the same group.

7 Case Study

The proposed approximate multipliers are evaluated using two image processing applications: image sharpening [37] and Canny edge detection [38]. Simulation is performed using MATLAB where exact multiplication is replaced by approximate multiplication. The results produced are used to analyze the performance of different approximate multipliers.

7.1 *Image Sharpening*

Structural similarity index (SSIM) and PSNR are used as a quality metric to quantify the output image. Consider a reference image a and a test image b, both of dimension X×Y then SSIM and PSNR are given by

$$SSIM(a, b) = l(a, b)c(a, b)s(a, b) \tag{2.10}$$

where $l(a, b)$, $c(a, b)$, $s(a, b)$ are the luminance, contrast, and structure comparisons, respectively.

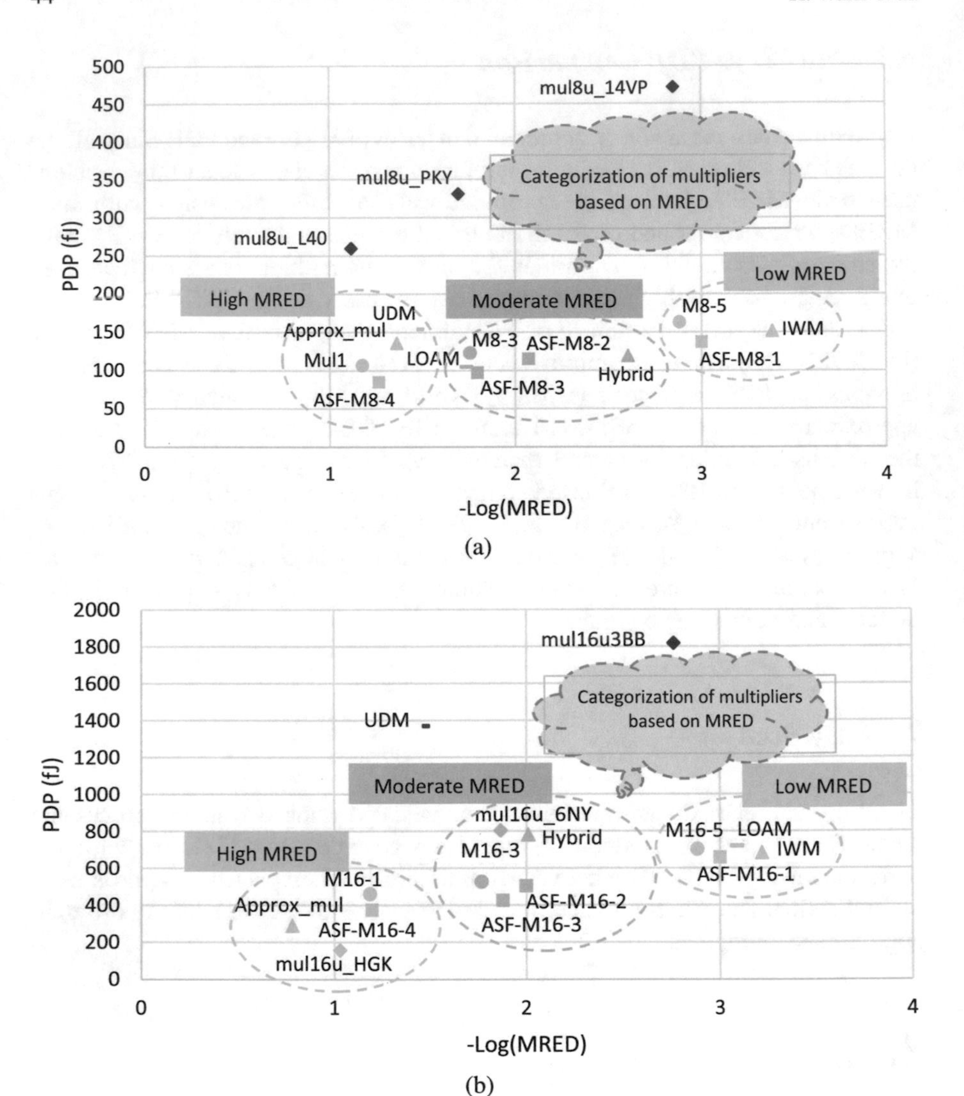

Fig. 2.7 A comparison of power-delay product and MRED for the approximate. (**a**) 8×8 multipliers. (**b**) 16×16 multipliers

$$MSE(a, b) = \frac{1}{XY} \sum_{n=1}^{X} \sum_{m=1}^{Y} (x_{n,m} - y_{n,m})^2 \tag{2.11}$$

$$PSNR(a, b) = 10log_{10}(Max^2/MSE(a, b)) \tag{2.12}$$

where Max represents the maximum value of each pixel. The image sharpening algorithm is given as

$$R(a,b) = 2I(a,b) - P(a,b) \tag{2.13}$$

where, I(a,b) is the input image and P(a,b) is the Gaussian smoothing

$$P(a,b) = 1/273 \sum_{n=-2}^{2} \sum_{m=-2}^{2} H(n+3, m+3) I(a-n)(b-m) \tag{2.14}$$

The Gaussian smoothing filter H is given by

$$H = \begin{bmatrix} 1 & 4 & 7 & 4 & 1 \\ 4 & 16 & 26 & 16 & 4 \\ 7 & 26 & 41 & 26 & 7 \\ 4 & 16 & 26 & 16 & 4 \\ 1 & 4 & 7 & 4 & 1 \end{bmatrix}$$

Blurred image (Fig. 2.8a) is given as an input to the algorithm, the addition and division involved in the sharpening process are kept accurate whereas exact multipliers are replaced by the approximate multipliers. The sharpened images produced by the ASF-16x designs are shown in Fig. 2.8b–e. Moreover, SSIM and the PSNR of the proposed approximate multipliers are given in Table 2.11. ASF-M16-3/4 have a higher error distance resulting in a less PSNR; however, ASF-M16-1/2 multipliers have a PSNR (close to 50 dB) that is acceptable for most of the applications.

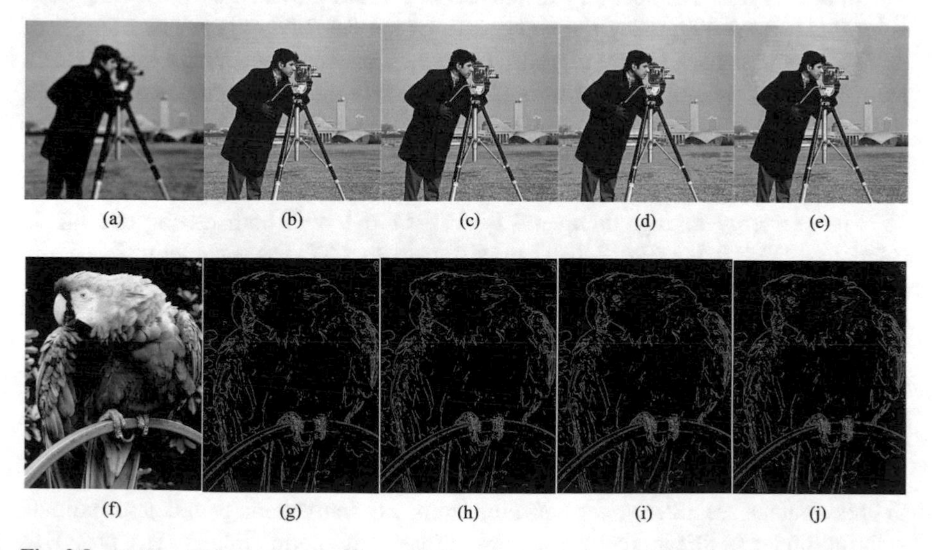

Fig. 2.8 (**a**) Blurred input image and result of image sharpening using (**b-e**) proposed ASF-M16-(1/2/3/4) multipliers. (**f**) Input image (16-bit) and output for Canny edge algorithm using (**g–j**) proposed ASF-M16-(1/2/3/4) multipliers

Table 2.11 Evaluation of ASF multipliers in image sharpening and Canny edge detection algorithms

	Image sharpening		Canny edge detection	
Approximate multipliers	SSIM	PSNR (dB)	Edges detected (%)	Energy (mJ)
Wallace$_{16x16}$	1	Inf.	100	3.21×10^{-3}
ASF-M16-1	0.97	53.06	96.23	2.53×10^{-3}
ASF-M16-2	0.92	48.21	93.48	1.95×10^{-3}
ASF M16-3	0.88	45.36	90.56	1.34×10^{-4}
ASF-M16-4	0.81	39.41	86.91	1.12×10^{-4}

7.2 Canny Edge Detection

The Canny algorithm is considered as a most effective edge detection method; it produces smooth and thin edges compared to other edge detection methods. It performs the following operations: Gaussian filter is applied to removing the noise; edge strength is calculated using the gradient of the image; local maxima are retained by applying a non-maximum suppression; potential edges are determined by thresholding; finally, all weak edges that are not adjacent to strong edges are suppressed. The size of the Gaussian kernel is inversely proportional to the detector's sensitivity to the noise; so smaller the size, the larger the detector's sensitivity to noise. Therefore, 5×5 Gaussian kernel with 16-bit fixed point arithmetic is utilized. Multiplication in the Gaussian filter is performed by an approximate 16×16 multiplier. A 16-bit (16 bits/pixel) gray-scale image is used as an input, as shown in Fig. 2.8f. The accuracy of the output images provided by the Canny edge algorithm is evaluated using the percentage of the detected edges. The number of multiplications in the Canny algorithm depends on the input image size. Except the multiplier all components in the algorithm deliver accurate results; therefore, the energy savings reported in Table 2.11 are due to the use of an approximate multiplier. ASF-M16-1 (Fig. 2.8g) offers a 21% energy reductions and detects 96.23% of the edges. ASF-M16-2 (Fig. 2.8h) detects 3% less edges, but it provides 18% more energy savings compared to ASF-M16-1 with both having acceptable PSNR. ASF-M16-3/4 (Fig. 2.8i, j) provides up to 65% improvements in energy reductions; therefore, it can be efficiently used in applications that have large error bounds and require reduced energy consumption.

8 Conclusion

In this chapter we have presented an automated framework called approximate solution finder (ASF) to synthesize approximate arithmetic circuits. The proposed methodology identified the erroneous outputs in a truth table and compensated them within allowed error bounds. Moreover, the approximate solutions can easily be incorporated with an FPGA/ASIC design flow. The recursive methodology was used to design large arithmetic designs. Four 8×8 and 16 × 16 multipliers have been proposed using different configurations of the basic 4×4 blocks. Image sharpening and Canny edge applications are used to evaluate the proposed multipliers. To facilitate further research and development in the area of approximate computing and for the sake of reproducibility of the results, we have made the libraries of both the proposed approximate multipliers open-source. The synthesizable Verilog files are provided as open-source libraries at https://sourceforge.net/projects/approxsolutionfinder/.

References

1. Huang P, Wang C, Liu W, Qiao F, Lombardi F. A hardware/software co-design methodology for adaptive approximate computing in clustering and ANN learning. IEEE Open J Comput Soc. 2021;2:38–52
2. Biasielli M, Bolchini C, Cassano L, Mazzeo A, Miele A. Approximation-based fault tolerance in image processing applications. IEEE Trans Emerg Top Comput. 2021. https://doi.ieeecomputersociety.org/10.1109/TETC.2021.3100623
3. Liu W, Lombardi F, Schulte M. A retrospective and prospective view of approximate computing. Proc IEEE (PIEEE). 2020;108(3):394–9
4. Xu Q, Mytkowicz T, Kim NS. Approximate computing: a survey. IEEE Des Test 2016;33(1):8–22
5. Yuan T, Liu W, Han J, Lombardi F. High performance CNN accelerators based on hardware and algorithm co-optimization. IEEE Trans Circ Syst I: Regul Pap. 2021;68(1):250–263
6. Venkataramani S, Chakradhar ST, Roy K, Raghunathan A. Approximate computing and the quest for computing efficiency. In: Proceedings of the 52nd annual design automation conference (DAC);2015. pp 1–6
7. Vahdat S, Kamal M, Afzali-Kusha A, Pedram M. TOSAM: an energy-efficient truncation- and rounding-based scalable approximate multiplier. IEEE Trans Very Large Scale Integr (VLSI) Syst. 2019;27(5):1161–73
8. Jiang H, Liu C, Lombardi F, Han J. Low-power approximate unsigned multipliers with configurable error recovery. IEEE Trans Circ Syst I: Regul Pap. 2019;66(1):189–202
9. Yin P, Wang C, Liu W, Waris H, Han Y, Lombardi F. Design and analysis of dynamic range approximate logarithmic multipliers (DR-ALMs) for machine learning applications. IEEE Trans Sustain Comput. 2020. https://doi.org/10.1109/TSUSC.2020.3004980
10. Leon V, Asimakopoulos K, Xydis S, Soudris D, Pekmestzi K. Cooperative arithmetic-aware approximation techniques for energy-efficient multipliers. In: Proceedings of the 56th ACM/IEEE design automation conference (DAC);2019. pp 1–6
11. Liu W, Qian L, Wang C, Jiang H, Han J, Lombardi F. Design of approximate radix-4 Booth multipliers for error-tolerant computing. IEEE Trans Comput. 2017;66(8):1435–1441
12. Ullah S, Schmidl H, Sahoo SS, Rehman S, Kumar A. Area-optimized accurate and approximate softcore signed multiplier architectures. IEEE Trans Comput. 2021;70(3):384–392
13. Shin D, Gupta SK. Approximate logic synthesis for error tolerant applications. In: Proceedings of the design, automation and test in Europe (DATE);2010. pp 957–960
14. Venkataramani S, Sabne A, Kozhikkottu V, Roy K, Raghunathan A. SALSA: systematic logic synthesis of approximate circuits. In: Proceedings of the 49th design automation conference (DAC);2012. pp 796–801
15. Lingamneni A, Enz C, Palem, Piguet C. Synthesizing parsimonious inexact circuits through probabilistic design techniques. ACM Trans Embedd Comput Syst (TECS) 2013;12(2):1–26

16. Samadi M, Lee J, Jamshidi DA, Hormati A, Mahlke S. SAGE: self-tuning approximation for graphics engines. In: Proceedings of the 46th IEEE/ACM International Symposium on Microarchitecture (MICRO);2013. pp 13–24
17. Miao J, Gerstlauer A, Orshansky M. Approximate logic synthesis under general error magnitude and frequency constraints. In: Proceedings of the international conference on computer-aided design (ICCAD);2013. pp 779–786
18. Ranjan A, Raha A, Venkataramani S, Roy K, Raghunathan A. ASLAN: synthesis of approximate sequential circuits. In: Proceedings of the design, automation and test in Europe (DATE);2014. p 364
19. Nepal K, Li Y, Bahar RI, Reda S. ABACUS: a technique for automated behavioral synthesis of approximate computing circuits. In: Proceedings of the design, automation and test in Europe (DATE);2014. pp 1–6
20. Nepal K, Hashemi S, Tann H, Bahar RI, Reda S. Automated high-level generation of low-power approximate computing circuits. IEEE Trans Emerg Top Comput 2019;7(1):18–30
21. Hashemi S, Reda S. Generalized matrix factorization techniques for approximate logic synthesis. In: Proceedings of the design, automation & test in Europe conference & exhibition (DATE);2019. pp 1289–92
22. Bernasconi A, Ciriani V, Villa T. Approximate logic synthesis by symmetrization. In: Proceedings of the design, automation & test in Europe conference & exhibition (DATE);2019. pp 1655–60
23. Venkataramani S, Kozhikkottu VJ, Sabne A, Roy K, Raghunathan A. Logic synthesis of approximate circuits. IEEE Trans Comput-Aid Des Integr Circ Syst. 2020;39(10):2503–15
24. Wu Y, Qian W. ALFANS: multilevel approximate logic synthesis framework by approximate node simplification. IEEE Trans Comput-Aid Des Integr Circ Syst. 2020;39(7):1470–83
25. Su S, Zou C, Kong W, Han J, Qian W. A novel heuristic search method for two-level approximate logic synthesis. IEEE Trans Comput-Aid Des Integr Circ Syst. 2020;39(3):654–69
26. Rabaey JM, Chandrakasan AP, Nikolic B. Digital integrated circuits, vol 2. Englewood Cliffs: Prentice Hall; 2002
27. Meng C, Qian W, Mishchenko A. ALSRAC: approximate logic synthesis by resubstitution with approximate care set. In: Proceedings of the 57th ACM/IEEE design automation conference (DAC); 2020. pp 1–6
28. Debnath D, Sasao T. A heuristic algorithm to design AND-OR-EXOR three-level networks. In: Proceedings of the Asia and South Pacific design automation conference (ASP-DAC);1998. pp 69–74
29. Brayton RK, Logic minimization algorithms for VLSI synthesis. Berlin/Heidelberg: Springer Science & Business Media; 1984
30. Kulkarni P, Gupta P, Ercegovac M. Trading accuracy for power with an underdesigned multiplier architecture. In: Proceedings of the international conference on VLSI design (VLSID);2011. pp 346–51
31. Lin C, Lin I, High accuracy approximate multiplier with error correction. In: Proceedings of the international conference on computer design (ICCD); 2013. pp 33–8
32. Ansari MS, Jiang H, Cockburn BF, Han J. Low-power approximate multipliers using encoded partial products and approximate compressors. IEEE J Emerg Select Top Circ Syst. 2018;8(3):404–416
33. Guo Y, Sun H, Kimura S, Design of power and area efficient lower-part-or approximate multiplier. In: Proceedings of the IEEE region 10 conference (TENCON); 2018. pp 2110–5
34. Strollo AGM, Napoli E, De Caro D, Petra N, Meo GD. Comparison and extension of approximate 4-2 compressors for low-power approximate multipliers. IEEE Trans Circ Syst I: Regul Pap. 2020;67(9):3021–34
35. Sabetzadeh F, Moaiyeri MH, Ahmadinejad M. A majority-based imprecise multiplier for ultra-efficient approximate image multiplication. IEEE Trans Circ Syst I: Regul Pap. 2019;66(11):4200–8

36. Mrazek V, Hrbacek R, Vasicek Z, Sekanina L. EvoApprox8b: library of approximate adders and multipliers for circuit design and benchmarking of approximation methods. In: Proceedings of design, automation & test in Europe conference & exhibition (DATE); 2017. pp 258–261
37. Waris H, Wang C, Liu W, Han J, Lombardi F. Hybrid partial product-based high-performance approximate recursive multipliers. IEEE Trans Emerg Top Comput. 2020. https://doi.org/10.1109/TETC.2020.3013977
38. Waris H, Wang C, Liu W, Lombardi F. AxBMs: high performance approximate radix-8 Booth multipliers for FPGA-based accelerators. IEEE Trans Circ Syst II: Brief Expr. 2021;68(5), 1566–1570

Chapter 3
Approximate Multiplier Design for Energy Efficiency: From Circuit to Algorithm

Ying Wu, Chuangtao Chen, Chenyi Wen, Weikang Qian, Xunzhao Yin, and Cheng Zhuo

1 Introduction

Thanks to the rapid growth of Artificial Intelligence (AI) and Internet-of-Things (IoT), energy efficiency has become a critical concern for IoT devices with constrained resources [1]. Among various efforts for energy efficiency optimization, approximate computing has emerged as a promising alternative for designers to trade computational accuracy with energy efficiency. This is especially applicable to human sensory or machine learning tasks where a small amount of inaccuracy is tolerable [2–5].

At the edge, IoT devices are designed to consume the minimum resource to achieve the desired accuracy. However, the conventional processors, such as CPU or GPU, can only conduct all the computations with pre-determined but sometimes unnecessary precisions, inevitably degrading their energy efficiency.

This work was partially supported by National Key R&D Program of China (Grant No. 2018YFE0126300) and National Natural Science Foundation of China (Grant No. 62034007 and 61974133).

Y. Wu · C. Wen · X. Yin · C. Zhuo (✉)
College of Information Science and Electronic Engineering, Zhejiang University, Hangzhou, China
e-mail: ying.wu@zju.edu.cn; wwency@zju.edu.cn; xzyin1@zju.edu.cn; czhuo@zju.edu.cn

C. Chen
College of Electrical Engineering, Zhejiang University, Hangzhou, China
e-mail: chtchen@zju.edu.cn

W. Qian
University of Michigan-Shanghai Jiao Tong University Joint Institute, Shanghai Jiao Tong University, Shanghai, China
e-mail: qianwk@sjtu.edu.cn

W. Liu, F. Lombardi (eds.), *Approximate Computing*,
https://doi.org/10.1007/978-3-030-98347-5_3

For example, when running data-intensive applications, e.g., streaming, neural network, and image processing, etc., multiplication is frequently invoked and consumes non-trivial energy. However, for a neural network, even with an inaccurate multiplier with limited precision, such inaccuracy may get cancelled out without impacting the inference accuracy [6]. In other words, when running inaccuracy-tolerable applications on the conventional processors, significant energy and time are actually spent on the multipliers computing highly accurate outputs that are not necessarily demanded. Thus, for the multiplication in IoT devices, there is a need to optimize its energy efficiency by providing sufficient instead of excessively accurate computational precisions.

As a common arithmetic component that has been studied for decades [7, 8], the past focus for the multiplier is mainly placed upon accuracy and performance. Recently, with awareness of the compromise between the stringent resource constraint and the accuracy tolerance for edge applications, there have been various research efforts for approximate multiplier design and optimization, ranging from algorithm, architecture, to circuit [9–23].

Since many prior designs happen to rely on hand-crafted structures or heuristics, it is then highly desired to systematically review and understand the pros and cons of how and where the inaccuracy can be introduced into the design. Thus, this chapter will review approximate multipliers from three different levels, i.e., architecture, algorithm, and circuit. The remainder of this chapter is organized as follows. In Sect. 2, we review the background of approximate multiplier. Sections 3–5 discuss the approximate multipliers at architecture, algorithm, and circuit levels, respectively, followed by the conclusions in Sect. 6.

2 Background

2.1 Fixed Point vs. Floating Point

Similar as many arithmetic functions implemented in hardware, the multiplier design can be categorized to fixed point and floating point implementations as a trade-off among accuracy, dynamic range, and cost. The major difference between fixed point and floating point numbers is whether the implementation has a specific number of digits reserved for the integer and fractional parts, respectively. In other words, fixed point numbers have a decimal point at a fixed position. Obviously, floating point may offer a wider dynamic range and higher precision than its fixed point counterpart, but at the cost of area, speed, and power consumption.

Figure 3.1 compares fixed point and floating point formats in the binary number system. The fixed point format consists of a sign bit, an integer part, and a fractional part, with a fixed binary point position. On the other hand, according to the IEEE 754 standard [24], which is a technical standard for floating point arithmetic, a floating point number consists of sign, exponent, and mantissa. The mantissa of a *normalized*

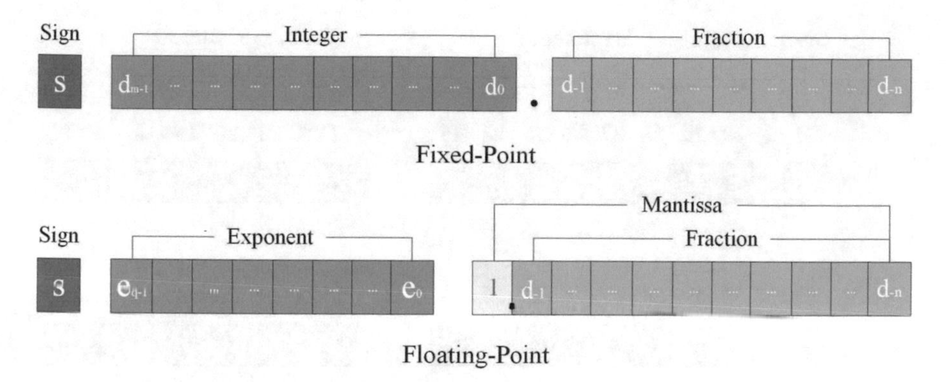

Fig. 3.1 Comparison between fixed point and floating point formats

floating point number is a fraction with its value between 1 and 2, where its first digit is fixed to 1 and the rest is the fraction in the range of [0,1).

Depending on the underlying number representations, designers may use either a fixed point multiplier or a floating point multiplier to conduct multiplication. For the floating point numbers, the multiplication procedure of 32-bit floating point numbers is demonstrated in Fig. 3.2. The sign bits are XORed together and the exponents are summed by an adder. Then, a bias of $2^{exponent\_width-1} - 1$ is subtracted from the sum to allow both negative and positive values for the exponent. Finally, the two mantissas are multiplied and shifted to the range of 1 and 2 to produce the normalized representation. The exponent will be adjusted if a shift happens. For a floating point multiplication, the mantissa part is much more energy- and delay-consuming than the other two parts, which is hence the focus of most research work [22, 23]. On the other hand, if no overflow, the fixed point multiplication is carried out as a regular multiplication with its fractional part truncated to the designed bit-width. However, the difference between the two multiplications is actually smaller than it seems. The multiplication of the mantissa parts for floating point numbers can be always viewed as a special case of fixed point multiplication, where the integer part is 1. Thus, the most critical operations in fixed and floating point multiplications can be considered as the same.

2.2 Binary Multiplier

Before we go into details of approximate multiplier design, we would like to introduce the basics of a binary multiplier, which is straightforward but sheds light on different approximation techniques introduced in the latter sections.

Due to the nature of dealing with only two digits, i.e., 0 and 1, binary multiplication actually can be considered as a process of addition and shifting. For example, assume we have two 4-bit operands of x and y, where x is a multiplicand

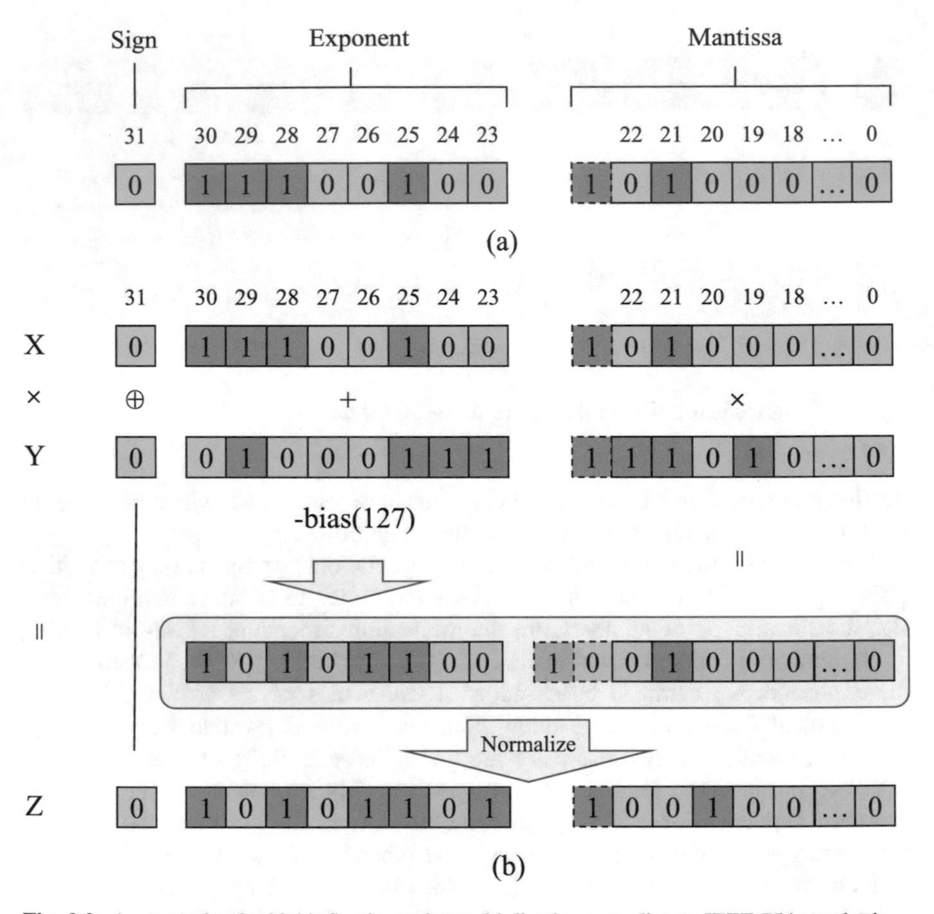

Fig. 3.2 An example of a 32-bit floating point multiplication according to IEEE 754 standard

and y is a multiplier. As shown in Fig. 3.3, similar as a decimal multiplication operation, the binary multiplication is carried out for each bit of the multiplier (i.e., y_i for $i = 0, 1, 2, 3$) and the multiplicand (i.e., x=$\{x_3, x_2, x_1, x_0\}$) to generate a partial product, e.g., $\{p_{3,0}, p_{2,0}, p_{1,0}, p_{0,0}\}$ for the first row. This process is then repeated for each bit of the multiplier y, with the partial product left-shifted by 1 bit. Finally, all the partial products are accumulated to obtain the multiplication result of $\{r_{7,0}, r_{6,0}, r_{5,0}, r_{4,0}, r_{3,0}, r_{2,0}, r_{1,0}, r_{0,0}\}$. Thus, the operation of a binary multiplier can be roughly divided into three stages, data input, partial product generation, and accumulation.

Since the binary product does not generate a carry, the bit-wise multiplication can be calculated with AND gates. Once all the partial products are generated, we can use an array of adders to accumulate partial products as shown in Fig. 3.4a, where HA refers to half adder and FA refers to full adder. Obviously, the critical path of such a structure is the carry propagation. For example, Fig. 3.4a demon-

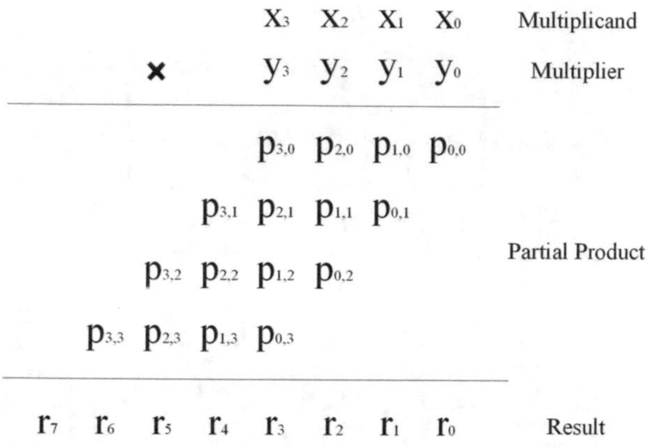

				x_3	x_2	x_1	x_0	Multiplicand
		×		y_3	y_2	y_1	y_0	Multiplier
				$p_{3,0}$	$p_{2,0}$	$p_{1,0}$	$p_{0,0}$	
			$p_{3,1}$	$p_{2,1}$	$p_{1,1}$	$p_{0,1}$		
		$p_{3,2}$	$p_{2,2}$	$p_{1,2}$	$p_{0,2}$			Partial Product
	$p_{3,3}$	$p_{2,3}$	$p_{1,3}$	$p_{0,3}$				
r_7	r_6	r_5	r_4	r_3	r_2	r_1	r_0	Result

Fig. 3.3 An example of 4-bit multiplication

strates a ripple-carry adder (RCA) based accumulator, with the carries propagated horizontally from right to left, while Fig. 3.4b plots a carry-save adder (CSA) based accumulator with carries propagated diagonally to achieve a shorter critical path for faster speed.

In order to further accelerate accumulation, C. S. Wallace proposed the Wallace Tree structure in 1964 [25]. As shown in Fig. 3.5, the Wallace Tree groups three partial products together column-wisely to generate two outputs, *i.e.*, a sum and a carry, thereby reducing the number of partial products by a factor of approximately 1.5. The operation is repeated until only two rows are left, i.e., 4 steps as in Fig. 3.5, which are then added up to obtain the final result. Parallel computation and partial product compression in each stage can be utilized to speed up the accumulation process [25].

2.3 *Approximate Multiplier*

Approximate arithmetic has been a popular research area in the past decade. Many prior work on approximate multiplier tackle the problem by introducing approximations at circuit, architecture, or algorithmic levels to reduce critical path delay or improve energy efficiency. For example, references [15, 26–32] propose to approximate K-map or prune out a few gates to simplify the gate netlist. Many work also focused on improving the conventional multiplier architecture with approximate components, such as adders, to speed up addition or partial product generation [15, 33–37]. Kulkarni et al. proposed to construct a new approximate multiplier architecture using a modified 2×2 multiply block [15]. From an even higher design level, Ahmed et al. proposed a pipelined log-based approximation using the classical Mitchell multiplier with an iterative procedure to improve the accuracy [38]. To speed up the iterative procedure, they proposed to truncate the

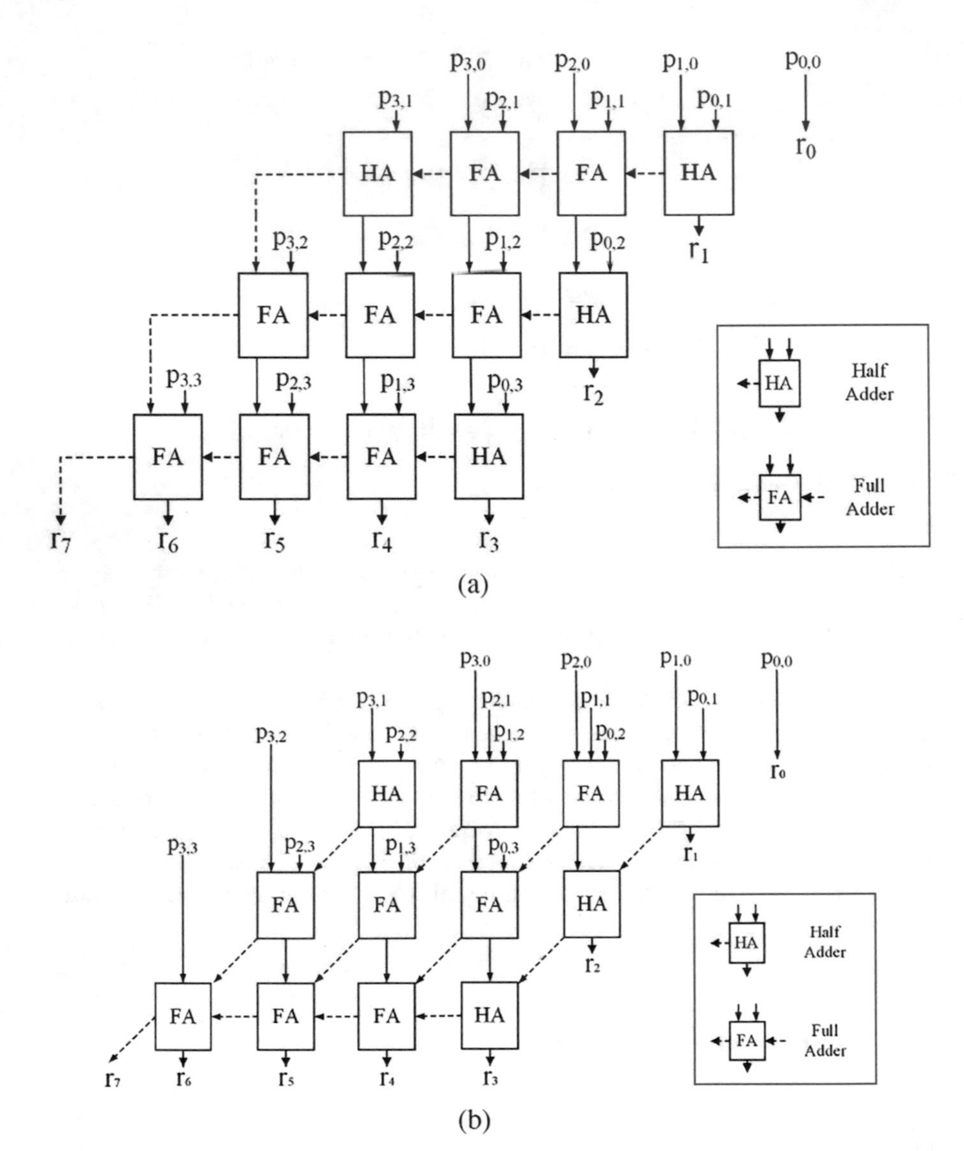

Fig. 3.4 (**a**) An example of RSA based accumulator; (**b**) an example of CSA based accumulator

bits after the leading one to save energy. To satisfy various accuracy requirement in different scenarios, another alternative is to utilize hybrid methods with both approximate and accurate multipliers to adjust the computational accuracy by selecting the appropriate multiplier, thereby trading off between accuracy and cost [39, 40].

For all the prior work with various approximation techniques, it is actually very challenging to precisely categorize the introduced approximation to a particular

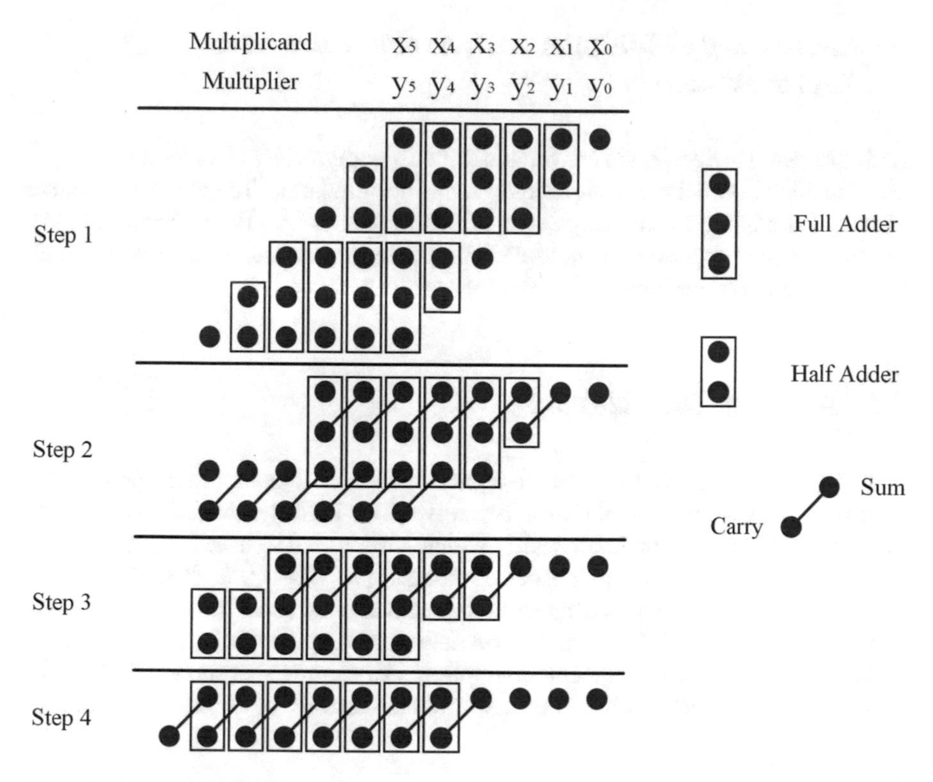

Fig. 3.5 An example of Wallace Tree based multiplier

design level, i.e., circuit, architecture, or algorithm. Many of them actually involve multiple design levels, as the high level approximation, e.g., algorithm, may always incur additional architecture changes [22, 23, 38–40].

In order to facilitate our review in the following sections, we would like to utilize the following rules for categorization:

- Architecture: With the binary multiplier architecture in Sect. 2.2 as a reference, the introduced approximation is intended to improve the efficiency of a particular stage in the reference architecture.
- Algorithm: The introduced approximation originates from a different algorithm to conduct multiplication.
- Circuit: The approximation technique is not limited to a particular multiplier architecture/algorithm and can be combined with the approximation techniques at other design levels.

3 Approximate Multiplier with Architecture Level Approximation

As is discussed in Sect. 2, the conventional multiplier typically involves three stages, i.e., data input, partial product generation, and accumulation. To reduce the number of partial products, an encoding stage can be included, e.g., Booth encoding [41]. When we design approximate multipliers based on such an architecture, approximations can be introduced into any of the four aforementioned stages.

3.1 Approximation at Input

It is simple yet effective to introduce approximation in data input for approximate multiplier design. For example, we can remove a few least significant bits (LSBs) of the input to reduce the input bit-width, which is supposed to have lower impact on the result than those most significant bits (MSBs) [34, 42–44]. In general, there are two types of data segmentation, dynamic segment method (DSM) and static segment method (SSM) [34]. DSM segments the data according to the leading one, while SSM is based on a given segmentation option. For example, as shown in Fig. 3.6a, DSM keeps k consecutive bits from the first non-zero bit of an unsigned number.

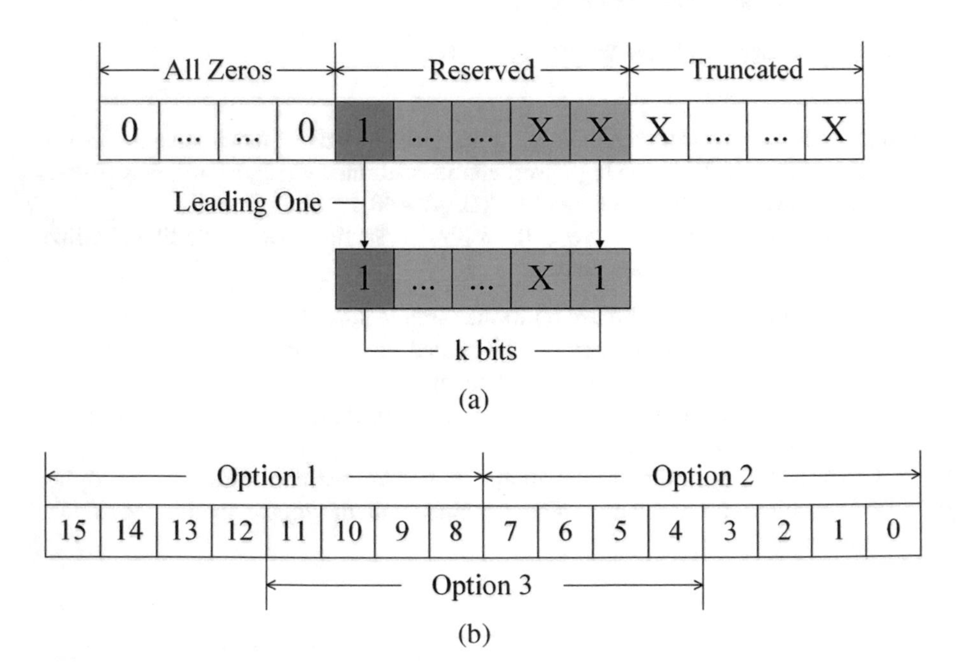

Fig. 3.6 (**a**) An example of DSM truncation; (**b**) an example of SSM truncation

Table 3.1 Comparison on 8-bit approximation multipliers with input approximations [43]

Multiplier	MRED	Power (μW)	Delay (ns)	Area (μm^2)	PDP (fJ)
Accurate	0	360	0.85	417	306
SSM [34]	N/A	68	N/A	75	N/A
DSM(3) [34]	0.1444	128	0.8	182	102.2
DSM(4) [34]	0.0680	205	1.08	233	221.46
DRUM(3) [42]	0.1260	104	0.7	143	72.73
DRUM(4) [42]	0.0640	172	1	208	172.16
LETAM(3) [44]	0.0290	270	1	310	270
TOSAM(1,5) [43]	0.0406	231	0.88	291	203.44

The parameter k determines the level of accuracy loss for approximate multiplier. On the other hand, SSM in Fig. 3.6b provides a few pre-determined (i.e., *static*) options when truncating the input data. The options can be like either leading k bits (option 1) or last k bits (option 2), as suggested in [34]. It is also possible to keep the bits in the middle (option 3) as a trade-off (Fig. 3.6b). Unlike DSM, SSM consumes less hardware resources but may include more redundant bits. In [42], the additional support for DSM requires 2 extra Leading One Detectors (LOD), 2 extra encoders and 1 extra barrel shifter.

Table 3.1 compares several approximate multipliers with approximation at input stage, the results of which are compiled from [34, 42–44]. The data is collected on 8-bit unsigned multiplication using 45-nm Nangate technology. Five approximate multipliers are included here and compared to an accurate multiplier [34, 42–44], where SSM [34] is the approximate multiplier using SSM to truncate the input data; DSM [34] and DRUM [42] both use DSM to truncate, while DRUM [42] always sets the last bit to 1 and DSM [34] leaves as it is; LETAM [44] and TOSAM [43] truncate both partial product and bit-width, while TOSAM uses 2 separate parameters for partial product and bit-width, respectively, and LETAM only uses 1 parameter for both. Parameter k is the number in the brackets for each multiplier. Five metrics are presented in the table, mean relative error distance (MRED), power, delay, area, and power-delay product (PDP) as the metric for energy efficiency. As shown in the table, SSM results in a smaller area and power while the other DSM based methods consume at least 1.5× larger area. Moreover, a larger bit-width or more complex control generally yields to a higher accuracy or smaller delay but at the cost of larger power and area consumption. Among all the DSM based multipliers, DRUM [42] provides a better trade-off between accuracy and energy efficiency.

3.2 *Approximation at Partial Product Generation*

Venkatachalam et al. propose an under-designed multiplier (UDM) architecture, which brings approximation into the partial product generation stage [15]. UDM

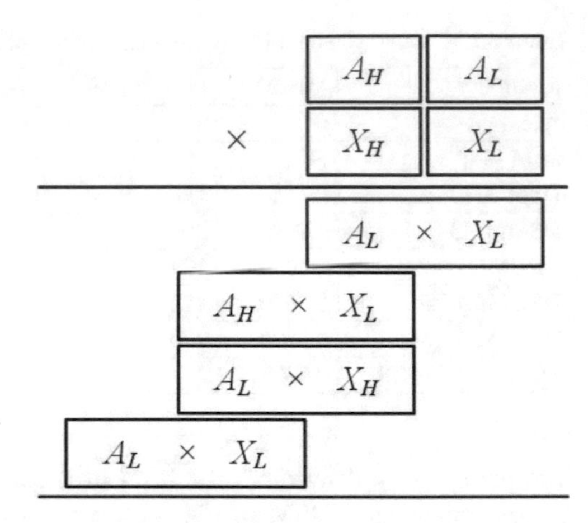

Fig. 3.7 An example of UDM in [15]

partitions both multiplier and multiplicand into 2 parts and then formulates a 2 × 2 multiplication. As shown in Fig. 3.7, each partial product can be produced with an approximate multiplier. Another alternative to partial product generation is to introduce an intermediate variable to replace the partial products (*a.k.a.* altered partial product (APP)) and then conduct approximations [45–47]. As discussed in Sect. 2, a partial product can be generated using AND gates:

$$pp_{m,n} = x_m \cdot y_n \ , \tag{3.1}$$

where x_m and y_n represent m^{th} and n^{th} bit of two inputs x and y, respectively. Similar as carry look-ahead adder, the propagate and generate signals can be defined as:

$$p_{m,n} = pp_{m,n} + pp_{n,m} \ , \tag{3.2}$$

$$g_{m,n} = pp_{m,n} \cdot pp_{n,m} \ . \tag{3.3}$$

Since the generate signals are possibly all 0's, they can then be compressed column-wisely using an OR gate. The propagate signals can be computed with approximate adders to achieve a more compact design than the original multiplier. Yang et al.. employ a similar idea of using two signals of approximate sum and error recovery vector to approximate the partial product [46]. Table 3.2 compares the impact of different partial product approximation methods with results compiled from [15, 45], where UDM refers to the method in Fig. 3.7 [15]; APP and APP_M refer to the approximate multiplier using altered partial products as in [45], while the most significant column is accurately computed without approximation in APP_M. The designs are compared on MRED, normalized mean error distance (NMED), power, delay, area, and PDP. It is expected that APP incurs larger error than APP_M with smaller area, delay, and power. UDM is more accurate than APP at the cost of more

Table 3.2 Comparison on 16-bit approximate multipliers using partial product approximation [45]

Multiplier	MRED	NMED	Power (μW)	Delay (ns)	Area (μm^2)	PDP (fJ)
Accurate	0	0	1776.49	0.68	4859.28	1208.01
UDM [15]	3.32e−2	1.39e−2	1318.51	0.67	3938	883.4
APP [45]	7.63e−2	1.78e−2	503.15	0.47	2158.56	236.48
APP_M [45]	2.44e−4	7.10e−6	1102.03	0.66	3319.2	727.34

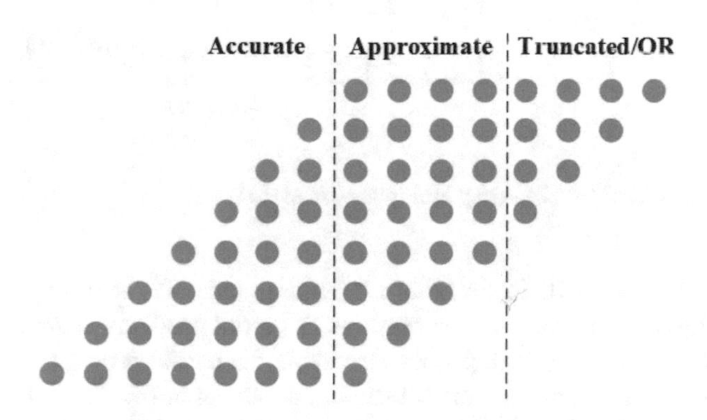

Fig. 3.8 An example of partial product array that is divided into three groups with different levels of approximation

than 2× power consumption, but inferior to APP_M almost across all the metrics. Thus, APP_M achieves a better trade-off between accuracy and energy efficiency.

3.3 *Approximation at Accumulation*

At accumulation stage, adders and compressors are the major computing modules. In addition to approximate adders, it is a natural idea to use approximate compressors to speed up accumulation [28, 36, 45–50]. For example, Venkatachalam et al. use approximate adders and approximate compressors to compress the partial product array to two rows, which are then added up through a ripple-carry adder [45]. Liu *et al.* further propose to ignore the carry signals in adders to reduce the critical path delay, which are then utilized later for error recovery [36].

Recently, many researchers also propose to separate the partial product array column-wisely to two or three groups, as shown in Fig. 3.8. As the leading bits may have larger impact on accuracy, each group can introduce different levels of approximation [28, 46, 47, 49, 50]. For example, OR gates can be deployed in the last group for LSBs to reduce the hardware cost. The accuracy of the approximate multiplier can be tuned by adjusting the control parameters [46]. References [28, 49] propose to use high-order approximate compressors with error recovery for the

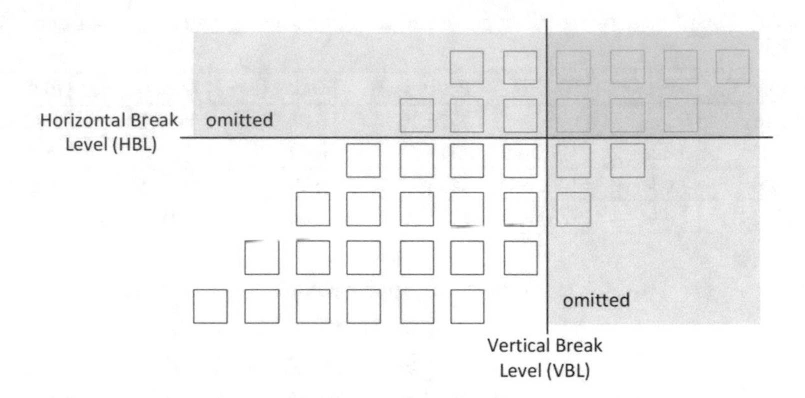

Fig. 3.9 An example of Broken-Array Multiplier (BAM) [51]

partial product groups with lower accuracy requirements. It is also straightforward to completely ignore the less important groups of partial products as more aggressive approximation. Mahdiani et al. propose to divide the partial product array into four groups through horizontal and vertical slicing, as shown in Fig. 3.9 [51]. The partial products on the right of Vertical Break Level (VBL) or above the Horizontal Break Level (HBL) are then ignored. In other words, only the partial products on the bottom left are used for calculation. Apparently, the approximation level can be adjusted by tuning VBL and HBL.

3.4 Approximate at Booth Encoding

Booth encoding is used to reduce the number of partial products, which can be generated in parallel at the cost of additional area. Radix-4 Booth algorithm is a common option deployed for high-bit-width multipliers [52]. Qian et al. propose an approximate Wallace-Booth multiplier with approximate modified Booth encoding (MBE), approximate 4-2 compressors, and approximate Wallace tree [29]. In addition to Radix-4 algorithm, Radix-8 Booth algorithm is also widely used to further reduce the number of partial products. However, Radix-8 algorithm demands odd multiples and hence needs additional adders. To reduce the increased partial product generation delay in Radix-8 algorithm, Jiang *et al.* suggest an approximate adder to generate the odd multiples for multiplication [53], which can reduce the delay of carry propagation as a trade-off between speed and accuracy.

4 Approximate Multiplier with Algorithm Level Approximation

Unlike the work in the last section that modify the reference multiplier architecture, some researchers propose to rebuild the multiplication operation from a higher level, i.e., algorithm, which naturally results in a new multiplier architecture. In this section we will review three different multiplier approximations at algorithm level: logarithm-based approximation, approximation with linearization, and hybrid approximation.

4.1 Logarithm-Based Approximation

With logarithmic transformation, the multiplication can be converted to addition, where the two operands are the logarithms of multiplicand and multiplier, respectively. The first logarithm-based multiplier (LM) was proposed by Mitchell et al. in 1962 [54]. For a multiplication of $A \times B$, we have:

$$A = 2^{k_1}(1 + x_1) , \tag{3.4}$$

$$\log_2(A) = k_1 + \log_2(1 + x_1) , \tag{3.5}$$

where A is the input operand, k_1 is the position of leading one, and x_1 is the fraction part that lies in $[0, 1)$. The same formulation can be applied to the other operand B with the parameters of k_2 and x_2. The logarithm of the multiplication can be written as:

$$\log_2(A \times B) = k_1 + k_2 + \log_2(1 + x_1) + \log_2(1 + x_2) . \tag{3.6}$$

According to Eq. (3.6), the implementation based on the Mitchell's algorithm [54] requires leading one detector (LOD), binary-logarithm converter (BLC), adder, and logarithm-binary converter (LBC). The procedure for the Mitchell's algorithm is demonstrated in Fig. 3.10 for a 16×16 multiplier. To reduce the implementation complexity, the logarithm computation in Eq. (3.6) can be approximated by:

$$\log_2(x + 1) \approx x, 0 \le x < 1 . \tag{3.7}$$

Then we have: $A \times B \approx 2^{k_1+k_2+x_1+x_2} = 2^{k_1+k_2} \times 2^{x_1+x_2}$. Based on the carry of $x_1 + x_2$, Eq. (3.7) can be further approximated as:

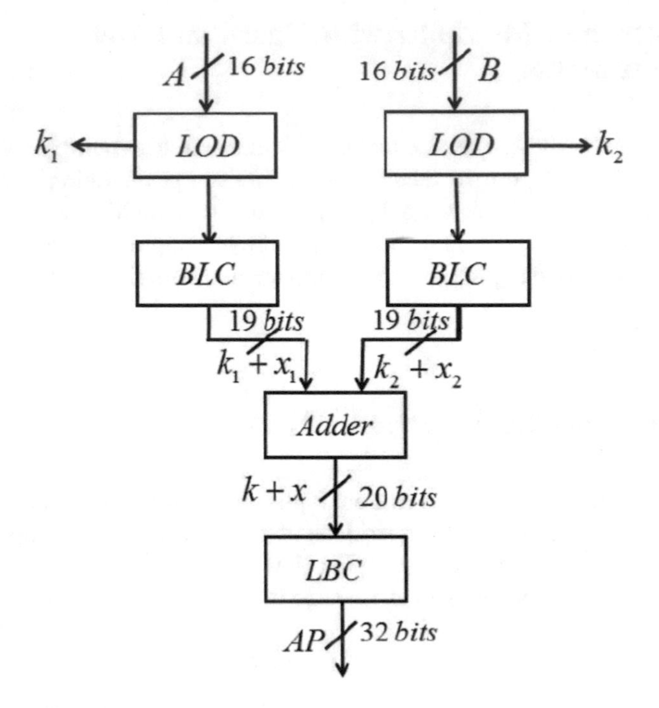

Fig. 3.10 Procedure for the Mitchell's algorithm [54]

$$A \times B \approx \begin{cases} 2^{k_1+k_2}(x_1 + x_2 + 1), & x_1 + x_2 < 1 \ , \\ 2^{k_1+k_2+1}(x_1 + x_2), & x_1 + x_2 \geq 1 \ . \end{cases} \tag{3.8}$$

Compared with the original multiplication, when $x_1 + x_2 < 1$, the error of Eq. (3.8) can be expressed as:

$$\begin{aligned} Error &= A \times B - 2^{k_1+k_2}(x_1 + x_2 + 1) \\ &= 2^{k_1+k_2}(1 + x_1)(1 + x_2) - 2^{k_1+k_2}(x_1 + x_2 + 1) \\ &= 2^{k_1+k_2}x_1x_2. \end{aligned} \tag{3.9}$$

It is noted that the error term of $2^{k_1+k_2}x_1x_2$ has the same structure as $A \times B = 2^{k_1+k_2}(1+x_1)(1+x_2)$. Then we can repeat the approximation procedure to compute $2^{k_1+k_2}x_1x_2$, which indicates an iterative process to achieve higher accuracy using logarithm-based approximation. In [38], the iterative approximation for $x_1 + x_2 \geq 1$ has been explored together with a truncation scheme. Liu et al. further investigate the logarithmic based approximate multipliers using different approximate adders and find that set-one-adder (SOA) can achieve a higher accuracy [55]. As shown in Fig. 3.11, an SOA consists of one approximate adder for the lower m bits and one exact adder for the higher $n - m$ bits. The approximate adder always sets the

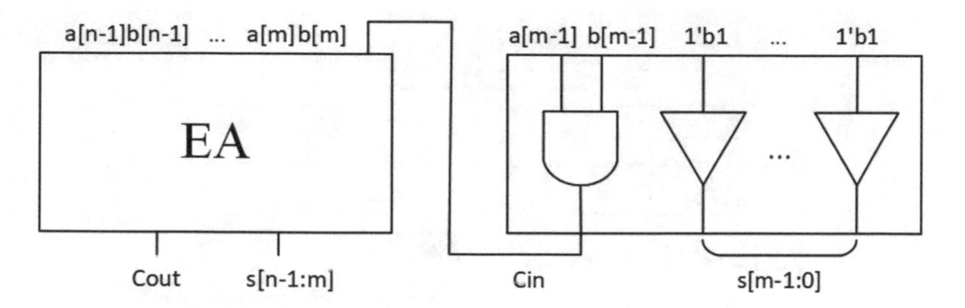

Fig. 3.11 Architecture of an n-bit set-one-adder [55]

Table 3.3 Comparison on 8-bit logarithm-based approximate multipliers [58]

Multipliers	MRED	NMED	Power (μW)	Delay (ns)	Area (μm^2)	PDP (fJ)
Accurate	0	0	99.3	1.06	235.9	105.2
Mitchell [54]	0.0368	0.0014	66.26	1.42	281.2	94.09
ALM-SOA-5 [55]	0.0396	0.0007	61.04	1.39	255.4	84.84
ILM-5 [58]	0.0951	0.001	50.37	1.64	255.3	82.61

lower m bits to logic 1 and hence results in over-estimation. Such an over-estimation is particularly designed to compensate for the accuracy loss of a logarithmic based approximate multiplier, as the Mitchell's algorithm always underestimates the multiplication result. Similar compensation schemes have been introduced in [56–58] to improve the average error introduced by the Mitchell's algorithm at the cost of area and power consumption.

Table 3.3 evaluates 8 × 8 logarithm-based approximate multipliers using ST Micro's 28 nm technology [58]. The results are compiled from [58] to compare the metrics of MRED, NMED, power, delay, area, and PDP. Three logarithm-based approximate multipliers are compared with an accurate Wallace tree based multiplier, where Mitchell [54] refers to the original Mitchell's algorithm; ALM-SOA-5 combines Mitchell's algorithm with SOA with $m = 5$ as in [55]; ILM-5 includes additional rounding of the inputs to the nearest power of 2 on top of the approximation techniques used in ALM-SOA-5. As shown in the table, all the logarithmic approximate multipliers can achieve good accuracy with 33.3–49.3% power reduction consumption, while the delay is increased by 31.1–54.7%. Among the three approximate multipliers, ALM-SOA-5 can achieve a better trade-off between accuracy and energy efficiency.

4.2 Approximation with Linearization

Multiplication is a nonlinear operation implemented with a few additions and compressions. In mathematics, it is a natural idea to approximate a nonlinear curve

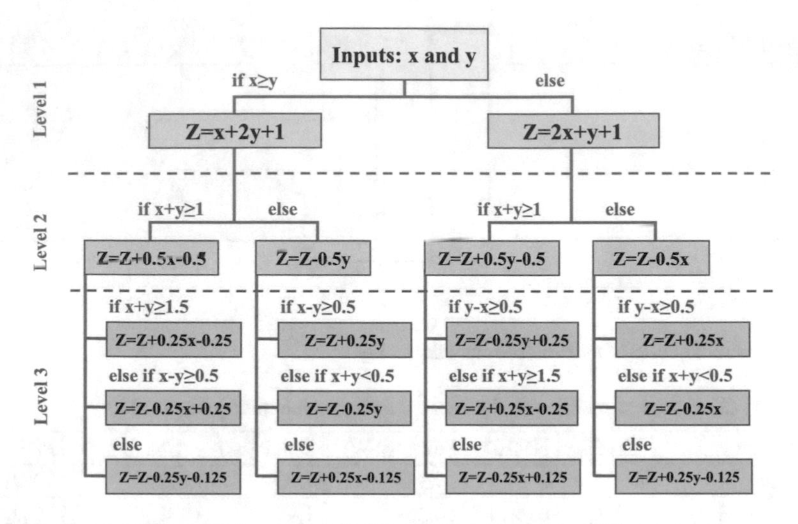

Fig. 3.12 Computation procedure using the linear approximation in [22]

with a piece-wise linear function. Thus, researchers have attempted to use linear arithmetic operations to approximate the nonlinear multiplication [22, 23]. It is noted that, while logarithm-based approximate designs are built on top of Eq. (3.8), it is actually a special case of linearization approximation.

Without loss of generality, the multiplication can be considered as a function of two variables, whose linear approximation can be always expressed as:

$$f = xy \approx f_{approx} = ax + by + c, \tag{3.10}$$

where x and y are the input operands; a, b, and c are the coefficients. In [22], an iterative linear approximation for floating point multiplication is proposed to approximate the multiplication according to Eq. (3.10). For the mantissas of normalized floating point numbers, the range is $[1, 2) \times [1, 2)$, which is a square domain. By appropriately partitioning the domain into smaller sub-domains and assigning a proper linear function to each, the original nonlinear surface for the multiplication can be approximated by a series of piece-wise linear functions, one for each sub-domain. Figure 3.12 summarizes the computation procedure called ApproxLP using the linear approximation in [22]. It is clear that the accuracy can be improved by partitioning more sub-domains, the number of which grows exponentially with the approximate level. Thus, the efficiency of ApproxLP in [22] actually quickly degrades with a larger approximation levels. Moreover, the comparators used for each level in Fig. 3.12 also introduce non-trivial delay overhead. Figure 3.13 plots the error distributions of ApproxLP for different approximation levels, which are symmetric over 0 and hence result in a zero average error.

To reduce the number of comparators, Chen et al. propose to partition the input domain into identical smaller square sub-domains [23]. For one level higher, each

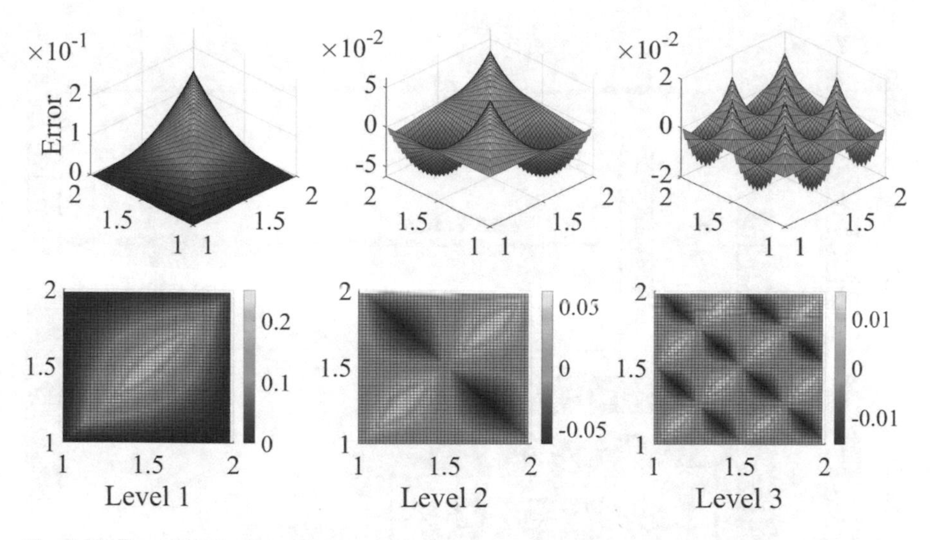

Fig. 3.13 Error distributions of ApproxLP at different approximation levels [22]

domain (or sub-domains) is further partitioned into four identical smaller ones. With such an iterative process, there are 4^n sub-domains for level n approximation. For a rectangular domain $[x_1, x_2] \times [y_1, y_2]$, the optimal coefficients to minimize the mean-square-error (MSE) between $f_{approx} = ax + by + c$ and $f = xy$ are [23]:

$$\begin{cases} a = \dfrac{y_1 + y_2}{2} \\ b = \dfrac{x_1 + x_2}{2} \\ c = -ab. \end{cases} \tag{3.11}$$

Figure 3.14 demonstrates the multi-level approximate multiplier architecture of OAM in [23]. In the figure, Level 0 is denoted as the basic approximation module, which provides an initial estimation f^0_{approx}, while the deeper levels act as error compensation to gradually improve the overall accuracy. Thus, the run-time configurability can be easily realized by specifying the desired depth. Unlike ApproxLP [22], the comparators are no longer needed for OAM [23]. Thus, the delay of OAM can be significantly improved when compared to ApproxLP even for a similar number of sub-domains.

Since the two coefficients of a and b are the middle points of the intervals where the operands belongs to, a circuit-friendly implementation can be achieved for the error compensation at each level as in Eq. (3.12) [23]:

$$\Delta f_n = \Big\{ \Big[\big(x[n]?(1) : (-1)\big) \times (y - \hat{y_{n-1}}) \Big]$$

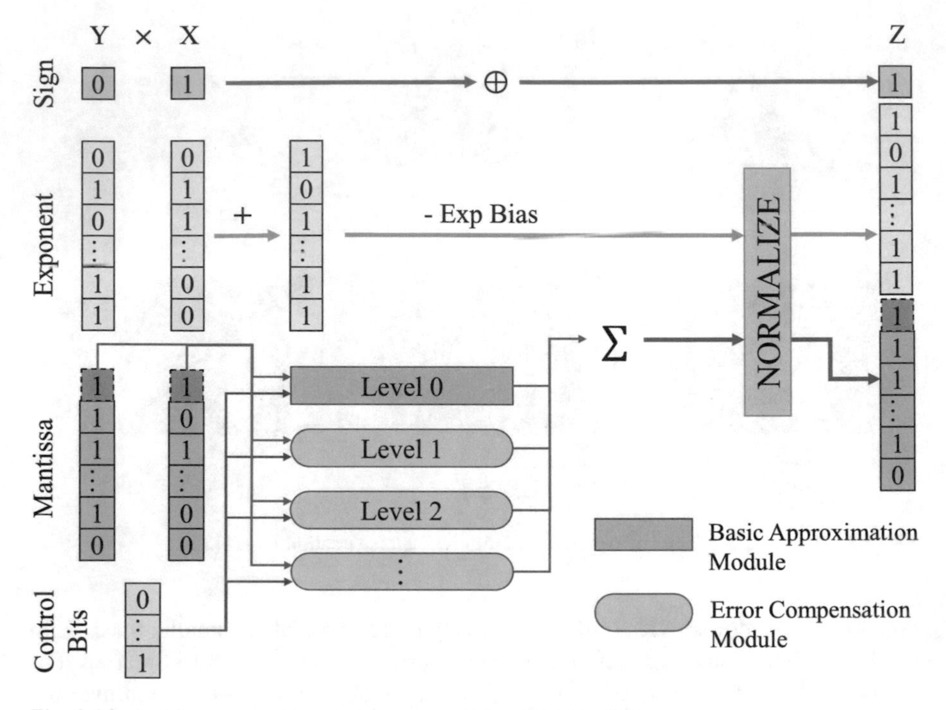

Fig. 3.14 Architecture of the approximate multiplier OAM in [23]

$$
\begin{aligned}
&+\Big[\big(y[n]?(1):(-1)\big)\times(x-x_{\hat{n-1}})\Big]\Big\}\gg(n+1) \qquad (3.12)\\
&+\Big[\big(x[n]\oplus y[n]\big)?(1):(-1)\Big]\gg(2n+2),
\end{aligned}
$$

where "? :" is the conditional operator and $x[n]$ is the nth bit of mantissa x; $\oplus$ is XOR operation; $x_{\hat{n-1}}$ is the $n-1$ bits truncation of mantissa x with an extra bit 1 at n^{th} position; $\gg$ represents right shift operation. Since the amount of right shift is pre-determined at each level, the right shift operation does not require additional circuits to implement. Thus, the number of operations at each level for OAM is reduced to 5, which results in a constant area complexity, while ApproxLP has an area complexity of $O(4^n)$ [23].

The errors of the approximate multiplier OAM [23] were reported as below for Maximum-Absolute-Error (MAE), Mean-Square-Error (MSE), and Mean-Absolute-Error (MeanAE) for approximation level n:

Table 3.4 Comparison on approximate multipliers using linearization-based approximation

Multiplier	MSE	Delay (ns)	Area (μm^2)	ADP (ns·μm^2)
Accurate	0	5.4	8219	44382.6
ApproxLP (0)	N/A	2.0	1446	2892
ApproxLP (1)	6.9e−04	2.7	2126	5740.2
ApproxLP (2)	4.3e−05	3.1	2890	8959
OAM (0)	N/A	1.9	1418	2694.2
OAM (1)	4.3e−04	2.3	2082	4788.6
OAM (2)	2.7e−05	2.7	2583	6974.1

$$\begin{cases} MAE = \dfrac{1}{4^{n+1}} \\ MSE = \dfrac{1}{9 \times 16^{n+1}} \\ MeanAE = \dfrac{1}{4^{n+2}}. \end{cases} \tag{3.13}$$

Similar as ApproxLP [22], OAM [23] has zero-mean error distribution, which is an appealing feature for applications with consecutive multiply-accumulate operations.

Table 3.4 compares two approximate multipliers using linearization-based approximations, where the accurate reference multiplier is a 32-bit floating point multiplier IP from the UMC 40 nm library, and the approximate multipliers are configured to different approximation levels as indicated by the numbers in the brackets. The results are compared on MSE, delay, area, and area-delay-product (ADP) as energy efficiency metric. It is found that, for the same approximation level, OAM [23] always performs better than ApproxLP [22]. When compared to the accurate multiplier IP, OAM [23] can achieve 68.6% area saving and 50% delay improvement at the cost of 2.7×10^{-5} MSE, with more than one order of magnitude energy efficiency improvement.

4.3 Hybrid Approximation

There are a few approximate designs that combine the multipliers with different precisions together to adapt to the varying accuracy requirements [39, 40], which are called hybrid approximation in this chapter. For example, reference [39] propose to combine accurate and approximate multipliers together to adjust the computational accuracy by selecting the appropriate multiplier. For the approximate multiplier, after detecting the number of consecutive 1's or 0's of the mantissa, the mantissa can then be rounded to 1 or 2, both of which make the multiplication as a shift operation. If a higher precision is required, the accurate multiplier is then invoked to conduct the calculation. Reference [40] uses the sum of two mantissas to approximate the multiplication. A tuning strategy is proposed to decide the working mode of the multiplier by detecting the number of the consecutive bits of the inputs. However, such methods heavily rely on an accurate or high-precision multiplier, which significantly increases the circuit area. Furthermore, it is difficult to predict whether approximate or accurate computation should be conducted.

5 Approximate Multiplier with Circuit Level Approximation

This chapter discusses a few general circuit level techniques for approximation, such as K-map modification, gate-level pruning, and voltage over-scaling (VOS), which are applicable to various architectures or algorithms.

5.1 K-Map Modification

Karnaugh map (K-map) is a common method for Boolean algebra expression simplification. The basic idea of K-map is to group the adjacent squares with the same logic values as much as possible. However, it is quite common in practice one or more squares cannot be grouped, causing additional logics and hence area. Thus, the approximation to K-map can be introduced to modify the adjacent square to the same value so as to group the squares and obtain a more compact representation. For example, the approximate multiplier UDM discussed in Sect. 3 is comprised of a 2×2 multiplication module [15], which can be designed through K-map modification. By modifying the K-map as in Fig. 3.15, the basic block can act as both a partial product generator and a compressor with an error rate of $1/16$ [15]. As shown in Fig. 3.16, when compared to the accurate logic implementation, the approximate implementation needs much fewer logic gates (37.5% reduction) with a shorter critical path.

The K-map modification can be applied to other arithmetic functions, such as adders [26], compressors [27, 28], and booth encoding modules [29, 30]. For example, Yin et al. use K-map modification to design an approximate modified Booth encoding (AMBE) module. With the modified K-map in Fig. 3.17, the original expression for modified Booth encoding algorithm:

$$PP_j = (X_{2i} \oplus X_{2i-1})(X_{2i+1} \oplus Y_j) + \overline{(X_{2i} \oplus X_{2i-1})}(X_{2i+1} \oplus X_i)(X_{2i+1} \oplus Y_{j-1}) \tag{3.14}$$

can be simplified to [30]:

$$PP_j^{'} = (X_{2i} \oplus X_{2i-1})(X_{2i+1} \oplus Y_j) . \tag{3.15}$$

A_1A_0 \ B_1B_0	00	01	11	10
00	000	000	000	000
01	000	001	011	010
11	000	011	111	110
10	000	010	110	100

Fig. 3.15 An example of modifying K-map to achieve more compact design [15]

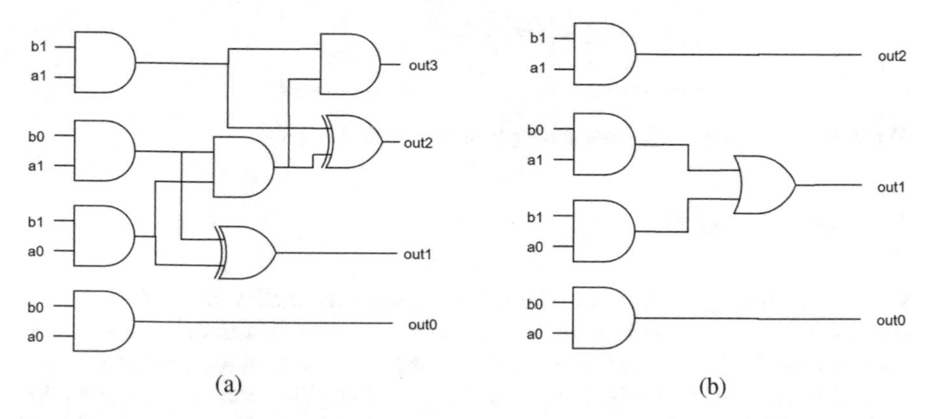

Fig. 3.16 Comparison on the implementations of the 2 × 2 multiplier module: (**a**) accurate logic implementation; (**b**) approximate logic implementation [15]

Y_jY_{j-1} \ $X_{2i+1}X_{2i}X_{2i-1}$	000	001	011	010	110	111	101	100
00	0	0	0	0	1	1	1	0
01	0	0	0	0	1	1	1	0
11	0	1	1	1	0	0	0	0
10	0	1	1	1	0	0	0	0

Fig. 3.17 A K-map example of AMBE [30]

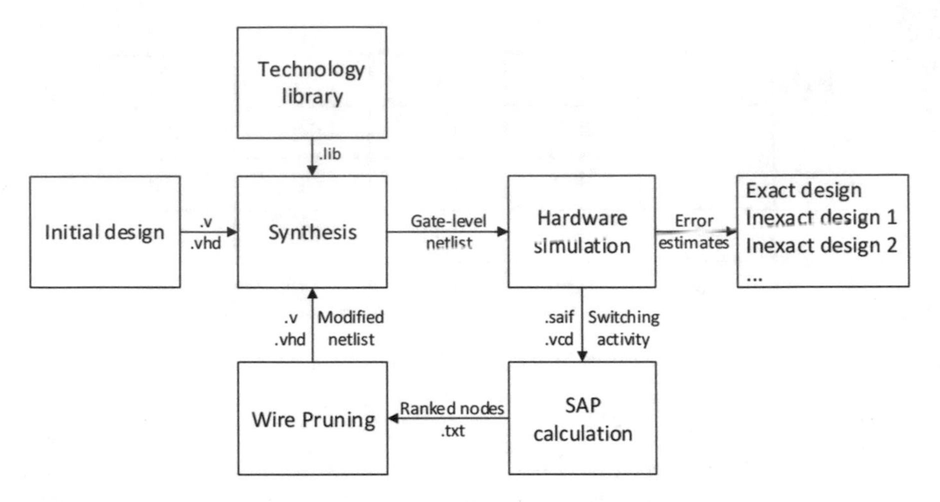

Fig. 3.18 An example of gate-level pruning in digital design flow [32]

5.2 Gate-Level Pruning

Gate-level pruning provides an alternative to simplify netlist. It is based on the probabilistic pruning, which prunes less active gates from a circuit with limited accuracy loss [31]. Jeremy et al. propose to transform the circuit to a graph and prune the nodes with the lowest significance-activity product (SAP) during synthesis [32]. The term "significance" indicates the importance of each node/gate, while "activity" refers to the toggling rate of the gate. The significance for the output nodes is user-defined and then backward propagated to calculate the significance of other gates. The activity of a node can be extracted from the .SAIF file (Switching Activity Interchange Format), which presents the toggle counts of wires. The digital design flow with gate-level pruning is presented in Fig. 3.18.

5.3 Voltage Over-Scaling

Voltage scaling is a common method used to reduce power consumption [20]. In general, the operating supply voltage needs to be higher than $V_{dd-crit}$, which is the minimum supply voltage to ensure the timing of the critical path [21, 59, 60]. While voltage over-scaling (VOS) reduces power effectively, error is introduced inevitably. Hence, the key idea of VOS is to reduce the errors introduced by timing violations due to VOS [60]. Since VOS mainly impacts the critical and near-critical paths, it is desired to adjust the architecture of each computation module to achieve a shorter critical path and alleviate the impact of low supply voltage [21]. Liu et al. further propose an analytical method to assess computation errors due to VOS, which can then select the corresponding architecture and setup [21].

6 Conclusions

In this chapter, we reviewed approximate multipliers from three levels, i.e., architecture, algorithm, and circuit. At architecture level, various approximation strategies were presented and discussed according to the stages of a conventional reference multiplier. At algorithm level, logarithm-based, linearization-based, and hybrid approximations were reviewed. Finally, at circuit level, we introduced three circuit level approximation techniques that can be applied together with any of the aforementioned approximation methods at architecture or algorithm levels. Detailed experimental results were presented in each section to help understand the pros and cons of different techniques at different design layers.

References

1. Atzori L, Iera A, Morabito G. The internet of things: a survey. Comput Netw. 2010;54(15):2787–2805
2. Gupta V, Mohapatra D, Park SP, Raghunathan A, Roy K. Impact: imprecise adders for low-power approximate computing. In: IEEE/ACM international symposium on low power electronics and design; 2011. pp 409–14
3. Han J, Orshansky M. Approximate computing: An emerging paradigm for energy-efficient design. In: 2013 18th IEEE European test symposium (ETS); 2013. pp 1–6
4. Imani M, Rahimi A, Rosing TS. Resistive configurable associative memory for approximate computing. In: Proceedings of the 2016 conference on design, automation & test in Europe. DATE '16. San Jose, CA: EDA Consortium; 2016. p 1327–32
5. Venkataramani S, Chippa VK, Chakradhar ST, Roy K, Raghunathan A. Quality programmable vector processors for approximate computing. In: Proceedings of the 46th annual IEEE/ACM international symposium on microarchitecture. MICRO-46. New York, NY: Association for Computing Machinery; 2013. pp 1–12 [Online]. Available: https://doi.org/10.1145/2540708.2540710
6. Jiang H, Santiago FJH, Mo H, Liu L, Han J. Approximate arithmetic circuits: a survey, characterization, and recent applications. Proc IEEE 2020;108(12):2108–2135
7. Govindu G, Zhuo L, Choi S, Prasanna V. Analysis of high-performance floating-point arithmetic on FPGAs. In: 18th international parallel and distributed processing symposium, 2004. proceedings; 2004. p 149
8. Yu RK, Zyner GB. 167 MHz radix-4 floating point multiplier. In: Proceedings of the 12th symposium on computer arithmetic; 1995. pp 149–54.
9. Courbariaux M, Bengio Y, David J. Low precision arithmetic for deep learning. In: Bengio Y, LeCun Y (eds.). 3rd international conference on learning representations, ICLR 2015, San Diego, CA, May 7–9, 2015. Workshop track proceedings; 2015. [Online]. Available: http://arxiv.org/abs/1412.7024
10. Deng J, Zhuo C. Energy efficient real-time UAV object detection on embedded platforms. IEEE Trans Comput-Aid Des Integr Circ Syst 2019;PP:1–1
11. He K, Gerstlauer A, Orshansky M. Circuit-level timing-error acceptance for design of energy-efficient DCT/IDCT-based systems. IEEE Trans Circ Syst Video Technol 2013;23:961–74
12. Imani M, Kim Y, Rahimi A, Rosing T. ACAM: approximate computing based on adaptive associative memory with online learning. In: Proceedings of the 2016 international symposium on low power electronics and design. ISLPED '16. New York, NY: Association for Computing Machinery; 2016. pp 162–7. [Online]. Available: https://doi.org/10.1145/2934583.2934595

13. Imani M, Patil S, Rosing TS. MASC: ultra-low energy multiple-access single-charge TCAM for approximate computing. In: Proceedings of the 2016 conference on design, automation & test in Europe. DATE '16. San Jose, CA: EDA Consortium; 2016. pp 373–8.
14. Imani M, Samragh M, Kim Y, Gupta S, Koushanfar F, Rosing T. RAPIDNN: in-memory deep neural network acceleration framework. CoRR abs/1806.05794; 2018. [Online]. Available: http://arxiv.org/abs/1806.05794
15. Kulkarni P, Gupta P, Ercegovac M. Trading accuracy for power with an underdesigned multiplier architecture. In: 2011 24th international conference on VLSI design. New York: IEEE; 2011. pp 346–51
16. Shafique M, Hafiz R, Rehman S, El-Harouni W, Henkel J. Invited: cross-layer approximate computing: from logic to architectures. In: 2016 53nd ACM/EDAC/IEEE design automation conference (DAC); 2016. pp. 1–6
17. Suhre A, Keskin F, Ersahin T, Cetin-Atalay R, Ansari R, Cetin AE. A multiplication-free framework for signal processing and applications in biomedical image analysis. In: 2013 IEEE international conference on acoustics, speech and signal processing. 2013; pp 1123–7
18. Vahdat S, Kamal M, Afzali-Kusha A, Pedram M, Navabi Z. Truncapp: a truncation-based approximate divider for energy efficient DSP applications. In: Design, automation test in Europe conference exhibition (DATE), 2017; 2017. pp 1635–8
19. Zhuo C, Unda K, Shi Y, Shih W-K. From layout to system: early stage power delivery and architecture co-exploration. IEEE Trans Comput-Aid Des Integr Circ Syst 2018;PP:1–1
20. Chandrakasan AP, Brodersen RW. Minimizing power consumption in digital CMOS circuits. Proc IEEE 1995;83(4):498–523
21. Liu Y, Zhang T, Parhi KK. Computation error analysis in digital signal processing systems with overscaled supply voltage. IEEE Trans Very Large Scale Integration (VLSI) Syst 2009;18(4):517–526
22. Imani M, Sokolova A, Garcia R, Huang A, Wu F, Aksanli B, Rosing T. ApproxLP: approximate multiplication with linearization and iterative error control. In: Proceedings of the 56th annual design automation conference 2019; 2019. pp 1–6
23. Chen C, Yang S, Qian W, Imani M, Yin X, Zhuo C. Optimally approximated and unbiased floating-point multiplier with runtime configurability. In: Proceedings of the 39th international conference on computer-aided design; 2020. pp 1–9
24. IEEE standard for floating-point arithmetic. IEEE Std 754-2019 (Revision of IEEE 754-2008); 2019. pp 1–84
25. Wallace CS. A suggestion for a fast multiplier. IEEE Trans Electron Comput 1964;1:14–17
26. Pabithra S, Nageswari S. Analysis of approximate multiplier using 15–4 compressor for error tolerant application. In: 2018 international conference on control, power, communication and computing technologies (ICCPCCT). New York: IEEE; 2018. pp 410–5
27. Van Toan N, Lee J-G. FPGA-based multi-level approximate multipliers for high-performance error-resilient applications. IEEE Access 2020;8:25481–25497
28. Ha M, Lee S. Multipliers with approximate 4–2 compressors and error recovery modules. IEEE Embed Syst Lett 2017;10(1):6–9
29. Qian L, Wang C, Liu W, Lombardi F, Han J. Design and evaluation of an approximate Wallace-Booth multiplier. In: 2016 IEEE international symposium on circuits and systems (ISCAS). New York: IEEE; 2016. pp 1974–7
30. Yin P, Wang C, Liu W, Swartzlander EE, Lombardi F. Designs of approximate floating-point multipliers with variable accuracy for error-tolerant applications. J Sign Process Syst. 2018;90(4):641–654
31. Lingamneni A, Enz C, Nagel J-L, Palem K, Piguet C. Energy parsimonious circuit design through probabilistic pruning. In: 2011 design, automation & test in Europe. New York: IEEE; 2011. pp 1–6
32. Schlachter J, Camus V, Enz C, Palem KV. Automatic generation of inexact digital circuits by gate-level pruning. In: 2015 IEEE international symposium on circuits and systems (ISCAS). New York: IEEE; 2015. pp 173–6

33. Camus V, Schlachter J, Enz C, Gautschi M, Gurkaynak FK. Approximate 32-bit floating-point unit design with 53% power-area product reduction. In: ESSCIRC conference 2016: 42nd European solid-state circuits conference; 2016. pp 465–8
34. Narayanamoorthy S, Moghaddam HA, Liu Z, Park T, Kim NS. Energy-efficient approximate multiplication for digital signal processing and classification applications. IEEE Trans Very Large Scale Integration (VLSI) Systems 2014;23(6):1180–4
35. Bhardwaj K, Mane PS, Henkel J. Power- and area-efficient approximate Wallace tree multiplier for error-resilient systems. In Fifteenth international symposium on quality electronic design; 2014. pp 263–9
36. Liu C, Han J, Lombardi F. A low-power, high-performance approximate multiplier with configurable partial error recovery. In 2014 design, automation & test in Europe conference & exhibition (DATE). New York: IEEE; 2014. pp 1–4
37. Lin C, Lin I. High accuracy approximate multiplier with error correction. In: 2013 IEEE 31st international conference on computer design (ICCD); 2013. pp 33–38
38. S. E. Ahmed, S. Kadam, and M. Srinivas, "An iterative logarithmic multiplier with improved precision," in *2016 IEEE 23nd symposium on computer arithmetic (ARITH)*. New York: IEEE; 2016. pp 104–111
39. Imani M, Peroni D, Rosing T. CFPU: configurable floating point multiplier for energy-efficient computing. In: 2017 54th ACM/EDAC/IEEE design automation conference (DAC). New York: IEEE; 2017, pp 1–6
40. Imani M, Garcia R, Gupta S, Rosing T. RMAC: runtime configurable floating point multiplier for approximate computing. In: Proceedings of the international symposium on low power electronics and design; 2018. pp 1–6
41. de Angel E, Swartzlander E. Low power parallel multipliers. In: VLSI Signal Processing, IX. New York: IEEE; 1996. pp 199–208
42. Hashemi S, Bahar RI, Reda S. Drum: a dynamic range unbiased multiplier for approximate applications. In: 2015 IEEE/ACM international conference on computer-aided design (ICCAD). New York: IEEE; 2015. pp 418–25
43. Vahdat S, Kamal M, Afzali-Kusha A, Pedram M. TOSAM: an energy-efficient truncation-and rounding-based scalable approximate multiplier. IEEE Trans Very Large Scale Integr (VLSI) Syst 2019;27(5), 1161–73
44. Vahdat S, Kamal M, Afzali-Kusha A, Pedram M. LETAM: a low energy truncation-based approximate multiplier. Comput Electr Eng 2017;63:1–17
45. Venkatachalam S, Ko S-B. Design of power and area efficient approximate multipliers. IEEE Trans Very Large Scale Integr (VLSI) Syst 2017;25(5), 1782–6
46. Yang T, Ukezono T, Sato T. A low-power high-speed accuracy-controllable approximate multiplier design. In: 2018 23rd Asia and South pacific design automation conference (ASP-DAC). New York: IEEE; 2018. pp 605–10
47. Yang T, Ukezono T, Sato T. Low-power and high-speed approximate multiplier design with a tree compressor. In: 2017 IEEE international conference on computer design (ICCD). New York: IEEE, 2017, pp 89–96
48. Momeni A, Han J, Montuschi P, Lombardi F. Design and analysis of approximate compressors for multiplication. IEEE Trans Comput 2014;64(4):984–994
49. Tung C-W, Huang S-H. Low-power high-accuracy approximate multiplier using approximate high-order compressors. In: 2019 2nd international conference on communication engineering and technology (ICCET). New York: IEEE; 2019, pp 163–7
50. Yang Z, Han J, Lombardi F. Approximate compressors for error-resilient multiplier design. In: 2015 IEEE international symposium on defect and fault tolerance in VLSI and nanotechnology systems (DFTS). New York: IEEE; 2015, pp 183–6
51. Mahdiani HR, Ahmadi A, Fakhraie SM, Lucas C. Bio-inspired imprecise computational blocks for efficient VLSI implementation of soft-computing applications. IEEE Trans Circ Syst I: Regul Pap 2009;57(4), 850–62
52. Lin H-L, Chang RC, Chan M-T, Design of a novel radix-4 booth multiplier. In: The 2004 IEEE Asia-Pacific conference on circuits and systems, vol 2. Citeseer; 2004. pp 837–40

53. Jiang H, Han J, Qiao F, Lombardi F. Approximate radix-8 booth multipliers for low-power and high-performance operation. IEEE Trans Comput 2015;65(8):2638–44
54. Mitchell JN. Computer multiplication and division using binary logarithms. IRE Trans Electron Comput 1962;4:512–7
55. Liu W, Xu J, Wang D, Wang C, Montuschi P, Lombardi F. Design and evaluation of approximate logarithmic multipliers for low power error-tolerant applications. IEEE Trans Circ Syst I: Regul Pap 2018;65(9):2856–68
56. Saadat H, Bokhari H, Parameswaran S. Minimally biased multipliers for approximate integer and floating-point multiplication. IEEE Trans Comput-Aid Des Integr Circ Syst. 2018;37(11):2623–35
57. Ansari MS, Cockburn BF, Han J. A hardware-efficient logarithmic multiplier with improved accuracy. In: 2019 design, automation & test in Europe conference & exhibition (DATE). New York: IEEE; 2019. pp 928–31
58. Ansari MS, Cockburn BF, Han J. An improved logarithmic multiplier for energy-efficient neural computing. IEEE Trans Comput 2020;70(4):614–25
59. Mohapatra D, Chippa VK, Raghunathan A, Roy K. Design of voltage-scalable meta-functions for approximate computing. In: 2011 design, automation & test in Europe. New York: IEEE; 2011. pp 1–6
60. Chen J, Hu J. Energy-efficient digital signal processing via voltage-overscaling-based residue number system. IEEE Trans Very Large Scale Integr (VLSI) Syst 2012;21(7):1322–32

Chapter 4
Low-Precision Floating-Point Formats: From General-Purpose to Application-Specific

Amir Sabbagh Molahosseini, Leonel Sousa, Azadeh Alsadat Emrani Zarandi, and Hans Vandierendonck

1 Introduction

Floating-point (FP) arithmetic [1] is one of the most prominent achievements of computer arithmetic that is ubiquitous in computing systems, from personal computers and smartphones to embedded processors, which are at the core of almost every electronic gadget used in our daily life [2]. The FP formats can provide high dynamic range as well as high precision within a fixed-size storage format [3]. This makes them prevalent number formats for a range of applications from embedded systems to scientific computations. The IEEE-754 standard [4] has been the mainstream floating-point number system that has been used in off-the-shelf processors for general-purpose use. However, application areas are experiencing a dramatic shift. Deep learning is gaining traction as a generic solution to numerous data-driven problems [5]. The explosion of digital devices connected to the Internet, dubbed the Internet-of-Things (IoT), requires energy-efficient processing of streams of sensor data [6]. However, FP operations are responsible for a significant amount of energy consumption of the system. For instance, in an embedded computing environment based on low-power microcontrollers, it has been shown that FP operations may consume over 30% of energy consumption and even an additional

A. Sabbagh Molahosseini · H. Vandierendonck
School of Electronics, Electrical Engineering and Computer Science, Queen's University Belfast, Belfast, UK
e-mail: a.sabbaghmolahosseini@qub.ac.uk; h.vandierendonck@qub.ac.uk

L. Sousa (✉)
INESC-ID, Instituto Superior Técnico (IST), University of Lisbon, Lisbon, Portugal
e-mail: las@inesc-id.pt

A. A. Emrani Zarandi
Department of Computer Engineering, Shahid Bahonar University of Kerman, Kerman, Iran
e-mail: a.emrani@uk.ac.ir

W. Liu, F. Lombardi (eds.), *Approximate Computing*,
https://doi.org/10.1007/978-3-030-98347-5_4

20% of energy consumed by the movement of FP operands between memory and registers [7]. Therefore, this shift in applications implies a renewed investigation of floating-point formats since they are larger than necessary for the majority of emerging applications, which results in needlessly high accuracy and energy waste [8].

There are various methods that aim to rethink conventional arithmetic to increase the performance efficiency of computing systems, such as unconventional computer arithmetic [9] and reduced-precision computing [10]. Data size shrinkage is one of the main targets of reduced-precision computing since it can decrease computation latency and memory bandwidth. This could be using the single precision, i.e., IEEE 32-bit floating-point format (FP32), instead of double-precision IEEE 64-bit floating-point format (FP64), in all or part of the computations that traditionally fully performed using double precision [11] or substituting FP32 with FP formats with less than 32 bits, namely *low-precision FP formats*. In this category, the only standard format is the general-purpose IEEE half-precision 16-bit floating-point format (FP16) [4]. Although FP16 has been widely used in various applications, from weather and climate modeling [12] to deep learning [13], in the narrow precision domain, even one bit can significantly change the range of the represented numbers. Therefore, the general-purpose FP16 cannot exploit the full potential of low-precision computing for all applications. This motivated researchers to design application-specific number formats to maximize performance while providing the required dynamic range and precision.

The industry demand and error resiliency of deep neural networks (DNNs) made deep learning as the driving application in the emerging era of application-specific number format design. Google derived the 16-bit brain floating-point format (BFloat16) [14] for particular use in its tensor processing units (TPUs) [15] for DNN inference and training [16–18]. This application-specific format was a successful example, being now supported on modern TPUs [19], NVidia graphics processing units (GPUs) [20], Intel central processing units (CPUs) [21, 22], and field-programmable gate arrays (FPGAs) [24] only within a few years after its proposal. Similar neural-optimized FP formats have been explored by other companies, such as IBM deep learning float (DLFloat) [25] and Entropy-coded Float (EFloat) [26], NVidia TensorFlow32 (TF32) [20, 27], Microsoft Floating-Point (MSFP) [28], and Intel Flexible FP format (Flexpoint) [29]. Additionally, there are some specific formats that target embedded machine learning (ML) applications, such as Flytes [30] and tunable floating-point (TFP) [31, 32]. Besides, specializing low-precision posit FP formats [33] for deep learning applications has attracted researchers [34–37]. Apart from ML, application-specific FP formats have been designed and applied in other areas such as customized precision mantissa segmentation (CPMS) FP formats for PageRank [38] and linear algebra applications [39], and Accuracy-Preserving Half-Precision (APHP) and Time-Optimizing Half-Precision (TOHP) for PageRank [40].

This chapter aims to focus the attention on the importance of low-precision FP formats for performance enhancement of modern computing workloads. In this regard, first, a collection of low-precision FP formats with their features and target applications are reviewed. Second, PageRank is studied as a case study to show how customized low-precision FP formats can provide a sufficient precision rather than considering the maximum possible precision available (e.g., FP32/FP64). Then, the effects of using low-precision FP formats on various aspects of performance including throughput, latency, and energy consumption based on different hardware platforms will be investigated. Finally, a step-by-step guideline will be presented to be used as a starting point for researchers and engineers to design customized FP formats given the algorithms and applications.

2 Low-Precision Floating-Point Formats

Representing always accurate real-valued numbers in computing systems with a limited number of bits is hard task. Number formats, either fixed point or floating point, are a way to approximate the original real values with a degree of error based on the selected rounding method [3]. There are two important parameters regarding numerical formats for representing real values: the dynamic range, which is the range of representable values, and the precision that indicates the resolution of encodable values [7]. Fixed-point formats suffer from a limited dynamic range, while floating point can provide high dynamic range in conjunction with high precision, which has resulted in the widespread use of FP number formats. This section analyzes various floating-point formats, from general-purpose to application-specific focusing on their structure and features. Later, in the next section, we will show how some of these formats are used in specific applications.

2.1 IEEE-Style FP Formats

This section first reviews the IEEE half-precision standard (i.e., FP16) as a general-purpose low-precision FP format and then analyzes other FP formats (Table 4.1) that are similar to IEEE standards, but with different exponents, mantissa sizes, and a different set of exception cases.

2.1.1 General-Purpose IEEE-Standard FP Formats

A real number can be represented in IEEE-754 format [4] by three components: sign (*S*), exponent (*E*), and mantissa (*M*) as follows:

$$(-1)^S \times 2^{E-Bias} \times 1.M, \tag{4.1}$$

where the bias value is $2^{(k-1)} - 1$, k is the number of exponent bits, and the mantissa is stored in normalized form. Consequently, the numbers below the minimum

Table 4.1 IEEE-style floating-point formats

Name	Size (bits)	Sign bit	Exponent (bits)	Mantissa (bits)	Hardware support	Application
FP64	64	✓	11	52	CPU/GPU	General-Purpose [4]
FP32	32	✓	8	23	CPU/GPU	General-Purpose [4]
Flyte24	24	✓	8	15	✗	Memory Format (General) [30]
TF32	19	✓	8	10	GPU	Deep Learning [20]
FP16	16	✓	5	10	GPU	General-Purpose [4, 41]
BFloat	16	✓	8	7	TPU/GPU/CPU	Deep Learning [16, 18, 20]
DLFloat	16	✓	6	9	ASIC	Deep Learning [25]
FP16alt	16	✓	8	7	SoC	Embedded Machine Learning [7, 42]
Flyte16	16	✓	8	7	✗	Memory Format (General) [30]
APHP	16	✗	3	13	✗	Memory Format (PageRank) [40]
TOHP	16	✗	6	10	✗	Memory Format (PageRank) [40]
CPMS16	16	✓	11	4	✗	Memory Format (PageRank) [38]
EFloat16	16	✓	3	12	✗	Memory Format (Deep Learning) [26]
MSFP9	9	✓	5	3	FPGA	Deep Learning [28]
MSFP8	8	✓	5	2	FPGA	Deep Learning [28]
FP8alt	8	✓	4	3	SoC	Embedded Machine Learning [7, 42]

normalized number are considered as subnormal. Full details and analysis about IEEE-standard FP format can be found in [1, 3, 4].

The size of the exponent and mantissa fields results in FP formats with different dynamic ranges and precisions. The IEEE standard [4] defines five binary FP formats with different sizes, with FP32 and FP64 having widespread hardware support nearly in all off-the-shelf general-purpose processors. The IEEE **FP16** format includes 5 exponent bits and 10 mantissa bits and follows the same arithmetic principles of FP32 and FP64 regarding rounding and subnormal support. However, mainstream commercial CPUs do not currently support FP16 operations, while GPUs do [41, 43]. It was the first half-precision format that was used for mixed-precision DNN training [13] since the use of FP16 instead of FP32 halves the memory storage and reduces data movement, particularly for big ML models. However, the narrow dynamic range of FP16 with only 5 exponent bits is not sufficient for DNN training. Due to this, mixed-precision DNN training using both FP16 and FP32 has been introduced [13]. Here, FP16 is used for weights, activations, and gradients, and a master copy of weights is shared in FP32. Moreover, FP16 knows other applications such as in high-performance computing (HPC) for iterative solvers [44] and spiking neural network simulations [45]. Additionally, FP16 is recently used for accelerating graph convolutional neural networks (GCNs) on GPUs [46] as well as low-precision implementation of reinforcement learning [47].

2.1.2 Application-Specific FP Formats

There is a recent research trend to derive specialized number formats for applications, mostly in the area of deep learning. The most important feature of number system design is the format, i.e., width of exponent and mantissa fields, according to the requirement of the application. Even one bit difference in exponent can significantly change the range of supported numbers; therefore, careful adjustment of exponent and mantissa sizes should be done to prevent accuracy loss. Similarly, the other related parameters such as subnormal support and rounding mode should be selected according to the application-level requirement. The IEEE-754 standard [4] covers all of these requirements, which increase the implementation complexity, as well as the lack of flexibility of adjusting the required precision. This subsection investigates some of the customized low-precision floating-point formats with focus on their structure as well as applications.

(a) BFloat16 FP32 plays an important role in mixed-precision computing, from DNN training [13, 48] to iterative graph processing algorithms [38, 49]. Therefore, conversion from low-precision FP formats to FP32 and vice versa is necessary and impacts performance. Exponent conversion between two different FP formats is complex since it can lead to overflow/underflow. Due to this, using the same representation for the exponent as FP32 simplifies the conversion process [50]. In this regard, Google proposed the BFloat16 [14] FP format with 8-bit exponent and 7-bit mantissa. The main purpose of BFloat16 is to provide the same dynamic range as FP32. It provides the dynamic range needed to represent DNN data, being easy to convert into FP32 without concerns regarding overflow and underflow. BFLoat16 does not support subnormals since its wide dynamic range can cover the required numbers for deep learning applications [18].

BFloat16 was first supported by TPUv2 [15] targeting DNN inference and training and then extended to other TPUs (i.e., TPUv3 and TPUv4i) [19]. The mixed-precision DNN training using BFloat16 and FP32 was introduced in [51]; convolutions are performed with BFloat16, and FP32 is used for non-convolutional operations such as batch normalization and gradient summation [51]. This structure is exactly in line with the hardware structure of the TPUv2 [15]. Most of arithmetic operations of DNNs are required in convolution layers, and therefore using BFloat16 instead of FP32 for convolution improves the performance due to reduced memory traffic as well as higher arithmetic speed [19, 48] while achieving the expected accuracy due to the use of FP32 for accumulation and normalization that prevents accuracy drop [51]. In addition to TPUs, NVidia and Intel support BFloat16 in their Ampere GPU architecture [20], and Cooper Lake CPU family [21], respectively. Although BFloat was designed especially for deep learning applications [16, 17], it has found other applications such as Monte Carlo simulation [52] and linear algebra [53]. Overall, it is expected with emerging native BFloat16 hardware support [23]; it will be applied to even more application areas.

(b) FP16alt and FP8alt A specific version of BFloat16 but with subnormal support and the same rounding and special cases handling like IEEE-754 was introduced in

[7]. Moreover, they design FP8alt with the same exponent bits as FP16, i.e., 5 bits, to ease conversion between FP8alt and FP16 [54]. These customized FP formats are designed for embedded machine learning applications such as K-Means and convolutional neural networks (CNNs) [55].

(c) DLFloat DLFloat is proposed by IBM researchers to provide a precision trade-off between FP16 and BFloat16 [25]. DLFloat includes 6 exponent bits and 8 mantissa bits, which leads to higher precising than BFloat and higher dynamic range than FP16 due to the one additional exponent bit. Apart from precision and dynamic range, the DLFloat designers tried to simplify the floating-point unit (FPU) of DLFloat while preserving the required dynamic range and precision suitable for deep learning applications. In this regard, DLFloat did not support subnormal representation (the subnormal support increases the complexity of the FPU but provides a higher presentation range). Moreover, it considers only round-nearest-up rounding mode for DLFloat [25]. Both techniques simplify the hardware logic of DLFloat FPUs.

(d) TF32 NVidia used TF32 as a part of its GPU Tensor Cores in the Ampere architecture [20] to provide the high dynamic range of BFloat and FP32 while preserving the suitable precision of FP16. In this structure, the FP32 operands are converted into TF32 that is a 19-bit FP format (a truncated version of FP32 ignoring the 13 least significant bits of FP32 mantissa), and then accumulation of 19-bit multiplications will be stored in FP32 format. This format is only used internally for Tensor Core computations, and therefore, the input and output operands are both FP32 [20]. The use of 8-bit exponents keeps the conversion to FP32 simple.

(e) MSFP MSFP is an 8-bit FP format with 5 exponent bits and 2 mantissa bits, which was designed for DNN inference [28]. DNN quantization, i.e., converting high-precision weights from FP32 into low-precision fixed-point and FP-reduced data types, is very prevalent since it can lead to efficient inference especially suitable for resource-constrained devices. However, the dynamic range of an 8-bit fixed-point or integer number is narrower than FP8. Therefore, FP8 can provide higher accuracy than those. For example, the MSFP8 results in the same accuracy as 16-bit fixed-point for DNN inference [28]. DNN training with this 8-bit FP structure has been developed in [56]. In this chapter, 8-bit FP format with the structure of MSFP8 is used for representation of weights, activations, and gradients of the model, while a 16-bit FP format with the structure of DLFloat is used for accumulation [56]. Apart from 8-bit version of MSFP, Microsoft has developed a 9-bit version, MSFP9, and used it for DNN inference in situations where higher precision than MSFP8 is needed [28].

(f) Flytes Flytes are a class of floating-point formats that aim to make a trade-off between precision and memory requirements [30]. In contrast to other formats like FP16 and BFloat16 that have direct hardware support, Flytes are designed to reduce memory storage and bandwidth. The main idea behind Flytes is to design various low-precision FP formats with less mantissa bits and the same exponent bits as the higher hardware-supported formats such as FP32 [30]. Therefore, Flytes16 and Flytes24 both have 8-bit exponent but with different mantissa sizes to cover the required precision needed by the application. It should be noted that Flytes are only stored in memory, and computations will be done using FP32/FP64. Therefore, conversions between Flytes and FP32/FP64 are required before each computation. They considered a customized single-instruction multiple-data (SIMD) implementation [57] paradigm based on AVX vectorization to overcome the overhead of conversion and read/write of unconventional number sizes such as 24-bit [30].

(g) CPMS Similar to [30], the idea of decoupling the arithmetic format from the memory format considered in [38, 39] but combined with the customized precision mantissa segmentation technique, i.e. CPMS. Their target is to design a communication reduction technique through the use of customized precision specified for iterative algorithms. In this scheme, the FP64 version of numbers is stored in memory but re-organized for efficient memory access. Therefore, some mantissa bits will be transferred to the processor in each iteration according to the required precision, which can lead to reduced memory traffic [38].

(h) APHP and TOHP Two completely customized FP formats are introduced in [40] by analyzing the required dynamic range and precision needed in each iteration of the PageRank algorithm. Following the half-precision formats with 16-bit width, [40] have derived two FP formats, the APHP format, with 3 exponent bits and consequently 13 mantissa bits, which achieve higher precision than any other available 16-bit format. Moreover, the TOHP format has 6 exponent bits and can provide less precision than APHP but a very simple conversion between FP32 than APHP [40]. Reference [40] was the first work that explored the possibility of removing the sign bit from the FP format when only unsigned numbers are used. It results in one free bit in the 16-bit frame that can be either added to the exponent to increase the dynamic range, or to the mantissa for precision improvement.

(i) EFloat and AdaptiveFloat EFloat is a variable-precision FP format that was designed especially for deep learning applications [26]. Similar to [40], they analyzed the exponent range of trained DNN models. When their range is narrow, it can be represented with a limited number of bits. Therefore, they proposed a FP format template that can be adjusted according to the weight range of DNN model to optimize the exponent and mantissa sizes [26]. At the present, there is no hardware available that implements EFloat, and they provide some hints for using it as a memory format. Similar to EFloat, AdaptiveFloat [58] also tries to dynamically capture the required range of DNN model weights. They customized the exponent bias to dynamically adjust the required exponent range to achieve efficient low-precision inference [58].

(j) TFP TFP [31] is a hardware-oriented scheme with the idea of designing variable-precision FP arithmetic units to enable low precision for a single arithmetic operation. TFP can cover various precisions from 5–8 exponent bits to 4–24 mantissa bits [31]. The TFP format is applied to some embedded machine learning applications such as Singular-Value Decomposition (SVD) [59] and neural networks [60]. Additionally, a half-precision version of TFP, i.e. TFP16, was developed in

[32] for embedded processing where the exponent can change from 5 to 8 bits, and consequently, the mantissa can vary from 10 to 7 bits. Variable-precision TFP-based arithmetic circuits, including adders, multipliers, and dividers, have been developed in [31, 32, 60, 61].

2.2 Block Floating-Point Formats

The block floating-point (BFP) format considers a block of FP numbers that have the same exponent value but with different values for mantissa. BFP was originally proposed for signal processing [62] where a shared exponent can be used for a block of numbers, which can lead to reduced data size. BFP has been used in many applications of low-precision implementation of DNN inference and training. In this regard, Intel researchers proposed the *FlexPoint* [29] format with 5 bits dedicated for the shared exponent and a block of 16-bit mantissas. A hybrid use of BFP and FP for DNN training is introduced in [63] to provide accuracy similar to FP32. In addition, the *Block Mini-Float* [64] scheme introduces another customization of BFP for DNN training. In this format, a block of FP numbers with less than 4-bit exponents shares the exponent's bias [64].

2.3 Posit Floating-Point Formats

Posit [33] is an unconventional FP number system that includes an additional field called *regime* to the conventional IEEE-style FP formats. This unique variable-size field (i.e., regime) can provide a trade-off between dynamic range and precision by decreasing exponent bits and increasing mantissa bits when a small exponent is needed. Consequently, when a large exponent is needed, fewer mantissa bits are available. Therefore, posit can provide higher precision for numbers near zero but less for numbers with high magnitude [35]. This feature attracted researchers for low-precision computation using small-size posit formats. On the other hand, the size variability of the regime field results in complex arithmetic circuits for posit. In this regard, Lu et al. [36] proposed an efficient hardware implementation of an 8-bit posit format for low-precision DNN training with competitive accuracy as FP32 but with higher performance. Recently, various customized low-precision posit formats with 8-bit size or less have been developed for DNN inference [34, 37].

3 Application-Specific FP Formats: Case Study

The previous section reviewed state-of-the-art low-precision FP formats with a focus on their mathematical structure and features. This section aims to show what motivates researchers to design application-specific FP formats, and how those formats can be used in different applications. As it can be seen from Table 4.1, PageRank (PR) has been the target of several low-precision FP formats since it relies on large datasets and is suitable to apply low-precision FP formats. This section considers it as a case study. Low-precision computing in graph processing algorithms recently has attracted attention [49]. PageRank [65] is the most well-known web page ranking algorithm that is at the basis of the Google search engine. The overall process starts from the behavior of a random surfer on the Internet who follows links from one web page to another. Therefore, the importance of a page can be defined in terms of other web pages that link to it [65]. In this regard, the first part of PR is translating the web page links to a directed graph without self and duplicate links, where the pages are nodes and their links are edges. A common solution for PR is the power iteration method that iteratively improves the accuracy of an initial estimate of the PR vector until the residual error, i.e., the absolute sum of the differences between PR values in current and previous iterations, converges to a preset accuracy [65]. This subsection investigates the use of customized FP formats based on [38, 40] to show how the transprecise computing paradigm [8] can be leveraged to dedicate sufficient, not extra precision to internal computations of an iterative algorithm like PageRank.

3.1 Mixed-Precision PageRank: CPMS Approach

The mixed-precision implementation of PR was developed in [38]. This work follows the trend of decoupling the data format from the arithmetic format [30] via the use of a customized precision mantissa segmentation mechanism. They considered FP64 as the baseline and then divided it into the same size segments. The segments are shared in memory in an interleaved format for efficient read/write. The 4-segment CPMS [38] includes four segments of 16-bit numbers where the first part includes 1 sign, 11 exponent, and 4 mantissa bits. The next three parts include the remaining 48 bits of the mantissa. It should be noted that all segment parts (totaling 64 bits) are always stored in memory in separate banks. Segments are retrieved only when the required precision demands it. The remaining bits up to 64 bits will be filled by zero to perform computations with double precision.

The main advantage of this method is its efficient conversion into double precision (i.e., FP64) for computations, since only the concatenation according to the required precision is needed to form the 64-bit number. They also used normalization to fix the stochastic problem of PR values due to the use of rounding toward zero via truncation. However, the experimental results presented in [38] show that only one iteration with 16-bit segment can be done, and most of the iterations used the 32- and 48-bit FP data formats. The main reason for this is the severe lack of precision since only 4 mantissa bits are considered in CPMS16. Hence, it is insufficient to reach convergence for a graph processing application like PR.

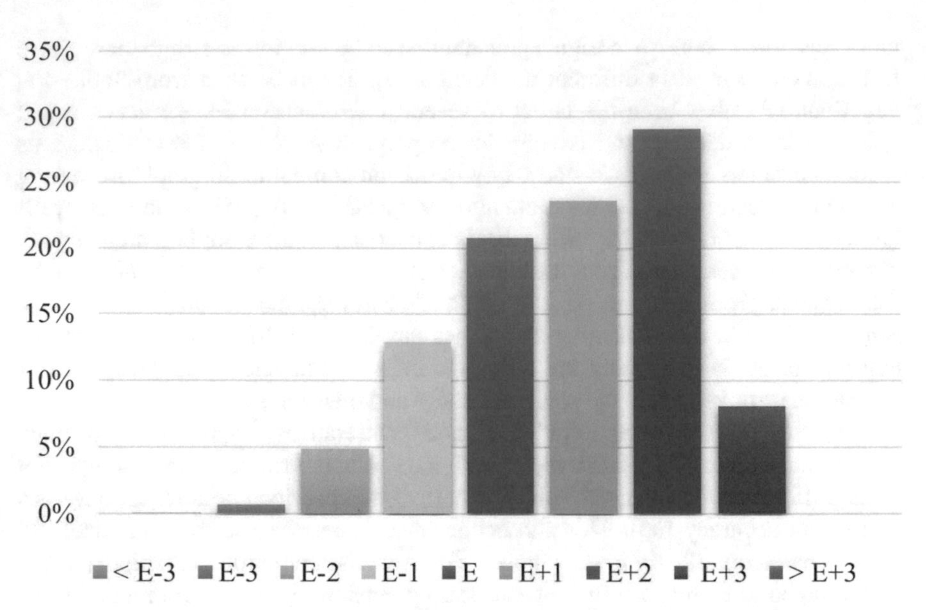

Fig. 4.1 Distribution of PageRank values' exponents for the Pokec dataset

3.2 *Low-Precision PageRank: Half-Precision with APHP*

Grützmacher et al. [38] made an attempt to low-precision implementation of PageRank by considering FP64 as the backbone computation format. However, the half-precision version of CPMS with 4 mantissa bits could not provide sufficient precision. In this regard, Molahosseini et al. [40] proposed a different approach to enable low-precision implementation of PR through defining fully customized FP formats for PR. Two formats were defined: APHP and TOHP. The following investigations have been done in [40] to derive APHP:

- First, Molahosseini et al. [40] developed a comprehensive analysis of PR values in each iteration based on FP32 and FP64 formats. Then, the residual error, the exponent and mantissa of each PageRank value in each iteration are analyzed.
- Second, an exponent-aware FP format is developed with the minimum possible exponent bits (i.e., 3 bits) to cover 99% of the exponents together with a separate mechanism to handle the remaining 1% exponents in a separate data structure. As an example, Fig. 4.1 shows the exponent distribution of PageRank values for the Pokec dataset [66] where *E* is the exponent of one divided by the number of nodes [40]. It can be seen that FP32 with 8 exponent bits is not required to cover this narrow range. Therefore, exponents can be coded with 3 bits as described in [40] and reserve more bits for the mantissa. This 3-bit exponent field together with the elimination of sign bits enables the APHP to be the only half-precision format that can provide high precision with 13-bit mantissas in a frame of 16 bit.

- The experimental results reported in [40] showed that APHP achieved the same accuracy as FP32 and FP64 at many initial iterations up to the residual error of 10^{-4}. As an example, Table 4.2 presents the residual error of PR for the Pokec dataset [66] based on three FP formats, APHP, FP32, and FP64. It can be seen that at the iteration 12, all formats at the same time reached the accuracy of 10^{-3}. In other words, up to iteration 12, PR is completely resilient to lower precision than FP64. Then, after iteration 12, the FP32 and FP64 formats converge to 10^{-4} faster than APHP. The APHP format finally achieves a residual error of 10^{-4} at iteration 23. After this point, the precision provided by APHP is not sufficient to achieve the residual error of 10^{-5} or less. In this case, if higher accuracy is needed, the algorithm should switch to FP32, and then finally for a high-precision requirement, switch to FP64 is necessary.

4 Low-Precision FP Formats: Performance Impact

This section investigates the impact of using low-precision FP formats in terms of energy, throughput, and speedup on different hardware platforms. In this regard, FP32, FP16, and BFloat16 were selected for analyzing performance. FP32 and FP16 can be considered as high- and low-precision general-purpose FP formats. The BFloat16 is selected since it can be considered as the flagship of application-specific FP formats, as it has found significant attention and was implemented in commercial hardware a few years after it was proposed. BFloat16 showed how adjusting the format structure according to the specific features of the target application can lead to performance efficiency [67]. The following subsections investigate the various performance aspects of these selected FP formats based on different hardware platforms. It should be noted that the bold numbers in Table 4.2 indicate the identical residual error digits between the three number formats.

4.1 Low Precision in CPU

The use of low-precision FP formats in off-the-shelf general-purpose processors is a challenging task, since most of the conventional CPUs are based on IEEE-standard high-precision formats of FP32 and FP64 to represent and perform operations on real numbers. Therefore, we have the following options to use low-precision FP formats in CPUs:

- Emulation of low-precision FP formats with integer operations.
- Use Intel Cooper Lake CPUs with BFloat16 support.
- Use low-precision FP format only for memory storage and data movement and perform computations with FP32/FP64.

Table 4.2 The PageRank residual error in each iteration for Pokec dataset based on different FP formats

Iteration number	Double precision (64 bit)	Single precision (32 bit)	APHP (16 bit)	Comment
1	**0.691**831174	**0.691**831172	**0.691**662908	
2	**0.216**602612	**0.216**602609	**0.216**580942	
3	**0.0848**39075	**0.0848**39083	**0.0848**36423	All formats reached 10^{-1}
4	**0.04308**3992	**0.04308**4003	**0.04308**0341	
5	**0.024934**474	**0.024934**459	**0.024934**081	
6	**0.01547**7497	**0.01547**7517	**0.01547**8806	All formats reached 10^{-2}
7	**0.0099**39473	**0.0099**39456	**0.0099**42646	
8	**0.0065**29166	**0.0065**2918	**0.0065**32167	
9	**0.00436**1450	**0.00436**1454	**0.00436**3786	
10	**0.0029**57729	**0.0029**57714	**0.0029**62223	
11	**0.00203**2458	**0.00203**2463	**0.00203**7544	
12	**0.0014**14251	**0.0014**14248	**0.0014**24945	**All formats reached** 10^{-3}
13	**0.000**996100	**0.000**996117	**0.00**1014317	
14	**0.0007**10398	**0.0007**10389	**0.0007**38321	
15	**0.0005**12790	**0.0005**12801	**0.0005**50668	
16	**0.000**374610	**0.000**374605	**0.000**426437	
17	**0.000**276766	**0.000**276760	**0.000**344576	
18	**0.000**206740	**0.000**206746	**0.000**291405	
19	**0.000**155997	**0.000**156002	**0.000**256525	FP32/FP64 reached 10^{-4}
20	**0.000**118869	**0.000**118906	**0.000**233005	
21	**0.0000**913745	**0.0000**913459	**0.000**217190	
22	**0.0000**708271	**0.0000**708319	**0.000**206075	
23	**0.0000**552993	**0.0000**552904	**0.000**198388	**APHP reached** 10^{-4}
24	**0.0000**434717	**0.0000**434689	**0.000**192726	
25	**0.0000**343761	**0.0000**343889	**0.000**188625	
30	**0.0000**114684	**0.0000**114520	**0.000**177869	FP32/FP64 reached 10^{-5}
31	**0.0000**0932617	**0.0000**0933541	**0.000**176823	
43	**0.0000**0104634	**0.0000**0112793	**0.000**172312	FP32/FP64 reached 10^{-6}
44	**0.000**000886029	**0.000**000915507	**0.000**172300	
57	**0.000**000106323	**0.000**000149782	**0.000**172263	F32/FP64 reached 10^{-7}
58	**0.0000**000903559	**0.000**000140276	**0.000**172274	
62	**0.0000**000471256	**0.000**000100569	**0.000**172274	
63	**0.0000**000400477	**0.000**000097057	**0.000**172267	
71	**0.000**000010895800	**0.000**000121643	**0.000**172266	FP64 reached 10^{-8}
72	**0.000**000009259900	**0.000**000090155	**0.000**172275	
85	**0.000**000001117020	**0.0000**001130620	**0.000**172265	FP64 reached 10^{-9}
86	**0.000**000000949327	**0.0000**000834129	**0.000**172274	
100	**0.000**000000097337	**0.0000**000828433	**0.000**172273	

The emulation of FP formats and their operations can be done using several integer operations on existing commercial processors [68, 69]. However, emulation needs handling several special cases such as NaN, infinity, and subnormal. It might lead to high latency that offsets the speed gain achieved by data movement reduction due to reduced-sized FP formats. Therefore, regular emulation of customized FP formats cannot provide better performance than native hardware-supported formats such as FP32 [30]. On the other hand, Intel followed the Google TPUs [19] and NVidida GPUs [20] by including BFloat16 in its new generation processors. The Intel CPUs based on Cooper Lake architecture with deep learning boost extensions recently came to the market. The impact of the embedded BFloat16 arithmetic unit in the performance has been analyzed [21, 22]. In this regard, [22] performed a detailed analysis of the impact of native BFloat16 FPU on the performance of DNN training. Their experiments showed that training using BFloat16 for both weights and activations based on SLIDE [70] DNN architecture, when applied on Amazon-670K and WikiLSH-325K datasets, can result in 1.28× and 1.39× higher speed than FP32, respectively [22]. Besides, a similar study for BERT's model training showed that BFloat16 can achieve 3×–4× speedup over FP32 while maintaining almost the same accuracy [21]. These results show the potential of these emerging CPUs for high-speed DNN training implementation. However, the CPUs with BFloat16 support mostly machine learning applications and situations where narrow precision can be used due to 7 mantissa bits of BFloat16. For instance, the PageRank algorithm that was investigated in previous section requires a narrow dynamic range but a high precision [40].

The third way to enable low-precision FP formats for applications running on CPUs is by decoupling the memory format from the computation format [39], with the aim of reducing memory storage and bandwidth requirement. Nowadays, the data communication dominates the FP arithmetic operations latency [71]. However, the memory format paradigm relies on conversion to/from machine-supported formats, such as FP32 and FP64. Flytes [30], CPMS [38], and APHP/TOHP [40] are examples of FP formats that use customized FP formats for data storage/movement and FP32/FP64 for computations. Anderson et al. [30], Grützmacher et al. [38], and Molahosseini and Vandierendonck [40] used the bit slice vectorization, mantissa segmentation, and truncated exponent, respectively, to overcome the conversion overhead.

4.2 Low Precision in TPUs/GPUs

TPUs [15] are accelerators designed specifically for deep learning applications [16]. The TPUv2 was the first hardware platform that embedded BFloat16 arithmetic units, which provides 1.5 times energy advantage in comparison to FP16 [15]. Figure 4.2 shows the energy per operation for different FP formats [19]. It can be seen that BFloat16 requires less energy than FP32 and FP16.

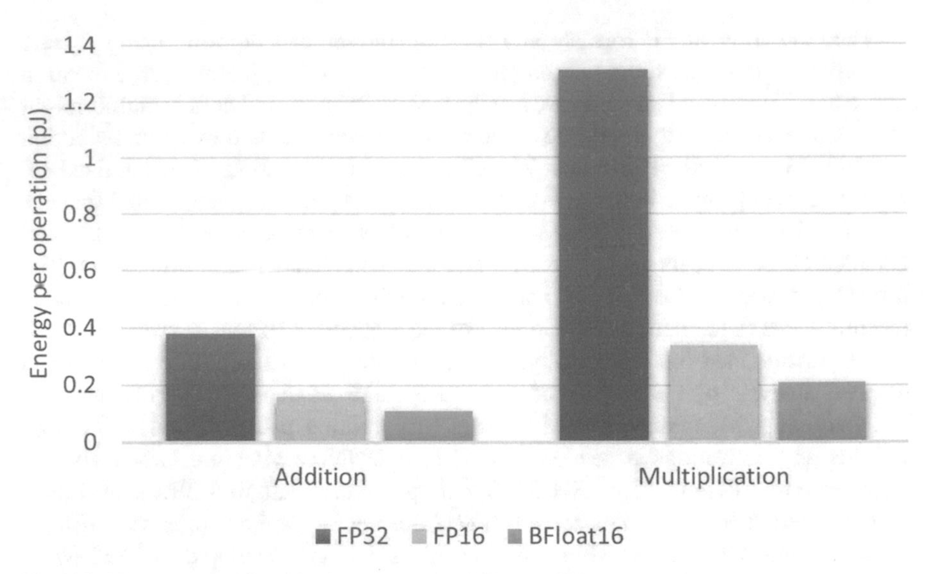

Fig. 4.2 Energy per addition/multiplication for different FP formats [19]

NVidia has supported FP16, first as a storage format in CUDA 7.5, and then as an arithmetic format with native hardware support starting with the Pascal GPUs [43]. They include a range of low-precision FP formats in the Tensor Cores (TCs) of the Ampere GPU architecture [20]. The number format can directly affect the computation's throughput and speedup. Figure 4.3 shows the maximum throughput of various FP formats in terms of trillion floating-point operations per second (TFLOPS) based on GPU-A100 architecture [20]. It can be seen that both FP16 and BFloat16 achieve higher throughput than FP32, i.e., 78 and 39 in comparison to 19.5 TFLOPS. Besides, using low-precision formats in TCs can significantly increase the throughput. Particularly, the TCs with FP16 or BFloat16 as inputs can result in 2× and 16× more throughput than the inputs with TF32 and FP32, respectively [27]. It should be noted that the accumulations in all TC's modes are performed using FP32.

4.3 *Low Precision in ASIC/FPGA*

The flexibility of customized hardware implementations such as application-specific integrated circuits (ASIC) and FPGA is ideal for customized FP realization. In this regard, Ref. [72] showed that reduced precision arithmetic can be exploited to increase FPGA parallelism. Moreover, Zoni et al. [73] provided a system-on-chip (SoC)-based design of a multi-precision FPU (Fig. 4.4) for application in embedded systems. They showed that for PolyBench benchmarks, which include various FP-intensive applications, the mix of both BFloat16 and FP32 can lead to

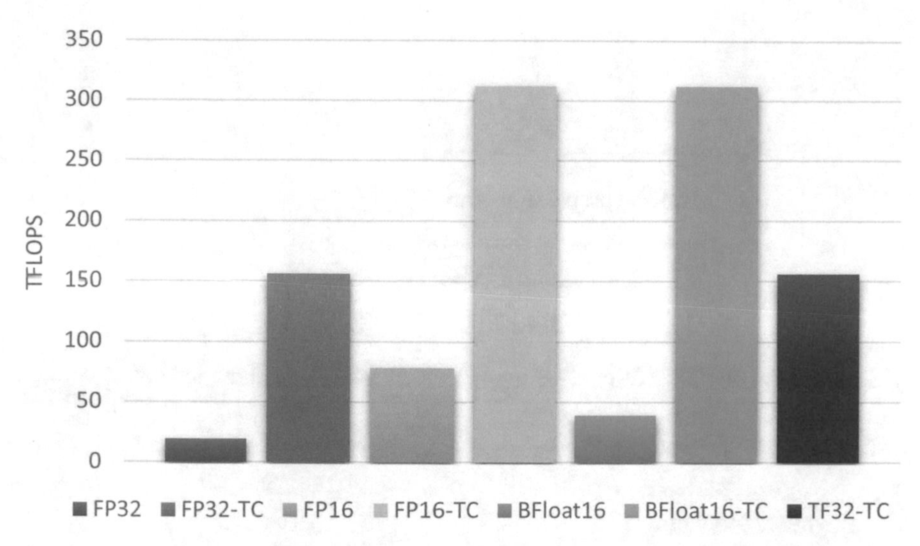

Fig. 4.3 Throughput of different FP formats on the GPU-A100 [20, 27]

19% improvement of energy-delay product with less than 2.5% average accuracy loss [73]. Additionally, Mach et al. [54] introduced the design of a full transprecision floating-point unit (TP-FPU) implemented using a single-core SoC platform. The TP-FPU supports both scalar and vector operations resulting in an efficient and flexible FPU. Figure 4.5 presents the energy of FP operations in different precisions based on both scalar and vectorized modes. It can be seen that low-precision formats, i.e., FP16, FP16alt, and FP8, result in significant energy savings in comparison to FP32. Moreover, the SIMD implementation with low-precision results in lower energy than scalar since vector operations apply more low-precision operands increasing the throughput per instruction [54]. It should be noted that FP16alt includes the same structure as BFloat16, but it follows IEEE-standard rules for special cases and subnormal support [7]. Overall, computing using the transprecision paradigm [55] based on the TP-FPU resulted in 1.67× speedup and decreased the energy consumption by 37% while maintaining end-to-end precision than FP32 [54].

5 Application-Specific Low-Precision FP Formats: Design Guidelines

This section presents a step-by-step simplified guideline that can be used as a starting point for engineers and researchers to investigate the possibility of using low-precision FP formats in their applications:

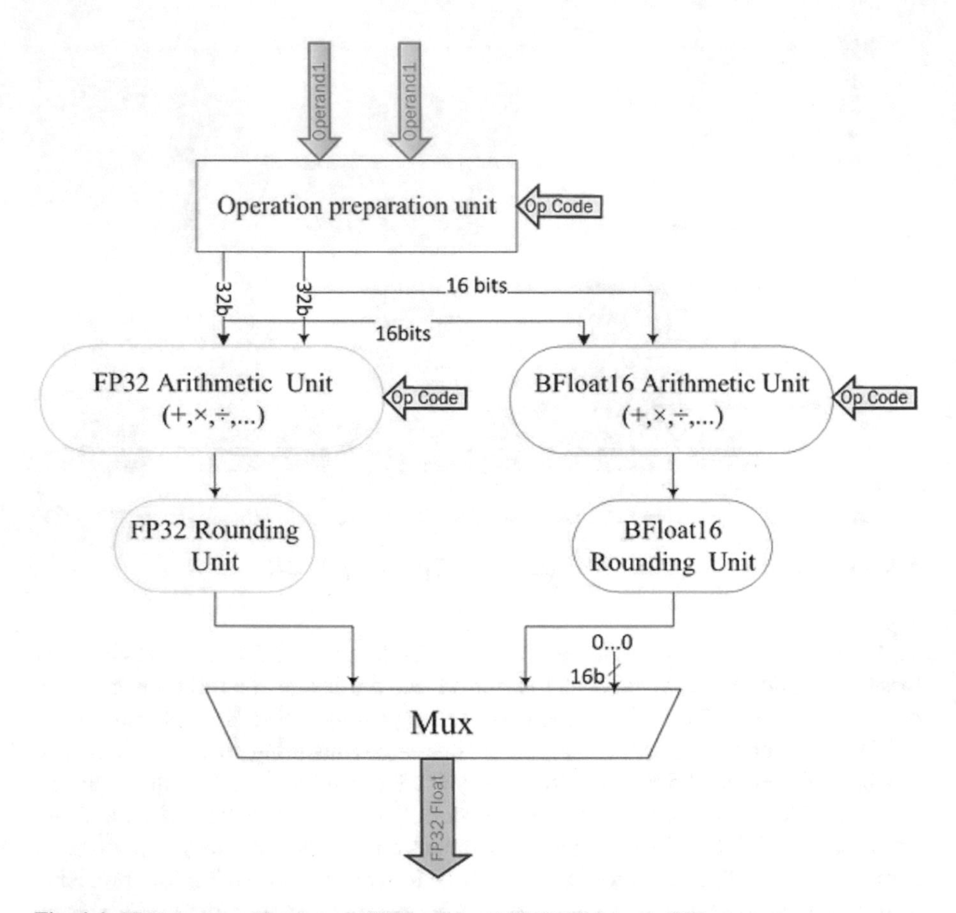

Fig. 4.4 FPU design with support of BFloat16 and FP32 FP formats [73]

- **Step 1**: The target algorithm/program should be analyzed to identify the required floating-point data types (FP32 or FP64) and the kind of operations such as addition and multiplication. In this stage, both the dynamic range (exponent) and precision (mantissa) of the required FP data should be checked to find the range and distribution of values. These analyses will experimentally show how many bits are required for each data. Besides, it will give a clue for determining the structure of the customized format including size of exponent and mantissa fields and a sign bit, if required.
- **Step 2**: The resiliency of the target algorithm against low precision can be tested by simulation with supported higher precision formats on off-the-shelf processors. In this regard, some low-precision simulation tools and libraries, such as [74] and [75], can be used. Alternatively, the behavior of the customized FP format can be emulated using FP32 by forcing the least significant bits of mantissa to zero for direct conversion, and considering appropriate rounding for reverse conversion [76]. In this stage, the engineer/researcher should test

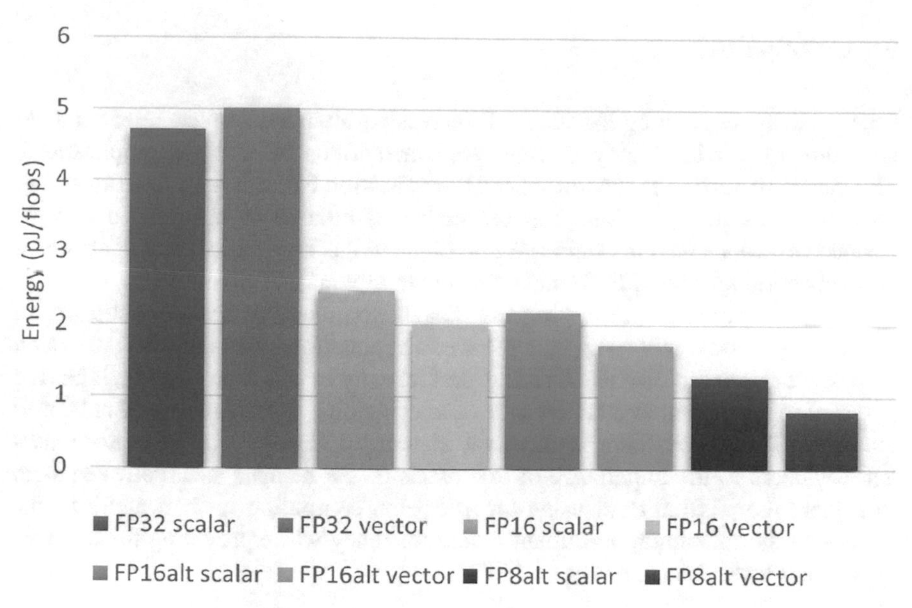

Fig. 4.5 Energy per operation of different FP formats based on SoC implementation [54]

the impact of different sizes of exponent and mantissa and different rounding schemes such as rounding to the nearest or stochastic rounding, on the accuracy of the program. This accuracy analysis can be started by substituting FP64 with FP32 to check the effect in accuracy. If the program achieves the expected accuracy/convergence with FP32 in all or some parts of the program (e.g., some iterations of an iterative algorithm), then further precision reduction can be tried by discarding more mantissa bits.

- **Step 3**: This stage aims to leverage the accuracy analysis results of the previous stage to determine the size of FP formats and its features. The first option is to use the existing low-precision FP format structures such as FP16 and BFloat16. Since FP16 and BFloat16 have hardware support on GPUs/CPUs, the potential of low-precision FP format in both bandwidth reduction and arithmetic speed can be assessed. Otherwise, a customized FP format with the required exponent and mantissa sizes can be considered, together with conversion routines to enable storing and transferring data between memory and processor. This transference is based on the customized FP format, but computations are performed based on the available hardware-supported FP formats, such as the FP32/FP64.
- **Step 4**: This step analyzes the impact of the low-precision FP format, determined in the previous step, on the performance, energy consumption, and other figures of merit. According to the results, the previous steps may need to be revisited for increasing the performance.

6 Conclusions

Low-precision computing has attracted increasing attention in recent years due to its ability to provide highly efficient implementations of emerging applications. This chapter investigates state-of-the-art low-precision FP arithmetic from emerging formats with native hardware support such as BFloat16 to customized formats designed to reduce memory traffic. We highlighted the importance and effectiveness of application-specific FP formats for modern workloads. However, the low-precision revolution is still in progress. Thanks to emerging hardware support, in the coming years, it will be applied in several applications. Although low precision is mostly explored in machine learning, and recently in graph processing, it has the potential to be applied to other areas for achieving high performance and enhanced efficiency. This chapter aims to attract the attention of non-FP specialist researchers and engineers to the importance of low-precision FP formats and motivates them to explore the possibility of using low-precision computing in their applications, to achieve the maximum performance and efficiency while providing the required application-level precision.

Acknowledgments This work is supported by FCT (Fundação para a Ciência e a Tecnologia, Portugal) through the Project UIDB/50021/2020, DiPET project (grant agreement EP/T022345/1 and CHIST-ERA Consortium of European Funding Agencies project no CHIST-ERA-18-SDCDN-002), and OPRECOMP project (European Union's H2020-EU.1.2.2.—FET Proactive research and innovation programme under grant agreement no. 732631), and Entrans project (EU Marie Curie Fellowship, grant agreement no 798209).

References

1. Goldberg D. What every computer scientist should know about floating-point arithmetic. ACM Comput Surv. 1991;23(1):5–48.
2. Molahosseini AS, Sousa L, Chang CH, editors. Embedded systems design with special arithmetic and number systems. Springer; 2017.
3. Parhami B. Computer arithmetic: algorithms and hardware designs. 2nd ed. New York: Oxford University Press; 2010.
4. IEEE Computer Society. IEEE standard for floating-point arithmetic. IEEE Std 754-2008:1–70; 2008.
5. Khan M, Bilal J, Haleem F. Deep learning: convergence to big data analytics. Springer; 2019.
6. Alioto M, editor. Enabling the Internet of Things: from integrated circuits to integrated systems. Springer; 2017.
7. Mach S. Floating-point architectures for energy-efficient transprecision computing. Doctoral Thesis, ETH ZURICH, Switzerland. 2021.
8. Malossi ACI, et al. The transprecision computing paradigm: Concept, design, and applications. In: Proc. of design, automation and test in Europe conference and exhibition (DATE). 2018.
9. Sousa L. Nonconventional computer arithmetic circuits, systems and applications. IEEE Circ Syst Mag. 2021;21(1):6–40.
10. Cherubin S, Agosta G. Tools for reduced precision computation: a survey. ACM Comput Surv. 2020;53(2):1.

11. Lee J, Vandierendonck H. Towards lower precision adaptive filters: facts from backward error analysis of RLS. IEEE Trans Signal Process. 2021;69:3446–3458.
12. Klöwer M, Düben PD, Palmer TN. Number formats, error mitigation, and scope for 16-bit arithmetics in weather and climate modeling analyzed with a shallow water model. J Adv Model Earth Syst. 2020;12(10):1–17.
13. Micikevicius P, Narang S, Alben J, Diamos G, Elsen E, Garcia D, Ginsburg B, Houston M, Kuchaiev O, Venkatesh G, Wu H, Mixed precision training. Preprint. arXiv:1710.03740. 2017.
14. Abadi M, Agarwal A, Barham P, Brevdo E, Chen Z, Citro C, Corrado GS, Davis A, Dean J, Devin M, Ghemawat S. Tensorflow: Large-scale machine learning on heterogeneous distributed systems. Preprint. arXiv:1603.04467. 2016.
15. Norrie T, et al. The design process for Google's training chips: TPUv2 and TPUv3. IEEE Micro. 2021;41(2):56–63.
16. Ying C, Kumar S, Chen D, Wang T, Cheng Y. Image classification at supercomputer scale. CoRR, abs/1811.06992. 2018.
17. Kalamkar D, et al. A study of BFLOAT16 for deep learning training. Preprint. arXiv:1905.12322. 2019.
18. Burgess N, Milanovic J, Stephens N, Monachopoulos K, Mansell D. Bfloat16 processing for neural networks. In: Proc. of IEEE 26th symposium on computer arithmetic (ARITH). 2019. pp. 88–91.
19. Jouppi NP, et al. Ten lessons from three generations shaped Google's TPUv4i: industrial product. In: Proc. of ACM/IEEE 48th annual international symposium on computer architecture (ISCA). 2021. pp. 1–14.
20. Choquette J, Gandhi W, Giroux O, Stam N, Krashinsky R. NVIDIA A100 Tensor Core GPU: performance and innovation. IEEE Micro. 2021;41(2):29–35.
21. Ozturk ME, Wang W, Szankin M, Shao L. Distributed BERT pre-training & fine-tuning with intel optimized TensorFlow on Intel Xeon scalable processors. In: Proc. of the ACM/IEEE international conference for high performance computing networking, storage, and analysis. 2020.
22. Daghaghi S, Nicholas M, Mengnan Z, Shrivastava A. Accelerating slide deep learning on modern CPUs: Vectorization, quantizations, memory optimizations, and more. In: Proc. of machine learning and systems conference. 2021.
23. bfloat16 – Hardware Numerics Definition. White Paper, Intel Corporation, USA, Nov. 2018.
24. Chromczak J, Wheeler M, Chiasson C, How D, Langhammer M, Vanderhoek T, Zgheib G, Ganusov I. Architectural enhancements in Intel Agilex FPGAs. In: Proc. of the ACM/SIGDA international symposium on field-programmable gate arrays (FPGA '20), New York, NY, USA. 2020. pp. 140–149.
25. Agrawal A, et al. DLFloat: A 16-b floating point format designed for deep learning training and inference. In Proc. of IEEE symposium on computer arithmetic (ARITH), Kyoto, Japan. 2019. pp. 92–5.
26. Bordawekar R, Abali B, Chen MH. EFloat: Entropy-coded floating point format for deep learning. arXiv:2102.02705. 2021.
27. Krashinsky R, Giroux O, Jones S, Stam N, Ramaswamy S. NVIDIA ampere architecture in-depth, NVidia Blog. https://developer.nvidia.com/blog/nvidia-ampere-architecture-in-depth/. Last Accessed: 30 Aug 2021.
28. Chung E, et al. Serving DNNs in real time at datacenter scale with project brainwave. IEEE Micro. 2018;38(2):8–20.
29. Köster U, et al. Flexpoint: an adaptive numerical format for efficient training of deep neural networks. In: Proc. of the 31st international conference on neural information processing systems, Red Hook, NY, USA. 2017. pp. 1740–50.
30. Anderson A, Muralidharan S, Gregg D. Efficient multibyte floating point data formats using vectorization. IEEE Trans Comput. 2017;66(12):2081–96.
31. Nannarelli A. Tunable floating-point for energy efficient accelerators. In: Proc. of IEEE symposium on computer arithmetic, Amherst, MA, USA. 2018. pp. 29–36.

32. Nannarelli A. Variable precision 16-bit floating-point vector unit for embedded processors. In: Proc. of IEEE symposium on computer arithmetic (ARITH), Portland, OR, USA. 2020. pp. 96–102.
33. Gustafson JL, Yonemoto I. Beating floating point at its own game: posit arithmetic. Supercomput Front Innov 2017;4(2):71–86.
34. Carmichael Z, Langroudi HF, Khazanov C, Lillie J, Gustafson JL, Kudithipudi D. Performance-efficiency trade-off of low-precision numerical formats in deep neural networks. In: Proc. of the conference for next generation arithmetic (CoNGA'19), New York, NY, USA. 2019. pp. 3, 1–9.
35. Montero RM. Leveraging posit arithmetic in deep neural networks. Master Thesis, Complutense University of Madrid. 2021.
36. Lu J, Fang C, Xu M, Lin J, Wang Z. Evaluations on deep neural networks training using posit number system. IEEE Trans Comput. 2021;70(2):174–87.
37. Langroudi HF, Carmichael Z, Gustafson JL, Kudithipudi D. PositNN framework: tapered precision deep learning inference for the edge. In: Proc. of IEEE space computing conference (SCC). 2019. pp. 53–9.
38. Grützmacher T, Cojean T, Flegar G, Anzt H, Quintana-Ortí ES. Acceleration of PageRank with customized precision based on mantissa segmentation. ACM Trans Parallel Comput. 2020;7(1):1.
39. Grützmacher T, Cojean T, Flegar G, Göbel F, Anzt H. A customized precision format based on mantissa segmentation for accelerating sparse linear algebra. Concurr Comput Pract Exp. 2020;32(15):1–12.
40. Molahosseini AS, Vandierendonck H. Half-precision floating-point formats for PageRank: opportunities and challenges. In: Proc. of IEEE high performance extreme computing conference (HPEC). 2020. pp. 1–7.
41. Markidis S, Chien SWD, Laure E, Peng IB, Vetter JS. Nvidia Tensor Core programmability, performance & precision. In Proc. of IEEE international parallel and distributed processing symposium workshops (IPDPSW). 2018. pp. 522–531.
42. Mach S, Rossi D, Tagliavini G, Marongiu A, Benini L. A transprecision floating-point architecture for energy-efficient embedded computing. In Proc. of IEEE international symposium on circuits and systems (ISCAS). 2018. pp. 1–5.
43. Ho N, Wong W. Exploiting half precision arithmetic in Nvidia GPUs. In: Proc. of IEEE high performance extreme computing conference (HPEC). 2017. pp. 1–7.
44. Haidar A, Tomov S, Dongarra J, Higham NJ. Harnessing GPU tensor cores for fast FP16 arithmetic to speed up mixed-precision iterative refinement solvers. In: Proc. of international conference for high performance computing, networking, storage and analysis. 2018. pp. 603–13.
45. Zambelli C, Ranhel J. Half-precision floating point on spiking neural networks simulations in FPGA. In: Proc. of international joint conference on neural networks (IJCNN). 2018. pp. 1–6.
46. Brennan J, Bonner S, Atapour-Abarghouei A, Jackson PT, Obara B, McGough AS. Not half bad: exploring half-precision in graph convolutional neural networks. In: Proc. of IEEE international conference on big data (Big Data). 2020. pp. 2725–34.
47. Björck J, Chen X, De Sa C, Gomes CP, Weinberger K. Low-precision reinforcement learning: running soft actor-critic in half precision. In: Proc. of the 38th international conference on machine learning. 2021. pp. 980–91.
48. Venkataramani S, et al. Efficient AI system design with cross-layer approximate computing. Proc IEEE. 2020;108(12):2232–50.
49. Firoz JS, Li A, Li J, Barker K. On the feasibility of using reduced-precision tensor core operations for graph analytics. In: Proc. of IEEE high performance extreme computing conference (HPEC), Waltham, MA, USA. 2020. pp. 1–7.
50. Carvalho A, Azevedo R. Towards a transprecision polymorphic floating-point unit for mixed-precision computing. In: Proc. of 31st international symposium on computer architecture and high performance computing (SBAC-PAD). 2019. pp. 56–63.

51. Xie S, Davidson S, Magaki I, Khazraee M, Vega L, Zhang L, Taylor MB. Extreme datacenter specialization for planet-scale computing: ASIC clouds. ACM SIGOPS Oper Syst Rev. 2018;52(1):96–108
52. Yang K, Chen YF, Roumpos G, Colby C, Anderson JR. High performance Monte Carlo simulation of Ising model on TPU clusters. CoRR, abs/1903.11714. 2019.
53. Henry G, Tang PTP, Heinecke A. Leveraging the bfloat16 artificial intelligence datatype for higher-precision computations. In: Proc. of IEEE 26th symposium on computer arithmetic (ARITH), Kyoto, Japan. 2019. pp. 69–76.
54. Mach S, Schuiki F, Zaruba F, Benini L. FPnew: an open-source multiformat floating-point unit architecture for energy-proportional transprecision computing. IEEE Trans Very Large Scale Integr (VLSI) Syst. 2021;29(4):774–87.
55. Tagliavini G, Marongiu A, Benini L. FlexFloat: a software library for transprecision computing. IEEE Trans Comput Aided Des Integr Circ Syst. 2020;39(1):145–56.
56. Wang N, Choi J, Brand D, Chen CY, Gopalakrishnan K. Training deep neural networks with 8-bit floating point numbers. In: Proc. of the 32nd international conference on neural information processing systems, Red Hook, NY, USA. 2018. pp. 7686–95.
57. Xu S, Gregg D. Bitslice vectors: a software approach to customizable data precision on processors with SIMD extensions. In: Proc. of 46th international conference on parallel processing (ICPP). 2017. pp. 442–51.
58. Tambe T, et al. Algorithm-hardware co-design of adaptive floating-point encodings for resilient deep learning inference. In: Proc. of 57th ACM/IEEE design automation conference (DAC). 2020. pp. 1–6.
59. Franceschi M, Nannarelli A, Valle M. Tunable floating-point for embedded machine learning algorithms implementation. In: Proc. of 15th international conference on synthesis, modeling, analysis and simulation methods and applications to circuit design (SMACD). 2018. pp. 89–92.
60. Franceschi M, Nannarelli A, Valle M. Tunable floating-point for artificial neural networks. In Proc. of 25th IEEE international conference on electronics, circuits and systems (ICECS). 2018. pp. 289–92.
61. Nannarelli A. Tunable floating-point adder. IEEE Trans Comput. 2019;68(10):1553–60.
62. Elam D, Iovescu C. A block floating point implementation for an N-Point FFT on the TMS320C55x DSP. Application Report, Texas Instruments. 2003.
63. Drumond M, Lin T, Jaggi M, Falsafi B. Training DNNs with hybrid block floating point. In: Proc. of the 32nd international conference on neural information processing systems (NIPS'18), Red Hook, NY, USA. 2018. pp. 451–61.
64. Fox S, Rasoulinezhad S, Faraone J, Leong P. A block minifloat representation for training deep neural networks. In: Proc. of international conference on learning representations. 2020.
65. Langville AN, Meyer CD. Google's PageRank and beyond: the science of search engine rankings. Princeton University Press; 2012.
66. Takac L, Zabovsky M. Data analysis in public social networks. In: Proc. of international scientific conference and international workshop present day trends of innovations, Lomza, Poland. 2012.
67. Jouppi NP, Yoon DH, Kurian G, Li S, Patil N, Laudon J, Young C, Patterson D. A domain-specific supercomputer for training deep neural networks. Commun ACM. 2020;63(7):67–78.
68. Hauser J. The SoftFloat and TestFloat validation suite for binary floating-point arithmetic. Technical Report, University of California, Berkeley. 1999.
69. Gerlach L, Payá-Vayá G, Blume H. Efficient emulation of floating-point arithmetic on fixed-point SIMD processors. In Proc. of IEEE international workshop on signal processing systems (SiPS). 2016. pp. 254–9.
70. Chen B, et al. Slide: In defense of smart algorithms over hardware acceleration for large-scale deep learning systems. arXiv preprint arXiv:1903.03129. 2019.
71. Mutlu O, Ghose S, Gómez-Luna J, Ausavarungnirun R. Processing data where it makes sense: Enabling in-memory computation. Microprocess Microsyst 2019;67:28–41.
72. Lee J, Peterson GD, Nikolopoulos DS, Vandierendonck H. AIR: Iterative refinement acceleration using arbitrary dynamic precision. Parallel Comput 2020;97:1–13.

73. Zoni D, Galimberti A, Fornaciari W. An FPU design template to optimize the accuracy-efficiency-area trade-off. Sustain Comput Inform Syst. 2021;29:1–10.
74. Higham NJ, Pranesh S. Simulating low precision floating-point arithmetic. SIAM J Sci Comput. 2019;41(5):585–602.
75. Fasi M, Mikaitis M. CPFloat: A C library for emulating low-precision arithmetic. Technical Report, The University of Manchester. 2020.
76. Romanov AY, et al. Analysis of Posit and Bfloat arithmetic of real numbers for machine learning. IEEE Access. 2021;9:82318–24.

Chapter 5
Spintronic Solutions for Approximate Computing

You Wang, Hao Cai, Kaili Zhang, Bo Wu, Bo Liu, Deming Zhang, and Weisheng Zhao

1 Introduction

With the fast development of the internet of things (IoT), cloud computing and artificial intelligence, multimedia applications have brought data boost to computing systems. As the conventional complementary metal oxide semiconductor (CMOS) technology continues to scale down, the variability and standby energy have been largely increased. As a result, a relatively high supply voltage or oversized transistors should be applied to ensure high reliability designs, which further increases the energy consumption or area overhead [1]. Thus, the requirement for computing methods with high energy efficiency becomes urgent. In this context, approximate computing is a promising solution to save energy and reduce circuit complexity by relaxing the accuracy of computations or data precision for applications with inherent error tolerance. For certain applications, such as multimedia, approximate outputs can be well interpreted by human senses despite being inaccurate [2].

The traditional approximate techniques at circuit level can be categorized into two main strategies: voltage over-scaling (VOS) and design complexity reduction (DCR) [3]. For VOS, a long range of supply voltage is applied to pursue high energy

The authors Y. Wang and H. Cai contributed equally to this work.

Y. Wang (✉)
Hefei Innovation Research Institute, Beihang University, Hefei, China
e-mail: youwang@buaa.edu.cn

H. Cai · B. Liu
School of Electronic Science and Engineering, Southeast University, Nanjing, China

K. Zhang · B. Wu · D. Zhang · W. Zhao
School of Integrated Circuit Science and Engineering, Beihang University, Beijing, China
e-mail: weisheng.zhao@buaa.edu.cn

W. Liu, F. Lombardi (eds.), *Approximate Computing*,
https://doi.org/10.1007/978-3-030-98347-5_5

efficiency with the cost of possible errors in the circuit functionality. Moreover, as the voltage applied on transistors is lowered, the aging effect can be weakened and the transistors may have a longer lifetime. The DCR strategies are usually realized by two methods: one is to implement part of the circuit extracted from the initial design; the other is to process the most significant bits accurately and accept potential errors in the least significant bits.

The early approximate computing solutions are mainly implemented in CMOS technology [4, 5]. As the physical limit is approaching, the silicon-based devices cannot continue to maintain performance boost as technology node scales down, especially in terms of energy efficiency. Moreover, the inconsistency between the operation speed of memory and that of the logic parts has become another bottleneck, which limits the further development of traditional CMOS-based computing [6]. Thus, new devices are urgently needed to meet the requirements of high energy-efficient computing.

Several emerging nonvolatile technologies have been intensively researched to address the increasing power issues induced by continuous scaling of CMOS technology node [7]. As the most commercialized spintronic device, spin transfer torque magnetic random access memory (STT-MRAM) is considered as the most suitable candidate for on-chip cache applications [8, 9]. Compared to other technologies, STT-MRAM is outstanding in terms of access speed, endurance, energy consumption, reliability and compatibility with CMOS technology [10]. STT-MRAM has the potentials to achieve the integration density of dynamic RAM (DRAM) and performance comparable to static RAM (SRAM). Moreover, the current-induced threshold operation provides a large design space to explore in approximate computing. Different from the STT switching method, MTJ with voltage-controlled magnetic anisotropy (VCMA) effect provides flipping of the magnetization upon a voltage pulse, irrespective of the initial state. Thus, this magnetoelectric random access memory (MeRAM) achieves less energy consumption and higher density, as well as the improved switching latency thanks to the very little charge flow required to operate [11, 12]. Another switching method denoted as the spin orbit torque (SOT) has been well developed recently with fast magnetization switching. Furthermore, the emerging Nand-like STT memory with SOT erase operation results in a high density and overcomes the asymmetric switching and source degeneration [13]. These spintronic devices with new switching mechanisms can also be used to explore innovative approximate techniques in computing chips.

A variety of previous works rely on additional CMOS circuits, which are sensitive to variability issues from both nonvolatile devices and CMOS transistors. The mixed precision processing in memory (PIM) concept is firstly proposed based on the phase change memory [14], which reveals the significance of the nonvolatile approximate PIM procedure. In this chapter, two categories of approximate techniques are applied in the PIM based on spintronics. In the first group, STT-MRAM is used in traditional full adders to achieve approximate computing by using the traditional methods of voltage over-scaling (VOS) and design complexity reduction (DCR). This group is presented to explore the benefits that the approximate techniques may bring to commercialized STT-MRAM-based circuits. In the second group, the joint switching mechanism of MTJ (VCMA and SOT) is proposed for approximate computing. Different from traditional PIM methods with sensing, processing and pass transistor logic-based logic-in-memory, this method only uses the MTJ writing operation to complete the PIM. This group is developed to show how emerging spintronic devices may improve the performance of the approximate processing in memory. The comprehensive study on performance evaluation shows that the energy consumption can be considerably reduced by these methods.

2 Basics of Spintronic Devices

The development of spintronic devices is originated from the discovery of the tunnel magnetoresistance (TMR) effect [15]. Magnetic tunnel junction (MTJ) was then invented based on this effect. The functionality of MTJ is mainly determined by three nanoscale films, i.e. two ferromagnetic layers sandwiched by an oxide barrier. The resistance value of MTJ depends on the relative magnetization orientation of the two ferromagnetic layers (R_p at parallel state and R_{ap} at antiparallel state). With different configurations, MTJ can be used to represent logic ‘0’ or ‘1’ in the memories and logic circuits. As logic value can be stored in MTJ by magnetic properties without voltage bias, it is an ideal candidate for memories.

The working mechanism of MTJ is demonstrated in Fig. 5.1a. The nanopillar resistance (R_p, R_{ap}) depends on the relative magnetization between RL and FL (Parallel (P) or Anti-Parallel (AP)). The resistance difference of P and AP state is characterized by tunnel magnetoresistance ratio: TMR $= (R_{ap} - R_p)/R_p$. With spin transfer torque (STT) mechanism, the configuration of MTJ can be changed between two states when a bidirectional current I higher than the threshold current

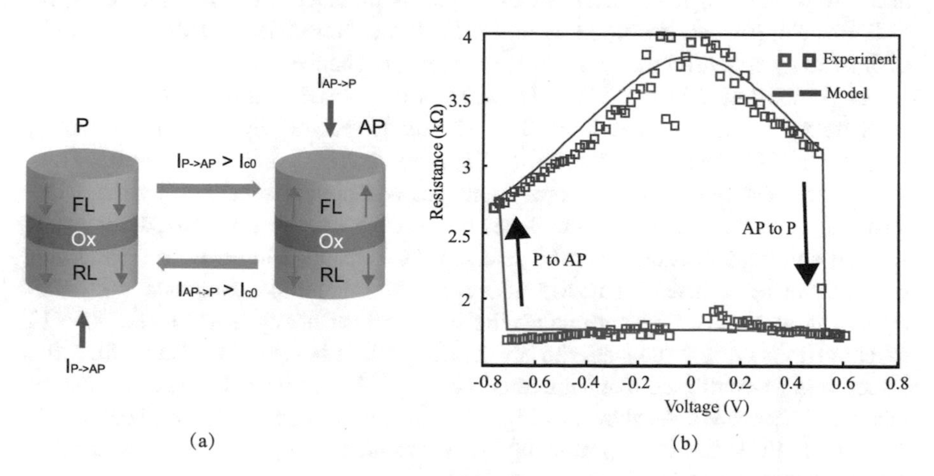

Fig. 5.1 (**a**) The main structure and working mechanism of MTJ. (**b**) The resistance transition according to the voltage applied across the STT-MTJ

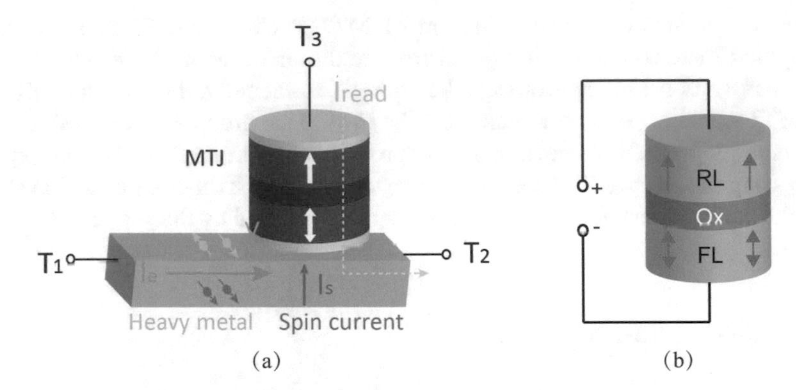

Fig. 5.2 (**a**) SOT switching approach: I_e is the writing current, I_s is the spin current generated by heavy metal and I_{read} is the reading current. (**b**) The structure of VCMA-MTJ: the oxide layer (Ox) is usually much thicker than that of STT-MTJ and SOT-MTJ

I_{c0} is injected. Figure 5.1b illustrates the resistance transition according to the voltage applied across the STT-MTJ. The red line signifies the simulation results of a compact model [16], while the blue rectangles represent the experimental results in [17].

TMR and the threshold current I_{c0} are two main characters of MTJ, which prompt the evolution of MTJ. Lower threshold current facilitates the writing of MTJ, while higher TMR ratio contributes to improved reading performance in terms of delay, energy and reliability. The switching method has evolved from field-induced magnetic switching (FIMS$\sim$10 mA), thermally assisted switching (TAS$\sim$1 mA) to the currently widely commercialized spin transfer torque switching (STT$\sim$100 uA). Without the need of magnetic field, STT makes it possible to achieve high density and low power magnetoresistive random access memory (MRAM). However, the STT switching speed is limited by an intrinsic incubation delay. This issue can be addressed by spin orbit torque (SOT) switching mechanism [14].

As demonstrated in Fig. 5.2a, SOT-MTJ is composed of a STT-MTJ deposited on a heavy metal (e.g. tantalum). The writing is realized by the spin current I_s across the MTJ generated by injecting a charge current I_e into heavy metal. Thus, the writing and sensing (I_{read}) operations are completely separated by the three-terminal configuration [13]. Therefore, low resistance can be realized for easier writing and high resistance can be realized for sensing. Moreover, the switching current can be reduced by nearly one order of magnitude compared with STT switching mechanism by optimizing the thickness of heavy metal layer. As STT mechanism is current induced, the self-heating effect is inevitable. To address this issue, voltage-controlled magnetic anisotropy (VCMA) assisted switching method has been discovered. As shown in Fig. 5.2b, instead of current, the utilization of a voltage via VCMA effect is promising to achieve ultra-low-power memory and PIM by significantly reducing Joule heating [18].

The resistance of p-MTJ is very sensitive to process variation due to the nanoscale size. The major source of process variation of MTJ arises from two aspects: (1) variable geometrical parameters due to surface roughness and inherent film variations (cross-sectional area $Area$, thickness of oxide barrier t_{ox} and thickness of free layer t_{sl}) and (2) imprecise magnetic properties due to inhomogeneity of materials induced by imperfect process (anisotropy field H_k, magnetization saturation M_s). These variations have severe impacts on the electrical properties of MTJ (resistance, TMR ratio, critical current I_{c0} and switching delay τ_{sw}) and further lead to performance degradation. The parameter variations are usually considered to follow approximately Gaussian distribution.

In reality, the switching of FM layers magnetization is usually not immediate but has an incubation period, denoted as switching delay τ_{sw}. The dynamic model is mainly composed of calculating the average switching delay τ_{sw} (with 50% of switching probability). Depending on the magnitude of switching current, the dynamic behaviour of MTJ can be divided into two regimes [19]: Sun model ($I > I_{c0}$) [20] and Neel–Brown model ($I < 0.8I_{c0}$) [21]. The former is also called precessional switching, which addresses fast switching (until sub 3 ns) but consumes more energy with high current density. Reversely, the latter consumes less energy with low current density but leads to slower switching, which is called thermally assisted switching. The two regimes are derived from the Landau–Lifshitz–Gilbert equation. τ_{sw} can be calculated as follows [20, 21]:

$$\tau_{sw} = \tau_0 \cdot \exp\left[\Delta\left(1 - \frac{I}{I_{c0}}\right)\right], \quad \text{when } I < 0.8I_{c0} \tag{5.1}$$

$$\frac{1}{\tau_{sw}} = \left[\frac{2}{C + \ln(\frac{\pi^2\Delta}{4})}\right] \frac{\mu_B P_{ref}(I - I_{c0})}{em_m(1 + P_{ref}P_{free})}, \text{ when } I > I_{c0} \tag{5.2}$$

where τ_0 is the attempt period, C is the Euler's constant, m_m is the magnetization moment and P is the tunnelling spin polarization. Usually, a high current ($I > I_{c0}$) is applied to guarantee fast writing in memory. Meanwhile, MTJ can also be erroneously switched by relatively low current ($I < 0.8I_{c0}$) during a long period of reading operation, determining the data retention time.

The thermal fluctuation of the environment introduces the randomness in the switching process, denoted as stochastic switching behavior. With injected current I, the switching probability can be described as follows [22]:

$$P_{sw} = \exp\left\{-\frac{t_{pulse}}{\tau_0}\exp\left[-\Delta\left(1 - \frac{I}{I_{c0}}\right)\right]\right\} \tag{5.3}$$

where t_{pulse} is the current pulse width. The stochastic switching is integrated into the model by using the random functions in Verilog-A language. The users are free to reconfigure the simulation conditions by choosing different types of statistical

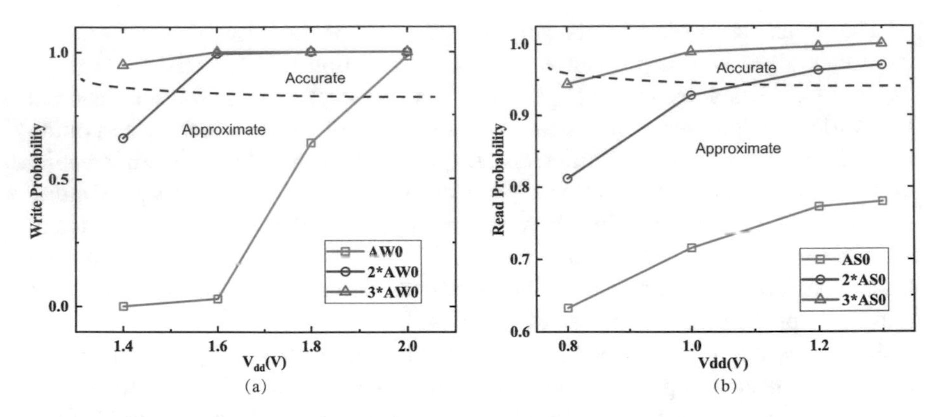

Fig. 5.3 (**a**) The writing success probability on function of supply voltage and the area. (**b**) The reading success probability on function of supply voltage and the area. AW0 and AS0 are the area of writing circuit and reading circuit with the minimum size of transistors

distributions for switching delay τ_{sw}. The variables used in the model are adjusted to achieve good agreement with the Eq. (5.3).

In the development of STT-MRAM, write failure induced by stochastic switching behavior and read failure induced by process variation are two major challenges for reliable operations [23]. Figure 5.3 demonstrates the effect of process variations and stochastic switching of MTJ on the success probability of MTJ writing and reading circuits. The writing circuit is the traditional 4T-2M circuit, while the reading circuit is the precharge sense amplifier [23]. The circuits are designed by using a 28 nm CMOS design kit and a 40 nm MTJ compact model [24]. It can be observed that the error rates decrease with the growth of supply voltage and area occupation. In realistic designs, high supply voltage and area overhead are usually used to guarantee the accurate results. Thus, much energy and area cost can be saved by applying the approximate techniques in STT-MRAM-based circuits.

3 STT-MRAM-Based Approximate Computing

There are two strategies to apply approximate techniques in the operations of spintronic devices: implementation of spintronic devices in traditional circuits, novel circuit designs based on spintronic devices. The details will be presented by taking the examples of STT-MRAM-based traditional approximate full adders and write-only bitwise approximate full adders based on SOT-MTJ and VCMA-MTJ.

3.1 STT-MRAM-Based Traditional Approximate Full Adders

At circuit level, the design strategies can be categorized into two groups: design complexity reduction DCR and voltage over-scaling (VOS). These two groups have been applied in [25] to realize approximate full adders based on the accurate ones proposed in [26]. As illustrated in Fig. 5.4, DCR is realized by eliminating the input carry of the full adder to get an approximation sum (AX-NV-FA1), while VOS is realized by insufficient write current of MTJ (AX-NV-FA2), which stores the data B for computation. With simplified logic, the former one has less area occupation and shorter delay. Without the pressure of successful writing, the latter one can work at accurate mode or at approximate mode under ultra-low-supply voltage, which consumes much less energy.

Figure 5.5 shows the timing waveform of the approximate adder with simplified logic. All possible combinations of inputs A, B (NV data stored in MTJs) and C_{in} are designated. Note that inexact output Sum is marked whereas output C_{in} is accurate. The total error distance is 4 in simplified logic-based approximate NV-FA. When V_{dd} is lower than 0.8 V, the current flowing through MTJ is lower than threshold MTJ switching current and MTJ states cannot be changed between P and AP. Thus, power consumption in both MTJ sensing and write operation is greatly reduced. Meanwhile, inexact output occurs at adder output (both Sum and C_{in}). The approximate mode can be executed over a supply ranging from 0.3 to 0.75 V. The transient simulation results of the dual-mode approximate adder are shown in Fig. 5.6. A 0.5 V supply is applied as an example in this NV-FA structure. The

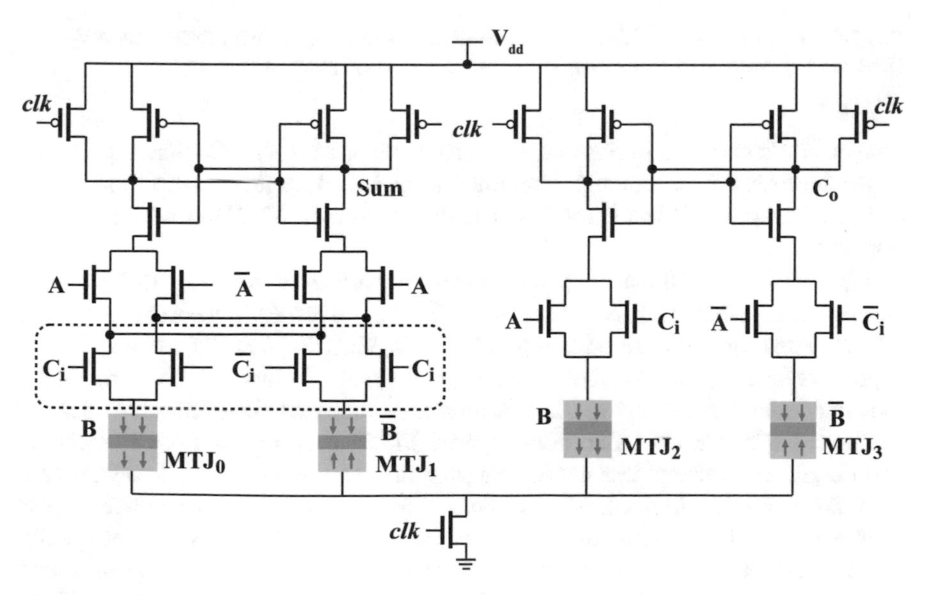

Fig. 5.4 Approximate NV-FA based on STT-MTJ: AX-NV-FA1 (without dashed line box) and AX-NV-FA2

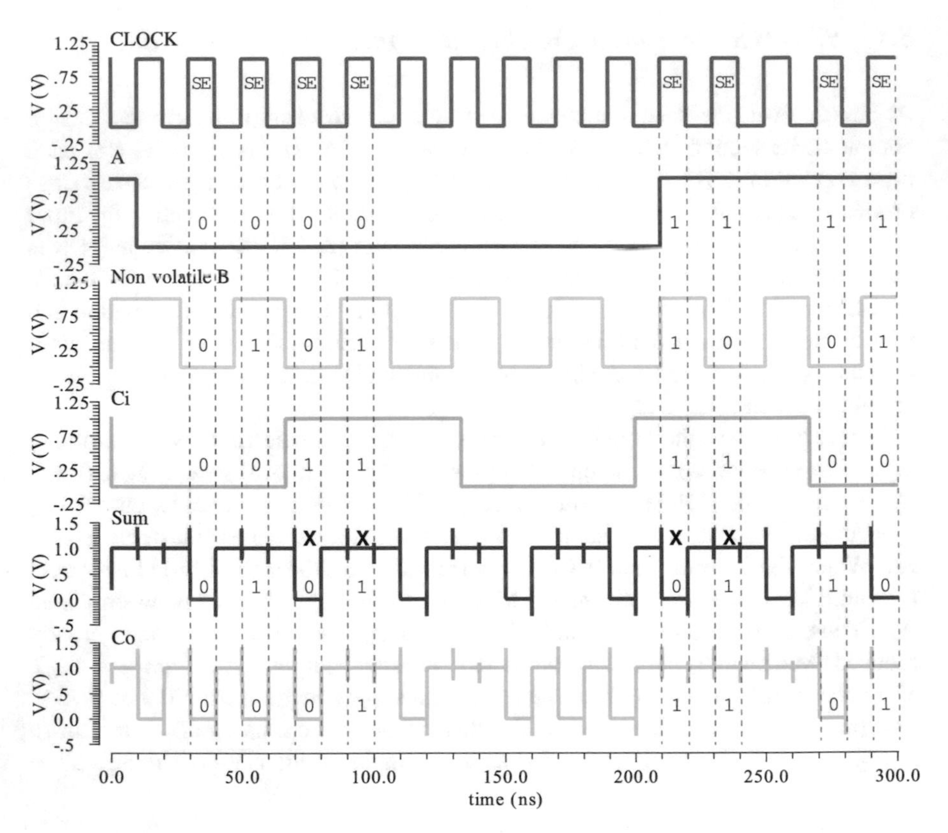

Fig. 5.5 The transition simulation waveforms of approximate adder with DCR (AX-NV-FA1). Output *Sum* is with errors distance 4. There is no error in C_o operation

total error distance is 6 in this simplified approximate NV-FA. Comparing the two approximate NV-FAs, although simplified logic-based implementation has 4 less transistor count and 2 less error distance, the dual-mode NV-FA is with a flexible implementation.

Figure 5.7 demonstrates the performance comparison of the STT-MRAM-based approximate adders (AX-NV-FA1, AX-NV-FA2), accurate adder (NV-FA) and CMOS-based approximate adders (CMOS-FA1, CMOS-FA2) [25]. Note that the supply voltage of AX-NV-FA2 is 0.5 V, while that of others is 1 V. The delay, dynamic/leakage accuracy and layout area are compared for each adder design. Compared with CMOS-based counterparts, MTJ-based approximate full adders feature higher accuracy, less energy consumption and larger area (the layout area includes magnetic flip-flop). Meanwhile, the approximate NV FAs consume less energy than their accurate counterparts at the expense of reduced output quality and enlarged delay. Compared to the DCR technique (AX-NV-FA1), the write and sensing power of the VOS technique (AX-NV-FA2) are reduced by nearly 70%

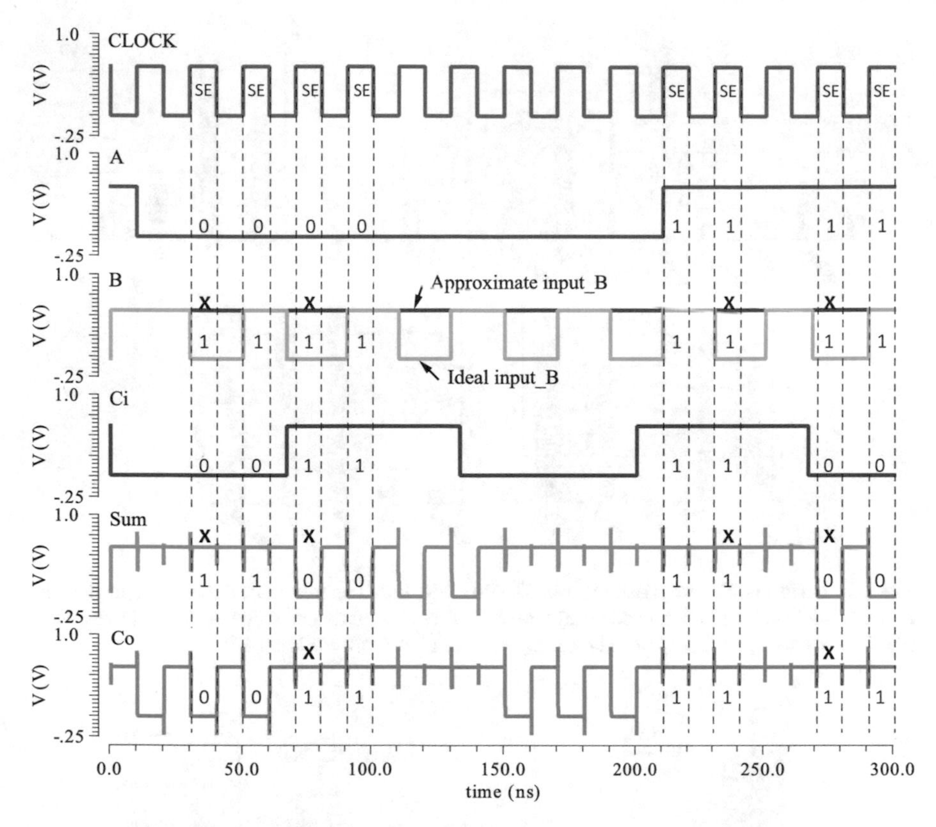

Fig. 5.6 The transient simulation waveforms of approximate adder by insufficient write current of MTJ (AX-NV-FA2, dual-mode NV-FA). The error distance is 4 in *Sum* and 2 in C_o

with the cost of increased sensing delay. For the VOS technique, the circuit can be modulated between accurate mode and approximate mode by modifying the supply voltage.

3.2 Write-Only Bitwise Approximate Full Adders Based on SOT-MTJ and VCMA-MTJ

Controlled by the voltage pulse, VC-MTJ can be at either parallel (logic '0') or anti-parallel (logic '1') states. Its writing operation is completed by switching the state of an MTJ. For the structure in Fig. 5.8a, the MTJ is switched by turning on the NMOS transistor (setting WL = '1') and adding a specific voltage pulse between BL and SL [27]. For example, a 1.2 V voltage pulse with a duration of 0.18 ns followed by a 0.6 V voltage pulse with a duration of 0.22 ns results in an MTJ switching; this is referred to as the STT-assisted precessional VCMA. On the other hand, a

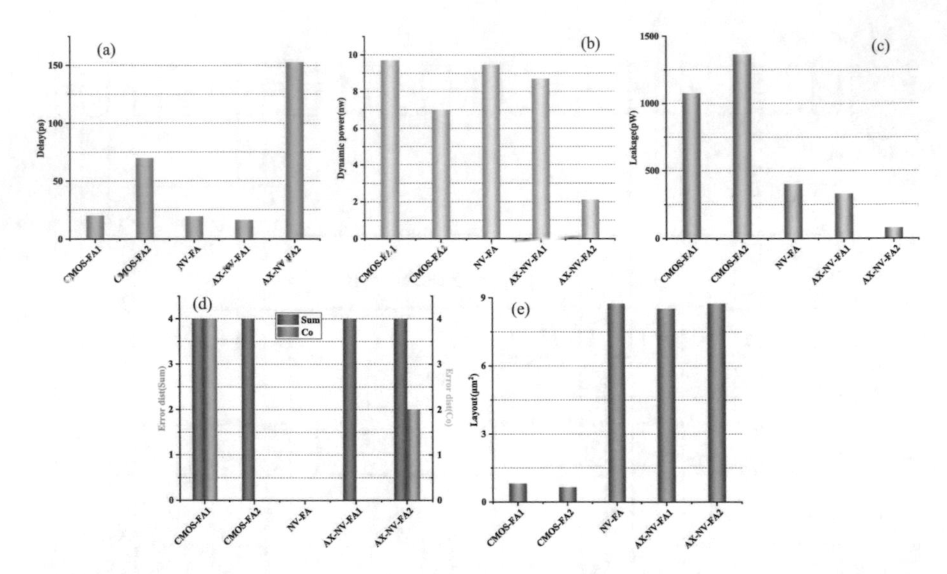

Fig. 5.7 Performance comparison of the STT-MRAM-based approximate adders, accurate adder and CMOS-based approximate adder in terms of (**a**) delay, (**b**) dynamic power, (**c**) leakage power, (**d**) error distance and (**e**) layout. The supply voltage of AX-NV-FA2 is 0.5 V

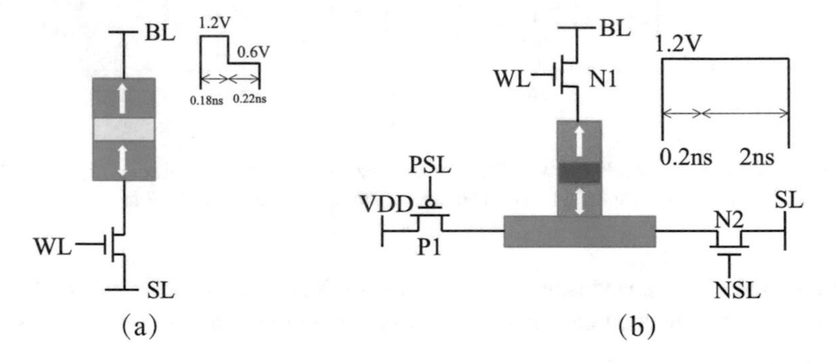

Fig. 5.8 Two typical interplay switching schemes for MTJ as writing/data storage process: (**a**) STT-assisted precessional VCMA and (**b**) SOT-STT interactions. Reproduced from [27]

0.55 ns voltage pulse of 1.1 V can also switch the MTJ, which is denoted as the precessional VCMA (P-VCMA). In Fig. 5.8b, an erasing operation is performed before each writing by turning on P1 and N2 and turning off N1, and the MTJ is at anti-parallel state. The MTJ switches to parallel state by a positive voltage pulse between VDD and BL to complete the writing of '0', when P1 and N1 are on, N2 is off. Otherwise, the MTJ stays at anti-parallel state.

Unlike traditional In-MRAM computing methods: first sensing then processing or pass transistor logic-based logic-in-memory, the proposed design relies on MTJ writing to complete the FA operations. Specifically, the input A of a FA is initially stored in an MTJ. The inputs B and C_{in} are consecutively fed into WL of a 1T-

1M bit-cell as demonstrated in Fig. 5.8a, whereas B is connected to NSL (or PSL) and C_{in} is connected to WL (or NSL) when using the structure in Fig. 5.8b. By controlling the voltage on the MTJ, the 1-bit addition is approximately implemented with state switches of the MTJ. Finally, four AxFAs are obtained by applying different switching mechanisms, which are referred to as Ax1 (SOT+STT), Ax2 (SOT), Ax3 (VCMA+STT) and Ax4 (P-VCMA). The truth table of the proposed AxFAs is listed in Table 5.1. Ax1 and Ax2 are implemented by the structure of Fig. 5.8b, and Ax3 and Ax4 are implemented by the bit-cell in Fig. 5.8a.

- **Ax1**: In this design, the inputs B and C_{in} are connected to NSL and WL (Fig. 5.8b), respectively. In step one, P1 is on and N1 is off, and B controls SOT current, i.e. the data in MTJ is '1' when B = '1$'$, otherwise it is A. In step two, P1 is on and N2 is off, and C_{in} controls STT current flow, i.e. the data stored in MTJ is ultimately '0' if C_{in} is '1'.
- **Ax2**: Unlike Ax1, the inputs B and C_{in} in Ax2 together control the SOT current, and hence, only one step is required to obtain the sum result. For Ax2, B is connected to PSL, and C_{in} is fed into NSL. To use the SOT mechanism, WL is always off. The data in MTJ is erased to '1' when B='0' and C_{in}='1', otherwise it stays as A.
- **Ax3**: In Ax3, the inputs B and C_{in} are consecutively input to WL (Fig. 5.8a). In this case, B and C_{in} control VCMA and STT effects, respectively. Thus, each high input signal generates a voltage pulse. The data stored in MTJ will be flipped due to the VCMA-STT switching when B = '1' and C_{in} = '1', otherwise it does not change.
- **Ax4**: As discussed in Section II, P-VCMA without the assistance of STT also results in a state switching of MTJ driven by a different voltage pulse. Thus, Ax4 uses two P-VCMA effects; the inputs B and C_{in} are consecutively fed into WL. Ax4 achieves the same function as Ax3.

As shown in Table 5.1, the signal stored in the MTJ after the required operations for each design is output as the Sum. The carry-out of the four designs C_o is taken from the input signal B to avoid the carry propagation delay when using the AxFAs in a multi-bit adder. The "X" following each digit indicates an incorrect result. E_{Axi} shows the errors for Axi considering both Sum_{Axi} and C_o. Table 5.1 shows that the probability of generating an error C_o is pretty low (25%). Also, the C_o results in a very low average error because both positive and negative errors can be produced.

In this section, the analysis is executed in Cadence Virtuoso with a VCMA-MTJ and a SOT-MTJ compact model, as well as a 28-nm CMOS technology [13, 18]. As discussed in Section II, the MTJ depends on the voltage/bias condition of the different bit-cell structures for MRAM. Table 5.2 lists the setup of the proposed four AxFAs.

Table 5.1 The truth table of the proposed joint switching approximate FA. The "X" indicates an incorrect output. Reproduced from [27]

FA inputs			SOT+STT		SOT		VCMA+STT		P-VCMA		
A (In MTJ)	B	C_{in}	Sum_{Ax1}	E_{Ax1}	Sum_{Ax2}	E_{Ax2}	Sum_{Ax3}	E_{Ax3}	Sum_{Ax4}	E_{Ax4}	C_o
0	0	0	0	0	0	0	0	0	0	0	0
0	0	1	0 (X)	−1	1	0	0 (X)	−1	1	0	0
0	1	0	1	+2	0 (X)	+1	0 (X)	+1	1	+2	1 (X)
0	1	1	0	0	0	0	1 (X)	−1	0	0	1
1	0	0	1	0	1	0	1	0	1	0	0
1	0	1	0	−2	1 (X)	−1	1 (X)	−1	0	−2	0 (X)
1	1	0	1 (X)	+1	1 (X)	+1	1 (X)	−1	0	0	1
1	1	1	0 (X)	−1	1	0	1	0	1	0	1

Table 5.2 Switching setup of the proposed four approximate FAs. Reproduced from [27]

	1st step			2nd step		
AxFA	Mechanism	Pulse duration (ns)	V_{dd} (V)	Mechanism	Pulse duration (ns)	V_{dd} (V)
Ax1	SOT	0.2	1.2	STT	2	1.2
Ax2	SOT	0.2	1.2	No operation		
Ax3	VCMA	0.18	1.2	STT	0.22	0.6
Ax4	VCMA	0.55	1.1	VCMA	0.55	1.1

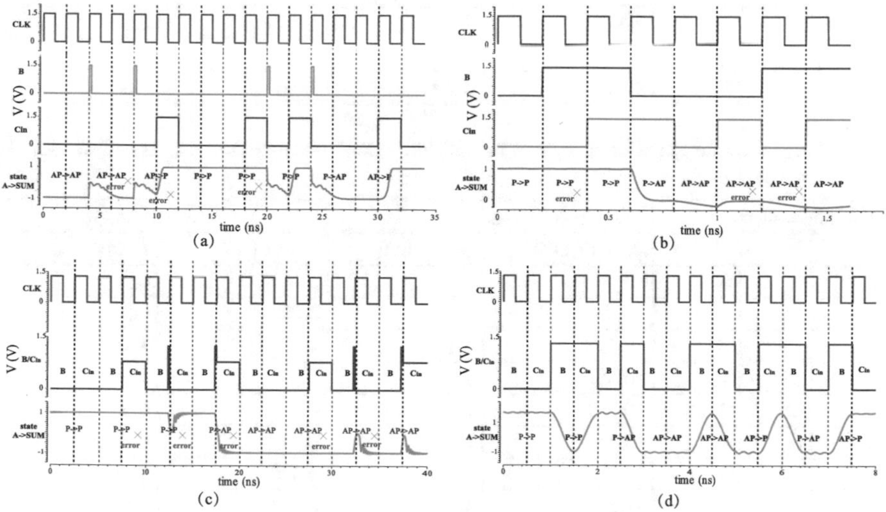

Fig. 5.9 Simulation waveform of different proposed approximate full adder, the SUM operations (**a**) interplay of SOT and STT, (**b**) SOT, (**c**) interplay of STT and VCMA and (**d**) precessional VCMA. Reproduced from [27]

3.2.1 AxFAs Energy-Delay Performance

Figure 5.9 shows thc simulation waveform of full adder Ax1 to Ax4. The error is highlighted according to the truth table. Table 5.3 shows the power consumption and latency of Ax1 and Ax2. Although Ax2 consumes a larger power than Ax1 when the state of MTJ switches, Ax2 has more input cases that no power is consumed. The average power of Ax1 is 186.7 μW, while it is 391.8 μW for Ax2. The maximum delay for both Ax1 and Ax2 is 2.56 ns.

The power and delay results for the eight different inputs of Ax3 and Ax4 are shown in Table 5.4. Compared with Ax3, Ax4 is more power efficient with a shorter maximum latency. The average power of Ax3 (5.547 μW) is roughly the twice that of Ax4. The maximum delay for Ax3 and Ax4 are 1.40 ns and 1.19 ns, respectively. Tables 5.3 and 5.4 illustrate that the VCMA-based AxFAs have smaller power dissipation and maximum delay than the SOT-based designs.

Table 5.3 Power-delay performance of SOT-STT and SOT-based AxFAs. Reproduced from [27]

Full adder input			SOT+STT (Ax1)		SOT (Ax2)	
A	B	C_{in}	Power (μW)	Delay (ns)	Power (μW)	Delay (ns)
0	0	0	0	2.2	0	0.2
0	0	1	238	2.2	1567	2.56
0	1	0	143.5	2.56	0	0.2
0	1	1	379.6	2.2	0	0.2
1	0	0	0	2.2	0	0.2
1	0	1	209.3	2.44	1567	2.56
1	1	0	143.5	2.56	0	0.2
1	1	1	379.6	2.2	0	0.2

Table 5.4 Power-delay performance of STT-assisted VCMA and P-VCMA-based AxFAs. Reproduced from [27]

Full adder input			VCMA+STT (Ax3)		P-VCMA (Ax4)	
A	B	C_{in}	Power (μW)	Delay (ns)	Power (μW)	Delay (ns)
0	0	0	0	1.1	0	0.5
0	0	1	5.63	1.15	0.086	0.5
0	1	0	5.575	1.1	4.931	1.15
0	1	1	10.98	1.4	6.719	1.19
1	0	0	0	1.1	0	0.5
1	0	1	5.458	1.15	0.087	0.5
1	1	0	5.525	1.1	4.378	1.18
1	1	1	11.21	1.4	6.094	1.09

Table 5.5 1-Bit energy comparison of proposed writing-based and sensing-based schemes. Reproduced from [27]

Energy consumption	Ax1	Ax2	Ax3	Ax4
Sensing-based (fJ)	900.28	900.28	93.68fJ	94.61
Writing-only (fJ)	742.89	210.41	16.25fJ	21.27

Comparing with the traditional sensing-based approach (sense the data from MTJ and then calculate with 28-transistor CMOS-FA), more than 80% energy reduction is obtained using the proposed write-only in-memory computing (see Table 5.5).

3.2.2 Accuracy Tradeoff

Assuming the inputs are equally likely to be '0' and '1', the ER, NMED, MRED and ED_{max} are calculated as shown in Table 5.6. Among the approximate designs for 1-bit addition, Ax2 is the most accurate in terms of NMED and MRED, whereas Ax1 has the poorest accuracy with the largest NMED and MRED. Ax4 has the lowest ERs but the largest maximum error distances.

Table 5.6 The error characteristics of the AxFAs. Reproduced from [27]

Error metric	Ax1	Ax2	Ax3	Ax4
ER (%)	62.5	37.5	62.5	**25.0**
NMED (%)	29.2	**12.5**	20.8	16.7
MRED (%)	60.4	**25.0**	41.7	37.5
ED_{max} (%)	66.7	**33.3**	**33.3**	66.7

The proposed approximate PIM scheme can be further applied to multi-bit adders. To maintain a high-accuracy, k LSBs of an n-bit adder are usually approximated ($k < n$), while the more significant bits are accurately computed. The k LSBs can be processed by cascading k AxFAs. To further assess the accuracy of the designs, the functions of 8-bit and 16-bit approximate adders consisting of AxFAs are implemented in MATLAB. An exhaustive unsigned input combinations are used as the inputs for 8-bit adders. Monte-Carlo simulations are performed for 16-bit adders, where the inputs are ten million input combinations in the range of [0, 65535] from a uniform distribution.

Figures 5.10 and 5.11 show the error characteristics of the 8-bit and 16-bit adders (using different approximate designs), respectively. These figures demonstrate that the comparison results with respect to ER, NMED, MRED and ED_{max} for the 8-bit and 16-bit approximate adders are the same. Figures 5.10a and 5.11a indicate that the adders based on Ax2 have the largest ERs Ax1 and Ax2 have the largest ERs, while the ones implemented by Ax4 show the smallest ERs.

In terms of NMED and MRED, the adder using Ax2 results in the smallest values; the one based on Ax2 has similar small results when the number of approximate LSBs is larger than 4. The adders using Ax1 have the largest NMEDs and MREDs. These comparison results are slightly different from the ones obtained from Table 5.6 due to the dependencies between adjacent AxFAs in a multibit adder. Being consistent with Table 5.6, the adders using Ax1 and Ax4 have the largest ED_{max}s. The adders based on Ax3 show medium values in all the four error metrics.

To assess the availability of the proposed approximate designs, the proposed AxFAs are evaluated in the image sharpening application. Figure 5.12 shows the image sharpening results by 16-bit approximate adders implemented by AxFAs with 9 approximate LSBs, where the inputs are 512 × 512 pixels in 8-bit grey scale. The peak signal-to-noise ratio (PSNR) shown below each image illustrates that the images sharpened using approximate 16-bit approximate adders consisting of nine Ax2, Ax3 and Ax4 have a similar quality as the accurate result. The PSNRs for the images sharpened by approximate adders with different k values are shown in Table 5.7. It shows that the image sharpening results are of good quality when the number of approximate LSBs is less than 10 for Ax2, Ax3 and Ax4. However, the maximum number of approximate LSBs is 7 for Ax1 to achieve a good enough image sharpening result.

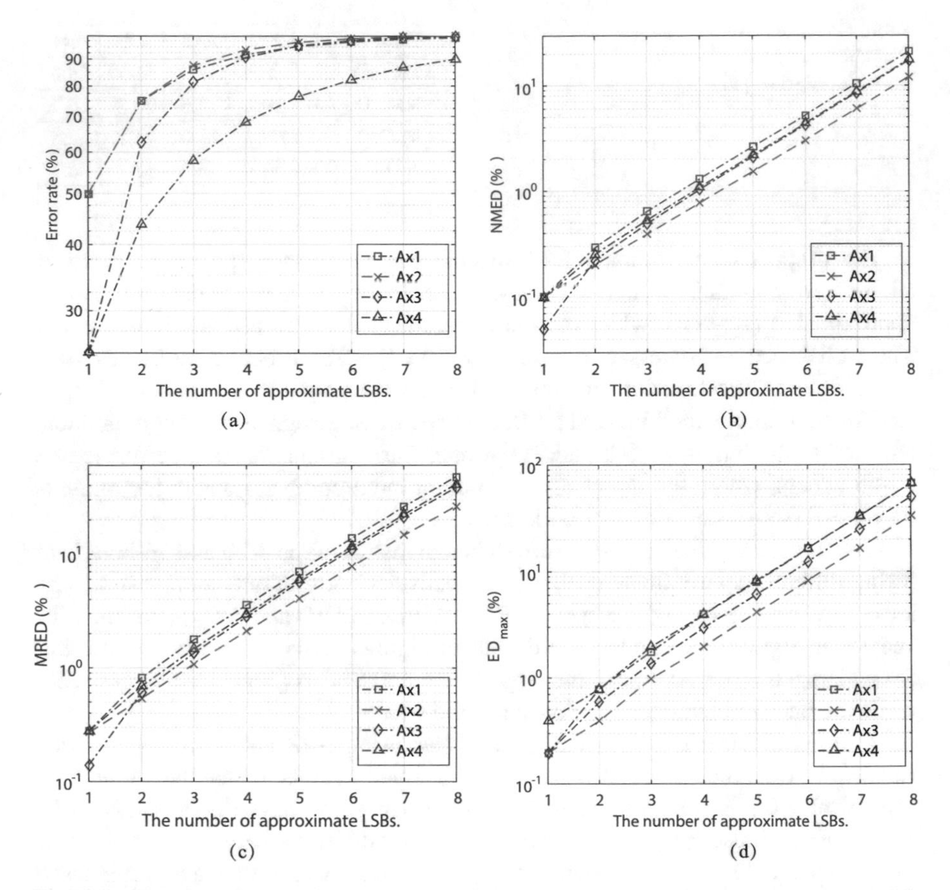

Fig. 5.10 The error characteristics of 8-bit adders with different number of approximate LSBs implemented by different approximate designs (**a**) Error Rate (**b**) NMED (**c**) MRED (**d**) EDmax. Reproduced from [27]

4 Summary

In this chapter, two design strategies are proposed for approximate computing based on spintronic devices. In the first one, the traditional approximate techniques of design complexity reduction and voltage over-scaling have been applied to STT-MRAM. The energy consumption has been drastically reduced with reduced reliability. Meanwhile, write-only bitwise approximate computing is used for processing in memory for the second one. The bit-wise addition has been approximately realized in the bit-cells without sensing/reading operation and extra peripheral circuits. The simulation results of controllable approximation operations have shown significant improvement in energy saving and design complexity reduction with negligible quality loss of the output.

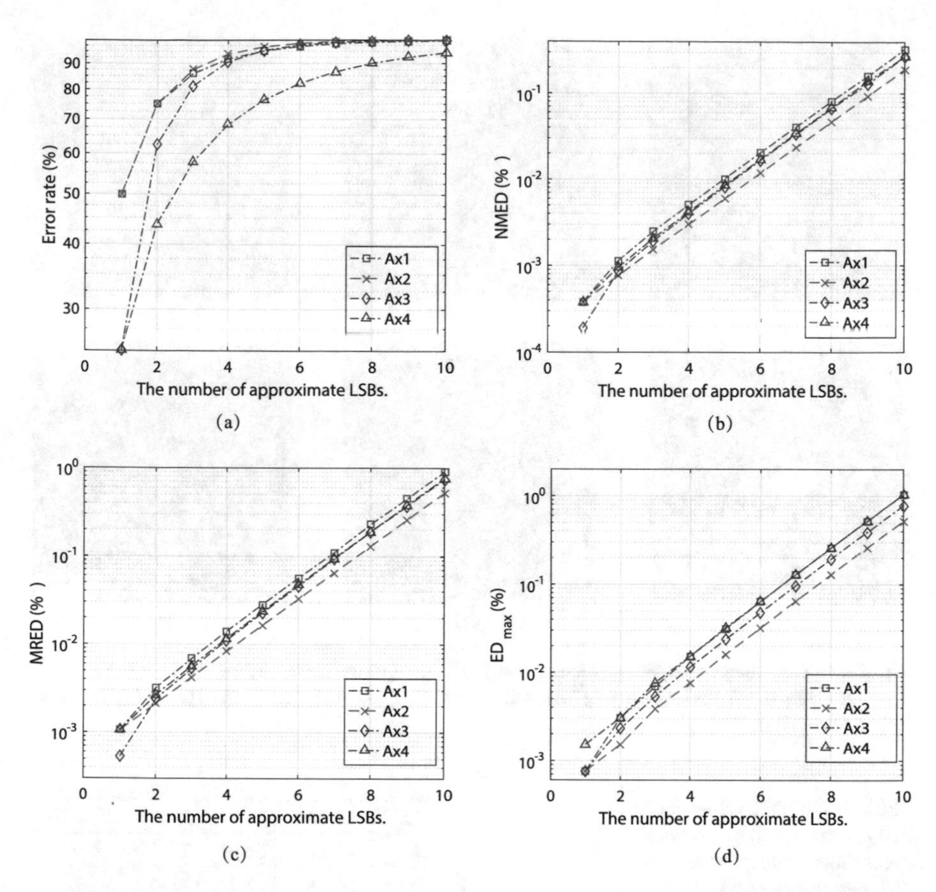

Fig. 5.11 The error characteristics of 16-bit adders with different number of approximate LSBs implemented by different approximate designs (**a**) Error Rate (**b**) NMED (**c**) MRED (**d**) EDmax. Reproduced from [27]

Fig. 5.12 The image sharpening results using 16-bit approximate adders (**a**) Input image (**b**) Accurate output (**c**) Ax1 (21.5 dB) (**d**) Ax2 (32.4 dB) (**e**) Ax3 (33.1 dB) (**f**) Ax4 (34.6 dB). Reproduced from [27]

Table 5.7 PSNR values for images sharpened by 16-bit approximate adders with different values of k (dB). Reproduced from [27]

AxFA	$k = 7$	$k = 8$	$k = 9$	$k = 10$
Ax1	33.0	26.9	21.5	16.0
Ax2	43.4	38.1	32.4	27.1
Ax3	44.8	39.1	33.1	29.0
Ax4	47.1	41.0	34.6	28.7

Acknowledgments The authors gratefully acknowledge the Beihang Hefei Innovation Research Institute Project BHKX-19-02, National Key R&D Program of China (2018YFB0407602) and Beijing Municipal Science and Technology Project under Grant Z201100004220002 for financial support of this work.

References

1. Liu W, Lombardi F, Shulte M. Proc IEEE 2020;108:2103–2107.
2. Angizi S, Jiang H, DeMara RF, Han J, Fan D. IEEE Trans Nanotechnol. 2018;17(4):795.
3. Amanollahi S, Kamal M, Afzali-Kusha A, Pedram M. Proc IEEE 2020;108(12):2150.
4. Chang CH, Gu J, Zhang M. IEEE Trans Circuits Syst I: Regul Pap. 2004;51(10):1985.

5. Gupta V, Mohapatra D, Park SP, Raghunathan A, Roy K. In: IEEE/ACM international symposium on low power electronics and design. 2011. p. 409–414.
6. Zhao W, Prenat G. Spintronics-based computing. Berlin: Springer; 2015.
7. Yu S, Chen PY. IEEE Solid-State Circuits Mag. 2016;8(2):43.
8. Wong HSP, Salahuddin S. Nat Nanotechnol. 2015;10:191. https://doi.org/10.1038/nnano.2015.29
9. Endoh T, Koike H, Ikeda S, Hanyu T, Ohno H. IEEE J Emerg Sel Top Circuits Syst. 2016;6(2):109.
10. Fong X, Kim Y, Venkatesan R, Choday SH, Raghunathan A, Roy K. Proc IEEE 2016;104(7):1449. https://doi.org/10.1109/JPROC.2016.2521712
11. Wang WG, Li M, Hageman S, Chien CL. Nat Mat. 2012;11:64.
12. Maruyama T, Shiota Y, Nozaki T, Ohta K, Toda N, Mizuguchi M, Tulapurkar A, Shinjo T, Shiraishi M, Mizukami S, Ando Y, Suzuki Y. Nat Nanotechnol. 2009;4:158.
13. Wang Z, Zhang L, Wang M, Wang Z, Zhu D, Zhang Y, Zhao W. IEEE Electron Device Lett. 2018;39(3):343.
14. Liu L, Pai CF, Li Y, Tseng HW, Ralph DC, Buhrman RA. Science 2012;336:555.
15. Julliere M. Phys Lett A 1975;54(3):225.
16. Wang Y, Cai H, Naviner LAdB, Zhang Y, Zhao X, Deng E, Klein JO, Zhao W. IEEE Trans Electron Dev. 2016;63(4):1762.
17. Wang M, Cai W, Cao K, Zhou J, Wrona J, Peng S, Yang H, Wei J, Kang W, Zhang Y, Langer J, Ocker B, Fert A, Zhao W. Nat Commun. 2018;9(671):1.
18. Wang L, Kang W, Ebrahimi F, Li X, Huang Y, Zhao C, Wang KL, Zhao W. IEEE Electron Dev Lett. 2018;39(3):440.
19. Koch RH, Katine JA, Sun JZ. Phys Rev Lett. 2004;92:088302.
20. Worledge DC, Hu G, Abraham DW, Sun JZ, Trouilloud PL, Nowak J, Brown S, Gaidis MC, O'Sullivan EJ, Robertazzi RP. Appl Phys Lett. 2011;98(2):022501.
21. Heindl R, Rippard WH, Russek SE, Pufall MR, Kos AB. J Appl Phys. 2011;109(7):073910.
22. Sato H, Yamanouchi M, Ikeda S, Fukami S, Matsukura F, Ohno H. Appl Phys Lett. 2012;101(2):022414. https://doi.org/10.1063/1.4736727
23. Zhao W, Zhang Y, Devolder T, Klein JO, Ravelosona D, Chappert C, Mazoyer P. Microelectron Reliab. 2012;52(9–10):1848.
24. Wang Y, Zhang Y, Deng E, Klein J, Naviner L, Zhao W. Microelectron Reliab. 2014;54(9–10):1774.
25. Cai H, Wang Y, De Barros Naviner LA, Zhao W. IEEE Trans Circuits Syst I Reg Pap. 2017;64(4):847.
26. Deng E, Zhang Y, Kang W, Dieny B, Klein JO, Prenat G, Zhao W. IEEE Trans. Circuits Syst I Reg Pap. 2015;62(7):1757.
27. Cai H, Jiang H, Zhou Y, Han M, Liu B. CCF Trans High Perf Comput. 2020;2(3):282.

Chapter 6
Majority Logic-Based Approximate Multipliers for Error-Tolerant Applications

Tingting Zhang, Honglan Jiang, Weiqiang Liu, Fabrizio Lombardi, Leibo Liu, Seok-Bum Ko, and Jie Han

1 Introduction

As the Dennard scaling is expected to come to an end, it becomes more difficult to obtain the reduction of power dissipation and the increase of clock frequency although the number of transistors on a chip increases as per Moore's Law [1]. Emerging nanotechnologies have been explored for digital circuit design due to their properties of high density and low power dissipation. Substantially different from conventional Boolean logic, these nanotechnologies often rely on majority logic (ML) [2], such as nanomagnetic logic (NML) [3], and magnetic tunnel junction (MTJ) devices [4].

T. Zhang · J. Han (✉)
University of Alberta, Edmonton, AB, Canada
e-mail: ttzhang@ualberta.ca; jhan8@ualberta.ca

H. Jiang
Shanghai Jiao Tong University, Shanghai, China
e-mail: honglan@sjtu.edu.cn

W. Liu
Nanjing University of Aeronautics and Astronautics, Nanjing, China
e-mail: liuweiqiang@nuaa.edu.cn

F. Lombardi
Northeastern University, Boston, MA, USA
e-mail: lombardi@ece.neu.edu

L. Liu
Tsinghua University, Beijing, China
e-mail: liulb@tsinghua.edu.cn

S.-B. Ko
University of Saskatchewan, Saskatoon, SK, Canada
e-mail: seokbum.ko@usask.ca

W. Liu, F. Lombardi (eds.), *Approximate Computing*,
https://doi.org/10.1007/978-3-030-98347-5_6

The ML gate performs an odd-input logic function, i.e., it outputs true (false) if more than half of the inputs are true (false). A three-input majority gate (voter, MV) implements the logic operation, $F = M(A, B, C) = AB + BC + AC$. Since the three-input majority voter is the most commonly used logic gate for all these ML-based nanotechnologies, the approximate designs employing multi-input majority voters [5, 6] are not reviewed in this chapter. The AND and OR functions in the conventional Boolean logic can be realized by setting one of three inputs to '0' or '1.' Due to the more extensive configurable ability of the majority gate, fewer gates are usually needed when building circuits based on ML.

Approximate computing has emerged as a new paradigm to improve the performance and the efficiency of circuits and systems at a limited loss in accuracy [7]. Although computational errors generally are not desirable, fully accurate computation is not always necessary for error-tolerant applications, such as image processing, and neural networks. A major hurdle in image processing or NNs is the energy overhead; so, approximate computing can be used, especially for multiplication, which is one of the most hardware-intensive arithmetic operations [8].

Approximate designs based on these ML-based nanotechnologies can be symbiotically utilized for low power and high performance [9]. Although approximate designs based on conventional Boolean logic can be directly interpreted to adapt into ML-based nanotechnologies, the naive mapping method usually cannot fully utilize the specific properties of ML. Therefore, approximate circuits need to be specially designed based on ML. The efficient ML-based implementations of approximate multiplier designs can achieve a significant improvement in hardware overhead if the majority gate can be cleverly manipulated.

Unlike approximate designs for CMOS circuits, approximate ML-based designs have only recently been pursued [9–13]. This chapter focuses on the recently developed ML-based approximate unsigned and signed multiplier designs. The circuit implementations using three-input majority gates and inverters as the basic logic gates are investigated. Specifically, they are classified depending on their approximation methodology adopted, and comprehensively analyzed in terms of hardware and error metrics. Case studies with these approximate unsigned and signed multiplier designs for error-tolerant applications are also provided as an assessment.

In this chapter, error metrics including the normalized mean error distance (NMED), the mean relative error distance (MRED), the root-mean-square error (RMSE), and the error rate (ER) are utilized to assess the error characteristics of approximate designs [14]. The circuit characteristics are evaluated by the number of utilized majority voters (MVs), the number of utilized inverters (INVs), the critical path delay (D), and area-delay product (ADP, given by $MV \times D$). The critical path delay is measured by the number of majority gates on the critical path since the delay for inverters is often very small for ML-based nanotechnologies [15]. The ADP is used to assess the overall hardware efficiency.

The remainder of this chapter is organized as follows. The classifications and evaluations for ML-based approximate unsigned and signed multipliers are shown in Sects. 2 and 3, respectively. Finally, Sect. 4 concludes the chapter.

2 ML-Based Approximate Unsigned Multipliers

In this section, designs of approximate unsigned multipliers are discussed from three methodologies, including the approximation in generating the PPs, the approximation in the PP array, and the approximation in PP accumulation.

2.1 Preliminaries

For an $n \times n$ unsigned multiplier, let $A = a_{n-1}a_{n-2}...a_2a_1a_0$ denote the multiplicand and $B = b_{n-1}b_{n-2}...b_2b_1b_0$ denote the multiplier, respectively. Firstly, the PPs are generated as $pp_{ij} = M(a_j, b_i, 0)$. The PPs are then accumulated based on different structures, such as the carry save adder array [9], the Wallace tree [16], and the Dadda tree [17] to reduce and compress the PP array into two rows. A multi-bit adder is employed to generate the final result. Figure 6.1 gives an example of the basic arithmetic process for the 8×8 unsigned multiplier using a Dadda tree structure and a Wallace tree structure. The black circle denotes a multiplier input bit, a partial product, or an output bit. The Dadda tree structure in Fig. 6.1a utilizes full adders (FAs), half adders (HAs), and 4:2 compressors to accumulate PPs efficiently. In the first stage, the PP array is compressed into four rows using two HAs, two FAs, and eight 4:2 compressors. Then, the second stage requires one FA and ten 4:2 compressors to further reduce the terms into two rows, which can be computed by using a 14-bit adder. The Wallace tree structure utilizes thirty-eight FAs and fifteen HAs in the four stages of compression in total (as per Fig. 6.1b).

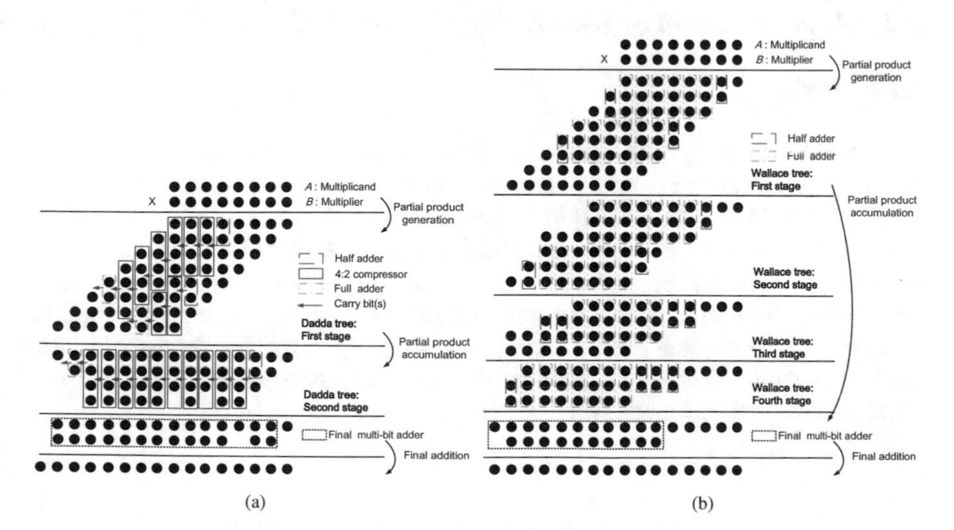

Fig. 6.1 The basic arithmetic process of an 8×8 unsigned multiplier: (**a**) using a Dadda tree structure [9], and (**b**) using a Wallace tree structure [16]

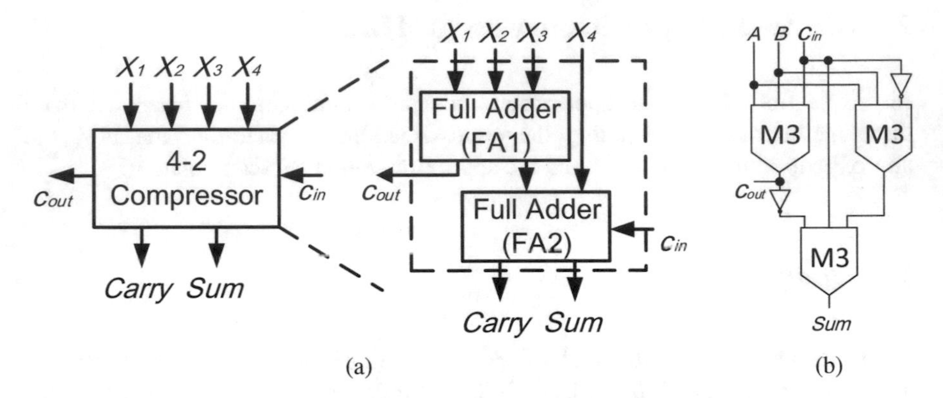

Fig. 6.2 A 4:2 compressor using two FAs: (**a**) the overall structure of a 4:2 compressor [18], and (**b**) the diagram of a full adder based on ML [19]

A 4:2 compressor has five inputs: four primary inputs (denoted as X_1, X_2, X_3, X_4) and one carry bit input from the lower position block (denoted as C_{in}) [18]. It produces three outputs: two primary outputs (denoted as $Carry$ and Sum) and one carry bit output to the higher position block (denoted as C_{out}). As shown in Fig. 6.2a, the 4-2 compressor is commonly implemented by cascading two full adders. For a full adder, let A, B, and C_{in} be the input operands; C_{out} and Sum be the outputs. The ML implementation of a full adder requires three majority gates and two inverters, as per Fig. 6.2b, where C_{out} is produced by using one majority gate [19].

2.2 *Approximation in Generating Partial Products*

2.2.1 Design

An approximate multiplier proposed in [11] utilized an approximate 2×2 multiplier design to construct larger multipliers. Consider the Boolean logic-based approximate 2×2 multiplier design [20]. It approximately produces the product "1001" by using "111" when the multiplier and multiplicand are both "11." Thus, the output can be approximately represented by using three bits with a saving of one output bit. Assume $A = a_1a_0$ denote the multiplicand, $B = b_1b_0$ denote the multiplier, and $out = out_2out_1out_0$ denote the approximate output. By directly mapping the design based on Boolean logic [20] into ML implementation as per (6.1)–(6.3), out_1 requires three majority gates, which is two more than out_0 and out_2.

$$out_0 = a_0b_0 = M(a_0, b_0, 0), \tag{6.1}$$

$$out_1 = a_1b_0 + a_0b_1 = M(M(a_1, b_0, 0), M(a_0, b_1, 0), 1), \tag{6.2}$$

$$out_2 = a_1b_1 = M(a_1, b_1, 0). \tag{6.3}$$

Kulkarni et al. [20] approximates out_1 by splitting (6.2) into two parts (i.e., (6.4) and (6.5)), such that one is employed as the out_1, the other is used as a compensation bit (denoted as Δ). Therefore, the equations of the approximate 2×2 multiplier design with three output bits are given by (6.1), (6.4), and (6.3), whereas (6.5) serves as the compensation bit.

$$out_1 = M(A_0, B_1, 0) \tag{6.4}$$

$$\Delta = M(A_1, B_0, 0). \tag{6.5}$$

2.2.2 Evaluation

An n-bit approximate multiplier can be constructed by using approximate 2×2 multiplier designs and compensation bits to control the error within a reasonable bound [11]. A tradeoff must be assessed when selecting an appropriate number of compensation bits for different sizes of multipliers.

Taken the 8×8 approximate multiplier as an example (as per in Fig. 6.3), let $A = a_7a_6...a_0$ be the multiplicand ($A_{H1} = a_7a_6$, $A_{H0} = a_5a_4$, $A_{L1} = a_3a_2$, and $A_{L0} = a_1a_0$) and $B = b_7b_6...b_0$ be the multiplier ($B_{H1} = b_7b_6$, $B_{H0} = b_5b_4$, $B_{L1} = b_3b_2$, and $B_{L0} = b_1b_0$). There are sixteen units of the 2×2 multiplier and each unit has a candidate compensation bit (denoted as ${C_{2^i}}^x$, where 2^i represents the weight (or significance) and x denotes the signed number of the complement bit if multiple complement bits exist under the current weight). The blocks in the same dotted diagonal line share the same significance.

Assume the number of ignored compensation bits be q. Figure 6.4 gives the error metrics (in NMED, MRED, RMSE, and ER) of the 8×8 approximate unsigned multiplier with different values of q. The accuracy decreases as q increases. With the increase of q from 1 to 16, the ER increases relatively slowly from 0.25 to 0.90; the NMED increases from 7.69×10^{-6} to 5.54×10^{-2}; the MRED increases from 2.96×10^{-4} to 0.45 and the RMSE increases from 1 to 5.71×10^3. When the weight of the compensation bits remains unchanged, the NMED, MRED, and RMSE will increase linearly; otherwise, these error metrics increase sharply. Moreover, with every increase in the value of q by one, one majority gate is saved, thus simplifying the PP reduction and compression.

2.3 *Approximation in Partial Product Accumulation*

Consider the full adders and 4-2 compressors are used in the PP reduction and compression (as per Fig. 6.1); this section reviews and evaluates the approximate full adder (AFA) designs and approximates 4-2 compressor (AC) designs.

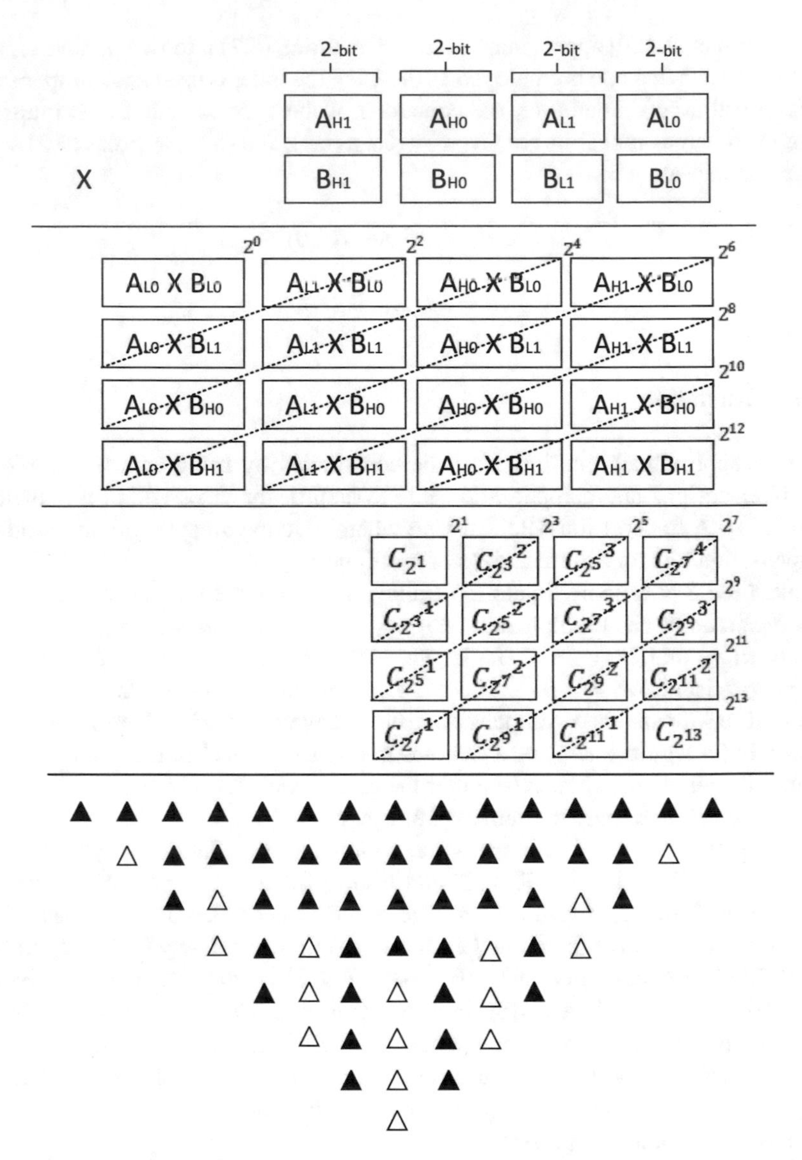

Fig. 6.3 PP generation and complement bit generation of an 8 × 8 approximate multiplier constructed by 2 × 2 multipliers [11]

2.3.1 Design

Approximate Full Adder

Figure 6.5 presents two recently proposed AFAs, denoted by AFA1 [10] and AFA2 [21]. Assume A, B, and C_{in} denote the input operands; S and C_{out} denote the output

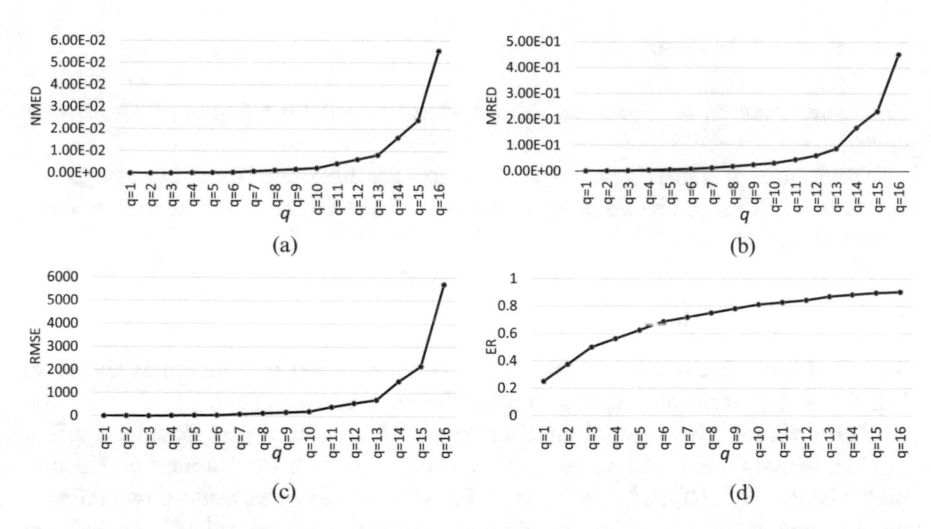

Fig. 6.4 Error measurements for an 8 × 8 approximate unsigned multiplier with the different values of q: (**a**) $NMED$ vs q, (**b**) $MRED$ vs q, (**c**) $RMSE$ vs q, and (**d**) ER vs q

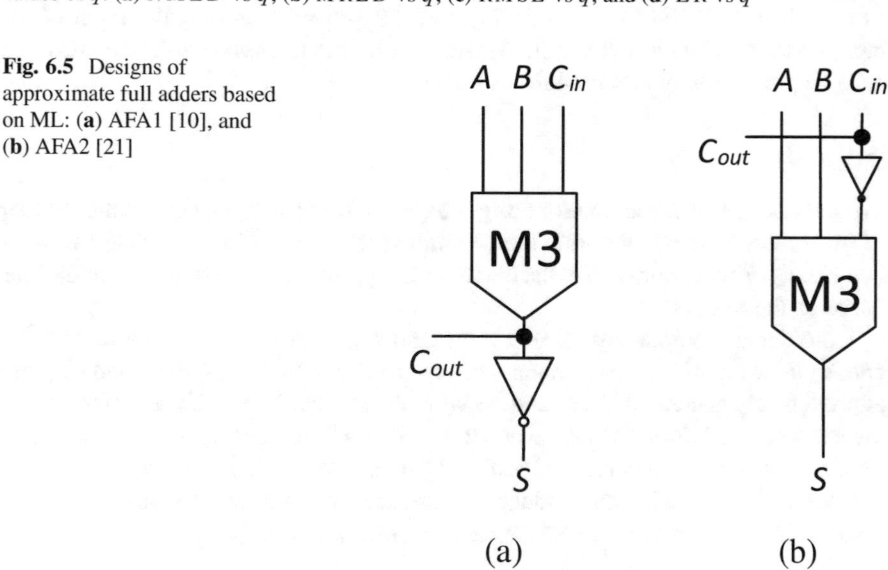

Fig. 6.5 Designs of approximate full adders based on ML: (**a**) AFA1 [10], and (**b**) AFA2 [21]

results of sum and carry out, respectively. The AFA1 [10] exactly produces the carry out C_{out}, while approximately produces the sum result, whereas the AFA2 [21] approximately generates both C_{out} and S.

Approximate 4-2 Compressor

The approximation is applied in the PP accumulation by using approximate 4-2 compressors.

The designs of approximate 4-2 compressors are discussed from three categories, i.e., the approximation based on approximate full adders, based on the truth table or output reduction.

Based on Approximate Full Adder

Since a 4-2 compressor is constructed by two connected full adders as shown in Fig. 6.2, AFAs were considered to replace the exact versions [11].

Figure 6.6a–f give designs of approximate 4-2 compressors based on AFAs in [11], namely from AC1 to AC6. They are constructed by different connection methods of AFA1[10] and AFA2 [21]. The AC1 is built by cascading two AFA1s; the AC2 and AC3 are built by cascading two AFA2s; the AC4 and AC5 are built by using AFA1 as module 1 and AFA2 as module 2; the AC6 is built by using AFA2 as module 1 and AFA1 as module 2. The AC2 and AC4 employ the input of the compressor as $Carry$ and use its negation to calculate Sum, while the AC3 and AC5 employ the output of module 1 as $Carry$.

Based on the Truth Table

The approximate 4-2 compressor designs based on the truth table change the number of true or false cases in the truth table by introducing errors to reduce the hardware complexity. Figure 6.6g–h give the diagrams of approximate 4-2 compressor designs based on the truth table.

Consider the output results of $Carry$ and C_{out} have the same logic with X_1 and X_2 in 24 out of 32 input assignments, respectively. In AC7, $Carry$ and C_{out} are approximately generated by X_1 and X_2 with no extra hardware. Then, by modifying the expression of Sum, the errors due to the approximate generation of $Carry$ and C_{out} can be partly compensated. AC8 is different from AC7 in generating the C_{out} and Sum. $Carry$ and Sum produce incorrect results in 8 and 16 cases out of 32 input assignments with the exact expression of C_{out}, respectively.

Based on Output Reduction

To further decrease the hardware complexity, two approximate 4-2 compressors are designed by reducing the number of the outputs, as shown in Fig. 6.6i,j. These designs use 2 outputs (rather than 3) for the final result (C_{out} is not employed). In AC9, the output result is first represented by using 2 outputs so that the binary '11' is employed to represent results larger than '11.' Then, the approximation is introduced to simplify the expressions. AC10 is specially designed for only four input operands with an assumption of $C_{in} = 0$. Sum is considered as a constant '1' without additional hardware.

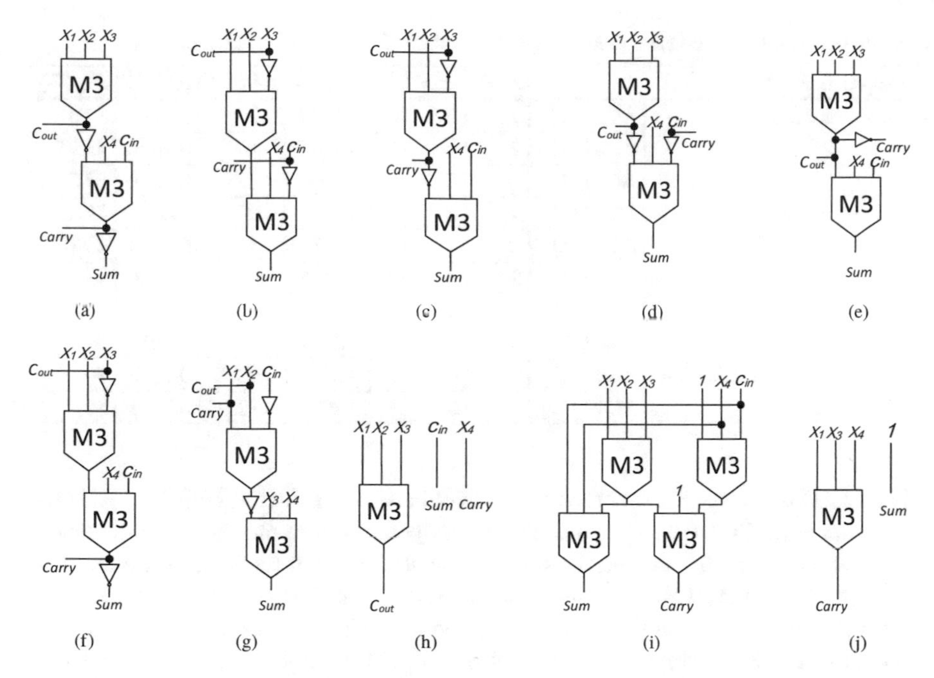

Fig. 6.6 Designs of approximate 4-2 compressors based on ML: (**a**) AC1 (AFA1+AFA1) [11], (**b**) AC2 (AFA2+AFA2) [11], (**c**) AC3 (AFA2+AFA2) [11], (**d**) AC4 (AFA1+AFA2) [11], (**e**) AC5 (AFA1+AFA2) [11], (**f**) AC6 (AFA2+AFA1) [11], (**g**) AC7 [22], (**h**) AC8 [23], (**i**) AC9 [11], and (**j**) AC10 [12]

Table 6.1 Hardware and error metrics of approximate full adders based on ML

AFAs	MV	INV	D	NMED	RMSE	ER (%)
AFA1 [10]	1	1	1	0.083	0.25	25
AFA2 [21]	1	1	1	0.083	0.25	25

2.3.2 Evaluation

Table 6.1 presents the hardware and error measurements of AFA1 and AFA2. They share the same error metrics with similar hardware complexity. The AFA1 approximately produces the sum result S with ED=1 when $A = B = C_{in} = 0$ and $A = B = C_{in} = 1$. The AFA2 introduces the errors for producing C_{out} and S, with ED=1 when $A = B = 0, C_{in} = 1$ and $A = B = 1, C_{in} = 0$. Although the approximate generation of C_{out} will lead to some catastrophic results when building the multi-bit adders by cascading AFA1s, the critical path can significantly be shortened.

A comprehensive comparison of approximate 4-2 compressor designs is provided in Table 6.2 in terms of hardware and error metrics. The designs based on 1-bit AFAs are similar; they utilize the same number of majority gates and share the same delay. Compared with the other AFA-based AC designs, the AC5 saves one inverter. The AC1, AC3, and AC4 show better performance in accuracy (MRED,

Table 6.2 Hardware and error metrics of approximate 4-2 compressors based on ML

Approximate designs		MV	INV	D	MRED	NMED	RMSE	ER (%)
Based on AFAs	AC1 [11]	2	2	2	0.1625	0.075	0.6124	37.5
	AC2 [11]	2	2	2	0.2344	0.1	0.7906	43.75
	AC3 [11]	2	2	2	0.1953	0.075	0.6124	37.5
	AC4 [11]	2	2	2	0.1495	0.075	0.6124	37.5
	AC5 [11]	2	1	2	0.2339	0.1	0.7906	43.75
	AC6 [11]	2	2	2	0.2146	0.1	0.7906	43.75
Based on the truth table	AC7 [22]	2	2	2	0.2344	0.1	0.7906	43.75
	AC8 [23]	1	0	1	0.2604	0.1	0.7071	50
Based on output reduction	AC9 [11]	4	0	2	0.2339	0.0875	0.7071	40.63
	AC10 [12]	1	0	1	0.2031	0.125	0.7071	50

NMED, RMSE, and ER); moreover, the AC4 has the smallest MRED. The AC2 and AC7 are essentially the same since the input operands have the same significance in the computation, thus sharing the same hardware and error metrics. Compared with AC7, although the AC8 has an increase of 11.1% in MRED and 14.2% in ER, it can improve the accuracy in RMSE by 11.3% with a reduction of one utilized majority gate, two utilized inverters, and one delay unit. Different from AC9, there are only 16 input combinations for AC10 since it only has four input operands. It utilizes only one majority gate with the MRED of 0.2031, the NMED of 0.125, the RMSE of 0.125, and the ER of 50%.

Therefore, among the AC designs with five input operands and three outputs, AC4 [11] and AC8 [23] are superior in terms of accuracy and hardware. AC9 [11] benefits from its characteristics of the fewer number of outputs, which can decrease the complexity of PP reduction and compression design. AC10 [12] with only two outputs is specially designed for the case considering four input operands.

2.4 Approximation in the Partial Product Tree

2.4.1 Design

Truncation as a commonly used approximation technique is considered in [12] to improve the performance and efficiency of the approximate multiplier. As shown in Fig. 6.7a, taken the approximate 8×8 multiplier design as an example, Sabetzadeh et al. [12] truncated the four least significant columns of the partial products. Moreover, the approximate compression is then employed by using approximate 4-2 compressors (AC10) when accumulating the six columns of PPs in the least signification after truncation. Therefore, three half adders, three full adders, twelve AC10s, and four 4-2 compressors are required before obtaining two rows of PPs.

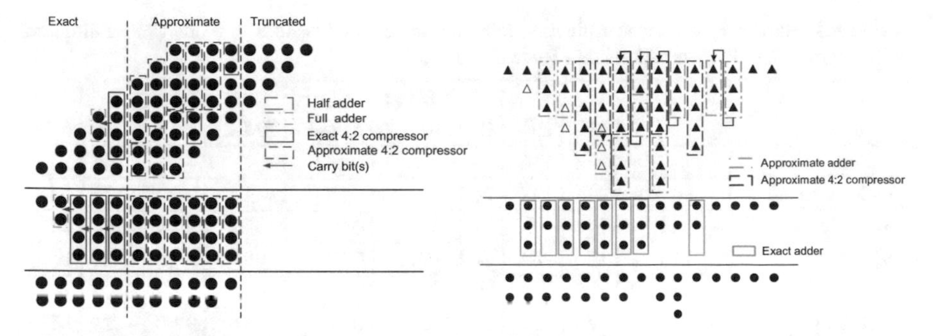

Fig. 6.7 The partial product reduction and compression design of an 8×8 unsigned multiplier using approximate 4-2 compressors with two outputs: (**a**) using AC10 [12], and (**b**) using AC9 [11]

2.4.2 Evaluation

The approximate multiplier design in [12] employs both the Boolean logic gates and majority logic gates. Since we only focus on the full majority logic-based circuit designs, the Boolean logic gates used in [12] are replaced by using majority gates and inverters for evaluation.

Assume the approximate factor p denote the number of columns with approximate compression from the least significance (2^0) and t denote the number of truncated columns from the least significance (2^0). If the approximate multiplier design employs both truncation and approximation, the columns of PP arrays with approximation can be truncated ($p \geq t$). For example, $p = 10$ and $t = 4$ are for the design in [12] as per Fig. 6.7a. For a fair comparison, the same values of p and t are considered for the approximate 8×8 unsigned multipliers with different approximation strategies.

The ACs with better performance (AC4 and AC8) and the AFAs (AFA1 and AFA2) are employed in the Dadda tree-based or the Wallace tree-based PP accumulation, respectively. Moreover, the approximate multiplier design with approximate PP generation and accumulation are also considered. Specifically, the AC9 based multiplier design with approximate PP generation ($q = 10$, $p = 14$) using the compression process specially designed for the ACs with two outputs [11] (as per Fig. 6.7b) is also considered. It also utilizes AFA2 on the first stage of compression.

Table 6.3 presents the hardware and error metrics for the exact and approximate unsigned multipliers. BW denotes the bit-width of the final results. Due to the approximate compression, the use of the truncation technique will increase the ER but without the significant decrease of the accuracy of the multiplier. Especially for the compression based on the Dadda tree using AC4 and the compression based on the Wallace tree using AFA2, combining with the truncation can achieve a slight improvement up to 1.13% in NMED, 1.5% in MRED, 1.0% in RMSE. Compared with the AC8 and AFA1-based designs, the AC4 and AFA2-based designs are superior in error metrics. The AFA2-based design has a smaller MRED and ER, and

Table 6.3 Hardware and error evaluation of exact and approximate 8 × 8 unsigned multipliers with approximate PP generation and PP accumulation

Multiplier	q	p	t	MV	INV	D	NMED (10^{-3})	MRED	RMSE	ER %	ADP	BW
Dadda based	–	–	–	232	116	24	–	–	–	–	5568	16
Wallace based	–	–	–	256	128	21	–	–	–	–	5376	16
AC10-based [12]	–	10	4	148	48	20	11.7	0.62	900	99.1	2960	12
AC4-based	–	10		176	84	24	8.8	0.63	695	98.5	4224	16
	–	10	4	155	70	20	8.7	0.62	688	99.9	3100	12
AC8-based	–	10	–	176	84	24	10.1	0.14	791	98.8	4224	16
	–	10	4	155	70	20	10.2	0.14	797	99.0	3100	12
AFA1-based	–	10	–	178	50	20	14.8	0.97	1110	99.1	3650	16
	–	10	4	160	36	16	14.8	0.97	1111	99.2	2560	12
AFA2-based	–	10	–	178	50	20	9.2	0.14	751	96.7	3650	16
	–	10	4	160	36	16	9.1	0.14	745	97.9	2560	12
AC9-based [11]	10	14	–	138	36	16	27.1	0.25	2423	98.7	2208	16
	10	14	4	123	32	15	27.1	0.25	2447	99.0	1845	12

the AC8-based design has a smaller NMED and RMSE. The ACs with two outputs can significantly reduce the hardware at the cost of accuracy. Compared with the AC10-based design, the improvement of ADP up to 40% and the improvement of MRED up to 60% can be obtained with the increase of 56% in NMED. Moreover, it can save up to 51.9% of the utilized majority gates and 66.8% in ADP compared with the exact designs.

2.5 *Application*

The 8 × 8 approximate multiplier is applied to image processing; they are used to multiply the same two images on a pixel basis so combining the two input images into a single output image. The structural similarity index measure (SSIM) and the peak signal-to-noise ratio (PSNR) are used to evaluate the quality of the image processing results.

As discussed in Sect. 2.4.2, the AC9 based design can significantly reduce the hardware cost but with a relatively large accuracy loss. Figure 6.8 gives the results of image multiplication using exact or approximate multipliers. Even using AC9-based approximate multipliers can still return a PSNR of 45.1 dB and an SSIM of 0.96.

Fig. 6.8 Results of image multiplication: (**a**) using exact multipliers and (**b**) using approximate multipliers based AC9 [11]

2.6 Discussion

The ML-based approximate unsigned multiplier designs have been investigated in this section. The PPs can be approximately generated using ML-based 2×2 approximate multiplier designs with so-called complement bits. More complement bits can lead to more accurate result. The appropriate number of compensation bits should be determined to balance the hardware cost and accuracy. The PP array can also be reduced by using the truncation technique, which is one of the most common approximate strategies. It can significantly decrease the complexity of PP compression designs.

As the extensively used arithmetic logic units during the compression of PPs, the approximate full adders and approximate 4-2 compressors have been considered to replace the exact counterparts. The designs of approximate 4-2 compressors can be classified into three categories: cascading approximate full adders, introducing errors in the truth table, and reducing the number of outputs. Especially, the designs based on the output reduction can effectively simplify the compression process since they will produce fewer terms but may with the problem of accuracy.

The comprehensive evaluations from hardware-level, accuracy-level, and application-level of the ML-based 8×8 approximate multiplier designs are performed to show the promise.

3 ML-Based Approximate Signed Radix-4 Booth Multipliers

The ML-based approximate signed multiplier designs have been recently studied using radix-4 Booth encoding [13, 24]. This section discusses the approximate radix-4 Booth multiplier designs based on ML from approximate PP generation, reduction, and compression.

3.1 Preliminaries

The ML implementations of different PP generation methods based on the modified Booth encoding (MBE) are firstly discussed in [13, 24]. The advantages and disadvantages of these PP generators (PPGs) are evaluated by analyzing their circuit characteristics in this section.

3.1.1 Radix-4 Booth Encoding

The modified Booth algorithm (also known as the radix-4 Booth algorithm) [25] has been widely utilized to solve the sign correction issue of signed multiplication and reduce the number of PPs. For an $n \times n$ signed multiplier, let $A = a_{n-1}a_{n-2}...a_2a_1a_0$ denote the multiplicand and $B = b_{n-1}b_{n-2}...b_2b_1b_0$ denote the multiplier. The most significant bits of A and B are the sign bits. In the MBE, the multiplier bits are first grouped into sets of three adjacent bits ($b_{2i+1}b_{2i}b_{2i-1}$) and the bits on the two sides overlap with the neighboring sets, except for the first set. Then, the multiplicand is encoded into $-2A$, $-A$, 0, A, or $2A$ to generate the PP array, as per Table 6.4 [25], where $0 \leq i \leq \frac{n}{2} - 1, 0 \leq j \leq n - 1$.

Let pp_{ij} be the PP bit in the ith row and the jth column. The negation operation is performed by inverting every bit of A (or $2A$, realized by pp_{ij}) and adding a '1' to the LSB (implemented by Neg_i) to get the 2's complement.

The Baugh Wooley algorithm is used to avoid the sign bit extension [28]. Figure 6.9 gives the PP and sign bit generation for an 8-bit radix-4 Booth multiplier

Table 6.4 Implementations of MBE scheme using MLCG [26], MLNG [27] and MLGA [24], MLGB [24]

		MLCG [26]		MLNG [27]		MLGA [24]		MLGB [24]	
$b_{2i+1}b_{2i}b_{2i-1}$	PP	pp_{ij}	Neg_i	pp_{ij}	Neg_i	pp_{ij}	Neg_i	pp_{ij}	Neg_i
000	0	0	0	0	0	1	1	1	1
001	+A	a_j	0	a_j	0	a_j	0	a_j	0
010	+A	a_j	0	a_j	0	a_j	0	a_j	0
011	+2A	a_{j-1}	0	a_{j-1}	0	a_{j-1}	0	a_{j-1}	0
100	−2A	$\overline{a}_{j-1}$	1	$\overline{a}_{j-1}$	1	$\overline{a}_{j-1}$	1	$\overline{a}_{j-1}$	1
101	−A	$\overline{a}_j$	1	$\overline{a}_j$	1	$\overline{a}_j$	1	$\overline{a}_j$	1
110	−A	$\overline{a}_j$	1	$\overline{a}_j$	1	$\overline{a}_j$	1	$\overline{a}_j$	1
111	0	0	0	1	1	0	0	1	1

15	14	13	12	11	10	9	8	7	6	5	4	3	2	1	0
					$\overline{pp_{08}}$	pp_{08}	pp_{08}	pp_{07}	pp_{06}	pp_{05}	pp_{04}	pp_{03}	pp_{02}	pp_{01}	pp_{00}
				1	$\overline{pp_{18}}$	pp_{17}	pp_{16}	pp_{15}	pp_{14}	pp_{13}	pp_{12}	pp_{11}	pp_{10}		Neg_0
		1	$\overline{pp_{28}}$	pp_{27}	pp_{26}	pp_{25}	pp_{24}	pp_{23}	pp_{22}	pp_{21}	pp_{20}		Neg_1		
1	$\overline{pp_{38}}$	pp_{37}	pp_{36}	pp_{35}	pp_{34}	pp_{33}	pp_{32}	pp_{31}	pp_{30}		Neg_2				
									Neg_3						

Fig. 6.9 PP and sign bit generation for an 8-bit radix-4 Booth multiplier [25]

as an example. All Booth PP generation methods discussed in this section are developed based on the same structure.

3.1.2 ML Implementations of Partial Product Generators

As shown in Table 6.4, different implementations of the MBE (i.e., PP generators, PPGs) have been discussed in [24] by leveraging the capabilities of ML, including the ML-based conventional PPG (namely MLCG [24]), the ML-based new PPG (namely MLNG [24]), and two recently proposed ML-based PP generators (namely MLGA [24] and MLGB [24]).

The conventional PPG in [26] provides a common implementation of the MBE using Boolean logic. Consider there is no effect on the final result if the negation operation is employed on "0." The new PPG based on Boolean logic performs the negation operation on the PP "0" when $b_{2i+1}b_{2i}b_{2i-1} = 111$: pp_{ij} is inverted to '1'and Neg_i is '1' [27]. The MLCG and MLNG give the efficient ML implementations of the conventional PPG [26] and the new PPG [27], respectively. The MLGA and the MLGB [24] were developed by modifying the PP "0" when $b_{2i+1}b_{2i}b_{2i-1} = 000$ and when both $b_{2i+1}b_{2i}b_{2i-1} = 000$ and 111. The MLGB can be considered as a combination of the MLNG and the MLGA.

Table 6.5 presents the ML-based equations and the required logic gates for different PP generation methods evaluated by MV and D. Due to the massive reuse of inverters among different PPs, it is unnecessary to evaluate the INVs for a pp_{ij}. All PP generation methods use the inversion of every input operand bit. Moreover, MLCG and MLGB use one more inverter when computing pp_{ij} due to the reuse of a majority gate.

For MLCG, when $j = 0$, $\overline{M}(b_{2i}, b_{2i-1}, a_j)$ in the equation of pp_{ij} is equal to $M(\overline{b}_{2i}, \overline{b}_{2i-1}, 1)$ that can be reused in the equation of Neg_i; similarly, for MLGB, $\overline{M}(b_{2i}, b_{2i-1}, \overline{a}_j)$ when $j = 0$ in the equation of pp_{ij} can be reused in the equation of Neg_i. Compared with MLCG and MLGB, MLNG requires one more majority gate and thus, it incurs into a longer critical path to generate pp_{ij}; however, it does

Table 6.5 Comparison of ML implementations for different partial product generation methods based on the MBE

PPGs	Equations	MV	D
MLCG	$pp_{ij} = M(M(M(b_{2i+1}, 0, \overline{M}(b_{2i}, b_{2i-1}, a_j)), M(b_{2i}, 1, M(b_{2i-1}, \overline{a}_{j-1}, 1)), 0), M(\overline{b}_{2i+1}, 0, M(b_{2i}, b_{2i-1}, a_j)), M(a_{j-1}, 1, M(\overline{b}_{2i-1}, \overline{b}_{2i}, 1)), 0), 1)$	10	4
	$Neg_i = M(b_{2i+1}, 0, M(\overline{b}_{2i}, \overline{b}_{2i-1}, 1))$	2	2
MLNG	$pp_{ij} = M(M(b_{2i+1}, M(M(b_{2i}, b_{2i-1}, \overline{a}_j), M(\overline{b}_{2i}, M(\overline{b}_{2i-1}, \overline{a}_{j-1}, 0), 0), 1), 0), M(M(\overline{b}_{2i+1}, M(b_{2i}, b_{2i-1}, a_j), 0), M(\overline{b}_{2i}, M(\overline{b}_{2i-1}, a_{j-1}, 1), 1), 0), 1)$	11	5
	$Neg_i = b_{2i+1}$	0	0
MLGA	$pp_{ij} = M(M(M(b_{2i+1}, 0, M(\overline{b}_{2i}, \overline{b}_{2i-1}, \overline{a}_j)), M(b_{2i}, 1, M(b_{2i-1}, \overline{a}_{j-1}, 1)), 0), M(M(\overline{b}_{2i+1}, 0, M(\overline{b}_{2i}, \overline{b}_{2i-1}, a_j)), M(M(\overline{b}_{2i+1}, b_{2i}, 0), M(b_{2i-1}, a_{j-1}, 0), 0), 1), 1)$	12	4
	$Neg_i = M(b_{2i+1}, \overline{b}_{2i}, \overline{b}_{2i-1})$	1	1
MLGB	$pp_{ij} = M(M(M(M(b_{2i}, b_{2i-1}, \overline{a}_j), b_{2i+1}, 0), M(\overline{M}(b_{2i}, b_{2i-1}, \overline{a}_j), \overline{b}_{2i+1}, 0), 1), M(M(\overline{a}_{j-1}, M(\overline{b}_{2i-1}, \overline{b}_{2i}, 0), 0), M(b_{2i}, M(b_{2i-1}, a_{j-1}, 0), 0), 1), 1)$	10	4
	$Neg_i = M(b_{2i+1}, 1, M(\overline{b}_{2i}, \overline{b}_{2i-1}, 0))$	2	2

not need any logic gate to implement Neg_i. MLGA uses two more majority gates to generate pp_{ij} and one fewer majority gate to generate Neg_i.

For an $n \times n$ multiplier, $\frac{n}{2}$ Neg_i and approximately $\frac{n^2}{2}$ PPs are required. Therefore, the efficiency to generate pp_{ij} has a more significant effect on the overall design of a multiplier. Compared with MLNG and MLGA, although MLCG and MLGB have higher complexity in generating Neg_i, they need fewer majority gates to implement pp_{ij} and a shorter critical path. Thus, MLCG and MLGB are more efficient for exact PP generation.

3.2 Approximation in Partial Product Generation

This section discusses ML-based approximate PPG designs with different types of errors, including positive or negative single-sided and double-sided errors. For approximate designs, the errors that make the approximate results always smaller or larger than the exact results are referred to as negative or positive single-sided errors. Otherwise, the errors are called double-sided errors.

Table 6.6 Approximate designs of pp_{ij} for Radix-4 booth PP generators in [13, 24]

	Approximate pp_{ij} based on MBE						
PP	Design	Type	Equation	MV	D	ER (%)	Errors
MLCG	APPG1 [13]	app_{ij}	$M(M(b_{2i+1}, \overline{M}(b_{2i}, b_{2i-1}, a_j), 0), M(\overline{b}_{2i+1}, M(b_{2i}, b_{2i-1}, a_j), 0), 1)$	4	3	12.5	Positive
	APPG2 [24]	app_{0j} $(j \neq 0)$	$M(M(\overline{b}_1, M(b_0, 0, a_j), 0), M(b_1, \overline{a}_j, 0), 1)$	4	3	12.5	Double
		app_{00}	$M(M(b_1, \overline{M}(b_0, 0, a_0), 0), M(\overline{b}_1, M(b_0, 0, a_0), 0), 1)$	4	3	0	
		other app_{ij}	$M(M(M(b_{2i+1}, \overline{a}_j, 0), \overline{M}(b_{2i}, b_{2i-1}, a_j), 0), M(M(\overline{b}_{2i+1}, a_j, 0), M(b_{2i}, b_{2i-1}, a_j), 0), 1)$	6	3	12.5	
MLNG	APPG3 [24]	app_{0j} $(j \neq 0)$	$M(M(\overline{b}_1, M(b_0, 0, a_j), 0), M(b_1, \overline{a}_j, 0), 1)$	4	3	12.5	Double
		app_{00}	$M(M(b_1, \overline{M}(b_0, 0, a_0), 0), M(\overline{b}_1, M(b_0, 0, a_0), 0), 1)$	4	3	0	
		other app_{ij}	$M(M(b_{2i+1}, M(b_{2i}, b_{2i-1}, \overline{a}_j), 0), M(\overline{b}_{2i+1}, M(b_{2i}, b_{2i-1}, a_j), 0), 1)$	5	3	12.5	
MLGA	APPG4 [24]	app_{0j} $(j \neq 0)$	$M(M(\overline{b}_1, M(\overline{b}_1, \overline{b}_0, a_j), 0), M(b_1, \overline{a}_j, 0), 1)$	4	3	12.5	Double
		app_{00}	$M(M(b_1, M(\overline{b}_0, 1, \overline{a}_0), 0), M(\overline{b}_1, M(\overline{b}_0, 1, a_0), 0), 1)$	5	3	0	
		other app_{ij}	$M(M(b_{2i+1}, M(\overline{b}_{2i}, \overline{b}_{2i-1}, \overline{a}_j), 0), M(\overline{b}_{2i+1}, M(\overline{b}_{2i}, \overline{b}_{2i-1}, a_j), 0), 1)$	5	3	12.5	
MLGB	APPG5 [24]	app_{ij}	$M(M(b_{2i+1}, M(b_{2i}, b_{2i-1}, \overline{a}_j), 0), M(\overline{b}_{2i+1}, \overline{M}(b_{2i}, b_{2i-1}, \overline{a}_j), 0), 1)$	4	3	12.5	Negative
	APPG6 [24]	app_{0j} $(j \neq 0)$	$M(M(\overline{b}_1, M(\overline{b}_1, \overline{b}_0, a_j), 0), M(b_1, \overline{a}_j, 0), 1)$	4	3	12.5	Double
		app_{00}	$M(M(b_1, M(\overline{b}_0, 1, \overline{a}_0), 0), M(\overline{b}_1, M(\overline{b}_0, 1, a_0), 0), 1)$	5	3	0	
		other app_{ij}	$M(M(b_{2i+1}, M(M(b_{2i}, b_{2i-1}, \overline{a}_j), \overline{a}_j, 1), 0), M(\overline{b}_{2i+1}, M(\overline{M}(b_{2i}, b_{2i-1}, \overline{a}_j), a_j, 1), 0), 1)$	6	4	12.5	

3.2.1 Design

Table 6.6 presents six approximate PP generator designs based on four different PP generation methods in Table 6.5 by introducing single-sided or double-sided errors. In Table 6.6, the first column presents the exact PP generation methods that approximate PPGs are designed based on; the second to fourth columns show the different PPs and the corresponding formulas for different APPGs in [24], respectively; The fifth to seventh columns present the required number of MVs, D and ER of each approximate pp_{ij} (app_{ij}); the eighth column reports the direction of the introduced errors (positive single-sided, negative single-sided or double-sided).

The approximate designs that produce double-sided errors (APPG2, APPG3, APPG4, and APPG6) leverage different characteristics of the input operands, i.e., app_{00}, $app_{0j}(j \neq 0)$ and the other common app_{ij}, whereas APPG1 and APPG5 use the same equations to generate all the app_{ij} by introducing positive or negative single-sided errors, respectively. The MLCG-based pp_{ij} includes relatively fewer true cases, thus APPG1 introduces positive single-sided errors. The MLGB encodes pp_{ij} with more true cases. The APPG5 does not include few true cases for simplification by introducing negative single-sided errors. However, if only single-sided errors are introduced, it leads to relatively large errors, especially used for building a large circuit. The APPG2 and APPG6 introduce double-sided errors based on MLCG and MLGB, respectively. Double-sided errors are also introduced in APPG3 and APPG4 for achieving the simplification of ML implementations since the MLNG- and MLGA-based pp_{ij}s have a moderate number of true cases.

3.2.2 Evaluation

As shown in Table 6.6, except for app_{00} with double-sided errors, all app_{ij}s share the same error rate of 12.5%. The APPGs based on MLCG and MLGB require a similar hardware complexity. Compared with APPG1 and APPG5, APPG2 and APPG6 introducing double-sided errors result in the increase of hardware complexity. Compared with the exact PP generation using MLCG and MLGB in Table 6.5, the generation of one PP using APPG1 and APPG5 can save six majority gates, whereas the saving of 40% in MVs can be obtained by using APPG2 and APPG6. With double-sided errors, APPG3 and APPG4 reduce the use of majority gates by two than APPG2 and APPG6. Compared with their exact counterparts in Table 6.5, APPG3 and APPG4 save 54% and 58% of utilized majority gates with unbiased errors and single-sided errors, respectively. An APPG6-based app_{ij} has the same critical path as the exact one. The other approximate PPG-based app_{ij}s can shorten the critical path to three units of delay.

Assume the approximation factor p denote the number of columns with approximate PPs in the PP array. The approximation is applied from the PPs of the least significance. The errors are analyzed for 8×8 Booth multipliers. Figure 6.10 presents the error metrics (in $NMED$, $MRED$, $RMSE$, and ER) of these six APPG designs with p (from 4 to 8). The single-sided errors introduced in one PP will lead to catastrophic effects on the final results with the increase in the values of p. Compared with the APPG1, the APPG5 has a worse performance in accuracy. The approximate MLCG- or MLGB-based multipliers with unbiased errors (APPG2 and APPG6) lead to the same error characteristics. Compared with their counterparts with single-sided errors (APPG1 and APPG5), with an increase of p, the improvement in errors increases up to 56.9% in NMED, 81.1% in MRED and 55.1% in RMSE; however, the APPG2 has a lower ER than APPG1 only when $p = 7$. The APPG4 shows better performance in accuracy than the APPG3, with improvements of 14.8% in NMED, 2.2% in MRED, 12.2% in RMSE, and 11.1% in ER when $p = 8$.

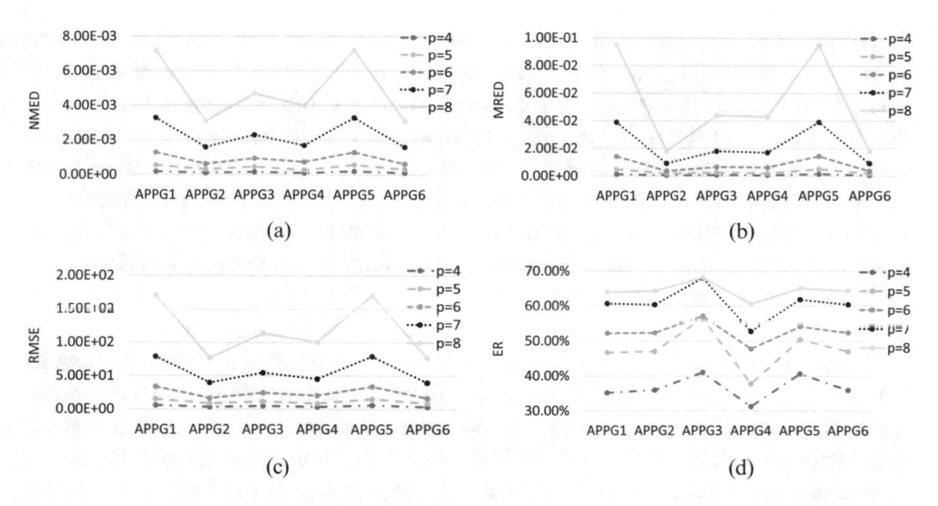

Fig. 6.10 Error measurements for an 8×8 Booth multiplier using different APPGs: (**a**) $NMED$ vs p, (**b**) $MRED$ vs p, (**c**) $RMSE$ vs p, and (**d**) ER vs p

The curves of the accuracy for these six PPG designs in NMED, MRED, RMSE in Fig. 6.10 have a similar tendency. For double-sided errors, when $p < 5$, the APPG4-based multiplier designs show a higher accuracy than other PPG-based multipliers; when $p > 5$, the APPG2- or APPG6-based designs are better choices for higher accuracy. When $p = 5$, the APPG2- and APPG6-based designs are superior in RMSE and MRED, respectively; but the APPG4-based multiplier design is more accurate in NMED. Moreover, in terms of ER, the APPG4-based designs always have the highest accuracy among the six PPG designs.

3.3 Approximation in PP Reduction and Compression

Depending on the different error characteristics introduced by approximate PPGs discussed in Sect. 3.2.1, the approximation in PP reduction and compression in [24] utilizes complementary strategies in the PP reduction based on a probability analysis and performs the compression depending on the characteristics of PP array. The 8×8 Booth multiplier is presented as a case study.

3.3.1 Design

PP Reduction

The APPG1-based design with positive single-sided errors results in larger results than the exact results. As a common approximation technique, truncation can significantly reduce hardware complexity to generate output values that are smaller than the exact results. Zhang et al. [24] considers the truncation as a complementary strategy for the positive errors due to the approximate PP generations.

Assume that the number of truncated columns from the least significant column in the PP array is $t + 1$ ($0 \leq t < p$, and $t = 0$ means that Column 0 including Neg_i is truncated). So, $p - t - 1$ columns remain using the approximate PPGs. A probabilistic analysis in [24] is performed to decide how many columns in the PP array are truncated under different values of p to maximize the effects of error compensation. Two probabilities are introduced in [24]: the probability of generating effective carries from the truncated columns and the probability of making the possible error compensation by introducing positive single sided errors by the remaining approximate PPs after truncation. The selection of t is determined to ensure the difference between these two probabilities is the smallest.

The APPG5-based design with negative single-sided errors makes the results smaller than the exact ones. To compensate for the errors resulting from the approximate encoding with negative single-sided errors, the PPs of lower significance are considered to be replaced by '1's in [24] to make the approximate results larger.

Assume the number of columns in which the PPs are set to '1's (from the least significance in the PP array) be $l + 1$ ($0 \leq l < p$, *e.g.* $l = 0$ means that the PP in Column 0 is set as '1's, except for Neg_0); then, the remaining $p - l - 1$ columns of PPs are generated by the approximate PPGs. A similar probabilistic analysis is then performed to decide the pairs of l and p.

Table 6.7 shows the analytical results of the values of t and l for p ($p \in [4, 8]$) and the error results for 8×8 multipliers (also verified by exhaustive simulation). It can be seen that excluding the exact PPs, it is better to keep two columns of higher significance for approximation while the other columns complement the errors using complementary strategies. The values of ER will increase if complementary strategies are employed. For the APPG1-based approximate multiplier design, combined with truncation, the improvement in accuracy increases with p, up to 27.7% in NMED, 8.4% in MRED, 36.8% in RMSE.

For the APPG5-based approximate multiplier design, the use of the complementary strategy achieves an improvement up to 27.7% in NMED, 9.4% in MRED, and 36.8% in RMSE. These two PP generation methods with complementary strategies result in multipliers with similar error characteristics. Evaluated by $NMED$ or $RMSE$, the APPG1-based design is relatively more accurate; the complementary strategy for an APPG5-based design is slightly more effective in terms of $MRED$ and ER. For an APPG1-based design with PP reduction, the complementary strategy (truncation) can significantly reduce the bit-width of the final results.

Truncation is also used as a complementary strategy for the other APPG-based designs with a relatively high error tolerance ($p \geq 5$) in [24]. Thus, the APPG2- and APPG6-based multipliers with double-sided errors are considered due to their smaller hardware overhead to generate exact PPs and the higher accuracy to generate approximate PPs when $p \geq 5$. From the experiments, although truncation increases the NMED, it can reduce the MRED. For an improvement in MRED and to maximize the advantages of truncation, APPG6-based multipliers with $t = 0, 1, 1$ for $p = 5, 6, 7$ are preferred; while the APPG2-based multiplier with $t = 2$ is preferred for $p = 8$ [24].

Table 6.7 Error measurements for the 8 × 8 APPG1 and APPG5-based multiplier designs using the PP reduction schemes with single-sided errors for PP generations [24]

PPGs	APPG1 (positive) [24]					APPG5 (negative) [24]				
p	t	$NMED$	$MRED$	$RMSE$	$ER(\%)$	l	$NMED$	$MRED$	$RMSE$	$ER(\%)$
4	1	1.9×10^{-4}	1.4×10^{-3}	5.1	69.5	1	2.0×10^{-4}	1.4×10^{-3}	5.1	67.1
5	2	4.9×10^{-4}	4.3×10^{-3}	10.8	86.6	2	5.0×10^{-4}	4.1×10^{-3}	11.5	83.6
6	3	1.1×10^{-3}	1.3×10^{-2}	23.7	91.1	3	1.1×10^{-3}	1.3×10^{-2}	24.4	89.8
7	4	2.4×10^{-3}	3.5×10^{-2}	50.7	96.1	4	2.5×10^{-3}	3.4×10^{-2}	53.0	95.1
8	5	5.2×10^{-3}	8.7×10^{-2}	108.5	97.6	5	5.2×10^{-3}	8.7×10^{-2}	110.8	96.9

PP Compression

After the PP generations and reduction, Zhang et al. [24] employs exact full adders for compression to preserve the accuracy; however, they can be replaced by approximate adders to further reduce hardware at a loss of accuracy.

Three accuracies at $p = 4$, $p = 6$, and $p = 8$ are considered. With double-sided errors, the APPG4-based design ($p = 4$), the APPG6-based design ($p = 6, t = 1$), and the APPG2-based design ($p = 8, t = 2$) are discussed in [24] since they have better performance in accuracy compared with other designs at the same value of p.

For APPG1- and APPG5-based multipliers, the PP array can be significantly reduced by using the complementary strategies, so that the hardware in the PP compression can be reduced, but with a high accuracy loss. Therefore, they are suitable for the low accuracy cases (i.e., $p = 6$ and 8) to reduce the hardware. From Table 6.7, the APPG1-based design (positive, $p = 6, t = 3$) and the APPG5-based design (negative, $p = 8, l = 5$) are considered. If the smallest bit-width of the output result is desired, the APPG1-based design is preferred.

The reduced PP array is compressed by leveraging the different characteristics of PPs, i.e., preprocessing the constants to reduce and regularize the PP array, generating and compressing the approximate PPs simultaneously with generating the exact PPs. The compression design for the APPG5 (negative, $p = 8, l = 5$)-based PP array requires thirteen full adders and a 9-bit ripple carry adder (RCA) [24]. It further introduces the approximation resulting form giving up the complementary bit in the least significance to reduce the hardware required for compression. The APPG1 (positive, $p = 6, t = 3$)-based PP array is compressed by using fifteen full adders and a 11-bit RCA. To simplify the compression process of the APPG4 (unbiased, $p = 4$)-based PP array, Neg_0 is discarded resulting in a 28% increase in NMED but a 2.8% decrease in MRED; thus, the compression requires sixteen full adders and a 12-bit RCA. The compression of the APPG6 (unbiased, $p = 6, t = 1$)-based PP array requires sixteen full adders and a 12-bit RCA; the compression of the APPG2 (unbiased, $p = 8, t = 2$)-based PP array requires seventeen full adders and a 10-bit RCA.

3.3.2 Evaluation

Table 6.8 presents the numbers of MVs and INVs, D, ADP, the error metrics and BW for the exact and approximate Booth multipliers in [24]. Similar with approximate multiplier designs, only full adders are applied in parallel prior to obtaining the two rows of PPs in the reduction of PPs for the exact multipliers. Thus, thirty full adders and a 13-bit RCA are required in the compression process.

For exact PP generations, the MLCG- and MLGB-based exact multipliers show good performance, reducing the number of majority gates by up to 11%, and the ADP by up to 12% compared with the MLNG- and MLGA-based exact schemes. Using approximate PP generation methods, the hardware decreases with a reduced accuracy. Compared with the MLCG-based exact multiplier, those using

Table 6.8 Hardware and error evaluation of exact and approximate 8 × 8 booth multipliers based on different PPGs in [24]

Multiplier		Errors	p	t	l	MV	INV	D	NMED (10^{-3})	MRED (10^{-3})	RMSE	ER %	ADP	BW
MLCG		–	–	–	–	442	138	24	0	0	0	0	10608	16
MLNG		–	–	–	–	479	106	25	0	0	0	0	11975	16
MLGA		–	–	–	–	497	106	24	0	0	0	0	11928	16
MLGB		–	–	–	–	444	138	24	0	0	0	0	10656	16
MLCG	APPG1	Positive	6	3	–	329	102	20	1.1	13	23.7	91.1	6580	12
	APPG2	Unbiased	8	2	–	325	106	20	3.3	17	76.9	92.2	6500	13
MLGA	APPG4	Unbiased	4	–	–	435	76	21	0.14	0.61	3.9	76.9	9135	16
MLGB	APPG5	Negative	8	–	5	258	84	19	5.2	86	111.1	99.8	4902	16
	APPG6	Unbiased	6	1	–	363	106	21	0.71	3.5	16.8	93.8	7623	14

approximate PPGs can obtain an improvement up to 53% in the ADP and 20% in delay. The MLGA-based approximate multiplier achieves only a reduction of 1.5% in the number of majority gates, 45% of the inverters, and 12% of the delay. The APPG4-based multiplier with negative single-sided errors ($p = 8$) reduces up to 41% of the majority gates, 39% of the inverters, 20% of the delay, and 53% of the ADP at a relatively large decrease in accuracy. Moreover, with truncation, the bit-width of the final result can be reduced by up to 12.

The approximate multiplier designs with positive single-sided and double-sided errors based on MLCG perform similarly in hardware metrics. Thus, the APPG1 (positive, $p = 6$, $t = 3$)-based multiplier design is likely to be preferred due to the higher accuracy, resulting in a 12-bit final result. Compared with exact multiplier designs, the APPG4-based design achieves at least a reduction of 13% in ADP. With a further decrease in accuracy, the APPG6 (unbiased, $p = 6$, $t = 1$)-based design reduces the ADP by 28%. Although the MLCG-based approximate multiplier designs have a similar ADP, the one with positive single-sided errors is superior in NMED. The APPG6 (negative, $p = 8$, $l = 5$)-based design saves 53% of ADP with a large accuracy loss.

Therefore, four levels of accuracy can be established for different requirements of applications: high accuracy (APPG4, unbiased, $p = 4$), good accuracy (APPG6, unbiased, $p = 6$, $t = 1$), moderate accuracy (APPG1, positive, $p = 6$, $t = 3$), and low accuracy (APPG6, negative, $p = 8$, $l = 5$).

3.4 Application

Several applications are considered for the approximate multipliers at four different accuracy levels: low and moderate accuracy for image multiplication and edge detection using the Sobel operator, good and high accuracy for a multilayer perceptron (MLP) for classification and a multi-task CNN for joint face detection and alignment, respectively.

3.4.1 Image Processing

Image multiplication and edge detection using the Sobel operator [19] are considered for the approximate multipliers with low accuracy (APPG5, negative, $p = 8$, $l = 5$) and the ones with moderate accuracy (APPG1, positive, $p = 6$, $t = 3$), respectively.

The experiments on ten different images return on average a PSNR of 59.34 dB and an SSIM of 0.9993 for the image multiplication, and a PSNR of 71.09 dB and an SSIM of 0.9998 for the edge detection, respectively.

3.4.2 Classification

The approximate 8 × 8 radix-4 Booth multipliers (APPG6, unbiased, $p = 6, t = 1$) with good accuracy are used in an MLP of three layers (an input layer of 784 neurons, one hidden layer of 100 neurons, and an output layer of 10 neurons) to classify the MNIST dataset. The performance of the MLP is assessed by the accuracy of the classification.

The 8×8 approximate Booth multipliers are utilized in the inference. The trained weight matrix, the inputs, and biases are mapped into $[-128, 127]$ for the inference. Using MATLAB, 94.41% and 94.34% are obtained as classification accuracies when using the exact and approximate multipliers, respectively. Moreover, compared with the exact design, the approximate multiplier design reduces the ADP of the multipliers in inference by up to 28% along with a reduced hardware for the accumulation operations due to the truncation in each multiplication.

3.4.3 Face Detection and Alignment

A multi-task CNN (MTCNN) is considered for joint face detection and alignment, using the Face Detection Data Set and Benchmark (FDDB) and Annotated Facial Landmarks in the Wild (AFLW) datasets. The MTCNN consists of three cascaded CNNs: the so-called proposal network (PNet), refine network (RNet), and output network (ONet) [8]. The true positive rate (TPR) and the normalized mean error (NME) indicate the accuracy in face detection and alignment, respectively. The hardware overhead is shown by the average number of multiply-and-accumulate (MAC) operations used for detecting faces in an image.

Due to the large number of multiplications required for the MTCNN, approximate 8 × 8 multipliers with a relatively high accuracy (APPG4, unbiased, $p = 4$) are used to replace the exact circuits. Figure 6.11 shows an example for the results of face detection and face alignment. Compared with exact multipliers, the use of approximate multipliers achieves a 5.1% reduction in the number of MACs (0.48 billion vs 0.51 billion), thus approximately saving 18.9% in ADP for MACs at a 7.4% accuracy loss with a TPR of 80.24%; however, the NME is increased from 3.58% to 11.01%.

3.5 *Discussion*

This section discusses the ML-based approximate Booth multiplier designs. Different PP generation methods (pp_{ij} and Neg_i) of the MBE have been developed for efficient ML implementations, i.e., MLCG, MLNG, MLGA, and MLGB. The MLNG and MLGA can achieve the simplification of Neg_i, but with the increase of hardware complexity of pp_{ij}. Thus, the MLCG and MLGB are superior in terms of exact PP generation. Based on these PP generation methods, approximate

(a) (b)

Fig. 6.11 Results of face detection and alignment using 8 × 8 approximate Booth multipliers with high accuracy (APPG4, unbiased, $p = 4$): (**a**) face alignment and (**b**) face detection

PPGs have been designed with different types of errors: positive single-sided, negative single-sided, double-sided errors. These approximate PPG designs with single-sided errors will lead to catastrophic results especially when multiple PPs employ approximate generation. Approximation in the PP reduction can serve as complementary strategies by mathematically analyzing the properties of the introduced errors to control accuracy, with the improvement of accuracy and the decrease of hardware complexity. Four approximate multipliers with different levels of accuracy can meet various design requirements in area, delay, and accuracy. A hardware evaluation at the gate level shows a saving of 28% in ADP compared with the exact design and very low NMED and MRED of 0.00071 and 0.0035.

4 Conclusion

In this chapter, designs of approximate multipliers based on ML are reviewed. Their hardware complexity and error characteristics are evaluated using theoretical analysis and functional simulation.

For the ML-based unsigned approximate multiplier design, the recent studies are investigated from the approximation in the PP generation, PP accumulation, and PP array. The recent ML-based 2 × 2 approximate multiplier design with the complementary bit can be employed as a unit to build the larger size of multipliers. The reduction of the number of PPs can be obtained with configurable accuracy by selecting different numbers of complementary bits. Moreover, the truncation technique is also an effective and efficient method to simplify the accumulation process. In the approximate PP accumulation, approximate compressors including full adders and 4-2 compressors can be extensively used. Specifically, approximate 4-2 compressors have been designed by cascading approximate full adders, introducing errors in the truth table, and reducing the number of outputs.

The approximate signed radix-4 Booth multiplier designs have been studied from the approximate PP generation, PP reduction, and compression. The PPs can be efficiently generated by four different ML-based implementation methods of the MBE. The approximate PPGs introduce single-sided or double-sided errors depending on the different characteristics of these PP generation methods. Double-sided errors can lead to more accurate PP generations than single-sided errors. The reduction and compression designs can further introduce positive (negative) single-sided errors to complement negative (positive) single-sided errors due to the approximate PP generation.

Case studies of image processing and neural networks show the great potential of the ML-based approximate multipliers designs.

Acknowledgments This work was supported by the Natural Sciences and Engineering Research Council (NSERC) of Canada (Project Number: RES0048688). T. Zhang is supported by a PhD scholarship from the China Scholarship Council (CSC).

References

1. Dennard RH, Cai J, Kumar A. A perspective on today's scaling challenges and possible future directions. Solid-State Electron. 2007;51(4):518–25.
2. Parhami B, Abedi D, Jaberipur G. Majority-logic, its applications, and atomic-scale embodiments. Comput Electr Eng. 2020;83:106562.
3. Vacca M, Graziano M. Zamboni M. Majority voter full characterization for nanomagnet logic circuits. IEEE Trans Nanotechnol 2012;11(5):940–7.
4. Jamshidi V. A vlsi majority-logic device based on spin transfer torque mechanism for brain-inspired computing architecture. IEEE Trans Very Large Scale Integr Syst. 2020;28(8):1858–66.
5. Angizi S, Jiang H, DeMara RF, Han J, Fan D. Majority-based spin-CMOS primitives for approximate computing. IEEE Trans Nanotechnol. 2018;17(4):795–806.
6. Jiang H, Angizi S, Fan D, Han J, Liu L. Non-volatile approximate arithmetic circuits using scalable hybrid spin-CMOS majority gates. IEEE Trans Circ Syst I Reg Pap. 2021;68(3):1217–30.
7. Liu W, Lombardi F, Shulte M. A retrospective and prospective view of approximate computing [point of view. Proc IEEE. 2020;108(3):394–9.
8. Jiang H, Santiago FJH, Mo H, Liu L, Han J. Approximate arithmetic circuits: a survey, characterization and recent applications. Proc IEEE. 2020;108:2108.
9. Behrooz P. Computer arithmetic: algorithms and hardware designs. Oxford University Press; 2010.
10. Labrado C, Thapliyal H, Lombardi F. Design of majority logic based approximate arithmetic circuits. In: 2017 IEEE international symposium on circuits and systems (ISCAS). 2017. pp. 1–4.
11. Liu W, Zhang T, McLarnon E, O'Neill M, Montuschi P, Lombardi F. Design and analysis of majority logic based approximate adders and multipliers. IEEE Trans Emerg Top Comput. 2019;9:1609.
12. Sabetzadeh F, Moaiyeri MH, Ahmadinejad M. A majority-based imprecise multiplier for ultra-efficient approximate image multiplication. IEEE Trans Circ Syst I Reg Pap. 2019;66(11):4200–08.
13. Zhang T, Liu W, Han J, Lombardi F. Design and analysis of majority logic based approximate radix-4 Booth encoders. In: 2019 IEEE/ACM international symposium on nanoscale architectures (NANOARCH). 2019. pp. 1–6.
14. Liang J, Han J, Lombardi F. New metrics for the reliability of approximate and probabilistic adders. IEEE Trans Comput. 2013;62(9):1760–71.

15. Pudi V, Sridharan K, Lombardi F. Majority logic formulations for parallel adder designs at reduced delay and circuit complexity. IEEE Trans Comput. 2017;66(10):1824–30.
16. Wallace CS. A suggestion for a fast multiplier. IEEE Trans Electron Comput 1964;1:14–7.
17. Dadda L. Some schemes for parallel multipliers. Alta Frequenza 1965;34:349–56
18. Chang C, Gu J, Zhang M. Ultra low-voltage low-power CMOS 4-2 and 5-2 compressors for fast arithmetic circuits. IEEE Trans Circ Syst I. 2004;51(10):1985–97.
19. Cho H, Swartzlander EE. Adder and multiplier designs in quantum dot cellular automata. IEEE Trans Comput. 2009;58(6):721–7.
20. Kulkarni P, Gupta P, Ercegovac M. Trading accuracy for power with an underdesigned multiplier architecture. In: Proc VLSI Design. 2011. pp. 346–51.
21. Zhang T, Liu W, McLarnon E, O'Neill M, Lombardi F. Design of majority logic (ML) based approximate full adders. In: 2018 IEEE international symposium on circuits and systems (ISCAS). 2018. pp. 1–5.
22. Moaiyeri MH, Sabetzadeh F, Angizi S. An efficient majority-based compressor for approximate computing in the nano era. Microsyst Technol. 2018;24(3):1589–1601.
23. Taheri M, Arasteh A, Mohammadyan S, Panahi A, Navi K. A novel majority based imprecise 4: 2 compressor with respect to the current and future VLSI industry. Microprocess Microsyst. 2020;73:102962.
24. Zhang T, Jiang H, Mo H, Liu W, Lombardi F, Liu L, Han J. Design of majority logic-based approximate Booth multipliers for error-tolerant applications. IEEE Trans Nanotechnol. 2022;21:81–89.
25. MacSorley O. High-speed arithmetic in binary computers. Proc. IRE. 1961;49:67–91.
26. Yeh W, Jen C. High-speed Booth encoded parallel multiplier design. IEEE Trans Comput. 2000;49(7):692–701
27. Cui X, Liu W, Swartzlander E, Lombardi F. A modified partial product generator for redundant binary multipliers. IEEE Trans Comput. 2016;65(4):1165–71.
28. Wang J, Kuang S, Liang S. High-accuracy fixed-width modified Booth multipliers for lossy applications. IEEE Trans Very Large Scale Integr. 2011;19(1):52–60.

Part II
Introduction: Design Automation and Test

Weiqiang Liu and Fabrizio Lombardi

In the second part of this book, automated tools and approaches for design automation and test are presented. The four chapters consider diverse technology platforms and metrics to design and test approximate computing systems. The first chapter is titled "Approximate Logic Synthesis for FPGA by Decomposition," authored by Zhiyuan Xiang, Niyiqiu Liu, Yue Yao, Fan Yang, Cheng Zhuo, and Weikang Qian. In the current technical literature, there are many approximate logic synthesis (ALS) methods, but few of them specifically target field programmable gate array (FPGA) designs. In this chapter, the authors propose an ALS method for FPGAs based on decomposition; this method resynthesizes some fanout-free cones in an FPGA with the fewest look-up tables by using the proposed approximate decomposition techniques. Experimental results show that the proposed method considerably reduces the required LUT count compared to previous ALS methods as applicable to FPGAs.

A different design aspect is treated in the next chapter "Design Techniques for Approximate Realization of Data-Flow Graphs" (by Marzieh Vaeztourshizi, Hassan Afzali-Kusha, Mehdi Kamal, and Massoud Pedram). This chapter deals with optimization techniques for the approximate realization of a high-level description of input applications. It includes a study and analysis of different design space exploration (DSE) methodologies, which find an approximate realization of an input application subject to given design constraints. Initially, the authors review the background and concepts related to power consumption calculation, approximate computing (AC) techniques and DSE methodologies, approximate high-level synthesis (HLS), and evaluation metrics for AC. Next, the authors explain and compare prior works in this field with emphasis on approximation methods used in an exact design and the DSE methodologies.

Data-flow graphs are also utilized in the next chapter, namely "Approximation on Data-Flow Graph Execution for Energy Efficiency" (by Qian Xu, Md Tanvir Arafin, and Gang Qu). In this chapter, the authors discuss approximation methods at different levels of a data-flow graph (DFG) to reduce energy consumption with a guaranteed set of results. Initially, the authors consider a probabilistic design framework that approximates an application by intentionally terminating specific DFG executions prior to reaching a deadline. Subsequently, the authors present a real-time estimation-and-recomputing approach that executes non-critical parts of the DFG in an approximate manner. Finally, the authors use the floating-point logarithmic operation as an example to show an approach that reduces and adjusts the computational data bit width based on their proposed DFG model.

The last chapter, “Test and Reliability of Approximate Hardware,” is authored by Marcello Traiola, Bastien Deveautour, Arnaud Virazel, Patrick Girard, and Alberto Bosio. As approximate computing is gaining increasing interest, important challenges, as well as opportunities, arise concerning the dependability of such systems. This chapter focuses on test and reliability issues related to approximate hardware systems; it covers problems and solutions concerning the impact of an approximation on hardware defect classification, test generation, and application. Moreover, the impact of the approximation on fault tolerance is also discussed, along with related design solutions to mitigate it.

Chapter 7
Approximate Logic Synthesis for FPGA by Decomposition

Zhiyuan Xiang, Niyiqiu Liu, Yue Yao, Fan Yang, Cheng Zhuo, and Weikang Qian

1 Introduction

With the breakdown of Dennard scaling, power consumption has become a bottleneck for circuit design [1]. Meanwhile, many useful applications are inherently error-tolerant. These include machine learning, pattern recognition, and image pro-

This work was supported by the State Key Laboratory of ASIC & System Open Research Grant 2019KF004. Zhiyuan Xiang and Niyiqiu Liu contributed equally.

Z. Xiang · N. Liu
University of Michigan-Shanghai Jiao Tong University Joint Institute, Shanghai Jiao Tong University, Shanghai, China
e-mail: xzy242215@sjtu.edu.cn; lnyq10@sjtu.edu.cn

Y. Yao
School of Computer Science, Carnegie Mellon University, Pittsburgh, PA, USA
e-mail: yueyao@cs.cmu.edu

F. Yang
State Key Laboratory of ASIC & System; Microelectronics Department, Fudan University, Shanghai, China
e-mail: yangfan@fudan.edu.cn

C. Zhuo
College of Information Science and Electronic Engineering, Zhejiang University, Hangzhou, China
e-mail: czhuo@zju.edu.cn

W. Qian (✉)
University of Michigan-Shanghai Jiao Tong University Joint Institute and MoE Key Laboratory of Artificial Intelligence, Shanghai Jiao Tong University, Shanghai, China

State Key Laboratory of ASIC & System, Fudan University, Shanghai, China
e-mail: qianwk@sjtu.edu.cn

W. Liu, F. Lombardi (eds.), *Approximate Computing*,
https://doi.org/10.1007/978-3-030-98347-5_7

cessing. Given these trends, approximate computing was proposed as a promising way to design low-power digital circuits for these error-tolerant applications [2]. It relaxes the stringent accuracy requirement to further reduce circuit area, delay, and power consumption. An important area in approximate computing is approximate logic synthesis (ALS), which automatically synthesizes a good approximate circuit under a given error constraint.

Many existing ALS methods are designed for application-specific integrated circuits (ASICs) [3–12], while only few target at field programmable gate arrays (FPGAs) [13, 14]. Modern FPGAs are implemented by a network of lookup tables (LUTs). A LUT of k inputs, known as k-LUT, can implement any k-input Boolean function. This causes fundamental difference between FPGAs and ASICs. The existing ALS methods for ASIC work on a different circuit representation than the LUT network representation. Some ALS methods cannot be applied to handle LUT networks [11], while the others can [5]. Although the former can still be applied in the technology independent synthesis phase for FPGA, they require an additional LUT mapping step to convert the intermediate design into the FPGA design, which leads to a weak control over the final hardware cost. For the latter, they cannot fully exploit the special features of LUT networks. Finally, the few ALS methods for FPGA still rely heavily on the existing LUT mapping tools [13, 14]. Thus, they also have a weak control over the final hardware cost. Therefore, in order to fully explore the power of ALS for FPGAs, it is imperative to develop a method that directly works on the LUT network representation and fully exploits the flexibility of the FPGA designs.

For this purpose, we propose a novel method to perform ALS for FPGAs in this work. Our method aims at reducing the LUT count. It directly works on the LUT network and exploits the reconfigurability of the LUTs to reduce its size.

The basic idea of our method is to approximately implement a LUT subnetwork by the minimum number of LUTs determined by the input size of the subnetwork. To illustrate our idea, consider an optimized LUT subnetwork of 6 inputs shown in the left part of Fig. 7.1. We assume that 3-LUTs are used. This optimal design is implemented by four 3-LUTs. Note that in theory, the minimum number of 3-LUTs needed to implement a 6-input function is 3.[1] However, there does not exist any LUT network of three 3-LUTs to implement the given function, as the LUT network in the left part of Fig. 7.1 is claimed to be optimal. Nevertheless, if we allow errors, it is possible to change the given function properly so that it can be implemented by only three 3-LUTs; one example is shown in the right part of Fig. 7.1. Indeed, there exist several different ways to connect three 3-LUTs. Furthermore, each 3-LUT can implement many different functions. Thus, the number of functions that can be implemented by a network of three 3-LUTs is enormous. On the one hand, this flexibility is helpful, since it is possible for us to find one function very close to the original one, thus with the minimum error introduced. On the other hand, given the large design space, how to efficiently find such a function is a big challenge. Since a

[1] With only two 3-LUTs, we can only realize a function with no more than 5 inputs.

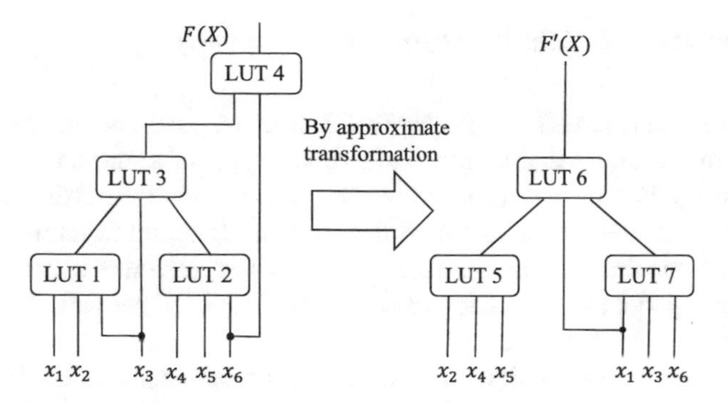

Fig. 7.1 An approximate transformation that reduces the LUT count to the minimum value determined by the input size of the function

LUT network naturally corresponds to a series of decompositions of a function, we propose a decomposition-based technique to solve the above challenge and derive a novel ALS method for FPGA.

Our main contributions are as follows:

- We propose a heuristic method to find an approximate disjoint decomposition for a given function with small error.
- We extend the above method and propose a heuristic method to find an approximate non-disjoint decomposition for a given function with small error.
- We propose an iterative decomposition algorithm that exploits both the approximate disjoint and non-disjoint decompositions to realize a Boolean function with the minimum number of LUTs determined by the input size of the function.
- We design an ALS flow for FPGA based on the iterative decomposition algorithm. Our experimental results showed that the proposed flow achieves more LUT count reduction than the previous state-of-the-art ALS methods for FPGAs.

The rest of the chapter is organized as follows. Section 2 discusses the related works. Section 3 provides the preliminaries. Section 4 presents our proposed methods. Section 5 shows the experimental results. Finally, Sect. 6 concludes the chapter.

2 Related Work

We describe some related works in this section. We first discuss the works related to ALS for FPGA, followed by works related to ALS based on decomposition.

2.1 Approximate Logic Synthesis for FPGA

Wu et al. proposed an ALS method for FPGA designs [13]. The method is based on a heuristic that assumes that by removing some inputs of a function, the final LUT count will drop. Following this heuristic, the method simultaneously removes some inputs of a local subnetwork and modifies its logic function to minimize the error introduced. Once the approximate function is derived, it further applies an existing FPGA mapping tool to the function to obtain the final LUT network. Liu and Zhang also proposed an ALS method and applied it to FPGA synthesis [14]. Their method works on the gate network first and then maps the network into the FPGA design through an FPGA mapping tool. Due to the use of the additional FPGA mapping tool, these previous methods have a weak control over the final hardware cost. In contrast, our method works on the LUT network representation directly to reduce the LUT count.

2.2 Approximate Logic Synthesis Based on Decomposition

Another direction of ALS is to perform approximate decomposition to some local circuits to simplify them. Hashemi et al. [10] proposed a method based on *Boolean matrix factorization*. It approximately factors the truth table of a multi-output function as the product of two Boolean matrices. It then synthesizes an approximate design based on the factorization. Yao et al. [15] proposed a method based on Boolean decomposition. It performs *approximate disjoint bi-decomposition* that recursively separates the input set of a function into two disjoint subsets. However, the method is based on the special disjoint bi-decomposition and thus can only generate networks of 2-input gates. In contrast, our method is based on a general disjoint decomposition and therefore can be applied to synthesize networks of k-LUTs with $k \geq 2$. Furthermore, beyond approximate disjoint decomposition, we also propose an approximate non-disjoint decomposition. Another related work is the proposal of a two-LUT architecture that approximately realizes a given function originally implemented by a single larger LUT [16]. The work also exploits Boolean decomposition to configure the LUTs. However, it targets at LUT-based computation, where a single large LUT implements the target function, which is different from LUT-based FPGA, where many small LUTs are interconnected to realize the target. Due to the architecture difference, the work only requires to perform the approximate disjoint decomposition once. In contrast, our work needs to perform multiple approximate disjoint decompositions recursively.

3 Preliminaries

In this section, we introduce the related preliminaries, including simple disjoint decomposition, fanout-free cone, error measurement, and Monte Carlo simulation.

3.1 Simple Disjoint Decomposition

Our proposed method is based on simple disjoint decomposition, which was pioneered by Ashenhurst [17] and Curtis [18]. We first introduce some definitions.

Definition 1 Let f be a logic function of n variables and $X = \{x_1, \ldots, x_n\}$ be its inputs. Let $\{A, B\}$ be a partition on X. The function f has a *simple disjoint decomposition* with *bound set A* and *free set B* if there exist functions ϕ and F such that $f(X) = F(\phi(A), B)$. The functions F and ϕ are called the *free-set function* and the *bound-set function*, respectively. If the function f has a simple disjoint decomposition, the function is said to be *decomposable*.

Not every logic function is decomposable. Ashenhurst gives a necessary and sufficient condition for the existence of a simple disjoint decomposition under a given partition on the input variables [17]. It is based on a *2-dimensional (2D) truth table* representation of the Boolean function, in which some variables define the columns and the remaining define the rows. An example of the 2D truth table is shown in Fig. 7.2. In what follows, we will also call this representation a *Boolean matrix*. The following theorem gives the necessary and sufficient condition.

Theorem 1 *Let $\{A, B\}$ be a partition on X. A logic function f is decomposable with bound set A and free set B if and only if the Boolean matrix with the variables in A and B defining the columns and the rows, respectively, has at most four distinct types of rows:*

1. *A pattern of all 0s*
2. *A pattern of all 1s*
3. *A fixed pattern p of 0's and 1's*
4. *The complement of the pattern p*

A proof to the above theorem can be found in [19]. We use the following example to illustrate how to obtain the simple disjoint decomposition once the condition in Theorem 1 is satisfied.

Example 1 Figure 7.2 shows a Boolean matrix of a Boolean function $f(x_1, x_2, x_3, x_4)$ with variables x_1 and x_2 defining the rows and variables x_3 and x_4 defining the columns. It satisfies the condition described in Theorem 1: row 1 falls into Type 3, rows 2 and 4 fall into Type 4, and row 3 falls into Type 2. Thus, function f is decomposable with free set as $\{x_1, x_2\}$ and bound set as $\{x_3, x_4\}$. We can set the truth table of the function $\phi(x_3, x_4)$ as the pattern in Type 3. For this example, the truth table is "0110," and correspondingly, $\phi(x_3, x_4) = \overline{x_3}x_4 + x_3\overline{x_4}$. Now, the first,

		x_3x_4				
		00	01	10	11	
x_1x_2	00	0	1	1	0	$\phi \wedge \overline{x_1} \wedge \overline{x_2}$
	01	1	0	0	1	$\bar{\phi} \wedge \overline{x_1} \wedge x_2$
	10	1	1	1	1	$x_1 \wedge \overline{x_2}$
	11	1	0	0	1	$\bar{\phi} \wedge x_1 \wedge x_2$

Fig. 7.2 A 2D truth table, or Boolean matrix, of a function f

second, third, and fourth rows of the Boolean matrix represent the functions $\phi\overline{x_1}\,\overline{x_2}$, $\overline{\phi}\overline{x_1}x_2$, $x_1\overline{x_2}$, and $\overline{\phi}x_1x_2$, respectively. Therefore, we obtain the final expression of f as

$$f = \phi\overline{x_1}\,\overline{x_2} + \overline{\phi}\overline{x_1}x_2 + x_1\overline{x_2} + \overline{\phi}x_1x_2 = F(\phi, x_1, x_2).\square$$

3.2 *Fanout-Free Cone (FFC)*

Our proposed method is based on simple disjoint decomposition. However, it is restricted to single-output Boolean functions. Given this restriction, we apply our method to a particular structure in a circuit called *fanout-free cone (FFC)*. We give the relevant definitions in this section.

We focus on combinational circuits implemented by FGPA. They can be viewed as a directed acyclic graph $N = (V, E)$, where V is the vertex set that contains all LUTs in the circuit and E is the edge set that represents the wire connection among all the LUTs. Given a node $v \in V$ in a graph $N(V, E)$, a *cone* of node v, denoted as C_v, is a subgraph of N consisting of node v and some of its predecessors such that any path from a node in C_v to v lies entirely in C_v [20]. The node v is called the root of C_v. A *fanout-free cone (FFC)* is a cone in which the fanouts of every node other than the root are in the cone [21].

3.3 *Error Rate and Monte Carlo Simulation*

There are two typical quantities to evaluate the error of an approximate circuit, *error rate (ER)* and *error magnitude (EM)*. ER is defined as the probability of an input pattern that gives an erroneous output for the approximate circuit. EM measures how much the output of the approximate circuit deviates from the correct output. It is typically used for arithmetic circuits. In this work, we focus on ER.

The number of input combinations of a circuit is exponential to the input size, and thus, it is impractical to enumerate them to calculate the exact ER. Instead, we perform Monte Carlo simulation to obtain an estimation of the ER of an approximate circuit, as is done in many other works [10, 11]. In our implementation, we chose the sample size as 10^5. Besides that, as we will show later, in order to determine proper approximate transformations for a sub-circuit, we need the occurrence probability for each combination of some internal signals. It is also calculated through Monte Carlo simulation. In this case, the simulation results for internal nodes are needed. To speed up our program, we store the simulation results for each node in the memory.

4 Methodology

In this section, we present the proposed method. We begin with an overview of the basic idea, followed by the technical details.

4.1 Basic Idea

Our method works on the FFCs in the given LUT network. An FFC implements a single-output function through a LUT subnetwork. In order to minimize the total LUT count, we minimize the number of LUTs needed to implement a selected FFC. Our method is based on the following observation: the minimum number of k-LUTs needed to implement an n-input function is $\lceil \frac{n-1}{k-1} \rceil$. For example, in order to implement a 7-input (respectively, 8-input) function with 4-LUTs, the minimum number of LUTs needed is 2 (respectively, 3). Although there exists flexibility in the LUT connection and configuration, it may be impossible to exactly realize the given FFC function with the minimum number of LUTs. However, if we introduce minor modification to the original function f, it is possible.

For this purpose, we develop a method to perform approximate disjoint decomposition for a given function. For an arbitrary function, it may not be decomposable. Approximate disjoint decomposition essentially finds a decomposable function close to it. In the context of FPGA synthesis, we repeatedly apply the approximate disjoint decomposition to the function implemented by an FFC. In each round, we derive an approximate disjoint decomposition with a bound set of size k. Assume the obtained bound-set function is ϕ. Then, we implement it by a k-LUT. With this, k inputs in the bound set are replaced by the single signal ϕ. Therefore, the input size of the function is reduced by $(k - 1)$. We repeat this process until the input size of the function is no more than k. At this moment, the final function can be implemented by a single k-LUT. However, for some cases, the number of inputs of the final function is fewer than k, making the last LUT not fully exploited. To make full use of the last LUT, we propose to connect some signals in the previous levels of the LUT network to the unused inputs of the last LUT. This requires to replace the final approximate disjoint decomposition by an *approximate non-disjoint decomposition*. We also develop a method for this.

Example 2 Consider a LUT network in the left part of Fig. 7.1. It implements a function f with 6 inputs, i.e., $X = \{x_1, \cdots, x_6\}$. If we want to implement f using the minimum number of 3-LUTs, we need two rounds of approximate disjoint decomposition. The intermediate steps are shown in Fig. 7.3a and b. In the first round, suppose that the approximate disjoint decomposition produces an approximation to f as $F_1(\phi_1(A_1), B_1)$ with the bound set $A_1 = \{x_2, x_4, x_5\}$ and the free set $B_1 = \{x_1, x_3, x_6\}$. The corresponding circuit is shown in Fig. 7.3a. After the first round, we find that the function F_1 has 4 inputs ϕ_1, x_1, x_3, x_6. Thus, it cannot be realized by one 3-LUT. Therefore, we continue to the second round of approximate disjoint decomposition. Now $X = \{\phi_1, x_1, x_3, x_6\}$. Suppose that the approximate disjoint decomposition produces an approximation to F_1 as $F_2(\phi_2(A_2), B_2)$ with the bound set $A_2 = \{x_1, x_3, x_6\}$ and the free set $B_2 = \{\phi_1\}$. The resulting circuit is shown in Fig. 7.3b. Now the function F_2 has only two inputs ϕ_1 and ϕ_2. Thus, it can be realized by a 3-LUT. However, the last LUT has an unused input. In this case, we perform an approximate non-disjoint decomposition to change the disjoint decomposition $F_2(\phi_2(x_1, x_3, x_6), \phi_1)$ into a non-disjoint one as $F_3(\phi_3(x_1, x_3, x_6), \phi_1, x_1)$. The final LUT network is shown in Fig. 7.3c. For this example, we reduce both the LUT count and the LUT network depth by one compared to the original LUT network shown in the left part of Fig. 7.1. □

We call the above iterative procedure *iterative approximate decomposition*. It requires $\left(\lceil \frac{n-1}{k-1} \rceil - 1\right) = \lceil \frac{n-k}{k-1} \rceil$ iterations. The number of k-LUTs in the resulting circuit is $\lceil \frac{n-1}{k-1} \rceil$, the minimum achievable value. However, some FFCs in the circuit may already contain the minimum number of LUTs. In this case, to further reduce their LUT counts, we need to remove some of their inputs. For this purpose, we also introduce an iterative input removal method. It can be treated as a special case of the iterative approximate decomposition.

In the following, we will describe details of the approximate disjoint and non-disjoint decompositions in Sects. 4.2 and 4.3, respectively. Then, we will describe the iterative approximate decomposition in Sect. 4.4, followed by the iterative input removal in Sect. 4.5. Finally, we will show the overall ALS flow in Sect. 4.6.

4.2 Approximate Disjoint Decomposition

In this section, we present the approximate disjoint decomposition for a fixed bound set and free set. We note that a similar method is presented in [16]. The problem solved by this technique is formally stated as follows: *given a Boolean function*

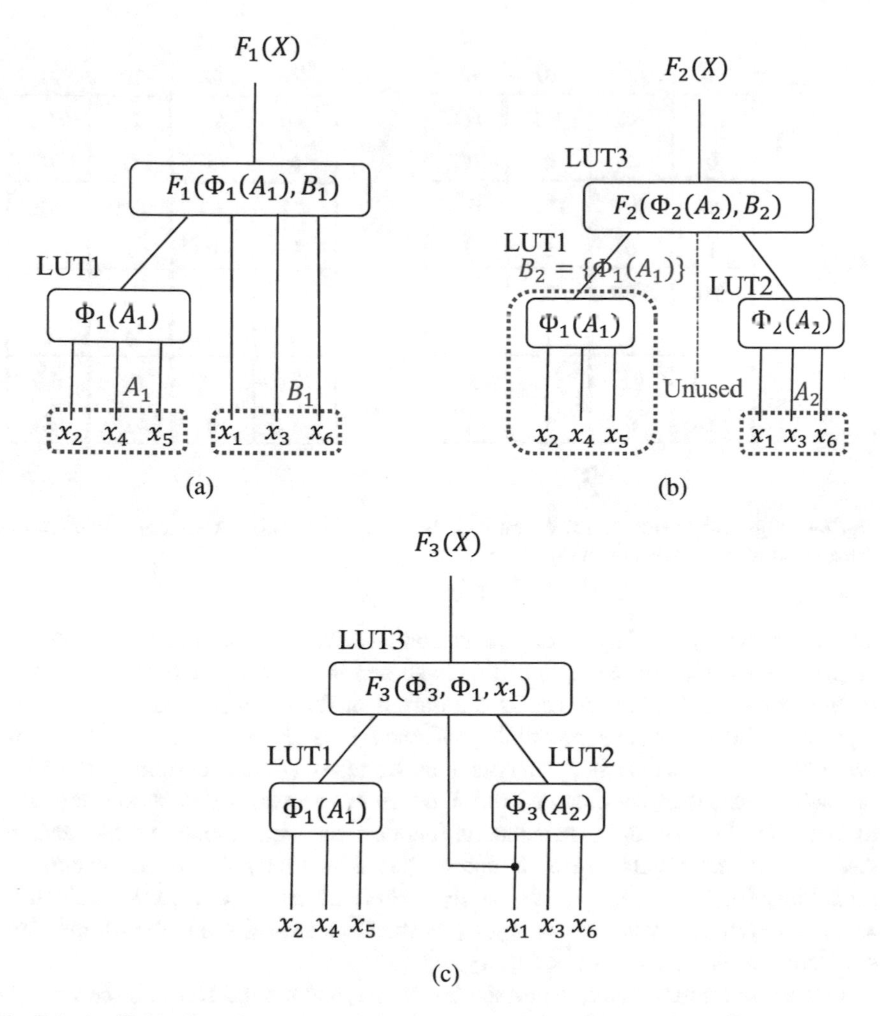

Fig. 7.3 An illustration of the proposed iterative approximate decomposition. (**a**) Applying the first round of approximate disjoint decomposition. (**b**) Applying the second round of approximate disjoint decomposition. (**c**) Applying an approximate non-disjoint decomposition

$f(X)$ and a partition (X_1, X_2) of the input set X, find a decomposable function $F(\phi(X_1), X_2)$ with the smallest ER over f. In what follows, if the bound set and the free set are clear from the context, we will also represent the decomposable function by a pair (F, ϕ) for simplicity.

The solution works on the Boolean matrix of the given function f with the variables in X_1 and X_2 defining the columns and rows, respectively. Besides that, in order to evaluate the ER of the approximate function over the original function, we also need to know the occurrence probability of each input pattern of the function f. For this purpose, we augment the Boolean matrix by including

x_1x_2 \ x_3x_4	00	01	10	11
00	3.0 0	6.7 1	0.5 0	11.0 0
01	8.7 1	2.2 0	12.9 0	6.3 1
10	0.0 0	17.0 0	2.2 1	1.3 0
11	12.0 1	9.5 0	0.8 1	5.9 1

(a)

	00	01	10	11
00	3.0 0	6.7 0	0.5 0	11.0 0
01	8.7 1	2.2 0	12.9 1	6.3 1
10	0.0 0	17.0 0	2.2 0	1.3 0
11	12.0 1	9.5 0	0.8 1	5.9 1

(b)

Fig. 7.4 Augmented matrices. (**a**) A non-decomposable function. (**b**) A decomposable function that approximates the function in (**a**)

the occurrence probability of each input pattern. We call the resulting matrix an *augmented matrix*. An example of this is shown in Fig. 7.4a. In each entry of the matrix, the binary value represents the output of the function and the real value represents the occurrence probability of each input pattern. For simplicity, we actually show the occurrence percentage in the matrix. The occurrence probability of each input pattern is obtained through the Monte Carlo simulation described in Sect. 3.3. For example, consider a function on local inputs x_1, x_2, and x_3. Assume that the Monte Carlo simulation has M samples. Then, the occurrence probability for $(x_1, x_2, x_3) = (a_1, a_2, a_3)$, where $a_1, a_2, a_3 \in \{0, 1\}$, is calculated as $C(a_1, a_2, a_3)/M$, where $C(a_1, a_2, a_3)$ is the number of times in the Monte Carlo simulation when $(x_1, x_2, x_3) = (a_1, a_2, a_3)$.

Given an arbitrary function, a fixed bound set, and a fixed free set, the function may not be decomposable with the bound set and the free set. However, if we make some proper changes to the entries in the Boolean matrix, we can construct a decomposable function by introducing some error. The ER is calculated as

$$\varepsilon = \sum_i \sum_j |B'[i][j] - B[i][j]| \cdot P[i][j], \tag{7.1}$$

where $B[i][j]$'s and $B'[i][j]$'s are the Boolean entries in the augmented matrices of the original function and the decomposable function, respectively, and $P[i][j]$'s are the probability entries in the augmented matrix of the original function. In other words, the ER is the sum of the occurrence probabilities of those input patterns with an output change.

Example 3 Given the function, the bound set, and the free set shown in Fig. 7.4a, the function is not decomposable with the bound set and the free set according to Theorem 1. However, by flipping the outputs of some input patterns, we can construct a decomposable function. A possible example is shown in Fig. 7.4b. The changed bits are labeled in red. By Eq. (7.1), the ER of the approximation is $\varepsilon = (6.7 + 12 + 2.2)\% = 20.9\%$. □

By Theorem 1, for a fixed bound set A and free set B, the Boolean matrix of a decomposable function is fully determined by two factors. The first is the fixed pattern p in Theorem 1, which we call *a pattern vector*. Note that $p \in \{0, 1\}^{2^{|A|}}$. As shown in Example 1, the pattern vector determines the bound-set function. The second is the collection of the type indices of all the rows. We represent it as a vector $r \in \{1, 2, 3, 4\}^{2^{|B|}}$, which we call *a row-type vector*. The ith entry in the row-type vector represents one of the four types that the ith row belongs to. For example, for the decomposable function shown in Fig. 7.4b, its pattern vector is $p = (1, 0, 1, 1)$ and its row-type vector is $r = (1, 3, 1, 3)$.

In order to determine the decomposable function for a fixed bound set and free set with the smallest ER over the original function, it is equivalent to finding the optimal pair of pattern vector and row-type vector (p, r). However, there are $2^{2^{|A|}} \cdot 4^{2^{|B|}}$ possible pairs in the solution space. A brute-force enumeration is prohibitive. Instead, we propose an algorithm to search for a good local optimal solution. The algorithm is based on the following two observations.

1. Once the pattern vector p is fixed, we can identify an optimal row-type vector r efficiently. To do this, we only need to decide each entry in the optimal row-type vector. To determine the ith entry, we compare the ith row of the original Boolean matrix with four choices, which are a pattern of all 0s, a pattern of all 1s, pattern p, and the complement of pattern p, and obtain the ER for each choice. The final best choice for the ith entry is just the one with the smallest ER.
2. Once the row-type vector r is fixed, we can identify an optimal pattern vector p efficiently. To do this, we only need to decide each entry in the optimal pattern vector. Consider the ith entry in the pattern vector. It has only two choices, 0 and 1. We enumerate these two choices. For each choice, since the row-type vector is fixed, we can obtain the ith column of the Boolean matrix B' of a decomposable function determined by the row-type vector r and the pattern vector p with the ith entry as that choice. We then obtain the ER of the ith column of B' over that of the original Boolean matrix. We compare the ERs for the two choices and select the choice giving a smaller ER.

Example 4 Consider the Boolean matrix shown in Fig. 7.4a. Suppose the fixed pattern vector $p = (1, 0, 1, 1)$. Now, we decide the entries in the optimal row-type vector r one by one. With $r[1]$ chosen as $1, 2, 3, 4$, the first row in the new Boolean matrix is $(0, 0, 0, 0)$, $(1, 1, 1, 1)$, $(1, 0, 1, 1)$, and $(0, 1, 0, 0)$, respectively. Comparing these four choices with the first row in the original Boolean matrix, we can obtain their ERs as 6.7%, 14.5%, 21.2%, and 0, respectively. Since the

---

Algorithm 1: Function *ApxDecomp* for finding a decomposable function with a given bound set and free set that has a small ER over the given Boolean function

---

Input : An augmented matrix M specifying the given Boolean function, the bound set, the free set, and the occurrence probability of each input pattern, and a parameter T.
Output : A decomposable function (F, ϕ).

```
1  IniPSet ⇐ ∅;
2  for each row p in M.B do
3  |   calculate the optimal row-type vector r with the pattern vector set as p;
4  |   p.error ⇐ ER of the decomposable function determined by p and r;
5  choose T distinct rows in M.B with the smallest ERs and add them into IniPSet;
6  OptimSet ⇐ ∅;
7  for each pattern vector p in IniPSet do
8  |   while p and r have been updated do
9  |   |   fix p and calculate the optimal r;
10 |   |   fix r and calculate the optimal p;
11 |   add the decomposable function (F, φ) determined by p and r into OptimSet;
12 return the decomposable function (F, φ) in OptimSet with the minimum ER;
```

---

last choice has the minimum ER, we choose $r[1]$ as 4. The other entries of r are determined similarly. The final optimal row-type vector is $r = (4, 3, 1, 3)$.

Now, suppose the fixed row-type vector $r = (4, 3, 1, 3)$. We decide the entries in the optimal pattern vector p one by one. We use $p[3]$ as an example. For $p[3] = 0, 1$, the third column in the new Boolean matrix is $(1, 0, 0, 0)^T$ and $(0, 1, 0, 1)^T$, respectively. Comparing these two choices with the third column in the original Boolean matrix, we can obtain their ERs as 3.5% and 15.1%, respectively. Thus, we choose $p[3] = 0$. The other entries are determined similarly. The final optimal pattern vector is $p = (1, 0, 0, 1)$. □

The above two observations lead to a method to search for a local optimal solution. Instead of searching for p and r simultaneously, we first fix p and optimize r. Then, we fix r and optimize p. In this way, we keep updating p and r until they do not change. Then, we reach a local optimal solution.

Based on the above idea, we propose an algorithm for finding a decomposable function with a given bound set and free set that has a small ER over the original function f. It is shown in Algorithm 1. The function takes an augmented matrix M as inputs. The matrix specifies the given Boolean function, the bound set, the free set, and the occurrence probability of each input pattern. The algorithm has two major parts. The first part creates multiple initial pattern vectors p's (see Lines 2–5). This is important because there may exist many local minima in which our basic optimization algorithm may get stuck. In order to improve the quality, one way is to choose multiple initial starting points. In our implementation, we choose the initial pattern vectors p's from the existing rows. For this purpose, we visit each row p in the Boolean matrix $M.B$ of the augmented matrix, obtain the associated optimal row-type vector r, and calculate the ER for this pair of p and r. We choose T distinct row patterns p's that give the smallest ERs as the initial pattern vectors p's.

The second part of the algorithm visits each initial p selected from the first part. For each p, it performs optimization on p and r alternatively until a local minimum is reached (see Lines 8–10). Then, it stores the corresponding decomposable function (F, ϕ) into the set *OptimSet* (see Line 11). Finally, it returns the decomposable function (F, ϕ) with the smallest ER in the set *OptimSet*.

4.3 Approximate Non-disjoint Decomposition

The approximate disjoint decomposition has the restriction that the bound and free sets are disjoint. Depending on the input size of the given FFC, some LUTs in the final LUT netlist obtained by the approximate disjoint decomposition may have some unused inputs. For instance, as shown in Example 2, if only using approximate disjoint decomposition, the last LUT *LUT3* has one input unused. *LUT3* together with *LUT2* implements the last approximate disjoint decomposition of the form $F_2(\phi_2(x_1, x_3, x_6), \phi_1)$. If we choose an input from *LUT2* and connect it to the unused input, as shown in Fig. 7.3c, the circuit area and depth do not change. By properly configuring the functions of these two LUTs, we can possibly reduce the ER. However, the resulting decomposition is not a disjoint decomposition anymore. It is a non-disjoint decomposition of the form $F(\phi(A'), B')$, where $A' \bigcap B' \neq \emptyset$. The number of possible non-disjoint decompositions is even larger than the number of disjoint ones. The search for an optimal one is time-consuming. We propose a fast solution for a good approximate non-disjoint decomposition based on an existing approximate disjoint decomposition.

Assume that an existing approximate disjoint decomposition has its bound set as A and free set as B. The original function has a corresponding augmented matrix with the variables in the sets A and B defining the columns and rows, respectively. To create a non-disjoint decomposition, a variable from the bound set A will be added to the free set B, creating a larger free set B'. Correspondingly, we first update the original augmented matrix. We call the resulting matrix the *updated augmented matrix*. We use the following example to illustrate how to obtain the updated augmented matrix.

Example 5 Suppose that the target function is $f(x_1, x_2, x_3)$ and the bound set A and the free set B of an existing approximate disjoint decomposition are $A = \{x_1, x_2\}$ and $B = \{x_3\}$. Figure 7.5a shows the augmented matrix of f with the variables in A defining the columns and the one in B defining the rows. We assume that the input combination (x_1, x_2, x_3) follows a uniform distribution. Thus, the occurrence probability of each input pattern is $1/8 = 12.5\%$.

Without loss of generality, suppose that we add x_2 into the free set $\{x_3\}$ to construct an approximate non-disjoint decomposition. We also let it be the first variable in the free set. With x_2 added into the free set, we modify the augmented matrix. The updated one is shown in Fig. 7.5c. The height of the augmented matrix is doubled. The upper half has $x_2 = 0$, while the lower half has $x_2 = 1$. The Boolean entries

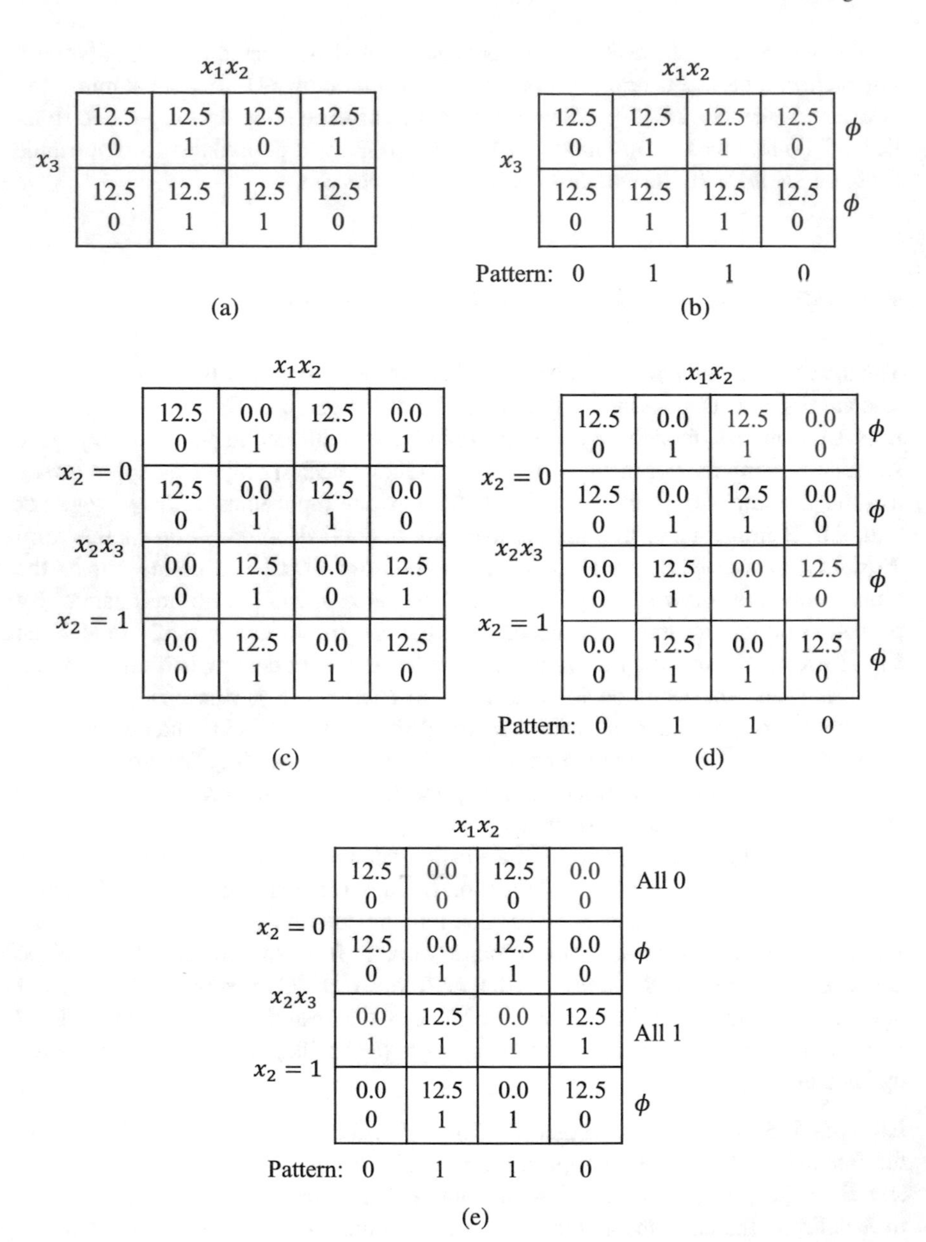

Fig. 7.5 An illustration of obtaining an approximate non-disjoint decomposition from an existing approximate disjoint decomposition. The approximate outputs different from the original ones are highlighted in red. (**a**) The augmented matrix for a function f. (**b**) The augmented matrix for the closest approximate disjoint decomposition to f. (**c**) The updated augmented matrix for the one in (**a**) with x_2 added into the free set. (**d**) The updated augmented matrix for the one in (**b**) with x_2 added into the free set. (**e**) The augmented matrix for the closest approximate non-disjoint decomposition to f

of the upper and lower half of the new matrix are the same as those of the original augmented matrix. However, the probability entries are changed. For the entries where the x_2's in the row and the column take different values, their probability values are 0, since these combinations can never occur, while for the entries where x_2's in the row and the column take the same value, their probability values are $1/8$. □

For the existing approximate disjoint decomposition of the original function, it also has a corresponding augmented matrix M_F. Suppose that the Boolean part of the augmented matrix is characterized by a pattern vector p and a row-type vector r. With a variable x from the bound set added into the free set, we can also obtain an update augmented matrix from M_F using the same method described above. The Boolean part of the updated augmented matrix is characterized by a pattern vector p' and a row-type vector r'. Assume that x is added as the first variable in the free set. Then, it can be easily seen that $p' = p$ and r' is two r's cascaded together.

Example 6 For the function shown in Fig. 7.5a, Fig. 7.5b shows the augmented matrix M_F of an approximate disjoint decomposition with the lowest ER. The Boolean part of the matrix is characterized by the pattern vector $p = (0, 1, 1, 0)$ and the row-type vector $r = (3, 3)$.The ER of this decomposition is $1/4$.

With x_2 added into the free set as its first variable, we can also obtain the updated augmented matrix as shown in Fig. 7.5d from M_F. The Boolean part of the updated augmented matrix is characterized by the pattern vector $p' = (0, 1, 1, 0)$ and the row-type vector $r' = (3, 3, 3, 3)$. Clearly, $p' = p$ and r' is two r's cascaded together. □

To derive an approximate non-disjoint decomposition, we work on the updated augmented matrix of the original function. Similar as the disjoint case, a non-disjoint decomposition can also be characterized by a pattern vector and a row-type vector. Thus, we only need to find an optimal pattern vector p^* and row-type vector r^*. By the above discussion, we can see that the existing approximate disjoint decomposition gives an initial solution for the non-disjoint decomposition with the pattern vector as p' and the row-type vector as r'. Then, by applying the alternative pattern and row-type vectors updating mechanism described in Sect. 4.2, we are guaranteed to find a non-disjoint decomposition with ER no more than that of the given approximate disjoint decomposition.

Example 7 For the function shown in Fig. 7.5a, the existing approximate disjoint decomposition gives an initial solution for the non-disjoint decomposition as shown in Fig. 7.5(d). By Example 6, the initial pattern vector is $p' = (0, 1, 1, 0)$ and the initial row-type vector is $r' = (3, 3, 3, 3)$. By applying the alternative pattern and row-type vectors updating mechanism, we can eventually derive a non-disjoint decomposition shown in Fig. 7.5e. Its pattern vector is $p = (0, 1, 1, 0)$ and its row-type vector is $r = (1, 3, 2, 3)$. From the pattern and row-type vectors, we can get the final approximate non-disjoint decomposition as $F(\phi, x_2, x_3) = x_2\overline{x_3} + x_3\phi$ with $\phi = \overline{x_1}x_2 + x_1\overline{x_2}$. Its ER is 0. □

Note that the above discussion assumes that only one variable from the bound set is added into the free set. However, the method can also be extended to add more than one variable from the bound set into the free set. Also, the discussion is on a special case where the input added into the free set is a direct fanin variable x_i of the bound-set function $\phi(x_1, \ldots, x_n)$. However, the method can also be extended when the input is a transitive fanin of the bound-set function. For this general case, the probability entries in the updated augmented matrix should be set as the occurrence probabilities of the corresponding input patterns, which can be obtained by the Monte Carlo simulation.

4.4 Iterative Approximate Decomposition

In this section, we present the details on how we obtain a structure with the fewest LUTs to implement a given function corresponding to an FFC in the circuit. It exploits both the approximate disjoint and non-disjoint decomposition techniques described above. It first builds a LUT network with the fewest LUT only using the approximate disjoint decomposition. However, the last LUT may have some unused inputs. Then, it applies the approximate non-disjoint decomposition to exploit the unused inputs to further reduce ER.

The proposed approximate disjoint decomposition method works under the assumption that the bound set and the free set are given. Before we apply it to find a good approximate decomposition, we need to decide the bound set. Suppose that the given function has n inputs. By the basic idea described in Sect. 4.1, we need to select a bound set of size k. Therefore, there are $\binom{n}{k}$ choices for the bound set in total. Different bound sets lead to different decomposable functions $F(\phi(X_1), X_2)$ and the associated ERs.

Furthermore, as we stated in Sect. 4.1, for a typical function, we need to do multiple rounds of approximate disjoint decomposition. This brings another problem. That is, the different choices made at the previous rounds influence the later choices. For example, in Example 2, in the first round, we choose bound set $A_1 = \{x_2, x_4, x_5\}$, and by applying the approximate disjoint decomposition, we obtain the decomposable function $F_1(\phi_1(A_1), B_1)$. Then, in the second round, the target function for the decomposition is F_1. In contrast, if we choose the bound set $A'_1 = \{x_1, x_3, x_5\}$ in the first round, then by applying the approximate disjoint decomposition, we obtain another decomposable function $F'_1(\phi'_1(A'_1), B'_1)$. Then, in the second round, the target function for the decomposition is F'_1. This will lead to a different final solution.

In order to address the above issues, we propose an iterative approximate decomposition algorithm based on local beam search [22]. The idea is that in each round of the decomposition loop, we always keep $m \geq 1$ promising decomposable functions

$$F_1(\phi_1(A_1), B_1), \ldots, F_m(\phi_m(A_m), B_m).$$

Algorithm 2: The proposed iterative approximate decomposition

Input: Simulation result *sim*, a given FFC C, and parameters T and m.
Output: An approximate LUT network with the minimum LUT count.

1 *TopChoice* $\Leftarrow \{C\}$;
2 *TopChoice*$[1].H \Leftarrow$ the Boolean function of C;
3 ▷ *TopChoice*[1] is the first element in the set *TopChoice*;
4 $n \Leftarrow$ the input size of C;
5 **for** $i = 1$ **to** $\lceil (n-k)/(k-1) \rceil$ **do**
6 *MaxHeap* $\Leftarrow \emptyset$;
7 ▷ *MaxHeap* stores the approximate circuits. Its key is the ER;
8 **for** $j = 1$ **to** |*TopChoice*| **do**
9 **for** *each partition (bSet, fSet) on the input set of TopChoice*$[j].H$ **do**
10 $M \Leftarrow$ *createMatrix(sim, bSet, fSet, TopChoice*[j]);
11 $(F, \phi) \Leftarrow$ *ApxDecomp*(M, T);
12 *NewCkt* $\Leftarrow$ *apply*(*TopChoice*[j], F, ϕ, *bSet*, *fSet*);
13 *NewCkt.calculateError(sim)*;
14 *MaxHeap.insert*(*NewCkt*);
15 **if** |*MaxHeap*| $> m$ **then**
16 *MaxHeap.deleteMax*();
17 *TopChoice* $\Leftarrow \emptyset$;
18 **for** $j = 1$ **to** |*MaxHeap*| **do**
19 insert the circuit corresponding to the free-set function of *MaxHeap*[j] into *TopChoice*;
20 **if** $(n-1)$ *is not a multiple of* $(k-1)$ **then**
21 **for** $j = 1$ **to** |*TopChoice*| **do**
22 *TopChoice*[j].*ApxNonDisjuncDecomp()*;
23 **return** *the approximate circuit in TopChoice with the smallest ER*;

In the next round, we obtain all approximate decomposable functions derived from $F_1, \ldots, F_m$ and keep the top m based on the smallest ERs.

The algorithm is shown in Algorithm 2. It takes an FFC as its input. Suppose the input size of the FFC is n. As we stated in Sect. 4.1, the iterative approximate decomposition needs $\lceil (n-k)/(k-1) \rceil$ rounds. In each round, there is a set of partial LUT networks obtained from the previous round. They are stored in the set *TopChoice*. Each element in *TopChoice* also has a data member H storing the latest function to be decomposed. We iterate over all elements in *TopChoice* (see Line 8). For the jth element *TopChoice*$[j]$, we further iterate over all pairs of bound set and free set partitioned from the input set of the function H of *TopChoice*$[j]$ (see Line 9). For each pair of bound set *bSet* and free set *fSet*, the function *createMatrix* prepares the augmented matrix M from the bound set, the free set, the simulation trace, and the current partial LUT network *TopChoice*$[j]$ (see Line 10). Then, the function *ApxDecomp* shown in Algorithm 1 is called to generate a good decomposable function that approximates the function H associated with *TopChoice*$[j]$ (see Line 11). After that, the obtained decomposable function is applied to the partial circuit *TopChoice*$[j]$ to derive a new circuit *NewCkt* (see Line 12). Then, the ER of the new circuit is calculated (see Line 13) before it is inserted into a max heap *MaxHeap* (see Line 14). The max heap is indexed on the

ER of the circuit. If its size is larger than m, then the element in it with the largest ER is removed (see Lines 15–16). This essentially ensures that in each round, at most m candidates with the smallest ERs are kept. After all the elements in the set *TopChoice* have been visited, the set *TopChoice* is first reset to an empty set (see Line 17), and then, the circuit corresponding to the free-set function of each element in *MaxHeap* is inserted into *TopChoice* (see Lines 18–19). This leads to the next round.

After the entire decomposition loop finishes, we obtain m LUT networks with the minimum LUT count only through the approximate disjoint decomposition. If $(n - 1)$ is not a multiple of $(k - 1)$, then the last k-LUT of each LUT network in the set *TopChoice* will have some unused inputs. In this case, for each LUT network, we apply the approximate non-disjoint decomposition to further reduce its ER, while keeping the circuit area and depth (see Lines 20–22). The non-disjoint decomposition is based on the disjoint decomposition for the last k-LUT. To reduce the search space, the additional inputs added into the free set are chosen as the primary inputs of the FFC that are different from the inputs in the current free set. We iterate over all possible input choices. For each choice, we build the updated augmented matrix with its probability entries obtained from the Monte Carlo simulation. Then, the solution described in Sect. 4.3 is applied to find an optimal non-disjoint decomposition for that input choice. Once all choices of the primary inputs are traversed, the non-disjoint decomposition with the lowest ER is selected. Finally, the approximate LUT network with the smallest ER in the set *TopChoice* is returned (see Line 23).

4.5 Iterative Input Removal

The proposed iterative approximate decomposition can effectively reduce the LUT count of an FFC if the LUT count of the original FFC is more than the minimum value $\lceil \frac{n-1}{k-1} \rceil$. However, some FFCs in a circuit may already contain the minimum number of LUTs. We call them *optimal-size FFCs*. For these FFCs, if we want to further reduce their LUT counts, we need to reduce the number of inputs. The way we reduce the number of inputs of FFCs is called *input removal*. One basic requirement for input removal is that the ER introduced should be minimal.

We can exploit the proposed approximate disjoint decomposition to remove any given set of inputs and obtain a function on the remaining inputs with the smallest ER over the original function. In the approximate disjoint decomposition, if we only allow each row to be either all 0s or all 1s, then the resulting function is independent of the bound-set inputs. For example, in a disjoint decomposition with the bound set as $\{x_1, x_2\}$ and the free set as $\{x_3, x_4\}$, if the first and fourth rows of the Boolean matrix are all 0 and the second and third rows are all -1, then the function given by the decomposition is $F(x_1, x_2, x_3, x_4) = \overline{x_3}x_4 + x_3\overline{x_4}$, independent of the variables x_1 and x_2.

Thus, if we want to remove a set of inputs, we can set them as the bound set and solve for a special case of the approximate disjoint decomposition where each entry of the row-type vector is only limited to either 1 or 2. This will give a function without these inputs and with the lowest ER. However, another problem is that we do not know which set of inputs we should remove to obtain the lowest ER. Theoretically, this can be determined by examining all input combinations, but it will lead to an exponential complexity. To improve the efficiency of the algorithm, instead, we do the input removal iteratively. In each iteration, the bound set size is fixed to 1 and the input choice with the lowest ER is obtained and removed. The iteration repeats until we cannot remove an input without letting the error accumulated so far exceed a given threshold ε, which is set as 0.1 times the given ER threshold in our implementation. This process is called *iterative input removal*. In the extreme case where all inputs of the FFC are removed, we essentially reduce the FFC to a constant. In our overall ALS flow, besides the optimal-size FFC, we also apply the iterative input removal to nonoptimal-size FFCs, which can also help reduce the LUT count.

4.6 Overall ALS Flow

In this section, we present the overall ALS flow integrating our proposed techniques. It is shown in Fig. 7.6a. During the process, we keep an ER margin, initialized as the given ER threshold (see Box 1), to help us ensure that the ER of the approximate circuit does not exceed the ER threshold. We perform multiple synthesis rounds. In each round, we first perform global Monte Carlo simulation (see Box 3), which will be used later to obtain the probability entries in the augmented matrices for various local FFCs. Then, we traverse the nodes in a topological order (see Box 4). For each node, we visit all of its candidate FFCs (see Box 5). In our study, we consider FPGA technology using 4-LUTs, that is, $k = 4$. Given this k value, we found that our proposed iterative approximate decomposition is runtime-efficient when the input size is no more than 12. Also, a nontrivial FFC in the LUT network should have at least $(k + 1)$ inputs. Thus, we limit the candidate FFCs to those with the input sizes from 5 to 12.

For each candidate FFC, we obtain a local approximate transformation for it (see Box 5). The details of this step are shown in Fig. 7.6b. It applies the proposed iterative input removal followed by the iterative approximate decomposition. For each node n, after all of its FFCs are visited, we first eliminate those FFCs with the local transformations that would increase the delay of n if the transformation is applied. This guarantees that our method does not increase the depth of the LUT network. We also eliminate those FFCs such that the ERs of their local transformations exceed the ER margin. This guarantees that the approximate transformation to be selected will not let the ER of the circuit exceed the ER threshold. For the remaining FFCs, we choose the FFC c with the largest score and replace c by its local transformation (see Box 6). The score for an FFC is defined as l/e, where l is the number of LUTs we can reduce by replacing the FFC with its local transformation and e is the ER of that transformation. By our scoring

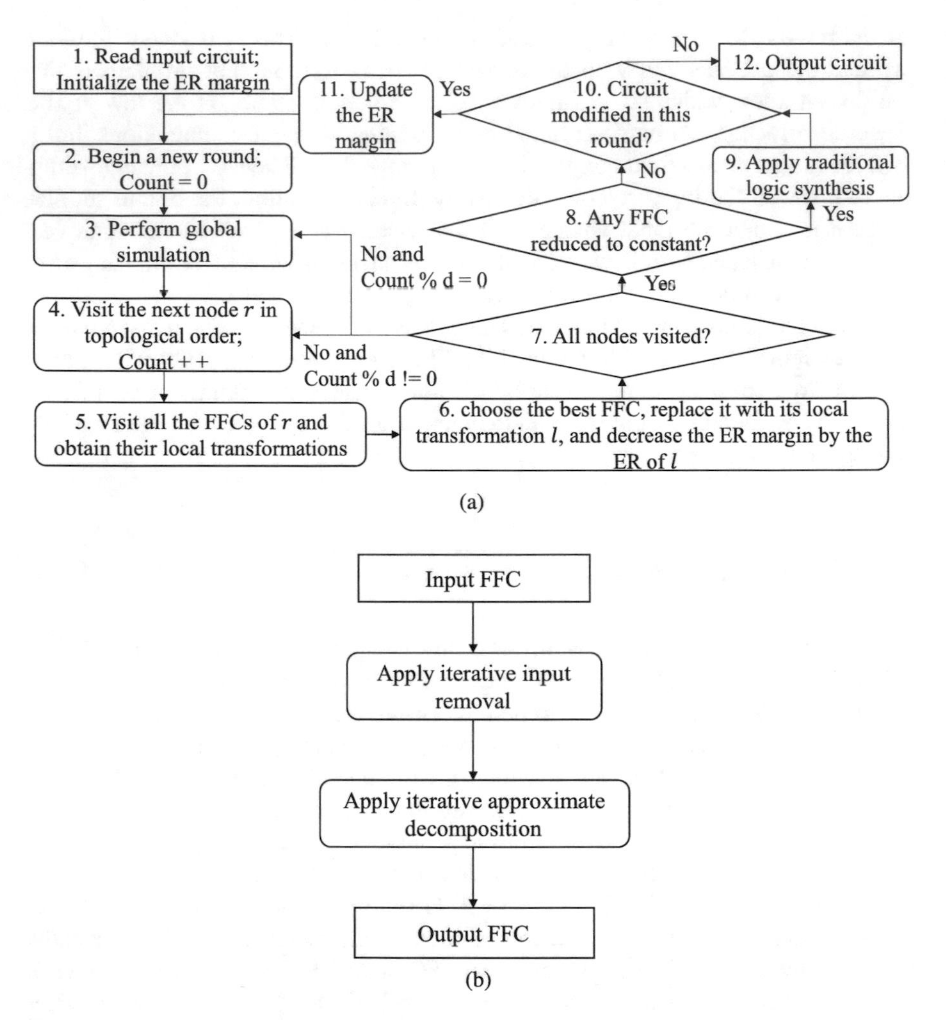

Fig. 7.6 The proposed approximate logic synthesis flow for FPGAs. (**a**) The proposed flow. (**b**) The flow to obtain a local approximate transformation

mechanism, this change maximizes the LUT count reduction per ER introduced. After the change, we decrease the ER margin by the ER of the selected local transformation (see Box 6). Then, we visit the next node in the topological order until all the nodes have been visited.

Within each round, if we re-simulate the whole circuit each time we change an FFC, it would take a long time. However, if we do not, for any FFC that overlaps with the modified one and has not been visited, the occurrence probabilities of its input combinations may not be accurate. To balance the runtime with the accuracy, we count the nodes we have visited in the current round. When d more nodes have been visited, we will perform global Monte Carlo simulation. In our implementation, we set d as 10. Also, for any FFC that overlaps with an FFC that has been modified after the last global simulation, we will skip it.

After all nodes in the LUT network are visited, we check whether any FFC has been modified to a constant value as a special case of the local approximate transformation (see Box 8). If so, we will apply traditional logic synthesis method to propagate the constant and further simplify the circuit (see Box 9). Then, we check whether in the current round, the circuit has been modified (see Box 10). If not, it means that there is no chance to further improve the circuit and we return the circuit as the final result (see Box 12). Otherwise, we begin a new round and update the ER margin by decreasing it by the actual ER (see Box 11). This update is because the ER of an approximate transformation is measured at the output of an FFC. However, due to the logic masking effect, an error at the output of an FFC may not be observed at the primary outputs of the circuit [11]. Thus, the actual ER could be smaller. Therefore, before the next round starts, we obtain the actual ER through simulation and update the ER margin as the given ER threshold minus the current ER.

5 Experimental Results

In this section, we present the experimental results. We implemented our algorithm in C++ and tested it on a computer with 3.4 GHz CPU and 8GB memory. The parameters T and m in Algorithm 2 were both set as 5.

Our benchmark circuits include all the random control circuits in the EPFL benchmark suite with the LUT count smaller than 1000. Besides them, in order to compare our method with the state-of-the-art ALS methods for FPGA [13, 14], we also included the same MCNC circuits used in [13] and [14]. The original circuits were mapped into networks of 4-LUTs using the logic synthesis tool ABC [23]. We executed the mapping command "if -K 4" multiple times to get the FPGA circuits with the minimum numbers of LUTs as the inputs to our algorithm.

5.1 *The Performance of Our Method*

Tables 7.1 and 7.2 show the synthesis results of our ALS method for the MCNC and EPFL benchmarks, respectively, under 5% ER constraint. The columns "size" and "depth" list the LUT network size, measured by the number of used LUTs, and the LUT network depth, respectively. The columns under "our baseline" list the sizes and depths of the well-optimized circuits as the inputs to our ALS method. The column "SRR" lists the size reduction ratio (SRR), calculated as the ratio of the LUT number reduction over the LUT number of the input circuit. The arithmetic mean SRRs for the two sets of benchmarks are 24.9% and 29.3%, respectively. For most benchmarks, our method can achieve more than 10% of size reduction.

Table 7.1 The results of our method and the previous methods [13, 14] for the MCNC benchmarks under the ER threshold of 5%

	Our baseline		Our method				Baseline [13, 14]		SCALS [14]			Wu [13]	
Circuit	Size	Depth	Size	Depth	SRR	Runtime (s)	Size	Depth	Size	Depth	SRR	Size	SRR
C432	68	11	61	11	0.103	175.4	97	10	55	10	**0.433**	79	0.186
C880	117	8	97	8	0.171	72	128	8	107	8	0.164	102	**0.203**
C1908	117	9	46	3	**0.607**	32	122	9	88	9	0.279	50	0.590
C2670	210	7	148	7	**0.295**	896.7	295	7	224	7	0.241	252	0.146
C3540	351	12	332	11	0.054	1829.3	346	12	305	11	**0.118**	325	0.061
C5315	465	8	450	8	0.032	2954.7	503	9	439	9	**0.127**	468	0.070
C7552	596	8	410	8	**0.312**	2566.6	593	8	440	8	0.258	486	0.180
Alu4	689	7	421	7	0.389	216	710	7	411	7	**0.421**	483	0.320
Alu2	198	10	150	8	**0.242**	442.3	160	12	135	11	0.156	136	0.150
Apex6	247	6	197	4	0.202	668.4	253	6	210	6	0.170	197	**0.221**
dalu	445	11	297	8	**0.333**	1100.6	425	11	329	11	0.226	349	0.179
Mean	318	8.8	237	7.5	**0.249**	995.8	330	9	249	8.8	0.236	266	0.210

Table 7.2 The result of our methods for the EPFL benchmarks under the ER threshold of 5%

	Our baseline		Our method			
Circuit	Size	Depth	Size	Depth	SRR	Runtime (s)
ALU ctrl	58	3	52	3	0.103	11.4
cavlc	299	6	248	6	0.171	365.5
Decoder	305	2	290	2	0.049	14.7
i2c	376	5	246	4	0.346	36.7
int2float	68	6	59	6	0.132	50.8
Priority	183	42	25	4	0.863	104.8
Router	72	5	31	1	0.569	3.2
Arbitor	650	23	286	13	0.560	36.7
mean	251	11.5	165	8.8	0.293	78.0

The effectiveness of our method is closely related to the number of nonoptimal-size FFCs in the input circuit. For some circuits like `decoder` in Table 7.2, they have few nonoptimal-size FFCs. With an insufficient number of such FFCs fed into the iterative approximate decomposition algorithm, our method cannot find approximate transformations with small ERs to further improve the circuit size. Thus, the size reduction for them is limited.

In our method, we need to re-simulate the whole circuit after applying some approximate transformations to the circuit. This process is repeated until the ER of the approximate circuit exceeds the error threshold. Before each round of re-simulation, we record the size of the current LUT network together with its ER. The plots of SRR versus ER are shown in Figs. 7.7 and 7.8 for the MCNC and the EPFL benchmarks, respectively. For all the circuits, the size reduction gradually increases as the ER increases to 5%. For our method, if we introduce $x\%$ error into the circuit, we can typically gain a $2.5\times\%$ to $4\times\%$ size reduction. For some circuits such as `priority` and `router`, the LUT network sizes reduce significantly with a small amount of error introduced. Furthermore, for some circuits (e.g., `priority` and `router`), the approximation process stops before the ER threshold 5% is reached because at these stopping points, our algorithm cannot find nonoptimal-size FFCs in the circuits or the remaining few nonoptimal-size FFCs only have local transformations with ERs larger than the remaining ER margin.

Tables 7.1 and 7.2 also list the running time of our method. As the running time depends not only on the size of the circuit but also on the number of iterations in the synthesis procedure, we only show the average running time of one iteration for each circuit. The average running time theoretically is proportional to the number of FFCs in the circuit, which is roughly proportional to the number of LUTs of the circuit. However, there are some exceptions, as they have too many or too few FFCs compared to their LUT sizes (e.g., `decoder`). Also, some circuits may only have small-size FFCs, which lead to extremely short running time compared to LUT sizes of the circuits, like `Alu4`, `i2c`, `router`, and `arbitor`.

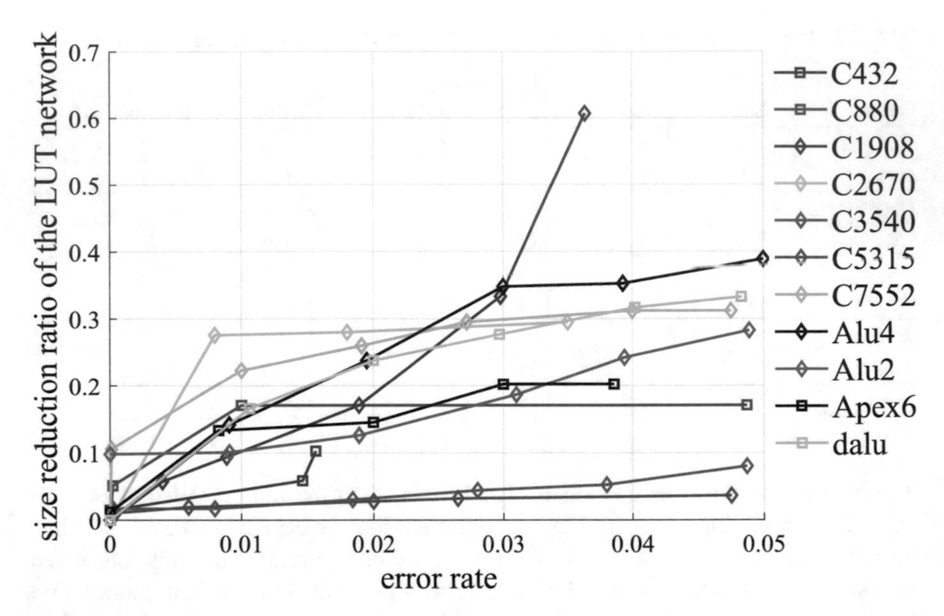

Fig. 7.7 Size reduction ratio vs. error rate plot for the MCNC benchmarks

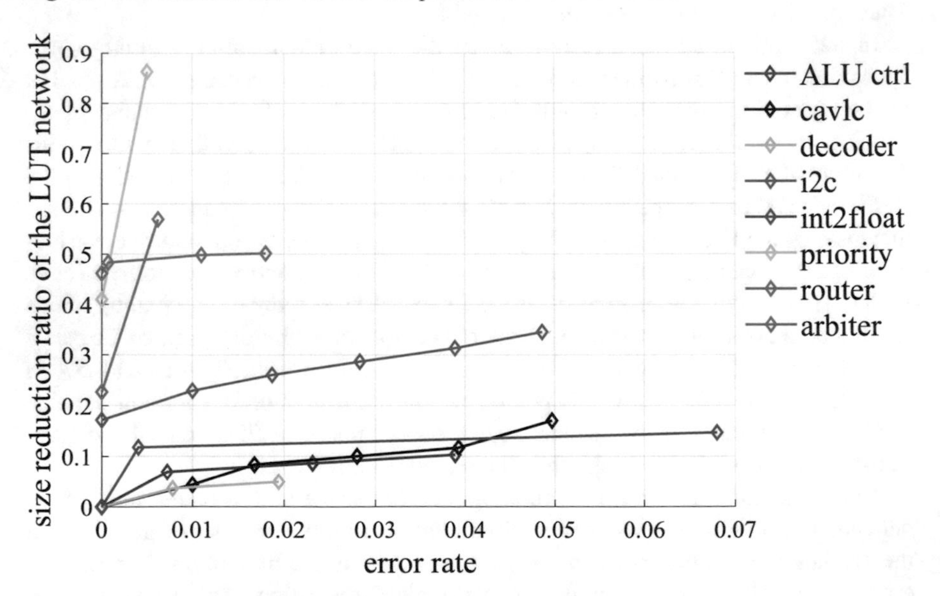

Fig. 7.8 Size reduction ratio vs. error rate plot for the EPFL benchmarks

5.2 *Comparison with the Existing ALS Methods for FPGA*

We compared our method with the state-of-the-art ALS methods for FPGA [13, 14] using the 11 MCNC circuits. Their synthesis results for the 11 circuits under 5%

ER constraint are also listed in Table 7.1. The columns under "baseline [13, 14]" show the sizes and depths of their baseline input circuits. By comparing the sizes and depths of our baseline inputs and theirs, we find that, on average, our baseline inputs are slightly better than theirs. We believe that it is caused by the different synthesis tools used in the initial FPGA optimization. Note that as our baselines are smaller and faster than theirs, it means that our synthesis task is more challenging than theirs.

Due to the difference in baseline inputs, the SRRs of the methods in [13] and [14] were calculated with their baseline results. The entries in bold highlight the cases where the method of the corresponding column is the best. We note that our method is not as good as the previous methods for some benchmarks, such as `C432`, `C3540`, and `C5315`. There are two reasons for this. One is that some benchmarks (e.g., `C3540`) have few FFCs, causing our method to yield limited improvement. The other is that our baseline inputs for some benchmarks (e.g., `C432` and `C5315`) are much smaller than theirs, making the improvement space of our method limited. Nevertheless, it can be seen that our proposed method achieves more best results than the other two. By comparing the arithmetic mean SRRs listed in Table 7.1, we can also conclude that our method is better than those in [13] and [14] in size reduction. Although our method is area-oriented, it can also effectively reduce the circuit depth. The average depth reduction ratio on the MCNC benchmarks by our method is 14.1%, while that by the method in [14] is 1.5%.

6 Conclusion

In this chapter, we proposed an ALS method for FPGAs. It is based on the novel approximate disjoint and non-disjoint decomposition techniques and the iterative approximate decomposition method, which transforms local LUT subnetworks into approximate ones with the minimum numbers of LUTs. Our experimental results showed that our proposed method is better than the state-of-the-art ALS methods for FPGA.

The current work only considers the error metric as ER. However, it is possible to extend it for other metric like average error magnitude (AEM) by further considering how the local error affects different POs. We will develop method for AEM in our future work.

References

1. Waldrop MM. The chips are down for Moore's law. Nature. 2016;530(7589):144–7.
2. Mittal S. A survey of techniques for approximate computing. ACM Comput Surv. 2016;48(4):62:1–62:33.

3. Shin D, Gupta SK. A new circuit simplification method for error tolerant applications. In: Design, automation & test in Europe conference & exhibition. 2011. pp. 1–6.
4. Venkataramani S, Sabne A, Kozhikkottu V, Roy K, Raghunathan A. SALSA: systematic logic synthesis of approximate circuits. In: Design automation conference. 2012. pp. 796–801.
5. Venkataramani S, Roy K, Raghunathan A. Substitute-and-simplify: A unified design paradigm for approximate and quality configurable circuits. In: Design, automation & test in Europe conference & exhibition. 2013. pp. 1367–1372.
6. Miao J, Gerstlauer A, Orshansky M. Multi-level approximate logic synthesis under general error constraints. In: International conference on computer-aided design. 2014. pp. 504–510.
7. Vasicek Z, Sekanina L. Evolutionary approach to approximate digital circuits design. IEEE Trans Evol Comput. 2015;19(3):432–44.
8. Chandrasekharan A, Soeken M, D. Große, Drechsler R. Approximation-aware rewriting of AIGs for error tolerant applications. In: International conference on computer-aided design. 2016. pp. 83:1–83:8.
9. Scarabottolo I, Ansaloni G, Pozzi L. Circuit carving: A methodology for the design of approximate hardware. In: Design, automation & test in Europe conference & exhibition. 2018. pp. 545–550.
10. Hashemi S, et al. BLASYS: Approximate logic synthesis using Boolean matrix factorization. In: Design automation conference. 2018. pp. 55:1–55:6.
11. Wu Y, Qian W. ALFANS: Multilevel approximate logic synthesis framework by approximate node simplification. IEEE Trans Comput Aided Des Integr Circ Syst. 2020;39(7):1470–83.
12. Meng C, Qian W, Mishchenko A. ALSRAC: Approximate logic synthesis by resubstitution with approximate care set. In: Design automation conference. 2020. pp. 187:1–187:6.
13. Wu Y, Shen C, Jia Y, Qian W. Approximate logic synthesis for FPGA by wire removal and local function change. In: Asia and South Pacific design automation conference. 2017. pp. 163–9.
14. Liu G, Zhang Z. Statistically certified approximate logic synthesis. In: International conference on computer-aided design. 2017. pp. 344–351.
15. Yao Y, Huang S, Wang C, Wu Y, Qian W. Approximate disjoint bi-decomposition and its application to approximate logic synthesis. In: International conference on computer design. 2017. pp. 517–24.
16. Meng C, Xiang Z, Liu N, Hu Y, Song J, Wang R, Huang R, Qian W. DALTA: A decomposition-based approximate lookup table architecture. In: International conference on computer-aided design. 2021. pp. 1–9.
17. Ashenhurst RL. The decompositions of switching functions. In: International symposium on the theory of switching functions. 1959. pp. 74–116.
18. Curtis HA. A new approach to the design of switching circuits. Van Nostrand; 1962.
19. Shen VS, McKellar AC. An algorithm for the disjunctive decomposition of switching functions. IEEE Trans Comput. 1970;100(3):239–48.
20. Cong J, Ding Y. Flowmap: an optimal technology mapping algorithm for delay optimization in lookup-table based FPGA designs. IEEE Trans Comput Aided Des Integr Circ Syst. 1994;13(1):1–12.
21. Cong J, Ding Y. Combinational logic synthesis for LUT based field programmable gate arrays. ACM Trans Des Autom Electron Syst. 1996;1(2):145–204.
22. Russell S, Norvig P. Artificial intelligence: a modern approach. Prentice Hall; 2009.
23. Mishchenko A, et al. ABC: A system for sequential synthesis and verification. 2007. [Online]. Available: http://www.eecs.berkeley.edu/~alanmi/abc/.

Chapter 8
Design Techniques for Approximate Realization of Data-Flow Graphs

Marzieh Vaeztourshizi, Hassan Afzali-Kusha, Mehdi Kamal, and Massoud Pedram

1 Introduction

Electronic systems including application-specific integrated circuits (ASICs) are designed and realized using very-large-scale integrated (VLSI) circuits. The design may be performed at different abstraction levels including register-transfer level (RTL). Due to the increasing number of devices in VLSI circuits, the RTL design process can be cumbersome and error prone. To overcome the problems, designers go to a higher abstraction level and utilize high-level synthesis (HLS) tools, whereby a behavioral description of an input application in a high-level language is mapped to a corresponding RTL realization. More specifically, at the behavioral level, an input application is typically represented by a data-flow graph (DFG) where HLS is exploited to map the DFG to some RTL realization. The resulting RTL code may utilize multiple modules, including arithmetic units, to perform the required functions of the DFG nodes. There may or may not exist a one-to-one correspondence between the arithmetic nodes in the RTL realization with the operations in the DFG.

The design of electronic systems, which mainly consists of digital circuits, is subject to different design constraints including speed, area, and energy/power consumption. The last constraint is critically important in embedded and mobile devices where there is a limited energy budget due to device size and form factor considerations as well as volumetric energy densities of current batteries. This has mandated energy efficiency as a major goal in the design of these computing

M. Vaeztourshizi (✉) · H. Afzali-Kusha · M. Pedram
University of Southern California, Los Angeles, CA, USA
e-mail: vaeztour@usc.edu

M. Kamal
University of Tehran, Tehran, Iran

W. Liu, F. Lombardi (eds.), *Approximate Computing*,
https://doi.org/10.1007/978-3-030-98347-5_8

platforms. The importance of achieving energy efficiency can be seen, for example, when we try to run machine learning (ML) applications (e.g., natural language processing, forecasting, and decision-making) on these energy-limited devices. The applications are both compute and data intensive, and their execution leads to large energy consumption. As a sidenote, the ubiquitous use of ML applications in data centers has resulted in a considerable increase of electrical energy consumption in data centers, with significant monetary cost implications for data center owners and operators (see, e.g., [1]).

The silver lining is the fact that many of these applications, including the ML ones, have a relatively high resiliency to the output error, which means approximate results at the outputs may be tolerable. The main reasons for this tolerance are the facts that (i) users of these applications, that is, humans, may not be able to detect small imperfections in the quality of the outputs (e.g., the human visual system does not perceive infrequent drop of video frames), (ii) there are no golden or unique solutions to offer and good enough solutions are still acceptable (e.g., movie recommendations on Netflix), and (iii) the input to the system itself has a lot of redundancy or noise (e.g., the data captured by a surveillance camera) [2].

This provide us with the opportunity to use an *approximate computing* (AC) paradigm whereby the exactness of the computations (and memory data operations) is abandoned in favor of improvements in the energy efficiency. Obviously, the limit of the approximation is set by the quality of service (QoS) constraint determined by the (end) user. Also, it should be noted that many approximate techniques, in addition to reducing the energy consumption, tend to increase the computational speed. Finally, notice that AC can be applied at different abstraction levels of the design, including behavioral (e.g., skipping a number of loop iterations [3]), architecture (e.g., selection of approximate custom instructions by the instruction set architecture [ISA] [4]), and circuit (e.g., simplifying the hardware of exact blocks such as arithmetic units [5]).

Obviously, effective utilization of AC as a design technique to optimize the system energy efficiency (and possibly increase the computational speed and reduce the layout area cost) of the electronic systems requires automating the generation of approximate circuits using a computer-aided design (CAD) tool. Such a tool would accept user-defined constraints (i.e., a quality of service) as an input parameter and map an exact high-level description of the input application to an approximate RTL representation. We call this tool an *approximate high-level synthesis* (AHLS) tool to distinguish it from its exact counterpart (known as the HLS tool). The AHLS tool considers different approximation techniques at different layers of the design abstraction and chooses a proper subset of these techniques to be deployed in the target system to optimize a combination of energy efficiency, computational speed, and area cost.

During the DFG design, one may apply some approximation techniques to the DFG implementation, which results in the generation of an *approximate data-flow graph* (ADFG). Examples include replacement of some of the exact arithmetic unit nodes with their approximate counterparts taken from a library of approximate arithmetic units. The approximate units may support several approximation degrees/levels, which may be determined statically (during the mapping process) or dynamically (at runtime). In addition to differing energy consumption values, the speed (delay) and area may vary from one configuration of the ADFG to another. Therefore, in applying approximation to the DFG nodes, one may use different levels of approximation for different nodes providing the designer with a huge number of ADFG choices/configurations. Obviously, each one leads to some specific energy consumption and quality of service for the ADFG. For example, a designer may want to choose an ADFG that results in the minimum energy consumption while satisfying some maximum output error constraint. This means that during design, the output quality for each configuration should be determined. Moreover, although finding the best approximation configuration through brute force search can provide the optimal design with respect to the objective functions and given constraints, this is an intractable problem that will require exponential time/space to solve. One should, therefore, resort to efficient solution methods to heuristically solve the said problem.

As mentioned above, to determine the optimal combinations of approximate vs. exact DFG nodes, a designer should search the design space. The operation, which is called design space exploration (DSE), is to optimize one or more circuit design metrics (e.g., energy consumption and output accuracy) while satisfying other circuit design metrics (e.g., average or maximum output error levels) as constraints. This may be formulated as a multi-objective optimization problem, where based on the application and user constraints, a set of non-dominated ADFGs are generated and presented to the user to choose from, for example, the Pareto-optimal frontier (PF) of all ADFGs in, say, a 2D space of energy efficiency vs. output accuracy. The set of non-dominated ADFGs represent the best trade-off between energy consumption (and/or other design metrics) and output accuracy (QoS). Figure 8.1a provides an example of a DFG with three operations, two addition and one multiplication, where each operation has three different implementations (i.e., one exact and two approximate implementations). The corresponding design space of approximate DFGs and the extracted PF are depicted in Fig. 8.1b where the provided tuples represent the degree of approximation for each DFG node (i.e., a higher value corresponds to a more aggressive approximation). The fully exact design refers to the original DFG, while the fully approximate circuit refers to the ADFG where all the DFG nodes are replaced by the approximate arithmetic unit (with the same operation type) from the library with the highest degree of approximation (i.e., the highest output error value).

The frontier may be used for maximizing the QoS given a maximum energy constraint or minimizing the energy consumption given a minimum QoS constraint. On the other hand, if a user is interested in a single ADFG, based on a given constraint, that is, QoS (or energy budget), a single optimal design that minimizes the energy consumption (or maximizes the output accuracy) will be provided. Obviously, the larger the design space is (e.g., approximate units with more configurable accuracy levels), the more challenging this exploration is. The design exploration problem for ADFGs is NP-hard [6], and therefore, many heuristic approaches have been employed to solve the problem either statically or dynamically.

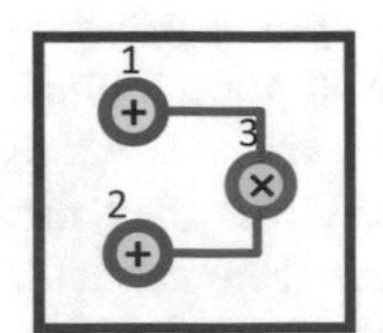

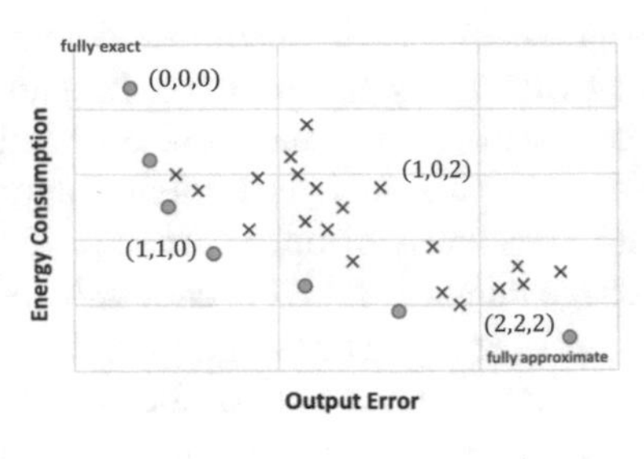

Fig. 8.1 (**a**) A DFG with two adder and one multiplier nodes, where each node has three different implementations and (**b**) the design space of all the ADFGs represented with red crosses and PF of the design space represented by blue dots

In this chapter, a survey of different optimization techniques for approximate realization of high-level description of an input application (including but not limited to DFGs and abstract synthesis trees [ASTs]), which use different DSE methodologies, is presented. The remainder of this chapter is organized as follows. In Sect. 2, we provide a brief review of some background materials pertaining to energy and power consumption, approximation methods and multi-objective optimization problems, approximate high-level synthesis, and error metrics utilized for evaluating approximate circuits. Sect. 3 deals with different DSE algorithms used for approximate high-level designs, and three works are discussed in more detail. Other works are also presented briefly. Finally, potential directions for future research in this area is discussed in Sect. 4.

2 Preliminaries

In this section, a review of some important concepts related to power consumption, approximate computing and multi-objective optimization problems, and AHLS are provided. Finally, a brief discussion of error metrics exploited to evaluate the quality at the output of the approximate circuits concludes this section.

2.1 *Energy and Power Consumption in Compute Units*

The two main components of the power consumption of a digital circuit are dynamic power and static (leakage) power where with a proper design, the power that is associated with the short circuit current is a small fraction of the dynamic power. The dynamic power may be expressed as

$$P_{\text{dynamic}} = \alpha\, C_{\text{t}} V^2 f \tag{8.1}$$

where α is the switching activity and represents the average zero to one transition activities of the nodes, C_t is the total capacitance value, V is the operating voltage, and f is the operating frequency. The reduction in any of the above parameters leads to a lower power consumption of the circuit. For instance, approximating the realization of a function (e.g., using a simpler circuit than that of the exact one) would reduce the C_t value, lowering the power consumption. The simplification may lower the delay of the circuit path of the approximated circuit too. This can be utilized to improve the overall computational performance of the design by increasing the operating frequency more than that of the exact circuit.

As another approximation approach, one may use a supply voltage lower than the one required for the exact operation of the circuit. In this approach, again, the power that is proportional to the square of supply voltage is reduced. The propagation delay of the circuit, which depends on the operating voltage, is approximated by (using the simple delay model of alpha-power law)

$$\text{Delay} \propto \frac{V}{(V - V_{\text{th}})^{\beta}} \tag{8.2}$$

where β is a parameter between 1 and 2 used for gate delay modeling in super-threshold regime and V_{th} is the threshold voltage of a short-channel (NMOS) transistor. Applying the same frequency as that of the nominal voltage when the supply voltage is reduced causes the violation of timing constraints (more specifically setup time violations), and hence, output errors can occur.

There are multiple sources for static (leakage) power consumption of the circuit where the main factor of it is the sub-threshold leakage current that occurs when the transistors are in the off state. The leakage power may be modeled as

$$P_{\text{leakage}} = I_0\, e^{\left(\frac{-V_{\text{th}} + \eta\, V}{m \vartheta_T}\right)} V \tag{8.3}$$

where I_0 is a fitting parameter, η is the drain-induced barrier lowering coefficient, m is a process-dependent parameter typically in the range of 1.2 to 1.35 for conventional CMOS devices, and ϑ_T is the thermal voltage (i.e., kT/q where k is the Boltzmann constant, T is the die temperature, and q is the electron charge). Also, when a circuit is approximated to lower the power consumption, the reduction in power consumption leads to a lower die temperature and exponential reduction in leakage. With the supply voltage reduction, again, an exponential reduction in leakage power is achieved. Additionally, with the circuit simplification, the leakage power that is linearly proportional to the number of devices is reduced [7].

Finally, it should be noted that one can make use of the obtained reduced critical path delay (positive timing slack) due to circuit simplification to lower the supply voltage to further reduce the power consumption. Here, the operating voltage reduction does not induce additional errors at the output [8].

2.2 Approximate Computing and DSE in Multi-Objective Optimization Problems

AC is an emerging design technique that is applicable to error-resilient applications and may be applied at different levels of the design abstraction. At the behavioral level, for example, some iterations of loops may be skipped, and a pre-computed value can be used instead [3]. At the architectural level, approximate custom instructions are selected together with exact ones by the instruction set architecture [4]. At the circuit level, one may employ functional approximation where an exact operation is substituted with an alternative simplified one to reduce the delay and power consumption of the circuit. The most important application of functional approximation is in designing arithmetic units (e.g., adders and multipliers) as they consume a large portion of the computing system power [9]. In fact, in digital signal processing (DSP) applications, arithmetic units are the main ones responsible for running the algorithm. There are different techniques for realizing approximate arithmetic functions. Breaking the carry chain or replacing the least significant bit (LSB) full adders (FAs) with OR gates in a multi-bit adder and ignoring some of the additions in the partial product accumulation stage of a multiplier are some examples of approximate arithmetic units [5]. Another AC technique at the circuit level is introducing timing errors, where the supply voltage of a component is reduced to a point that the critical paths of the design are violated, while power consumption is reduced. This technique is called voltage overscaling (VOS) and can be applied at runtime to the circuit (dynamic accuracy configurability) [10].

The design metrics (energy, delay, area, and output accuracy) of a circuit are functions of different options (parameters), which are used for the design of the circuit. The options include, for example, sizes of the gates and/or threshold voltages of the transistors. More specifically, differently sized gates have different input/output capacitances causing different power consumption values, current drive capabilities (speeds), and layout areas. These design metrics of power consumption and latency compete with each other, which means that improving one often leads to the degradation of the other. Introducing AC as a new design paradigm has added another conflicting design metric (i.e., output accuracy) to the design process of VLSI circuits, which is determined by the amount of the output error induced by, for example, approximating functions. One can obtain a design with smaller power consumption (or higher speed) by using more approximation, which leads to a larger inaccuracy at the output. As an example, applying different degrees/levels of approximation (e.g., using a different number of approximate LSB bits in a multi-bit adder) results in approximate solutions with different accuracy and energy dissipation values. These different designs may be included in a library of approximate arithmetic units to be utilized during the AHLS design of target circuits.

The conflicting behaviors of the objective functions of interest set the framework of the optimization in a way that the goal of the designer becomes finding designs that optimize the key circuit metric(s) while not violating the constraints related to other circuit metric(s). In the case that one is dealing with a single objective function (or a set of non-conflicting objective functions), the problem may be formulized as a single objective optimization problem. The optimization of different (conflicting) objective functions at the same time with the goal of reaching a solution that is best regarding all the objective functions is called a multi-objective optimization problem (MOP) [11]. The set of all the combinations of the design options/parameters from the circuit components (i.e., all possible designs) and their corresponding circuit metrics of interest create the design space with different objective functions as the dimensions of the space. This design space needs to be judiciously explored to find the set of best combinations of the component design options/parameters (i.e., best design points) that optimize the design metrics(s) of interest while considering the constraints. The process is called design space exploration (DSE), which is generally an NP-hard problem [6].

In an MOP with conflicting objective functions, one may not find a solution that optimizes all the objective functions at the same time. In fact, the set of non-dominated design points in the design space, which exhibit the best trade-off between the objective functions of interest, constitute the Pareto frontier (PF) of the design space [12]. The design A with its specific values for the objective functions dominates design B if (i) A is better than B in all the objective function values or (ii) A is the same as B in some but better than B in at least one objective function value. We refer to the two mentioned conditions as dominating conditions. In Fig. 8.2, an example of the design space of an MOP with two objective functions (F_1 and F_2) is depicted. The goal of the optimization problem is to find the best designs that minimize both the objective functions. The points represented by blue dots are the PF of the design space from the design points depicted by red crosses. The points emphasized in the figure are representing the dominating conditions. For example, for the design point C, both the objective function values of D are smaller, and hence, C is dominated by D based on the first condition. For the design point A (E), the design point B (F) has similar values for F_2 (F_1) while larger F_1 (F_2) values. Therefore, B (F) dominates A (E) based on the second condition.

Having the PF of the design space for two objective functions, a single operating point can be obtained given a user-specified constraint on one of the objective metrics and an optimization goal on the other one. In Fig. 8.2, a user-specified maximum threshold value for F_2 is represented by a red dashed line. If the user is interested in finding the design that minimizes F_1 while meeting the constraint for F_2, the design point in the green circle is the one to choose. Having the PF available during the runtime, if the constraint changes, the system could migrate to a new operating point that meets the new condition dynamically.

2.3 *Approximate High-Level Synthesis*

Moving the design of VLSI circuits to higher abstraction levels and utilizing HLS to automatically generate an efficient RTL (Verilog or VHDL) code from an untimed

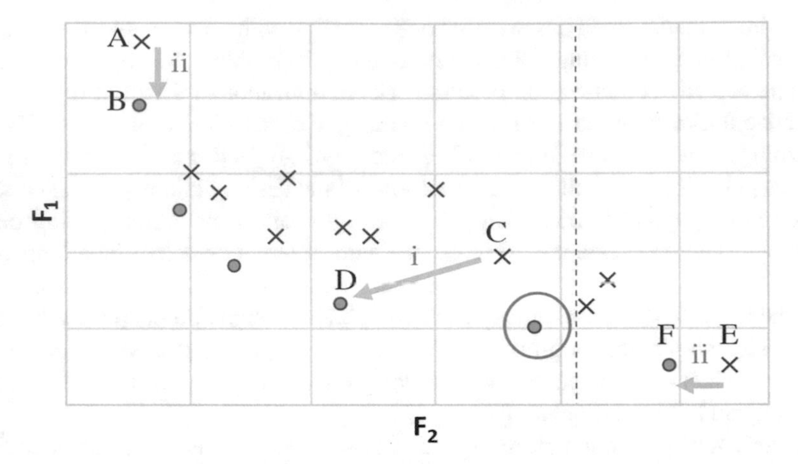

Fig. 8.2 An example of a design space for a MOP with two objective functions. The arrows from the represented dominating design points determine the dominating condition of that point in the design space. The design point inside the green circle is the design that meets the user-defined maximum threshold value for F_2 (denoted by dashed line) with the minimum value for F_1

behavioral input application lead to a shorter time to market, easier verification steps, and smaller effort to fix the design errors. In fact, the design at lower abstraction levels is very tedious owing to the significant complexity of (digital) electronic systems with a very large number of transistors. In the HLS domain, utilizing different synthesis directives (i.e., design options/parameters) leads to designs with different circuit design metrics including power, speed, and area [13]. As an example, loop unrolling is one common HLS directive where fully unrolling the loop by duplicating the hardware to perform the tasks in the loop provides the smallest latency for the loop execution. The required replication of the hardware resources (parallelism) results in higher power consumption and area of the design. Figure 8.3 depicts the design points (represented by red crosses) of the design space of different implementations of *Autocorr* function [14] (obtained by HLS tool) with the objective functions of area and latency [15]. The points located on the PF (represented by blue dots) have the best trade-off between latency and area.

The major steps of HLS include allocation, scheduling, and binding where among them, scheduling and component binding are the most significant ones. The scheduling step assigns starting time (or clock cycles) to the components of the design, while binding assigns hardware modules (from a given pre-characterized RTL module library) to the components. Along the scheduling and binding steps, the allocation step, which obtains the number of resources and their types, is also executed [13].

In the design at this level, first, the input application in the behavioral level is compiled into an intermediate representation (IR) [16]. There are different representations of IR, which most of them use graphs. Abstract synthesis tree (AST) is a directed acyclic graph where the nodes represent occurring actions

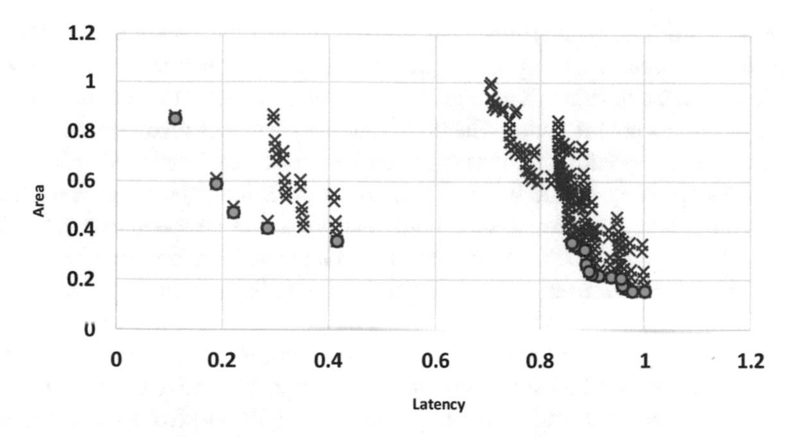

Fig. 8.3 The whole design space (red crosses) of the *Autocorr* function with the PF of the design space (red squares). (Redrawn from [15])

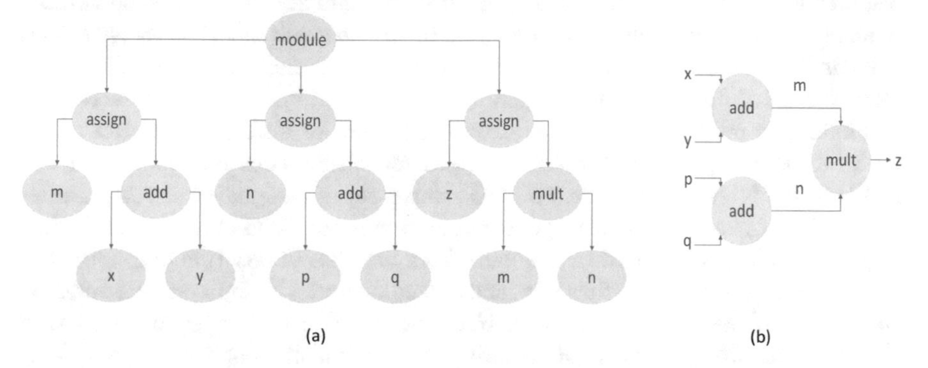

Fig. 8.4 (**a**) The AST and (**b**) the DFG representations of $z = (x + y) \times (p + q)$

in the behavioral code and edges represent the logical dependencies between the actions. ASTs contain information about the data flow in the input application and the abstract structures of the code [17, 18]. To represent only the data movements in a code, a data-flow graph (DFG) is defined. A DFG is a directed graph where nodes represent operations, and the data dependencies between the operations are captured by edges. The inputs to the graph are the set of primary inputs (PIs), while the outputs of the graph are the set of primary outputs (POs). An example of an AST and the corresponding DFG is represented in Fig. 8.4. Also, it should be mentioned that the control flow graph (CFG), which represents the control flow dependencies in the code, is not the focus of this chapter and, hence, not included here. Given the compiled representation of an input application and a pre-characterized library including RTL modules, the next step in HLS is to assign the design components to time steps and physical modules in the library. In this section, we concentrate on DFG as the high-level representation of the input design.

The scheduling step determines the starting time of the operations in the DFG. To obtain a minimum latency scheduling without any constraints, the as soon as possible (ASAP) algorithm schedules an operation as soon as the inputs to the node are ready from its parent nodes. The node that has its inputs ready is called a ready node. Thus, this algorithm represents the earliest time possible for scheduling (t_S). The as late as possible (ALAP) algorithm, on the other hand, performs a scheduling with a given latency constraint where the operations can be scheduled at the latest time possible (t_L). In a DFG, an operation can be scheduled anytime between (t_S) and (t_L), based on the user-defined constraints. The list scheduling algorithm can heuristically minimize the resource usage (i.e., number of modules taken from the RTL library) subject to a latency constraint or minimize the latency subject to a resource constraint. In the list scheduling algorithm, to minimize the resource usage subject to a latency constraint, at each time step t (starting from zero), the ready operations that have $t_L = t$ (critical operations) should be scheduled to that time step and bound to available resources (with the same type). If any proper required resource is not available, the resource is allocated, and the operation is bound to it in that time step. Also, the non-critical (i.e., operations with $t_L > t$) ready operations that can be assigned to any idle resources are also scheduled and bound. Therefore, along the scheduling and binding operations, the allocation is also performed [19].

Finding an optimized RTL code based on the use of different directives of the HLS tool is an intractable problem where a set of heuristics is required to solve it. When the HLS tool generates a DFG from a set of designated directives, after the scheduling and binding steps, every operation in the DFG is assigned to a time step (scheduling) and a hardware module from the RTL library (binding) [20]. The required design metrics are then obtained and compared with the given user-defined constraints. If the design is unacceptable, a new design should be generated with a new set of directives without requiring the modification of the original code. Using approximation methods in the HLS steps opens the new directions toward approximate HLS, making the problem even further complicated.

Before developing CAD tools for automatic generation of approximate circuits, one needed to resort to manual design approaches. The manual approaches normally have their limitations and may result in sub-optimal designs. With the increasing complexity of electronic circuits and the need for efficacy improvement of approximate circuits, development and employment of automatic techniques to obtain approximate instances of any general complex design are inevitable. Similar to conventional design tools, the tools that take user-defined constraints (i.e., a quality of service) as their inputs and automatically generate approximate circuits at the gate level may be called approximate logic synthesis (ALS) tools. Likewise, the tool that automatically maps the high-level description of an input application to the approximate RTL representation is called approximate high-level synthesis (AHLS) tool. The AHLS tools should employ various approximation techniques applicable at different layers of the design abstraction.

An approximate data-flow graph (ADFG) can be generated from applying approximation techniques to different operation nodes of a DFG. As an example, invoking ADFG where a subset of the arithmetic units in a DFG are replaced by their approximate counterparts from the library of approximate arithmetic modules provides the designer with lower energy consumption and possibly improvements in circuit delay and area. More specifically, this replacement along with the corresponding configurations (i.e., approximate degrees of the modules assigned to DFG nodes) generates the ADFG for the design. The improvement in the design parameters is accompanied with some degradation in the output quality. For a given DFG with approximate arithmetic units substituting the exact operation nodes, the output error is a function of the number of approximate units used in the design and the approximate levels of each approximate units when configurability exists. The replacement of exact components of a DFG with the approximate modules taken from the library of approximate modules resembles the binding step of the HLS tool [20].

The combination of different approximation units (and/or methods) along with their corresponding degrees of approximation for the DFG nodes as well as the output accuracy and other circuit design metrics (e.g., energy consumption) generates the design space of ADFGs. The design optimization here includes finding the best approximation combination for the DFG nodes such that the energy consumption (and/or possibly other design metrics) and output error are minimized while the minimum output quality constraint is satisfied. The challenge of the design technique is to determine the output quality for each combination. While estimating the output error for each approximate unit (e.g., approximate adder) can be performed more readily, say, by its simulation using many inputs, the estimation becomes very time consuming as the number of the DFG nodes increases. It should be also noted that depending on the application, the output quality metric may be different (we will discuss different error evaluation metrics later in this section). Therefore, estimating the quality of the output for different ADFG configurations is a challenge when design space of an application is explored, especially given the fact that potentially, there are many configurations for the ADFGs.

More specifically, let us consider a DFG with P nodes, which can be approximated. Also, assume that each approximate unit has Q possible implementations (one exact and $Q-1$ approximate ones). The design space for this ADFG will have Q^P design points with different output accuracy and energy consumption (and/or other design metric) values to be explored. In fact, exploring all possible design configurations exhaustively is not feasible due to combinatorial explosion, even for DFGs with small number of nodes. As an example, when $P=8$ and $Q=3$, we end up with 6,561 configurations. For each configuration, many inputs should be applied to the DFG simulator to estimate the error of the output, which requires excessively large computation time. Obviously, the larger the design space is (the larger Q and P are), the more time consuming this exploration will be. Therefore, heuristic approaches have been proposed [18, 24, 26–29, 31] to determine the (near-)optimal set of approximate vs. exact DFG nodes along with their corresponding levels of approximation to optimize the output accuracy and/or energy consumption, given a user-defined constraint. In the context of a MOP, the output of these approaches may be a PF that can be used to dynamically choose a single design point minimizing the energy consumption (maximizing the output accuracy) with respect to an accuracy

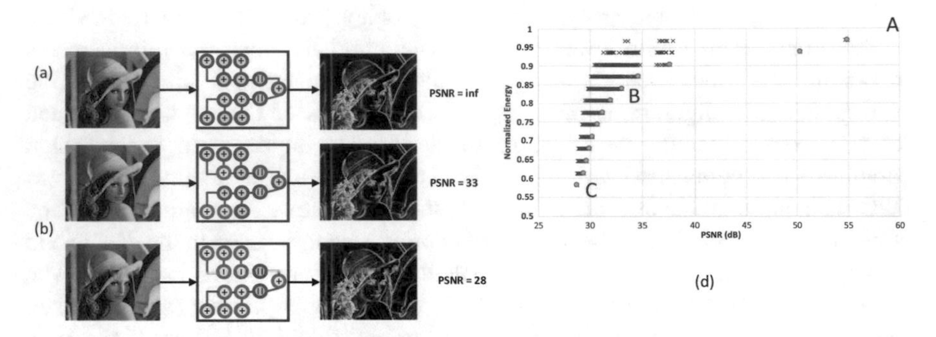

Fig. 8.5 The (**a**) exact and (**b**, **c**) two approximate DFG representations of the Sobel edge detector benchmark with an input sample [22] and the corresponding output after applying the filter. The design space of the ADFGs is depicted in (**d**) with the corresponding PF of the trade-off between energy consumption and PSNR

(energy budget) constraint. In the next part of this chapter, the works published in the literature dealing with heuristic approaches will be reviewed.

Figure 8.5a–c shows the exact Sobel edge detector DFG and its two possible ADFGs. In the ADFGs, some of the exact arithmetic units were substituted with their approximate counterparts taken from a library of approximate modules. For this example, the arithmetic units had one exact and one approximate implementation. The green nodes are 11-bit exact additions (ADDs) and the blue nodes are absolute value operators (ABSs), which consist of a multiplexer and an adder. The yellow nodes are approximate 11-bit lower-part OR adders (LOAs) with the four least significant bits approximated [21]. Therefore, with 13 ADD nodes, each having two different implementations, the design space contains $2^{13} = 8{,}192$ different ADFGs. Figure 8.5d depicts the whole design space of the ADFGs and the obtained PF of the accuracy vs. energy consumption trade-off. The energy consumption of the ADFGs has been normalized to that of the exact implementation. To obtain the accuracy, the peak signal-to-noise ratio (*PSNR*, defined in the next subsection) was defined based on the difference between the image generated by the exact and the approximate Sobel operators. The design points marked on (d) correspond to the DFG implementations of (a–c). Note that the exact DFG (a) has *PSNR* of infinity that cannot be represented in the figure and its energy value is only depicted.

2.4 Error Evaluation Metrics

In this subsection, we provide some of the employed error metrics in the AC design paradigm to assess the quality of the output when approximation methods are exploited in the design. To obtain the error characteristics of a design, a set of input values are applied to the original and approximate circuit. To calculate the exact error metrics, the set of all input vectors should be applied. For designs with large input dimensions, however, this is not feasible, and a set of random input vectors can be chosen to estimate the output error. In the following, we discuss the error metrics of error distance (ED), mean error distance (MED), mean relative error distance (MRED), signal-to-noise ratio (SNR), PSNR, mean squared error distance (MSED), and mean structural similarity index metric (MSSIM).

For approximate arithmetic units, error distance (ED) is defined as the numerical difference between the approximate output ($O^{'}$) and the exact output (O):

$$\text{ED} = \mid O' - O \mid \tag{8.4}$$

With the use of ED on a set of input vectors, mean error distance (MED), mean relative error distance (MRED), and mean squared error distance (MSED) are defined. MED is expressed as

$$\text{MED} = \left(\frac{1}{N}\right) \times \sum_{i=1}^{N} \text{ED}_i \tag{8.5}$$

where N is the number of applied input vectors and ED_i is the ED of the ith input vector. The normalized error distance with respect to the exact output values is averaged in MRED and formulated as

$$\text{MRED} = \left(\frac{1}{N}\right) \times \sum_{i=1}^{N} \frac{\text{ED}_i}{O_i} \tag{8.6}$$

where O_i is the exact output of the ith input vector.

For the applications in the domain of digital signal processing (DSP), some of the accuracy metrics from this domain are used to evaluate these applications with applied AC techniques. More specifically, for image processing applications, PSNR and mean structural similarity index metric (MSSIM) are used to reflect the quality of the output image when approximate circuits are used. The signal-to-noise ratio (SNR) metric is another metric in the DSP domain defined as

$$\text{SNR} = 10 \times \log_{10}\left(\frac{P_{\text{signal}}}{P_{\text{noise}}}\right) \tag{8.7}$$

where $P_{\text{signal}}(P_{\text{noise}})$ refers to the power of the output (noise) signal.

The metric PSNR is defined as

$$\text{PSNR} = 10 \times \log_{10}\left(\frac{\text{max}_{\text{pixel}}^2}{\text{MSED}}\right) \tag{8.8}$$

where $\text{max}_{\text{pixel}}$ is the maximum possible value for pixels of an image and MSED is the mean squared error distance obtained from

$$\text{MSED} = \left(\frac{1}{N}\right) \times \sum_{i=1}^{N} (\text{ED}_i)^2 \tag{8.9}$$

When perception by the human visual system is of importance, the parameter MSSIM is a better metric. The parameter is defined between two $N \times N$ blocks (one is the errorless one and the other is the approximate one) of x and y as

$$\text{SSIM}(x, y) = \frac{\left(2\mu_x\mu_y + C_1\right)\left(2\sigma_{xy} + C_2\right)}{\left(\mu_x^2 + \mu_y^2 + C_1\right)\left(\sigma_x^2 + \sigma_y^2 + C_2\right)} \tag{8.10}$$

Here, $\mu_x(\mu_y)$ and $\sigma_x(\sigma_y)$ represent the mean and standard deviation of the pixels in the considered block, and C_1 and C_2 denote constant values obtained (but not equal to) from the dynamic range of the pixel values. Also, σ_{xy} is obtained from the pixel and the μ values [7]. By averaging the extracted SSIM values, one obtains the MSSIM parameter.

3 DSE Algorithms in Optimization of Approximate Circuits

As discussed before, the DSE problem in the design optimization of approximate circuits is a difficult problem where heuristic approaches are preferably employed to obtain a heuristic solution in a reasonable time. There are approaches proposed in the literature to obtain the set of best designs that optimize the objective functions (e.g., power consumption and/or output accuracy) with a user-defined constraint (e.g., minimum output accuracy level). Generally, the goal in most of the works in this area has been the optimization of power consumption (and/or other design metrics) subject to satisfying a threshold (minimum value) of the output accuracy. For some platforms (e.g., field-programmable gate array [FPGA]) the available resources are frequently known a priori, and therefore, an optimization problem to maximize the output accuracy and/or energy efficiency subject to resource constraints is considered [23].

In the search for the best approximate configurations, most of the proposed heuristic approaches make use of iterations where in each iteration of the algorithm, a set of approximate designs are generated and evaluated. The evaluation step mostly involves simulation and synthesis. In these approaches, simulation tools are used to extract the quality of an approximate design point through comparing the exact results of a given set of test vectors with the approximate ones. On the other hand, synthesis tools are invoked to obtain the other circuit design metrics (e.g., power consumption). A major part of the DSE runtime belongs to the simulation and the synthesis steps [24]. In addition, it should be emphasized that the output quality of an approximate circuit is strongly input-dependent. Some works have proposed analytical [25] or ML-based models [26, 27] to estimate the quality of an approximate circuit. For the other design metrics such as power consumption, ML-based models have also been proposed to eliminate the need to invoke the synthesis tool [27]. From the perspective of exploring the design space, the DSE algorithms presented in the literature may be categorized into two groups: local search and evolutionary.

In the next subsections, we introduce the works in each group while providing more details about three of the works in the groups. Also, at the end of this chapter, key features of some of the works in the context of AHLS are compared.

3.1 First Category Local Search Approaches

In the first category, in each iteration of the iterative DSE algorithm, the design space is expanded from one design point (called a parent point) to other points in the space. From the parent point, either one or multiple new design points in the design space (approximate configurations) are generated. The generation of the new points is performed by changing some of the design options/parameters in the parent design (e.g., add inaccuracies to the exact components in the parent design). Regardless of the number of new points, only one design point is selected as the parent point for the subsequent iteration. Heuristics are mostly utilized to choose this parent design point for the subsequent iteration from the set of newly generated approximate designs and/or the original parent point. The DSE algorithm can generate the set of new approximate configurations to be explored in the design space either stochastically [27] or deterministically [24].

3.1.1 Automated Behavioral Approximate Circuit Synthesis (ABACUS) Technique

We review the ABACUS (**a**utomated **b**ehavioral **a**pproximate **circu**it **s**ynthesis) technique [18], which belongs to the first category of DSE algorithms. It is the first work addressing the DSE problem of approximate designs using the behavioral hardware description of any arbitrary input application. The ABACUS tool automatically generates approximate circuits (still represented at the behavioral level) that are located on the PF of the output accuracy vs. power consumption trade-off curve while meeting a user-specified quality constraint. ABACUS can be employed for the designs to be implemented on ASICs or FPGAs. The input to ABACUS, represented in RTL or behavioral hardware description, is captured by an AST structure. On the AST of the exact design, a set of *transformation operators* (TOs) are applied to generate new approximate ASTs. The modified ASTs are written back to readable behavioral code in order to be suitable for the subsequent synthesis and evaluation steps. The set of TOs utilized by ABACUS are listed below:

(i) Data-type simplification where the intermediate signals are truncated. The truncation operation in ABACUS is performed in two ways of (a) resetting a number of LSBs of a signal in the design to zero and (b) performing the operation only on the most significant bits (MSBs) of the operands and shifting the result.
(ii) Operation transformations where the exact arithmetic operations in the AST are replaced by their approximate counterparts with smaller power consumption from a given library of approximate arithmetic units.
(iii) Arithmetic expression simplification where the structure of an arithmetic statement is approximated to a near-similar one with the goal of increasing operation/operand sharing. For instance, $w_i \times x_i + w_j \times x_j$ can be approximated with $w_i \times (x_i + x_j)$ by substitution of w_j with w_i, which can save one multiplication operator.
(iv) Variable-to-constant (V2C) substitution where signals that have small variability (small standard deviation around the average [mean] value) are replaced with their average value. The information about the signal values is obtained from the simulation results of the exact design where the values of all intermediate signals are reported and evaluated. With the substitution of a variable with a constant value, the part of the circuit calculating the substituted signal can be pruned out.
(v) Loop transformation where the loops in the program are unrolled to increase the chances for other circuit simplifications. Moreover, some iterations in a loop may be skipped or replaced by pre-computed values from the previous iterations.

The aforementioned TOs can be applied at any location of the AST of the input application. In addition, in a chain of transformations, any modified AST can be considered as the starting design to which other TOs can be applied. As stated before, the design space of approximate designs has an excessively large number of design points, which make the exhaustive exploration impractical. In ABACUS, a stochastic greedy heuristic is proposed to explore the design space. The algorithm consists of *N* iterations, whereby each iteration generates *M* approximate new designs from the parent AST of that iteration. The parent AST of each iteration is generated at the end of a previous iteration with the very first iteration starting with the exact design as the parent design point. The general overview of ABACUS is depicted in Fig. 8.6.

In the ith iteration of the heuristic, *M* random transformation operators are separately applied to random locations of the AST of the ith parent design point to generate *M* independent modified ASTs. Since the TOs and their applied locations on the parent AST are chosen randomly with a probability, ABACUS is categorized as a stochastic heuristic. For each newly generated AST, the behavioral code is also written from its AST. The output accuracy of each generated approximate circuit is evaluated via simulations, and if the accuracy is within the user-specified threshold, it will be marked as a valid design point (i.e., added to the pool of approximate circuits). For the accuracy evaluation, a set of testbenches are simulated with the

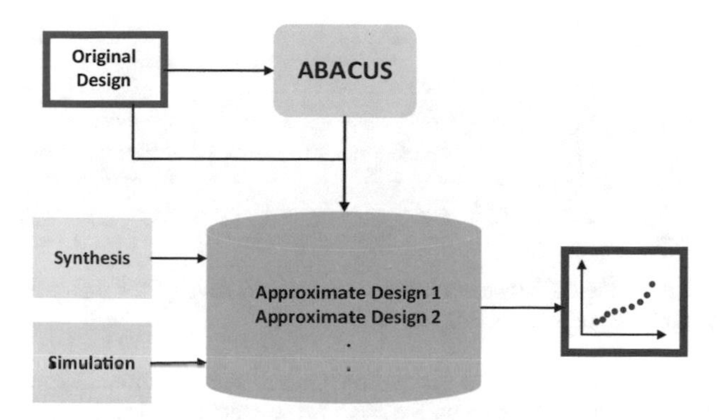

Fig. 8.6 General overview of the ABACUS framework [18]

Mentor Graphics ModelSim. Only if the design point is valid, its other circuit design metrics are obtained via the synthesis tool. In the case of ASIC designs, a standard cell library and, in the case of FPGA platform, look-up table and architecture clusters are provided to the synthesis tool. Once the generated approximate circuits are evaluated and synthesized, a *linear* figure of merit (FOM) for each approximate circuit is defined with respect to the accuracy, area saving, and power saving values of the design as

$$\mathrm{FOM} = \alpha_1 \times \mathrm{accuracy} + \alpha_2 \times \mathrm{power}_{\mathrm{saving}} + \alpha_3 \times \mathrm{area}_{\mathrm{saving}} \tag{8.11}$$

where $0 \leq \alpha_i \leq 1$ and $\sum_{i=1}^{3} \alpha_i = 1$.

The value of α_i is set based on the importance of the corresponding circuit design metrics, meaning that the highest value of α corresponds to the metric with the highest significance for the user. Among the generated approximate circuits, the one with the highest FOM is chosen as the starting parent design point for the subsequent iteration. The heuristic continues until the search budget (i.e., number of available iterations) is exhausted. At the end of the final iteration, the generated approximate design that has the highest FOM represents the design with the closest accuracy to the user-specified constraint while having the best other circuit metrics (of importance to the user). From the set of all the approximate circuits in the pool, the PF of the design space may be extracted.

The ABACUS tool was applied to three application benchmarks including perceptron classifier, block matcher, and finite impulse response (FIR) filter to show its efficacy compared to the case of the conventional method, where for a set of adder and multiplier operations in the designs, three LSBs are truncated. The PFs extracted with the ABACUS tool for the design space of different benchmarks are presented in Fig. 8.7. The green square in each figure represents the original design. From the extracted PFs and a given accuracy threshold (8% accuracy drop), the designs that have the most power saving are given in Table 8.1.

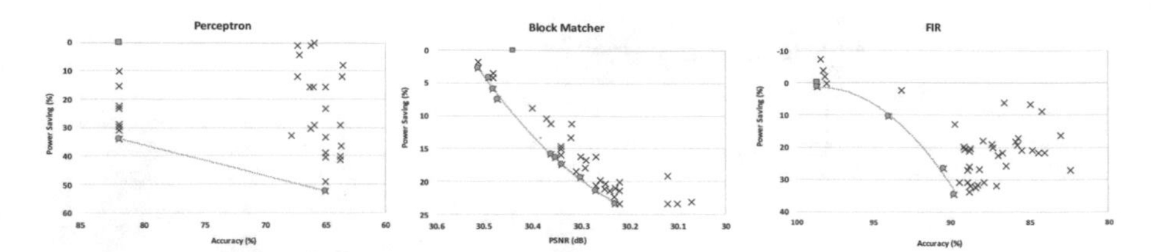

Fig. 8.7 PFs of the trade-off between accuracy and power savings for the design space of the different benchmarks extracted by ABACUS [18]

Table 8.1 Obtained designs with the best power savings for the different benchmarks with a maximum of 8% accuracy degradation with respect to the exact design [18]

Design	# Iterations	Accuracy threshold	Accuracy achieved	Power saving	Area saving
Perceptron	10	76.2%	82.9%	33.2%	38.3%
Block matcher	15	28.0 dB	30.0 dB	23.0%	19.4%
FIR	10	90.9%	93.9%	10.4%	15.8%

If in the first iteration of the algorithm, all newly generated approximate circuits are transferred to the next iteration as the parent points (instead of choosing a single parent design point with the highest FOM), and this process continues in the subsequent iterations, then obviously the algorithm will suffer from an exponential approximation configuration size increase as [18]

$$M^0 + M^1 + M^2 + \cdots + M^N = \frac{1 - M^{(N+1)}}{(1 - M)} \tag{8.12}$$

The tool ABACUS, however, limits itself to only $M \times N$ points in the design space, and as a result, it significantly speeds up the DSE process. Note that utilizing a linear FOM as a ranking metric that chooses the next parent design based on the local M new design points in each iteration may result in getting stuck in the saddle point of the design space. This drawback is mitigated in the work of [8], which will be explained further later.

3.1.2 Other Works in the First Category

In this part, we review other works that are grouped in the first category of DSE algorithms for the optimization of approximate circuits. In [24], for the operations in the design that can be approximated (called candidates), the impact factors (IFs) for each candidate (including area and quality IFs) are pre-calculated. To calculate the area IF (AIF), the output of each candidate in the exact design is tied to zero separately, and the area reduction compared to the exact design represents the AIF of that candidate. The AIF determines the impact of approximating a candidate on the overall area of the design, and approximating a candidate with a higher AIF is preferred. On the other hand, to calculate the quality IF (QIF), the least absolute shrinkage and selection operator (LASSO) is exploited as a variable selection technique to determine the impact of approximating a candidate on the quality of the output of the design (primary output). To utilize LASSO technique, the primary output of the original design (output parameter) is represented by a linear combination of the values of the outputs of the candidates (input parameters) with regularization. The absolute values of the obtained coefficients through LASSO determine the contribution of each candidate to the output of the design. A larger coefficient means that the candidate has higher impact on the primary output, and therefore, approximating that candidate is not preferable. In the proposed framework called Jump Search, the design search starts from the exact design at the root of a search tree. In each iteration (i.e., level of the search tree), a change in one of the candidates of the current tree node (i.e., parent design) creates different new designs as the children of the parent design point. Multiple approximation techniques may be performed to obtain new designs. Among all the possible designs, the design with the highest defined FOM (i.e., highest AIF and lowest QIF) is added to the tree. This design is also selected as the parent of the next iteration. In the last step, the obtained path in the design space tree (called promising path), which is identified by the choice of the selected approximate designs, is used to extract the best optimal design. The method performs a binary search on the promising path to choose a design that has the highest number of approximations (and, therefore, likely has the lowest area cost) while meeting a user-specified quality threshold (via simulation and synthesis tools). The output of the Jump Search algorithm is always a single specific design, classifying it as a deterministic algorithm.

In [26], a learning-based search method is proposed for the DSE problem, which works in two steps. In each iteration, starting from the exact synthesized circuit at the root of the search tree, for each tree level, the tree node that has the highest improvement factor (IMF) is selected to lie on a path that will reach to a leaf node (x) for further explorations. The IMF of the node i of the tree depends on the ratio of the summation of the improving metric values (e.g., area improvement) up to this point and the number of visits for this node in the search process. The number of node visits is considered to fairly distribute the search budget into different regions of the search space. On the selected leaf node x, an exact arithmetic unit is randomly chosen and approximated from a module in the given library to generate a new child node to x. The quality evaluation is performed via a pre-trained random forest (RF) model. If the child node satisfies the quality constraint, positive feedback (normalized area improvement of the child node estimated from the area values of the library components) is added to the IMF of all the parent nodes on that path from the root to the node x, and the node is added as the child to x. Otherwise, negative feedback (i.e., -1) is added to the IMFs. This way, in the next iterations, the new IMFs are utilized to select the new leaf node. When the search budget is exhausted, the nodes that have the highest IMF values are selected. The simulation and synthesis tools are invoked only on the final selected nodes where the optimal design is obtained.

The autoAx tool presented in [27] utilizes a large library (e.g., containing 6000 approximate implementations for an 8-bit adder) of approximate arithmetic units to generate approximate versions of a given input application. To choose the best set of approximate designs, which are located on the PF of the design space of area and *MSSIM*, instead of invoking simulation and synthesis tools for the first phase of the algorithm, autoAx utilizes pre-trained ML models (separate ML models for the quality and area metrics) to estimate the design metrics. In a pre-processing step, first, a set of suitable modules from the given library for each arithmetic unit operation are selected. The suitable modules for each circuit operation are the ones located on the PF of the error vs. area trade-off when the operation is replaced by approximate modules from the library. To obtain the error, the probability mass function (PMF) of the inputs to an operation is obtained through simulations of the original design on the benchmark data. With the use of input PMF and the ED for each input combination from PMF, the weighted MED is obtained. Next, a modified stochastic hill-climbing algorithm is performed to choose a pseudo-PF (PPF) set of the design space. Starting from a random design as the parent (a design where some arithmetic units are randomly replaced by approximate units from the library) and an empty PPF set, in each iteration of the algorithm, the parent design is modified (only in one random node of the circuit) to create a new design (C). The design metrics of C are estimated from the models, and if C dominates the designs previously in the PPF, it is added to the PPF and assigned as the parent for the next iteration. If the initial point is not changed after k successive iterations, the initial point is replaced by a new random design point from the PPF set to avoid getting stuck in a saddle point. After the termination condition is reached (e.g., reaching maximum number of iterations), simulation and synthesis tools are invoked on the designs located in the PPF set to extract the final PF set of the design space. The quality of the final PF depends on the performance of the pre-computed ML models for the design metrics.

In [32], the clang-Chimera tool (mutation engine for C/C++ codes) is used to integrate AC techniques to a given source code represented in C/C++. Starting from the exact design located at the root of the search tree, the branch and bound (B&B) algorithm is used to create other design points (i.e., tree nodes) in the search tree. The generated tree is constructed recursively (starting from the root) in a depth-first search manner. Therefore, in each recursion from a node in the tree, for each operation of the design corresponding to that tree node, the algorithm generates a new approximate design from the approximate version of the operation. In this work, the approximation method removes the LSBs of the operands, reducing the accuracy of the operands and, in turn, the operation results. If the obtained approximate design is within the user-specified threshold, a child for the evaluated tree node is added to the tree and the algorithm continues (repeats) from the child node. The accuracy evaluation is performed via software (SW) simulations. Once the search tree cannot be expanded in any directions, the set of the leaf nodes constitute the PF of the design as they are the designs that cannot be approximated anymore. The set of leaves are then implemented in the hardware and synthesized to obtain the other design metrics of interest. From the synthesized designs, the one that has the best design metrics can be selected. In the context of this work, PF refers to the configurations that have a quality as close as possible to the error threshold. This definition is not the same as the one used by other researchers.

The work in [31] proposed a multi-level method where different approximate techniques are applied at different levels of the design. In addition, different input distributions were used when generating approximate circuits to evaluate the robustness of the design regarding different input distributions. The proposed method has four phases, where each phase refers to applying approximations to a specific level of the design. At the software (SW) level, loop unrolling, code pruning, and variable-to-constant (V2C) and variable-to-variable (V2V) techniques are applied to simplify the input code for the next steps. The HLS level, which has the longest runtime, assigns the approximate modules in the given library to the operations in the design. In the first step of the HLS phase, separately for each operation of the code, all the applicable approximate modules from the library are assigned, and the generated code is synthesized with the HLS tool and simulated for error evaluations. At the end of this step, a list (D) of all the valid designs (i.e., designs with output error smaller than user-specified threshold) with their circuit metrics (i.e., output error and area) is generated. The next step iteratively chooses the combination of designs from D with a greedy method. First, a linear cost function ($\alpha_1 \times$ error $+ \alpha_2 \times$ area) is assigned to each design in D. In each iteration, the design with single approximation operation from D with the lowest cost function value is chosen as the starting parent design. Next, a new operation is approximated based on the cost function values. The new designs are simulated to evaluate the output error. For the area of the new designs, the difference between the area of the approximate module and the exact one estimates the area modification. The set of non-dominated approximate designs are selected and fully synthesized, and the RTL code for them is generated to be utilized in the next phase. In the RTL phase, for all the generated approximate RTL codes, the V2C and V2V optimization techniques are again applied since now all the internal signals are visible. On the signal bits of the designs, replacing the bits by "0" or "1" is also evaluated. Finally, on the gate-level netlist of the designs, the stability analysis is performed, where a list of five different input distributions is applied to the approximate circuits and the range of the output accuracy is extracted to be reported as the accuracy range for the approximated generated circuits instead of a single value.

3.2 Second Category Evolutionary Approaches

The second category of DSE algorithms for optimizing approximate circuits expands the design space from multiple design points in each iteration of the algorithm [8, 20, 28, 29]. In each iteration, in contrast to the algorithms of the first category, a set of parent design points in the design space (called a population) are used to generate a set of new approximate configurations. These algorithms can be stochastic in nature where different rounds of the algorithm with the same inputs result in different explorations of the design space.

3.2.1 Exploring the Trade-Off Between Accuracy and Energy Efficiency (EGAN) Framework

In [28], the goal is to find approximate circuits with their corresponding configurations (i.e., degree of approximation of each circuit component), which are located on the PF of the trade-off space of accuracy vs. energy consumption. The input to the framework, which is named EGAN, is the DFG of the input application. Replacing the arithmetic nodes in the DFG with their approximate counterparts from a given library results in an ADFG that has some level of inaccuracy at its outputs but potentially consumes less energy.

The configuration of a given ADFG is determined through the type and degree of the approximation of each node. To efficiently extract the PF of the space of the ADFGs, an iterative heuristic, which considerably reduces the exploration runtime compared to exhaustive search of the design space, is used. The steps of the EGAN framework are depicted in Fig. 8.8. The heuristic starts with a random initial set of M configurations, denoted as S, including the fully exact and fully approximate DFGs. The fully exact design point has the highest accuracy and energy consumption values, while the fully approximate one has the lowest accuracy and energy consumption. The corresponding energy consumption and output accuracy of all the ADFGs in the set S are obtained, and the sampling space (the space containing the design points) is generated. In a pre-processing step, the energy consumption of the arithmetic units in the library with different approximation degrees is pre-calculated (in a specific technology) and used to obtain the energy consumption of a given ADFG. Ignoring the scheduling step in the HLS process, the energy costs of the arithmetic modules assigned to an ADFG are accumulated to estimate the total energy consumption of the design. In the case of a design that does resource sharing, a synthesis tool should be utilized to obtain an accurate value for the energy consumption of the ADFG. Obviously, utilizing a synthesis tool increases the runtime, although it does not affect the efficacy of the proposed framework for extracting near-optimal PF of the design space. To obtain the output accuracy of a design with a specific configuration, the gate-level representation of the ADFG with the instantiated (exact or approximate) modules for each ADFG node (i.e., obtained from the configuration) is simulated with a set of test vectors.

The subsequent steps of EGAN are iterative. First, the design points in the initial sampling space, which contain information about the design configuration, energy consumption, and output accuracy values, are clustered by employing the k-means clustering algorithm [34] into (γ) clusters based on the similarities (their hamming distance) between the three mentioned parameters. In Fig. 8.8, the initial design points (gray dots) are clustered into two clusters represented by purple and blue dots. After that, in each iteration of EGAN, a set of new configurations are generated. The generation of new design points is based on the integer representation of the configurations, that is, a vector of n-bit integer numbers, represented by C. In the configuration vector C, n is the number of arithmetic units of the DFG, and each element in the vector represents the degree of approximation of the corresponding DFG node. Zero represents the exact implementation of the DFG node, and higher

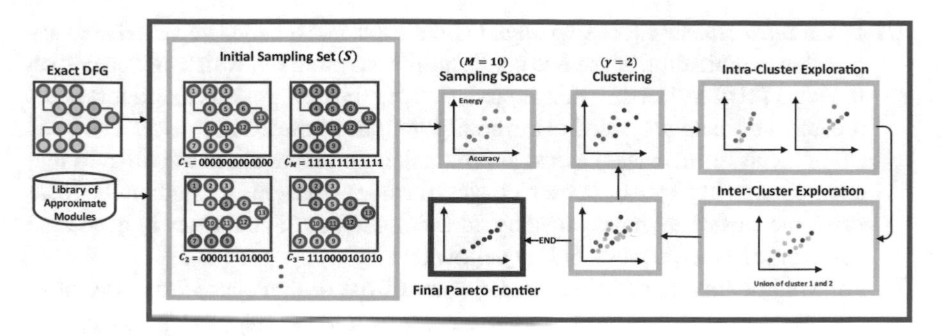

Fig. 8.8 Overall steps of the EGAN framework [28]

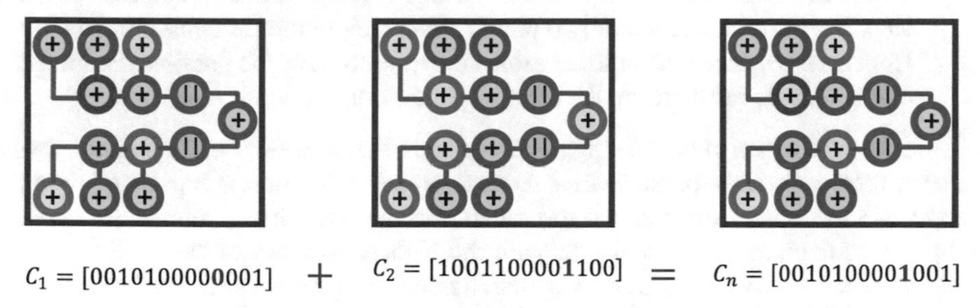

Fig. 8.9 An example of generating a new configuration (C_n) from two configurations C_1 and C_2 where C_1 has higher accuracy and energy consumption values than C_2. Comparing the configuration vectors of C_1 and C_2, there are four DFG nodes that can be approximated in C_1 where one of them is represented by a red border line. The new configuration has a similar vector than C_1 with the red border line DFG node modified

values correspond to approximate implementations with higher output error and energy saving values. Here, only adders and multipliers are approximated. To generate a new design point, two existing parent points with the configuration vectors of C_1 and C_2 (where C_1 has higher output accuracy and energy consumption values) are combined. Any vector element i in C_1 (i.e., $C_1[i]$) that has a lower value than the corresponding element in C_2 can be incremented to a number between $C_1[i]$ and $C_2[i]$. This is analogous to adding inaccuracies to a DFG node in C_1, which has a higher accuracy than the same DFG node in C_2. This way, a new configuration C_n is generated (with one DFG node implemented differently than C_1), which has a lower energy consumption than C_1 with a possibly lower output accuracy. An example of a combination of two possible configurations from the Sobel DFG is depicted in Fig. 8.9. There are only two implementations for a DFG node in this example, exact and one approximate implementation with a degree of one.

With the use of the mentioned combination method, the generation of new design points for design space expansion in EGAN has two steps:

(i) In the *intra-cluster expansion step*, each cluster is expanded separately. Every possible combination of the two configurations C_1 and C_2, which are located on the local PF of a cluster, is considered. Among the new possible configurations, α percent of them are added to the configurations of the design space. The local PF of a cluster includes the design points that dominate all other points in that same cluster. In Fig. 8.8, in each cluster, the configurations located on the local PF of the cluster are represented by dark-colored dots, and the newly generated configurations are represented by yellow dots.

(ii) To avoid getting stuck in the saddle point of the design space from the intra-cluster expansion step, in the *inter-cluster expansion step*, the union of two different clusters is generated. The combination of the configurations located on the local PF of the union of two clusters expands the design space in this step. For the combination of two points, one configuration is chosen separately from each cluster in the union. In this step, β percent of the possible combined configurations are added to the design space configurations.

In the generation of new configurations within the intra-/inter-cluster expansion steps, there are many possible new configurations that, considering all of them, increases the framework runtime too much. On the other hand, selecting a small number of them may not provide us with the highest accuracy of the final PF. The parameters α and β are tuned in a way that a good trade-off between the framework runtime and the accuracy of the final PF obtained can be achieved.

For all the new configurations, the energy consumption and output accuracy are obtained by using the procedure explained above. At the end of each iteration, the PF of all the points in the design space (including the initial set and the new generated points) is extracted. If the extracted PF is similar to the PF of the previous iteration or a total compute budget is reached (i.e., reaching the maximum number of design points to be explored), EGAN returns the PF of the last iteration as the final PF (represented by red dots in the design space of Fig. 8.8). If the heuristic is not terminated, the set of all design points are considered as the initial sampling set for the subsequent iteration.

To assess the efficacy of EGAN, the PF of the design space of the three benchmarks including Sobel edge detector, FIR filter, and discrete cosine transform (DCT) is extracted where the results are presented in Fig. 8.10. For the Sobel benchmark, each DFG node has three possible implementations, while for FIR and DCT, each node has two levels of implementations. For all the benchmarks, extracting the exact PF is infeasible. It can be observed that changing the initial set size or number of clusters leads to a slightly better PF. Also, to present the quality of the PF generated by EGAN, in Fig. 8.11, for a smaller design space of Sobel benchmark (each adder has two approximate degrees of one exact and one approximate implementations), the PF extracted by exhaustive simulations and EGAN framework is compared. It can be seen that the PF extracted by EGAN almost matches the PF obtained by the exhaustive search of the design space while having a smaller runtime (98% runtime reduction).

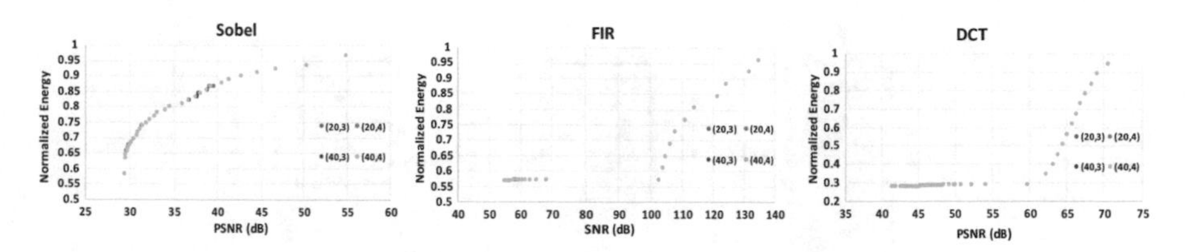

Fig. 8.10 PF of the design space for different benchmarks determined by EGAN for different values of M and γ represented by a tuple (M, γ) for each curve [28]

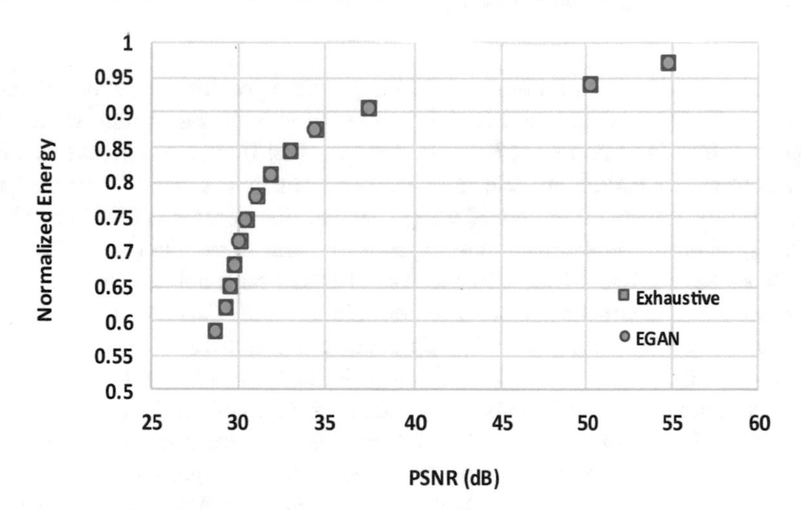

Fig. 8.11 Extracted PF by the EGAN framework and exhaustive simulations [28]

3.2.2 Modified ABACUS Work

The ABACUS tool was improved in [8] by changing the DSE algorithm. In the modified tool, instead of choosing a single AST with the highest defined FOM among the M-generated approximate ASTs to be the starting AST of the subsequent iteration, multiple (N_{sel}) parent ASTs are chosen with the use of NSGA-II algorithm. NSGA-II is a multi-objective evolutionary algorithm [30] that is a popular technique in solving problems with multiple conflicting objective functions instead of converting them to a single-objective optimization problem. The algorithm has a complexity of $O(KH^2)$ where K represents the number of objective functions and H represents the population size [30]. NSGA-II has a set of operators that generate new design points from the initial population.

The NSGA-II algorithm starts from a randomly generated initial set of P with the size of H. The set P is combined with a set of H offspring population from P (named Q) to generate a new set R. The R population (with the size of $2H$) is sorted into multiple PFs, where the PF of the original population R with respect to the objective functions represents the first PF (PF_1), and by removing PF_1 set

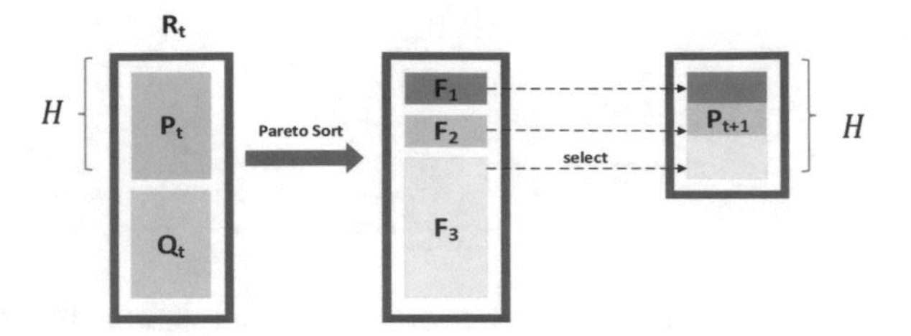

Fig. 8.12 Choosing step of the NSGA-II algorithm [30]

from the population, the PF of the remaining points represents the second PF (PF_2) and so on. To each design point in the population R, a fitness or rank value (F) is assigned, which is equal to its PF level. Now, starting from the members of PF_1, a set of H members from R are selected based on their fitness values to be the initial population for the next iteration. If $|PF_1| < H$, all the members of PF_1 are chosen and the remaining members are chosen from PF_2 and so on. On the other hand, if $|PF_1| > H$, the members from PF_1 are sorted based on a defined metric, and the best H members create the next iteration population [30]. This procedure for the tth iteration of the NSGA-II algorithm is represented in Fig. 8.12.

While in the modified ABACUS, the algorithm still has N iterations, instead of one parent point, there are N_{sel} initial designs in each iteration. For each N_{sel} parent point, M newly designs are generated and validated for accuracy satisfaction. Any valid design is synthesized and added to the local pool of approximate circuits for each parent. The global pool of ADFGs is updated to include the ADFGs stored in all the local pools. Then, a hybrid method, instead of a separate linear [18] or NSGA-II-based method, is utilized to choose the parent points for the subsequent iteration. The reason that a hybrid method is preferable is that due to the iterative nature of ABACUS, NSGA-II alone cannot provide a good solution. In fact, in this case, only the designs that globally dominate other designs in the global pool will be repeatedly selected as the parents for the next iterations, which limits the search space. With the linear method of choosing a single parent from local pools [18], also the search can get stuck in a saddle point of the design space. Therefore, for the designs in the global pool (i.e., the set of approximate circuits), the NSGA-II FOM function and, for the designs in the local pools, the linear FOM function are invoked. The best $N_{sel} - 1$ designs based on the NSGA-II method from the global pool and a single best design based on the linear method from the designs in the local pools are selected as the next N_{sel} parent points for the subsequent iteration. The modified ABACUS prioritizes approximations on the critical paths of the circuit to generate positive timing slacks used for VOS to increase the power reduction of the design. The generation of the approximate designs is also parallelized because of the independent nature of the ABACUS iterations to reduce the runtime of the search algorithm. The use of Verilator, a fast C-based simulator, instead of simulation tools used in [18], is another method, in addition to parallelization to increase the speed here [8]. The results of the ABACUS tool with different parent selection

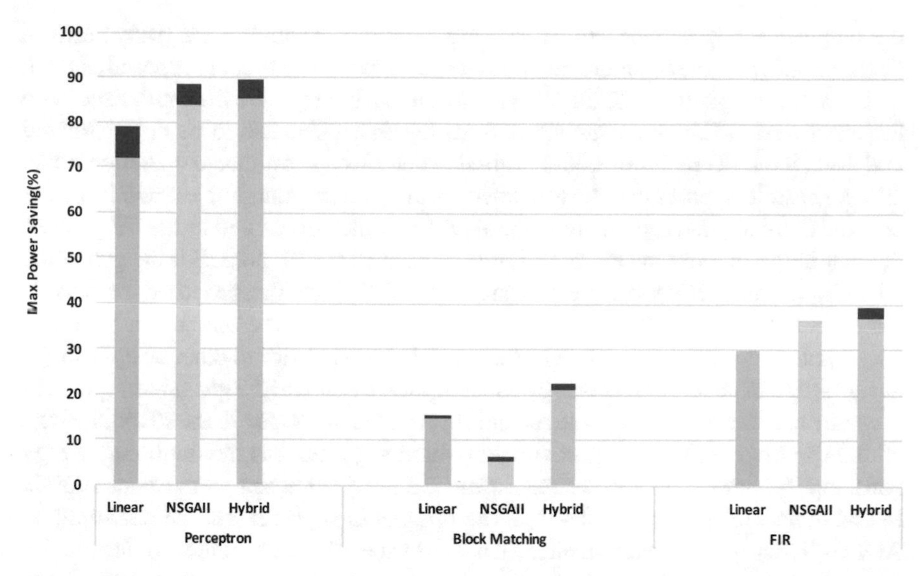

Fig. 8.13 Comparison of power saving for different designs of the modified ABACUS tool with the different parent selection methodologies and additional VOS technique applied [8]

methodologies and the additional VOS technique (the purple part) for different benchmarks are presented in Fig. 8.13. The hybrid method with the use of VOS technique achieves more reduction in power for all the designs.

3.2.3 Other Works in the Second Category

Other works in the second category of DSE algorithms for the optimization of approximate circuits are briefly discussed here. The work in [20] combines Cartesian generic programming (CGP) and NSGA-II algorithm [30] to create a library of approximate arithmetic units generated from a set of conventional functions (such as FAs and inverters) as the initial population. With the obtained library and the DFG of a given application, the optimization problem is defined based on minimizing the area, delay, and output error by replacing the exact DFG nodes with modules from the library. Based on the user preferences, the solution that optimizes the desired objective function can be chosen as the final solution. NSGA-II algorithm is utilized to solve the multi-objective optimization problem with the two mutation and crossover operators to create the offspring population from the parent approximate designs. In the mutation operation, an approximate unit is randomly assigned per DFG nodes in the ADFG, and in the crossover operation, an offspring is created from two parent points by combining them together. For the objective functions, the area and delay values are estimated from the summation of the area and delay values for all the corresponding nodes in the DFG where the output error is obtained by simulating a random sampling set (with uniform distribution) from the input domain and the variance-based model proposed in [25].

In [29], for a given DFG, the PF pool of the design space is first initialized with the exact design. The heuristic starts from the first operation to be approximated and visits the DFG nodes in a breadth-first order. For any evaluated operation node, all the possible approximations (obtained from different rounding bit widths of the operands) are applied separately to the ADFGs, which are stored in the PF pool. In the first iteration, only the exact design is present in the PF pool. For the generated ADFGs, the circuit design metrics are estimated from the proposed models in the literature. For the accuracy evaluation, for example, the method in [33] was employed. Among the new ADFGs, the ones that are inferior to other designs in the design space of accuracy and energy consumption or violate the given user-specified accuracy are discarded, while the remaining ADFGs are added to the PF pool. After all the DFG operations are approximated, the design that has the minimum energy consumption among all the designs in the PF pool is chosen as the final ADFG. In addition to the final (considered as the optimal design) design, the near-optimal ADFGs that have a distance smaller than a user-specified threshold from the quality and energy consumption of the optimal design are also kept as a near-optimal set. The optimal ADFG is scheduled to find the design with the minimum latency under the given resource constraints. Then, a final optimization step is performed to check if there is any better ADFG in the near-optimal set that has a smaller critical path delay than that of the optimal design. If that is the case, that design will be chosen as the final ADFG, and with the use of the voltage overscaling, the energy consumption will be reduced further. The voltage is scaled only to the point that there will not be any timing violation.

3.3 Comparison of the Works

Now let us compare the key features of some of the works in the area of DSE of approximate circuits in the context of AHLS. The features that are given in Table 8.2 include the category of the DSE algorithm, the input application type used in the proposed framework, the approximation methods invoked, the type of DSE heuristics, the output of the search process for the designer, the HLS steps considered in the work, and the estimation techniques utilized in determining the circuit design metrics, if there are any.

In the category column, the category of the DSE algorithms is presented. The DSE technique of [25] falls out of the two categories of DSE algorithms. The reason is that the optimization problem to find the optimal approximate circuit (minimizing leakage power while meeting a user-specified accuracy threshold) is formulated as a knapsack problem and solved with an ILP solver.

In the input application column, different works have considered different representations of the high-level input application. HW (SW) stands for hardware (software) representation of the input application. In the AC method column, the

Table 8.2 Features of the works in the AHLS domain

Works	Category	Input application	AC method	DSE algorithm	Output	HLS	Model
[18]	1	Behavioral Verilog	Multiple	Stochastic greedy	PF	b	×
[28]	2	DFG	Arithmetic unit replacement	Evolutionary approach	PF	b	×
[8]	2	Behavioral Verilog	Multiple + VOS	Stochastic greedy + NSGA-II	PF	b	×
[27]	1	HW (Verilog) + SW (C++) codes	Arithmetic unit replacement	Stochastic hill climbing	PF	b	ML
[20]	2	DFG	Arithmetic unit replacement	NSGA-II	PF	b	×
[32]	1	C/C++ code	Precision scaling	Branch and bound	PF*	b	×
[24]	1	Verilog code	Multiple	Deterministic greedy	Single	b	×
[26]	1	SystemC code	Arithmetic unit replacement	Stochastic greedy	Single	b	ML
[25]	×	DFG	Precision scaling	Knapsack problem	Single	s + b	AN
[33]	2	C code	Precision scaling + VOS	Evolutionary approach	PF	s + b	AN
[31]	1	C code	Multiple in different levels	Greedy	PF	s + b	×

approximation techniques applied to the input applications are listed. Precision scaling refers to the reduction of the bit widths of the operations or operators in a design subject to a user-defined accuracy threshold. Some works invoke multiple techniques. For example, in [31], multiple techniques (combination of approximation methods) are applied to the original circuits at different abstraction levels. ABACUS also used multiple approximation techniques to generate new approximate designs [18]. In the DSE algorithm column, the DSE heuristics are presented. Most of the works have proposed greedy methods to solve the optimization problems. Some works generate deterministic outputs for the DSE algorithm, while others have stochastic nature, where different rounds of the same algorithm result in different explorations paths on the design space. The outputs of the DSE heuristic are presented in output column. Some of them generate the PF of the design space and some extract the optimal design (single) based on the objective function and user-defined constraints. In the work of [32], the PF* refers to the configurations that have a quality as close as possible to the user-specified error threshold, and hence, its PF definition is different from the definition of other researchers.

For the HLS steps considered in each work, represented in HLS column, the scheduling (s) and binding (b) are considered. Most of the papers have not mentioned the HLS steps explicitly. The binding step refers to the assignment of approximate modules from the RTL library to the input design. In some works, the scheduling step is also mentioned. For example, in [25], a modified list scheduling algorithm is proposed, which is invoked after (and within) the precision optimization step that incorporates the approximate modules in the scheduling algorithm. Finally, to improve the runtime of the approximate design process, there are a limited number of works that propose estimation models to predict the design metrics. As an example, the application-level accuracy (*MSSIM*) of an approximate design and other circuit metrics are estimated with two separate pre-trained ML models to be incorporated in the DSE algorithms (see, e.g., [27]).

4 Future Directions

The main challenges in the DSE problem of optimizing approximate circuits are to have fast and efficient heuristics. As mentioned in [24], a major part of the DSE runtime belongs to the simulation and synthesis steps. Therefore, one important direction in the AC paradigm is to propose and/or improve analytical or computational models that accurately yet efficiently estimate the effect of approximation in a given input application without the need for using simulations tools. Also, incorporating the scheduling step in the approximation design flow is another research path in making approximate HLS more complete. Moreover, as mentioned in this chapter, the output quality of an approximate circuit is strongly input dependent. Therefore, obtaining optimal approximate designs from simulations of explored design points in the design space on a set of input vectors (i.e., static methods) cannot be generalized to all input workloads. It is possible that a design that is optimal for a set of input vectors violates the quality threshold with another set of input vectors. Therefore, runtime approximate computing techniques are required to solve the mentioned issue [35].

References

1. Kouzes RT, Anderson GA, Elbert ST, Gorton I, Gracio DK. The changing paradigm of data-intensive computing. Computer. 2009;42(1):26–34. https://doi.org/10.1109/MC.2009.26.
2. Mittal S. A survey of techniques for approximate computing. ACM Comput Surv. 2016;48(4):62:1–62:33.
3. Sidiroglou-Douskos, et al. Managing performance vs. accuracy tradeoffs with loop perforation. In: Proceedings of ACM SIGSOFT symposium and European conference on foundations of software engineering (ESEC/FSE); 2011. p. 124–34.
4. Kamal M, Ghasemazar A, Afzali-Kusha A, Pedram M. Improving efficiency of extensible processors by using approximate custom instructions. In: 2014 design, automation & test in Europe conference & exhibition (DATE); 2014. p. 1–4. https://doi.org/10.7873/DATE.2014.238.
5. Jiang H, Liu C, Liu L, Lombardi F, Han J. A review, classification, and comparative evaluation of approximate arithmetic circuits. J Emerg Technol Comput Syst. 2017;13(4):Article 60, 34 pages. https://doi.org/10.1145/3094124
6. Sengupta D, Snigdha FS, Jiang H, Sapatnekar SS. SABER: Selection of approximate bits for the design of error tolerant circuits. In: 2017 54th ACM/EDAC/IEEE design automation conference (DAC); 2017. p. 1–6. https://doi.org/10.1145/3061639.3062314.
7. Amanollahi S, Kamal M, Afzali-Kusha A, Pedram M. Circuit-level techniques for logic and memory blocks in approximate computing systems. Proc IEEE. 2020;108:2150–77. https://doi.org/10.1109/JPROC.2020.3020792.
8. Nepal K, Hashemi S, Tann H, Bahar RI, Reda S. Automated high-level generation of low-power approximate computing circuits. IEEE Trans Emerg Top Comput. 2019;7(1):18–30. https://doi.org/10.1109/TETC.2016.2598283.
9. Zendegani R, Kamal M, Fayyazi A, Afzali-Kusha A, Safari S, Pedram M. SEERAD: A high speed yet energy-efficient rounding-based approximate divider. In: 2016 design, automation & Test in Europe conference & exhibition (DATE); 2016. p. 1481–4.
10. Nakhaee F, Kamal M, Afzali-Kusha A, Pedram M, Fakhraie SM, Dorosti H. Lifetime improvement by exploiting aggressive voltage scaling during runtime of error-resilient applications. In: Proc. Integr, vol. 61; 2018. p. 29–38.
11. Kashfi F, Hatami S, Pedram M. Multi-objective optimization techniques for VLSI circuits. In: 2011 12th international symposium on quality electronic design; 2011. p. 1–8. https://doi.org/10.1109/ISQED.2011.5770720.
12. Miettinen K. Nonlinear multiobjective optimization. Netherlands: Springer US; 1999.
13. Coussy P, Gajski DD, Meredith M, Takach A. An introduction to high-level synthesis. IEEE Des Test Comput. 2009;26(4):8–17. https://doi.org/10.1109/MDT.2009.69.
14. Hara Y, Tomiyama H, Honda S, Takada H, Ishii K. CHStone: a benchmark program suite for practical c-based high-level synthesis. In: Proceedings of the 2008 IEEE international symposium on circuits and systems. IEEE; 2008. p. 1192–5.
15. Ferretti L, Ansaloni G, Pozzi L. Cluster-based heuristic for high level synthesis design space exploration. IEEE Trans Emerg Topic Comput. 2021;9(1):35–43. https://doi.org/10.1109/TETC.2018.2794068.
16. Camposano R. From behavior to structure: high-level synthesis. IEEE Des Test Comput. 1990;7(5):8–19. https://doi.org/10.1109/54.60603.

17. Muchnick SS. Advanced compiler design & implementation. Burlington: Morgan Kaufmann; 2003.
18. Nepal K, Li Y, Bahar RI, Reda S. ABACUS: A technique for automated behavioral synthesis of approximate computing circuits. In: 2014 design, automation & test in Europe conference & exhibition (DATE); 2014. p. 1–6. https://doi.org/10.7873/DATE.2014.374.
19. Hwang C-T, Lee J-H, Hsu Y-C. A formal approach to the scheduling problem in high level synthesis. IEEE Trans Computer-Aided Des Integr Circuits Syst. 1991;10(4):464–75.
20. Vaverka F, Hrbacek R, Sekanina L. Evolving component library for approximate high-level synthesis. IEEE Symp Series Comput Intell. 2016;2016:1–8. https://doi.org/10.1109/SSCI.2016.7850168.
21. Mahdiani HR, Ahmadi A, Fakhraie SM, Lucas C. Bio-inspired imprecise computational blocks for efficient VLSI implementation of soft-computing applications. IEEE Trans Circuits Syst I: Regular Papers. 2010;57(4):850–62.
22. The USC-SIPI Image Database [Online]. Available: http://sipi.usc.edu/database.
23. Leipnitz MT, Nazar GL. High-level synthesis of resource-oriented approximate designs for FPGAs. In: 2019 56th ACM/IEEE design automation conference (DAC); 2019. p. 1–6.
24. Witschen L, Mohammadi HG, Artmann M, Platzner M. Jump search: a fast technique for the synthesis of approximate circuits. In: Proceedings of the 2019 on Great Lakes symposium on VLSI (GLSVLSI '19). New York: Association for Computing Machinery; 2019. p. 153–8. https://doi.org/10.1145/3299874.3317998.
25. Li C, Wei Luo SS, Sapatnekar, Hu J. Joint precision optimization and high-level synthesis for approximate computing. In: 2015 52nd ACM/EDAC/IEEE design automation conference (DAC); 2015. p. 1–6. https://doi.org/10.1145/2744769.2744863.
26. Awais M, Mohammadi HG, Platzner M. LDAX: a learning-based fast design space exploration framework for approximate circuit synthesis. In: Proceedings of the 2021 on Great Lakes symposium on VLSI (GLSVLSI '21). New York: Association for Computing Machinery. p. 27–32. https://doi.org/10.1145/3453688.3461506.
27. Mrazek V, Hanif MA, Vasicek Z, Sekanina L, Shafique M. autoAx: an automatic design space exploration and circuit building methodology utilizing libraries of approximate components. In: 2019 56th ACM/IEEE design automation conference (DAC); 2019. p. 1–6.
28. Vaeztourshizi M, Kamal M, Pedram M. EGAN: a framework for exploring the accuracy vs. energy efficiency trade-off in hardware implementation of error resilient applications. In: 2020 21st international symposium on quality electronic design (ISQED); 2020. p. 438–43. https://doi.org/10.1109/ISQED48828.2020.9137041.
29. Lee S, John LK, Gerstlauer A. High-level synthesis of approximate hardware under joint precision and voltage scaling. In: Design, automation & test in Europe conference & exhibition (DATE), vol. 2017; 2017. p. 187–92. https://doi.org/10.23919/DATE.2017.7926980.
30. Deb K, Pratap A, Agarwal S, Meyarivan T. A fast and elitist multiobjective genetic algorithm: NSGA-II. IEEE Trans Evol Comput. 2002;6(2):182–97. https://doi.org/10.1109/4235.996017.
31. Xu S, Schafer BC. Exposing approximate computing optimizations at different levels: from behavioral to gate-level. IEEE Trans Very Large Scale Integr Syst. 2017;25(11):3077–88. https://doi.org/10.1109/TVLSI.2017.2735299.
32. Barbareschi M, Iannucci F, Mazzeo A. Automatic design space exploration of approximate algorithms for big data applications. In: 2016 30th international conference on advanced information networking and applications workshops (WAINA); 2016. p. 40–5. https://doi.org/10.1109/WAINA.2016.172.
33. Lee S, Lee D, Han K, Shriver E, John LK, Gerstlauer A. Statistical quality modeling of approximate hardware. In: 2016 17th international symposium on quality electronic design (ISQED); 2016. p. 163–8. https://doi.org/10.1109/ISQED.2016.7479194.
34. Elkan C. Using the triangle inequality to accelerate k-means. In: Fawcett T, Mishra N, editors. ICML. AAAI Press; 2003. p. 147–53.
35. Tabatabaei-Nikkhah S, Zahedi M, Kamal M, Afzali-Kusha A, Pedram M. ACHILLES: accuracy-aware high-level synthesis considering online quality management. IEEE Trans Computer-Aided Des Integr Circuits Syst. 2019;38(8):1452–65. https://doi.org/10.1109/TCAD.2018.2846625.

Chapter 9
Approximation on Data Flow Graph Execution for Energy Efficiency

Qian Xu, Md Tanvir Arafin, and Gang Qu

1 Introduction

Advances in hardware design methodology and semiconductor fabrication technology continue to enable the rapid growth of computer systems with high computation speed and low power consumption, despite the fact that we are getting to the end of Moore's Law that predicts the shrinking of transistor size and the increasing of the number of transistors per chip area. Meanwhile, with the explosion of data and the prevalence of the Internet, there is still a high demand to produce low-power, high-speed end devices, especially for the Internet of Things (IoT). This desire inspires plenty of research focusing on exploiting approximation to improve the system's overall performance by balancing the amount of computation, which determines power and energy, and the quality of the computation, which can be evaluated by metrics such as result accuracy [16].

It should be noted that approximation is not suitable for all applications or hardware systems. For example, when a system requires high precision and energy is not the primary concern, accuracy should not be traded for power and energy. Therefore, most of the reported approximate computing approaches are mainly designed for applications such as multimedia processing and machine learning. The former relates to the human cognition system, which accepts (and, to some extent, cannot detect) minor errors. The latter are examples of error-resilient systems where small computational errors will not impact the outcome. In short, before applying

Q. Xu · G. Qu (✉)
University of Maryland, College Park, MD, USA
e-mail: qxu1234@umd.edu; gangqu@umd.edu

Md. T. Arafin
Morgan State University, Baltimore, MD, USA
e-mail: mdtanvir.arafin@morgan.edu

W. Liu, F. Lombardi (eds.), *Approximate Computing*,
https://doi.org/10.1007/978-3-030-98347-5_9

approximation methods to a system, one needs to consider whether the specific system can tolerate the errors introduced by approximation.

As a basic building block of any computer system, hardware components that perform arithmetic operations are the foundation of computation. Their high occurrence in a system makes them one of the most popular research topics in approximate computing. Various types of approximate adders [11], approximate multipliers [14, 15], and approximate dividers [6] have been proposed and demonstrated to be effective in reducing power and energy consumption. Different data formats, such as integers, fixed-point numbers, floating-point numbers, and new data formats [4] have also been explored for approximate computation.

Data flow graph (DFG) is a popular model for computation at various abstract levels, from the system level and application level to individual operations such as addition, subtraction, and multiplication. It has been extensively used in the low power implementation of software [9, 10, 13]. In a DFG, the nodes represent computations, the edges represent the flow of data or data dependencies between the computations, and the inputs to the DFG are variables or constants. The aforementioned approximation approaches have considered the input data (approximation at data level) and the nodes (approximation at basic arithmetic units) of a DFG for optimization. Such local optimization methods fail to utilize the synergy among the data and computation in a DFG. Many applications and programs require many iterations of the same DFG. Therefore, we believe they will not achieve approximate computing's full potential in reducing power and energy. In this chapter, we demonstrate this through several projects.

2 Chapter Overview

This chapter aims to provide a comprehensive view of the approximate computing on DFG execution for energy efficiency. This will be quite different from the existing approximation approaches on single computation units or input data. We will consider the entire DFG or the underlying application/program represented by the DFG.

In Sect. 3, we first describe the main structure and components of DFGs. We then introduce the three main research questions for any DFG approximation scheme. We summarize the key ideas of the three DFG approximation algorithms that will be elaborated in the remainder of this chapter.

In Sect. 4, we consider the iterative execution nature of the DFGs, such as those used in multimedia applications and present a probabilistic design method to implement DFGs [9, 10]. Unlike the traditional approach that attempts to complete each iteration of the DFG execution, we intentionally terminate certain iterations before they miss the execution deadline to save energy. Such early termination may not guarantee the maximal number of completed iterations of the DFG. Still, it can provide us significant energy savings if the program does not require the highest completion ratio. In short, this method trades the program's completion ratio for energy efficiency.

In Sect. 5, we discuss another approximation technique on the single execution of a DFG. We identify the critical and non-critical branches in the DFG and perform accurate and approximate computing on them, respectively, to reduce energy consumption without large compromise to the precision of the computation. This approach is referred to as *estimate and recompute* [3].

In Sect. 6, we consider how approximation inside the operations can affect the overall performance in DFGs and discuss the significance of accommodating specific processes in the whole program. Using the logarithmic function as an example, we show how to decide the approximation level for each operation so that the best performance can be achieved through the cooperation of multiple processes [28].

Section 7 summarizes this chapter of discussion on approximation on DFG execution for energy efficiency.

3 Approximation on DFG Execution: Challenges and Solutions

3.1 Data Flow Graph Basics

A Data Flow Graph (DFG) is a directed acyclic graph where each node represents a computation (such as addition and multiplication), and an edge from node u to node v indicates that the computation result of node u will be an input for the computation at node v. Edges without a source node are input to the DFG and edges without destination node are outputs of the DFG. For example, consider a basic algebraic operation, the dot product, of two 1-D two-element vectors $a = [a_0, a_1]$ and $b = [b_0, b_1]$. Let c be the result of the dot product, then c can be calculated following Eq. 9.1:

$$c = a_0 * b_0 + a_1 * b_1 \tag{9.1}$$

In Eq. 9.1, there are two multiplication operators and one addition operator. Suppose we denote the first product by t_0 and the second product by t_1. It is easy to obtain the data dependencies that the addition at node c depends on the two intermediate products t_0 and t_1, while the two multiplications at nodes t_0 and t_1 depend on the availability of their operands, (a_0,b_0) and (a_1, b_1). Figure 9.1 is the DFG constructed based on Eq. 9.1, the dot product.

For any sequence of operations without conditionals, a DFG can be generated. Addition and multiplication, division, and even some complex operators like square root or logarithm can be represented as nodes. The numbers in all formats, including integer, fixed-point, floating-point, and even user-defined format, can be represented

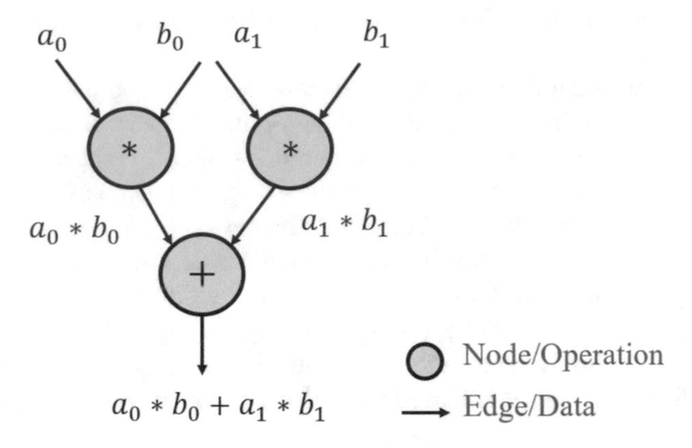

Fig. 9.1 DFG example of dot product on two 1-D two-element vectors

as edges. As discussed earlier, approximate computing on DFGs has been done on both the nodes (in the form of an approximate hardware implementation of the operations) and edges (on data approximation with different formats). But there is little work on how to consider the DFG systematically for approximation. In this chapter, we discuss how to fill this gap with approximate computing techniques focusing on the execution of the DFG instead of its nodes or edges.

3.2 *Main Questions for Approximation on DFGs and Their Execution*

Among the many considerations, answers to the following three questions are crucial for developing any approximate computing method for an application modeled as a DFG and its execution:

- **What is approximable:** whether the entire application is error resilient or which portions can tolerate errors? The answer to this question determines the targets for approximation.
- **How to approximate the designs:** which specific approximation techniques are applicable for the given application? The answer to this question focuses on the details of approximation.
- **How to ensure the output quality:** to which extent can we approximate the application and its execution? The answer to this question will be application-specific, and an optimal solution will balance the trade-off between energy efficiency and output quality.

Discovering the approximable portions in a DFG is not easy. First of all, this is application specific, and a well understanding of the DFG-represented application and its execution characteristics is necessary. At a high level, when we consider the entire DFG, applications like multimedia signal processing are approximable. Take video decoding, for example, failing to decode one frame will most likely be acceptable. At a low level, when we consider the nodes and edges in the DFG, there are many opportunities for approximation. One can use approximate adders and multipliers as the basic building blocks of DFGs to approximate the corresponding operations. One can also choose to approximate the input data to the DFG or certain nodes of the DFG to reduce the amount of computation and achieve energy efficiency.

After the approximation targets in the DFG are determined, there are still many different approximation techniques to choose from. For example, more than a dozen approximate adders and approximate multipliers have been proposed in the past. As a result, the options for approximation on a DFG will be exponential to the number of operations units such as adder and multiplier that the DFG has. As another example, the impact of each node on the output's precision may not be the same. Hence, one can approximate the computation on those non-critical nodes or approximate the input data to those nodes. However, whether a node is critical to the output might be input dependent and varies from one execution to another, making this a challenging problem.

Any form of approximate computing will result in degradation in the quality of the computation result. That is why we have stated earlier that approximate computing techniques can only be applied to systems that can tolerate errors. Suppose the error caused by approximation goes beyond the level that the system can tolerate. In that case, such an approximation method cannot be used because it fails to deliver the required quality of service. Therefore, when considering approximate computing for energy-efficient design and implementation, it will be vital to guarantee that a user-specified quality of execution (or the level of errors during execution) is maintained. For example, when running a machine learning model for a classification task, the model or the execution could be approximated as long as the correct class label is assigned.

Designing energy-efficient systems with an approximation is a complicated process. The three questions mentioned above are among the main challenges. Failing to address them adequately would most likely end up in a failed design. As one can see, solutions to these questions are highly related to the applications that the system will perform. Hence there does not exist a one-fit-all solution. In the following, we highlight the key ideas of three sample approximation techniques for DFGs. The technical details will be elaborated in the subsequent sections.

3.3 Approximation Techniques

In this subsection, we introduce three approximation techniques for DFGs, emphasizing their main ideas and how they answer the three previously discussed questions.

The first technique is proposed for systems such as multimedia data processing [9, 10]. It is based on the assumption that the same DFG will be executed periodically (e.g., the frame decoding task when playing a video), and successful completion for every iteration is not required. In other words, it is acceptable for the DFG execution to produce inaccurate (or even incorrect) outputs for some iterations. Thus, the authors proposed taking advantage of this error tolerance and the uncertainties in the execution time by terminating certain iterations early to save energy. These iterations are selected based on the real-time execution time information to guarantee that the overall completion ratio will meet the user's statistical performance requirement. In terms of the three questions, such probabilistic design (1) approximates the entire DFG on selective iterations, (2) uses early termination of the execution once a long execution time is predicted, and (3) provides probabilistic guarantees on the number of completed iterations to ensure performance.

The second technique considers a general DFG and is built on the observation that different inputs to the same node in a DFG may have significantly different impacts on the accuracy of the node's output [3]. That is to say, if one can identify this impact for each piece of data in a low-cost manner, the value of the "unimportant" data does not need to be as accurate as those "important" data; thus, the operations for generating these "unimportant" data can be approximated. The authors proposed a couple of low-cost runtime methods based on converting data to the logarithmic domain. The goal of the computation in the log domain is to distinguish, during the execution of a DFG, the critical subgraphs that produce "important" data and the non-critical subgraphs. In this estimate-and-recompute method, (1) the non-critical subgraphs in the DFG become approximable; (2) converting the expensive multiplication operation to the low-cost addition operation in the logarithmic domain is the proposed approximation method; and (3) threshold values are set to provide the required computation accuracy.

The last technique is designed for logarithms and other complicated operations (square root is another example). These operations are widely used and are in general implemented through iterations of the basic operations such as addition and multiplication [28]. Comparing to the basic operation nodes in DFG, such complicated operations consume much more energy, but their impact on the output accuracy may be the same as basic operations. For example, in the simple operation $a+b$, a and b will impact the output's accuracy even if a is the output of an adder and b comes from a logarithm operation. Determining the energy and accuracy trade-off of these operations will provide a guideline for approximation. We perform a thorough analysis of these complex operators' error and energy models and deduce their impact on the DFG outputs. More specifically, in this approach, (1) the computation inside the implementation of complex operators is approximable; (2) the number of loops or iterations in the complex operators can be reduced to save power and energy; and (3) the thorough error analysis guarantees that results from the approximated complex operators give the required accuracy while achieving a maximal reduction of power and energy.

4 Probabilistic Design for Multimedia Systems

For the current multimedia systems, most present techniques are based on worst- or average-case scenarios [2, 7, 17]. The worst-case performance requirement usually leads to overdesigning because each single data point should be considered and taken good care of to ensure the successful completion under all the cases. However, the average-case performance requirement usually overly relaxes the hardware constraint and cannot guarantee the completion of half the cases. Given the disadvantages of these two requirement settings, designers start to consider providing a statistical completion ratio guarantee for a multimedia system [8, 12, 25]. Based on this new requirement, a probabilistic design is proposed to relax the hardware constraints and avoid the overdesigning [9, 10]. The technique utilizes the minimum effort to quickly check whether the current task can be completed on time and decide whether the remaining steps are worth running early.

This section is organized as follows. We will use a toy example to explain why probabilistic designs are needed or what benefits can be achieved via such designs. Next, we will talk about the architecture for the probabilistic design and discuss the whole workflow and every part of it. Finally, we will show the design details for the two main processes inside the workflow, probabilistic timing performance profiling, and estimation, and offline and online resource management.

4.1 An Illustrative Example

Consider an application that should be executed repetitively under different data points but could tolerate a certain degree of execution failures. A probabilistic design is proposed for such an application. For example, a target application that needs to be approximated consists of three sequential nodes, A, B, and C, as shown in Fig. 9.2. It should be noted that Fig. 9.2 is not a complete representation of a DFG because no specific operations are provided. Still, it is sufficient to calculate the overall execution time since the computation sequence is provided. Suppose each task has two possible execution times: one for the best-case execution time (BCET) and the other for the worst-case execution time (WCET). It is easy to know that there are eight possible execution times for the execution of all three tasks when the sum for any combination differs from each other. All the eight scenarios are listed in Fig. 9.2. Besides, the real execution time (RET) and the probability of occurring this scenario are provided for each scenario.

Suppose that we set the execution time deadline to be 200 cycles and repetitively execute the application described above. In that case, all the iterations will be finished on time since the longest execution time for this design is 200 cycles. However, if only 70% of iterations are required to be executed on time, there is no need to set the deadline 200 cycles, and 100 cycles suffice. That means termination can be decided at 100 cycles when the execution fails to finish. Besides, termination

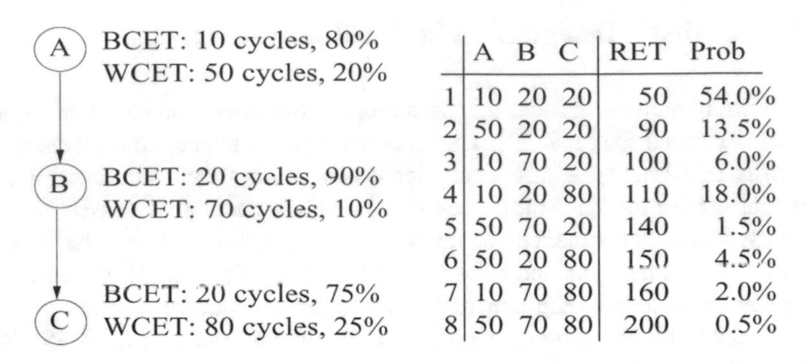

	A	B	C	RET	Prob
1	10	20	20	50	54.0%
2	50	20	20	90	13.5%
3	10	70	20	100	6.0%
4	10	20	80	110	18.0%
5	50	70	20	140	1.5%
6	50	20	80	150	4.5%
7	10	70	80	160	2.0%
8	50	70	80	200	0.5%

Fig. 9.2 An application with three sequential tasks and all the execution time scenarios (source of the figure: [10])

decision can be made even earlier. For example, if the total execution time for node A and node B is more than 80 cycles, it is not possible to finish all the three tasks on time. Therefore, if A and B cannot finish in 80 cycles, we could stop the current iteration early.

Based on the analysis of the toy example, we could obtain two basic conclusions. The first conclusion is that the execution time can be shortened if we do not need 100% completion ratio for the current application. The second conclusion is that the application can be stopped at early stages to avoid unnecessarily execution of the following nodes when time is insufficient.

4.2 *Probabilistic Design Overview*

The probabilistic design methodology is illustrated in Fig. 9.3. Similar to the toy example, designers should start with the application, performance requirements, and a list of target system architectures. The application can be represented by a DFG, which consists of a set of nodes and the edges between them where nodes and edges represent operations and data dependencies, respectively. For the performance requirements, two main metrics should be paid attention to: one is the statistical completion ratio and the other is timing constraints. With the profiling tools, we discover all the scenarios with different execution times and calculate the probabilities of their occurrence. After estimating the probabilistic timing performance, we can determine whether the current setting can meet the completion ratio constraint or not. If the answer is no, we need to change either the software optimization or the hardware configuration and repeat the profiling analysis with the new setting. However, if the answer is yes, we could step forward to the resource management process. Two specific techniques are utilized in this process. The first technique is static, allocating minimum resources to each node offline, and the second technique is dynamic, allocating resources at runtime by developing

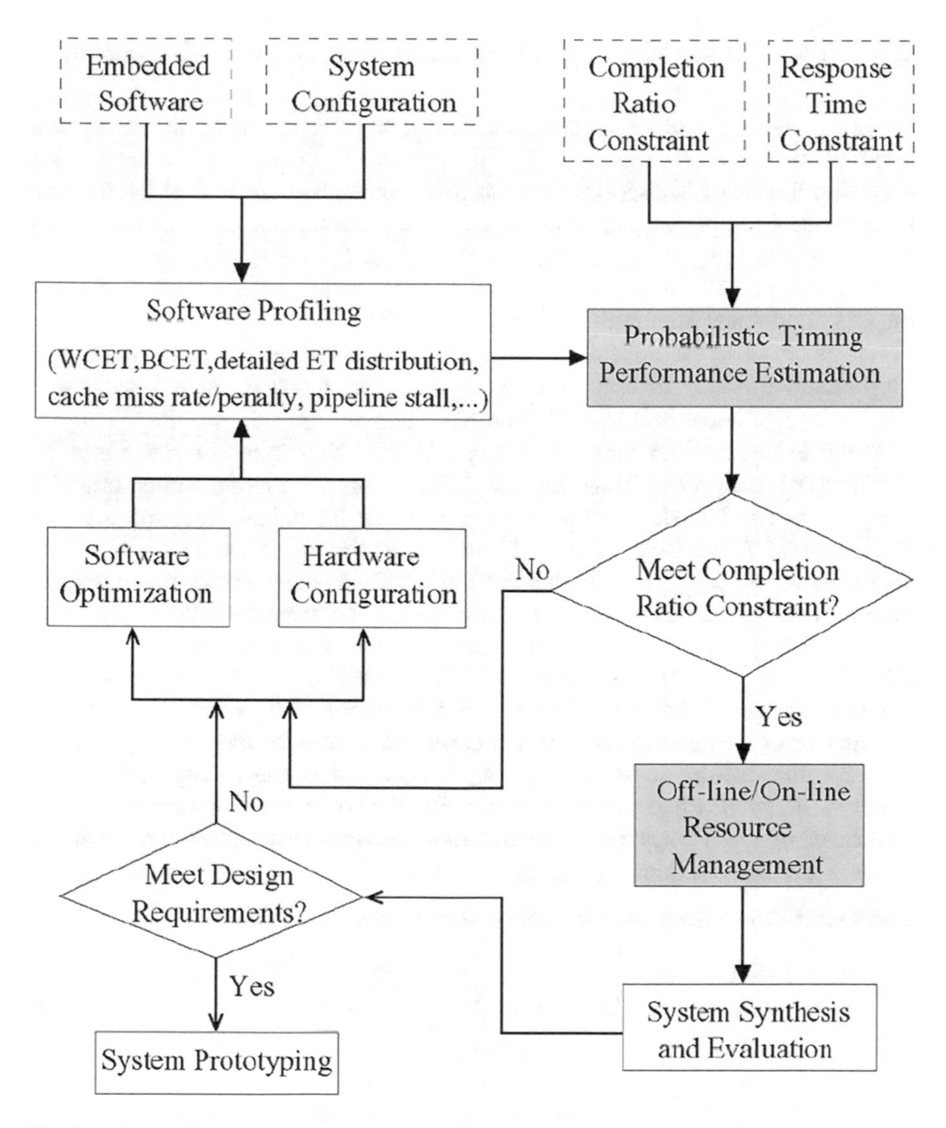

Fig. 9.3 Design flow in the probabilistic design methodology (source of the figure: [10])

real-time schedulers. Finally, system synthesis and evaluation will be done on the whole application to ensure all the performance requirements will be met under the current hardware, software, and scheduler settings. After the evaluation process, the application can be put into use. To summarize, probabilistic design mainly consists of probabilistic timing performance profiling and estimation and offline and online resource management, which will be discussed in detail in the following subsections.

4.3 Probabilistic Timing Performance Profiling and Estimation

There are several papers on the probabilistic timing performance estimation for soft real-time system designs [8, 12, 25]. The general assumption is that each task's execution time can be described by a discrete probability density function that can be obtained by applying path analysis and system utilization analysis techniques [18]. Following the previous works, we consider the following scenario.

Consider a data flow graph $G = (V, E)$ given an application, where V denotes the set of nodes and E denotes the set of edges. The execution time for each node is described by a discrete probability density function as explained in the toy example. To be more specific, for each node v_i, it can be executed for k_i different times $t_{i,1}, t_{i,2}, ..., t_{i,k_i}$ under corresponding probabilities $p_{i,1}, p_{i,2}, ..., p_{i,k_i} | \sum_{l=1}^{k_i} p_{i,l} = 1$. That is to say, the probability of node v_i taking $t_{i,j}$ time to be completed is $p_{i,j}$.

The total completion time for the DFG G under the execution order $< v_1 v_2 ... v_n >$ can be calculated by summing up the individual execution time for each node e_i. If we denote the total completion time by $C(< v_1 v_2 ... v_n >)$, it should be equal to $\sum_{i=1}^{n} e_i$. M denotes the deadline constraint, or the maximum time provided to complete the application. G will be executed repetitively under different data points. If the execution time is longer than the deadline constraint $C(< v_1 v_2 ... v_n >) > M$, we consider the current iteration a failure. Otherwise, the iteration has been completed. Given $N >> 1$ iterations, suppose that K can denote the number of completed iterations, the completion ratio for the given application with deadline constraint M is $Q = K/N$. Other than the timing performance requirement, another performance requirement, the completion ratio constraint, can be denoted by Q_0. We say the completion ratio constraint satisfies over N iterations under M timing constraint if $Q \geq Q_0$.

Theorem 1 *The maximum achievable completion ratio is given by:*

$$Q^{max} = \sum_{\sum_{i=1}^{n} t'_{i,j_i} \leq M} \prod_{i=1}^{n} p_{i,j_i} \tag{9.2}$$

where the sum is taken over the execution time combinations that meet the deadline constraint M and the product computes the probability that each combination occurs.

Based on the theorem above, $Q^{max} < Q_0$ implies the fact that the completion ratio constraint requirement can never be achieved under the current hardware and software settings. Thus designers should consider modifying the system settings and calculate the maximum achievable completion ratio again.

However, estimating the maximum completion ratio as shown in Eq. 9.2 is computationally expensive since there are multiple nodes in a DFG and there are multiple execution times for a specific node. Even if there are only two possible execution times, BCET and WCET, the number of execution time combinations is 2^n, which increases exponentially with the number of nodes. Therefore, the following polynomial heuristic has been proposed to achieve a fast estimate of the completion ratio.

The execution times for a specific node can be sorted as $t_{i,1} < t_{i,2} < ... < t_{i,k_i}$ and the prefix sum or the occurrence probability can be calculated by:

$$P_{i,l_i} = \sum_{j=1}^{l_i} p_{i,j} \tag{9.3}$$

which represents the probability when the execution time at node v_i does not exceed t_{i,l_i}. For each node v_i, a specific time such as t_{i,l_i} can be allocated. Based on this setting, the completion ratio can be given by:

$$Q = \prod_{i=1}^{n} P_{i,l_i} = \prod_{i=1}^{n}\sum_{j=1}^{l_i} p_{i,j} \tag{9.4}$$

A greedy algorithm is utilized for fast estimation of the completion ratio. First, the allocated execution time for each node could be its WCET, which makes sure $Q = 1$ and any completion ratio constraint can be achieved since $Q_0 \leq Q = 1$. If the allocated time for node v_i is reduced from t_{i,l_i} to $t_{i,(l_i-1)}$, based on Eq. 9.4, the completion ratio will be reduced by $\frac{P_{i,(l_i-1)}}{P_{i,l_i}}$. Besides, the total allocated time will be reduced by $t_{i,l_i} - t_{i,(l_i-1)}$. Therefore, we iteratively cut the time slot for node v_j, which yields the largest $(t_{j,l_j} - t_{j,(l_j-1)})\frac{P_{j,(l_j-1)}}{P_{j,l_j}}$ as long as the completion ratio is greater than Q_0. By cutting the time via this greedy-selection algorithm, the required completion ratio constraint can be achieved. If the total allocated time can reach the deadline constraint M, we say that the polynomial heuristic can conclude that the required Q_0 is achievable under the current hardware and software settings.

4.4 *Offline and Online Resource Management*

Determining that the current hardware and software setting can lead to a satisfying approximate design for the application under M deadline constraint and the Q_0 completion ratio constraint, and the next step is to carefully consider managing the resources to save the most cost. We will first discuss a naive approach and then propose another improved scheduling algorithm.

Let us first consider *a naive best-effort approach*. It is very straightforward to execute all the nodes at the highest voltage and the highest speed until the deadline M is reached. This approach can reach the maximum achievable completion ratio, but it does not help save more energy costs; thus, we call it the naive best-effort approach. For this approach, we need to compute the energy consumption for comparison with other approaches. Among N iterations of execution for the whole DFG, K of them can be completed before the deadline. Suppose the completion time can be denoted by $C_i (1 \leq i \leq K)$. If the power dissipation at the reference voltage is P_{ref}, the energy consumption over N iterations can be calculated by:

$$E = P_{\text{ref}} \left(\sum_{i=1}^{K} C_i + (N - K)M \right) \tag{9.5}$$

Another approach, *online best-effort energy minimization algorithm*, is proposed for systems whose supply voltage can be switched at runtime. The system can switch the voltage at runtime to save more energy consumption while satisfying the same completion ratio. The energy can be saved on two occasions. On the one hand, if the completion occurs earlier than expected, we could allocate more time for each node by running the application at the lower voltage. On the other hand, if the application cannot be finished as expected, we could terminate the application earlier to avoid much power waste.

Before the algorithm details are introduced, let us start with the notations. For each node v_i in the execution sequence $< v_1 v_2 ... v_n >$, we define its latest completion time T_{l_i} and earliest completion time T_{e_i}. For the last node v_n in the application, both the latest and earliest completion times are set to be M. For other nodes before v_n, their T_{l_i} and T_{e_i} depend on the shortest execution time and the longest execution time for the following nodes, respectively, which can be represented by the equations below:

$$T_{l_n} = T_{e_n} = M \tag{9.6}$$

$$T_{l_i} = T_{l_{i+1}} - t_{i+1,1} \tag{9.7}$$

$$T_{e_i} = T_{e_{i+1}} - t_{i+1,k_{i+1}} \tag{9.8}$$

We will discuss the online best-effort energy minimization algorithm with the notations mentioned above. The current node under execution is v_i and its actual execution time at the reference voltage in jth iteration is $e_{i,j,ref}$. If the current node cannot be finished on time, which can be represented by $t + e_{i,j,ref} > T_{l_i}$, the current iteration can be terminated early to save the energy. If the current node can be finished earlier than expected, which can be represented by $t + e_{i,j,ref} < T_{e_i}$, the voltage can be scaled so that the execution of the current node can be finished at T_{e_i}, which still leaves sufficient time for the following nodes.

5 Estimate and Recompute During DFG Execution

One method to fulfill the approximate computing for data flow graphs is to identify the non-critical computations by analyzing their impacts on the output accuracy. If the outputs are not sensitive to some nodes, these nodes do not need to be calculated accurately. Based on this idea, several previous works [20–22] try to target the non-critical parts by using the training data or by data range tuning and interval arithmetic. However, all these methods are working offline regardless of every input, as we call "static." Therefore, a runtime approximation paradigm is proposed in [3]. The proposed framework quickly estimates the impact of each input to output accuracy by converting the floating-point numbers to logarithmic numbers. With the fast estimate results, two algorithms are proposed to decide whether specific nodes need to be computed accurately or approximately.

5.1 Conversion from Floating Point to Logarithmic

Two different types of floating-point formats are widely used in multiple applications: single-precision floating point and double-precision floating point. Double precision is more likely to be applied to systems with high requirements on the precision, which are not suitable for approximate computing. Therefore, we mainly focus on the single-precision floating-point format. Based on the IEEE 754 standard, single-precision numbers are composed of 32 bits, a sign bit, 8 exponent bits, and 23 mantissa bits. The value can be computed by:

$$\text{num} = \text{sign} * (1 + \text{mantissa}) * 2^{(\text{exponent}-127)} \tag{9.9}$$

When ignoring the sign bit, Eq. 9.9 can be rewritten as $x = 1.m * 2^e$, where the dot represents the radix point. Let x_l be the logarithmic representation value of x and the value for x_l can be computed by:

$$x_l = \log_2(x) = \log_2(1.m * 2^e) = \log_2(1.m) + e \approx m + e = e.m \tag{9.10}$$

Based on Eq. 9.10, we can obtain the logarithmic value x_l by directly exploiting the bits in single-precision floating-point data format and truncating the least significant bits in mantissa. At the same time, if we would like to recover the floating-point format from the logarithmic representation, we could pad "0"s to the end. The conversion process between logarithmic representation and the floating-point data format has been illustrated in Fig. 9.4.

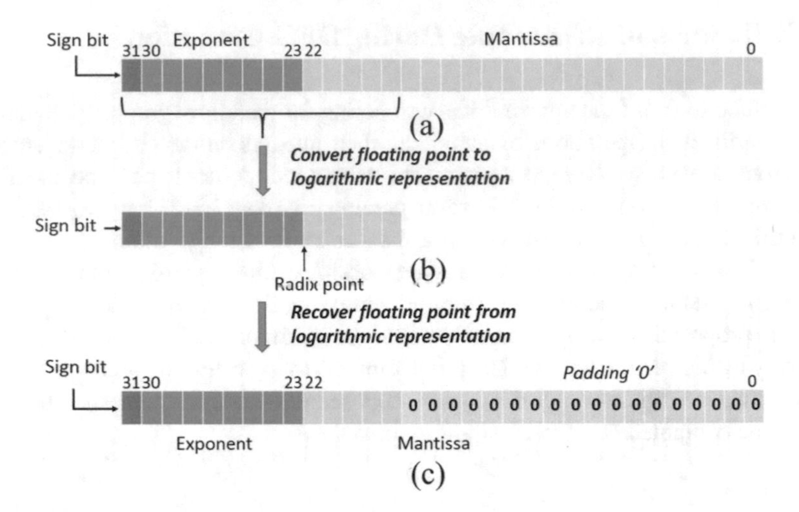

Fig. 9.4 Conversion of floating-point formation between linear domain and log domain (source of the figure: [3])

5.2 *Arithmetic Operations in Logarithmic Representations*

When numbers are represented in the logarithmic domain, the arithmetic operations should change accordingly. Therefore, we will discuss how the converted data can contribute the arithmetic operations. Intuitively, multiplication and division become easier to compute in the log domain, while addition and subtraction become more complicated. Based on this intuition, the operations can be divided into accurate and approximate conversion.

Accurate conversion refers to the operations that can be processed directly without providing any error compensation. The common operations of this type include multiplication, division, square root, and power.

- For multiplication where $S = A * B$, the logarithmic value of the product can be calculated by $S_l = \log_2 S = \log_2(A * B) = \log_2 A + \log_2 B = A_l + B + l$. That is, the product in log domain can be obtained by adding the logarithmic operands.
- For division where $S = A/B$, it is easy to get $S_l = A_l - B_l$. That is, the quotient in log domain can be obtained by subtraction of the logarithmic operands.
- For square root where $S = \sqrt{A}$, $S_l = \log_2 S = \log_2 \sqrt{A} = \frac{1}{2} \log_2 A = \frac{1}{2} A_l = A_l >> 1$. That is, square root can be transformed into shifting.
- For power where $S = A^n$, $S_l = \log_2 A^n = n \log_2 A = n A_l$. That is, power can be transformed into one multiplication with integers.

The common characteristic for the operations mentioned above is that their counterpart computation in log domain is much cheaper than their original version in floating-point domain.

Approximate conversion refers to the much more complicated operations in the log domain, such as addition and subtraction. Since the goal of the log domain is to identify the critical subgraph and the non-critical subgraph via fast estimation, there is no need to generate accurate results as long as the goal task can be finished as expected. Therefore, we introduce a novel addition and subtraction estimation with an error compensation technique. Consider addition as an example and suppose both the operands A and B are positive ($A > B$). The sum can be calculated by:

$$\begin{aligned} S &= A + B = 2^{A_l} + 2^{B_l} = 2^{A_l}(1 + 2^{B_l - A_l}) \\ &\approx 2^{A_l}(1 + 2^{\text{round}(B_l - A_l)}) = 2^{A_l}(1 + 2^{-ed}) \end{aligned} \tag{9.11}$$

where ed denotes the exponential difference. Therefore, the logarithmic value can be calculated by:

$$\begin{aligned} S_l &= \log_2 S = \log_2 2^{A_l}(1 + 2^{-ed}) = \log_2 2^{A_l} + \log_2(1 + 2^{-ed}) \\ &\approx A_l + 2^{-ed} \end{aligned} \tag{9.12}$$

It should be noted that A_l has only 5 bits in the fraction part. Therefore, if $ed > 5$, $S_l = A_l$, and if $ed \leq 5$, $S_l = A_l + 1 >> ed$. Based on this technique, the results of addition and subtraction under log domain can be estimated quickly.

5.3 *Noncriticality Truncation*

Error Resilient and Sensitive Operations For a DFG, we classify the operations into two types: (1) the error-sensitive operations and (2) the error-resilient operations. Since each node in the DFG represents one operation, the nodes with error-sensitive operations are defined as error-sensitive nodes, denoted by n^s. Error-sensitive operations include multiplication, division, and exponentiation because small changes on the operands can cause a significant difference in the output. Besides, the nodes with error-resilient operations are defined as error-resilient nodes, denoted by n^r. Error-resilient operations include addition, subtraction, and comparison because the dominant operand with a larger absolute value has a greater impact on the output. The dominant input and the minor input are denoted by I_d and I_m, respectively. More difference between the two operands means more resiliency for the operand with a smaller absolute value. This error-resilient feature is utilized to identify the non-critical inputs as well as their branches.

Noncriticality Definition and Classification The non-critical input only occurs for error-resilient nodes. Since all the inputs greatly impact the output precision, they are all critical inputs for error-sensitive nodes. For error-resilient nodes, an input I_m is a noncritical input in log domain if and only if $f(I_d - I_m) \geq \delta$, where

$$f(x) = \begin{cases} x & \text{for } \text{add}_l/\text{sub}_l \\ \text{abs}(x) & \text{for comparative operations} \end{cases}$$

δ represents the threshold designers use to control the computational quality. For example, $\delta = 1$ means that the dominant input is twice as large as the minor input. The larger δ means more difference between the two operands and less criticality for the I_m. After the threshold for the DFG is determined, the corresponding non-critical operands can be identified as well.

Truncation and Recomputation After identifying the non-critical input given an error-resilient node n_i^r in the log domain, the steps to produce the approximate results are listed below. The key idea is to replace the non-critical operand with the estimated value in the log domain and recompute critical operands accurately in the linear domain.

- Cut off n_i^r's noncritical parent branch.
- Replace the noncritical operand with the estimated value I_m in log domain.
- Convert I_m in log domain to the value in linear domain.
- Recompute n_i^r's critical parent branch in linear domain accurately.
- Compute node n_i^r accurately.

Figure 9.5 presents a motivation example of this truncation and recomputation step.

Error Analysis Next, we deduce a theoretical error analysis of the truncation and recomputation steps. Consider addition as an example and denote the accurate input values for n_i^r by v_m and v_d. Since l_m and l_d are their logarithmic representations, we can get $v_m = 2^{I_m}$ and $v_d = 2^{I_d}$. Suppose the estimated error for I_m and I_d are ϵ_m and ϵ_d. The recovered value for the minor input is $2^{I_m+\epsilon_m}$. Therefore, the error rate of node n_i^r is given by:

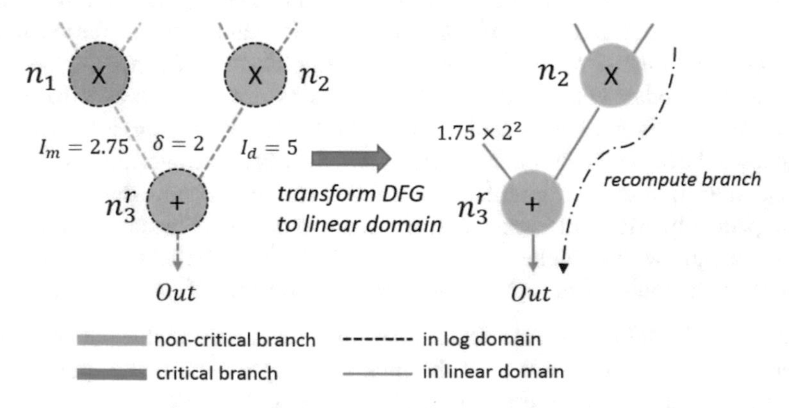

Fig. 9.5 Example of truncation and recomputation (source of the figure: [3])

$$
\begin{aligned}
er &= \frac{O_{\text{apx}} - O_{\text{acc}}}{O_{\text{acc}}} = \frac{(2^{I_m+\epsilon_m} + v_d) - (v_m + v_d)}{v_m + v_d} \\
&= \frac{2^{I_m+\epsilon_m} - v_m}{v_m + v_d} = \frac{2^{I_m+\epsilon_m} - 2^{I_m}}{2^{I_m} + 2^{I_d}} \\
&\leq \frac{2^{\epsilon_m} - 1}{2^{\delta - \epsilon_d + \epsilon_m} + 1}
\end{aligned}
\tag{9.13}
$$

where O_{apx} and O_{acc} are the approximate and accurate outputs. Assume $\epsilon_m - \epsilon_d \approx 0$, then

$$
er \leq \frac{2^{\epsilon_m} - 1}{2^{\delta} + 1} \tag{9.14}
$$

With a proper threshold value, the error rate can be controlled within a small range.

5.4 *Runtime DFG Approximation Algorithms*

Although we could determine the criticality of the parent branches of a given node n_i^r, we cannot remove all the non-critical nodes directly because the node n_k might be in the non-critical branch of n_i^r and the critical branch of n_j^r at the same time. Simply removing the node n_k will not cause a significant difference in the output of n_i^r but is likely to cause unacceptable errors in the output of n_j^r. Therefore, considering the scenario described above, we proposed two graph truncation algorithms.

The *GlobalCut* algorithm transforms the problem of removing the non-critical node to minimally reserving the critical nodes on the path from the primary inputs to the primary outputs. The algorithm starts from the output nodes and finds all the critical nodes in the path back to the input nodes. To traverse the graph, we could initialize a queue with ith output node O_i. For the first node in the queue, do the following steps recursively until the queue is empty: (1) figure out all the critical parent nodes and (2) mark them as executable nodes. After finishing the algorithm, the minimally reserved critical nodes are discovered. Approximate output can be calculated by simply executing these critical nodes in the linear domain.

Unlike the GlobalCut, which directly ignores the non-critical nodes, *LocalCut* algorithm considers them because the errors can go up when being propagated to the outputs of the graph. Therefore, LocalCut Algorithm starts from the input nodes and goes forward. First of all, estimate the graph in the log domain until meeting the next error-resilient node n_i^r. Then, do the truncation and recomputing steps described in the previous section. The output for node n_i^r can be obtained from its recomputed results. The key idea behind this is to recompute the critical nodes accurately in the linear domain while replacing the remaining non-critical nodes with their estimated values in the log domain.

6 Approximate Logarithms by Bit-Width Optimization

Various truncation methods have been proposed for basic operations such as addition [19] and multiplication [27], but its application to logarithm is not well explored. The logarithm is more highly energy hungry than these basic operations because it requires many iterations of the basic operations to achieve high accuracy. For example, a common way to implement logarithm is to use iterations, which we will elaborate later. Besides, the logarithm is a fundamental operation in statistical models and machine learning tasks such as variational inference, Bayesian methods, and neural networks on many resource-constrained systems. Therefore, implementing energy-efficient logarithm calculation is highly desirable. The recent explosion of data size, computation cost, and the approximative nature of these tasks provides an ideal environment for approximate computing. We will explore the feasibility of using approximate computing for energy-efficient logarithm operation and demonstrate the energy savings it achieves with little accuracy loss [28].

The critical challenge for approximate computing in trading off performance degradation for energy saving is estimating errors in the logarithmic operation and how it propagates and amplifies in the final results of the entire program or application. Truncation, an approximation approach whose introduced error can be expressed as a function of the number of approximate bits, provides a sound basis for error analysis and hence the quality control of the results. For instance, the bit-width optimization problem has been formulated to analyze the error behavior of truncation and to determine the optimal number of approximate bits for each operand in a program without causing significant accuracy degradation to the result. This problem has been well studied for basic operations such as addition and multiplication. Therefore, we will investigate how to achieve energy efficiency and quality control of a program with logarithm operations through bit-width optimization.

6.1 *Error Analysis for Logarithms*

We consider IEEE 754, the standard format of floating-point arithmetic, as the way of representing data. As shown in Fig. 9.6, the 32-bit IEEE 754 format for a floating-point data consists of three parts: a sign bit s_a, an 8-bit exponent e_a, and a 23-bit mantissa m_a.

The error introduced to the data comprises three parts: truncation error, propagation error, and calculation error. Truncation error is due to the direct modification of data, while propagation error is due to the propagation of truncation error through each operation in the program. Calculation errors are those from the imperfection of hardware or software implementation and are considered rare.

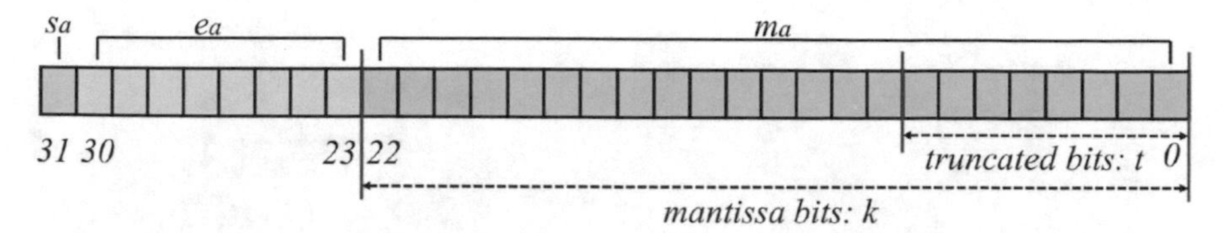

Fig. 9.6 IEEE 754 format with the last t bits to be truncated during the approximate computation (source of the figure: [28])

Before studying how it affects calculation results for an operation like logarithm, we first analyze how truncation affects operand's value or truncation error. Assume that the least significant t bits are truncated from the total k bits in the mantissa; the induced error range can be given by:

$$x_a \in \left[-(\frac{1}{2^{k-t}} - \frac{1}{2^k}) * 2^{e_a}, (\frac{1}{2^{k-t}} - \frac{1}{2^k}) * 2^{e_a}\right] \tag{9.15}$$

Since k is a fixed number such as 32 or 64, the range of truncation error x_a only depends on the number of bits to be truncated. However, the exact value of truncation error varies as the input value varies. In order to compute error variance, we assume that the value of last t bits in operand a is uniformly distributed and denotes the distribution probability as p_a. Thus, the variance of truncation error can be computed by:

$$\sigma_a^2 = \sum x_a^2 p_a = 2^{2(e_a - k)} * \frac{(2^t - 1)2^t}{3} \tag{9.16}$$

The next step is to analyze how this error will be transmitted to calculation results or propagation errors. Since some prior works [5, 23, 24] have already deduced propagation error caused by addition, multiplication, and shift operations, what we focus on in this chapter is the logarithm operation. Suppose the output for logarithm is denoted by out, and its variance is represented by σ_{out}^2. The logarithm output variance, as a function of operand variance σ_a^2, is given by $\sigma_{\text{out}}^2 = (\partial out/\partial a)^2 \sigma_a^2$. After substituting σ_a^2 with 9.16, the final truncation and propagation error variance for logarithm is shown below:

$$\sigma_{\text{out}}^2 = \frac{1}{a^2} * \sigma_a^2 = \frac{1}{(m_a * 2^k)^2} * \frac{(2^t - 1)2^t}{3} \tag{9.17}$$

which suggests that the maximal error variance is achieved when mantissa $m_a = 1$ for any given value of t.

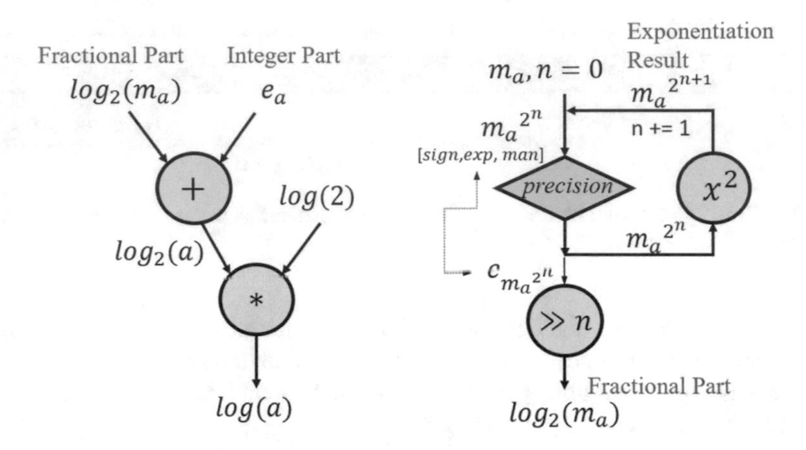

Fig. 9.7 Logarithm calculation flow chart (source of the figure: [28])

6.2 Energy Analysis for Logarithms

In this subsection, we discuss the impact of data truncation on energy savings for the logarithm operation. As shown in Fig. 9.7, the logarithm is implemented with iterations of basic arithmetic units. Therefore, energy savings can be achieved by low-precision computation units (e.g., adder or multiplier) and fewer iterations required for a specific error budget.

First, we illustrate the implementation details for the logarithm. Reference [26] proposes an approach to calculate logarithm by exploiting floating-point data exponent. As shown in Fig. 9.7a, the problem of computing $\log a$ (base e) can be easily transformed into computing $\log_2 a$. On the one hand, the integer part e_a can be directly extracted from IEEE 754 format. On the other hand, the fractional part $\log_2 m_a$ can be computed as illustrated in Fig. 9.7b. The core algorithm is to apply repeated square operations to m_a for exponentiation result $m_a^{2^n}$, where n denotes the number of loops executed. Since $m_a^{2^n}$ is also a floating-point number, its logarithm base 2 can be approximated by the exponent value $e_{m_a^{2^n}}$. Thus, the fractional part is given below, and it is precise up to $1/2^n$ because the first n bits are accurate.

$$\log_2 m_a = \frac{1}{2^n} * \log_2 m_a{}^{2^n} \approx \frac{1}{2^n} * e_{m_a^{2^n}} = e_{m_a^{2^n}} >> n \tag{9.18}$$

Next, we analyze how the number of bits to be truncated, t, affects the number of loops executed, n, for the logarithm operation given an error budget. Since the first n bits in the fractional part are accurate, calculation error x_{calc} ranges from 0 to $1/2^n$. Given the uniform error distribution, calculation error variance is given by:

$$\sigma_{\text{calc}}^2 = \int x_{\text{calc}}^2 p_{\text{calc}} - \mu^2 = \frac{1}{12} * \left(\frac{1}{2^n}\right)^2 \tag{9.19}$$

To stick to our prior assumption that calculation error caused by hardware or software is insignificant compared with truncation and propagation error, we choose n such that calculation error variance is restricted, as shown in 9.19 where δ is the hyperparameter, which the user can define.

$$\sigma_{\text{calc}}^2 \leq \delta\sigma_{\text{tNp}}^2 \tag{9.20}$$

After substituting σ_{calc}^2 and σ_{tNp}^2 with Eqs. 9.19 and 9.17, respectively, we obtain Eq. 9.21. It shows that at least $k - 1/2 * \log_2 4\delta$ loops are required to fully exploit all the significant digits when no bit is truncated. When t LSB bits are truncated, t loops can be omitted without introducing noticeable calculation error.

$$n \geq k - t - \frac{1}{2}\log_2 4\delta \tag{9.21}$$

Finally, we consider the effect of truncation on the overall energy consumption of the logarithm. The minimal number of iterations required for computing an acceptable logarithm result is determined in Eq. 9.21. Given the total number of loops n, the energy consumption for logarithm operation is linearly dependent on n. Besides, as shown in Fig. 9.7b, each iteration includes several arithmetic operations. Due to the reduced data length, some operations can be simplified, which also creates energy savings. After combing the above two effects, the total energy consumption for the truncated logarithm operation is given in 9.22, where E_{m_t}, E_{s_t}, and E_{add} denote energy for truncated floating-point multiplier, shifter, and integer adder, respectively.

$$E_{\text{log\_t}} \geq (k - t - \frac{1}{2}\log_2 4\delta) * (E_{\text{m\_t}} + E_{\text{s\_t}} + E_{\text{add}}) \tag{9.22}$$

6.3 BWOLF System Structure

We propose the BWOLF (Bit-Width Optimization for Logarithmic Function) system for programs with logarithm and other basic arithmetic operations. BWOLF considers a program in the popular data flow graph model and utilizes sequential quadratic programming to determine the optimal number of approximate bits for each operand to minimize energy consumption.

The proposed BWOLF system is illustrated in Fig. 9.8. The system requires three inputs: a target program, the input range, and an upper bound of error. The given program should have a fixed number of inputs and outputs, exclude conditionals, and be deterministic where output values will not change given the same input values. The data range for input is needed for the estimation of maximal output error variance. For example, the maximal error variance for a truncated input depends exponentially on the maximal value of exponent (see Eq. 9.16). The error budget

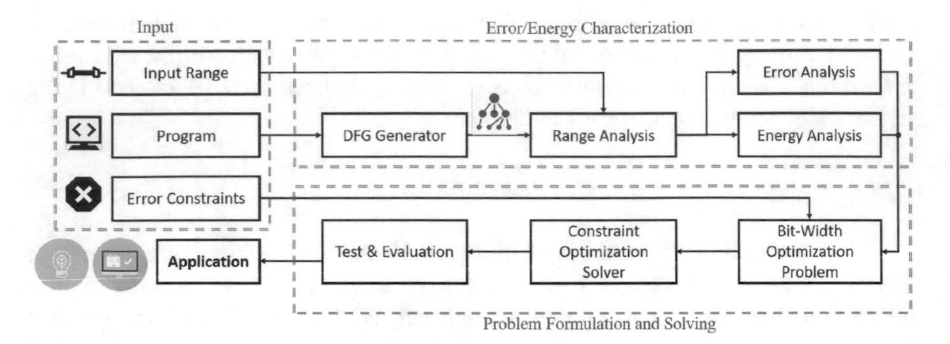

Fig. 9.8 Overview of BWOLF system structure (source of the figure: [28])

is an upper bound for the output error deviation σ (the square root of the error variance σ^2).

The *characterization stage* figures out error and energy models for a program. Suppose the number of approximate bits for each data is given. In that case, an error model can determine the maximal output error variance when inputs change, while an energy model can determine the energy required to produce results. There are three steps to obtain them. First, based on the program, the data flow graph (DFG) generator produces a directed graph whose nodes represent operators and edges represent data. The next step is to conduct a range analysis on the DFG to obtain the minimum and maximum value for each data, which provides the basis for further investigation on maximal error variance for outputs. Finally, based on the topological order in the DFG and the range for each data, error, and energy analyses will determine the relationship between the number of approximate bits for each data and the maximal output error variance and total energy consumption.

To obtain error and energy models for the whole program, we utilize approximate adder [19] and multiplier [27] whose error and energy behavior has been thoroughly analyzed. The error and energy behavior for the logarithm node is illustrated in the previous sections.

Let us consider a data flow graph that is composed of addition, subtraction, multiplication, and logarithm operations as shown in Fig. 9.9. Each node in the graph represents an operation, and each edge represents a value passed between operations. For a directed graph G, there are D edges and V nodes that are denoted as $D_i (1 \leq i \leq D)$ and $V_j (1 \leq j \leq V)$. The edges are sorted so that the input edge of a node has a smaller index than its output edge. The number of bits to be truncated for each edge is represented as $T_i (1 \leq i \leq D)$.

Since each edge in DFG can be truncated, truncation error σ_t^2 will be introduced to the whole flow. This error will then be propagated from a node's input edge(s) to its output edge. Such propagation transforms the error into truncation and propagation error, which is denoted by σ_{tNp}^2 (tNp: truncation for input edges and propagation for nodes). For example, σ_a^2 and σ_{out}^2 given in Eqs. 9.16 and 9.17 are indeed σ_t^2 and σ_{tNp}^2 for logarithm node. It should be noted that the outputs for prior

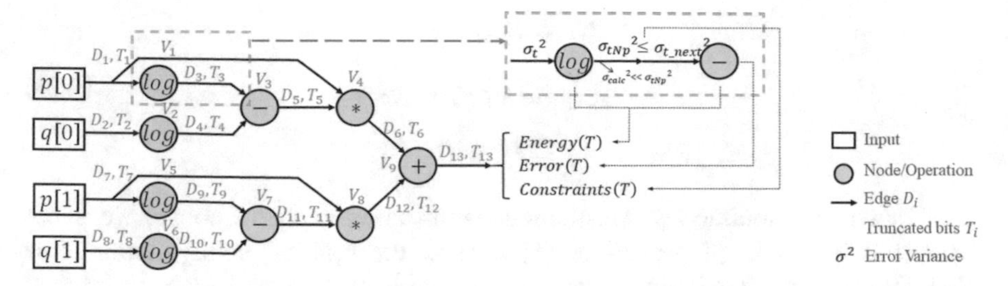

Fig. 9.9 DFG example of KL divergence for two Bernoulli distributions (source of the figure: [28])

nodes are the inputs for the following nodes, which are also eligible for truncation. As shown in the zoomed-in box in Fig. 9.9, we denote the error variance caused by truncation of a node's output edges as σ^2_{next} . More precisely, it is denoted as σ^2_{next} for logarithm but denoted as σ_t^2 for subtraction operation.

To obtain error model or maximal error variance for graph's outputs (i.e., D_{13} in Fig. 9.9), the first step is to calculate the truncation error for graph's inputs (i.e., D_1,D_2,D_7,D_8 in Fig. 9.9). Next, for each node in the topological order in DFG, its maximal truncation and propagation error can be obtained based on node operation, operands' error variance, and range. The node's output data, as well as error variance, will then be passed to the next node. However, if the number of approximate bits for the node's output edge (e.g., T_3 for logarithm node V_1's output edge D_3) is quite small, σ^2_{next} will be much smaller than σ^2_{tNp}. In this case, the precision for current data (i.e., D_3) is still controlled by σ^2_{tNp} when extra energy is required for higher-precision subtraction operation. Therefore, we select the number of approximate bits (i.e., T_3) by ensuring σ^2_{next} is compatible with or even larger than σ^2_{tNp}. This inequation is added to Constraints(T). After going through all nodes in the graph, maximal truncation and propagation error variance for all outputs in the graph can be obtained. If there are multiple output nodes, we sum up the error variance of each node for final error variance Error(T). Besides, we calculate energy consumption for each node based on the number of approximate bits in its input edges. Summing up energy consumption for all nodes in the overall energy Energy(T) required for finishing computing the graph.

The *problem formulation and solving stage* aims at selecting the best approximate configuration, which minimizes the energy consumption while keeping the maximal output error variance under constraint. The problem of bit-width optimization is seeking the best number of bits to be truncated for each data to minimize energy consumption when the error is within the user-defined range. As illustrated in the previous subsection, we have obtained error model Error(T), energy model Energy(T), and constraint functions Constraints(T). All of them depend on T, which represents the approximate configuration. Finding the best approximate configuration is formulated below, where *ec* represents the user's fixed error budget for σ.

$$\begin{aligned} \min_{T} \quad & \text{Energy}(T) \\ \text{s.t.} \quad & \text{Error}(T) \le ec^2 \\ \text{and} \quad & \text{Constraints}(T) \end{aligned} \tag{9.23}$$

To solve the nonlinearly constrained optimization problem above, we adopt sequential quadratic programming [1] to find the optimal configuration. This algorithm utilizes a quadratic programming subproblem to approximate the original complex nonlinear problem and exploits the solution to the simplified problem to help generate a better approximate quadratic subproblem. The solutions for the sequence of subproblems finally converge to the optimal point for the original nonlinear problem.

7 Summary

With the increasing demand for resource-constrained end devices such as the Internet of Things, low-power system design remains an active and challenging research topic. Approximate computing is a promising technique to save power and energy for systems that are error tolerant or error resilient. This chapter provides the motivation of approximate computing on data flow graphs (DFGs) and discusses the main concerns in approximating DFGs. Since both the basic arithmetic units such as adders and multipliers and the input data of a DFG have been well-studied for approximation, we focus on several novel approaches that target the entire DFG or the underlying applications for approximation. The first technique, probabilistic design of multimedia systems, approximates the DFG execution by terminating the iterations with predicted long execution time early to save energy under the completion ratio constraint. The second technique, estimate-and-recompute, approximates floating-point multiplications by addition in the logarithm domain and then recomputes for the accurate value only for operations on the critical subgraphs. The last example demonstrates how the logarithm operation can be represented as a DFG and computed approximately by optimizing the bit-width of the input. Our goal in this chapter is to inspire novel approximate computing approaches that investigate the intrinsic nature of the specific application to achieve the optimal approximation strategy to trade computation accuracy for energy reduction.

References

1. Boggs PT, Tolle JW. Sequential quadratic programming. Acta Numer. 1995;4:1–51.
2. Eikerling HJ, Hardt W, Gerlach J, Rosenstiel W. A methodology for rapid analysis and optimization of embedded systems. In: Proceedings IEEE symposium and workshop on engineering of computer-based systems; 1996. p. 252–9.

3. Gao M, Qu G. Estimate and recompute: a novel paradigm for approximate computing on data flow graphs. IEEE Trans Comput-Aid Des Integr Circuits Syst. 2020;39 2:335–45.
4. Gao M, Wang Q, Nagendra ASK, Qu G. A novel data format for approximate arithmetic computing. In: 2017 22nd Asia and South Pacific design automation conference (ASP-DAC); 2017. p. 390–5.
5. Gao M, Wang Q, Qu G. Energy and error reduction using variable bit-width optimization on dynamic fixed point format. In: 2019 IEEE computer society annual symposium on VLSI (ISVLSI); 2019. p. 152–7.
6. Hashemi S, Bahar RI, Reda S. A low-power dynamic divider for approximate applications. In: 2016 53nd ACM/EDAC/IEEE design automation conference (DAC); 2016. p. 1–6.
7. Henkel J, Ernst R. High-level estimation techniques for usage in hardware/software co-design. In: Proceedings of 1998 Asia and South Pacific design automation conference; 1998. p. 353–60.
8. Hu X, Zhou T, Sha EM. Estimating probabilistic timing performance for real-time embedded systems. IEEE Trans Very Large Scale Integr Syst. 2001;9 6:833–844
9. Hua S, Qu G, Bhattacharyya SS. Energy reduction techniques for multimedia applications with tolerance to deadline misses. In: Proceedings of the 40th annual design automation conference (DAC'03); 2003. p. 131–6.
10. Hua S, Qu G, Bhattacharyya SS. Probabilistic design of multimedia embedded systems. ACM Trans Embed Comput Syst. 2007;6 3:15–es.
11. Jiang H, Han J, Lombardi F. A comparative review and evaluation of approximate adders. In: Proceedings of the 25th edition on great lakes symposium on VLSI (GLSVLSI'15). New York: Association for Computing Machinery; 2015. p. 343–8.
12. Kalavade A, Moghé P. A tool for performance estimation of networked embedded end-systems. In: Proceedings of the 35th annual design automation conference (DAC'98). New York: Association for Computing Machinery; 1998. p. 257–62.
13. Kianzad V, Bhattacharyya SS, Qu G. Casper: an integrated energy-driven approach for task graph scheduling on distributed embedded systems. In: 2005 IEEE international conference on application-specific systems, architecture processors (ASAP'05); 2005. p. 191–7.
14. Liu C, Han J, Lombardi F. A low-power, high-performance approximate multiplier with configurable partial error recovery. In: 2014 design, automation test in europe conference exhibition (DATE); 2014. p. 1–4.
15. Liu W, Qian L, Wang C, Jiang H, Han J, Lombardi F. Design of approximate radix-4 booth multipliers for error-tolerant computing. IEEE Trans Comput. 2017;66 8:1435–41.
16. Liu W, Lombardi F, Shulte M. A retrospective and prospective view of approximate computing [point of view]. Proc IEEE 2020;108 3: 394–9.
17. Madsen J, Grode J, Knudsen PV, Petersen ME, Haxthausen A. LYCOS: the lyngby co-synthesis system. Des Autom Embed Syst. 1997;2 2:195–235.
18. Malik S, Martonosi M, Li YTS. Static timing analysis of embedded software. In: Proceedings of the 34th annual design automation conference (DAC'97). New York: Association for Computing Machinery; 1997. p. 147–52.
19. Nannarelli A. Tunable floating-point adder. IEEE Trans Comput. 2019;68 10:1553–60.
20. Nepal K, Li Y, Bahar RI, Reda S. Abacus: a technique for automated behavioral synthesis of approximate computing circuits. In: 2014 design, automation test in Europe conference exhibition (DATE); 2014. p. 1–6.
21. Riehme J, Naumann U. Significance analysis for numerical models. In: 1st workshop on approximate computing (WAPCO); 2015. p. 0278–0070.
22. Roy P, Ray R, Wang C, Wong WF. ASAC: automatic sensitivity analysis for approximate computing. In: Proceedings of the 2014 SIGPLAN/SIGBED conference on languages, compilers and tools for embedded systems (LCTES'14). New York: Association for Computing Machinery; 2014. p. 95–104.
23. Sengupta D, Snigdha FS, Hu J, Sapatnekar SS. Saber: selection of approximate bits for the design of error tolerant circuits. In: 2017 54th ACM/EDAC/IEEE design automation conference (DAC); 2017. p. 1–6.

24. Snigdha FS, Sengupta D, Hu J, Sapatnekar SS. Optimal design of jpeg hardware under the approximate computing paradigm. In: 2016 53nd ACM/EDAC/IEEE design automation conference (DAC); 2016. p. 1–6.
25. Tia TS, Deng Z, Shankar M, Storch M, Sun J, Wu LC, Liu JS. Probabilistic performance guarantee for real-time tasks with varying computation times. In: Proceedings real-time technology and applications symposium; 1995. p. 164–73.
26. TMS320 DSP development support reference guide. Tech. Rep., Texas Instruments; 1998.
27. Wires KE, Schulte MJ, McCarley D. FPGA resource reduction through truncated multiplication. In: Proceedings of the 11th international conference on field-programmable logic and applications (FPL'01). Berlin: Springer; 2001. p. 574–83.
28. Xu Q, Sun G, Qu G. BWOLF: bit-width optimization for statistical divergence with -logarithmic functions. In: 2020 IEEE 31st international conference on application-specific systems, architectures and processors (ASAP); 2020. p. 165–72.

Chapter 10
Test and Reliability of Approximate Hardware

Marcello Traiola, Bastien Deveautour, Alberto Bosio, Patrick Girard, and Arnaud Virazel

1 Introduction and Background

Since the early 1970s, the demand in electronic components and the necessity to push the limits of manufactured circuits for increased performance and transistor density has never stopped. Consequently, each new technology node suffers from reliability issues due to manufacturing defects, variability, interference, and wear-out. These well-known drawbacks lead to the occurrence of faults that can finally cause system failures in Integrated Circuits (ICs). Several approaches, such as test and fault tolerance, allow effectively improving the reliability of ICs, making them to work as intended.

Approximate Computing (AxC) is a paradigm that is extensively used to improve the circuit efficiency, by intentionally degrading the computation result quality. From the hardware standpoint, AxC enables the creation of circuits whose output values may differ from the original circuit for a certain set of input values [45]. The degradation introduced through the approximation must be judicious, in order

M. Traiola
University of Rennes, Inria, CNRS, IRISA, UMR 6074, Institut de Recherche en Informatique et Systèmes Aléatoires (IRISA), Rennes, France
e-mail: marcello.traiola@inria.fr

B. Deveautour · A. Bosio
University of Lyon, Ecole Centrale Lyon, INSA Lyon, CNRS, UCBL, CPE Lyon, Lyon Institute of Nanotechnology (INL), Ecully, France
e-mail: bastien.deveautour@cpe.fr; alberto.bosio@ec-lyon.fr

P. Girard (✉) · A. Virazel
Laboratory of Computer Science, Robotics and Microelectronics of Montpellier (LIRMM), University of Montpellier, French National Centre for Scientific Research (CNRS), Montpellier, France
e-mail: patrick.girard@lirmm.fr; arnaud.virazel@lirmm.fr

W. Liu, F. Lombardi (eds.), *Approximate Computing*,
https://doi.org/10.1007/978-3-030-98347-5_10

to preserve quality loss below a certain threshold. Depending on the application scenario and the related reliability figure, such quality threshold may be more or less stringent.

In this chapter, we review the issues related to the test and the fault tolerance of approximate hardware, along with the existing solutions.

1.1 Reliability Issues in Nanometer Technology

The reliability of digital circuits and systems is kept high owing to several methods. These methods ensure that the designs perform their function under defined conditions and during their estimated lifespan. They cover different aspects related to the manufacturing and the in-field functioning of electronics. For instance, clean rooms control impurities; industrial control systems achieve production consistency; burn-in and testing, before and after packaging, ensure the detection of design weaknesses and manufacturing defects after stressing the circuits. All these methods are necessary before introducing the semiconductors to the market, but they are not foolproof. Even though miniaturization offers many advantages, each new CMOS node faces reliability issues, as the trend is rapidly approaching the physical limits of operation and manufacturing [34].

Digital systems can experience failures during the three phases of their lifespan depicted in the bathtub curve in Fig. 10.1 [39]. Early failures are labeled as infant mortality; random failures occur during the working life and wear-out failures happen at the end of the circuit's lifespan. Different kinds of faults may occur to ICs, due to different phenomena. Faults can be classified according to their duration in three categories: *transient*, *intermittent*, and *permanent*.

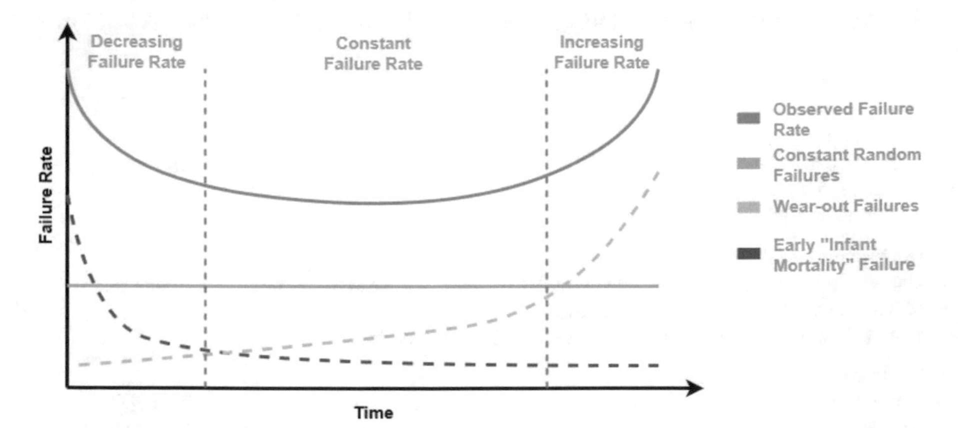

Fig. 10.1 The bathtub curve describing the failure rate of electronic circuits over their lifespan [39]

Transient Faults Transient faults randomly affect the correct functioning of the IC for a short time window. After this period, the device returns to a normal behavior. Variability and interferences are the main causes of transient faults.

Intermittent Faults Intermittent faults occur randomly like transient faults, but they never really disappear. In fact, their occurrence often precedes the occurrence of a permanent fault. Aging is the primary cause of intermittent faults.

Permanent Faults Permanent faults are irreversible and are mostly due to manufacturing defects. They can also appear at the end of the circuit's lifetime due to extreme wear-out effects.

As mentioned above, faults may have different causes: manufacturing defects, variability, interferences, and aging (wear-out).

Manufacturing Defects and Variability Early failures during the infant mortality phase are mainly due to manufacturing imperfections, possibly leading to permanent defects in a chip. Moreover, the variability of transistor characteristics has always been an issue in IC design [82]. This phenomenon prevents the circuit from functioning correctly even though each individual transistor behaves correctly [51]. Furthermore, while the number of transistors doubles every 2–3 years, according to past microprocessor data, the die size remains relatively constant [33]. This scaling inevitably leads to a chip power density increase leading to inadequate heat sinking, thus to hot spots. In turn, this alters the timing characteristics of circuits [37]. Inevitably, the increasingly complex logic of ICs entails the emergence of defects more and more hard to detect [60, 61]. Therefore, thoroughly testing the manufactured chips is crucial before releasing them to the market.

Interferences Variability generated by manufacturing imperfections may generate unexpected circuit behaviors during operation. Furthermore, as transistors become smaller, their supply voltage (Vdd) decreases. These conditions are favorable for the occurrence of temporary effects, such as transient or intermittent faults. These temporary effects can be the result of electromagnetic influences, alpha-particle emission, or cosmic radiations. They are responsible for the greatest part of digital malfunctions, and more than 90% of the total maintenance costs are credited to them [57]. Internal interferences can also be a cause of temporary malfunctioning. With the scaling of components, the scaling of the interconnect line thickness (width and separation) must also follow. In these conditions, a high crosstalk noise is becoming a major issue due to larger capacitive couplings between interconnects in a polluted environment. Additionally, supply voltage scaling lowers the noise sensitivity threshold and increases the transient fault sensibility of new technology nodes due to high-energy particle from environment or within the packaging.

Wear-Out Although area scaling followed an exponential trend, supply voltage had a way slower scaling pace. There are two main reasons behind that: (1) The need to keep up with the competitive frequency growth (2) The need to retain the basic noise immunity and cell stability [67] As a result, the discrepancy between area and voltage scaling leads to high power density and elevated temperatures. This, in turn, first causes modifications in the timing characteristics of the design and, eventually, wears out the chip's metal lines. Wear-out failures appear in-field after a certain period of use and limit both performances and lifetime of modern electronics [66]. This is especially critical for applications that demand high throughput (e.g., data centers) or for which technical support is expensive (e.g., space equipment).

When a fault propagates through the logic, it can be captured by a flip-flop (or memory cell) and stored as faulty value.

Depending on their nature, faults can become hard or soft errors. If the error reaches the service interface, this may cause a subsequent failure and alters the service [4].

Soft Errors Soft errors occur when particles, such as high-energy neutrons from cosmic rays or alpha particles generated from impurities in the packaging, strike a sensitive zone of the microelectronic device. This causes voltage glitches that propagate through combinational logic parts of the IC. Those events are referred to as *single-event transients (SETs)* [5, 15, 22]. We refer to *single-event upset (SEU)* as the phenomenon for which an SET propagates to some memory element of the circuit and its value is captured.

Hard Errors Hard errors are a consequence of permanent silicon defects (due to manufacturing imperfections or to the wear-out). In the last decades, as transistor density increased, the likelihood of getting more hard errors in a given core kept increasing as well. In addition, the high frequencies increase the switching activity rate that accelerate material aging due to temperature and voltage stress [12].

1.2 Reliability Improvement Approaches

To improve the reliability, a thorough testing is crucial after the IC manufacturing and during the IC lifetime. Testing of digital circuits is performed thanks to binary test patterns applied to circuit's inputs [18]. If the test result and the expected one (or *golden*) do not match, the circuit is declared as *faulty*. Testing can be classified depending on the goal it is intended to serve:

Production Testing After chip manufacturing, the production testing determines whether the actual manufacturing process produced correct devices or not. This process is performed by the device manufacturer that owns full details about the internal structure of the manufactured system and usually exploits Automated Test Equipments (ATEs) for performing the tests.

In-Field Testing Conversely, when the device is already in the field and under certain operative conditions, it requires to be tested during its normal operational life. Thus, a periodic *in-field* testing strategy is implemented. In this case, the test is carried out through the test mechanisms embedded in the device itself. Today, industrial standards—such as ISO26262, for automotive, and DO254, for avionics—provide the necessary guidelines to implement these test strategies in the context of different safety-critical applications.

Two types of tests are usually performed on VLSI chips:

Functional Test The test is performed through the device functional inputs and observing only the functional outputs. This leads to test the device functionality rather than the faults.

Structural Test The test is designed by taking into account the device's structural information, thanks to its netlist (i.e., the topological distribution of its logic gates). Very sophisticated algorithms, implemented within *Automatic Test Pattern Generators (ATPGs)*, exist to generate efficient structural tests.

The ATPG produces input sequences (referred to as *test set*) to efficiently detect possible faults in digital circuits. It is based on very advanced algorithms, such as the FAN [24]. Once the test set is available, by simulating it with the fault-free circuit, the fault-free output (or expected output) is obtained. In the test phase, if a fault occurs, the circuit outputs will be different from expected, when the test set is applied. In this way, we are able to detect the faults affecting the circuit.

The above reported concepts are not intended to be exhaustive; for an extensive dissertation about them, readers may refer to [9].

Furthermore, other reliability improvement practices can be adopted. They can be basically characterized into three categories, namely, *fault avoidance*, *fault removal*, and *fault tolerance* [32].

Fault Avoidance It aims at minimizing the sensibility of ICs to faults. To do so, specific tools and techniques assist the designers to specify, design, and manufacture systems by addressing the source of the mechanisms causing the failures [31, 56, 63, 87].

Fault Removal It includes approaches whose function is to detect and eliminate existing faults during specification and design. Fault removal also refers to removing faulty components during production and operational phases [63].

Fault Tolerance It aims at guaranteeing the service provided by the product despite the presence or appearance of faults [32]. This chapter focuses specifically on fault tolerance.

While improving the manufacturing process and thoroughly testing the manufactured circuits certainly help to cope with permanent faults, they do not solve random failures. Even the best efforts and investments to avoid or remove faults cannot prevent them from appearing in any operational system. However, as depicted in Fig. 10.2, it is possible to prevent those faults by using hardware fault-tolerance techniques. Some of them, such as *fault masking*, are classified as static, whereas some others, such as *error correcting codes*, as dynamic [10]. Some fault-tolerant designs are destined to resilient application, which are intrinsically able to provide and maintain an acceptable level of service despite faults occurring in the process. In these cases, the fault-tolerance mechanism aims at keeping the faults under an established level of impact. All the fault-tolerance techniques are usually referred to as *redundancy*, a principle introduced by John von Neumann in 1950s [50]. The

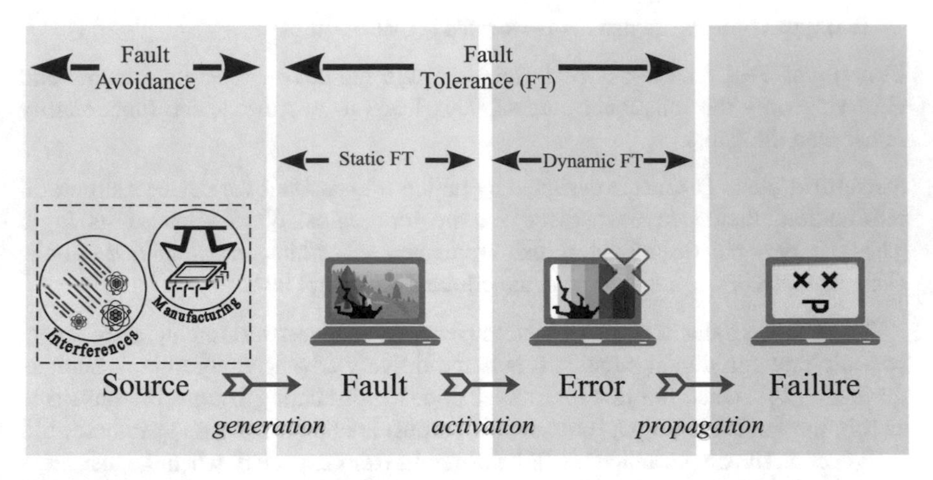

Fig. 10.2 Reliability improvement approaches across the fault-to-failure life cycle [10]

basic idea is to improve the system reliability by adding redundant information. Redundancy can be classified as *spatial*, *temporal*, or *of data* according to Mathew et al. [43].

Spatial Redundancy It refers to techniques employing extra hardware to process the same information multiple times in parallel. Even if some hardware parts are affected by a transient or permanent fault, a logic voting scheme can produce a single correct output thanks to the redundant computations. On the other hand, the extra hardware resources used to improve reliability lead to area and power overheads [41].

Temporal Redundancy It is based on repeating a computation or a transmission to detect (and sometimes correct) possible temporary errors [16]. In some cases, spending extra computation/transmission time (thus sacrificing performance) to tolerate faults is preferred to spatial redundancy, which implies additional area and power costs.

Data Redundancy It is particularly employed in memory devices, and it is based on detection and correction codes stored along with the data [65]. This extra information is generated from the original data to effectively identify the presence of one or more transient or permanent faults and possibly correct them.

The detection of errors during the lifetime functional operation of a system is called *on-line detection* [27]. Examples of commonly used fault detection technique are *duplication with comparison* and *error detecting codes*. Error detecting codes are based on using redundant information to detect possible errors in stored/transmitted data. The data are guaranteed to be correct if some characteristics are respected, e.g., parity, checksum, and cyclic redundancy check (CRC). Those are generally computed from the data and compared to the expected ones, stored/transmitted along with the data. Error detecting codes can provide protection against SEUs

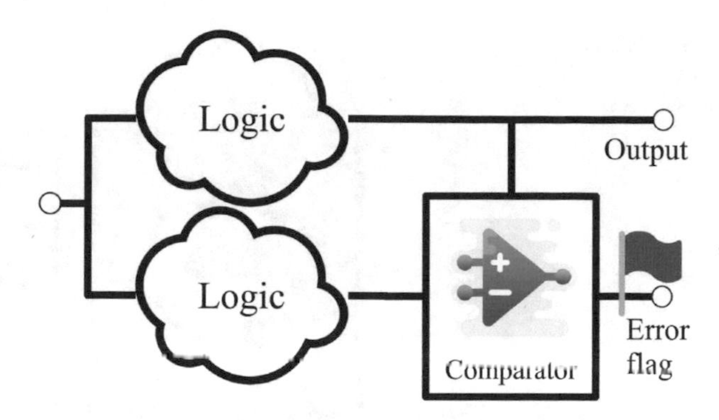

Fig. 10.3 Generic structure of the *duplication with comparison* architecture

and permanent faults [53] and are mainly used in memories, owing to their regular structure [17]. Error detection codes can be used in logic circuits to detect malfunctioning; however, their efficiency relies on ad-hoc designs [27]. Duplication with comparison, sketched in Fig. 10.3, employs two identical copies (at least functionally) of a system whose outputs are compared. A possible error affecting one of the copies is detected by comparing the two results. Its implementation, simplicity, and its ability to detect a wide variety of faults–permanent, transient, and timing—are the main reasons of its popularity.

Once a fault has been detected, a proper method to recover from its effect is needed. The *recovery* consists in restoring the last error-free state of the system. This approach is referred to as *rollback*. The rollback consists in repeating the last operation(s) before the fault detection. Usually, architectures based on rollback detect faults through spatial redundancy and correct them by applying temporal redundancy. Periodical or occasional checkpoints save the state of the system; when an error is detected, these checkpoints are restored and the next operations are recomputed. The rollbacks can restore the system to several thousand previous states or simply to the previous cycle [44, 76]. Fault tolerance achieved through fault detection and rollback implies lower area and power costs than using spatial redundancy to prevent a fault from propagating and eventually causing an error. However, in case of a recovery, power and timing costs are proportional to the quantity of computing cycles that are recovered.

1.2.1 Fault-Tolerant Architectures

To deal with reliability issues in ICs, in the last decades, several fault-tolerant architectures have been proposed, such as Pair-and-A-Spare [48], Razor [19], Soft and Timing Error Mitigation (STEM) [3], Conjoined Pipeline (CPipe) [68], Triple Modular Redundancy (TMR) [40], and Dynamic Adaptive Redundant Architecture

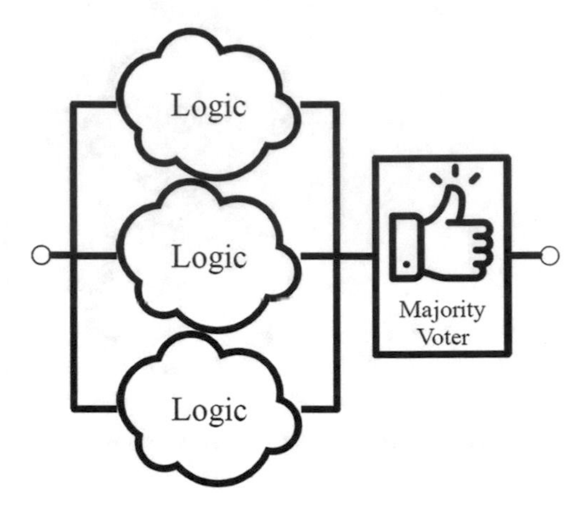

Fig. 10.4 Generic structure of the Triple Modular Redundancy architecture

TMR [84]. Since, in this chapter, we focus on the generic TMR, in the following we briefly summarize its structure and functioning.

Triple Modular Redundancy (TMR) Triple Modular Redundancy (TMR) is one of the most popular fault-tolerant architectures, based on spatial redundancy. Its first application to computing systems dates back to 1962 [40]. Its generic structure is sketched in Fig. 10.4. The TMR exists in different versions; the simplest is the one where three exact copies of the combinational logic (CL) of a circuit are connected to a voting circuit. The latter *masks* a possible faulty output that differs from the other two, by performing a majority vote. This configuration can mask single faults occurring in the CL. However, any fault located in the input or output register causes a system failure. In general, a fault occurring in any *unprotected* (i.e., non-triplicated) part of the circuit would result in a common-mode failure, thus undetectable for the approach. TMR variants triplicating the entire circuit—flip-flops included—and also the voter exist and mask any single fault in the circuit. On the other hand, the costs associated to the TMR structure are considerably high. Since the resources are triplicated, area and power overhead costs are at least 200%. The majority voter adds extra cost, also in terms of additional timing.

1.2.2 Selective Hardening Approaches

As discussed, all hardening methods rely on some variants of redundancy. Along the error recovery mechanisms, redundancy demands considerable resources to tolerate faults [36]. Moreover, often fault-tolerant solutions have limited application scenarios (e.g., ECC are primarily adapted for memory circuits). On the other hand, approaches most effectively dealing with a wide spectrum of failures often lead to massive hardware redundancy (e.g., TMR entails more than 200% area and power overheads [20]).

Researchers addressed this concern by proposing the *selective hardening* approach. The underlying idea is straightforward: if hardening the whole circuit is too resource-demanding, then only some chosen parts of the circuit are hardened. The choice of the parts needing hardening depends on two main factors, i.e., their particular exposure to failure and the criticality of their role in the system functioning. While the selective hardening greatly reduces the overall error rate and the redundancy costs, the fault coverage decreases. Thus, the approach trades off some reliability to gain resources and tries to find "sweet spots" where the costs are greatly reduced and the system reliability is not too much degraded. Selective hardening techniques first perform a *vulnerability assessment* and then use fault-tolerance approaches to protect the most vulnerable parts. Examples of work in this direction can be found in [8, 21, 42, 46, 52, 54, 89].

1.3 Toward Approximate-Computing-Based Reliable Hardware

As reported in [26], the continuous scaling of CMOS technologies into the nanometer range has increased the effect of variability and degradation mechanisms on the yield and reliability of CMOS circuits and systems. In fact, the normal lifetime of miniaturized ICs is more and more reduced [30]. Approximate Computing aims at transforming this problem into an opportunity [2]. The basic idea is to "*embrace*" errors as an intrinsic property of ICs and systematically design optimized approximate circuits functioning regardless of errors.

To achieve such a goal, on the one hand, test procedures have to be redesigned to be aware of the introduced approximation. On the other hand, the relaxation of non-critical computational constraints typical of the AxC [45] must be judicious—similarly to Selective Hardening—in order to keep a satisfactory reliability level. Therefore, we need to consider how AxC impacts on the role of testing and fault tolerance.

From a test perspective, the concept of *fault* changes. Indeed, the circuit is allowed to produce an *acceptable error*, by design. The maximum allowed error is defined by error metrics [38]. In the same way, during the test, the impact of a fault needs to be measured with such metrics. If the obtained measure is higher than the acceptable threshold, then the circuit has to be rejected; otherwise, it has to pass the test. Thus, test procedures have a twofold role: (1) Reject circuits whose error is greater than the threshold. (2) Avoid rejecting *acceptably faulty* circuits. This ultimately leads to yield increase and possibly to the test cost reduction (i.e., the fewer the faults to test, the fewer the required test vectors). As a result of this consideration, test procedures have to be carefully redesigned.

Concerning the fault tolerance, the application of AxC techniques has to satisfy different constraints, depending on the system criticality. In detail, if the target system is composed of modules having different criticality, the approximation cannot be applied indiscriminately; rather it has to be adapted. The parts of the system that are less critical have to be first identified; then, selective—possibly approximation-based—hardening techniques can be applied. Previous works [77] proposed a very fast and low-computationally intensive method that helps to select the most sensitive parts of a logic design. This allows identifying the necessary hardening degree to fulfill the soft error reliability constraints and reduce the design cost, in terms of area and power [80]. Based on this analysis, called *structural susceptibility analysis*, a selective hardening technique using the *Hybrid Transient Fault-Tolerant (HyTFT)* architecture is proposed in [78]: by reducing the number of output nodes of the CL and comparing it with a full version of the circuit, this selective hardening approach not only reduces the size of the comparator but also significantly reduces the size of the duplicated CL copy in a vulnerability-aware manner. The use of the structural susceptibility analysis employed in the HyTFT architecture has proven to be more efficient in terms of area and power consumption with respect to a full duplication scheme. However, this analysis does not consider any error metrics as usually done in AxC (e.g., Worst-Case Error for error magnitude). In this context, fault-tolerant architectures bringing an AxC-based partial protection have been emerging. It is important to make sure that they are a good alternative to the classical selective hardening architectures and to be aware of the challenges that they raise. Finally, all of this begs the question: is it possible to apply AxC techniques also to the most critical parts of the system, despite the imprecise nature of AxC-based circuits?

While Sect. 3 addresses the approximation-based fault tolerance, the next section discusses the aspects related to the test of approximate circuits.

2 Testing Approximate Hardware

The application of AxC in hardware leads to systems widely referred to as *Approximate Integrated Circuits (AxICs)*. An extensively used method to design those circuits is *functional approximation* [55]. This section specifically focuses on the test of functionally approximate circuits. Unlike *approximate test* [79], where the constraints of the test process itself are relaxed, in this context, we focus on how effectively test approximate circuits. As already mentioned, test techniques must be revisited to be applied to AxICs, since the approximation changes their functional behavior. As a matter of fact, extending to AxICs, the conventional testing concepts are not trivial. In particular, in the context of AxICs, even if a fault leads to exhibit a different behavior than expected, it may still be acceptable; thus the AxIC should not be discarded. Mastering these mechanisms may lead to increase the production process yield and/or reduce the test cost.

This section reviews the *approximation-aware* test flow that deals with such aspects. The flow is composed of three main steps: (1) Fault classification (2) Test pattern generation (3) Test set application Faults producing critical effects on the circuit behavior are divided from those producing acceptable effects, in the fault classification phase. Test stimuli covering all the critical faults and simultaneously leaving undetected as much acceptable faults as possible are produced in the *test pattern generation*. Finally, in the *test set application*, AxICs under test are classified as critically faulty—thus discarded—or as acceptably faulty, or fault-free—thus they pass the test. In turn, this reduces the number of AxICs discarded due to acceptable faults; thus it increases the manufacturing yield.

2.1 Approximation-Aware Fault Classification

Fault classification is the first step of the approximation-aware testing. It divides acceptable faults from critical ones [11, 25, 71, 72], under the single-fault assumption [9]. Moreover, the fault classification process produces the *expected yield increase* value of the approximation-aware test w.r.t. conventional test [75]. It is defined as follows:

$$\text{Expected yield increase} = \frac{\text{Acceptable faults}}{\text{Total faults}}. \tag{10.1}$$

The purpose of such a metric is to establish an upper bound to the achievable yield gain.

Measuring the output deviations of AxICs is a crucial task for a successful classification. Different error metrics have been proposed in the literature to measure arithmetic AxIC output deviations [38]. For instance, a widely accepted metric is the Worst-Case Error (WCE), defined as follows:

$$\text{WCE} = \underset{\forall i \in \mathcal{I}}{max} \left| O_i^{\text{approx}} - O_i^{\text{precise}} \right|, \tag{10.2}$$

where

- $i \in \mathcal{I}$ is the input value within the set of all possible inputs $\mathcal{I}$,
- $O_i^{precise}$ is the precise output integer representation, for input i,
- O_i^{approx} is the approximate output integer representation, for input i.

The complexity of evaluating the impact of a fault depends on the considered error metric [70]. Metrics based on the calculation of a mean entail a higher computational effort to carry out the fault classification compared to other metrics. Indeed, calculating the mean error of an arithmetic circuit requires the error measure for all the possible circuit input values, which is a $O(2^n)$ complexity task (where n is the number of input bits).

Existing techniques to deal with fault classification are based on the idea of masking the effects of acceptable faults by using a filter [11, 25, 70, 72], as shown in Fig. 10.5. In detail, at design time, both the original precise netlist (i.e., not approximate) and the AxIC netlist are put together with an *evaluation module* to form a new netlist (*classifier structure*, in the figure). The evaluation module compares the two circuit outputs with respect to the chosen error metric(s). It produces an error only if the AxIC produces an output outside the defined error

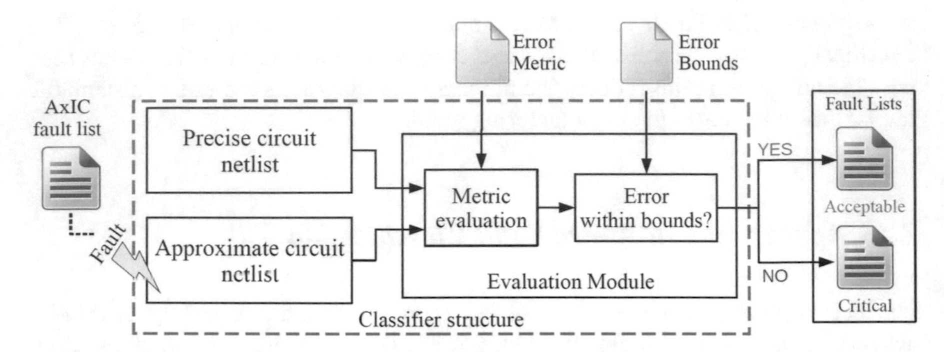

Fig. 10.5 Approximation-aware fault classification

bounds. The netlist of the so-built classifier is elaborated with conventional test approaches, such as conventional ATPG approaches or a SAT-based ones, to classify the AxIC possible faults. The underlying idea is using the evaluation module to filter acceptable fault effects. Therefore, only critical faults generate an error condition. For instance, when using the ATPG, if the generation procedure finds a test for a given fault, this means that the fault is critical; conversely, no tests are produced for acceptable faults, since their effect is masked.

Figure 10.6 reports the results of the approximation-aware fault classification applied to approximate arithmetic circuits [35, 47, 62, 85, 86, 88], expressed in terms of average *expected yield increase* (Eq. (10.1)). Results are extracted from [11, 25, 72], and the analyzed circuits are 8- and 16-bit adders [35, 47, 62, 86, 88], 8-, 16-, and 32-bit multipliers [47], single-precision IEEE-754 standard floating-point circuits [85], and fixed-point multipliers and dividers [85]. We report the results by organizing the circuits in groups based on the allowed percentage error. The reference metric is the WCE, defined in Eq. (10.2).

Adders and multipliers, on average, show an expected yield increase above 50%, except for 8-bit adders showing an average around 19% for circuits allowing WCE greater than 0% and lower than 10%. Fixed- and floating-point units show results ranging between 13% and 78%, on average. As foreseen, the expected yield increase gets higher when a larger error is allowed; this happens as the effect of more faults is masked; thus, they are considered as acceptable. More details on the approaches are reported in [11, 25, 70, 72, 75].

2.2 *Approximation-Aware Test Pattern Generation*

Test pattern generation is the second step of the approximation-aware testing. In conventional test, the higher the *fault coverage* a test generation methodology is able to achieve, the more efficient it is considered to be. This slightly changes, when it comes to AxICs. Indeed, on the one hand, test patterns must detect all critical

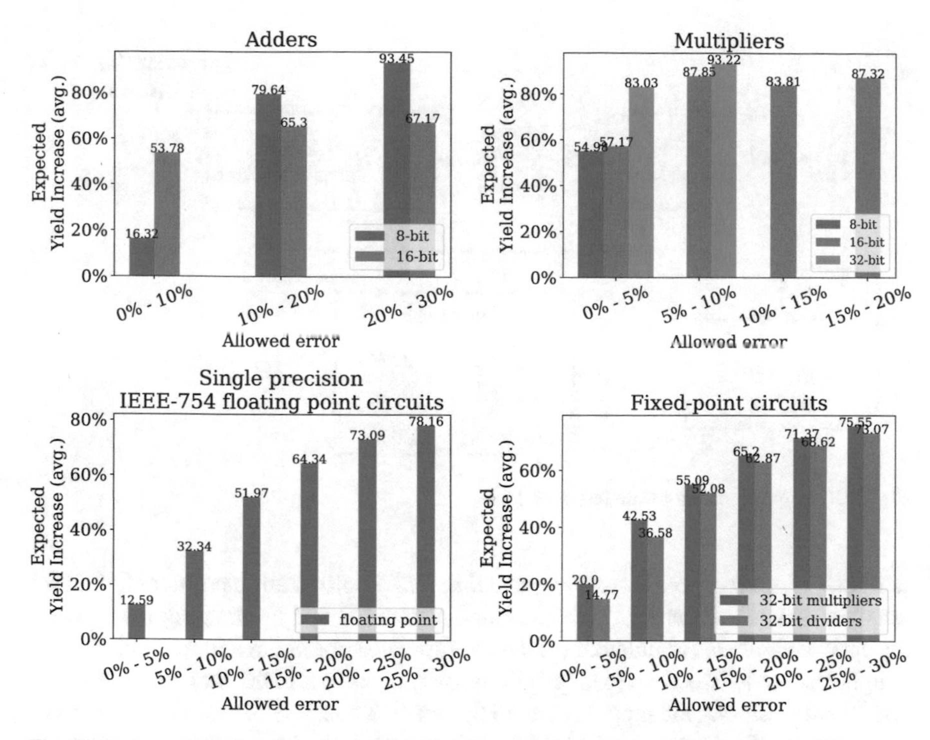

Fig. 10.6 Approximation-aware fault classification experimental results from [11, 25, 72]

faults; on the other hand, they should detect as few as possible acceptable faults. In turn, this allows identifying and rejecting all AxICs affected by critical defects and, simultaneously, preventing the elimination of those affected by acceptable defects.

Thus, we have to revisit the *fault coverage (FC)* definition, by extending it into two subclasses: *acceptable FC* and *critical FC*, defined below:

$$\text{Acceptable FC} = \frac{\text{Acceptable faults detected}}{\text{Total acceptable faults}} \tag{10.3}$$

$$\text{Critical FC} = \frac{\text{Critical faults detected}}{\text{Total critical faults}}. \tag{10.4}$$

In the AxIC context, an ideal test set should lead to 100% critical FC and 0% acceptable FC [73, 75].

Efforts toward achieving such test set have been spent. In [73], a methodology searching for the smallest subset of tests detecting all the critical faults and minimizing the detected acceptable faults has been developed. It is sketched in Fig. 10.7: (1) The input vector (sub-)set $\mathcal{S}$ is generated for the AxIC. (2) The coverage of both AxIC's critical and acceptable faults is measured by using fault simulation (i.e., circuit simulation by applying the test set and with the fault introduced in the

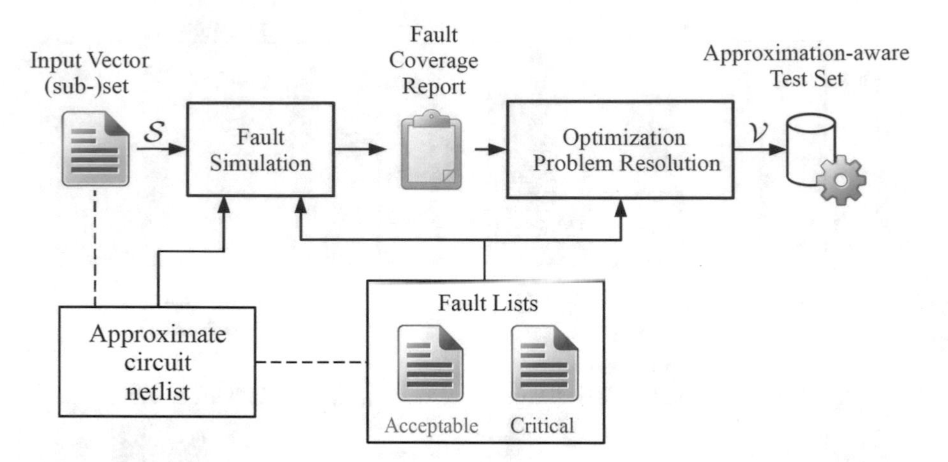

Fig. 10.7 Approximation-aware test generation

netlist). (3) An *Integer Linear Programming (ILP)* optimization problem to find the smallest test subset detecting 100% critical faults FC and minimizing the detected acceptable faults is formulated. (4) The optimization problem solution delivers the sought test set $\mathcal{V}$. The described technique guarantees to find the best possible vector combination among the input vectors in the set $\mathcal{S}$. Thus, if $\mathcal{S}$ is the exhaustive AxIC input set, the final test set will be the *global optimum*. The described approach is independent of the specific fault classification technique and of the chosen error metrics. The reader can refer to [73] for further details.

Figure 10.8 reports the results of the approximation-aware test pattern generation applied to approximate arithmetic circuits from [35, 47, 62, 86, 88], compared to the conventional generation (i.e., obtained with conventional ATPG tools). As shown, the conventional generation technique exhibits an average acceptable fault coverage between 65% and 92%. Significant lower (thus better) acceptable fault coverage values were achieved by using the approximation-aware technique from [73]. Indeed, acceptable fault coverage values between 33% and 76% were achieved. Furthermore, the average number of necessary tests is reduced when using the approximation-aware generation, thus reducing the necessary time to test the circuit. However, concerning the execution time, the approximation-aware test pattern generation technique entails an overhead due to the ILP problem intrinsic complexity.

Finally, due to the internal structure of the circuits, the ideal outcomes (i.e., 100% covered critical faults and 0% covered acceptable faults) are far from being achieved. Thus, in the last phase of the test, a methodology to distinguish acceptable from critical fault, after the test application, is needed. We show this in the next subsection.

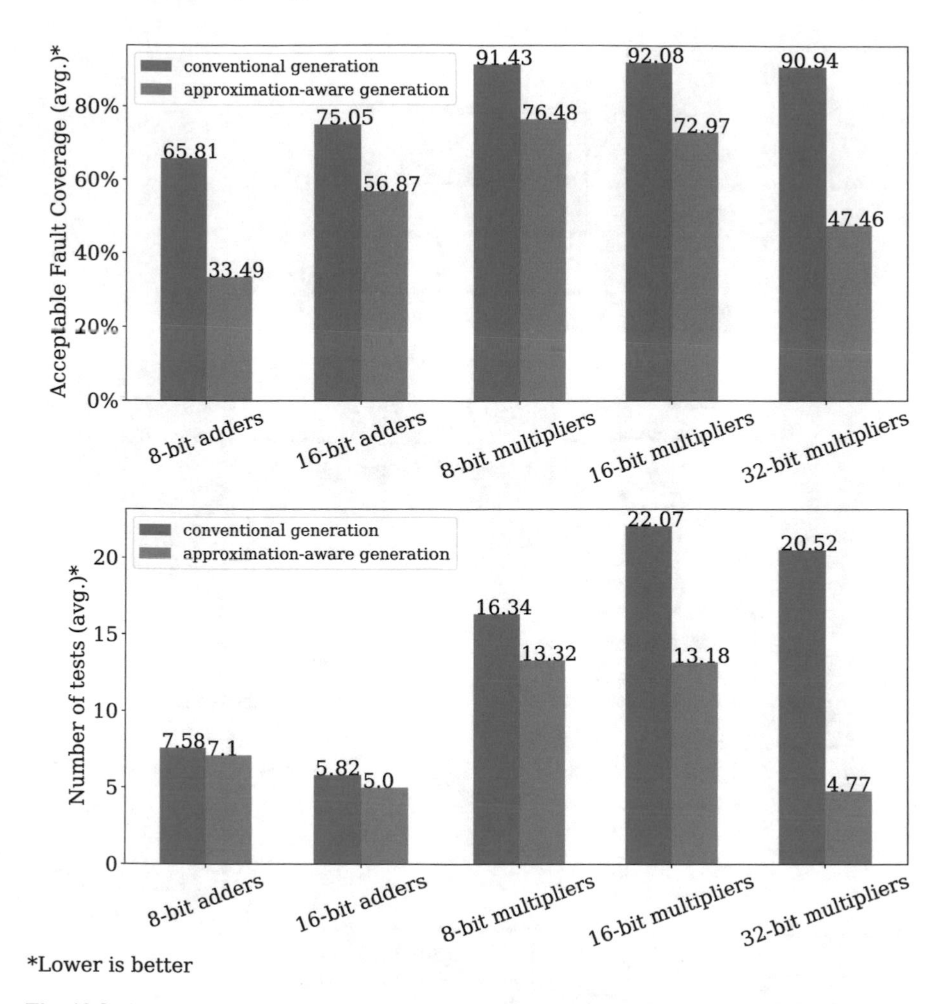

Fig. 10.8 Approximation-aware test pattern generation experimental results from [73, 75]

2.3 *Approximation-Aware Test Set Application*

Test pattern application is the last step of approximation-aware testing. When AxICs are involved, observing a different response than expected not always means that the circuit has to be rejected. Indeed, if a fault causing the issue was acceptable (i.e., its effects are within error bounds), the AxIC has to pass the test. Unfortunately, as shown in the previous subsection, often tests built to detect critical faults also detect acceptable ones. Thus, a methodology to determine whether the detected fault is acceptable or not is needed.

The well-known *signature analysis* concept [23] can help in this context. Briefly, it compacts test responses of a Circuit Under Test (CUT) into a *signature* and compares it with the expected one (produced by the fault-free circuit). If they match,

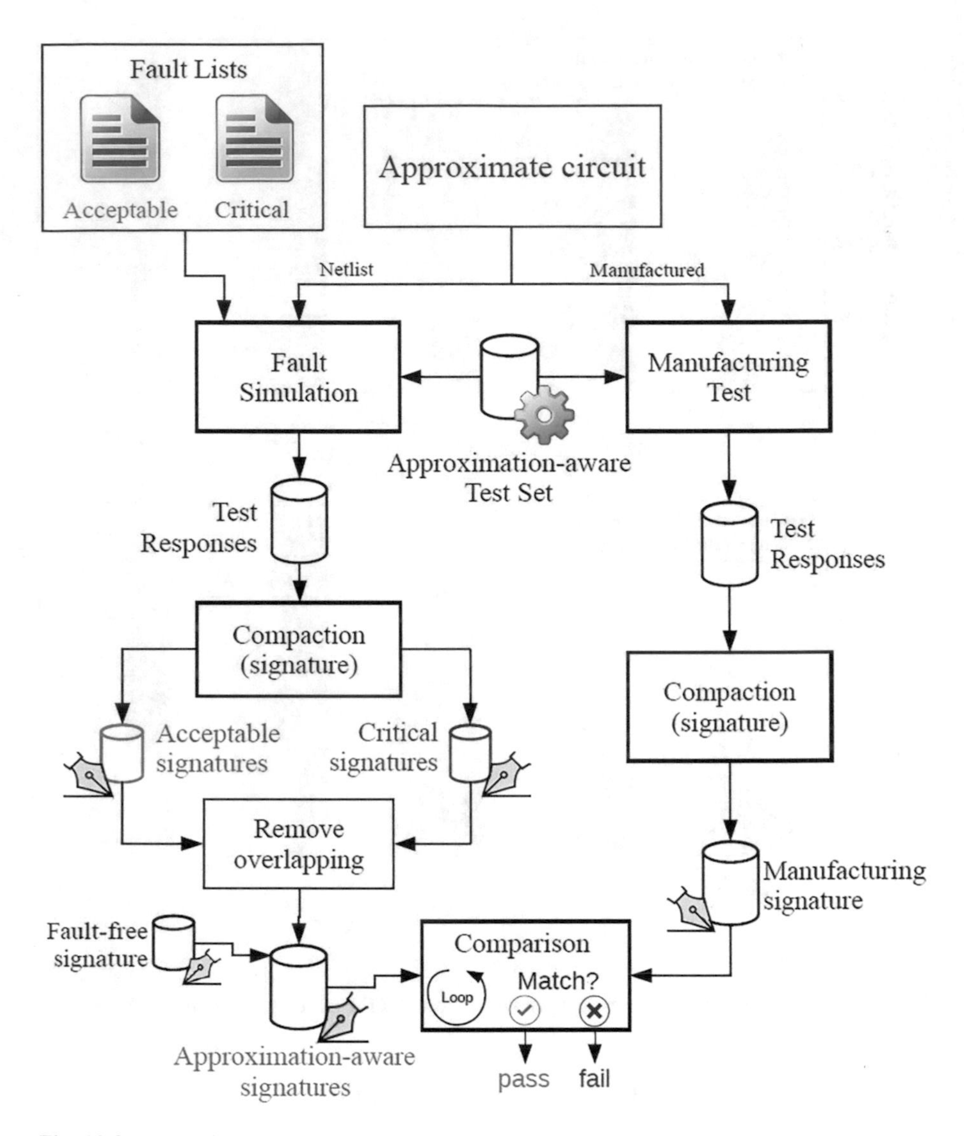

Fig. 10.9 Approximation-aware test application

the CUT passes the test. In [74], this concept is adapted to approximation-aware test, as depicted in Fig. 10.9. It is divided into two phases:

1. At design time (left side, in the figure), test patterns are *fault-simulated* with the AxIC's fault lists. For each fault, a signature is generated. If any, acceptable signatures overlapping with critical ones are removed. The golden signature (i.e., fault-free) is added to the acceptable ones to obtain the *approximation-aware signatures*.

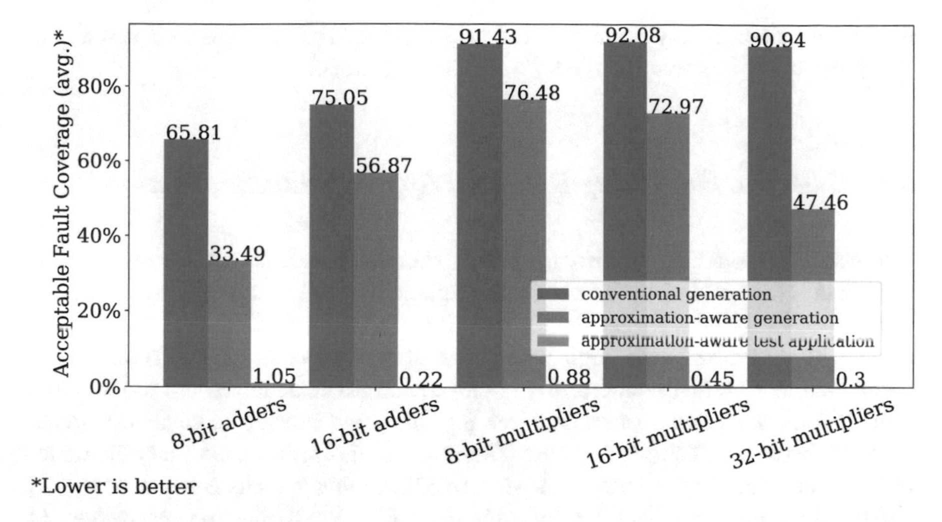

Fig. 10.10 Approximation-aware test set application experimental results from [69, 74]

2. At test time (right side in the figure), the test responses of the manufactured AxIC generate a signature; all the approximation-aware signatures are compared with it; if at least one match is obtained, then the AxIC passes the test.

Results obtained with the discussed technique are reported in Fig. 10.10, in terms of acceptable fault coverage. In general, the technique delivered 100% critical fault coverage and 0.58% acceptable fault coverage on average, which are very close to the ideal ones. For further details, the reader can refer to [74].

3 Fault-Tolerant Architectures Based on Approximate Computing

AxC has already been used in the literature in the context of fault-tolerant architectures. In [28] and [59], the authors presented the *Approximate TMR (ATMR)* and its extension, the *Full ATMR (FATMR)*. Just as the TMR, the ATMR scheme employs three CL copies, two AxICs and a precise one, and the FATMR uses three AxICs. In these architectures, only one AxIC delivers an erroneous response for a given input vector. The idea is that each approximate module has its own unique *domain of approximation*. Since the three modules together always deliver at least two correct outputs, the (F)ATMR can mask any approximate responses coming from one of the AxICs. However, in case of a fault, the so-designed architecture can only protect the circuit for a set of input vectors defined by the designer. Other proposals of a low-cost TMR based on approximate computing were presented in [64] and more recently in [58] and [28, 29]. Finally, authors in [1] show the interest of AxC for fault tolerance in arithmetic circuits. They proposed a configurable-accuracy approximated adder embedding a correction technique. While effective, this solution is also workload-dependent.

3.1 Selective Hardening Based on Approximate Duplication

The selective hardening philosophy, briefly discussed in Sect. 1.2.2, aims at minimizing the cost entailed by fault-tolerant architectures while trying to minimize the related reliability loss. As previously mentioned, AxC goes in a similar direction, aiming at minimizing the logic area cost at the expense of precision in the computation. This makes interesting to employ an AxIC as redundant module in a *duplication and comparison* scheme (see Fig. 10.3) and assess its contribution to the trade-off between reliability and cost. Therefore, in this section, we study the trade-off between reliability and cost of the selective hardening technique proposed in [78] and briefly discussed in Sect. 1.3, by comparing different duplication approaches. In detail, we explore four duplication scenarios:

1. A full duplication scheme
2. A reduced duplication scheme based on the structural susceptibility analysis presented in [80]
3. A reduced duplication scheme based on the logical weights of the arithmetic circuit outputs
4. A reduced duplication scheme based on an approximate circuit from a public benchmark suite [47], composed of arithmetic AxICs

Most of the conceived AxICs are arithmetic circuits, as their precision loss is easily measurable [38]. In light of this, we can assess the precision reduction of the duplication schemes with metrics commonly used for arithmetic circuits. Among the common ones, we resort to the Worst-Case Error (WCE) metric (Eq. (10.2)). Finally, the four considered scenarios are workload-independent.

Next subsection introduces the *structural susceptibility analysis*, and then we describe the different considered duplication scenarios.

3.1.1 Structural Susceptibility Analysis

The structural susceptibility analysis methodology proposed in [80] is based on the fact that not all outputs of a CL block have the same susceptibility to single-event transient (SET) effects and assumes that their susceptibility is a function of the number of nodes in their fan-in logic cone. It exploits the structural properties of the output fan-in cone to get their relative susceptibility estimates. The outputs are ranked on the basis of their relative susceptibility, and the most susceptible ones are selected for error detection. The susceptibility of each fan-in cone is calculated as the sum of all the logic cell weights in the corresponding fan-in cone. The cell weight is calculated as the number of inputs and outputs of that cell.

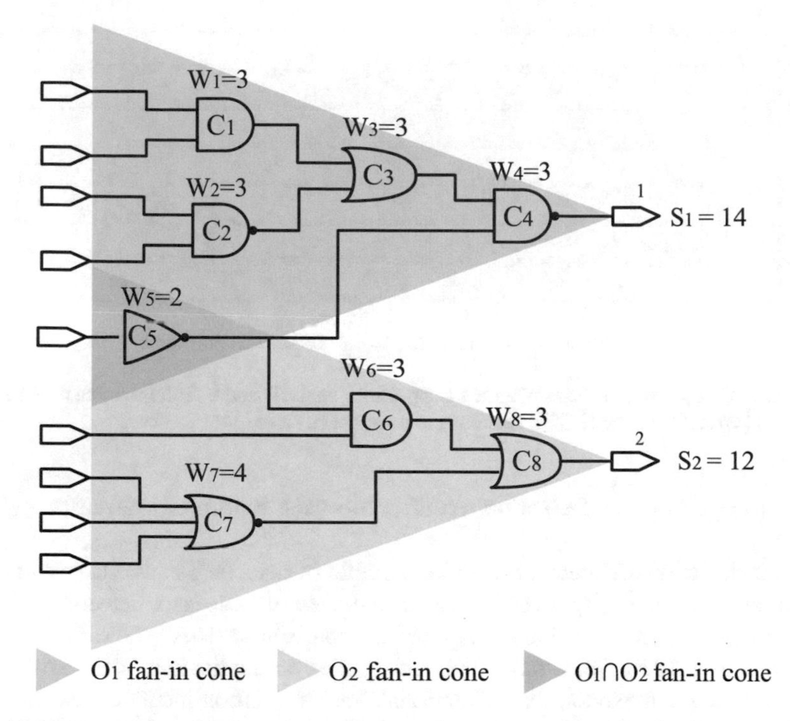

Fig. 10.11 Example of structural susceptibility analysis application to a simple circuit

Figure 10.11 sketches the structural susceptibility analysis application to a simple example circuit. The shaded regions mark the boundaries of the two output fan-in cones. For each gate, W_i indicates the respective weight. The sum of all the weights in the cone gives the preliminary fan-in cone susceptibility value (S_j). In this example, $S_1 = 14$ and $S_2 = 12$ are the preliminary fan-in cone susceptibility values for O_1 and O_2, respectively. Thus, O_1 is more susceptible to propagate an error than O_2. In other words, providing an error detection mechanism on the output O_1 can better improve the reliability of the circuit, compared to having it placed on O_2. Finally, to produce the final susceptibility values, the technique assigns the weight of the shared cells (i.e., belonging to multiple-output fan-in cones) only to the most susceptible output.

In order to provide an example of its effectiveness, we report in Fig. 10.12 (taken from [14]) the comparison of the structural susceptibility analysis results and a fault injection experiment, for the b03 circuit from the ITC'99 benchmark suite. The red line represents the normalized distribution of the structural susceptibility (S_j) for each circuit output; the blue line represents the distribution of the average number of soft error failures observed at the circuit outputs after a fault injection campaign. Further analyses and validations of the structural susceptibility analysis can be found in [80].

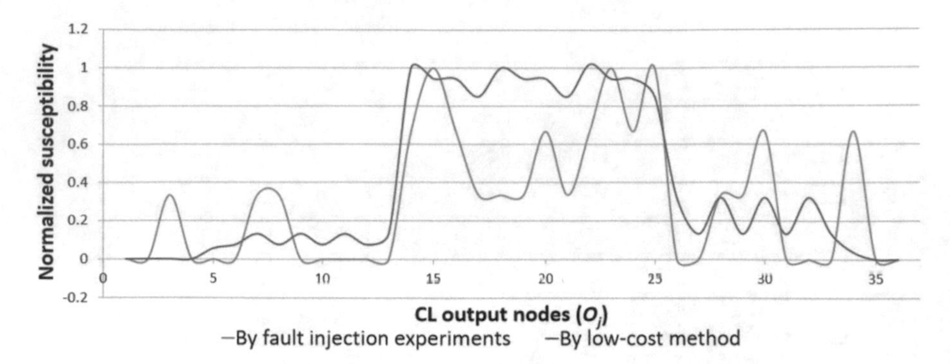

Fig. 10.12 Comparison of the structural susceptibility analysis and a fault injection experiment, for the b03 circuit from the ITC'99 benchmark suite. Taken from [14]

3.1.2 Selective Error Detection Architectures for Arithmetic Circuits

An error detection architecture must be capable of detecting transient, permanent, and timing faults that may occur in an arithmetic circuit. The error detection scheme that we evaluate employs duplication and comparison to detect the occurrence of faults (see Fig. 10.3). Since the architecture relies on duplication of the arithmetic logic and on a comparator, its conventional implementation incurs an overhead of more than 100% in terms of area and power. Cleverly selecting the functions to be duplicated is a practicable way to allow the designer to control the trade-off between the area/power overhead and the reliability improvement. Below, we briefly describe the different duplication scenarios that we compare.

Full Duplication Scheme In this scenario, the architecture is able to detect all faults (transient, permanent, and timing faults) that may occur in the arithmetic circuit. A full comparator circuit is used in this case.

Reduced Duplication Scheme Based on Structural Susceptibility Analysis To obtain multiple circuits having different structural susceptibilities and overheads, we use the structural susceptibility analysis, described in Sect. 3.1.1. As illustrated in Fig. 10.13, each copy is created by selecting a set of outputs ranked according to the analysis. In this scenario, the comparator is reduced since the obtained copies have fewer outputs. This duplication scheme is able to detect only faults affecting the common area between the original circuit and the reduced circuit. Consequently, we have to assess the impact of the undetected faults by using the aforementioned error metrics. In this scenario, the WCE is defined as follows:

$$WCE = \sum_{i \in B} 2^i, \tag{10.5}$$

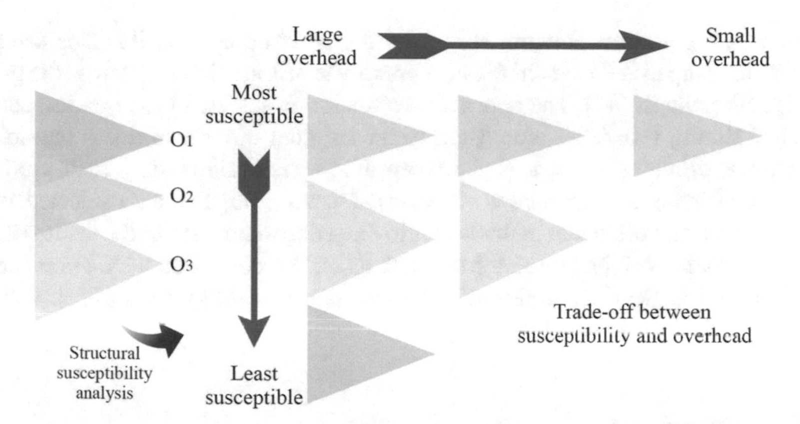

Fig. 10.13 Reduced duplication scheme based on structural susceptibility analysis

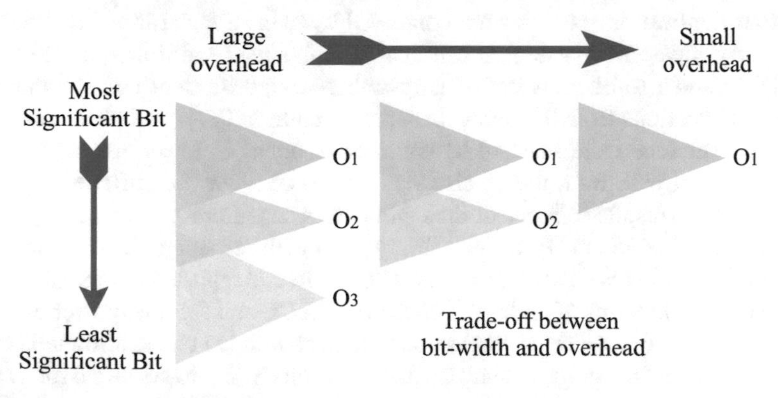

Fig. 10.14 Reduced duplication scheme based on output logic weights

where B indicates the set containing the positions in the original circuit of the outputs that are truncated in the reduced circuit. For example, if the removed outputs are in positions 0 and 2 in the original circuit, then the WCE will be $2^0 + 2^2 = 5$.

Reduced Duplication Scheme Based on Output Logic Weights In this scenario, we consider the possibility to duplicate the arithmetic circuit by using a functional metric. Indeed, we consider that the outputs of the arithmetic circuit can be ranked from Least Significant Bit (LSB) to Most Significant Bit (MSB). This partial duplication scheme may be considered as an Unequal Error Protection (UEP) scheme [6]. As shown in Fig. 10.14, the reduced arithmetic circuits are obtained by eliminating fan-in cones from the one driving the LSB up to the one driving the MSB. In this case, the reduced circuit entailing the smallest overhead corresponds to the logic cone driving the MSB output. Also in this case, we assess the impact of the undetected faults by calculating the error metric (in this case, the WCE in Eq. (10.5)).

Reduced Duplication Scheme Based on Approximate Circuits This scenario consists in using as reduced circuit an approximate arithmetic circuit from the public benchmark suite in [47]. The approximate version is selected based on its reduced area and timing properties, compared to the original circuit. In this scenario, the comparator provides an error signal when the precise arithmetic circuit produces a response having a difference w.r.t. the AxC version larger than a selected WCE value. It is worth noting that in this scenario the reduced circuit has the same number of outputs as the original circuit. More details on the design of such a comparator can be found in Sect. 2.1 (referred to as *evaluation module*) and in [72]. For this scenario, WCE is calculated by using Eq. (10.2).

3.1.3 Comparisons

The four duplication scenarios are compared by using four arithmetic circuits: an 8-bit carry look-ahead adder, an 8-bit carry look-ahead multiplier, a 12-bit array multiplier, and a 16-bit array multiplier. For these four arithmetic circuits, more than 1100 AxC versions from the public benchmark suite in [47] have been considered. The AxC versions were selected by considering equal or lower area and equal or shorter critical path w.r.t. the precise version. To compare the different scenarios, we present the results in terms of area and power consumption overhead (obtained with the NanGate 45 nm Open Cell Library [49]) with respect to the full duplication scenario, as well as WCE metric values. The results are reported in Fig. 10.15. First, they show that in terms of trade-off between the WCE and the area/power overhead, using the reduced duplication based on the logical weights of the arithmetic circuit outputs (LW in the graphs) leads to better results (i.e., lower area for a given WCE) compared to the structural susceptibility analysis (SSA in the graphs). More importantly, when the reduced duplication based on approximate arithmetic circuits is used, an even better trade-off is achieved.

Experimental results demonstrate the interest of using approximate structures as duplication scheme since both area overhead and power consumption are reduced compared to a full duplication scheme, while maintaining good levels on error metrics. The reader can refer to [14] for a more extensive discussion and additional experimental results.

3.2 *Ensuring Fault Tolerance Through Approximate Redundancy*

During the lifespan of a system used in harsh (e.g., radiative) environment, its hardware is subject to various physical phenomena that may alter its performance or provoke errors [81]. Moreover, some systems demand a high level of reliability since failures would imply catastrophic outcomes. Aerospace systems, submarine

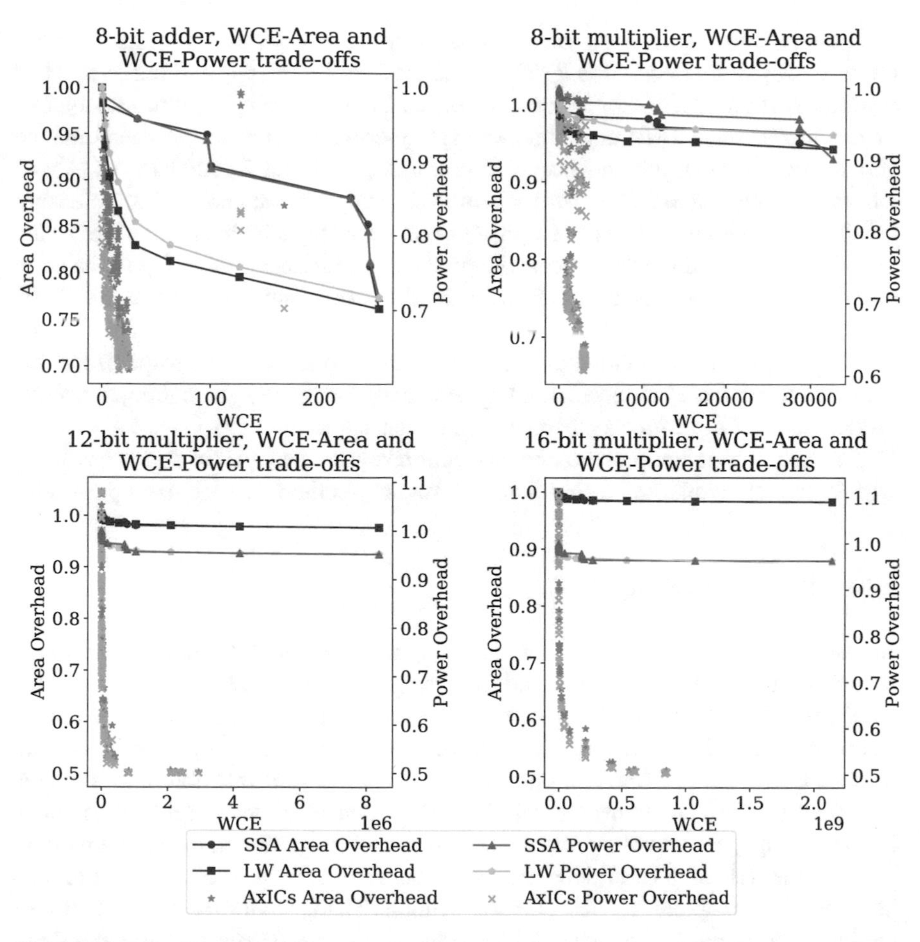

Fig. 10.15 Experimental results comparing the four scenarios: reduced duplication based on the structural susceptibility analysis (SSA); reduced duplication based on the logical weights of the arithmetic circuit outputs (LW); reduced duplication scheme based on an approximate arithmetic circuits from [47] (AxICs)

telecom, or even medical instruments cannot risk particle strikes, wear-out, or aging. However, high levels of reliability usually require heavy fault-tolerant designs to reach such high requirements. Several structures have been designed to maintain the accuracy of these safety-critical applications. A well-known existing structure capable of tolerating soft and hard errors is the Triple Modular Redundancy (TMR), briefly introduced in Sect. 1.2.1. A triplication of the circuit with a majority voter ensures logic error masking at a cost of a 200% area and power overhead. Indeed, a TMR masks (i.e., tolerates) permanent or transient faults occurring in one module—or in several modules, provided that they do not impact the same outputs if several modules are faulty—for any vector applied to its inputs. AxC philosophy may not seem compatible with safety-critical applications. AxC has been applied to applications where an approximate result is sufficient for their purpose [58]. While AxC was applied to TMR [28, 29, 58, 64] and led to reduced overheads, it entailed also a reduced error-masking capability. Unfortunately, this makes approximate TMR not suitable in safety-critical scenarios. To overcome the above issue, in [13], the Quadruple Approximate Modular Redundancy (QAMR) was introduced. QAMR is a novel scheme using approximate computing to ensure full logic masking (tolerance) of transient and permanent faults. As the TMR, the QAMR masks all single faults occurring in the logic replicas, delivering to the voter a majority of correct responses. It achieves the same fault tolerance as the TMR while still benefiting from the advantages of approximate computing. The QAMR uses four approximate circuit replicas. The essential condition to be respected is that the four approximated replicas are complementary, i.e., at any given time, they must produce at least three precise responses (i.e., non-approximated).

The next subsection presents the motivation behind the QAMR approach, along with its fundamentals, and a circuit approximation method adapted to its purpose.

3.2.1 The QAMR Scheme

As already mentioned, several proposals (known as *ATMR*) have been made to reduce the TMR area overhead by using AxC [28, 29, 58, 64]. Unfortunately, the ATMR suffers from severe limitations in terms of reliability. Let us resort to Fig. 10.16 to illustrate the above-mentioned issue. Let f be a generic multi-output function, whose input domain D can be split into four sub-domains $D1$, $D2$, $D3$, $D4$. The conventional TMR approach uses three identical modules implementing f and a voting scheme to ensure fault tolerance in the case a module incurs some defective conditions. In such case, the defective module will produce incorrect or incomplete response, for some inputs. Thanks to the other two fault-free modules, the correct response can be produced. Figure 10.16a sketches the three system copies used in TMR approach. All three copies produce correct results for all the function's domains (D1–D4), when no errors occur. The ATMR approach, on the other hand, proposes to use three reduced (i.e., approximate) modules, as sketched in Fig. 10.16b. The three modules produce correct results only for some of the function's domains, when no errors occur. In the figure, two modules out of three produce the correct results for domains $D2$, $D3$, and $D4$. Using three approximate modules instead of the three original copies surely enables the opportunity to achieve efficiency gains but also exposes the computation to some errors in case of faults. Defects impacting $D2$, $D3$, and $D4$ will not be tolerated. For instance, if a defect impacts the domain $D2$ of the second module, only the third module produces the right output and a correct vote might not be possible. To use the ATMR as fault-tolerant solution for safety-critical applications, only the protected parts of the system (i.e., $D1$ in the example is correct for all the modules) can be critical. Such design constraint can be challenging and not always achievable, even for resilient applications. Therefore, the ATMR is not suitable as fault-tolerant solution for safety-critical applications.

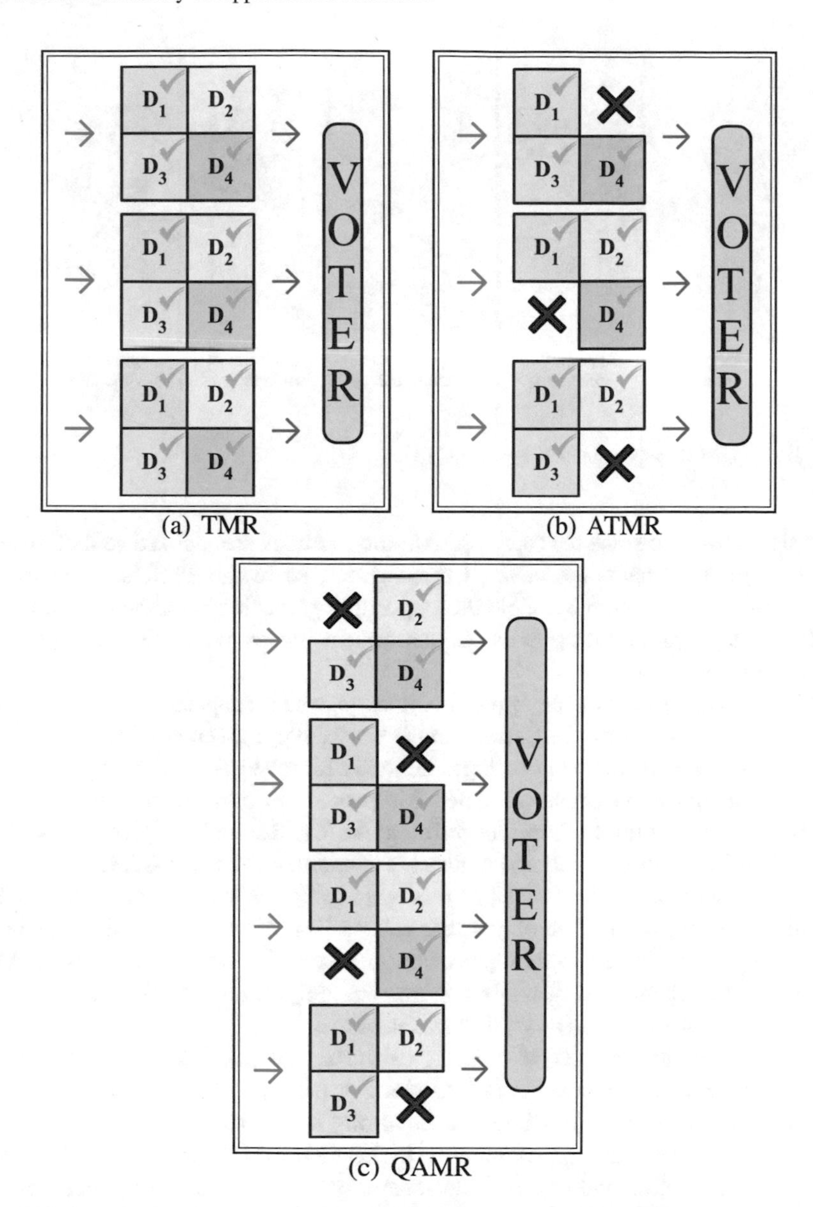

Fig. 10.16 (**a**): Three identical copies are used in TMR: all produce the correct results for all the function's domains (D1–D4) when no errors occur. (**b**): In ATMR approach, the three approximate modules produce exact results only for three out of four function's domains; some domains are not covered three times as in the TMR scheme. (**c**) QAMR approach: four approximate modules cover each function's domain three times, thus obtaining the same coverage as in the TMR scheme

In the light of this, the goal of QAMR is to achieve the TMR reliability level while still profiting from Approximate Computing advantages. To show the principle of the QAMR scheme, we resort to Fig. 10.16c. The QAMR offers a

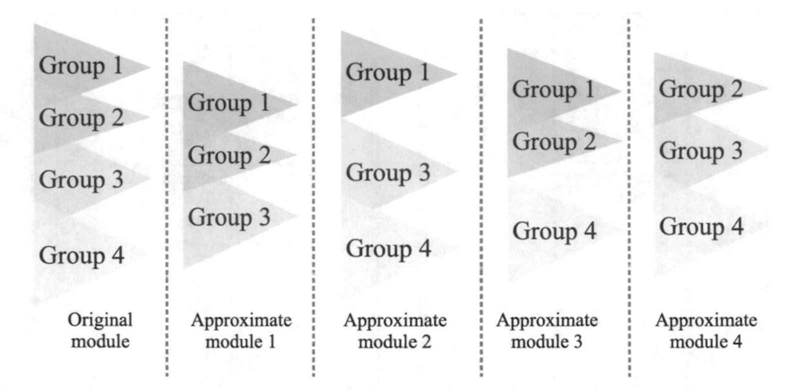

Fig. 10.17 QAMR approximation approach [13]

complete coverage since the four approximate modules are realized so that, overall, the function's domains are covered three times, as in the TMR scheme. At the same time, the four AxICs enable the opportunity to achieve efficiency gains. The underlying insight is that a good AxC technique achieves more gains than it reduces the system's accuracy.

Figure 10.17 sketches the approximation approach proposed in [13] to realize the modules required by the QAMR architecture. For a given circuit, one group of outputs, along with their fan-in logic cone, is removed from the original circuit to obtain a first approximate module; this process is performed four times with different output groups to form four different AxICs. Removing a given output from *only one* of the four approximate replicas is fundamental for the QAMR to work as intended. By realizing the QAMR in this way, using the same voter as the TMR is possible. Indeed, for a given output, the voter still receives three replicas as input. For better clarity, Fig. 10.18 shows an example of a 4-bit output circuit in the QAMR scheme. Since each AxIC has only one missing output, the voter is able to execute the majority vote just as in a classic TMR scheme.

With circuit having lots of outputs, exploring all possible ways of grouping their to-be-removed outputs is not feasible. For example, for a circuit having 245 outputs, the number of possible combinations is $\approx 10^{142}$. Therefore, in [13], the complementary groups of outputs to be removed were generated pseudo-randomly, in an iterative experimental campaign. Experiments have been carried out on combinational circuits from the public LGSynth'91 benchmark suite [83]. For each circuit, the conventional TMR (i.e., three precise replicas and the voter) was compared with the QAMR versions obtained in the experimental campaign (i.e., four approximate modules and the same voter). For further details on the experimental setup, the reader can refer to [13]. Figure 10.19 summarizes the results of the pseudo-random exploration. In the graphs, relative area gain (y-axis) and relative timing gain (x-axis) w.r.t. the original TMR are reported. The TMR is reported in the origin, symbol ★. Both area and timing gains are calculated as follows:

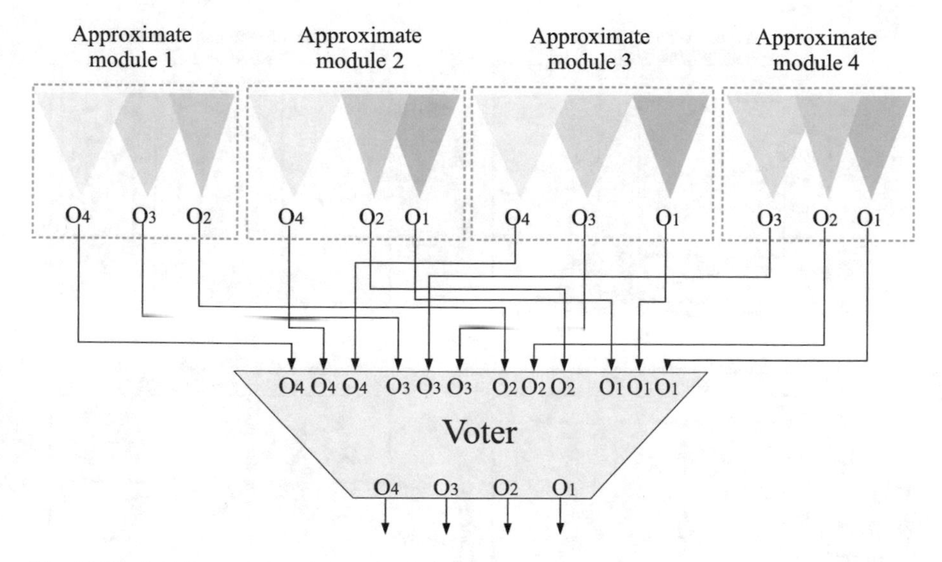

Fig. 10.18 QAMR example with conventional voting scheme

$$\alpha_{\text{gain}}\% = \frac{\alpha_{\text{TMR}} - \alpha_{\text{QAMR}}}{\alpha_{\text{TMR}}} \cdot 100, \tag{10.6}$$

where α represents either the area or the timing of the two architectures.

For each circuit, the graphs show the non-dominated solutions found with the pseudo-random exploration, i.e., the solutions having either better or at least equal area or delay w.r.t. each other solution in the set. They are organized as follows:

1. Circuits presenting at least one non-dominated solution achieving gains in terms of **both area and delay** (*Q1* quadrant, $x > 0 \wedge y > 0$) are reported in the first graph.
2. Circuits—which are not in Q1—presenting at least one non-dominated solution achieving gains only in terms of delay (*Q2* quadrant, $x \leq 0 \wedge y > 0$) are reported in the second graph.
3. Circuits—which are not in Q1—presenting at least one non-dominated solution achieving gains only in terms of area (*Q4* quadrant, $x > 0 \wedge y \leq 0$) are reported in the third graph.
4. Circuits—which are not in Q1—presenting non-dominated solutions either in terms of delay or area are reported in the fourth graph.
5. Circuits presenting all non-dominated solutions not achieving any gains (*Q3* quadrant, $x \leq 0 \wedge y \leq 0$) are reported in the last graph.

To help the reader, in the first and second graphs, we added a $\approx 2\times$ zoom of the densest parts. The graphs highlight that while for some circuits the exploration did not find any solutions achieving gains (e.g., the *decod* circuit, symbol ⋈, quadrant Q3); in general, it is possible to obtain superior QAMR implementations w.r.t. the

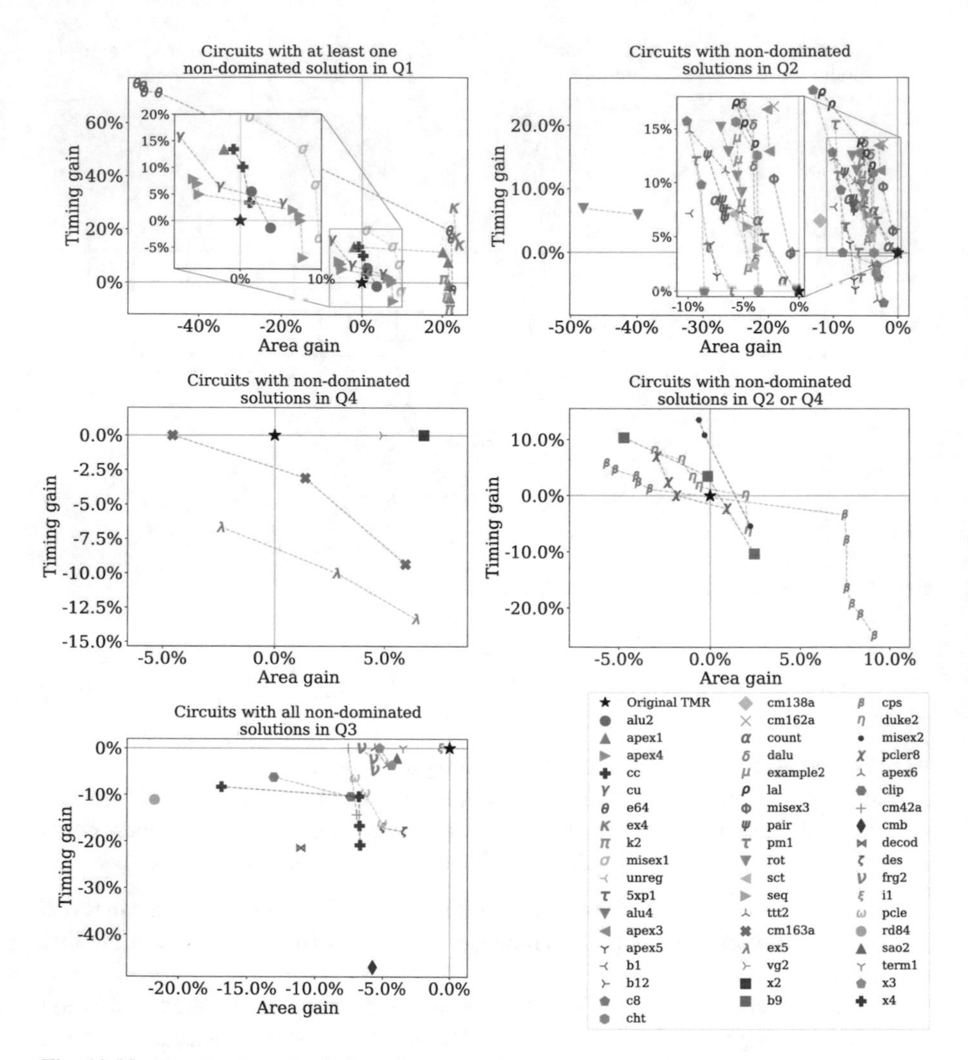

Fig. 10.19 Non-dominated solutions, in terms of relative area and timing gain, obtained with a pseudo-random exploration [13]

original TMR, in terms of timing (e.g., *alu4*, symbol ▼, quadrant Q2), area (e.g., *x2*, symbol ■, quadrant Q4), or both (e.g., *ex4*, symbol *K*, quadrant Q1).

The presented data clearly indicate that QAMR offers a cheaper alternative to the standard TMR scheme for safety-critical applications. The results show that QAMR is feasible and demonstrate that there is a real interest in using AxC to realize more efficient fault-tolerant architectures for safety-critical applications. Since the exploration results are far from being optimal, as also shown in [7], further studies are needed to provide enhanced approximation techniques fully exploiting AxC opportunities in safety-critical scenarios.

4 Conclusion

Regardless of the field of application, i.e., trading, health care, satellite telecommunications, civilian transports, military equipment, data centers, etc., there is an increasing demand of electronics able to perform complex and resource-intensive operations. Most of these operations demand a high degree of reliability, availability, and safety. However, the increasing vulnerability of transistors and interconnects requires electronic system designs to overcome reliability issues, which intensify for every new emergent CMOS technology generation. Furthermore, the complexity of modern electronic systems makes difficult to realize error detection, recovery, masking, etc. Moreover, also area and power limitations, as well as high-performance demands, are compelling requirements to satisfy. These requirements force the industry to limit the overheads in reliability enhancements or come up with more adaptive designs that respond to the reliability problematic of the targeted field of application. Among the work of the last two decades, Approximate Computing brought multiple opportunities to different extents. The fundamental goal is to improve the system efficiency (time/area/energy) by relaxing result's accuracy requirements. This also brought along new challenges, concerning the reliability of electronic chips. In particular, in this chapter, we focused on test and fault-tolerance issues related to approximate hardware. We first showed how Approximate Computing changes the conventional concepts of digital circuit testing and reviewed the approximate-aware test flow, able to deal with such scenario. Then, we discussed how Approximate Computing can be used to reduce the cost of fault detection and fault-tolerance mechanisms and showed how the state-of-the-art methodologies achieve this goal.

Finally, we deem interesting and valuable for the scientific community to move toward the study of new *Approximation-for-Reliability* principles, allowing the development of enhanced approximation methodologies that take into account also reliability aspects.

References

1. Al-Maaitah K, Qiqieh I, Soltan A, Yakovlev A. Configurable-accuracy approximate adder design with light-weight fast convergence error recovery circuit. In: 2017 IEEE Jordan conference on applied electrical engineering and computing technologies (AEECT). 2017. pp. 1–6. https://doi.org/10.1109/AEECT.2017.8257753.
2. Anghel L, Benabdenbi M, Bosio A, Traiola M, Vatajelu EI. Test and reliability in approximate computing. J Electron Test. 2018;34(4):375–87. https://doi.org/10.1007/s10836-018-5734-9.
3. Avirneni NDP, Somani A. Low overhead soft error mitigation techniques for high-performance and aggressive designs. IEEE Trans Comput. 2012;61(4):488–501. https://doi.org/10.1109/TC.2011.31
4. Avizienis A, Laprie JC, Randell B. Fundamental concepts of dependability. 2001. http://www.cs.ncl.ac.uk/publications/trs/papers/739.pdf.

5. Benedetto J, Eaton P, Avery K, Mavis D, Gadlage M, Turflinger T, Dodd P, Vizkelethyd G. Heavy ion-induced digital single-event transients in deep submicron processes. IEEE Trans Nuclear Sci. 2004;51(6):3480–85. https://doi.org/10.1109/TNS.2004.839173.
6. Borade S, Nakiboğlu B, Zheng L. Unequal error protection: An information-theoretic perspective. IEEE Trans Inf Theory. 2009;55(12):5511–39. https://doi.org/10.1109/TIT.2009.2032819
7. Bosio A, O'Connor I, Traiola M, Echavarria J, Teich J, Hanif MA, Shafique M, Hamdioui S, Deveautour B, Girard P, Virazel A, Bertels K. Emerging computing devices: Challenges and opportunities for test and reliability*. In: 2021 IEEE European test symposium (ETS). 2021. pp. 1–10. https://doi.org/10.1109/ETS50041.2021.9465409.
8. Bottoni C, Coeffic B, Daveau JM, Naviner L, Roche P. Partial triplication of a SPARC-V8 microprocessor using fault injection. In: 2015 IEEE 6th Latin American symposium on circuits systems (LASCAS). 2015. pp. 1–4. https://doi.org/10.1109/LASCAS.2015.7250415
9. Bushnell M, Agrawal, V.: Essentials of electronic testing for digital, memory, and mixed-signal VLSI circuits. 2000. https://doi.org/10.1007/b117406
10. Castano V, Schagaev I. Resilient computer system design. Cham: Springer International Publishing; 2015. https://link.springer.com/book/10.1007/978-3-319-15069-7. OCLC: 1194524751.
11. Chandrasekharan A, Eggersglüß S, Große D, Drechsler R. Approximation-aware testing for approximate circuits. In: 2018 23rd Asia and South Pacific design automation conference (ASP-DAC). 2018. pp. 239–244. https://doi.org/10.1109/ASPDAC.2018.8297312.
12. Chen CC, Milor L. Microprocessor aging analysis and reliability modeling due to back-end wearout mechanisms. IEEE Trans Very Large Scale Integr (VLSI) Syst. 2015;23(10):2065–76. https://doi.org/10.1109/TVLSI.2014.2357756.
13. Deveautour B, Traiola M, Virazel A, Girard P. QAMR: an approximation-based fully reliable TMR alternative for area overhead reduction. In: 2020 IEEE European test symposium (ETS). 2020. pp. 1–6. https://doi.org/10.1109/ETS48528.2020.9131574.
14. Deveautour B, Virazel A, Girard P, Gherman V. On using approximate computing to build an error detection scheme for arithmetic circuits. J Electron Test. 2020;36(1):33–46. https://doi.org/10.1007/s10836-020-05858-5. http://link.springer.com/10.1007/s10836-020-05858-5.
15. Dodd P, Shaneyfelt M, Felix J, Schwank J. Production and propagation of single-event transients in high-speed digital logic ICs. IEEE Trans Nuclear Sci. 2004;51(6):3278–84. https://doi.org/10.1109/TNS.2004.839172.
16. Dubrova E. Fault-tolerant design. New York: Springer; 2013. https://doi.org/10.1007/978-1-4614-2113-9. http://link.springer.com/10.1007/978-1-4614-2113-9.
17. Dutta A, Jas A. Combinational logic circuit protection using customized error detecting and correcting codes. In: 9th international symposium on quality electronic design (ISQED 2008). 2008. pp. 68–73. https://doi.org/10.1109/ISQED.2008.4479700. ISSN: 1948-3295.
18. Eldred RD. Test routines based on symbolic logical statements. J. ACM. 1959;6(1):33–37. https://doi.org/10.1145/320954.320957
19. Ernst D, Kim NS, Das S, Pant S, Rao R, Pham T, Ziesler C, Blaauw D, Austin T, Flautner K, Mudge T. Razor: a low-power pipeline based on circuit-level timing speculation. In: Proceedings. 36th annual IEEE/ACM international symposium on microarchitecture, 2003. MICRO-36. 2003. pp. 7–18. https://doi.org/10.1109/MICRO.2003.1253179.
20. Fazeli M, Miremadi S, Ejlali A, Patooghy A. Low energy single event upset/single event transient-tolerant latch for deep subMicron technologies. IET Comput Digit Techniques. 2009;3(3):289. https://doi.org/10.1049/iet-cdt.2008.0099. https://digital-library.theiet.org/content/journals/10.1049/iet-cdt.2008.0099.
21. Fazeli M, Ahmadian SN, Miremadi SG, Asadi H, Tahoori MB. Soft error rate estimation of digital circuits in the presence of Multiple Event Transients (METs). In: 2011 design, automation test in Europe. 2011. pp. 1–6. https://doi.org/10.1109/DATE.2011.5763020. ISSN: 1558-1101.
22. Ferlet-Cavrois V, Massengill LW, Gouker P. Single event transients in digital CMOS—a review. IEEE Trans Nuclear Sci. 2013;60(3):1767–90. https://doi.org/10.1109/TNS.2013.2255624.

23. Frohwerk RA. Signature analysis: a new digital field service method. Hewlett-Packard Journal. 1977;28(9):2–8.
24. Fujiwara H. FAN: A fanout-oriented test pattern generation algorithm. In: The IEEE international symposium on circuits and systems (ISCAS). 1985. https://www.researchgate.net/publication/234044505_FAN_A_fanout\discretionary-oriented_test_pattern_generation_algorithm.
25. Gebregiorgis A, Tahoori MB. Test pattern generation for approximate circuits based on Boolean satisfiability. In: 2019 Design, automation test in Europe conference exhibition (DATE). 2019. pp. 1028–1033. https://doi.org/10.23919/DATE.2019.8714898.
26. Gielen G, Wit PD, Maricau E, Loeckx J, Martin-Martinez J, Kaczer B, Groeseneken G, Rodriguez R, Nafria M. Emerging yield and reliability challenges in nanometer CMOS technologies. In: Design, Automation and Test in Europe (DATE). 2008. pp. 1322–1327. https://doi.org/10.1109/DATE.2008.4484862.
27. Göessel M, Ocheretny V, Sogomonyan E, Marienfeld D. New methods of concurrent checking, frontiers in electronic testing, vol. 42. Dordrecht: Springer Netherlands; 2008. https://doi.org/10.1007/978-1-4020-8420-1. http://link.springer.com/10.1007/978-1-4020-8420-1.
28. Gomes IA, Martins MG, Reis AI, Kastensmidt FL. Exploring the use of approximate TMR to mask transient faults in logic with low area overhead. Microelectron Reliab. 2015;55(9):2072–2076. https://doi.org/10.1016/j.microrel.2015.06.125. https://www.sciencedirect.com/science/article/pii/S0026271415300676. Proceedings of the 26th European Symposium on Reliability of Electron Devices, Failure Physics and Analysis.
29. Gomes IAC, Martins M, Reis A, Kastensmidt FL. Using only redundant modules with approximate logic to reduce drastically area overhead in TMR. In: 2015 16th Latin-American test symposium (LATS). 2015. pp. 1–6. https://doi.org/10.1109/LATW.2015.7102522.
30. Hamdioui S. Electronics and computing in nano-era: The good, the bad and the challenging. In: 2015 10th international conference on design technology of integrated systems in nanoscale era (DTIS). 2015. pp. 1–1. https://doi.org/10.1109/DTIS.2015.7127342
31. Hareland S, Maiz J, Alavi M, Mistry K, Walsta S, Dai C. Impact of CMOS process scaling and SOI on the soft error rates of logic processes. In: 2001 Symposium on VLSI technology. Digest of technical papers (IEEE Cat. No.01 CH37184). 2001. pp. 73–74. https://doi.org/10.1109/VLSIT.2001.934953.
32. Heimerdinger W, Weinstock C. A conceptual framework for system fault tolerance. Tech. Rep. CMU/SEI-92-TR-033, Software Engineering Institute, Carnegie Mellon University, Pittsburgh, PA. 1992. http://resources.sei.cmu.edu/library/asset-view.cfm?AssetID=11747.
33. Huang W, Stan MR, Gurumurthi S, Ribando RJ, Skadron K. Interaction of scaling trends in processor architecture and cooling. In: 2010 26th Annual IEEE semiconductor thermal measurement and management symposium (SEMI-THERM). 2010. pp. 198–204. https://doi.org/10.1109/STHERM.2010.5444290.
34. International Roadmap for Devices and Systems (IRDS™) 2020 Edition - IEEE IRDS™. https://irds.ieee.org/editions/2020.
35. Kahng AB, Kang S. Accuracy-configurable adder for approximate arithmetic designs. In: DAC design automation conference 2012. 2012. pp. 820–5. https://doi.org/10.1145/2228360.2228509.
36. Koren I, Krishna CM. Fault-tolerant systems. San Francisco (CA): Morgan Kaufmann; 2021. https://doi.org/10.1016/C2018-0-02160-X. www.sciencedirect.com/book/9780128181058/fault-tolerant-systems.
37. Kumar R. Temperature adaptive and variation tolerant CMOS circuits. Madison: University of Wisconsin; 2008.
38. Liang J, Han J, Lombardi F. New metrics for the reliability of approximate and probabilistic adders. IEEE Trans Comput. 2013;62(9):1760–71. https://doi.org/10.1109/TC.2012.146
39. Lienig, J, Bruemmer, H.: Reliability Analysis. In: Fundamentals of Electronic Systems Design, pp. 45–73. Springer International Publishing, Cham (2017). https://doi.org/10.1007/978-3-319-55840-0_4. http://link.springer.com/10.1007/978-3-319-55840-0_4.

40. Lyons RE, Vanderkulk W. The use of triple-modular redundancy to improve computer reliability. IBM J Res Devel. 1962;6(2):200–9. https://doi.org/10.1147/rd.62.0200.
41. Maheshwari A, Burleson W, Tessier R. Trading off transient fault tolerance and power consumption in deep submicron (DSM) VLSI circuits. IEEE Trans Very Large Scale Integr (VLSI) Syst. 2004;12(3):299–311. https://doi.org/10.1109/TVLSI.2004.824302. http://ieeexplore.ieee.org/document/1281801/
42. Maniatakos M, Makris Y. Workload-driven selective hardening of control state elements in modern microprocessors. In: 2010 28th VLSI test symposium (VTS). 2010. pp. 159–64. https://doi.org/10.1109/VTS.2010.5469589. ISSN: 2375-1053.
43. Mathew J, Shafik RA, Pradhan DK, editors. Energy-efficient fault-tolerant systems. New York: Springer; 2014. https://doi.org/10.1007/978-1-4614-4193-9. http://link.springer.com/10.1007/978-1-4614-4193-9.
44. Mehrara M, Attariyan M, Shyam S, Constantinides K, Bertacco V, Austin T. Low-cost protection for SER upsets and silicon defects. In: 2007 Design, automation test in Europe conference exhibition. 2007. pp. 1–6. https://doi.org/10.1109/DATE.2007.364449.
45. Mittal S. A survey of techniques for approximate computing. ACM Comput Surv. 2016;48(4):62:1–62:33. https://doi.org/10.1145/2893356.
46. Mohanram K, Touba N. Cost-effective approach for reducing soft error failure rate in logic circuits. In: International test conference, 2003. Proceedings. ITC 2003. vol. 1. 2003. pp. 893–901. https://doi.org/10.1109/TEST.2003.1271075. ISSN: 1089-3539.
47. Mrazek V, Hrbacek R, Vasicek Z, Sekanina L. EvoApprox8b: Library of approximate adders and multipliers for circuit design and benchmarking of approximation methods. In: Design, automation test in Europe conference exhibition (DATE). 2017. pp. 258–61. https://doi.org/10.23919/DATE.2017.7926993.
48. Naeimi H, DeHon A. Fault-tolerant sub-lithographic design with rollback recovery. Nanotechnology. 2008;19(11):115708. https://doi.org/10.1088/0957-4484/19/11/115708. https://iopscience.iop.org/article/10.1088/0957-4484/19/11/115708.
49. NanGate: Nangate 45nm open cell library. http://www.nangate.com/?pageid=2325.
50. Neumann Jv. Probabilistic logics and the synthesis of reliable organisms from unreliable components. In: Shannon CE, McCarthy J, editors. Automata studies. (AM-34). 1956. pp. 43–98. Princeton University Press. https://doi.org/10.1515/9781400882618-003. https://www.degruyter.com/document/doi/10.1515/9781400882618-003/html.
51. Oda S, Ferry DK, editors. Nanoscale silicon devices, 0 edn. CRC Press; 2018. https://doi.org/10.1201/b19251. https://www.taylorfrancis.com/books/9781482228687.
52. Pagliarini SN, Naviner LAdB, Naviner JF. Selective hardening methodology for combinational logic. In: 2012 13th Latin American test workshop (LATW). 2012. pp. 1–6. https://doi.org/10.1109/LATW.2012.6261262. ISSN: 2373-0862.
53. Peterson WW, Weldon EJ. Error-correcting codes, 2nd ed. Cambridge: MIT Press; 1972.
54. Polian I, Reddy SM, Becker B. Scalable calculation of logical masking effects for selective hardening against soft errors. In: 2008 IEEE Computer Society annual symposium on VLSI. 2008. pp. 257–262. https://doi.org/10.1109/ISVLSI.2008.22. ISSN: 2159-3477.
55. Rehman S, Prabakaran BS, El-Harouni W, Shafique M, Henkel J. Heterogeneous approximate multipliers: architectures and design methodologies. 2019. pp. 45–66. Springer. https://doi.org/10.1007/978-3-319-99322-5_3.
56. Rushby J. Formal methods and their role in the certification of critical systems. In: Shaw R, editor. Safety and reliability of software based systems. London: Springer; 1997. pp. 1–42.
57. Sachdev M. Defect oriented testing for CMOS analog and digital circuits, Frontiers in electronic testing. vol. 10. Boston: Springer US; 1999. https://doi.org/10.1007/978-1-4757-4926-7. http://link.springer.com/10.1007/978-1-4757-4926-7.
58. Sánchez-Clemente A, Entrena L, García-Valderas M, López-Ongil C. Logic masking for set mitigation using approximate logic circuits. In: 2012 IEEE 18th international on-line testing symposium (IOLTS). 2012. pp. 176–181. https://doi.org/10.1109/IOLTS.2012.6313868.
59. Sanchez-Clemente AJ, Entrena L, Hrbacek R, Sekanina L. Error mitigation using approximate logic circuits: A comparison of probabilistic and evolutionary approaches. IEEE Trans Reliab. 2016;65(4):1871–83. https://doi.org/10.1109/TR.2016.2604918.

60. Santoro M. New methodologies for eliminating no trouble found, no fault found and other non repeatable failures in depot settings. In: 2008 IEEE AUTOTESTCON. 2008. pp. 336–40. https://doi.org/10.1109/AUTEST.2008.4662636. ISSN: 1558-4550.
61. Segura J, Hawkins CF. CMOS electronics: how it works, how it fails. New York: IEEE Press/Wiley-Interscience; 2004. OCLC: ocm53192483.
62. Shafique M, Ahmad W, Hafiz R, Henkel J. A low latency generic accuracy configurable adder. In: 2015 52nd ACM/EDAC/IEEE design automation conference (DAC). 2015. pp. 1–6. https://doi.org/10.1145/2744769.2744778.
63. Shivakumar P. Techniques to improve the hard and soft error reliability of distributed architectures. Thesis. 2007. https://repositories.lib.utexas.edu/handle/2152/3304.
64. Sierawski BD, Bhuva BL, Massengill LW. Reducing soft error rate in logic circuits through approximate logic functions. IEEE Trans Nuclear Sci. 2006;53(6):3417–21. https://doi.org/10.1109/TNS.2006.884352.
65. Sosnowski J. Transient fault tolerance in digital systems. IEEE Micro. 1994;14(1):24–35. https://doi.org/10.1109/40.259897. http://ieeexplore.ieee.org/document/259897/.
66. Srinivasan J, Adve S, Bose P, Rivers J. The case for lifetime reliability-aware microprocessors. In: Proceedings. 31st Annual international symposium on computer architecture. 2004. pp. 276–287. https://doi.org/10.1109/ISCA.2004.1310781.
67. Srinivasan J, Adve S, Bose P, Rivers J. The impact of technology scaling on lifetime reliability. In: International conference on dependable systems and networks. 2004. pp. 177–186. https://doi.org/10.1109/DSN.2004.1311888.
68. Subramanian V, Somani AK. Conjoined pipeline: Enhancing hardware reliability and performance through organized pipeline redundancy. In: 2008 14th IEEE Pacific Rim international symposium on dependable computing. 2008. pp. 9–16. https://doi.org/10.1109/PRDC.2008.54.
69. Traiola M. Test techniques for approximate digital circuits. PhD thesis, Université Montpellier. 2019. https://tel.archives-ouvertes.fr/tel-02485781.
70. Traiola M, Virazel A, Girard P, Barbareschi M, Bosio A. Investigation of mean-error metrics for testing approximate integrated circuits. In: 2018 IEEE international symposium on defect and fault tolerance in VLSI and nanotechnology systems (DFT). 2018. pp. 1–6. https://doi.org/10.1109/DFT.2018.8602939.
71. Traiola M, Virazel A, Girard P, Barbareschi M, Bosio A. On the comparison of different ATPG approaches for approximate integrated circuits. In: IEEE 21st international symposium on design and diagnostics of electronic circuits systems. 2018. pp. 85–90. https://doi.org/10.1109/DDECS.2018.00022.
72. Traiola M, Virazel A, Girard P, Barbareschi M, Bosio A. Testing approximate digital circuits: Challenges and opportunities. In: 2018 IEEE 19th Latin-American test symposium (LATS). 2018. pp. 1–6. https://doi.org/10.1109/LATW.2018.8349681.
73. Traiola M, Virazel A, Girard P, Barbareschi M, Bosio A. A test pattern generation technique for approximate circuits based on an ILP-formulated pattern selection procedure. IEEE Trans Nanotechnol. 2019. p. 1. https://doi.org/10.1109/TNANO.2019.2923040.
74. Traiola M, Virazel A, Girard P, Barbareschi M, Bosio A. Maximizing yield for approximate integrated circuits. In: 2020 design, automation test in Europe conference exhibition (DATE). 2020.
75. Traiola M, Virazel A, Girard P, Barbareschi M, Bosio A. A survey of testing techniques for approximate integrated circuits. Proc IEEE. 2020;108(12):2178–94. https://doi.org/10.1109/JPROC.2020.2999613.
76. Tran D, Virazel A, Bosio A, Dilillo L, Girard P, Pravossoudovitch S, Wunderlich HJ. A hybrid fault tolerant architecture for robustness improvement of digital circuits. In: 2011 Asian test symposium. 2011. pp. 136–141. https://doi.org/10.1109/ATS.2011.89. ISSN: 2377-5386.
77. Wali I. Circuit and system fault tolerance techniques. PhD thesis, Université Montpellier. 2016. https://tel.archives-ouvertes.fr/tel-01807927.

78. Wali I, Deveautour B, Virazel A, Bosio A, Girard P, Sonza Reorda M. A low-cost reliability vs. cost trade-off methodology to selectively harden logic circuits. J Electron Test. 2017;33(1):25–36. https://doi.org/10.1007/s10836-017-5640-6. https://doi.org/10.1007/s10836-017-5640-6.
79. Wali I, Traiola M, Virazel A, Girard P, Barbareschi M, Bosio A. Towards approximation during test of integrated circuits. In: 2017 IEEE 20th international symposium on design and diagnostics of electronic circuits systems (DDECS). 2017. pp. 28–33. https://doi.org/10.1109/DDECS.2017.7934574.
80. Wali I, Virazel A, Bosio A, Dilillo L, Girard P. An effective hybrid fault-tolerant architecture for pipelined cores. In: 2015 20th IEEE European test symposium (ETS). 2015. pp. 1–6. https://doi.org/10.1109/ETS.2015.7138733.
81. Weide-Zaage K, Chrzanowska-Jeske M. Semiconductor devices in harsh conditions. https://www.routledge.com/Semiconductor-Devices-in-Harsh-Conditions/Weide-Zaage-Chrzanowska-Jeske/p/book/9780367656362.
82. Wirnshofer M. Variation-aware adaptive voltage scaling for digital CMOS circuits. Springer series in advanced microelectronics. Springer Netherlands; 2013. https://doi.org/10.1007/978-94-007-6196-4. https://www.springer.com/gp/book/9789400761957.
83. Yang S. Logic synthesis and optimization benchmarks user guide version 3.0; 1991. https://doi.org/10.1.1.49.591.
84. Yao J, Okada S, Masuda M, Kobayashi K, Nakashima Y. DARA: A low-cost reliable architecture based on unhardened devices and its case study of radiation stress test. IEEE Trans Nuclear Sci. 2012;59(6):2852–8. https://doi.org/10.1109/TNS.2012.2223715.
85. Yazdanbakhsh A, Mahajan D, Esmaeilzadeh H, Lotfi-Kamran P. AxBench: A multiplatform benchmark suite for approximate computing. IEEE Des Test. 2017;34(2):60–8. https://doi.org/10.1109/MDAT.2016.2630270.
86. Ye R, Wang T, Yuan F, Kumar R, Xu Q. On reconfiguration-oriented approximate adder design and its application. In: 2013 IEEE/ACM international conference on computer-aided design (ICCAD). 2013. pp. 48–54. https://doi.org/10.1109/ICCAD.2013.6691096.
87. Zhou Q, Mohanram K. Gate sizing to radiation harden combinational logic. IEEE Trans Comput Aided Des Integr Circ Syst. 2006;25(1):155–66. https://doi.org/10.1109/TCAD.2005.853696.
88. Zhu N, Goh WL, Yeo KS. An enhanced low-power high-speed adder for error-tolerant application. In: Proceedings of the 2009 12th international symposium on integrated circuits. 2009. pp. 69–72.
89. Zoellin CG, Wunderlich HJ, Polian I, Becker B. Selective hardening in early design steps. In: 2008 13th European test symposium. 2008. pp. 185–190. https://doi.org/10.1109/ETS.2008.30. ISSN: 1558-1780.

Part III
Introduction: Security

Weiqiang Liu and Fabrizio Lombardi

The third part deals with a specific topic, namely security; recently, security of approximate circuits and systems has grown in importance, and numerous research activities have been reported. The first chapter is titled "Security Vulnerabilities in Approximate Circuits," authored by Chongyan Gu, Yuqin Dou, Weiqiang Liu, and Máire O'Neill. Approximate computing (AC) introduces security vulnerabilities, because uncertain and unpredictable errors during an approximate execution may be indistinguishable from malicious modifications of the input data, the execution process, and the results. In this chapter, a comprehensive analysis of security vulnerabilities in approximate computing is presented; specifically, the security threats in approximate circuits and approximate testing techniques are investigated and analyzed. A countermeasure and a self-detection technique to tampering an attack are also proposed, and an experiment is pursued to evaluate the effectiveness of the proposed method. Experimental results show that the proposed method is effective for the tampering attack.

A physical-based technique is employed in the next chapter "Voltage Overscaling Techniques for Security Applications" by Md Tanvir Arafin, Qian Xu, and Gang Qu. Voltage overscaling (VOS) has emerged as a promising solution for low-power approximate computation. However, aggressive scaling of the operating voltage may result in faults/errors and introduce analog fingerprints of circuit components. For example, VOS has been utilized as a fault-inducing technique in attacks (such as clock screw), in which voltage regulators in system-on-chip processors are exploited to generate errors in the cryptographic computation. As errors due to aggressive VOS reveal a significant amount of information about the underlying hardware's physical variations, this chapter details the status of research and development for VOS in security applications, from authenticating devices to securing machine learning models in main memory. The objective is to demonstrate the security promises and pitfalls of VOS to engineers designing the next generation of low-power voltage overscaled systems.

The application of approximate computing to cryptography is the topic of the next chapter by Dur-e-Shahwar Kundi, Ayesha Khalid, Song Bian, and Weiqiang

Liu. Approximation has also the potential to be utilized for area and power efficiency in the domain of information security. This chapter surveys the practicality of the deployment of approximate computing for cryptographic primitives and applications, together with the implications on correctness as well as security-level reduction.

The last chapter of this part is "Towards Securing Approximate Computing Systems: Security Threats and Attack Mitigation," authored by Qiaoyan Yu, Pruthvy Yellu, and Landon Buell. Recent literature has shown that some AC mechanisms can be exploited by attackers to implement new attack surfaces. In this chapter, unique attacks that are applicable to AC systems are introduced and examples of practical attacks are provided. Furthermore, general principles of defense mechanism designs are proposed to strengthen the resilience of AC systems against emerging attacks. Several countermeasures are examined in applications, such as digital signal processing and artificial neural networks.

Chapter 11
Security Vulnerabilities and Countermeasures for Approximate Circuits

Chongyan Gu, Yuqin Dou, Weiqiang Liu, and Máire O'Neill

1 Introduction

In the last decades, various advanced computing systems, including supercomputers, ubiquitous computing centers, and servers have been developed and widely deployed. However, due to the limit of Moore's law [8], conventional computing techniques are lack of providing further performance improvement. Therefore, new nanoscale computing paradigms are urgently required for low power and high performance computing systems. Implementing a fully correct computation design will dramatically increase the cost due to the low power supply and higher integration density at nanoscale. The International Technology Roadmap for Semiconductors (ITRS) reported that fault tolerance can significantly reduce the cost of manufacturing testing and verification for devices [9]. Hence, appropriate reduction of computational accuracy could effectively improve the performance of computing systems without sacrificing functionality and perception.

Approximate computing, which accepts errors during computation without affecting the results of certain human perception and recognition, has been introduced. It has been widely employed to many applications, such as artificial intelligence (AI), machine learning (ML), signal processing, etc., in which noisy data and erroneous results are acceptable for the computation. Research in this area has attracted a number of interests from both academia and industry [7, 11, 18, 26].

C. Gu (✉) · M. O'Neill
Centre for Secure Information Technologies (CSIT), Queen's University Belfast, Belfast, UK
e-mail: c.gu@qub.ac.uk; m.oneill@ecit.qub.ac.uk

Y. Dou · W. Liu
College of Electronic and Information Engineering, Nanjing University of Aeronautics and Astronautics, Nanjing, China
e-mail: douyuqin@nuaa.edu.cn; liuweiqiang@nuaa.edu.cn

W. Liu, F. Lombardi (eds.), *Approximate Computing*,
https://doi.org/10.1007/978-3-030-98347-5_11

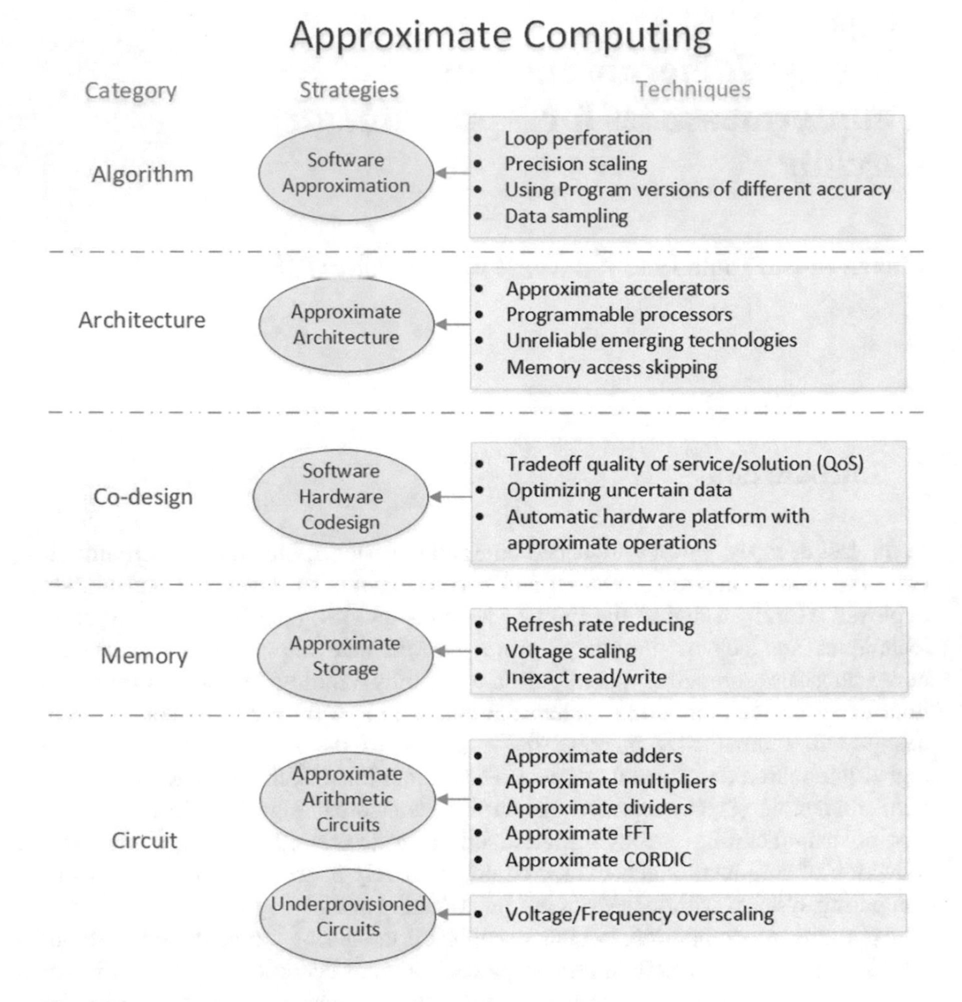

Fig. 11.1 Approximate computing strategies and techniques

Approximate computing, as shown in Fig. 11.1, can be applied to different categories, in hardware and software and in different layers of systems.

Approximate computing techniques are crucial for energy-efficient systems and are being considered for high speed and low power nanoscale integrated circuit (IC) designs [5, 12]. Google's deep learning (DL) chip, which embedded tensor processing unit (TPU), has applied common approximate computing techniques, such as precision scaling [12], to enhance performance. IBM research also launched a project for building on-chip AI accelerators with approximate computing techniques, such as precision scaling and approximate arithmetic units, which achieve a 4-200x speedup over existing methods [5].

Previous research focused on efficiently improving performance with acceptable loss of accuracy [19, 26]. However, the security implications of approximate computing have not been well studied. Recent research has shown that approximate computing may introduce security vulnerabilities due to the uncertainty and unpredictability of intrinsic errors during approximate processing, which is indistinguishable from malicious manipulations [16]. If approximate computing has potential security threats, its applications will undoubtedly be affected. To ensure the security of an approximate design, it is important to understand how to differentiate errors which my be introduced by the approximate design or malicious manipulations.

Yellu et al. [29] has investigated several potential threat models and countermeasures for approximate circuits. [20] has presented that the approximate circuit may have the threat of leaking information, but no specific design was provided. hardware Trojan (HT), as one of the most harmful hardware attacks, has been widely investigated. Approximate circuits could lead to HTs being more easily inserted into systems. Approximate circuits are typically controlled by an additional signal that does not belong to the original design. Hence, it could be easily exploited to trigger a HT. Recent research in [4] has pointed out that approximate adder circuits may be vulnerable to hardware Trojans than exact adder circuits. In this book chapter, a comprehensive analysis of the security vulnerabilities of approximate circuits, including the impact of HT attacks to approximate arithmetic circuits, will be investigated.

2 Approximate Arithmetic Circuits

Approximate computing can improve computing performance by losing the accuracy of calculation results. It has attracted many research efforts, especially for approximate arithmetic circuits, which simplify circuit designs to achieve approximate operations of the desired function, such as addition, multiplication, and division. The main approximate arithmetic circuits include approximate adder [3, 6], approximate multipliers [13–15], and approximate dividers [2].

Arithmetic circuits play an important role in the processor [25], in which arithmetic circuits directly affect the performance and power consumption of the whole computing system. Hence, it is expected to achieve higher speed and lower power consumption as well as satisfied the use of fault-tolerant applications. Most of the approximate computing circuits are based on the approach of logic reduction, which includes logic gate reduction or replacement with different gate elements. Although the approximate circuit is computationally more efficient and consumes less energy, the current development is mostly reflected in the design level, while ignoring its security issue.

Among the approximate arithmetic units, the study of approximate adders has attracted a lot of attention. Specifically, faster and more energy-efficient adders, achieved by designing shorter carry chains or using specific bits, have been widely

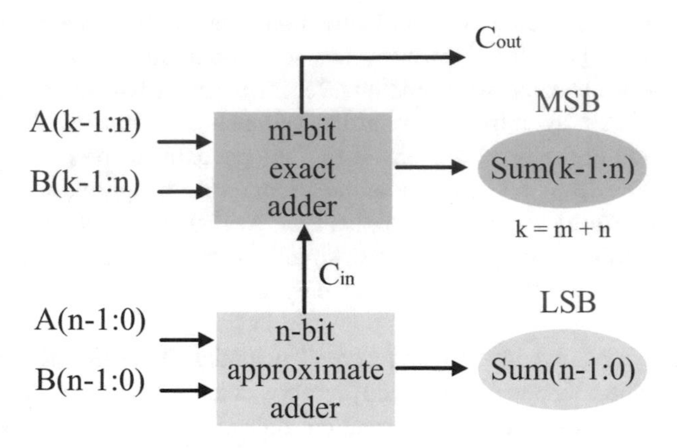

Fig. 11.2 The revised LOA adder structure [17]

developed. An overview and classification of current adders is presented in [17]. A survey [11] has also reviewed different types of approximate adders.

A low-part-or-adder (LOA) design is chosen as an example to illustrate the design of approximate adders. The idea of LOA is to execute exact computation from the most significant bits in an output and inaccurate computation for the least significant bits in the output. As shown in Fig. 11.2, a k-bit LOA consists of two parts: an m-bit exact adder and an n-bit inaccurate adder. The revised LOA will be employed as an example to present the security analysis of approximate circuits. The exact component is composed of n exact 1-bit adders connected in parallel, and the approximate component is executed using m approximate 1-bit adders. For a k-bit adder, $k = n + m$, where k is the k-bit adder, n represents the n-bit accurate bits and m is for the m-bit approximate bits.

3 Hardware Trojans

Resulting from the globalization of the semiconductor supply chain, the design and fabrication of ICs are now distributed worldwide. It brings great benefit to IC companies, leading to a lower design cost and a shorter time-to-market window. However, it also raises serious concern about IC trustworthiness triggered by the use of third party vendors. As a result, it is becoming very difficult to ensure the integrity and authenticity of devices. The aim of a HT [27] is to modify or add a circuit for malicious purpose. Common malicious goals of HT include leaking sensitive information, changing or controlling the functionality of the circuit, and reducing circuit reliability. A HT can be inserted into IC products at any untrusted phase of the IC production chain by third party vendors or adversaries with an ulterior motive. Most of the HTs are triggered by rare event or signal such that they will not be discovered easily. Figure 11.3 shows an example of hardware Trojan inserted

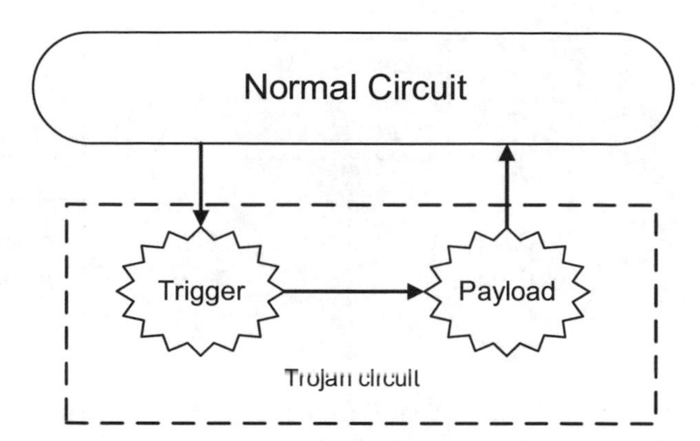

Fig. 11.3 An example of the Trojan-infected

circuit. The trigger circuit activates the subsequent payload circuit when trigger circuit receives a specified signal, including a specified input or internal status. The play load circuit can leak information or destroy circuit after activation. [10] indicated that the payload circuit could be connected to the output of an approximate circuit so that the error of the result exceeds an acceptable error range for cognitive applications. HT detection and prevention is still very challenging for trusted IC designs.

In a logic circuit, the importance of each path is different. For example, the path of the most significant bit, which includes the most important information, is normally considered as an important path. Inserting a hardware Trojan to this path is more destructive than that of the less important path. Therefore, in order to analyze the vulnerability of a circuit after approximation, transition probability is a commonly utilized metric. A testing flow was designed in [4] to analyze HTs in an approximate circuit is shown in Fig. 11.4. At first, hardware description languages, such as VHDL or Verilog, are utilized to implement both approximate and exact circuits. EDA tools, such as Synopsys, are used to generate the netlist of both circuits. The physical design tool calculates the transition probability of each node in the synthesized netlist for the circuits. Finally, the transition probability is calculated to investigate the security vulnerabilities of approximate circuits by analyzing both the approximate circuit architecture and critical path.

4 Security Vulnerabilities in Circuit Designs

Approximate circuits may be vulnerable for HT insertion compared with exact circuits [16]. Approximate circuits are typically controlled by an additional signal that does not belong to the original circuit design. This could be easily utilized as a trigger signal to excite a HT or modified by malicious modifications. Moreover, some approximate circuits may require an error correction code (ECC) or other error

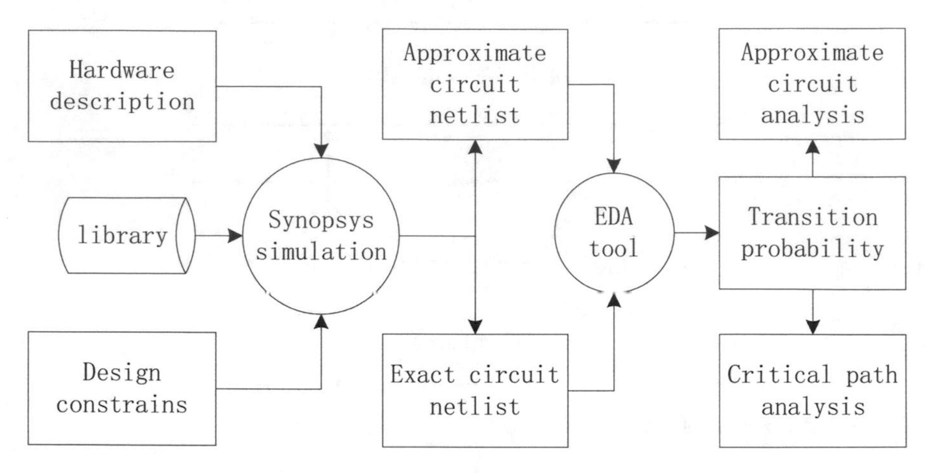

Fig. 11.4 A flow for the analysis of an approximate circuit

correction techniques to derive an acceptable result. These ECC circuits become another target for HT attacks. Recent research shows that signals in an approximate circuit and its corresponding exact different transition probability [4].

Two examples, including a malicious modification of an approximate circuit and a HT insertion to an approximate circuit, are presented as follows.

4.1 Malicious Modification

As an example, the inputs of the LOA as shown in Fig. 11.2 can be deliberately manipulated. An attacker can force the LOA to continuously generate erroneous outputs. The ECC or fault-tolerant process will be activated more than usual. [28] presented that the power consumption of the approximate adder (when 50% of errors is acceptable) is higher than the same adder (when 25% of errors is acceptable) when this malicious attack is undertaken.

Figure 11.5 presents an example of the images of a potential malicious modification attack on an approximate logarithmic multiplier (ALM), which utilizes truncation scheme on a certain number of least significant bits using a truncation parameter t, which represents the number of the remaining most significant bits is t. If the original number of full bits is s, the number of the truncated bits is $s - t$. The truncation parameter is normally stored in a register. A malicious attack is applied to maliciously tamper the truncation parameter t to an unacceptable value when it is read out from the register.

An exact result derived from the exact logarithmic multiplier circuit with 8-bit input and 8-bit output is shown in Fig. 11.5a. Figure 11.5b–d present the results of the same image generated by the ALM with different truncation parameters (t = 6, 4, and 2, respectively). An adversary may maliciously modify the value of t from

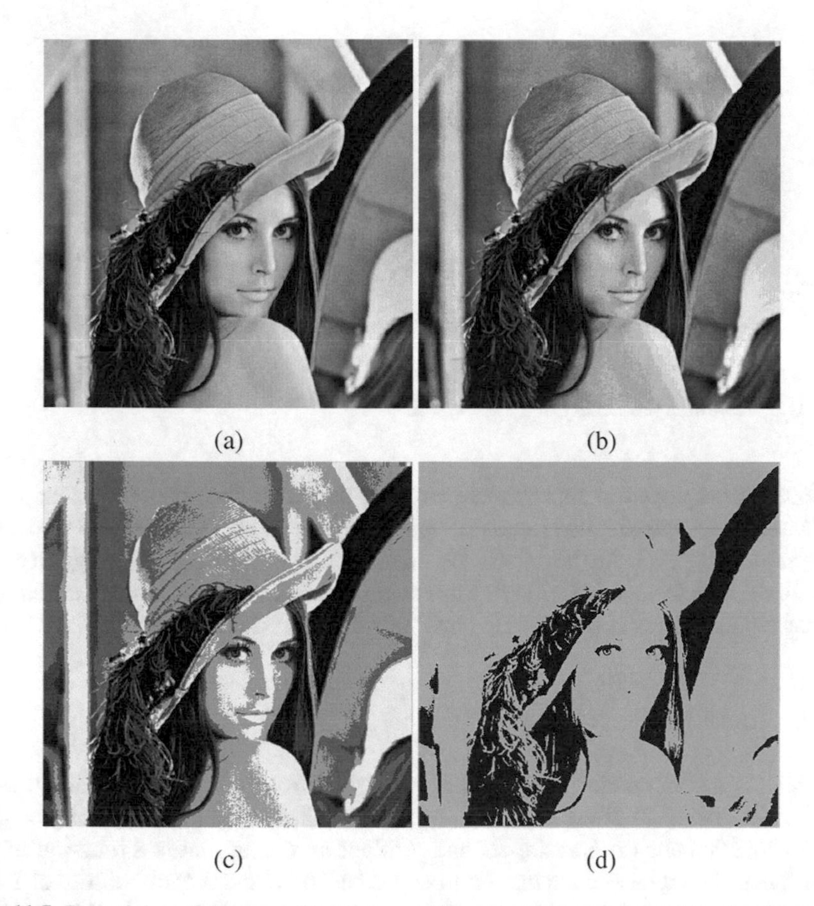

Fig. 11.5 Potential malicious modifications on the truncation parameter (t) of approximate logarithmic multiplier (ALM): (**a**) original result with 8-bit input, (**b**) $t = 6$, (**c**) $t = 4$, and (**d**) $t = 2$

6 ($s = 8$) to 2, which generates the image result in Fig. 11.5. It has a low quality and may not be acceptable to the user. The ECC mechanisms may misunderstand the calculation of $t = 6$ is unacceptable. Finally, the user will either not use approximate computing designs or have to active the ECC. However, the overuse of ECCs will consume extra power, which could result in the approximate computing designs invalid. Interestingly, the attacker achieves this without modifying the value of the inputs.

4.2 Hardware Trojans Attacks

As above-mentioned, approximate computing has been shown more vulnerable to HT insertion compared with exact circuits [4]. If an ECC or other error detection

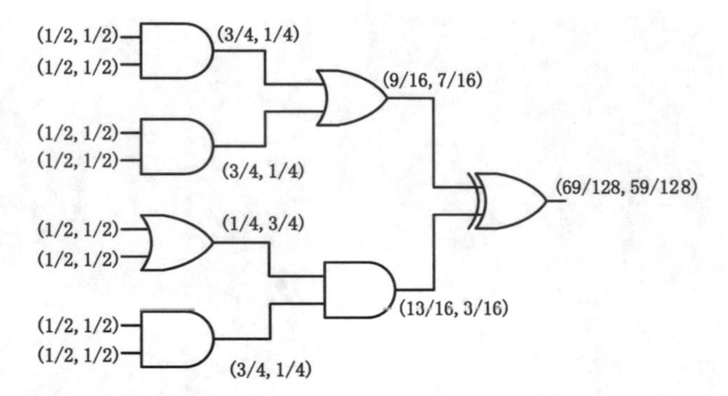

Fig. 11.6 Transition probability of a logic circuit

circuits are employed to approximate circuits, they may become another target for HT insertion. Approximate circuits normally require an additional signal to control its approximate level. As the control signal is not part of the original design, it could be utilized to activate a HT circuit. Moreover, signals in an approximate circuit and its corresponding exact circuit may have very different transition probability [4].

4.2.1 Transition Probability

A circuit is a combination of a number of logic gates, the most basic digital units having the ability to allow or block the flow of digital signals. Typically, a gate circuit contains one or more inputs that produce an output that is a function of the input value. The state of each input port could be 0 or 1, and the probability of 1 and 0 is equal. The state of the output could also be 0 or 1 but different probabilities. The probability calculation requires the truth tables of gate-level circuits. For example, for an XOR gate, the input ports are X and Y, and the output is O. Only if X and Y inputs are opposite, the output O is 1. For signals x and y, the probability of $(x = 0 andy = 1)$ is $(p_{x0}, p_{y1}) = (1/2, 1/2)$, which leads to the probabilities of the output signal $(o = 0)$ and $(o = 1)$ are $p_{o0} = 1/2$ and $p_{o1} = 1 - p_{o0} = 1/2$, respectively. (11.1) defines a switching probability between 1 and 0.

$$p_{tran} = p_{o1} * p_{o0} \tag{11.1}$$

As an example, Fig. 11.6 shows the transition probability of a HT-based computational logic circuit. The input probabilities for both 0 and 1 are assumed as $1/2$. The probability of the final output can be calculated since the state probability of each node in the circuit can be computed using (11.1). The flip probability of the final gate is calculated as $69/128 * (59/128) = 0.2484$. A geometric distribution can be utilized to model the flip using the topology of the circuit. The flip probability can be expressed as follows:

$$P(n) = p * (1 - p)^n (n = 0, 1 \ldots) \tag{11.2}$$

It indicates that the node will flip at the $(n + 1)$th clock cycle. (p_0, p_{o1}) are the probabilities of a circuit node when the value of the node is 0 or 1, respectively. Note, the probability calculation should follow three operations. First, nodes are independent to each other. Second, it should obey the rule, $p_0 + p_1 = 1$. Third, the probability of the input nodes is $(1/2, 1/2)$.

Transition probability plays an important role in HT insertion. Generally, in an approximate circuit, some modules are approximated, and some modules remain as exact. In logic circuits, the importance of each path is different. For example, in the arithmetic circuit, the most significant bit derives from an important path since it has the highest weight of contribution to the final result. Inserting a HT to the more important path is more destructive than that of the less important path. Hence, in order to investigate the vulnerability of the circuit after approximation, transition probability is utilized.

4.2.2 Approximate Circuit Analysis

We still use the LOA as an example. The security vulnerabilities of the LOA circuit will be investigated under different parameter settings. The transition probabilities of both an exact adder and approximate LOA adders are calculation and compared. To obtain the transition probability of the approximate circuit, an EDA tool is utilized and MATLAB is applied to analyze the performance. Figure 11.7 presents a scatter diagram of the transition probabilities of the comparison between exact LOA adders and approximate LOA circuits, respectively, in different parameters (n), which indicates the number of inaccurate bits. They are also grouped and compared in two categories, 8-bit and 16-bit LOA circuits, respectively. The x-axis, nodes from start to output, represents the nodes from the input gates to the output gates in circuits. The y-axis, transition probability, presents the transition probability of each node. Figure 11.7a, b, and c show the transition probabilities of an 8-bit approximate LOA adder in different approximate parameters, $n = 2, 4$, and 6, respectively, compared to that of an 8-bit exact LOA adder. The number of transition probabilities with high approximate parameters (n) in the approximate adders is increased compared with that of the exact adders. It reveals that the approximate adders are more likely to insert HTs when the number of inaccurate components is increased. Figure 11.7d, e, and f present the transition probabilities of an 16-bit approximate LOA adder in different approximate parameters, $n = 2, 4$, and 6, respectively, compared to that of an 16-bit exact LOA adder. Similar to the results of the comparison between the 8-bit exact adder and approximate adders, the transition probabilities with high parameters (n) in the 16-bit approximate adders are also increased compared with that of the exact adders. Moreover, the more the number of approximate modules (the larger the value of n), the higher the possibility of HT insertion. Particularly in Fig. 11.7f, the number of low transition probabilities in the approximate adder with the approximate parameter of 12 is much more than that of

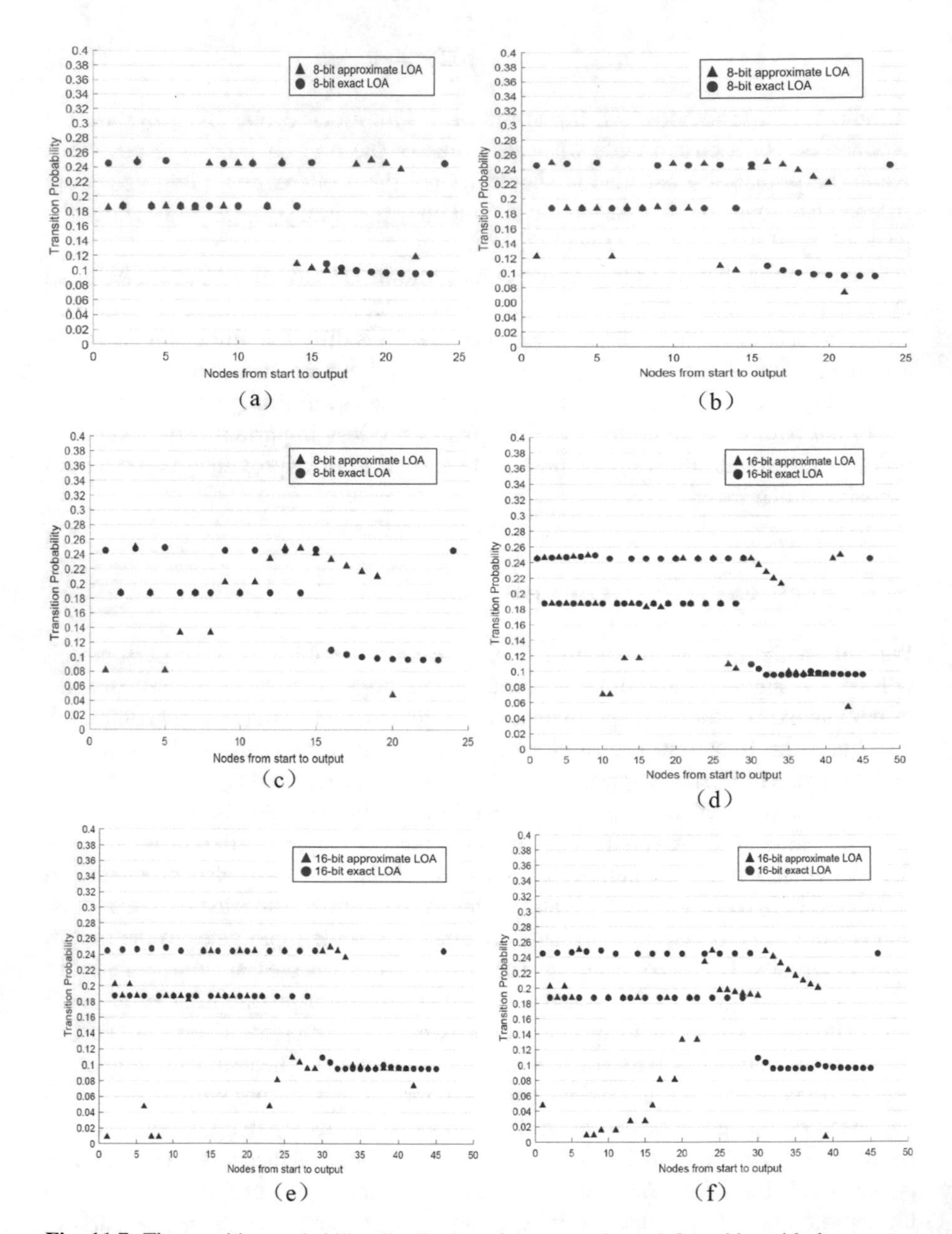

Fig. 11.7 The transition probability distribution of the approximate LOA adder with the parameters of (**a**) 8-bit, k = 8, n = 2, (**b**) 8-bit, k = 8, n = 4, (**c**) 8-bit, k = 8, n = 6, (**d**) 16-bit, k = 16, n = 4, (**e**) 16-bit, k = 16, n = 8, (**f**) 16-bit, k = 16, n = 12, respectively [4]

the exact adder. Some values of the transition probabilities in the approximate adders are close to 0, which indicates a high risk of the approximate adders to HT attacks. The transition probability of the last node in approximate adders is lower than that of exact adders, which also shows a potential security vulnerability to HT insertions. From the experimental analysis, it alerts the security threats in approximate circuits. It also provides an instruction to circuit designers to avoid such vulnerabilities in the future designs.

5 Security Threats in Approximate Testing Techniques

As the use of approximate circuits in practical applications, the testing techniques in approximate designs are different with conventional testing approaches. Hence, the investigation of security vulnerabilities in approximate testing techniques is also needed.

5.1 Approximate Testing Techniques

A post-production test is commonly utilized to ensure the functionality of a design and detect physical faults caused by factory manufacturing process variations. Due to the error tolerance of approximate circuits, the conventional testing techniques used for exact circuits may not be applicable to approximate circuits due to the fact that the acceptable faults in approximate designs may be mistakenly classified as unacceptable faults. Several approximation-aware testing techniques [1, 22, 23] have been proposed for approximate computing designs to date. However, the use of approximate-aware testing processes may also introduce security vulnerabilities.

The approximation-aware testing technique includes two phases. One is fault classification process and the other one is test vector generation. Figure 11.8 presents a flow of an approximate-aware testing process. At first, a test data (fault) is input into both the exact circuit and approximate circuit. The difference (δ) between the outputs of both circuits is calculated and compared with a threshold (t), which is derived from a predefined error metric. The fault will be determined to a non-acceptable fault or an acceptable fault. The fault will be added to the non-acceptable fault list when the difference δ is greater than the threshold t. Otherwise, the fault goes to the acceptable fault list. Until now, the fault classification process is completed. The next step is test vector generation. The non-acceptable fault will be fed again to the approximate circuit to generate the test vector. The faults that are not affected by the error metric will not be detected by the approximation-aware method. The approximation-aware testing process is completed.

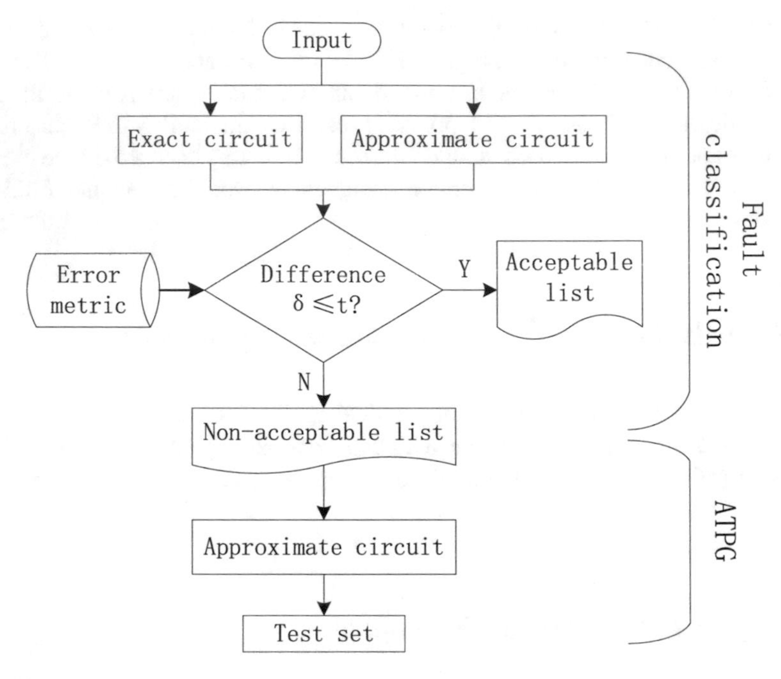

Fig. 11.8 Flow of approximation-aware testing technique

5.2 *Malicious Tampering Attack*

The introduction of new testing methods may potentially bring new attack opportunities. It is necessary to investigate the security vulnerabilities in the approximation-aware testing techniques. As introduced in Fig. 11.8, fault classification is the core process in approximation-aware testing techniques, and it requires a reliable error metric. However, the error metric has to be stored in memory and if it is not well protected, it can be easily accessed and tampered by attackers. The error metric is utilized to determine the threshold (t) for the comparison between the exact and approximate results in the fault classification process. The fault classification result could be affected when the threshold (t) is maliciously modified.

5.3 *Countermeasures*

In order to address the above-mentioned security threat, a Self-detection test flow is proposed. The fault classification is carried out after the generation of a test pattern. The proposed classification process is similar to the fault simulation (FS) method proposed in [24], which analyzed the classification process of FS and approximation-aware testing technique. The analysis results show that the classification of both methods varies according to the value of the error metric WCE, which is the worst error metric, *i.e.*, the maximum arithmetic distance between the full-precise and approximate output. It is adopted as the error metric. The WCE is defined as follows:

$$WCE = \max_{\forall i} |O^{(i)}_{approx} - O^{(i)}_{prec}| \tag{11.3}$$

where $O^{(i)}_{approx}$ represents the approximate result, $O^{(i)}_{prec}$ represents the exact result, and WCE calculates the largest difference between them. The testing threshold $t = WCE$ is in the range of $(8, 256)$. In this experiment, $t = WCE$ is set to 8 as an example. Generally, if the value of a WCE is small, the classification of FS is better than that of approximation-aware testing technique. If the value of the WCE is large, then the classification of approximation-aware testing is better than FS. Since the classification of the proposed self-detection test is the same as the FS in the [24], the classification result should be the same as the FS. The proposed self-detection test technique includes three steps: (1) test pattern generation, (2) error metric detection, and (3) fault classification. The steps (2) and (3) are presented in Fig. 11.9.

Test Pattern Generation

The test pattern generation process is the same as the traditional test. A fault is injected into an approximate circuit to generate a test netlist. Then, the netlist is fed into ATPG for test pattern generation. This process can be completed by EDA tools.

Error Metric Detection

The generated test pattern from the above step and the given metric are fed into perform comparison. Two types of results can be derived. One is the approximate result of the approximate circuit under test, and the other is the relevant exact result. The test pattern of the exact circuit is easy to obtain. For an arithmetic unit, the exact test vectors of the exact circuit can be obtained through conventional computing platforms. The difference between the exact result and the approximate result is utilized to detect abnormal metric output. This process can be described by Algorithm 1, which includes two steps. First, the exact result (O_i^E) and approximate result (O_i^A) of the test pattern is calculated, and the maximum difference between them (D) is derived. Second, the maximum difference is compared with the metric (m). If it is greater than the flag threshold (t) , a warning signal $(flag = 1)$ is output. It is assumed that an attacker applies an attack, such as tampering with the metric to an unacceptable value. To counter this, the self-detection test only needs to define a threshold with a smaller range to determine the metric result.

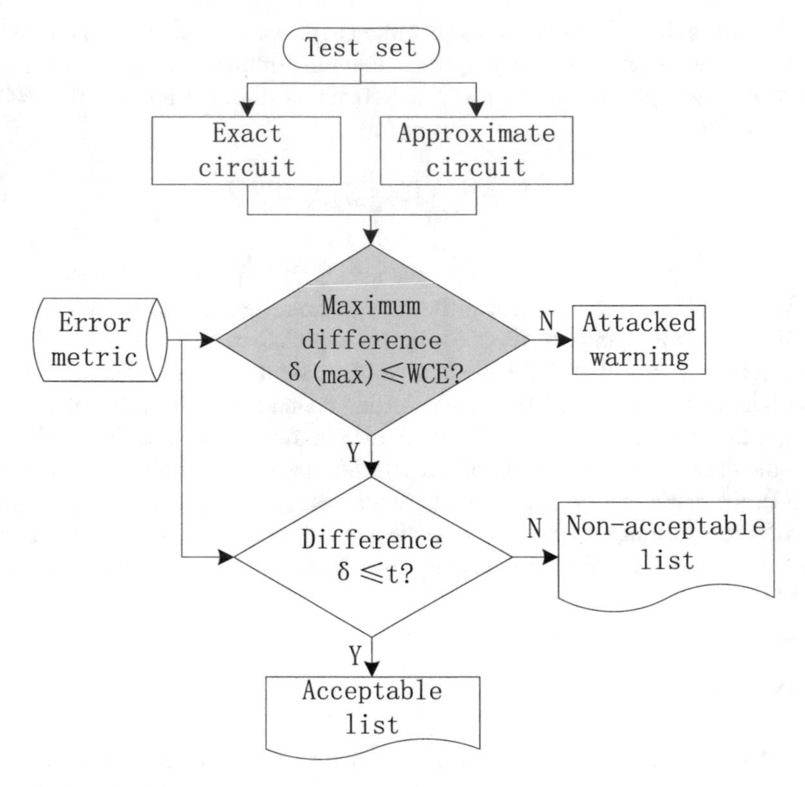

Fig. 11.9 Proposed tampering attack detection method

Fault Classification

After completing the metric detection, a fault classification process will be carried out. At the fault classification stage, a fault is first inserted into the approximate circuit netlist to generate the test pattern, used as an input to obtain an approximate result with the fault. Then the approximate result with fault is compared with the corresponding exact result to derive the difference. If the difference result is larger than the threshold t obtained by the metric, the test pattern corresponding to this result will be classified as non-acceptable, otherwise the test vector is considered as acceptable.

5.4 *Experimental Analysis*

To verify the effectiveness of the proposed self-detection method, we consider a tampering attack that aims at the error metric in approximation-aware testing technique.

Algorithm 1: Metric difference detection

```
Input: approximate output(O_i^A), Exact output(O_i^E), threshold (t), error metric (m), Number
       of test vectors (N)
Output: flag (1: abnormal, 0: normal)
1  Define two intermediate variables s and d for the comparison with the final error metric (m);
2  Initialize S = 0, d = 0;
3  for i < N do
4  |   s =|O_i^A − O_i^E| ;
5  |   if d < s then
6  |   |   d = s;
7  |   end
8  |   i++;
9  end
10 if |d − m| > t then
11 |   flag = 1;
12 else
13 |   flag = 0;
14 end
```

5.4.1 Experimental Setup

Table 11.1 introduces the experimental setup for the evaluation of the proposed self-detection method. The experiment is carried out using Matlab simulation. A 16-bit LOA approximate adder is utilized to evaluate the proposed self-detection method. An example of error metric is set as $WCE = 2'b111$ and a tampered value is set as $WCE = 2'b11111$. The alert threshold is defined as $t = 2'b101$.

5.4.2 Experimental Results

Figure 11.10 presents the experimental results of the evaluation of the proposed self-detection technique on the tampering attack as shown in Table 11.1. $S(i)$ represents the difference between an exact result and approximate result of a test pattern. The *Alert Value* $(D + m)$ is an untampered WCE value, which is the maximum error value of an approximate circuit. The *WCE (TEM)* shows a warning line generated from the proposed self-detection technique. It can be seen that the tampered line is much larger than the warning line, which means that when the indicator is detected, a warning signal will be output.

Table 11.1 Experimental setup

Simulation software	ATPG tool [21] , MATLAB
Object circuit	16-bit LOA approximate adder
Error metric	$WCE = 2'b111$
Attack method	Tamper with $WCE = 2'b11111$
Alert threshold	$t = 2'b101$

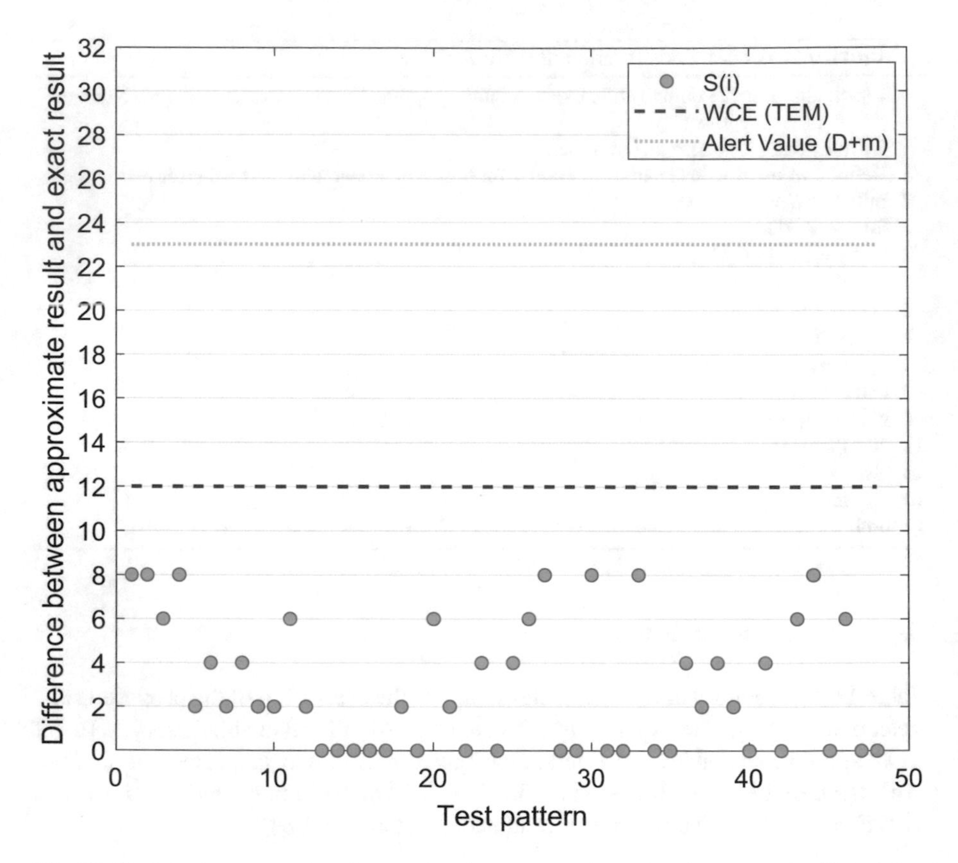

Fig. 11.10 Experimental results of approximation-aware testing using the proposed self-detection technique

6 Conclusion

Security vulnerabilities have been shown in approximate computing systems due to the fact that the uncertain intrinsic errors during approximate execution may be indistinguishable from malicious modification of the input data. This book chapter presented a comprehensive and deep study of security vulnerabilities in approximate computing systems, including HT insertion to approximate circuits and tampering attacks on approximation-aware techniques. The experimental analysis presented that the approximate adder is more vulnerable to HT attack than the exact adder circuit. The tampering attack, which targets the error metric of an approximation-aware testing technique, in particular affects the fault classification of approximation-aware testing. A self-detection testing method to tackle the challenge of tampering attack is proposed to thwart the security vulnerabilities in approximation-aware testing technique.

References

1. Chandrasekharan A, Eggersglüß S, Große D, Drechsler R. Approximation-aware testing for approximate circuits. In: Proceedings of the 23rd Asia and South pacific design automation conference (ASP-DAC); 2018. pp 239–44
2. Chen L, Han J, Liu W, Lombardi F. Design of approximate unsigned integer non-restoring divider for inexact computing. In: Proceedings of the ACM 25th edition on Great Lakes symposium on VLSI (GLSVLSI); 2015. pp 51–6
3. Chen L, Lombardi F, Montuschi P, Han J, Liu W. Design of approximate high-radix dividers by inexact binary signed-digit addition. In: Proceedings of the great lakes symposium on VLSI (GLSVLSI); 2017. pp 293–298
4. Dou Y, Yu S, Gu C, O'Neill M, Wang C, Liu W. Security analysis of hardware trojans on approximate circuits. In: Proceedings of the 2020 on great lakes symposium on VLSI. GLSVLSI '20. New York, NY: Association for Computing Machinery; 2020. pp 315–320
5. Fleischer B, Shukla S, Ziegler M, Silberman J, Oh J, Srinivasan V, Choi J, Mueller S, Agrawal A, Babinsky T, Cao N, Chen C, Chuang P, Fox T, Gristede G, Guillorn M, Haynie H, Klaiber M, Lee D, Lo S, Maier G, Scheuermann M, Venkataramani S, Vezyrtzis C, Wang N, Yee F, Zhou C, Lu P, Curran B, Chang L, Gopalakrishnan K. A scalable multi-teraOPS deep learning processor core for AI trainina and inference. In: 2018 IEEE symposium on VLSI circuits; 2018. pp 35–6
6. Gupta V, Mohapatra D, Raghunathan A, Roy K. Low-power digital signal processing using approximate adders. IEEE Transact Comput-Aid Des Integr Circ Syst. 2013; 32(1):124–137
7. Han J, Orshansky M. Approximate computing: An emerging paradigm for energy-efficient design. In: Proceedings of the 18th IEEE European test symposium (ETS), May 2013; pp 1–6
8. Hruska J. Nvidia's CEO Declares Moore's Law Dead; 2017
9. International Technology Roadmap for Semiconductors. Last accessed 16 January 2018
10. Islam SA. On the (IN) security of approximate computing synthesis (2019). Preprint. arXiv:1912.01209
11. Jiang H, Liu C, Liu L, Lombardi F, Han J. A review, classification, and comparative evaluation of approximate arithmetic circuits. ACM J Emerg Technol Comput Syst 2017;13(4):60:1–60:34
12. Jouppi N, Young C, Patil N, Patterson D, Agrawal G, Bajwa R, Bates S, Bhatia S, Boden N, Borchers A. In-datacenter performance analysis of a tensor processing unit. In: Proceedings of the 44th annual international symposium on computer architecture (ISCA); 2017. pp 1–12
13. Liu W, Qian L, Wang C, Jiang H, Han J, Lombardi F. Design of approximate radix-4 Booth multipliers for error-tolerant computing. IEEE Trans Comput 2017;66(8):1435–41
14. Liu W, Xu J, Wang D, Wang C, Montuschi P, Lombardi F. Design and evaluation of approximate logarithmic multipliers for low power error-tolerant applications. IEEE Trans Circ Syst I: Regul Pap 2018;65(9):2856–68
15. Liu W, Cao T, Yin P, Zhu Y, Wang C, Swartzlander EE Jr, Lombardi F. Design and analysis of approximate redundant binary multipliers. IEEE Trans Comput 2019;68(6):804–819
16. Liu W, Gu C, O'Neill M, Qu G, Montuschi P, Lombardi F. Security in approximate computing and approximate computing for security: challenges and opportunities. Proc IEEE 2020;108(12):2214–2231
17. Liu W, Gu C, Qu G, O'Neill M. Approximate computing and its application to hardware security. New York: Springer; 2018. pp 43–67
18. Liu W, Lombardi F, Shulte M. A retrospective and prospective view of approximate computing [point of view. In: Proceedings of the IEEE; 2020
19. Mittal S. A survey of techniques for approximate computing. ACM Comput Surv 2016;48(4):62:1–62:33
20. Regazzoni F, Alippi C, Polian I. Security: the dark side of approximate computing? In: Proceedings of the IEEE/ACM international conference on computer-aided design (ICCAD); 2018. pp 1–6

21. Tetramax (2018). https://www.synopsys.com
22. Traiola M, Virazel A, Girard P, Barbareschi M, Bosio A. On the comparison of different ATPG approaches for approximate integrated circuits. In: IEEE 21st international symposium on design and diagnostics of electronic circuits systems (DDECS); 2018. pp 85–90.
23. Traiola M, Virazel A, Girard P, Barbareschi M, Bosio A. Maximizing yield for approximate integrated circuits. In: Design, automation test in Europe conference exhibition (DATE); 2020. pp 810–815
24. Traiola M, Virazel A, Girard P, Barbareschi M, Bosio A. On the comparison of different ATPG approaches for approximate integrated circuits. In: 2018 IEEE 21st international symposium on design and diagnostics of electronic circuits & systems (DDECS). New York: IEEE; 2018. pp 85–90
25. Venkataramani S, Chippa VK, Chakradhar ST, Roy K, Raghunathan A. Quality programmable vector processors for approximate computing. In: Proceedings of the 46th annual IEEE/ACM international symposium on microarchitecture (MICRO); 2013. pp. 1–12.
26. Xu Q, Mytkowicz T, Kim N. Approximate computing: a survey. IEEE Des Test 2016;33(1):8–22
27. Xue M, Gu C, Liu W, Yu S, O'Neill M. Ten years of hardware trojans: a survey from the attacker's perspective. IET Comput Digit Tech. 2020;14(6):231–46
28. Yellu P, Boskov N, Kinsy MA, Yu Q. Security threats in approximate computing systems. In: Proceedings of the great lakes symposium on VLSI (GLSVLSI); 2019. pp 387–392
29. Yellu P, Monjur MR, Kammerer T, Xu D, Yu Q. Security threats and countermeasures for approximate arithmetic computing. In: 2020 25th Asia and South pacific design automation conference (ASP-DAC); 2020

Chapter 12
Voltage Overscaling Techniques for Security Applications

Md Tanvir Arafin, Qian Xu, and Gang Qu

1 Introduction

With the advent of the Internet-of-Things (IoT), low-power design techniques are becoming essential for designing and implementing resource-constrained IoT nodes. These design techniques include aggressive voltage overscaling scaling (VOS), dynamic voltage and frequency scaling (DVFS), clock gating, retention power gating, approximate computing etc., which can provide significant improvement on the power budget of a digital system. Unfortunately, these techniques can suffer from unintentional side effects arising from the budget compromises. For example, attacks such as CLKScrew [20], PlunderVolt [12], and VoltJockey [14, 15] have demonstrated that energy management oblivious designs lead to security concerns in modern commodity devices. Therefore, system designers and architects wishing to explore low-power and approximate computing paradigms must examine side effects and security vulnerabilities resulting from these power-saving techniques.

On the other hand, approximate computing techniques also offer designs for novel security primitives. For example, Arafin et al. [1] have demonstrated that voltage overscaling in a processor design creates physically unclonable hardware signatures for device fingerprinting. Later, Zhang et al. [24] have shown how voltage overscaling-based primitives can be helpful to design strong physically unclonable functions (PUFs). Additional works such as approximate adder-based information hiding [4], acceleration of side-channel analysis using approximate computing

Md. T. Arafin
Morgan State University, Baltimore, MD, USA
e-mail: mdtanvir.arafin@morgan.edu

Q. Xu · G. Qu (✉)
University of Maryland, College Park, MD, USA
e-mail: qxu1234@umd.edu; gangqu@umd.edu

W. Liu, F. Lombardi (eds.), *Approximate Computing*,
https://doi.org/10.1007/978-3-030-98347-5_12

[5, 8], and approximate DRAM PUFs [16, 21] show the promises and opportunities of approximate computing in securing next generation of lightweight devices and systems.

Voltage overscaling-based security primitives and vulnerabilities are examples of hardware security problems. Hardware security is an expanding branch of computer engineering that studies the security impact of the physical implementation of a given computing system. The dynamic nature of hardware design and the constant evolution of computer architecture require a careful introspect of the security and privacy footprint of the emerging devices and systems from both the hardware and software perspectives. Traditional software security practices leave the hardware side vulnerabilities out of scope, eventually making the underlying hardware the weakest link of the system security. However, if hardware security issues are resolved and hardware properties are utilized, hardware can act as the most vital link in the system. Therefore, new system design and architecture should explore the hardware security-based opportunities for the system. Hence, this chapter engages in studying the VOS-based to generate low-power and lightweight hardware security primitives for the next generation of IoT devices.

In this chapter, we explore the security opportunities arising from voltage overscaling—one of the common practices of approximate computing techniques. In Sect. 2, we introduce the readers to the basics of voltage overscaling techniques. Section 3 discusses voltage overscaling in computing systems such as processors and provides an example of VOS in a simple 8-bit ripple carry adder. Section 4 explores voltage overscaling in the main memory, i.e., DRAM systems. Based on the discussions in the previous sections, Sect. 5 presents two security applications of voltage overscaling—one in authentication and the other in securing machine learning models from model inversion attacks. Finally, Sect. 6 discusses some vulnerabilities introduced by voltage and frequency overscaling. The chapter ends in Sect. 7.

2 Voltage Overscaling

Common voltage overscaling technique can be defined as follows:

Definition 2.1 Voltage overscaling is a low-power operating condition of a digital system where the system is operated under the nominal voltage that guarantees correct output for all input conditions at a given frequency.

The definition shows that a system running in a voltage-overscaled condition cannot guarantee correct output for all the given inputs. Thus, VOS is an error-inducing technique. These errors are generated due to the timing inaccuracies in the critical path of the system. Hence, the timing faults due to VOS are the artifact of aggressive undervoting. Overclocking also induces VOS-related defects since severe overclocking forces the system to operate over the critical error-free margin for a given supply voltage. Both of these cases are related to the dynamic nature of voltage and frequency scaling.

The origin of VOS-induced timing errors can be explained by the essential operation of CMOS-based digital systems. As the operating voltage reduces, it takes longer to charge and discharge the transistors. Therefore, a processor needs to be operated at a lower frequency to provide sufficient charging/discharging operation time. Overclocking pushes the operating limit by increasing the frequency as long as the system can generate the acceptable output at a given voltage. At the same time, undervolting reduces the operating voltage at a given frequency to minimize power consumption. Modern processors, memories, and graphical processing units (GPUs) employ kernel-level power management and dynamic voltage and frequency scaling (DVFS) techniques to support these power-saving and throughput-increasing techniques. DVFS avoids generating timing faults due to unmet path-delay requirements by the circuit. However, voltage and frequency overscaling techniques sacrifice the error-free operation to gain additional power savings. The goal of aggressive VOS is to ensure that the errors remain under a critical level so that the final output is acceptable.

Interestingly, voltage overscaling-based error can reveal information regarding the fabrication variation of a digital integrated circuit (IC). Fabrication variations in an IC come from (1) imperfection of the manufacturing process, (2) random dopant fluctuation, and (3) variation in the gate oxide thickness. At nominal operating voltage and frequency, it may be difficult to observe these variations. With voltage overscaling, fabrication variations could have a larger impact on the execution and become observable in the forms of errors or imprecise outputs.

The effect of voltage overscaling is prominent on the processor and the memory devices of the system. Therefore, in this chapter, we will explore the voltage overscaling issues in CPU components and dynamic random access memory (DRAM) components.

3 VOS in Processing Elements

VOS effects in logic components of a digital system arise from the timing properties of the transistors. On the other hand, the standard deviation of threshold voltage variation increases with shrinking transistor size. This relationship is given by Hong et al. [7]

$$\sigma_{\Delta V_t} = A_{\Delta V_t}/\sqrt{WL}, \tag{12.1}$$

where $A_{\Delta V_t}$ is the characterizing matching parameter for any given process. This variation in V_t will have a direct consequence in the delay of a CMOS gate that can be approximated using the following equation [7]:

$$d_{gate} \propto \frac{V_{DD}}{\beta(V_{DD} - V_t)^{\alpha}}, \tag{12.2}$$

where α and β are the fitting parameters for a gate and the given process. Therefore, static timing analysis (STA) of a given design is performed on the process corners to maintain timing correctness. Scaling V_{DD} from the predefined critical operating voltage at a given frequency or increasing frequency from the critical operating frequency at a given voltage creates significant timing errors and degrades the output quality.

From Eq. 12.2, it is evident that with voltage overscaling and transistor shrinking, underlying device signature due to process variation (e.g., V_t variation) manifests itself more prominently in the delay output. If a proper correction mechanism is not applied, this variation will cause timing errors in the output for a given computation. Since manufacturing variation is a random process, one can generate a unique device signature for the processor by profiling this error. Furthermore, these errors are input/computation-dependent, and therefore, one can extract challenge–response pairs based on hardware-dependent authentication keys from these timing errors. To understand the issue, let us explore a simple example in the next section.

3.1 VOS Example

To understand the effects of voltage overscaling on processed data, we have simulated a simple image processing application using a ripple carry adder (RCA). This experiment helps us analyze the impact of process variations, voltage variations, and temperature on a digital system that is operated under the nominal voltage that guarantees the timing correctness of the critical paths.

Ripple carry adders are a good candidate to experiment with the effect of timing failures on a critical path. Venkatesan et al. have provided process variation-independent error profiles for ripple carry adders, carry look-ahead adders (CLA), and Han–Carlson Adder (HCA) [22]. It was found that the error probability increases as the number of critical paths that fail to meet the timing constraint increases [22]. Therefore, with the presence of randomness in the manufacturing process, the variations in the transistors in the critical paths will have a significant contribution to the errors produced by the adders.

For this example, we design the adder in Verilog and synthesize it using the Cadence Virtuoso RT compiler. The synthesized design is then converted into a netlist with standard cells randomly chosen from our modified library. We create a modified standard cell library for the FreePDK 45 nm libraries [18]. We use 200 modified NMOS and PMOS models with variable threshold voltages to introduce

Table 12.1 Parameters used for the adder simulations

Parameter Name	Value(s)
Supply voltage (VDD)	0.4 V/0.45 V/1 V
NMOS threshold voltage (Vtn)	0.322 ± 0.02415 V
PMOS threshold voltage (Vtp)	-0.302 ± 0.02265 V
Operating temperature (T)	25 °C
Clock Period (T_{clk})	1 ns

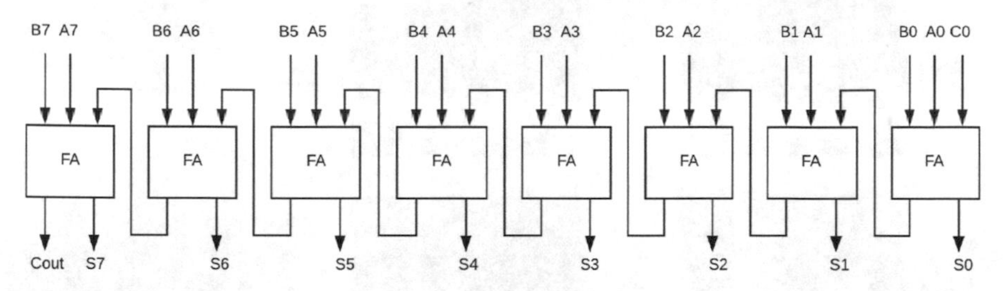

Fig. 12.1 8-bit ripple carry adder using 1-bit full adders. The long carry chain can propagate errors in the least significant bits to create a cascade of errors in higher significant bits

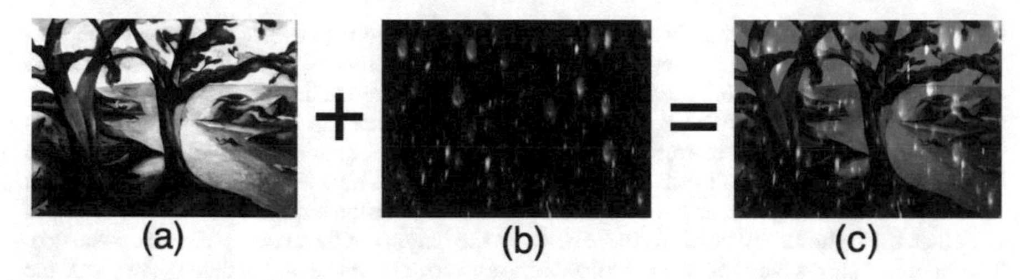

Fig. 12.2 An example of superimposing two images. We have used two grayscale images trees (**a**) and snowflakes (**b**) from MATLAB library to generate the superimposed image snowfall (**c**)

process variation. We assume a Gaussian distribution with a ±7.5% standard deviation for the variation of the threshold voltages. These modified NMOS and PMOS transistor models are randomly chosen to build 100 different versions of each standard cell in the FreePDK 45 nm library. These cells are used to replace similar logic cells in the netlist. Finally, we simulate the netlist using the HSpice platform. The parameters used for the simulation are given in Table 12.1.

We present a simple image processing application based on the superimposition of two images under general operating conditions and compare their results. The image processing application (i.e., superimposition) reads the 8-bit values stored at every pixel location for any two given image and adds the values. The processing is first done on an accurate adder and then on two voltage overscaled ripple carry adders (as shown in Fig. 12.1) with process variations (Fig. 12.2).

From visual observation of Fig. 12.3a, b, and c, one can clearly notice the effect of voltage overscaling in this simple image processing application. If we plot the histogram of the Euclidean distances of the right figures of the Fig. 12.3a, b and a, c (as done in Fig. 12.4), then we can see some interesting results on the error pattern.

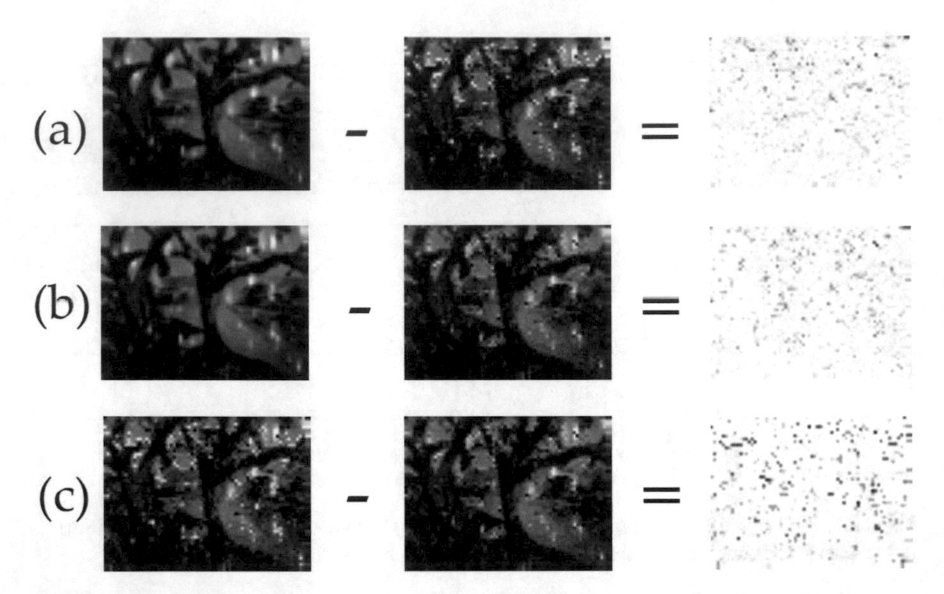

Fig. 12.3 An example of the effect of process variations in voltage overscaling-based computation. In (**a**), the difference between two snowfall images is presented. The left one is the accurate result for the superimposition, and the middle image is computed under voltage overscaling using $v_{dd} = 0.4\,\text{V}$. The error profile is presented on the right image. In (**b**), a similar adder that is identical to the adder in (**a**) in every aspect, except the process variation of the transistors, is used for computing the same result with supply voltage $v_{dd} = 0.4\,\text{V}$. Here the correct result is shown on the left, the output for the voltage-overscaled adder is shown in the middle, and the error profile is shown on the right. In (**c**), the differences between the outputs of these two adders are presented. The left figure shows the result from the first voltage-overscaled adder, the middle figure shows the output of the second adder, and the right figure shows the difference between the two approximate images. The source images were downsized to 52×40 pixels for reducing computation time

It can be noted that the peaks of this histograms mainly appear at 2, 4, 8, 16, 32, 64..., which suggests that most of these errors in the approximated results come from a single bit errors. Furthermore, the peaks at 1, 2, 4, 8, 16 are higher in most cases, revealing the fact that for these two adders most of the errors are in the LSBs.

3.2 *Errors in an Approximated Circuit*

To understand the effect of process variation in voltage overscaling, we have studied the error profiles generated by adders. If a given circuit is operated with a clock period that is less than the maximum delay produced by the circuit, then the circuit output becomes a function of current and previous input values [22]. Therefore, the output data in a circuit under VOS not only is a function of process variations but also a function of the input values applied to the circuit. Hence, for a combinational adder with two operands $\mathbf{x}$ and $\mathbf{y}$, we can write the current output $\mathbf{z_i}$ of a voltage

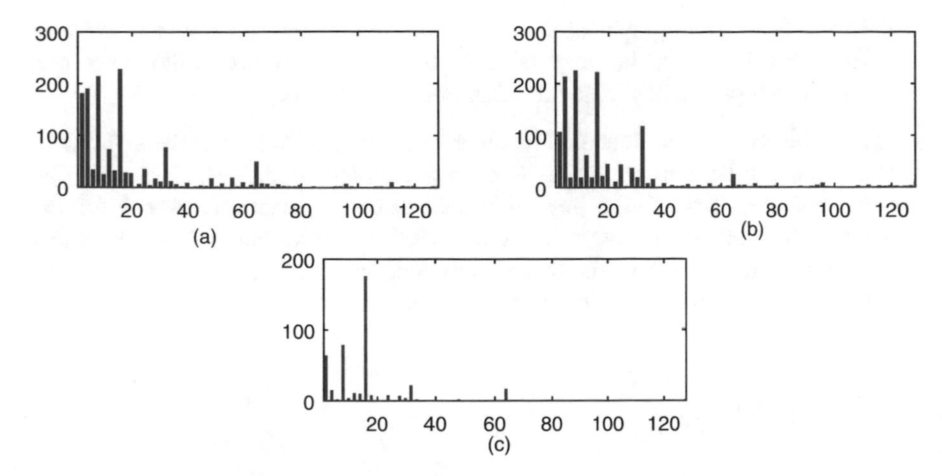

Fig. 12.4 A histogram of the Euclidian distances for the rightmost figures of Fig. 12.3. The x-axis in the sub-figures represents bins of all the possible distance values from 1 to 128, and the y-axis in the sub-figures represents the number of pixels (in the two corresponding computation of Snowfall images) that share the same distance values

overscaled adder as a function of current inputs $\mathbf{x_i}$, $\mathbf{y_i}$, and previous inputs $\mathbf{x_{i-1}}$, $\mathbf{y_{i-1}}$. Therefore,

$$\mathbf{z_i} = f(\mathbf{x_i}, \mathbf{y_i}, \mathbf{x_{i-1}}, \mathbf{y_{i-1}}), \tag{12.3}$$

where $f()$ defines a process variation-dependent addition. This dependence on previous inputs can cause cascading errors in the output. Therefore, to correctly predict the output of a voltage overscaled adder, one can take the following measures:

1. Save the output data for the set of input patterns that will be used on that circuit. For example, if and only if a set S_I containing n-input pattern will be used for processing in a given adder, then Alice needs to save outputs for all the combination of the patterns resulting from S_I. This would reveal the partial behavior of the circuit for a subset of input data.
2. Profile the adder for all possible input patterns. For profiling the adder, one not only needs to consider all possible current input values but also requires previous input values. Therefore, a correct profile of an n-bit voltage overscaled adder would consist of a table of entries comprised all possible current input value times all possible input values in the previous step. This would amount to $2^{2n} \times 2^{2n} = 2^{4n}$ entries of the input values and the corresponding output values. Since all the additions are not incorrect and dependent on the previous values, one can significantly reduce strongly the size of the profile by only the cases where the adders provide inaccurate results.

3. Use a delay-based graphical model to learn the properties of the adder. Since Alice has the adder, she can profile the device and use this profile to create a conditional probability table for a Bayesian network.

From this experiment, it should be clear to the reader that the process variation information can be extracted using voltage-overscaled conditions. Extracting this information does not require any additional circuit components except for the ones used for voltage regulation in a chip. The process variation details manifest themselves in the error patterns of the generated output. This is the key idea for using voltage overscaling to extract entropy from physical hardware.

4 VOS in Memory

The effect of voltage overscaling is also observed in memory modules. For example, reducing the operating voltage of the main memory module under the nominal operating voltage for error-free operation leads to physical variation-dependent error in the memory storage. To present a detailed overview of VOS in memory, in this section, we illustrate the overscaling effects in dynamic random access memory (DRAM) systems.

4.1 Basics of Dynamic Random Access Memory (DRAM) System

A DRAM memory system consists of DRAM cells that utilize storage capacitors and access transistors to store bits and control read/write operations. In Fig. 12.5, a unit 1-transistor-1-capacitor (1T1C) DRAM cell is depicted. Fundamental DRAM operation can be divided into four phases: (1) ACTIVATE, (2) READ, (3) WRITE, and (4) PRECHARGE. Reduced supply voltage lengthens the latency of DRAM operations, leading to insufficient time to fully complete them and causing errors. To understand how voltage overscaling leads to errors in a DRAM, we need to examine the effects of reduced voltage operation on the timing parameters during reading and writing a DRAM cell.

For reading a bit from a DRAM row, the row is activated with the ACTIVATE signal. The equilibrate signal (V_{eq}), which ensures that the $V_{bitline}$ is at the $V_{supply}/2$, precedes the activation phase as shown in Fig. 12.6. During the read operation, $V_{bitline}$ moves to either V_{supply} or 0 V depending on the charge stored in the cell capacitor. As the row is activated, charge sharing starts between the cell capacitor and the bit line, which increases/decreases the bit line voltage. A sense amplifier (SA) senses the change in the bit line voltage, and based on the change, SA starts driving the bit line to V_{supply} or 0 V. The latency of activation operation t_{RCD} is defined as the minimum time before the row buffers read the data. In this

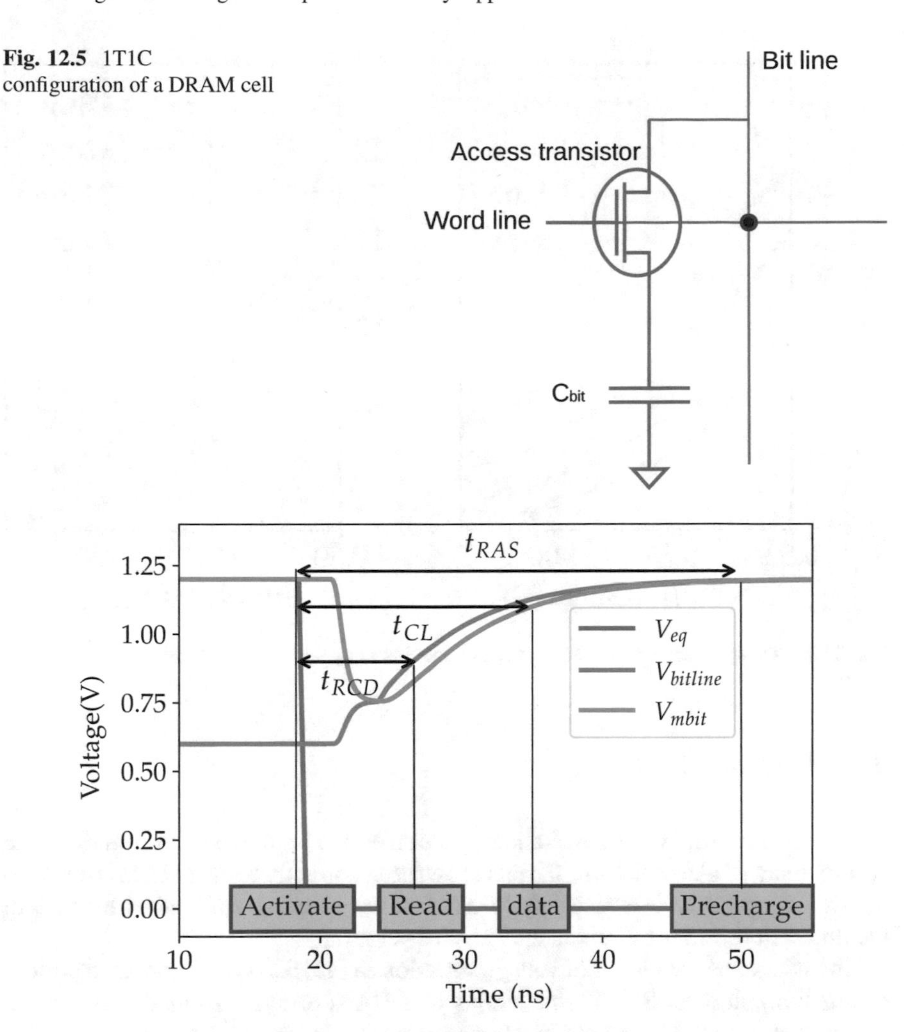

Fig. 12.5 1T1C configuration of a DRAM cell

Fig. 12.6 Timing parameters, bit line voltage ($V_{bitline}$), equilibrate signal (V_{eq}), and cell-capacitor voltage (V_{mbit}) during a READ operation of a DRAM cell

chapter, we measure this latency as the time it takes for $V_{bitline}$ to reach 75% of the supply voltage after the issue of the ACTIVATE signal.

Once the data is ready for the READ operation, column access time (t_{CL}) defines the time required to read the cache line. This is an internal DRAM timing parameter that is determined by the internal clock in the DRAM. Finally, SAs force both the bit line and the cell-capacitor voltage (V_{mbit}) to propagate toward the V_{supply} or 0 V. The timing parameter t_{RAS} ensures that the cell-capacitor charge is fully restored before another cycle of operation can begin. Figure 12.6 demonstrates the READ from a DRAM cell and the corresponding timing parameters for this operation.

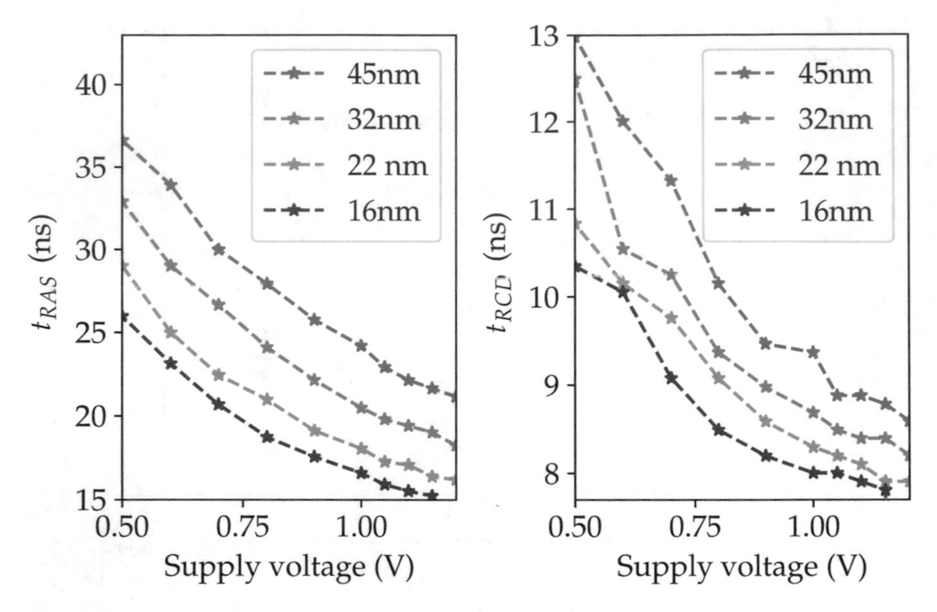

Fig. 12.7 Supply voltage dependence of read latencies t_{RAS} and t_{RCD} with different technology nodes

4.2 VOS in DRAM

Let us assume that WRITE operations are performed with sufficient latency in the approximate DRAM cells, *i.e.,* the bits are written correctly to the DRAM. However, the READ operation is overscaled. As a result, the bit errors suffered by the system would result due to the errors in the READ operation.

To understand the effects of voltage variation in the READ operation, we present detailed circuit-level SPICE simulations of DRAM arrays. We model the circuit components using 22 nm DRAM components reported in [6]. The sense amplifier and the access transistors are modeled using the Predictive Technology Model (PTM)'s 22 nm models [25]. Extrapolation toward 45, 32, and 16 nm designs is obtained based on the ITRS roadmap [6, 9, 17]. Figure 12.7 provides the timing variations in read operation at different supply voltages for other technology nodes.

When reducing supply voltage, fabrication variation and device can make a significant difference in the introduced errors to DRAM cells. Given a DRAM chip with megabytes or even gigabytes, the electric components' properties, e.g., w/l for transistors and capacitance for capacitors, are different from those in other cells. To understand the effect of this, we have examined the variations in the READ operation timings using Monte Carlo simulations for the 45 nm technology node. We perform the Monte Carlo simulations with 1000 iterations with 5% variations in the circuit parameters. The results of the Monte Carlo simulation are presented in Fig. 12.8. From the simulation, we find the values for t_{RAS} and t_{RCD} should be

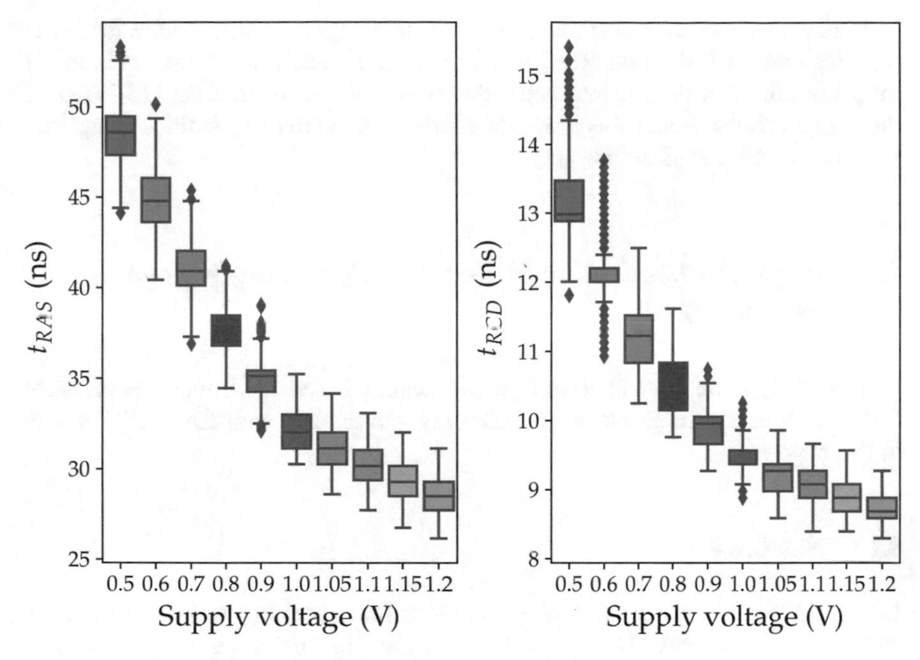

Fig. 12.8 Supply voltage dependence of t_{RAS} and t_{RCD} for 45 nm technology

in the range of >30 ns and 10ns, respectively, to avoid fabrication variation-related errors. This matched the reported values $t_{RAS} = 35$ ns and $t_{RCD} = 13$ ns for typical DDR3L designs.

Moreover, devices are likely to be designed in various structures or manufactured by different vendors, thus demonstrating distinctive error patterns. To summarize, fault can be introduced to the DRAM by scaling voltage and the cells that would be affected depending on fabrication variation and device. Detailed discussions on intentional DRAM fault can be found in [2, 13].

5 Security Applications of Voltage Overscaling

Timing-error-based security applications are not prevalent in practice due to two factors: (a) errors containing hardware-dependent information are prevalent for recent technology nodes such as the sub-45 nm devices, (b) DVFS techniques are designed with performance consideration that ensures correct computation in cases of voltage/frequency scaling. Thus, standard theoretical DVFS practice hinders the security applications that require device profiling or overscaling based on timing errors. However, careful implementation of DVFS to support VOS can open up new avenues for security applications in modern low-power IoT devices. In this section, we discuss two security applications of VOS techniques. The first one is

VOS-based authentication [1] that demonstrates the applicability of VOS for device profiling and authentication. The second application details a defense again model inversion attack in deep neural networks using voltage overscaling [23]. Both of these applications should encourage the reader to explore opportunities using VOS for novel security applications [11].

5.1 Simple Authentication Protocol Design Using Voltage Overscaling

In this section, we present a voltage overscaling-based authentication protocol. Before describing the protocol, we introduce the readers with the notations used in this chapter.

5.1.1 Notations

In this chapter, we use the following notations for describing different algorithms and security protocols. The set of integers modulo an integer $q \geq 1$ will be denoted by $\mathbb{Z}_q$. Matrices, vectors single elements over $\mathbb{Z}_q$ will be represented by consecutively upper case bold letter, lower case bold letters, and lower case letters such as **X**, **x**, and x. For a vector **x**, the length of the vector is denoted by $|\mathbf{x}|$, i^{th} element is represented by $\mathbf{x}[i]$, and $wt(\mathbf{x})$ denotes the Hamming weight (i.e., the number of indices for which $\mathbf{x}[i] \neq 0$) of the vector **x**. The Hamming distance between two binary matrices is denoted by $hd(\mathbf{A}, \mathbf{B})$ (i.e., the number of index pairs (i,j) for which $\mathbf{A}[i][j] \neq \mathbf{B}[i][j]$). Concatenation of two vectors is represented with || symbol. $c \xleftarrow{\$} \{\mathbf{x} \in \mathbb{Z}\}$ represents a random sampling of **x**. We denote *probabilistic polynomial time* (PPT) algorithms with upper case calligraphic alphabets such as $\mathcal{A}$. Therefore, if an algorithm $\mathcal{A}$ is probabilistic, then for any input $\mathbf{x} \in \{0, 1\}^*$, there exists a polynomial $p(.)$ such that the computation of a terminates in at most $p(|\mathbf{x}|)$ steps.

5.1.2 Description of the Protocol

First, we construct an authentication protocol using process-dependent delay information. In this application, delay information is used for generating responses to random challenges. We assume an interactive protocol between a single prover $\mathcal{P}$ and a verifier $\mathcal{V}$. Both the prover and the verifier have some knowledge about a shared secret **x**. The secret is generated through a key-generation procedure $KeyGenVOLtA(1^\lambda)$, where λ is a security parameter. The authentication protocol responds with the outputs accept or reject after a successful run of the protocol.

Table 12.2 Single-round interactive authentication VOLtA

Prover($\mathbf{x_1}, \mathbf{x_2}, H$)		Verifier($M, \mathbf{x_1}, \mathbf{x_2}, \epsilon$)
		$\mathbf{R} \xleftarrow{\$} \mathbb{Z}_p^{\ell \times n}$
	$\xleftarrow{\mathbf{R}}$	
Calculate $\mathbf{L} = \mathbf{H}(\mathbf{R}, \mathbf{x_1}) = \mathbf{R} + \mathbf{x_1}$ using the adder and then calculate $\mathbf{z} = \mathbf{L} \oplus \mathbf{x_2} = (\mathbf{R} + \mathbf{x_1}) \oplus \mathbf{x_2}$		
	$\xrightarrow{\mathbf{z}}$	
		Calculate $\mathbf{z}' = M(\mathbf{R}, \mathbf{x_1}) \oplus x_2$. If distance $(\mathbf{z}', \mathbf{z}) \leq \epsilon$ accept.

This authentication protocol assumes that the prover has a voltage overscaled computation unit (H) that generates process-dependent errors. The verifier either knows the correct model or profile (M) to simulate the computation unit.

Algorithm 4 RNG-based key generation for VOS-based authentication

1: **procedure** $(\mathbf{x_1}, \mathbf{x_2}) \leftarrow KeyGenVOLtA(1^\lambda)$
2: Sample $\mathbf{x_1} \xleftarrow{\$} \mathbb{Z}_p^\ell$
3: Sample $\mathbf{x_2} \xleftarrow{\$} \mathbb{Z}_p^\ell$
4: **return** $\mathbf{x_1}, \mathbf{x_2}$

Enrollment The prover and the verifier use the key-generation procedure $KeyGenVOLtA$ to generate secrets $\mathbf{x_1}, \mathbf{x_2}$. ϵ is the predetermined error threshold for the authentication (Table 12.2).

Note that the distance in the protocol can be measured by standard distance measurement functions such as Hamming distance or Euclidean distance. Also, with multiple keys, the verifier can authenticate numerous users using the same device. Moreover, the verifier can verify the prover over different devices, given that the verifier knows the correct model of those devices.

5.1.3 Evaluation of the Protocol

For our discussions on threat models and attacks of this simple protocol, let us assume that a verifier named Alice tries to authenticate a prover, Bob, over an untrusted channel where an attacker, Malice, performs the following attacks: obtain the security keys or being erroneously recognized as Bob.

The protocol presented in this section is a simple two-factor authentication scheme. The key idea is that the protocol requires knowledge of both the secret was known (i.e., key $\mathbf{x_1}, \mathbf{x_2}$) and the secret possessed (i.e., properties of the voltage overscaled adder). To prove the protocol's effectiveness, we start by analyzing the potential weakness for the case when we have a perfect adder. If the adder is perfect when calculating $\mathbf{L}$, this protocol is not secure. Assume the following scenario: the malicious attacker Malice is pretending to be Alice, and she wants to resolve Bob's key $\mathbf{x_1}, \mathbf{x_2}$ by sending some messages $\mathbf{R}$ and receiving the corresponding $\mathbf{z}$ from Bob. Then she will apply eavesdropping and bit manipulation techniques to recover the key.

However, when applying the voltage overscaling approach, the addition will become non-deterministic because the physical variations will affect the arithmetic result as discussed in Sect. 3.2. Therefore, the result of $\mathbf{M}(\mathbf{R}, \mathbf{x_1})$ cannot be accurately predicted. As a result, the bitwise attacking scenario fails. Overall, the uncertainty of the calculation in the arithmetic function needs to be guaranteed for the security of this protocol. Given the lower process nodes can demonstrate this uncertainty in the calculation due to the process variation in the device, the security of this simple protocol against weak attacks can be guaranteed.

There can also be random guessing attacks, eavesdropping attacks, man-in-the-middle attacks, compromised keys, learning-based attacks, and side-channel attacks on this protocol. A detailed discussion of these attacks and their countermeasures can be found in [1]. Although, this simple protocol cannot guarantee security against a strong attack such as machine-learning-based modeling attacks as discussed in [19, 24]. A voltage overscaling-based stronger authentication protocol that is resistant to ML-based attacks can be found in [24]. Hence, this authentication example should be used as a starting point for designing security algorithms using voltage overscaling. Additionally, reliability issues of these designs due to the physical operation of devices must be taken into account to ensure the robustness of the derived security protocols. Let us explore some of the reliability issues in the next section.

5.1.4 Reliability Issues in VOS-Based Designs

Uniqueness and the Effect of Process Variation

To understand the effect of process variation of VOS-based algorithms, we use eight adders generated from a process variation-aware 45 nm process as discussed in Sect. 3. As shown in Sect. 3, the results from the RCA adders at VDD = 0.4 V contain errors. To understand the uniqueness of each approximate adder regarding these errors, we evaluate the variations using the following metrics:

1. The pairwise Hamming distance of results between adder i and adder j. We used a 3670-byte random input sequence related to the image processing application discussed in Sect. 3. For each adder, we collect the results on each clock cycle and concatenate all the results to create the complete output bit-stream generated by the adder. Then, we calculate the pairwise Hamming distance of these output bit-streams. We divide the Hamming distance with the length of the bit-stream and report the result in percent in Table 12.3.

Table 12.3 Pairwise hamming distance (in percent)

	A1	A2	A3	A4	A5	A6	A7	A8
A1	0	18.82	18.24	18.0	19.4	18.4	18.3	17.52
A2	18.82	0	5.36	5.21	5.67	5.65	3.89	5.39
A3	18.24	5.36	0	4.62	5.98	5.11	5	6.79
A4	18.0	5.21	4.62	0	5.73	3.53	4.13	6.44
A5	19.4	5.67	5.98	5.73	0	6.04	5.59	6.28
A6	18.4	5.65	5.11	3.53	6.04	0	4.96	6.64
A7	18.3	3.89	5	4.13	5.59	4.96	0	5.41
A8	17.52	5.39	6.79	6.44	6.28	6.64	5.41	0

Table 12.4 Pairwise average numerical difference between the output from the devices at 0.4 V

	A1	A2	A3	A4	A5	A6	A7	A8
A1	0	12.01	14.64	16.11	12.30	14.13	12.13	13.25
A2	12.01	0	9.04	12.30	8.03	9.19	8.08	8.97
A3	14.64	9.04	0	8.43	9.03	7.24	9.07	11.13
A4	16.11	12.30	8.43	0	10.68	8.58	8.88	11.13
A5	12.30	8.03	9.03	10.68	0	6.01	6.51	9.45
A6	14.13	9.19	7.24	8.58	6.01	0	8.11	8.89
A7	12.13	8.08	9.07	8.88	6.51	8.11	0	8.47
A8	13.25	8.97	11.13	11.13	9.45	8.89	8.47	0

2. The pairwise average numerical difference of adder i and adder j is given by

$$avg_d = \sum_{l=1}^{N} \frac{|result(i,l) - result(j,l)|}{N}, \tag{12.4}$$

where N is the size of the output in bytes. The result is shown in Table 12.4.

The Hamming distance is widely used to measure the difference between two binary bit-streams. However, it omits the bit orders. For example, the Hamming distance between the pair (00001, 00011) and a pair (00001,10001) is all 1. Therefore, we introduce the second metric to have a measure of the average numerical difference between the values represented by 8-bit outputs of every two adders. Tables 12.3 and 12.4 are symmetric since for our measures distance(i,j) = distance(j,i) [1].

In Tables 12.3 and 12.4, A1 represents the results from an exact adder. From the first rows of Table 12.3, we can notice that there are about 20% bit flips in the outputs of the voltage overscaled adders. From the rows and columns of Table 12.3, it is evident that there is a significant difference in the output of two different voltages overscaled adders. Moreover, from Table 12.4, it is evident that there is a significant average difference between the numerical results produced from each adder.

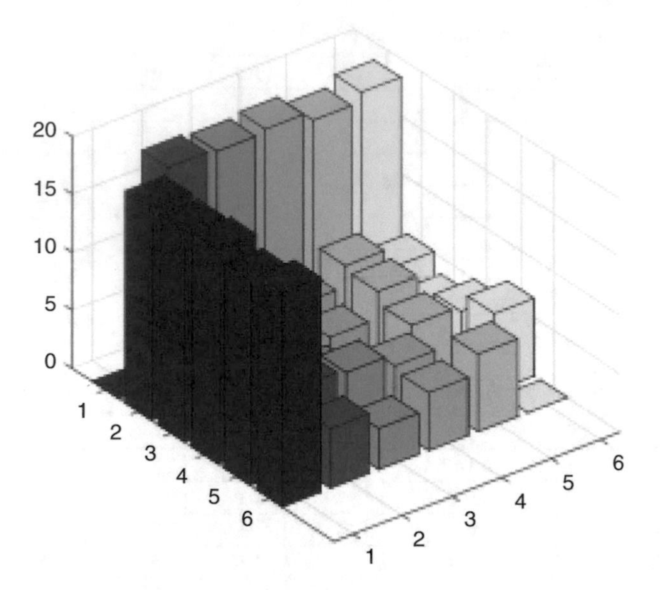

Fig. 12.9 Hamming distance (in percent) between devices at 0.45 V. Here, 1 represents a correct adder (A1), and 2–6 represent voltage overscaled approximate adders (A2–A6)

Effect of Variations in Supply Voltage
Variations in supply voltage will also cause reliability issues for the security protocols designed using VOS-based overscaled devices and processors. Therefore, care must be taken to ensure voltage supply with a minimal amount of noise for proper implementation of the security protocols. In Fig. 12.9, we have plotted the response of voltage overscaled adders at 0.45 V. This is a bit higher voltage than the one used for the results reported in Table 12.3. It can be noted that as the voltage increases the overall Hamming distance decreases, which represents the eventual convergence of all the adders to the correct output at sufficiently higher voltage.

Effect of Variations in Temperature Variations in operating temperature can also affect the security protocols. To understand the effect of temperature, we have calculated the percentage Hamming distance between the results of the same adder at different temperatures as shown in Fig. 12.10.

We find that ± 5 degree variations result in less than 1% bit flips. Therefore, by carefully calibrating the threshold of error tolerance ϵ of the protocols, one can negate the effects of minor temperature variations.

5.1.5 Mitigating Reliability Issues

Noise due to environmental variations can cause deviation from the expected result in a voltage-overscaled circuit. However, taking majority votes in multiple reading

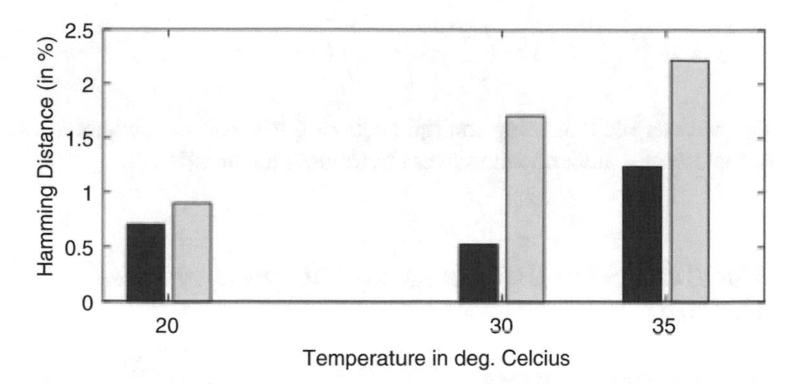

Fig. 12.10 Temperature-dependent bit flips for two different adders. The distance is calculated from the results produced at $T = 25\,^{\circ}\mathrm{C}$. The blue (dark) line represents the temperature-dependent bit flip for adder A2, and the yellow (light) line represents the adder A3

with the same challenge can reduce error due to environmental variations. Hence in the small error case, suppose that the voltage-overscaled device $\mathcal{A}$ produces the expected response x with probability 0.9 over the choice of a challenge c. Then one can amplify this probability to $1/(10n)$ by making $10logn$ repetitions and take the majority vote. In this way, one can correctly recover all of the bits of x with high probability.

Another way of mitigating reliability issues due to environmental variation is to use multiple challenge–response pairs (CRP) for security applications. For example, assume that there is an interchip variation τ exists between two given adders. τ is defined to be the probability that for a given random challenge **c**, the response of the two adders at VOS limit would be different. Furthermore, let us quantify the noise as μ, where μ represents the probability that for a given challenge the response would be different for the δ change in environmental condition. Then, the probability that at least $2t + 1$ out of k reference responses differ between two chips is given by Lee et al. [10]

$$a = 1 - \sum_{i=0}^{2t} \binom{k}{i} \tau^i (1 - \tau)^{k-i}. \tag{12.5}$$

For a single adder, the probability that at most t out of k responses differ is given by Lee et al. [10]

$$b = 1 - \sum_{j=t+1}^{k} \binom{k}{j} \mu^j (1 - \mu)^{k-j}. \tag{12.6}$$

Therefore, the probability of identification of N chips using a set of k challenges is at least [10]

$$p = a^{\binom{N}{2}} b^N \approx \left(1 - \binom{N}{2}(1-a)\right)(1 - N(1-b)). \tag{12.7}$$

Therefore, we can see that using multiple sets of CRPs for authentication, we can guarantee successful authentication over environmental variations.

5.2 *Model Inversion Defense Using Voltage-Overscaled Memory*

5.2.1 Model Inversion Attacks

Model inversion attacks (MIA) expose private training information by abusing the open access-control policy of machine learning algorithms. Inversion attacks are successful because the ML model itself already contains a lot of information about the training data. Since a model is usually trained to obtain a very high prediction accuracy, it tends to "memorize" information provided in the training data set and is likely to be overfitting. Under this scenario, it will show great confidence or high prediction accuracy when fed with some input images, which are very similar to those in the training set.

In general, the higher accuracy a model shows under overfitting, the more vulnerable it is regarding model inversion attacks. If we consider the model as a function f whose input is x and the output is $y = f(x)$, MIA finds the approximate inverse function $g \approx f^{-1}$. Suppose x_0 is an image in the training set, and the overfitting model produces the exactly expected label $y_0 = f(x_0)$ with high confidence. Then, when an approximate inverse function is applied to y_0, the result (i.e., $g(y_0)$) will be very close to x_0. Moreover, if the adversary can access all the original model parameters, further details about the input data can be derived with the existing algorithms, such as the autoencoder mentioned above [3]. Based on this idea, [3] successfully reconstruct the facial image based on a given identity of a person by building and training the decoder, as shown in Fig. 12.11.

5.2.2 Hardware-Oriented Defense Against MIA

Based on this understanding of the vulnerability of the neural networks to MIA, defense mechanisms have been proposed that add some uncertainty to the prediction results to help mitigate a trained model's overfitting [3]. More specifically, if given x_0, the model generates $y_0 + \Delta y$ instead of y_0, the exact representation of the label. When applying an approximate inverse function to this label, $g(y_0)$ would not directly produce x_0. It needs to be mentioned that x_0 and y_0 represent an input/image vector and an output/label vector, respectively. Unfortunately, this solution leads to the permanent degradation of the model.

Fig. 12.11 An example of model inversion attack. The left image represents a collection of data for forty individuals used for training a machine learning model. The image on the right demonstrates the information extracted using model inversion attacks on the trained model. In this example, we have inverted a differential autoencoder (DAE) model trained on AT&T Laboratories Cambridge database of faces using the algorithm given in [3]

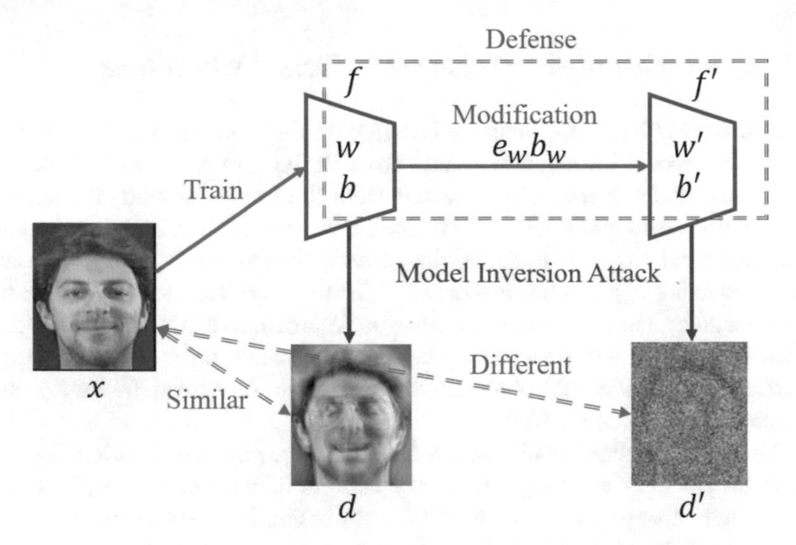

Fig. 12.12 Proposed solution for defending against model inversion attack

To avoid damaging the model parameters, here we demonstrate a run-time degradation of the model by utilizing the random bit errors generated in a voltage-overscaled approximate memory system. We exploit the DRAM's intrinsic error introduction mechanism when applying voltage overscaling-based READ operation as discussed in Sect. 4. Figure 12.12 illustrates the defense mechanism against MIA.

Suppose a neural network model f is trained over input data x, referring to face images, and predicts labels y, referring to a person's identity. As explained in the previous section, a model inversion attack can be conducted to the model for an approximate inverse function, denoted by M_f. This approximate inverse function can be fed with all the labels to obtain the corresponding recovered data given by

$d = M_f(y)$. In the specific case of face recognition classifiers, the attack targets recovering the face image given a person's identity. To defend against this attack, instead of using the original vulnerable neural network, we create a modified version of neural network f' using voltage overscaling, which does the classification task. Note that this updated model is auto-generated. When a ML model is loaded into the voltage-overscaled adder, the READ operation that brings the model parameters from DRAM to cache introduces random errors. Therefore, the processor uses this modified model for the classification task.

For this new model, the recovered data can be represented by $d' = M_{f'}(y)$. If the model f includes parameters w and b, the minor changes made to these parameters are denoted by e_w and e_b. In this case, the modified model f' includes parameters $w' = w + e_w$ and $b' = b + e_b$. Therefore, voltage overscaling introduces e_w and e_w to the model during the execution of the ML algorithm. However, for proper operation of the ML algorithm, one needs to ensure both satisfactory classification accuracy and security.

5.2.3 Accuracy and Security Issues for VOS-Based MIA Defense

To measure the effect of approximate DRAM, we first train the neural network and attack it with model inversion as usual. Then all the parameters in an 8-bit fixed-point representation format are extracted from the trained system. We apply the reconstruction attack proposed in [3] to each individual in the data set and compare the reconstructed images with the original ones in the data set to measure the MIA's power in stealing confidential information. The two critical metrics we are using are test accuracy and the Pearson Correlation Coefficient (PCC). The test accuracy is used to evaluate the network's overall performance on solving the original classification task under the voltage scaling technique. A detailed discussion on this experiment can be found in [23].

After applying voltage scaling to DRAM chips, the physical variation-dependent random bit error percent ranges from 3×10^{-8} to 0.89 according to [2]. If a low supply voltage that generates large bit flips is chosen, the network itself might not be working on the original task. In contrast, a supply voltage close to the standard setting produces small, or zero bit flips, and in that case, the defense will not be robust. Therefore, we focus on a random bit flip percentage, ranging from 10^{-4} to 0.2.

We first analyze how the physical variation randomness affects the network's functionality and security given a specific bit flip rate. When the bit error rate is set to be 0.01, we conduct the defense as well as model inversion attack for 100 times and then measure the test accuracy and PCC similarity for each experiment. The density results follow a Gaussian distribution, as shown in Fig. 12.13. In spite of the physical variation randomness, the average test accuracy decreases from 93% without defense to 91% after defense, and the average PCC similarity between the retrieved images and the training set images decreases from 0.54 without defense to 0.27 after defense.

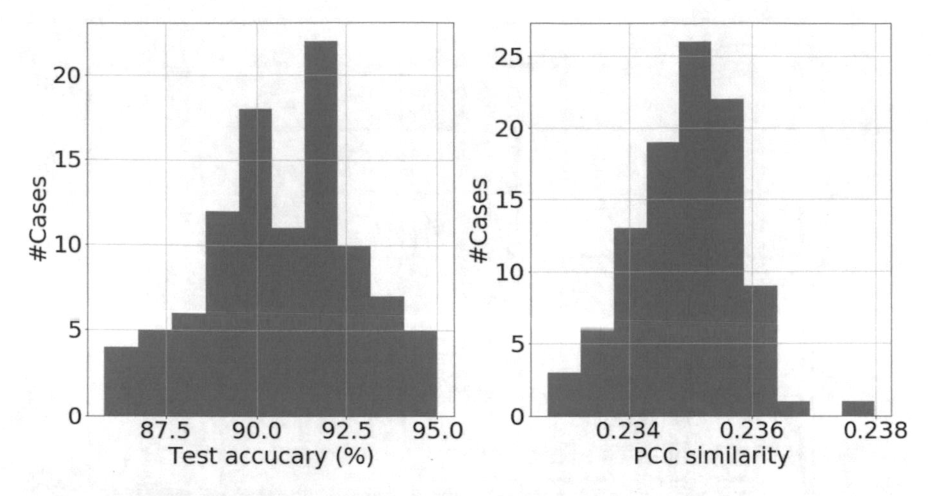

Fig. 12.13 Effect of physical variation randomness on test set classification accuracy and PCC similarity between retrieved images after defense and the original training images when applying reduced voltage with 0.01 bit error rate

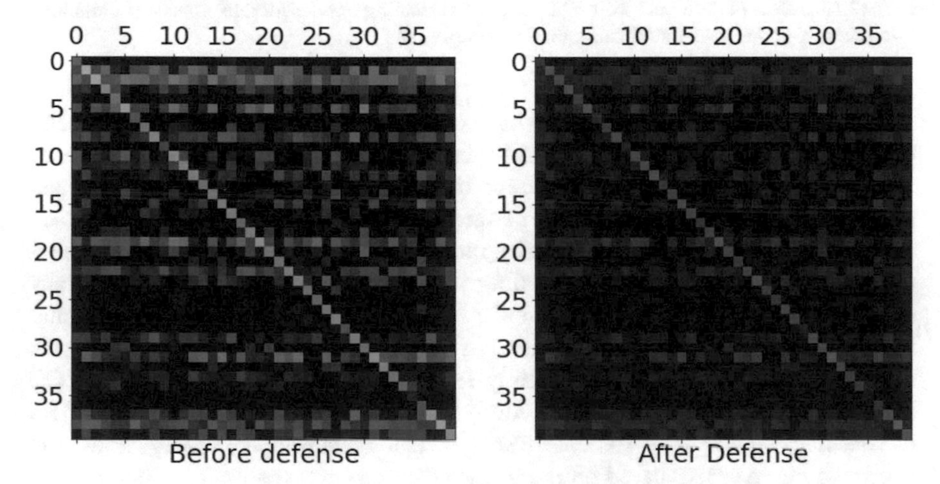

Fig. 12.14 PCC similarity matrix between retrieved images of MIA and original training images for 40 individuals before and after the MIDAS defense

To understand how the defense works for all the 40 individuals, we compute the PCC similarity between each individual's retrieved image and the training images for all people. The generated PCC matrices without/with defense are shown in Fig. 12.14. Based on the diagonal lines in the two matrices, we notice that our algorithm successfully exploits the DRAM voltage overscaling-based defense in reducing the PCC similarity for all individuals. Besides, our proposed method blends the retrieved images with the context, making it harder to identify the individual's identity based on the retrieved image.

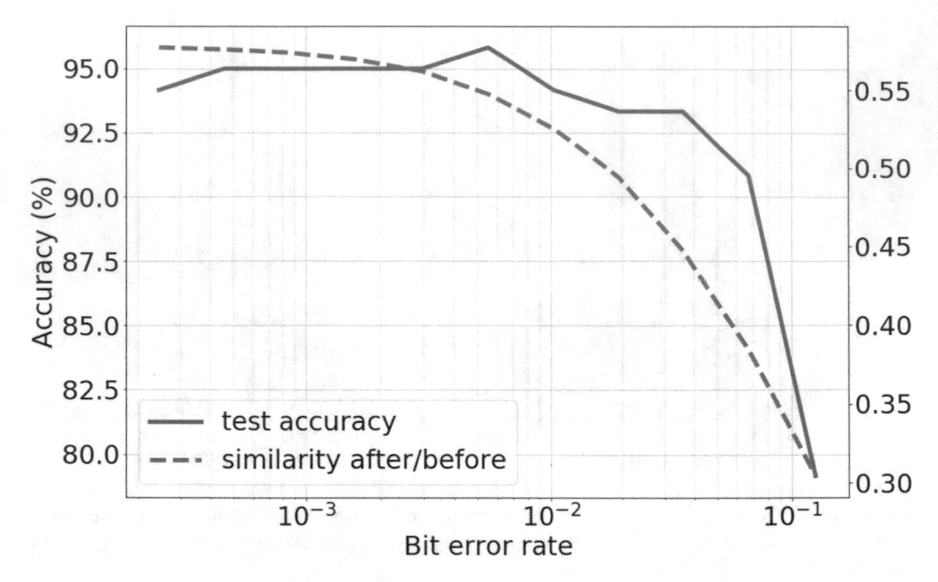

Fig. 12.15 Effect of different settings of voltage overscaling or bit error rate on test set classification accuracy and after/before defense PCC similarity ratio

Another significant question is what is the best voltage overscaling or bit flip rate setting using approximate DRAM memory systems. To answer this question, we select some settings within the range mentioned above, repeat the defense and attack process over 50 times for each setting, and calculate the average test accuracy and after/before defense PCC similarity ratio. As shown in Fig. 12.15, when the bit error rate caused by voltage overscaling is smaller than 0.005, the test accuracy does not change much, and the defense has shown its effectiveness in reducing PCC similarity by 45%. With the bit error rate being set between 0.005 and 0.06, the average test accuracy drops slightly from 95% to around 92%, and the PCC similarity decreases by 55%. Thus, more significant bit error rates tend to affect test accuracy significantly. Hence, we suggest that the developers wisely choose the operating bit error rate based on the accuracy-vs-security trade-off, as depicted in Fig. 12.15.

6 Voltage Overscaling and Security Vulnerabilities

VOS is a fault-inducing operation, i.e., the effect of voltage overscaling is translated into errors generated during a computation. This is useful for fault-injection attacks such as CLKscrew, where voltage regulators of commodity ARM/Android devices are exploited for undervolting/overclocking the SoC processors to induce faults in cryptographic computation [20]. Moreover, errors due to aggressive VOS on sub-45 nm technology node can reveal a significant amount of information about the physical variations in the underlying hardware [1]. Practical DVFS implementations on CPUs are hardware oblivious and software controlled, where the OS kernel supervises the power management IC that controls the voltage and clock generators that generate the operating frequency. This is a *security-oblivious* design since an attacker with kernel privileges can undervolt a processor while keeping the operating frequency same and thus creating timing errors. Recent attacks such as CLKscrew [20] have documented experimental techniques for achieving such faults using the flaws in conventional DVFS implementation. The software-controlled VoltJockey attack has successfully broken ARM TrustZone [14] and Intel SGX [15]. Similar methods can also be employed to extract unique fabrication variation-dependent path errors from modern sub-22 nm processor cores and identify the computing elements.

On the other hand, voltage and frequency scaling are supported by the commodity GPUs, and they are usually exposed to the user through vendor-specific kernel drivers. Reverse engineering efforts for the software drivers for GPUs such as Nouveau are continually working on volting and clocking improvement of open-source GPU drivers. These works enable research on the performance and error-tolerance issues in modern GPUs as well as provide experimental tools to generate voltage and frequency overscaling-based timing faults in the hardware.

It should be noted that by profiling the input patterns of a voltage overscaled circuit, one could easily deanonymize the approximate circuit later based on the physical variations. This exposes a critical flaw in voltage-overscaling-based approximate computing. Therefore, the anonymity of the approximate devices should also be thoroughly studied. The discussions in this chapter can be considered a step toward such analysis.

7 Conclusions

Securing the Internet requires creating trust between the interconnected devices. The general overview of the hardware component from a computation point is that all integrated circuits of the same circuit components are identical, which is at the farthest from the physical truth. Works on physically unclonable functions (PUFs) over the last decade have demonstrated that each physical IC is unique, and one can use such uniqueness for user and device authentication, preventing IC counterfeiting, and provide hardware metering techniques. However, traditional PUF designs require additional components to be added in the IC that inhibits its widespread use. The overscaling techniques discussed in this chapter are an attempt to bridge this gap where the physical variation of the existing commodity hardware will be extracted using kernel modification without introducing new hardware components for the end users.

Hardware variation-dependent error will be more prominent in the next generation of devices where device dimensions are reduced further to accommodate more transistors per area. Standard digital design practices hide this entropy treasure trove from the users through operating over a critical margin. This chapter demonstrates how to securely harness these errors that are reproducible, unique, hardware-dependent analog signatures. We hope that this chapter will inspire further research in the security application of voltage overscaling in the next generation of low-power devices and systems.

References

1. Arafin MT, Gao M, Qu G. VOLtA: Voltage over-scaling based lightweight authentication for IoT applications. In: 2017 22nd Asia and South Pacific design automation conference (ASP-DAC). Piscataway: IEEE; 2017. p. 336–341.
2. Chang KK, Yağlıkçı, AG, Ghose S, Agrawal A, Chatterjee N, Kashyap A, Lee D, O'Connor M, Hassan H, Mutlu O. Understanding reduced-voltage operation in modern DRAM devices: experimental characterization, analysis, and mechanisms. Proc ACM Meas Analy Comput Syst 2017;1(1):1–42.
3. Fredrikson M, Jha S, Ristenpart T. Model inversion attacks that exploit confidence information and basic countermeasures. In: Proceedings of the 22nd ACM SIGSAC conference on computer and communications security. 2015. p. 1322–1333.
4. Gao M, Wang Q, Arafin MT, Lyu Y, Qu G. Approximate computing for low power and security in the internet of things. Computer 2017;50(6):27–34.
5. Gilmore R, Hanley N, O'Neill M. Neural network based attack on a masked implementation of AES. In: 2015 IEEE international symposium on hardware oriented security and trust (HOST). Piscataway: IEEE; 2015. p. 106–111.
6. Hassan H, Patel M, Kim JS, Yaglikci AG, Vijaykumar N, Ghiasi NM, Ghose S, Mutlu O. CROW: a low-cost substrate for improving DRAM performance, energy efficiency, and reliability. In: 2019 ACM/IEEE 46th annual international symposium on computer architecture (ISCA). 2019. p. 129–142.
7. Hong I, Kirovski D, Qu G, Potkonjak M, Srivastava MB. Power optimization of variable-voltage core-based systems. IEEE Trans Comput-Aided Des Integr Circuits Syst. 1999;18(12):1702–1714.
8. Hospodar G, Gierlichs B, De Mulder E, Verbauwhede I, Vandewalle J. Machine learning in side-channel analysis: a first study. J Cryptograp Eng. 2011;1(4):293.
9. ITRS Reports
10. Lee JW, Lim D, Gassend B, Suh GE, Van Dijk M, Devadas S. A technique to build a secret key in integrated circuits for identification and authentication applications. In: 2004 symposium on VLSI circuits, 2004. Digest of technical papers. Piscataway: IEEE; 2004. p. 176–179.
11. Liu W, Gu C, O'Neill M, Qu G, Montuschi P, Lombardi F. Security in approximate computing and approximate computing for security: challenges and opportunities. Proc IEEE 2020;108(12):2214–2231.
12. Murdock K, Oswald D, Garcia FD, Van Bulck J, Gruss D, Piessens F. Plundervolt: software-based fault injection attacks against Intel SGX. In: 2020 IEEE symposium on security and privacy (SP). 2020.
13. Patel M, Kim JS, Mutlu O. The reach profiler (REAPER): enabling the mitigation of DRAM retention failures via profiling at aggressive conditions. ACM SIGARCH Comput Archit News 2017;45(2):255–268.
14. Qiu P, Wang D, Lyu Y, Qu G. VoltJockey: breaching TrustZone by software-controlled voltage manipulation over multi-core frequencies. In: Proceedings of the 2019 ACM SIGSAC conference on computer and communications security. New York, NY, 2019, CCS '19, Association for Computing Machinery. 2019. p. 195–209.

15. Qiu P, Wang D, Lyu Y, Qu G. VoltJockey: breaking SGX by software-controlled voltage-induced hardware faults. In: 2019 Asian hardware oriented security and trust symposium (AsianHOST). Piscataway: IEEE; 2019. p. 1–6.
16. Rahmati A, Hicks M, Holcomb DE, Fu K. Probable cause: the deanonymizing effects of approximate DRAM. In: Proceedings of the 42nd annual international symposium on computer architecture. 2015. p. 604–615.
17. Son YH, Seongil O, Ro Y, Lee JW, Ahn JH. Reducing memory access latency with asymmetric DRAM bank organizations. In: Proceedings of the 40th annual international symposium on computer architecture. 2013. p. 380–391.
18. Stine JE, Castellanos I, Wood M, Henson J, Love F, Davis WR, Franzon PD, Bucher M, Basavarajaiah S, Oh J, et al. FreePDK: an open-source variation-aware design kit. In: 2007 IEEE international conference on microelectronic systems education (MSE'07). Piscataway: IEEE; 2007. p. 173–174.
19. Su H, Zhang J. Machine learning attacks on voltage over-scaling-based lightweight authentication. In: 2018 Asian hardware oriented security and trust symposium (AsianHOST). Piscataway: IEEE; 2018. p. 50–55.
20. Tang A, Sethumadhavan S, Stolfo S. {CLKSCREW}: exposing the perils of security-oblivious energy management. In: 26th {USENIX} security symposium ({USENIX} security 17). 2017. p. 1057–1074.
21. Tehranipoor F, Karimian N, Xiao K, Chandy J. DRAM based intrinsic physical unclonable functions for system level security. In: Proceedings of the 25th edition on great lakes symposium on VlSI. 2015. p. 15–20.
22. Venkatesan R, Agarwal A, Roy K, Raghunathan A. MACACO: Modeling and analysis of circuits for approximate computing. In: Proceedings of the international conference on computer-aided design. Piscataway: IEEE Press; 2011. p. 667–673.
23. Xu Q, Arafin MT, Qu G. MIDAS: model inversion defenses using an approximate memory system. In: 2020 Asian hardware oriented security and trust symposium (AsianHOST). Piscataway: IEEE; 2020. p. 1–4.
24. Zhang J, Shen C, Su H, Arafin MT, Qu G. Voltage over-scaling-based lightweight authentication for IoT security. IEEE Trans Comp. 2021;71:323–336.
25. Zhao W, Cao Y. Predictive technology model for nano-CMOS design exploration. ACM J Emerg Technolog Comput Syst. 2007;3(1):1–es.

Chapter 13
Approximate Computing for Cryptography

Dur-e-Shahwar Kundi, Ayesha Khalid, Song Bian, and Weiqiang Liu

1 Introduction

Approximate computing [1, 2] is one of the promising energy-efficient hardware technique that provides significant efficiency gains at the cost of permissible quality degradation. Right now, it has been employed in a number of error-tolerant applications such as digital signal processing (DSP), image processing, machine learning, pattern recognition, etc. [3–5], due to the fact that these applications are driven by the human perception and have a significant margin in which numerical exactness can be relaxed. However, many information security primitives/applications find their suitability to adopt approximate computing techniques either because of their inherent approximate nature or the characteristics of the application utilizing it. In case of information security, the cryptographic primitives/applications have different set of design parameters, including security, area, power, etc. The security here represents the quality metric similar to accuracy in case of multimedia and DSP applications. However, a naive degradation of computational accuracy might not always be possible and sensible in case of every cryptographic primitive, primarily because the errors due to approximation may be difficult to distinguish

D.-S. Kundi (✉) · A. Khalid
Centre for Secure Information Technologies (CSIT), Queen's University Belfast, Belfast, UK
e-mail: d.kundi@qub.ac.uk; a.khalid@qub.ac.uk

S. Bian
Department of Communications and Computer Engineering, Kyoto University, Kyoto, Japan
e-mail: sbian@easter.kuee.kyoto-u.ac.jp

W. Liu
College of Integrated Circuits (CIC), Nanjing University of Aeronautics and Astronautics, Nanjing, China
e-mail: liuweiqiang@nuaa.edu.cn

W. Liu, F. Lombardi (eds.), *Approximate Computing*,
https://doi.org/10.1007/978-3-030-98347-5_13

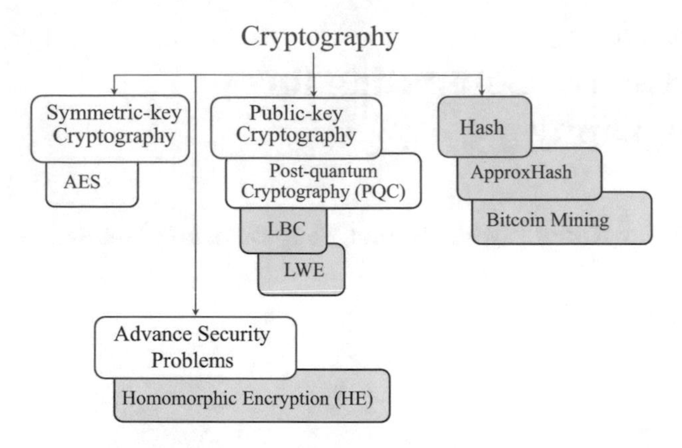

Fig. 13.1 Cryptographic primitives

from the malicious falsification from the adversary. Additionally, approximation can introduce new threats especially in case of embedded devices.

A brief survey is presented in [6], pointing the use of approximate computing for information security, such as cryptography, hardware security, and machine learning-based security approaches. However, this work takes up a comprehensive view of various cryptographic primitives and applications, having the potential for the approximation. It is worth mentioning here that not every cryptographic category is able to adapt the approximation techniques, the possible solutions are shown by the gray color code in Fig. 13.1.

The first potential approximation primitive is the cryptographic hash function as discussed in detail in Sect. 2, which plays an important role in numerous security-based applications. It was designed with the aim for a set of random outputs with some specific properties for a correlated set of inputs. Therefore, intermediate computations in conventional hash function can be executed approximately subjected to the output sets fulfilling the designed criterion. Based on this, the approximate hash function [7] finds its further application in the Bitcoin cryptocurrency [8] provided in Sect. 2.1.

The second one is the Learning with Errors (LWE) problem; that is, a building block of lattice-based cryptography (LBC) is inherently approximate in nature [9]. It is undertaken as an alternate of classical public-key cryptography (PKC) (i.e., ECC and RSA) used today, under the threat of a viable quantum computer. Approximate versions of LWE specially the Ring-LWE (R-LWE) explored in Sect. 3 are especially attractive to IoT applications with end-node devices that are limited in terms of computational power [10]. The more advanced security problem, i.e., homomorphic encryption (HE) based on LWE can also be accelerated using the approximation techniques [11], which is explained in Sect. 3.1.

2 Approximate Hash

Cryptography hash functions are the work horse of cryptography and are used in many information security applications such as digital signatures, message authentication codes (MACs), key derivation, RNG, etc. in order to provide essential security services. It is a deterministic procedure that computes a fixed length of output for an arbitrary length of input data such that either the contents or length of input data is impossible to recovered. The data to be hashed is called the message M, while the hash value is called as message digest or simply digest H and is generated by a hash function h of the form $H = h(M)$.

There are two major families of hashes, one is the Addition-Rotation-XOR (ARX) family, i.e., Secure Hash Algorithm-1 (SHA-1), SHA-2 with variants SHA-224, SHA-256, SHA-384, etc. as specified in [12] as Secure Hash Standard (SHS). The other one belongs to the family of SPONGE construction [13], i.e., the SHA-3 which is a new hash algorithm being standardized by National Institute of Standards and Technology (NIST) in 2015 [14], to supplement the previous SHS algorithms. The main difference between the two families of hashes is the type of operations being involved in it, as summarized in Table 13.1. The most costly operation of the ARX family of hash is the 32-/64-bit modular adders that determines the overall delay, power, and area of their designs. Therefore, these modular adders can be replaced with the approximate version by taking in account the cryptographic properties of the hash functions. However, these adders are eliminated in the new standardized SHA-3 but still have the potential to be explored for the approximation based on the performance/delays of individual bitwise operations. For example, XOR gate is the expensive one, which can be approximated with AND/OR gate.

The cryptographic hash functions do not require a particular output, and hence, there is no concept of error. However, they must satisfy the following three major properties [15]:

- Pre-Image Resistance: It is a measure of difficulty to find an input message from which a given hash value is computed. That is to find any pre-image M such that $h(M) = H$ when given any H for which a corresponding input is not known.

Table 13.1 Cryptographic hash functions

Hash family	Algorithm	Variants	State size	Rounds	Operations
ARX	SHA-1	—	160	80	Modular 32-/64-bit Addition
	SHA-2	SHA-224	256	64	AND, OR, XOR, ROT
		SHA-256			
		SHA-384	512	80	
		SHA-512			
SPONGE function	SHA-3	SHA3-224	1600	24	AND, OR, XOR, NOT, ROT
		SHA3-256			
		SHA3-384			
		SHA3-512			

This means that hash functions are one-way functions and functions that lack this property are vulnerable to pre-image attack.

- Second Pre-Image Resistance: Given an input message, it is a measure of difficulty to find another input message such that both messages have the same hash value. That is given M to find a second pre-image $M \neq M'$ such that $h(M) \neq h(M')$. This property is also referred to as a weak collision resistance and functions that lack this property are vulnerable to the second pre-image attack.
- Collision Resistance: Collision resistance is a measure of difficulty to find two different messages that have the same hash value. To find any M, M_0 which has the same hash such that $h(M) = h(M_0)$. This property is also referred to as strong collision resistance.

To examine all these hash function criterion, NIST has provided the statistical test suit [16]. Out of all the commonly used tests are (i) avalanche test, (ii) Maurer's universal statistical test, (iii) statistical Mobius analysis, and (iv) near-collision test. Avalanche test counts the number of bits changed in digest with one bit change in the input. The hash function is said to passed avalanche test, if it satisfies Strict Avalanche Criterion (SAC) [17], which requires 50% probability of changing each bit of hash digest with single bit change in the input. The randomness is measured using Maurer's universal statistical test that is carried out for significance levels of 0.1 and 0.01 [18]. Furthermore, statistical Mobius test analyzes any exploitable bias due to design flaw, which is done at a significance level of 0.05. These ranges of measurement levels provide us the opportunity to replace the accurate circuits with the approximate ones, which will be faster and more compact and consume less power, while maintaining the acceptable range.

The same concept has been demonstrated in [7] using the SHA-1. The SHA-1 has 80 rounds, where the basic round function comprises a nonlinear function, shifters, and 32-bit modular adders as shown in Fig. 13.2. In order to reduce delay, power, and area of SHA-1 architecture, the costly 32-bit modular adders are replaced by the approximate version of Ripple Carry Adder (RCA) that can be designed using Approximate Full Adders (AFAs) or Approximate Mirror Adders (AMAs), etc. A detailed overview about the approximate adders is provided in [19].

In [7], the authors investigated the level of approximation across the 80 rounds of SHA-1 (named the design as ApproxSHA-1) that is permitted according to hash function criterion. Hence, one can select the appropriate ApproxSHA-1 design with N stages of approximation as shown in Fig. 13.3, according to his application's hardware as well as security strength requirements.

2.1 Approximate Bitcoin Mining

Bitcoin is a digital cryptocurrency, mainly created to simplify transaction processes without needing a third party, to increase the speed of cross-border transactions, and

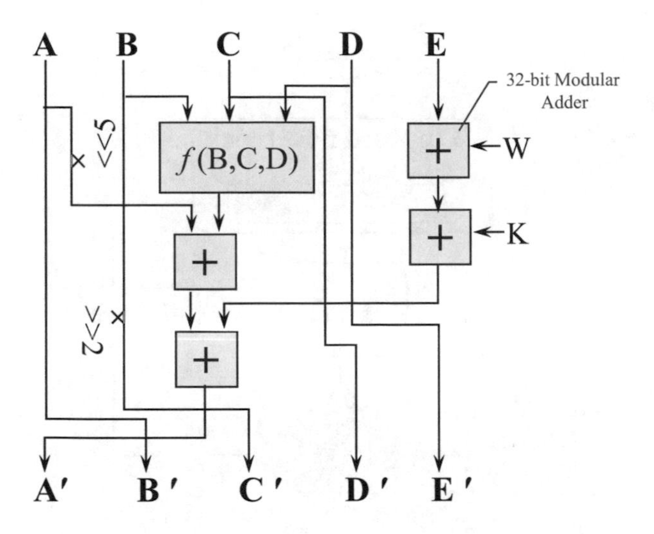

Fig. 13.2 SHA-1 round function

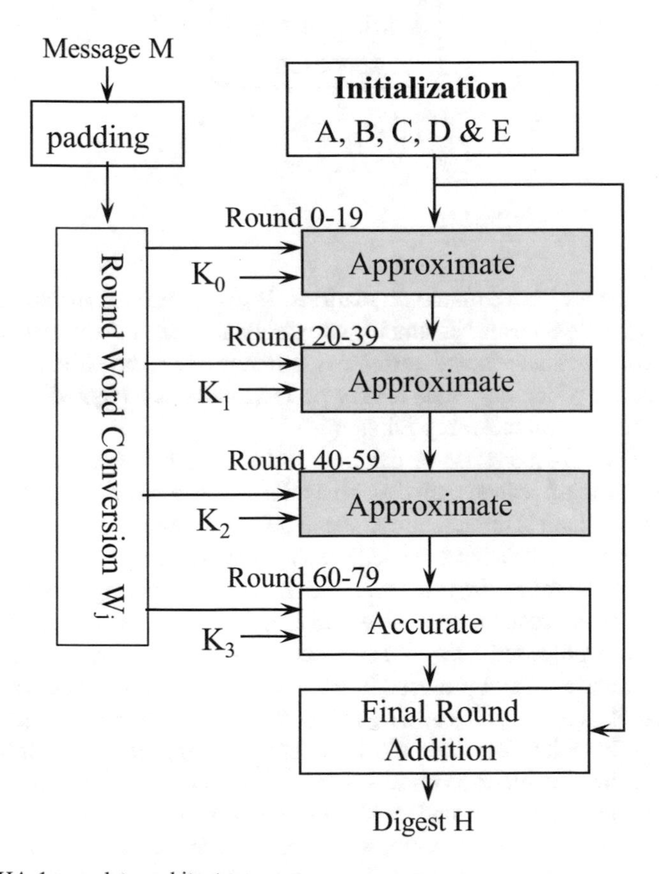

Fig. 13.3 SHA-1 complete architecture

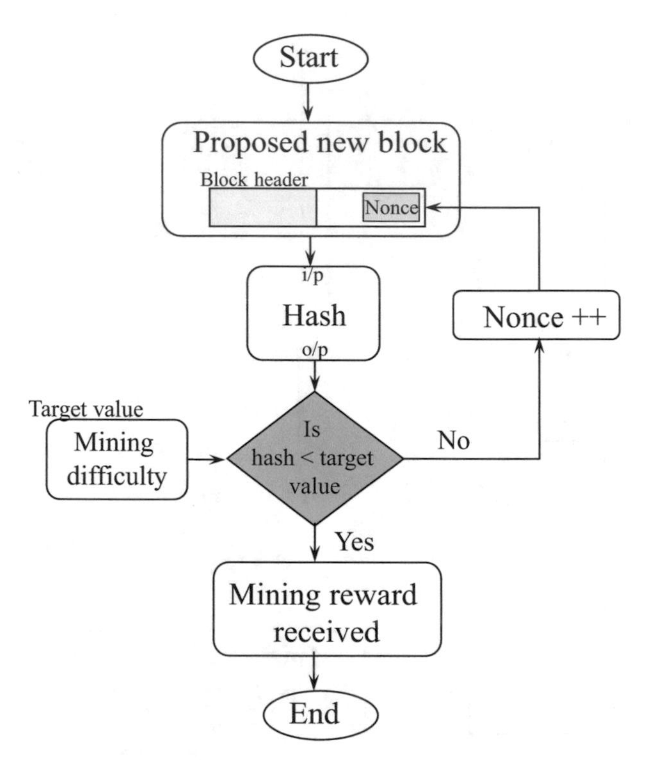

Fig. 13.4 Bitcoin mining process

to be independent of government regulations. It was presented in January 2009 by the pseudonymous Satoshi Nakamoto, whose identity till date remains a mystery [20]. Unlike government-issued currencies, Bitcoin offers the attractive promise of decentralization, which has made it very popular and has triggered the creation of many other currencies, including Ethereum.

Bitcoin mining is a process of creating and adding transactions in the form of bitcoin to the Bitcoin ledger, called as blockchain. An attractive and unique feature offered by blockchain is immutability, which makes it impossible for any third party to interfere with Blockchains the continuously growing chained data. The Bitcoin mining process is shown in Fig. 13.4, and it requires solving a computationally hard block to be discovered as a *proposed new block*, which will be added to the blockchain. The proposed new block contains the hash of that block, transaction data, cryptographic hash of previous block, and a timestamp. When someone joins the network, it gets the full copy of the blockchain for the nodes to verify that everything is in order. The new block is sent to everyone in the network; each node then verifies the block to make sure they have not been tampered with. After successful verification, each node adds this block in to their blockchain. All the nodes in this network create *consensus*. They agree on which blocks are valid and which are not. This validation is performed by the nodes that are connected in the network using specific consensus mechanism to ensure the new member is authentic. This process of validating the transaction is commonly known as mining and miner is the node performing this validation.

As multiple miners are trying to mine a new block, the problem is embarrassingly parallel. Miners propose a potential Bitcoin block of transactions and use this for an input to the ledger. The proposed block is hashed via the SHA256 hash function, yielding a fixed-sized hash. If the hash value is lower than the Bitcoin network difficulty, then the miner who proposed the block wins. Otherwise, the miner continues trying to compute more hashes. The successful miner's block is then added to the blockchain, and the miner is rewarded with newly issued Bitcoin for their work. As the number of miners in the network increases, the mining difficulty goes high. The Bitcoin has a limited circulating supply of 21 million coins.

Bitcoin mining, based on complex computation, is inherently error-tolerant. Since the cost (e.g., electricity) for Bitcoin mining is very high, hence, low-power strategies are important for Bitcoin mining. To address this, approximate computing has been considered to improve the efficiency of Bitcoin mining. Approximate circuits can be built to reduce delay and area consumption but trading off reliability.

The mining algorithm is shown in Algorithm 1. There are two inputs to the function: D(t) that is the difficulty of the network that is adjusted regularly, and the other input is the header of the new block proposed. As can be seen from the line number 4 of Algorithm 1, the process of mining, in essence, is to search for a nonce value that results after a double SHA-256 hash digests less than a given threshold. The process continues after incrementing the nonce, till such a value is found. There is incentive given to miners for successfully mining new blocks; without miners, new transactions cannot be added to the public Bitcoin ledger. Consequently, the miners compete against each other to get higher hash rate in order to maximize the chances of getting the reward, called *block reward*. Back in 2009, this reward was originally 50 Bitcoins, and however, every four years this reward halves. Since Bitcoin has a strict circulation supply of 21 million bitcoins and at the time of writing this chapter more than 83% of them have already been mined, the job of a miner is getting harder. By the year 2031, 83% of the bitcoins will be mined.

Algorithm 1 Bitcoin mining process

Require: D(t), header
1: nonce=0;
2: **while** (nonce $< 2^{32}$) **do**
3: threshold = $((2^{16}1) << 208)/D(t)$;
4: digest = SHA-256(SHA-256(header));
5: **if** digest $<$ threshold **then**
6: **return** (nonce);
7: **else**
8: nonce=nonce+1;
9: **end if**
10: **end while**

2.2 The Economics of Bitcoin Mining

With the increasing hashing difficulty in Bitcoin mining process, the miners must weigh the cost of special purpose equipment plus the operating expenses they need vs. the bitcoin reward. The evolution of the mining hardware, across several generations, from early GPU-based machines to today's application-specific integrated circuits (ASICs) is discussed in [21].

1. First generation, CPU-based mining: the so-called first generation of mining comprised of simple C code (opensource, available on GitHub) that can run on simple laptop/desktop. Several operations that the existing high-performance SHA-256 hashing libraries offered were employed for speed optimizations.
2. Second generation, GPU-based mining: GPU-based mining software was released about a decade ago that successfully enabled leveraging GPUs for Bitcoin mining enthusiasts to generate profits. Scaling up with rigs with multiple GPUs however increased the power dissipation far too high to need specialized warehouse spaces with economical cooling solutions.
3. Third generation, FPGA-based mining: FPGA-based Bitcoin miner implementations were introduced soon after the GPU-based mining setups. The implementations employed parallelism and exhaustive loop unrolling to generate one nonce trial (hash) per clock cycle on low-power FPGA families (e.g., Xilinx Spartan).
4. Fourth generation, ASIC-based mining: soon after the FPGA-based mining was successfully introduced, several ASIC-based mining solutions were introduced. Several companies made their solutions commercially available with a drastic improvement over the hash rate costs, throughput performance and energy efficiency compared to GPU-based solutions.
5. Fifth generation, The ASIC War: as ASICs proved their performance in Bitcoin mining, many venture capitalists jumped into the war of better solutions primarily relying on better architectures and more advanced process nodes.
6. Sixth generation, The ASIC Victors: the current generation of Bitcoin miners are the companies that survived The ASIC War competition and were successful in advancing to better technology nodes successfully. BitFury and Bitmain both producing 16-nm chips are two big names in this context. These companies have created their own ASIC-based cloud data centers in places where low-cost energy is not a problem.

2.3 Mining Approximation

In terms of monetary evaluation, the price of a single Bitcoin touched an all-time high of around $65K in April 2021. Moreover, Bitcoin mining is highly criticized for being power hungry and consequently too expensive. Consequently, there is high interest in making the mining problem fast. Approximation can be used to as a technique to reduce delay and area of Bitcoin mining circuits, thereby increasing profits [22]. Since hashing is an embarrassingly parallel process, the propagation of approximation errors is limited. Furthermore, in case of an error, the hash of any new block is verified by all the rest of nodes in the network to reject any invalid nodes. As mentioned in the [22], SHA-256 Pipeline data path employs several different types of adders, including the carry-save adder (CSA), carry-propagate adder (CPA), carry-lookahead adder (CLA), and ripple carry adder (RCA). Adders are a basic component of digital circuits and their approximation has been extensively studied [23]. The authors in [22] undertake the following two approximate adder designs in the mining process and analyze it for profit maximization:

1. A gracefully decaying adder (GDA) [24]: a GDA rearranges a carry-lookahead adder logic with a configurability for graceful decay. GDA(1,4) with a 16-bit carry chain is taken up for analysis in mining hardware as its error rate is low.
2. The Kogge–Stone adder (KSA) [25]: KSA is a variable latency speculative adder that minimizes delay at the expense of area. A KSA_{32} design is considered as it minimizes the hashing pipeline's delay-area product.

The approximate Bitcoin mining work [22] concludes that profits can be maximum if the approximation can reduce the delay of the adders at the price of using more area resources. Operating hash cores on clock rates that are Better Than Worst-Case (BTWC) also helps achieve better profits. The KSA adder using functional and operational approximation was able to generate a substantial 30% improvement in the Bitcoin mining profits.

3 Approximate Learning with Error (LWE)

The *Learning with Errors* (LWE) problem was first introduced by Regev in 2005 [26] and since then became the foundation of many lattice-based cryptography (LBC) primitives. The LBC has emerged as one of the most viable alternatives to the classical PKC schemes, due to its security, versatility, and efficiency. The security proofs of LBC are based on worst case hardness of lattice problems, which are up till now resistant to attacks by the quantum computers. It provides a diverse set of primitives to tackle with the challenges of traditional security problems, i.e., Public-Key Encryption (PKC) [27, 28]/Key Exchange Management (KEM) [29, 30] and digital signature [31, 32] as well as emerging security problems, i.e., Fully Homomorphic Encryption (FHE) [31, 33–36], and many more.

What remains common however is the fact that these quantum-resilient algorithms are much more complex than the currently deployed PKC techniques, i.e., larger key sizes, higher computational complexity, making them impractical for low-cost devices specially in the *Internet of Things* (IoT) scenario. IoT today plays a pivotal role in connecting and exchanging information. An increasing number of heterogeneous devices are getting connectivity, and the number of IoT devices connected to Internet are predicted by Cisco to easily cross 500 billion by 2030 [37]. The IoT revolution is a double-edged sword, driving the promise of digital transformation on the one hand and opening a plethora of potential security and privacy vulnerabilities on the other. In 2020, the literature statistics of SCOPUS database, for a 10th straight year, reported the highest number of IoT attacks and threats related articles [38]. The computational capabilities of these ubiquitous IoT devices depend on the layer in which they are deployed, i.e., server, edge, or the end-nodes. Out of all, the end-node entities have limited area/power capabilities that constitute the main bottleneck in the realization of secure quantum-resilient hardware. However, these devices have lower recommended classical security level of at least 112 bits [39].

The original LWE problem was based on the *Standard Lattices* involves high computational and spatial cost restricting performance. To tackle that, LWE was modified to the Ring-LWE (R-LWE) problem by Lyubashevsky et al. in 2010 [40] based on *Ideal/Ring Lattices*, eliminating the computations with large matrices saving memory and also results in small key sizes. Lightweight R-LWE coprocessors on FPGAs have been actively researched [41–45], proving their feasibility for the practical deployment on server/edge-based IoT devices. However, very limited Application-Specific Integrated Circuit (ASIC) implementations of R-LWE have been reported in [46, 47] utilizing 776k GE and 106k GE, respectively, while the lightweight hardware implementation of ECC processor [48] requires only 20k GEs to 29.7k GEs for GF(2^{163}) to GF(2^{233}) (having respective classical security levels of 80 bits to 112 bits), hence highly infeasible for resource-constrained IoT end-node devices.

However, if an R-LWE problem is analyzed, it is inherently approximate in nature. The decryption process involved a threshold decoding to recover the final message from the added Gaussian noise. Generally, the R-LWE-based PKE cryptosystem comprises three stages: key generation, encryption, and decryption, as illustrated in Fig. 13.5. All the operations are based on polynomials in the ring $\mathbb{R}_q$, U represents the uniform distribution and χ is a discrete Gaussian (DG) distribution with a standard deviation σ and mean $\mu = 0$, whereas "×" and "+" represent the modular multiplication and addition operations, respectively. Observing the decryption equation $\mathbf{c}_1 \times \mathbf{r}_2 + \mathbf{c}_2$, it translates to

$$\begin{aligned} &= (\mathbf{a} \times \mathbf{e}_1 + \mathbf{e}_2)\mathbf{r}_2 + \mathbf{p} \times \mathbf{e}_1 + \mathbf{e}_3 + m \cdot \lfloor q/2 \rfloor \\ &= (\mathbf{a} \times \mathbf{e}_1 + \mathbf{e}_2)\mathbf{r}_2 + (\mathbf{r}_1 - \mathbf{a} \times \mathbf{r}_2)\mathbf{e}_1 + \mathbf{e}_3 + m \cdot \lfloor q/2 \rfloor \\ &= \mathbf{e}_2\mathbf{r}_2 + \mathbf{e}_1\mathbf{r}_1 + \mathbf{e}_3 + m \cdot \lfloor q/2 \rfloor . \end{aligned} \tag{13.1}$$

It is noticed that as long as $|\mathbf{e}_2\mathbf{r}_2 + \mathbf{e}_1\mathbf{r}_1 + e_3|$ (denoted as $|e|$) is less than $\lfloor q/4 \rfloor$, $-\lfloor q/4 \rfloor < m \cdot \lfloor q/2 \rfloor + e < \lfloor q/4 \rfloor$ if $m = 0$, and otherwise if $m = 1$. Since e_3 is small compared to the products, and the products are obtained from independently drawn Gaussian samples, so can be written as $|e| = |\langle \mathbf{e}, \mathbf{r} \rangle|$, where $\mathbf{e}, \mathbf{r} \leftarrow \chi_\sigma$. The decryption will fail if error threshold of decoder, i.e., $\langle \mathbf{e}, \mathbf{r} \rangle > \lfloor q/4 \rfloor$. This is known as per-symbol error probability as denoted by δ, which in turn depends on error distribution according to Lemma 3.1 of correctness criteria [49].

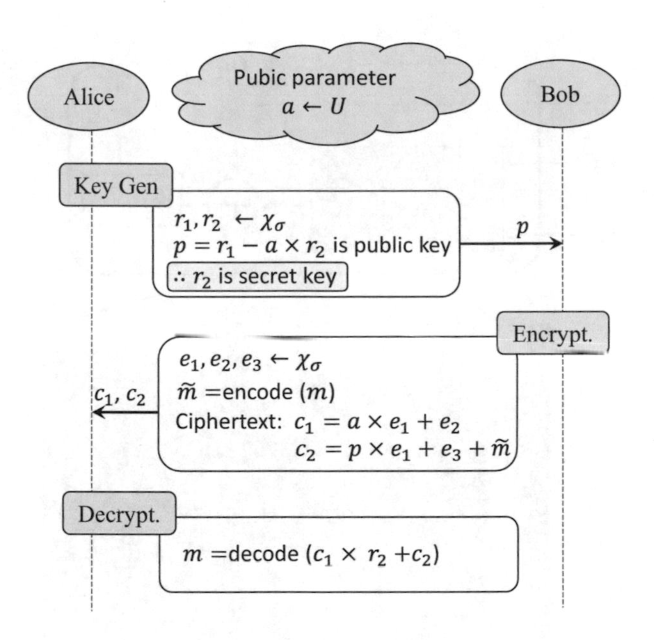

Fig. 13.5 R-LWE-based PKE cryptosystem

Table 13.2 Hardware results [9]

Designs	Delay (ns)	Area (GE[a])	Power (μW)
Exact 12-bit	7.0	2912	33.3
DRUM9_12 [50]	5.3	1413	26.5
App. multiplier [9]	4.0	698	14.4

[a] Gate equivalent (NAND)

Hence, this approximate nature of R-LWE has been exploited in [9, 10] to use the approximate computing techniques, to reach even lowered power and compact designs, better suited to highly constrained IoT devices, at the cost of lowered security levels. In 2018, Bian et al. [9] designed an optimized version of approximate Dynamic Range Unbiased Multiplier (DRUM) [50] instead of the exact one for the polynomial multiplication in R-LWE decryption, thereby satisfying the per-symbol error probability of 2% with overwhelming confidence. The 12×12-bit exact multiplier is approximated to 9×9-bit as shown in Fig. 13.6. The architecture on a 65 nm low-power process node achieves a $2\times$ area reduction and $1.84\times$ power reduction than a conventional DRUM [50] as shown in Table 13.2.

However, Kundi et al. [10] demonstrated the effect of approximation via operand truncation strategy [50–52], by exploiting the fact that one of the inputs to the multiplier always comes from the DG distribution, i.e., $\mathbf{r}_2$, $\mathbf{e}_1$ as shown in Fig. 13.5. Based on the statistical properties of DG noise (i.e., 99.7% of values lie within interval $[\mu - 3\sigma, \mu + 3\sigma]$ are 4-bit, whereas only 0.2% of values are greater than this), truncate the DG noise 1-bit in addition to the technique proposed in

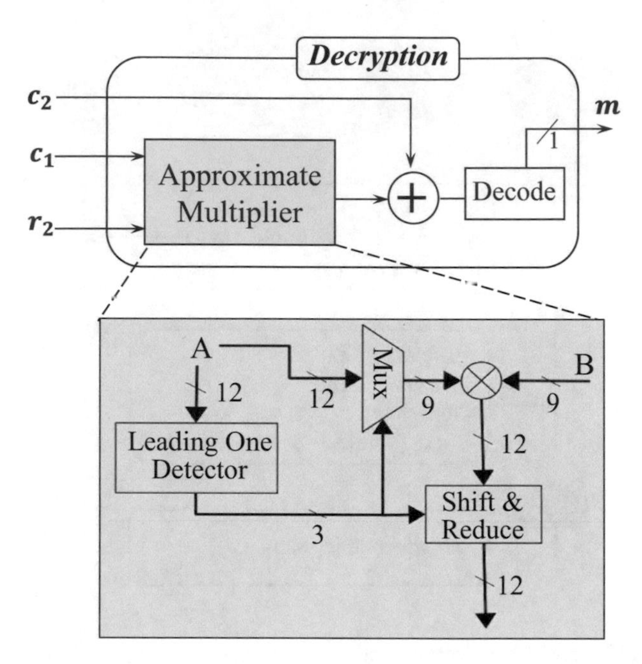

Fig. 13.6 LWE approximate multiplier [9]

[45]. Hence, transform the 12×12-bit unsigned exact multiplier to 12×4-bit signed multiplier. The proposed approximate modular multiplier (referred to as AxMM) is shown in Fig. 13.7, comprising of an AxMult followed by an approximate modular reduction (AxMR) circuitry. The architecture can be employed in the whole R-LWE cryptosystem. Using 45 nm fast process technology achieves the area and power reduction by 35% and 23%, respectively, as compared to the smallest exact R-LWE multiplier design [45].

Hence, the application of AxC techniques to the ingredients of R-LWE provides us the benefit to greatly reduce the hardware as well as power consumption, thereby making it an ideal choice for resource-constrained devices. However, there will be corresponding reduction in the security level because of the smaller size of error samples but sufficient for meeting the lowest 112-bit security level of IoT end-node devices. On the other hand, the per-symbol error probability of R-LWE scheme will be improved as stated in [41] and the same is proved in Table 1 of [53] (Table 13.3).

3.1 Homomorphic Encryption

LWE-based homomorphic encryption (HE) schemes are also known to be efficient in constructing secure multi-party computation (MPC) protocols. In particular, as suggested in [54], HE schemes based on LWE (in most cases, R-LWE) are

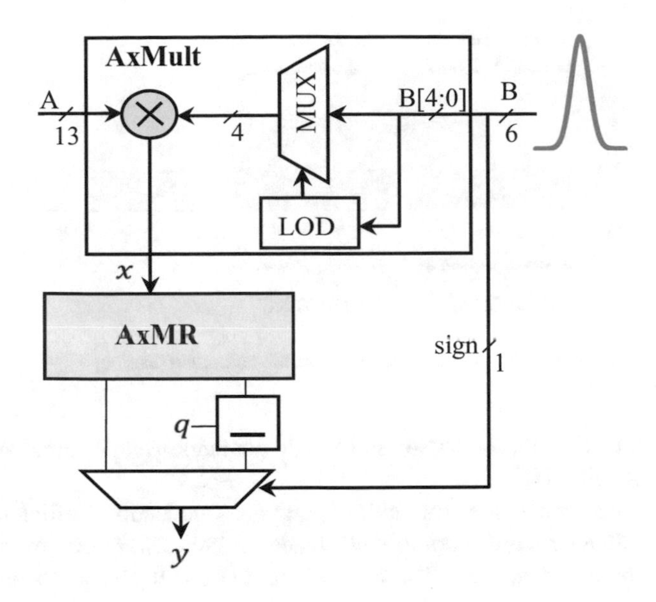

Fig. 13.7 Approximate modular multiplier [10]

Table 13.3 Hardware results [10]

Designs	Area (GE)	Delay (ns)	Power (μW)
Exact 13-bit [44]	2628	1.70	496.85
App. multiplier [9]	2421	1.60	310.46
Exact 5-bit [45]	1431	1.29	179.71
AxMM [10]	920	0.96	137.98

highly efficient in processing linear operations (e.g., matrix-vector products) with minimum communication costs when compared to other MPC protocols such as secret sharing [55, 56] or Yao's garbled circuit [57, 58]. Hence, LWE-based HE schemes are actively deployed in the field of privacy-preserving machine learning [59, 60], encrypted database [61], secure computation offloading [62], secure genome-wide association [63], and many more.

As HE schemes based on the LWE problem are also inherently approximate, exploring how approximate computing can be appropriately applied to HE is under research. The main difficulty in adopting approximate computing in LWE-based HE is that, because errors accumulate extremely fast during homomorphic evaluations, error estimation is rigorously performed to ensure that the designed parameter sets are minimized while being able to contain the errors. Furthermore, in most HE applications, number theoretic transform (NTT) is applied to accelerate polynomial multiplications for the underlying R-LWE ciphertexts. Unfortunately, since NTT is applied over finite rings, the computations are exact and cannot tolerate any error in the NTT-transformed domain. Hence, most existing protocols as well as

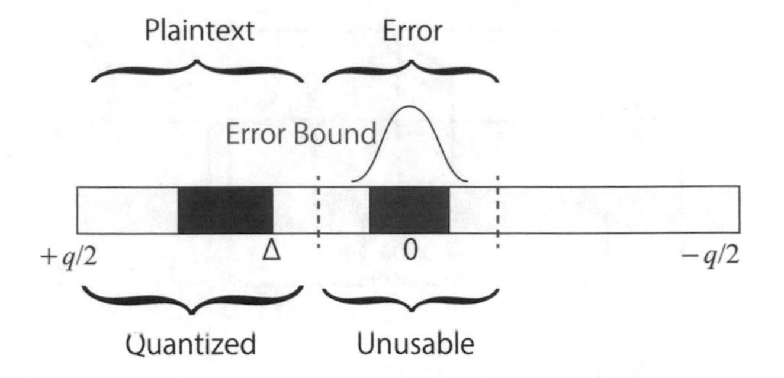

Fig. 13.8 A graphical depiction of one single slot in the ciphertext array $\mathbf{c}_1$

hardware accelerators for R-LWE-based HE do not employ approximate computing techniques [54, 59, 64].

In order to maximize the computational and communicational efficiency for HE evaluations with approximate computing, a new technique for homomorphic linear transform is proposed in [11]. The authors in [11] target on the specific case of privacy-preserving neural network inference and realize that NTT can be avoided in such case without lowered computational efficiency. Specifically, let $\mathbb{Z}$ denote the set of integers. Let $m \in \mathbb{Z}$ be the lattice dimension, $n = \phi(m)$, where ϕ is Euler's totient function, $p \in \mathbb{Z}$ be the plaintext modulus, and $q \in \mathbb{Z}$ be the ciphertext modulus. Consider the standard R-LWE ciphertext $\mathbf{c} = (\mathbf{c}_0, \mathbf{c}_1)$ encrypted as follows:

$$\mathbf{c}_0 = -\mathbf{a} \tag{13.2}$$

$$\mathbf{c}_1 = \mathbf{a} * \mathbf{s} + \Delta \cdot \mathbf{u} + \mathbf{e} \bmod q, \Phi_m, \tag{13.3}$$

where $\mathbf{a}, \mathbf{s}, \mathbf{e} \in \mathbb{Z}_q^n$, $\mathbf{u} \in \mathbb{Z}_p^n$, and Φ_m is the mth order cyclotomic polynomial. A single encrypted plaintext of $\mathbf{c}_1$ is illustrated in Fig. 13.8, where the ciphertext space (i.e., $\mathbb{Z}_q$) is divided into two sections: (i) the error section and (ii) the plaintext section.

As pointed out in [11], the polynomial multiplications involved in R-LWE cryptosystems are simply convolutions, with the additional operation of negacyclic reduction. As a result, it is observed that ciphertext–plaintext convolution can be performed without NTT. Specifically, consider the following operation between a R-LWE ciphertext $\mathbf{c} = (\mathbf{c}_0, \mathbf{c}_1) \in \mathbb{Z}_q^n$ and a plaintext vector $\mathbf{w} \in \mathbb{Z}_p^n$:

$$\mathbf{r}_0 = \mathbf{c}_0 * \mathbf{w} \bmod q, \Phi_m, \tag{13.4}$$

$$\mathbf{r}_1 = \mathbf{c}_1 * \mathbf{w} \bmod q, \Phi_m. \tag{13.5}$$

For the decryption of the resulting ciphertext, $\mathbf{r} = (\mathbf{r}_0, \mathbf{r}_1)$ can compute

$$\left\lfloor \frac{1}{\Delta}(\mathbf{r}_0 * \mathbf{s} + \mathbf{r}_1) \right\rceil$$
$$= \left\lfloor \frac{1}{\Delta}(\mathbf{c}_0 * \mathbf{w} * \mathbf{s} + \mathbf{c}_1 * \mathbf{w}) \right\rceil$$
$$= \left\lfloor \frac{1}{\Delta}\left(-\mathbf{a} * \mathbf{w} * \mathbf{s} + \left((\mathbf{a} * \mathbf{s} + \Delta \cdot \mathbf{u} + \mathbf{e}) * \mathbf{w}\right)\right) \right\rceil$$
$$= \left\lfloor \frac{1}{\Delta}(\Delta \cdot \mathbf{u} * \mathbf{w} + \mathbf{e} * \mathbf{w}) \right\rceil$$
$$= \left\lfloor \mathbf{u} * \mathbf{w} + \frac{\mathbf{e} * \mathbf{w}}{\Delta} \right\rceil. \tag{13.6}$$

In order to correctly decrypt $\mathbf{u} * \mathbf{w}$, a certain error bound needs to be guaranteed, e.g., the L_2 norm of the resulting error, and $||\frac{\mathbf{e}*\mathbf{w}}{\Delta}||$ needs to be less than 0.5. Notice that the convolution here can be carried out using the schoolbook algorithm [44, 45, 65] and does not require NTT. While the schoolbook convolution algorithm has a time complexity of $O(n)^2$ when convolving two length-n vectors, the authors in [11] show that in most neural networks, the size of the filter, i.e., $\mathbf{w}$, is extremely short. In many cases, the length of $\mathbf{w}$ is less than $\log n$, where n is the length of the ciphertext.

The main observation made in [11] is that approximating the ciphertext $\mathbf{c}$ does not induce large decryption failures. Consider the approximated ciphertext $\overline{\mathbf{c}} = \mathbf{c} + \boldsymbol{\varepsilon}_{\mathbf{c}}$, where $\boldsymbol{\varepsilon}_{\mathbf{c}} = (\boldsymbol{\varepsilon}_{\mathbf{c}_0}, \boldsymbol{\varepsilon}_{\mathbf{c}_1})$ is some approximation error on $\mathbf{c}$. Then, the decryption function in Eq (13.2) evaluates to

$$= \left\lfloor \frac{1}{\Delta}\left(-\mathbf{a} * \mathbf{w} * \mathbf{s} + \boldsymbol{\varepsilon}_{\mathbf{c}_0} * \mathbf{s} + \left((\mathbf{a} * \mathbf{s} + \Delta \cdot \mathbf{u} + \mathbf{e}) * \mathbf{w} + \boldsymbol{\varepsilon}_{\mathbf{c}_1}\right)\right) \right\rceil$$
$$= \left\lfloor \mathbf{u} * \mathbf{w} + \frac{\mathbf{e} * \mathbf{w} + \boldsymbol{\varepsilon}_{\mathbf{c}_0} * \mathbf{s} + \boldsymbol{\varepsilon}_{\mathbf{c}_1}}{\Delta} \right\rceil. \tag{13.7}$$

Here, two additional terms are added, namely, $\boldsymbol{\varepsilon}_{\mathbf{c}_0} * \mathbf{s}$ and $\boldsymbol{\varepsilon}_{\mathbf{c}_1}$, which are the errors that stem from approximate decryption. Comparing Eq. (13.7) with Eq. (13.2), it is clear that as long as the proper bounds for the approximation error terms can be derived, the decryption can be correctly carried out on approximated ciphertexts. By tightly estimating the approximation errors generated in neural network-based privacy-preserving inference, the authors in [11] show that as much as 4 bits out of the 27-bit operands can be approximated without significantly affecting the accuracy of the inference. Experiment results show that the proposed convolution technique combined with the application-specific hardware can significantly reduce the energy consumption of the homomorphic convolution operation in privacy-preserving inference.

4 Conclusions

This chapter analyzes the possibility of approximating several classes of cryptography to take advantage of hardware efficient technique, i.e., Approximate Computing (AxC). We explored an approximation in cryptographic hash function based on its important cryptographic properties and outlined its possible application in bitcoin mining. The bitcoin mining application is in turn exploited due to its inherent tolerance to inaccuracy for gaining better mining profits. We have taken up the approximation of Learning with Errors (LWE) problem, which is a building block of lattice-based cryptography (LBC). Its approximate versions such as R-LWE are especially attractive to IoT end-node applications, which are limited in terms of computational power. Lastly, the acceleration of advanced security problem, i.e., homomorphic encryption, is also investigated for the use of AxC techniques.

References

1. Chippa VK, Chakradhar ST, Roy K, Raghunathan A. Analysis and characterization of inherent application resilience for approximate computing. In: Proceedings of the 50th annual design automation conference, ser. DAC'13. New York: Association for Computing Machinery; 2013. https://doi.org/10.1145/2463209.2488873.
2. Liu W, Lombardi F, Shulte M. A retrospective and prospective view of approximate computing. Proc IEEE. 2020;108:394–9.
3. Gupta V, Mohapatra D, Raghunathan A, Roy K. Low-power digital signal processing using approximate adders. IEEE Trans Comput Aided Des Integr Circuits Syst. 2013;32(1):124–37.
4. Liu W, Qian L, Wang C, Jiang H, Han J, Lombardi F. Design of approximate radix-4 booth multipliers for error-tolerant computing. IEEE Trans Comput. 2017;66:1435–41.
5. Waris H, Wang C, Liu W. Hybrid low radix encoding based approximate booth multipliers. IEEE Trans Circuits Syst II Express Briefs. 2020. https://doi.org/10.1109/tcsii.2020.2975094.
6. Liu W, Gu C, O'Neill M, Qu G, Montuschi P, Lombardi F. Security in approximate computing and approximate computing for security: challenges and opportunities. Proc. IEEE. 2020;108(12):2214–31.
7. Dutt S, Paul B, Chauhan A, Nandi S, Trivedi G. Approxhash: delay, power and area optimized approximate hash functions for cryptography applications. In: Proceedings of the 10th international conference on security of information and networks, ser. SIN '17. New York, NY: Association for Computing Machinery; 2017. p. 291–4.
8. Vilim M, Duwe H, Kumar R. Approximate bitcoin mining. In: 2016 53nd ACM/EDAC/IEEE design automation conference (DAC); 2016. p. 1–6.
9. Bian S, Hiromoto M, Sato T. DWE: decrypting learning with errors with errors. In: Proc. 55th ACM/ESDA/IEEE design automation conference (DAC); June 2018. p. 1–6.
10. Kundi D-S, Bian S, Khalid A, Wang C, O'Neill M, Liu W. AxMM: area and power efficient approximate modular multiplier for R-LWE cryptosystem. In: Proc. IEEE international symposium on circuits and systems (ISCAS); 2020. p. 1–5.
11. Bian S, Kundi D-S, Hirozawa K, Liu W, Sato T. APAS: application-specific accelerators for RLWE-based homomorphic linear transformations. IEEE Trans Inf Forensics Secur. 2021;1. https://doi.org/10.1109/TIFS.2021.3114032.
12. FIPS PUB 180-3, Secure Hash Standard (SHS), National Institute of Standards and Technology (NIST), Std.; Oct 2008. http://csrc.nist.gov/publications/fips/fips180-3/fips180-3_final.pdf.

13. Bertoni G, Daemen J, Peeters M, Assche GV. Cryptographic sponge functions, version 1. Tech. Rep.; Jan 2011. http://sponge.noekeon.org/CSF-0.1.pdf.
14. FIPS PUB 202, SHA-3 standard: permutation-based hash and extendable-output functions. National Institute of Standards and Technology (NIST), Std.; August 2015. http://csrc.nist.gov/publications/drafts/fips-202/fips_202_draft.pdf.
15. Paar C, Pelzl J. Hash functions. In: Understanding cryptography. Berlin: Springer; 2010.
16. Bassham L, Rukhin A, Soto J, Nechvatal J, Smid M, Leigh S, Levenson M, Vangel M, Heckert N, Banks D. A statistical test suite for random and pseudorandom number generators for cryptographic applications. Tech. Rep.; 2010.
17. Forrié R. The strict avalanche criterion: spectral properties of Boolean functions and an extended definition. In: Goldwasser S, editor. Advances in cryptology – CRYPTO' 88. New York: Springer; 1990. p. 450–68.
18. Maurer UM. A universal statistical test for random bit generators. J Cryptol. 1992;5:89–105.
19. Jiang H, Liu C, Liu L, Lombardi F, Han J. A review, classification, and comparative evaluation of approximate arithmetic circuits. ACM J Emerg Technol Comput Syst. 2017;13:60:1–3.
20. Nakamoto S. Bitcoin: a peer-to-peer electronic cash system. In: Decentralized business review; 2008. p. 21260.
21. Taylor MB. The evolution of bitcoin hardware. Computer. 2017;50(9):58–66.
22. Vilim M, Duwe H, Kumar R. Approximate bitcoin mining. In: 2016 53nd ACM/EDAC/IEEE design automation conference (DAC). Piscataway: IEEE; 2016. p. 1–6.
23. Verma AK, Brisk P, Ienne P. Variable latency speculative addition: a new paradigm for arithmetic circuit design. In: Proceedings of the conference on design, automation and test in Europe; 2008. p. 1250–5.
24. Ye R, Wang T, Yuan F, Kumar R, Xu Q. On reconfiguration-oriented approximate adder design and its application. In: 2013 IEEE/ACM international conference on computer-aided design (ICCAD). Piscataway: IEEE; 2013. p. 48–54.
25. Esposito D, De Caro D, Napoli E, Petra N, Strollo AGM. Variable latency speculative Han-Carlson adder. IEEE Trans Circuits Syst I Regul Pap. 2015;62(5):1353–61.
26. Regev O. On lattices, learning with errors, random linear codes, and cryptography. In: Proc. 37th Annual ACM symposium on theory of computing (STOC); May 2005. p. 84–93.
27. Regev O. On lattices, learning with errors, random linear codes, and cryptography. J. ACM. 2009;56(6):34.
28. Lindner R, Peikert C. Better key sizes (and attacks) for LWE-based encryption. In: Cryptographers' track at the RSA conference. Berlin: Springer; 2011. p. 319–39.
29. Bos J, Costello C, Ducas L, Mironov I, Naehrig M, Nikolaenko V, Raghunathan A, Stebila D. Frodo: take off the ring! practical, quantum-secure key exchange from LWE. In: Proceedings of the 2016 ACM SIGSAC conference on computer and communications security. New York: ACM; 2016. p. 1006–18.
30. Cheon JH, Kim D, Lee J, Song YS. Lizard: cut off the tail! Practical post-quantum public-key encryption from LWE and LWR. IACR Cryptol ePrint Arch. 2016;2016:1126.
31. Gentry C, Peikert C, Vaikuntanathan V. Trapdoors for hard lattices and new cryptographic constructions. In: Proceedings of the fortieth annual ACM symposium on theory of computing. New York: ACM; 2008. p. 197–206.
32. Lyubashevsky V. Lattice signatures without trapdoors. In: Annual international conference on the theory and applications of cryptographic techniques. Berlin: Springer; 2012. p. 738–55.
33. Brakerski Z, Vaikuntanathan V. Efficient fully homomorphic encryption from (standard) LWE. SIAM J Comput. 2014;43(2):831–71.
34. Brakerski Z, Gentry C, Vaikuntanathan V. (Leveled) fully homomorphic encryption without bootstrapping. ACM Trans Comput Theory. 2014;6(3):13.
35. Gentry C, Sahai A, Waters B. Homomorphic encryption from learning with errors: Conceptually-simpler, asymptotically-faster, attribute-based. In: Advances in cryptology–CRYPTO 2013. Berlin: Springer; 2013. p. 75–92.
36. Khedr A, Gulak G, Vaikuntanathan V. SHIELD: scalable homomorphic implementation of encrypted data-classifiers. IEEE Trans Comput. 2016;65(9):2848–58.

37. Cisco. Internet of things (IoT). The Washington Post, July 2015. http://www.cisco.com/web/solutions/trends/iot/portfolio.html.
38. Krishna RR, Priyadarshini A, Jha AV, Appasani B, Srinivasulu A, Bizon N. State-of-the-art review on IoT threats and attacks: taxonomy, challenges and solutions. Sustainability. 2021;13(16). https://www.mdpi.com/2071-1050/13/16/9463.
39. McKay KA, Bassham L, Turan MS, Mouha N. Report on lightweight cryptography. National Institute of Standards and Technology (NIST), Tech. Rep. NISTIR 8114, March 2017. https://doi.org/10.6028/NIST.IR.8114.
40. Lyubashevsky V, Peikert C, Regev O. On ideal lattices and learning with errors over rings. In: Gilbert H, editor. Advances in cryptology – EUROCRYPT. Berlin: Springer; 2010. p. 1–23.
41. Pöppelmann T, Güneysu T. Towards efficient arithmetic for lattice-based cryptography on reconfigurable hardware. In: Proc. international conference on cryptology and information security in Latin America; 2012. p. 139–58.
42. Pöppelmann T, Güneysu T. Towards practical lattice-based public-key encryption on reconfigurable hardware. In: Proc. international conference on selected areas in cryptography; 2013. p. 68–85.
43. Poppelmann T, Guneysu T. Area optimization of lightweight lattice-based encryption on reconfigurable hardware. In: 2014 IEEE international symposium on circuits and systems (ISCAS). Piscataway: IEEE; 2014. p. 2796–9.
44. Fan S, Liu W, Howe J, Khalid A, O'Neill M. Lightweight hardware implementation of R-LWE lattice-based cryptography. In: Proc. IEEE Asia Pacific conference on circuits and systems (APCCAS); 2018. p. 403–6.
45. Liu W, Fan S, Khalid A, Rafferty C, O'Neill M. Optimized schoolbook polynomial multiplication for compact lattice-based cryptography on FPGA. IEEE Trans Very Large Scale Integr Syst. 2019. https://doi.org/10.1109/TVLSI.2019.2922999.
46. Song S, Tang W, Chen T, Zhang Z. LEIA: a 2.05mm^2 140mw lattice encryption instruction accelerator in 40nm CMOS. In: Proc. IEEE custom integrated circuits conference (CICC); 2018. p. 1–4.
47. Banerjee U, Ukyab TS, Chandrakasan AP. Sapphire: a configurable crypto-processor for post-quantum lattice-based protocols. IACR Trans Cryptogr Hardw Embed Syst. 2019;4:17–61.
48. Salarifard R, Bayat-Sarmadi S, Mosanaei-Boorani H. A low-latency and low-complexity point-multiplication in ECC. IEEE Trans Circuits Syst I Regul Pap. 2018;65(9):2869–77.
49. Lindner R, Peikert C. Better key sizes (and attacks) for LWE-based encryption. In: Kiayias A, editor. Topics in cryptology – CT-RSA. Berlin: Springer; 2011. p. 319–39.
50. Hashemi S, Bahar RI, Reda S. DRUM: a dynamic range unbiased multiplier for approximate applications. In: Proc. IEEE/ACM international conference on computer-aided design (ICCAD); Nov 2015. p. 418–25.
51. Vahdat S, Kamal M, Afzali-Kusha A, Pedram M. TOSAM: an energy-efficient truncation- and rounding-based scalable approximate multiplier. IEEE Trans Very Large Scale Integr Syst. 2019;27(5):1161–73.
52. Vahdat S, Kamal M, Afzali-Kusha A, Pedram M. LETAM: a low energy truncation-based approximate multiplier. Comput Electrical Eng. 2017;63:1–17. http://www.sciencedirect.com/science/article/pii/S0045790616306310.
53. Gøttert N, Feller T, Schneider M, Buchmann J, Huss S. On the design of hardware building blocks for modern lattice-based encryption schemes. In: Cryptographic hardware and embedded systems (CHES). Berlin: Springer; 2012. p. 512–29.
54. Juvekar C, Vaikuntanathan V, Chandrakasan A. Gazelle: a low latency framework for secure neural network inference. arXiv:1801.05507 [Preprint]. 2018.
55. Keller M, Orsini E, Scholl P. MASCOT: faster malicious arithmetic secure computation with oblivious transfer. In: Proc. 2016 conference on computer and communications security. New York: ACM; 2016. p. 830–42.
56. Keller M. MP-SPDZ: a versatile framework for multi-party computation. In: Proceedings of the 2020 ACM SIGSAC conference on computer and communications security; 2020. p. 1575–90.

57. Yao AC. Protocols for secure computations. In: 23rd Annual symposium on foundations of computer science, 1982. SFCS'08. Piscataway: IEEE; 1982. p. 160–4.
58. Riazi MS, Samragh M, Chen H, Laine K, Lauter KE, Koushanfar F. XONN: XNOR-based oblivious deep neural network inference. IACR Cryptol ePrint Arch. 2019;2019:171.
59. Mishra P, Lehmkuhl R, Srinivasan A, Zheng W, Popa RA. Delphi: a cryptographic inference service for neural networks. In: 29th USENIX security symposium (USENIX Security 20); 2020. p. 2505–22.
60. Bian S, Wang T, Hiromoto M, Shi Y, Sato T. ENSEI: efficient secure inference via frequency-domain homomorphic convolution for privacy-preserving visual recognition; 2020.
61. Hackenjos T, Hahn F, Kerschbaum F. SAGMA: secure aggregation grouped by multiple attributes. In: Proceedings of the 2020 ACM SIGMOD international conference on management of data; 2020. p. 587–601.
62. Matsuoka K, Banno R, Matsumoto N, Sato T, Bian S. Virtual secure platform: a five-stage pipeline processor over {TFHE}. In: 30th USENIX security symposium (USENIX Security 21); 2021.
63. Kuo T-T, Jiang X, Tang H, Wang X, Bath T, Bu D, Wang L, Harmanci A, Zhang S, Zhi D, et al. iDASH secure genome analysis competition 2018: blockchain genomic data access logging, homomorphic encryption on GWAS, and DNA segment searching; 2020.
64. Roy SS, et al. Hardware assisted fully homomorphic function evaluation and encrypted search. IEEE Trans Comput. 2017;66(9):1562–72.
65. Zhang Y, Wang C, Kundi D-S, Khalid A, O'Neill M, Liu W. An efficient and parallel R-LWE cryptoprocessor. IEEE Trans Circuits Syst II Express Briefs. 2020;67(5):886–90.

Chapter 14
Towards Securing Approximate Computing Systems: Security Threats and Attack Mitigation

Qiaoyan Yu, Pruthvy Yellu, and Landon Buell

1 Introduction

There is an increasing demand for high-performance and energy-efficient computing systems, especially in the computational intensive applications such as data mining, machine learning, and artificial intelligence [8, 36]. As we are approaching the physical limit of silicon, the continuous improvement on performance and energy efficiency cannot be achieved by technology scaling only. Approximate Computing (AC) emerges as an alternative pathway to go and it represents a paradigm shift from conventional precise processing to inexact computation but still satisfying the system requirement on accuracy [7, 34, 36].

The deployment of AC techniques in a computing system is a synergy of software and hardware support. Figure 14.1 depicts the approximate modules in a computing system. As can be seen, approximation techniques can be applied at various levels of the computing stack. In the work [44], we categorize the approximation mechanisms into four categories: approximate system, approximate storage, approximate circuits, and software-level approximation. More specifically, approximate computing at the system level is realized by modifying architecture, in which approximate accelerators and programmable processors are typically adopted [21, 37, 42]. In software-level approximation, power consumption is reduced by skipping the execution of the functions in a predicted part of the code [16, 29], by relaxing the constraints on synchronization timing and hand-shaking [26], by leveraging domain specific knowledge to simplify application algorithms [38], or by alternating the implementation algorithms for a specific function [9]. Representable techniques for approximate storage include applying

Q. Yu (✉) · P. Yellu · L. Buell
University of New Hampshire, Durham, NH, USA
e-mail: qiaoyan.yu@unh.edu; py1007@wildcats.unh.edu; lhb1007@wildcats.unh.edu

W. Liu, F. Lombardi (eds.), *Approximate Computing*,
https://doi.org/10.1007/978-3-030-98347-5_14

different refresh rates on memory blocks [30], loosening guard band of multi-level memory cells [33], and selective voltage scaling [10]. Circuit-level approximation techniques reduce the computation resolution or shorten the critical delay path to improve performance and energy efficiency [14, 20, 49].

In spite of the advantages on performance improvement and power saving, AC techniques also bring in new security vulnerabilities to the computing system. The diversity of approximation techniques gives the attackers a wide exploration space to develop attacks at various levels. The work [20] implies that the utilization of approximate adders will make it easier for reverse engineers to identify the critical path. The work [32] briefly analyzes whether approximate techniques could facilitate reverse engineering, integrated circuit piracy, side-channel analysis, and hardware Trojan attacks. Security vulnerabilities on approximate storage (DRAM, SRAM, and Phase Change Memory) are discussed in the visionary work [44]. Multiple attack models for approximate arithmetic circuits are introduced in the work [46]. Comprehensive attack models together with elaborated examples are presented in the survey [47].

The rest of the chapter is organized as follows. Section 2 describes the security threat models applicable in approximate computing systems. The difference between the attacks conducted on conventional computing and approximate computing are discussed in detail in Sect. 3. Section 4 introduces general guidelines for the countermeasure designs against the unique attacks in AC systems. Sections 5 and 6 present the existing attack detection mechanisms and pro-active defense methods, respectively. This chapter is concluded in Sect. 6. Future research directions on the topic of securing AC systems are highlighted in Sect. 6, as well.

2 Security Threat Models in Approximate Computing

The primary focus of the existing research efforts on approximate computing is on developing different approximation mechanisms. There lacks good understanding on how AC techniques can be exploited by attackers to harm computing systems. Unfortunately, new security threats could indeed emerge in approximate computing systems. For example, a hybrid precise/approximate computing system could implicitly leave attackers a clue indicating which portion of the system is more vulnerable to attacks. The loosely protected portion will be a gateway for attackers to intrude the computing system. Attackers could also compromise the error management module for error detection, error correction, and precision adjustment in approximate computing systems. Consequently, fault attacks will be leveraged to leak confidential information, cause intermittent denial of service, and facilitate design piracy. Furthermore, the reduced testing efforts on approximate arithmetic and storage elements only offer a limited capability to differentiate whether the behavior deviation is caused by the approximation mechanism or a well-designed attack that disguises the attack effect as a benign error. The aforementioned attack

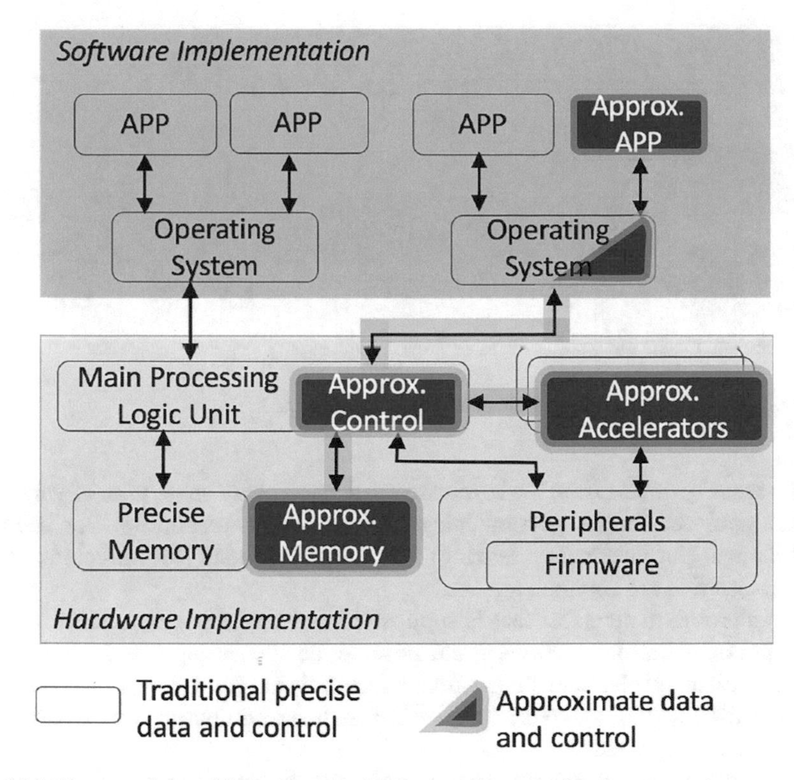

Fig. 14.1 Extra modules added to a computing system for approximation computing

examples drive us to study the new security threats posed on the different design and implementation stages of approximate computing systems.

The new data and control paths induced by the approximation operations highlighted in Fig. 14.1 are possible to be exploited by adversaries to perform various attacks. In this section, we zoom in the design flow of AC systems and summarize the new security threats that we envision in practical applications. As shown in Fig. 14.2, most of the design phases (software and hardware) could be breached. Attackers could come from hardware design groups, software developers, system integration teams, and application users.

1. In the phase of approximate computing accelerator design, the integrity of error tolerance mechanism could be stealthily interrupted such that the deviation from precise computation or storage is not caused by the approximation mechanism. Instead, the loss of accuracy is an effect of design tampering.
2. During the process of integrating precise and approximate hardware modules, the control unit that tunes the configuration parameters to obtain different accuracy goals could be a target of attacks. The burst errors caused by such attacks are likely to exceed the maximum degree that the system can tolerate.

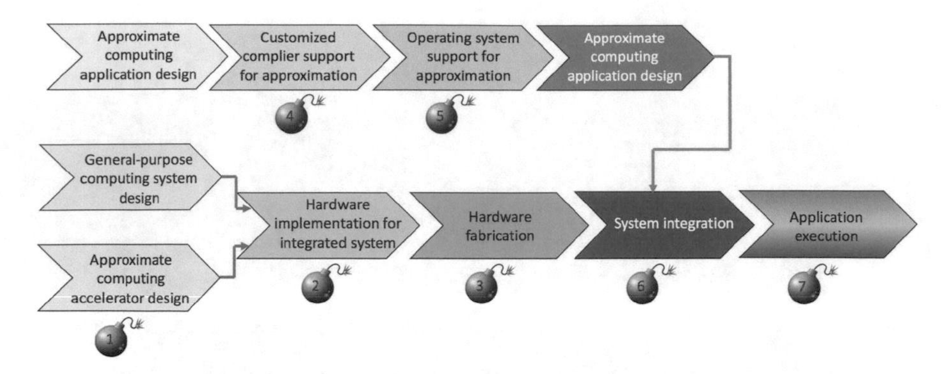

Fig. 14.2 Possible attack stages in the design flow of AC systems

3. The security threats from the hardware fabrication phase are similar to what have been analyzed for the general integrated circuit manufacturing. For instance, backdoors (for information leaking) and hardware Trojan (for malicious behaviors) insertion are common attacks.
4. A compromised compiler that is supposed to support the newly added approximate computing instructions could remove the special tag that is designed to differentiate precise and approximate instructions. As a result, the predetermined accuracy for the hybrid precise/approximate computing system cannot be achieved as expected.
5. If the operating system for approximate computing is tampered, the access could be granted to the unauthorized user (i.e., attacker) to read/write the well-protected precise memory banks, or terminate the current computation iteration earlier than what is required to achieve the aimed precision.
6. The attacks conducted in the stage of system integration will be the synergy effort to ensure the success of all the attacks mentioned in the previous phases. As both hardware and software components are available, high-level attacks (e.g., covert channel implementation) could be implemented.
7. A malicious application user is more likely to perform side-channel analysis and design piracy attacks. By measuring the error pattern in approximate memory [31], attackers are able to recognize the user's identity.

3 Hardware Attacks in Conventional Computing vs in Approximate Computing Systems

3.1 Hardware Trojans

Hardware Trojans (HT) are the extra circuitry that is employed to carry out malicious functions. The malicious modification is added either during the IC design phase or fabrication phase [4, 41]. There are many hardware Trojan based attacks reported in the literature [15, 17, 27, 40]. The common goals of HTs include leaking information, changing functionality, and causing denial of service. The design of HT trigger logic and payload logic plays an important role in carrying out stealthy HT-based attacks. Generally, the trigger logic of a hardware Trojan is activated rarely. For instance, the don't care conditions are leveraged to design a trigger circuit [4]. Once the Trojan is triggered, the payload executes the malicious behavior, such as generating intentional side-channel signals (e.g., power dissipation or delay) to leak information [23].

Approximate computing systems use non-deterministic techniques like voltage over-scaling or precision scaling to make the system behavior unpredictable [36]. The uncertainty of the systems output can be exploited to design the stealthy HT trigger logic. For the exact implementation of a computing system, a given input stimulus will result in the same system output. In contrast, the output of an approximate computing system will deviate from the exact computation result but the average accuracy of the overall computation is satisfied by the deployed application. If a rarely triggered HT induces errors occasionally, the average accuracy could stay close to the desired accuracy level. Consequently, it becomes difficult to differentiate the abnormal behaviors caused by HT attacks or the approximation mechanism. Thus, the HT in AC systems is stealthy in nature. Moreover, the degraded accuracy due to the inexact hardware implementation could be accumulated to form a malicious event [46].

A large group of approximation techniques leverage the inherent or add-on error resilience mechanisms available in computing systems to trade precision/accuracy with enhanced performance and energy efficiency. However, if the error resilience mechanism does not have a proper protection, attackers could exaggerate the approximation effect to the degree that the system cannot tolerate the approximation errors any more. For example, AC systems use some additional circuitry to tune the overall accuracy dynamically. The integrity of the fault tolerant circuit or error correction logic may become the attack surface of HTs [32, 47].

Example We use a 16-bit ripple carry adder shown in Fig. 14.3 to illustrate the HT attack in an AC system. The adder is considered as an approximate adder when the 8 least significant bits of the inputs are calculated using the approximate technique. Alternatively, when all the bits are calculated using a precise adder logic, the design performs conventional computing. For both the designs, the HT is triggered when there is a carry propagated either from submodule1 or submodule2 (the rightmost two submodules in the Fig. 14.3) as indicated in the blue dashed lines. The payload is designed such that the four least significant bits of the output are driven as 1 (irrespective of the inputs), when HT is triggered. Although the same HT trigger and payload logic are designed for approximate adder and precise adder, the attack consequence differs because of different Trojan trigger probabilities [25]. It is observed that the accuracy is decreased to 75% when using the precise adder for HT activation. There is a 50% degradation of accuracy for the same HT activation in the case of the approximate adder. This example shows that the HT implemented

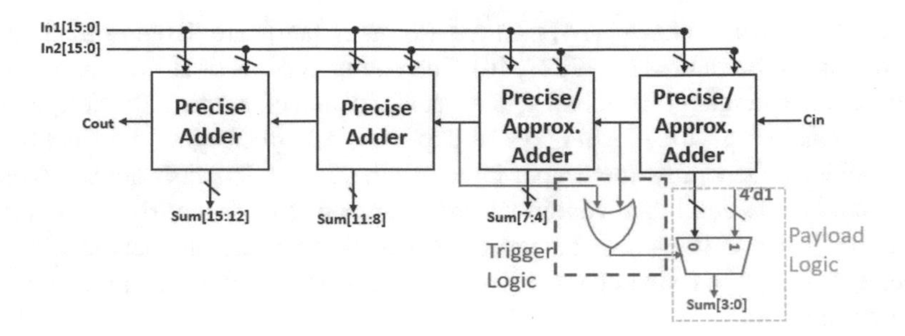

Fig. 14.3 Hardware Trojan payload and trigger logic example for a conventional adder and an approximate adder

at the approximate adder are better in hiding attack consequence compared to at the conventional computing systems.

3.2 *Fault Attacks*

A fault attack is an intentional manipulation of the device with the aim to provoke errors inside the device causing security and functionality failure [19]. Fault attacks are generally triggered by external sources to change the physical characteristics of the hardware, such as leading to bit flips in the data stored at memory cells, flip flops, and logic gates. Since the attack causes deliberate bit-flips, this step of the attack is commonly named as fault injection phase. Then, the injected faults are propagated from the micro-architectural level and to the software level.

The fault attack at the circuit level is mainly performed by fault injection techniques that alter physical operating conditions [12]. More specifically, the common sources of fault injections include:

- Creating clock glitches by shortening the clock signal period.
- Operating the device beyond the maximum tolerable temperature.
- Using laser and electromagnetic pulse to change the environment temperature.
- Decreasing the supply voltage deliberately to cause timing errors in logic gates.

The approximate computing techniques already use faulty hardware and inaccurate computing modules to gain better performance [36]. Changing the physical characteristics of the inaccurate hardware becomes easier, since slight variations to the physical device can make the system prone to the increased amount of erroneous results. However, in the case of conventional computing, the changes on the physical device for fault attack involve little extra effort. The other advantage of fault attacks in AC systems is that the accuracy due to attack could be treated as an inherent approximation effect.

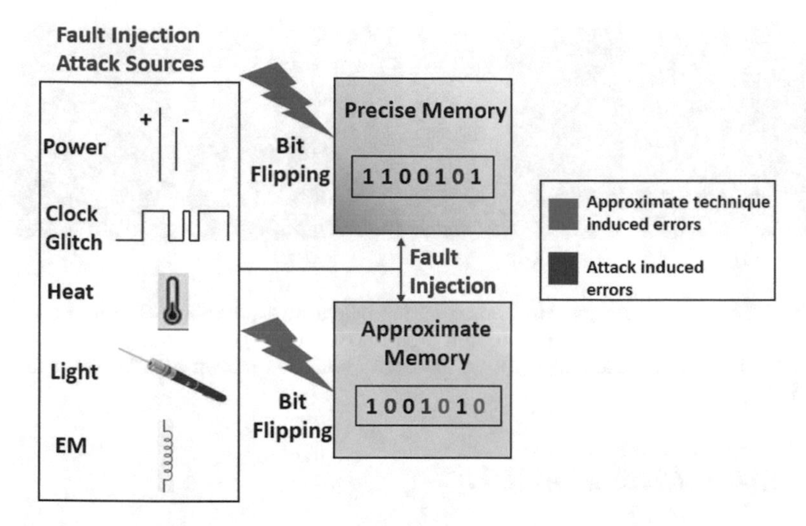

Fig. 14.4 Fault injection attack example using precise and approximate memories

Example We use approximate DRAM as an example to illustrate the fault attack in AC systems. Assume that "1100111" was stored at the precise memory before the attack. As the approximate DRAM has a lower refresh rate (due to the approximation principle), the data stored in the approximate memory location is changed to "1100010." That means the approximation leads to 2 erroneous bits as indicated in blue color in Fig. 14.4. The fault injection at the memory can be initiated by one of the sources shown in Fig. 14.4. The fault injection makes the data stored in the memories subject to bit flips and thus causes the data stored in precise memory to incur one bit change and the data in the approximate memory to have changes on four bits. This is indicated by the red color in Fig. 14.4. There is 87.5% degradation in accuracy for precise memory data due to fault attack and 50% degradation for approximate memory data due to approximate technique induced errors and fault attack induced errors. This simple example demonstrates that the attack consequence has more severe consequence for the approximate computing system compared to the conventional computing system.

Consider Fig. 14.5a is the image stored in the DRAM and it will be the same image when read from precise DRAM. However, the image read from the approximate DRAM is shown in Fig. 14.5b. Now, if the temperature is varied near the DRAM cell for conducting fault attacks, the number of errors increases in case of approximate DRAM compared to precise DRAM. Thus leading to more corrupted data. The Fig. 14.5c, d are the examples to show the effect of stuck at 0 and bit-flip errors caused by the fault attack triggered on approximate memory. If the image is considered for authentication, the authentication process fails because of the blurred image.

Fig. 14.5 Impact of incorrect refresh rates on the output of approximate DRAM. (**a**) original, (**b**) approximation output with 60 s refresh time interval, (**c**) regional attack with 60 s refresh time interval (bit-flip model), and (**d**) regional attack with 60 s refresh time interval (stuck-at-0 model) [44]

3.3 Covert Channel Attacks

A covert channel is a kind of attack that forms a communication channel secretly between two processes that are not supposed to communicate [5]. As the existing covert channel does not affect the normal functionality, covert channel attacks are considered as a stealthy attack. The main motivation of the covert channel attack is to leak the sensitive information such as secret keys or the critical parameters of a functional unit design.

The covert channel is mostly formed by utilizing the side-channel signals, e.g., power dissipation or electromagnetic emissions that modulate the information to be leaked. Then, the side-channel signals are decrypted by the attacker to interpret the leaked information. Based on how the data is transmitted, covert channels are classified as storage channel or timing channel [28]. In the storage-based covert channel, it is assumed that one process/module has access to the internal memory, where the secret data is stored by the target process/module (the data that has to be leaked). Furthermore, the receiver process/module does not gain the access to the internal memory of the target process/module but can reach the common memory shared among different processes/modules. In order to leak the information from target process/module, the other process/module that has access to the secret information will modify the value to reflect the data being transmitted. The receiver process/module will read the data.

In the timing based covert channel, the sender is able to invoke the receiver at any time, but it is based on the value of a commonly observed hardware-based quantity. In this sense, the value presented in the hardware at a particular time is important. If the hardware-based value increases with a predictable pattern, attackers could divide that value with a predetermined amount to lower the noise at the cost of transfer rate reduction.

The approximate techniques are commonly employed in non-critical functional modules. However, there could be some input or control signals shared between the critical modules and approximate modules [32]. The covert channel can be formed

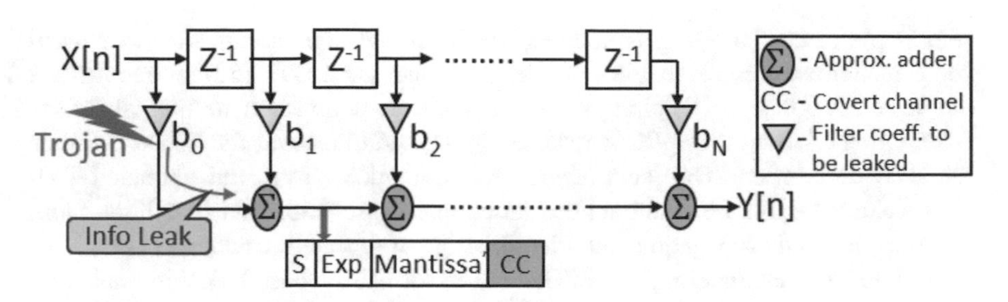

Fig. 14.6 Covert channel attack example using FIR filter

between the critical modules and approximate modules, with the help of shared control signals [47]. Since approximate modules do not require 100% verification, the adversaries can take control of approximate modules to hide the leaked information. The inaccuracy caused by the attack on the approximate modules could be confused with the effect of approximation. The information leaked via the side-channel signals may be misinterpreted or may not be properly captured. However, leaking the information within the system without modulation can help the attacker with reduced or no decrypting effort. Considering the aforementioned points, covert channel attacks for leaking the information within the same IC is more advantageous and stealthy in approximate computing systems compared to conventional computing systems.

Example Let us consider an example of an FIR filter. The filter coefficients are one of the important design parameters in the filter design. The leakage of coefficients can help the attacker to reconstruct the design for achieving counterfeiting and realize the IP piracy. Since the FIR filter is computational intensive, approximate adders can be utilized to reduce the area and power consumption [35]. The approximation principle could be misused to form a covert channel attack, which will leak the filter coefficients through the LSB bits of the approximate output (labeled as CC shown in the Fig. 14.6). To make the attack more stealthy, a Trojan trigger logic can be used here.

3.4 Reverse Engineering Attacks

The process of understanding the design implementation in the format of layout or netlist is regarded as reverse engineering attack. Hardware reverse engineering can be generally divided into FPGA and ASIC-oriented techniques [39]. Targeting at the FPGA based design, the reverse engineering process starts with scanning the bitstream to recover the netlist. The netlist is further analyzed to recover high-level RTL information. Existing literature uses several FPGA tool-chains to reverse engineer the bitstream and netlist [2, 48]. After the reverse engineering step is performed, the target modules are identified and the attacker can insert some malicious circuits to the modules by changing LUTs [50] or reconstruct the same RTL design (IP piracy) from the reverse engineered netlist [6]. In the ASIC-oriented design, chip/IC is reverse engineered to recover the netlist and then the RTL description. The identification of submodules from the extracted RTL uses techniques like FSM extraction, functional aggregation and matching, word identification and propagation, and identification of repeated structures [11].

The reverse engineering of FPGA based or ASIC based design that uses conventional computing techniques is not straightforward. The complexity of the approach could vary if the chip utilizes approximate computing techniques in the design [32]. Let us consider a scenario where the netlist of the design is successfully recovered using some reverse engineering tools. As the approximate modules produce erroneous results, the functional aggregation and matching that rely on the functional testing for identification of the module may fail to identify the module of interest (from attackers perspective). The FSM extraction technique may also be not successful. This is because the control flow for the approximate modules is not as same as the precise modules, and the overall flow in AC systems is more complex for reverse engineering analysis than conventional computing systems. However, identification of the repeated structure technique could be successful since the approximate circuits are generally accompanied by some additional control units. Attackers could use those control units as a clue to tell the location of the approximate units. In short, reverse engineering attack will succeed in certain AC cases.

4 General Guideline of Countermeasure Designs for AC Systems

To strengthen AC systems' resilience against the attacks mentioned in the previous sections, we introduce four general principles for the defense mechanism designs protecting AC systems: *randomization*, *access prohibition*, *noise injection*, and *design obfuscation*. As shown in Fig. 14.7, these four principles can be realized by diverse approaches to assure the confidentiality, availability, integrity, and anti-piracy capability of AC systems.

4.1 Randomization

The principle of randomization means making the approximate data or operations as unpredictable as possible from the attackers' point of view. The allocation of approximate memory should be not directly visible at the application level. Oblivious memory accessing can be a feasible way to randomize approximate

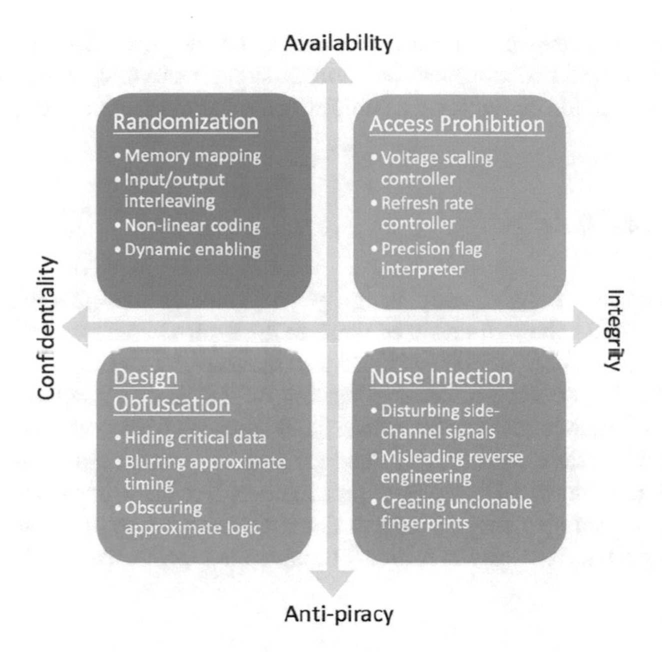

Fig. 14.7 Categories of general countermeasures design

memory. For approximate arithmetic operations, the inputs could be pre-processed so that the precision of approximate computing is not disclosed. Interleaving (or scrambling) controlled by a secret key will raise the bar for adversary to gain the full knowledge on the approximation mechanism adopted in the system. Non-linear coding is one step further than interleaving. If the average accuracy is the goal of approximation, randomization can be applied to dynamically enable the approximation mechanism.

4.2 Access Prohibition

The key idea of access prohibition is to add a protection barrier to stop attackers from intruding the approximation zone. Essentially, the techniques that prohibit unauthorized access turn the approximation zone into a Trusted Platform Module (TPM), which inherently has high resilience against intrusion attacks. The controller for configurable voltage/frequency scaling should be particularly isolated; the parameter adjustment in the controller needs a strictly verified procedure. If an attacker obtains full access, the circuit that manages memory refreshing could also be in danger. Thus, either circuit or architectural level access monitor should be in place to sense any abnormal behaviors in the refresh rate controller. A precision flag interpreter decodes the indication sign for approximation in an operation instruction.

If the interpreter is transparent to all users except the interpreter designer, it would be beneficial to thwart malicious modification. A strong restriction on the access entry to the approximation controller is a straightforward and effective path to pursue.

4.3 Noise Injection

For the purpose of maintaining integrity and enhance anti-piracy capability, a defender can inject the well-designed noise to the approximate system. The injected noise is benign to AC systems but will disturb the side-channel signals that attackers could utilize to perform reverse engineering on AC design itself or confidential information. Noise-induced delay, power, and thermal effects will alter the regular pattern of AC systems and thus mislead an adversary when the approximation is active and how much impact the approximation has been made to the system. Moreover, we can also consider to use the signature (e.g., on-current and power) created by the injected noise to form unclonable fingerprints like PUFs.

4.4 Design Obfuscation

Extensive research efforts have demonstrated that design obfuscation techniques are effective to strengthen the confidentiality and anti-piracy capability of a hardware design. Obfuscation is also applicable in AC systems, in which the implementation of approximate algorithms/operations/logic circuits should be obscured to defer the successful attacks. Two obfuscation strategies—*hiding critical information in the unused bits* and *blurring boundaries between AC and non-AC modules*—are particularly unique for AC systems. As not all operand bits are utilized in approximate computing, the unused bits can be used to carry signature/authentication code for the purpose of security assurance. If the mechanism of hiding critical bits is not disclosed in public, this technique will be low cost. The boundary blurring is a promising defense strategy for general AC system designs. The key outcome of blurring is to minimize the number of explicit attack surfaces.

5 Example of Attack Detection Methods

In this section, we introduce the state-of-the-art defense methods [43, 46] against various attacks performed in AC systems. We use the metric based detection method and logic function based detection method as examples to discuss how to implement the countermeasure.

5.1 Metric Based Attack Detection

5.1.1 Metrics

The main principle of approximation computing relies on the trade-off between accuracy and improved performance. The degraded accuracy is being calculated using different error metrics like error distance (ED), relative error distance (RED), normalized error distance (NED), mean relative error distance (MRED) [3, 18, 22, 24]. Some of the definitions used for calculating accuracy as indicated from the existing literature are summarized as below. Equation (14.1) [1, 18, 49] defines the mean value of accuracy ACC_{amp} to quantize to what extent the imprecise output R_e deviates from the precise output R_c.

$$ACC_{amp} = 1 - \frac{|R_c - R_e|}{R_c} \tag{14.1}$$

In which, R_e and R_c are the results of the imprecise and precise outputs being translated to a decimal number, respectively. A higher ACC_{amp} means that the imprecise output is closer to the precise result.

Hamming distance in Eq. (14.2) is another metric to measure accuracy [1, 18]. Different from ACC_{amp}, ACC_{hm} ignores the weight of the incorrect bit carried in the output; instead, the ratio of the number of mismatched bits B_e between the precise and imprecise outputs over the bit width of the output B_w is important in the evaluation.

$$ACC_{hm} = 1 - \frac{B_e}{B_w} \tag{14.2}$$

The metric ACC_{apx} in Eq. (14.3) pays attention to the bits from precise computation submodules (for MSB) and ignores those bits produced by the approximate submodules (for LSB). This metric assesses the accuracy of the critical portion of interest.

$$ACC_{apx} = \frac{T_{apx}}{T_{total}} \tag{14.3}$$

In which, T_{apx} is the total number of correct cases after we neglect the bits computed by the approximate submodules, and T_{total} stands for the total number of test cases.

Different than ACC_{apx}, the accuracy ACC_{gen} in Eq.(14.4) is defined as the ratio of the total number of completely correct cases T_c over the total number of test cases T_{total}. This metric equally considers all the mismatch bits between the imprecise and precise outputs. ACC_{gen} is generally referred as pass rate [1].

$$ACC_{gen} = \frac{T_c}{T_{total}} \tag{14.4}$$

Table 14.1 Comparison of accuracy measured with different metrics applied to 8-bit hybrid (precise and approximate) adders [43]

No. precise (approx.) units	ACC_{amp}	ACC_{hm}	ACC_{apx}	ACC_{gen}
1 (7)	90.28%	65.15%	82.95%	16.46%
2 (6)	95.04%	68.39%	82.57%	19.34%
3 (5)	97.45%	71.87%	81.84%	22.66%
4 (4)	98.64%	75.48%	80.47%	26.56%
5 (3)	99.22%	79.24%	78.13%	31.25%
6 (2)	99.52%	83.38%	75%	37.50%
7 (1)	99.72%	88.91%	75%	50.00%
8 (0)	100%	100%	100%	100%

The accuracy metrics above will have different sensitivities to errors. In the work [43], we used a 8-bit adder, composed of different numbers of approximate 1-bit full adders, to quantitatively compare the accuracy metrics defined by Eqs. (14.1)–(14.4). As shown in Table 14.1, it is observed that different metrics for accuracy make large difference on the measured values for the same configuration of adder. In addition, the speed of accuracy dropping with respect to the number of approximate units varies in different accuracy metrics. The accuracy of ACC_{amp} is above 90% even when seven 1-bit full adders are approximate. In contrast, the metric ACC_{gen} is the most sensitive to the utilization of approximate units (it could yield an accuracy rate as low as 16.46%). The two main observations led to the motivation of using the metrics for detection of attack.

5.1.2 Precise Approximate Unit Under Test for Security Examination Method

A differential metric is introduced to indicate if the approximation unit is compromised or not. Assume the measurement from the precise functional unit is P, the measurement from the original approximate functional unit is A, and the test result from the tampered approximate functional unit is A'. In a typical verification, one will compare the delay or power of the untampered and tampered approximate unit in a way expressed in Eq. (14.5).

$$DIFF_{scs} = \frac{A' - A}{A} \tag{14.5}$$

where $DIFF_{scs}$ stands for the differential side-channel signal (e.g., delay and power). Since the hardware Trojan logics are designed to be stealthy, the Trojan-induced increase on delay and power is typically insubstantial. Thus, the metric $DIFF_{scs}$ is not suitable for attack detection. However, the sensitivity of the Trojan detection can be increased by incorporating the side-channel signals obtained from the precise implementation to the differential metric expressed in Eq. (14.6).

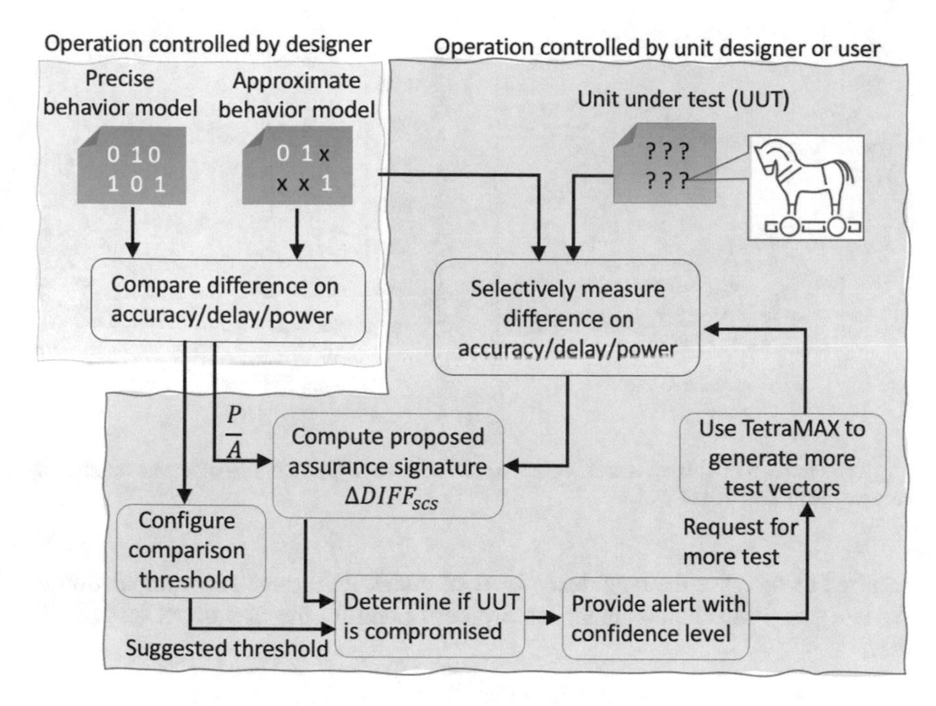

Fig. 14.8 Attack detection flow for approximate computing systems

$$\Delta_{DIFF_{scs}} = \frac{P(A' - A)}{A * A'} = \frac{P}{A} * \frac{A' - A}{A'} \tag{14.6}$$

The ratio $\frac{P}{A}$ in Eq. (14.6) is a constant enlargement factor. By multiplying the $\frac{P}{A}$ factor, the tampering induced difference on the side-channel signals of approximate functional units will be increased, thus reducing the false negative detection rate. We incorporate the differential metrics in a unique Trojan detection method designed for approximate computing systems. Figure 14.8 depicts the key idea of the Trojan dctection method, which exploits difference among *P*recise-*A*pproximate-*U*nit under test for *S*ecurity *E*xamination (PAUSE). The PAUSE method does not only rely on the difference between the output/delay/power of the tamper-free approximate system and the unit under test, but also leverages the difference that are intentionally introduced in the phase of designing the approximation algorithm. The approximate system designer typically provides performance improvement and trades-off accuracy. The ratio $\frac{P}{A}$ is available at design time, since $\frac{P}{A}$ is usually reported by the designer. A small number of input patterns will be used to measure the output accuracy, delay, and power of the behavioral approximate computation unit and its hard/firm macro. Next, the metric $\Delta_{DIFF_{scs}}$ will be computed accordingly. We compare that metric with the threshold ξ suggested by the approximate system designer. If $\Delta_{DIFF_{scs}}$ exceeds the threshold, something abnormal occurs (likely induced by a Trojan attack). Then more evaluation is

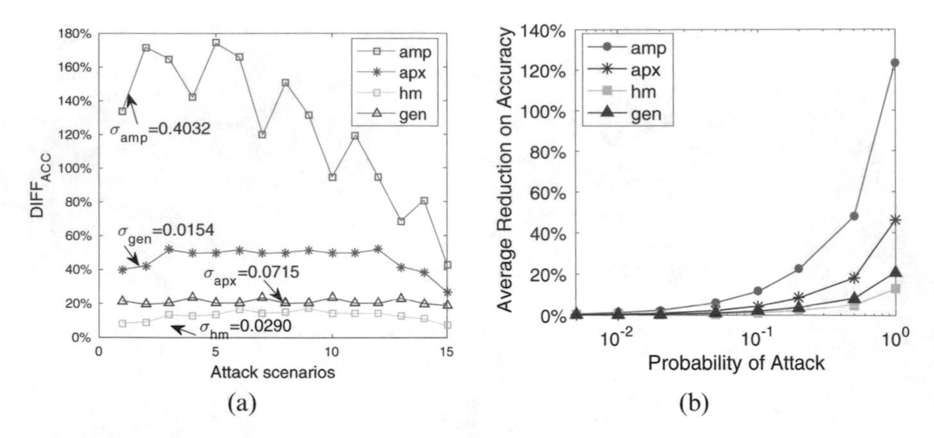

Fig. 14.9 Impact of diverse attack scenarios on accuracy. (**a**) Differential accuracy and (**b**) accuracy reduction due to Trojan

required to increase the confidence level of attack detection. The computation of $\Delta_{DIFF_{scs}}$ and threshold comparison are performed by the unit under test (UUT) user.

5.1.3 Attack Success Rate Assessment Using PAUSE Attack Detection

We continue to use the 8-bit adder to examine the difference on accuracy $DIFF_{ACC}$ induced by the Trojan attack. Figure 14.9a shows that the reduction on accuracy varies with different accuracy metrics. The use of ACC_{amp} leads to the most significant variation on $DIFF_{ACC}$ if the attack is injected in different locations. In contrast, ACC_{hm} is the least sensitive to the attack location. The standard deviation of the $DIFF_{ACC}$ for each accuracy metric is labeled in Fig. 14.9a. Because the Trojan attack is stealthy and is not always-on, we emulate the Trojan attack with different triggering probabilities. When the probability of attack decreases to 10^{-2}, the average reduction on accuracy is almost close to zero (shown in Fig. 14.9b). This means, it is challenging to detect the presence of Trojans based on the variation on accuracy.

5.2 *Logic Function Based Attack Detection*

5.2.1 Integrity Check and Exclusive Logic Based Attack Detection

As introduced in Sect. 3, the approximate techniques are exploited by the attackers to execute their malicious intentions. One of such exploitations is swapping the input and output interconnects to degrade the accuracy to achieve denial of service

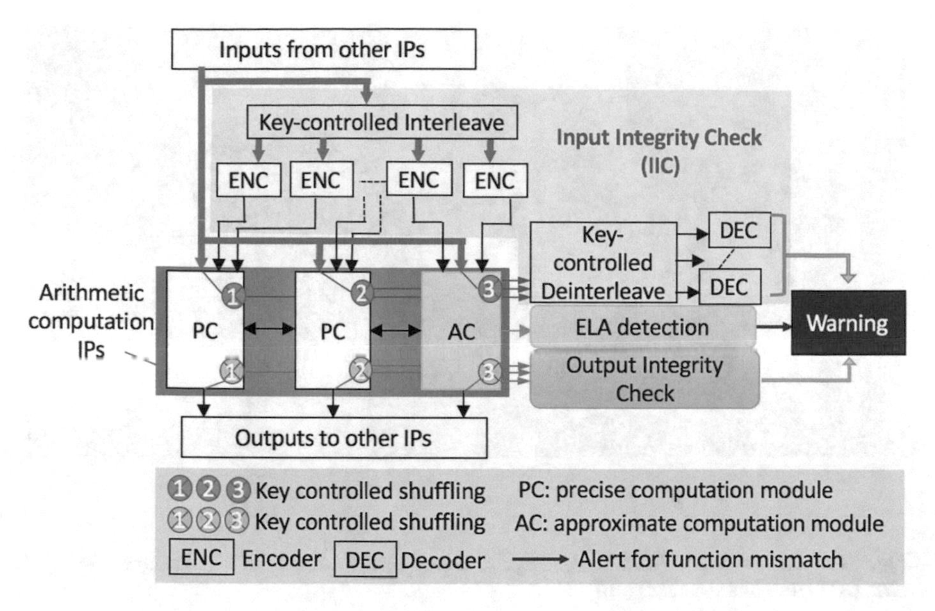

Fig. 14.10 Overview diagram of logic function based attack detection method [46]

or for triggering a HT logic. An encoding based detection mechanism called *Input Integrity Check* (IIC) and *Output Integrity Check* is designed in the work [46] to detect the attacks originating because of interconnect tampering. Other attack that is possible in AC systems is by altering the approximation function, as the malicious changes can be misinterpreted to be a predefined approximation. To thwart such attacks, an exclusive logic based attack detection mechanism was designed [46]. In the exclusive logic based attack detection module (*ELA Detection*), the outputs of the approximate function are selectively examined to generate an alert signal to indicate hardware attacks on AC IPs. The overall flow of the attack detection method is shown in Fig. 14.10.

Input or Output Integrity Check Method The encoding stage of integrity check detection method is designed in three steps. In the first step the inputs to the arithmetic computation IPs are divided into sub-groups. Key-controlled interleaving is the next step. The user key will determine how to interleave the inputs. In the last step even parity check code is adopted to generate a check bit for each group of interleaved inputs, which are fed to the arithmetic computation IPs. Before the actual arithmetic computation, the encoded inputs are de-shuffled and check bits are calculated. If the check bits transferred from the PC and AC IPs do not match the newly calculated ones, the interconnect tampering attack is detected and an alert signal turns on the warning system.

Exclusive Logic Based Attack Detection Method In the ELA method, the input patterns that will result in different outputs from the PC and AC IPs are scanned. Next, an input pattern that gives different outputs when driven by precise and

Fig. 14.11 Images in DCT-IDCT. (**a**) Original image before normal DCT, (**b**) image sabotaged after DCT (four bits swapped) [46]

approximate module is chosen to generate the alert logic. If attackers modify the logic defined in the AC IP, the alert logic is set to notify the occurrence of the attack.

To illustrate the development of ELA alert logic, an approximate adder design [13] is considered as an example in the work [46]. The input pattern of '0 1 0' (for A, B, and Cin, respectively) was observed to be one of the combinations that lead to different carry bits for the precise and approximate adders. Thus, the Boolean logic expressed in Eq. (14.7) is formed to generate the alert signal.

$$Alert = A + \overline{B} + Cin + Cout \tag{14.7}$$

The detection success rate was found to be further improved in the work [46], when the other output *Sum* is also considered for forming an alert logic equation as in Eq. (14.8).

$$Alert = A + \overline{B} + Cin + Cout + \overline{Sum} \tag{14.8}$$

5.2.2 Assessment of Attack Detection Rate

Assessment on IIC

The work [46] used DCT-IDCT algorithm to process the *cameraman* picture shown in Fig. 14.11a as a case study. To emulate the interconnect attack, a hardware Trojan is designed such that the LSB of the exponent and the MSB on the mantissa of the

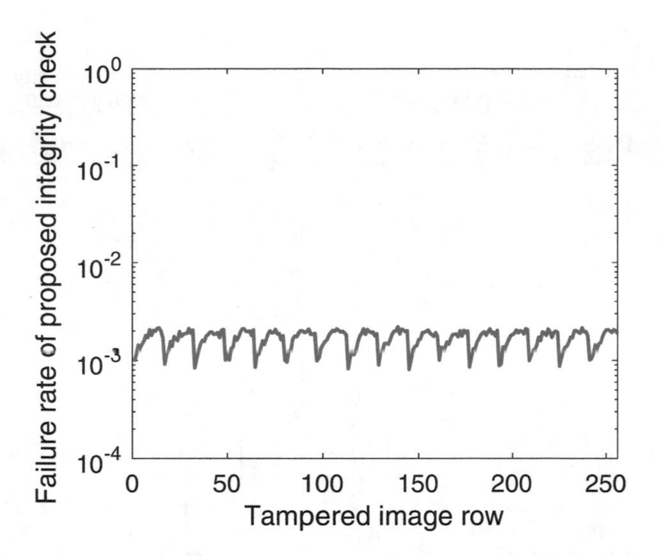

Fig. 14.12 Attack detection failure rate for all possible attack locations [46]

floating-point number representing the pixel content are swapped and the resultant image is shown in Fig. 14.11b. Now to examine the effectiveness of the IIC detection method, the image row under attack was swept from 1 to 256 (last row). As shown in Fig. 14.12, the attack detection failure rate is in the range of $2.2 * 10^{-3}$ and $8.087 * 10^{-4}$ and it is data independent.

Assessment on ELA Detection A 64-bit ripple carry adder is used in the work [46] to evaluate the ELA detection method. Hamming distance was adopted to compare the number of different output bits between the precise and approximate adders (w/o our protection method) experiencing under-propagation (up) and over-propagation (op) tampering attacks. Hamming distance for all test cases is shown in Fig. 14.13. For the under-propagation case, the alert logic will reduce the Hamming distance by 74.6%. If both sum and carry are used (Eq. 14.8) to detect attacks, the method successfully reduces Hamming distance to zero (i.e., 100% attack detection rate).

6 Pro-Active Defense Mechanisms

The pro-active defense for approximate systems is to make the boundary between precise and approximate modules indistinct ideally. The Fig. 14.14 depicts the concept of defense mechanism that changes the boundaries of AC components to different shapes and obfuscates them by mingling AC and non-AC modules. The dashed line areas are the obfuscated boundaries, which impede attackers from locating the AC components and implementing attacks.

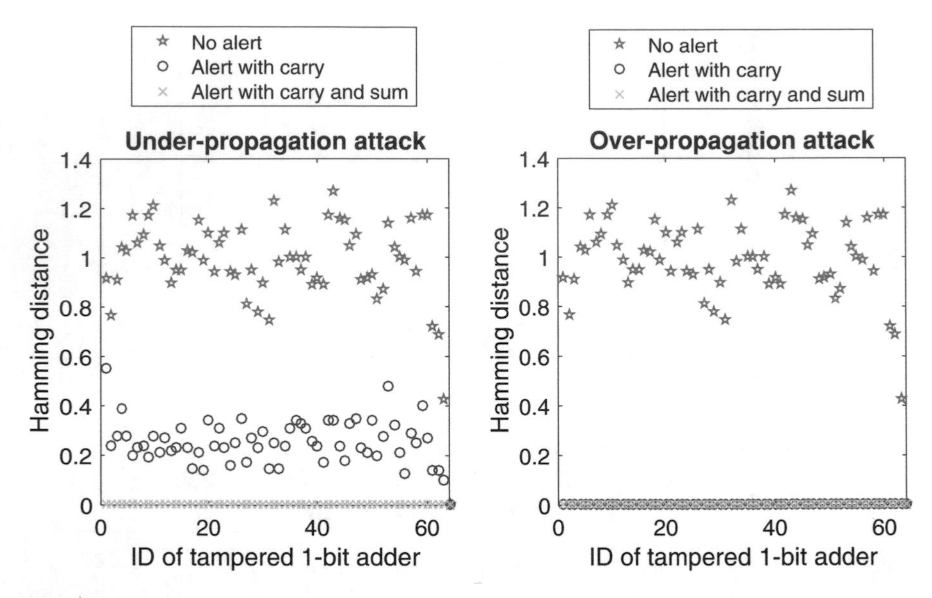

Fig. 14.13 Hamming distance improved by ELA detection method in a 64-bit adder [46]

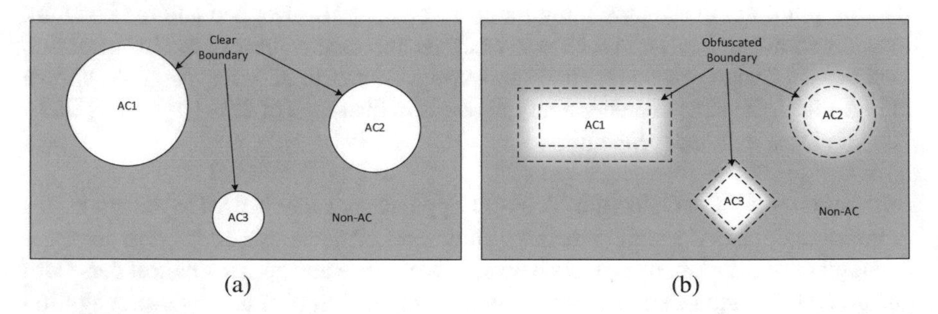

Fig. 14.14 The comparison of a AC system before and after boundary blurring protection. (**a**) A system including AC components (white parts) and non-AC (gray parts). The boundary between AC and non-AC components is easy to identify. (**b**) The system protected by boundary blurring method

In general, the obfuscation of approximate computing system can be derived from three aspects: *WHEN to approximate (WNA)*, *WHAT to approximate (WTA)*, and *HOW to approximate (HWA)*. Figures 14.15, 14.16, and 14.17 depict the general principle for how to explore the aspects above to implement the obfuscation function.

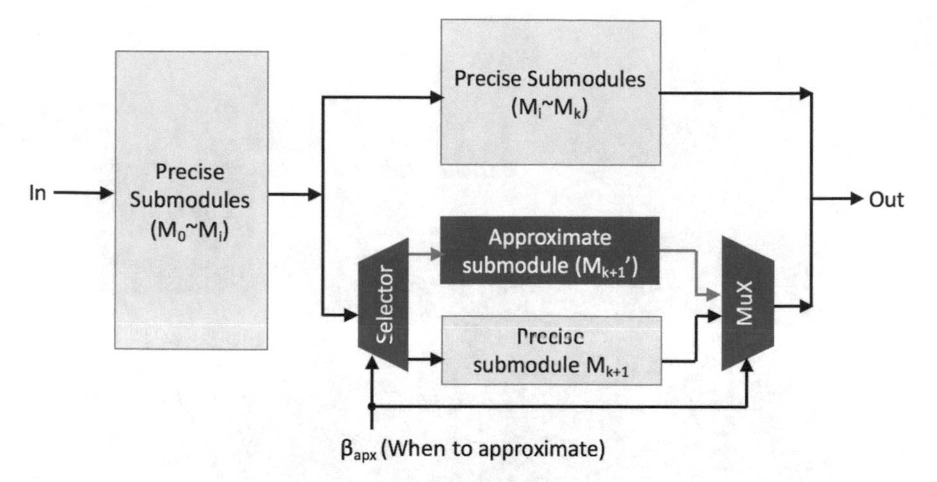

Fig. 14.15 Blurring approximate boundary with WNA scheme

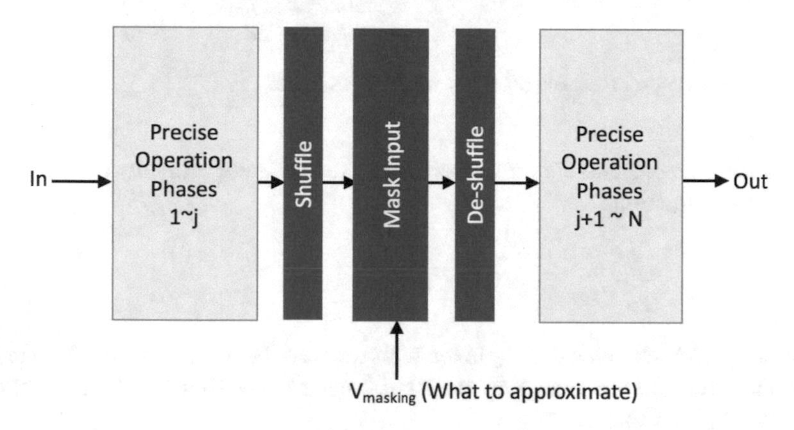

Fig. 14.16 Blurring approximate boundary with WTA scheme

6.1 Obfuscate WHEN to Approximate (WNA)

To save the power consumption, a submodule for precise operations in a computing systems is replaced with an approximate version M_{k+1}, which simplifies the functional logic to enable more optimization. However, the submodule M_{k+1} could be exploited as an attack surface. As shown in Fig. 14.15, a redundant copy of M_{k+1} is implemented with the original precise logic (i.e., M_{k+1}'), and then use the control signal β_{apx} to enable the toggling between M_{k+1} and M_{k+1}'. As long as β_{apx} is secretly controlled by a legitimate user, obfuscation on *WHEN to approximate* will effectively counteract attacks on the approximate submodule.

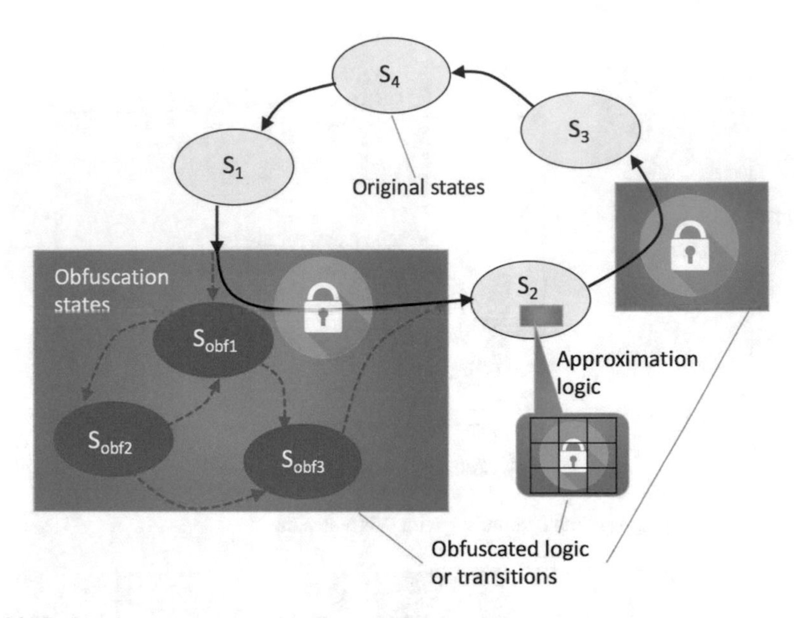

Fig. 14.17 Blurring approximate boundary with HWA scheme

Assume P_o represents for the probability of success tampering on the unprotected approximate function.

$$P_o = \frac{1}{L_C} \tag{14.9}$$

Equation (14.9) indicates that P_o is the reciprocal of the total number of the entries L_C to the protected function. Now the attack success rate after WNA obfuscation is expressed in Eq. (14.10).

$$L_C = (1 - \eta)x + \eta x \tag{14.10}$$

In which, η is the approximate rate ($<$ than 1), and x represents the entire logic function. Now, the probability that an attack can successfully reach the approximate submodule M_{k+1} is expressed in Eq. (14.11).

$$\Gamma_{apox} = \frac{\eta x}{L_C} = \eta \tag{14.11}$$

After WNA obfuscation, the approximate submodule M_{k+1} is utilized with a rate of β_{apx} (i.e., obfuscation rate). Thus, the new attack success rate is reduced to the expression in Eq. (14.12).

$$\Gamma_{whenobf} = Obfuscation\ Rate \cdot \Gamma_{apox} = \beta_{apx} \cdot \eta \tag{14.12}$$

As shown, the attack success rate depends on the approximation rate η adopted in the implementation of the approximate computing system and the utilization rate of the approximation submodule.

6.2 *Obfuscate WHAT to Approximate (WTA)*

The object of obfuscation includes (1) preliminary inputs, (2) internal signals, (3) processing logic, and (4) the combination of three above. The WTA method focus on obfuscating which signals in the data flow. Assume the entire approximate computing systems take *N* phases to produce the final output. Some internal signals are considered to obscure as opposed to the primary inputs. Due to the visibility, it is easier to bypass the obfuscation applied to primary inputs than those applied to internal signals.

Figure 14.16 shows one scheme that first shuffles the internal signals and then employs a masking vector to partially mute the internal signals before de-shuffling. Due to the muting operation, a portion of the internal signals remains constant and thus the corresponding precise submodules for those muted signals will not consume dynamic power. The masking vector V_{masking} is considered as an obfuscated input, since the vector does not straightforwardly direct one-to-one muting. Logic cones involved by the masking vector will affect the success rate of the attack on approximate computing systems.

In this type of approximation, attackers could manipulate the original muted bits to spread the malicious inputs to the other precise submodules. Assume the fan-out coefficient of the intended logic cone is ϕ. Equation (14.13) describes how WHAT to approximate scheme will affect the attack success rate.

$$\Gamma_{whatobf} = \eta \cdot \left(\sum_{i=1}^{V_{masking}} \phi_i - \sum_{i=1,j=1}^{V_{masking}} \phi_{ij} \right) \tag{14.13}$$

A larger selected ϕ will lead to a higher attack success rate. In contrast, a wider masking factor will lead to a less chance of a success attack.

6.3 *Obfuscate HOW to Approximate (HWA)*

The general approximation rules include (1) increasing the number of don't care cases to enlarge the space of optimization (2) dedicating precise and approximate operation zones, (3) separating the timing for enabling precise and approximate function, and (4) differentiating the controls of environmental supplies (e.g., temperature and voltage). In the HWA scheme shown in Fig. 14.17, the state transitions (e.g., from S1 to S2 and from S2 to S3) are obfuscated with a key vector, preventing

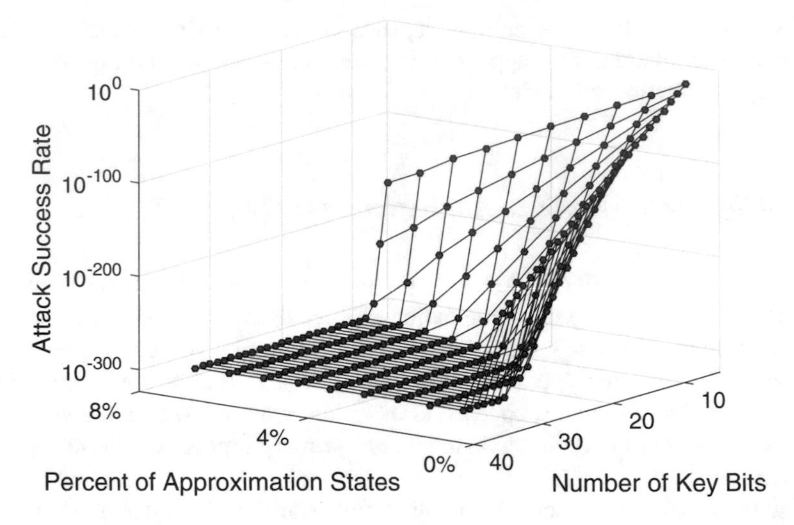

Fig. 14.18 Projected attack success rate reduced by HWA scheme

unauthorized users to access the state handling approximate operations. Incorrect key vectors will result in the system entering obfuscation states, which cause system malfunctions. In addition, the truth table for approximation logic (e.g., for S2) can be re-designed by integrating a key vector. This type of obfuscation extends the logic map by adding a unique "nonce" to the original approximation and thus introducing user-controlled "dummy logic."

Compared to WNA and WTA schemes, HWA is the most powerful obfuscation. Assume that the approximate system utilizes N_{Sorig} original states, among which $\eta \cdot N_{Sorig}$ are assigned to be the states for approximation. If the size of the applied key vector is K and the number of obfuscation states is N_{Sobf}, the attack success rate after HWA blurring boundary will be as expressed in Eq. (14.14).

$$\Gamma_{howobf} = \prod_{j=1}^{\eta \cdot N_{Sorig}} \left(\prod_{i=0}^{N_{Sobf}} \left(\frac{1}{2^{K-i}} \cdot \gamma_{logic_i} \right) \right) \tag{14.14}$$

In which, the term γ_{logic_i} is the coefficient of logic masking associated with the approximate state. This parameter will vary with the specific logic function obfuscated by the key vector. More logic masking effect achieved by the encryption key will yield a lower attack success rate. The increasing key size and the number of obfuscation states collaboratively contribute to reduce the success rate of attacks in a quasi-exponential fashion. The numerical plot shown in Fig. 14.18 projects the trend described in Eq. (14.14).

6.4 *Case Study : Design Obfuscation Using Image Classification Application*

As mentioned in Sect. 3, the AC systems can be exploited to conduct stealthy attacks. If such malicious approximate techniques are utilized in CNN, there could be decrease in the overall accuracy but could maintain the other metrics unchanged making it difficult to be detected during testing [45]. To protect the CNN from such attacks, the work [47] applied the WTA scheme [45] to blur the boundary between approximate and precise operations such that the utilization of the approximation module is not explicit and uncertain from the attacker's point of view. The CiFar-10 dataset that contains of $32 \times 32 \times 3$ shaped images were used for training and testing the model. The first hidden layer of the CNN model used approximate additions in the matrix computation to accelerate the processing speed at the neural nodes. The WTA scheme was deployed in the first hidden layer. The arrangement made by WTA increases the uncertainty on the error characteristic, which can be used to detect abnormal behaviors induced by the ECA attack.

The three classification metrics for the CNN are shown in Fig. 14.19. The toggling rate is set to 1/3 to the control signal β_{AP} so that the approximate module will be muted and the equivalent precise operations will be active only for one third of the entire training period. As the approximation operation is randomly enabled, the success rate of ECA is reduced. As shown in Fig. 14.19a, the loss function value for 33% blurred (i.e., WTA scheme) for small pixel grouping size is 7× higher than the baseline, original approximated, and ECA cases. As the size of the pixel group increases, the WTA method gradually loses its effectiveness. The precise and recall scores shown in Fig. 14.19b, c both confirm that the WTA will help in detecting the attacks easily during verification.

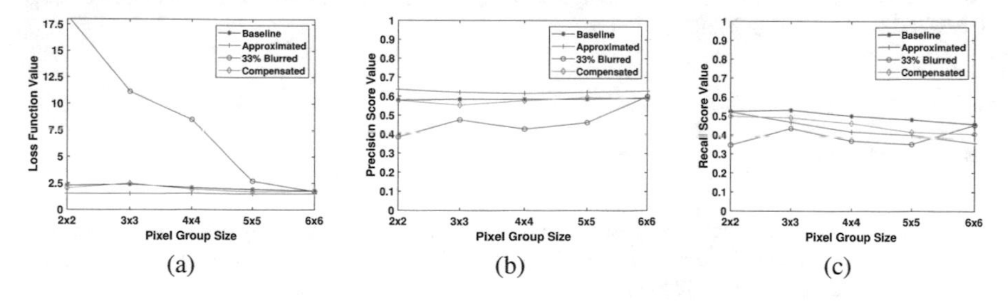

Fig. 14.19 Impact of WTA obfuscation based defense method on the classification rate of CNN. (**a**) Loss function, (**b**) precision score, and (**c**) recall score [47]

7 Summary and Future Research Directions

Although AC systems are promising for big data storage and computation-intensive analysis, the utilization of approximate computing techniques may open new exploration space for attackers to compromise the emerging computing systems. In this chapter, we summarize the state-of-the-art efforts that examine the security vulnerabilities associated with approximate computing techniques. We also compare the attacks in conventional computing systems and approximate computing systems and highlight the challenges on attack detection and mitigation for AC systems. However, the study of the security issues in approximate computing systems is still in its youth. More research efforts are needed to disclose new security threats through attack examples in real-world applications. We suggest researchers investigating the new security vulnerabilities originated from the devices, circuits, microarchitecture, protocols, compilers, and operation kernel applied in AC systems. In future, more advanced attacks may be performed as a combination effort of multiple levels.

Since the attacks on AC systems are diverse, we envision that attack detection and mitigation methods can be classified into four categories: randomization, access prohibition, noise injection, and design obfuscation. More extensive and in-depth research in each category will be needed. Hardware cost and performance overhead induced by those defense mechanisms shall be assessed via either application-oriented case studies or based on a systematic framework. The latter one will be more challenging to conduct. In the framework, we will need to seek a proper metric, which enables us to differentiate the errors caused by security attacks from the approximation errors. We encourage researchers to propose more security metrics in future to minimize the false positive and false negative attack detection rate.

Acknowledgments This work is partially supported by the National Science Foundation awards CNS-1652474 and CNS-2022279.

References

1. Benara V, Purini S, Accurus: a fast convergence technique for accuracy configurable approximate adder circuits. In: Proceedings of 2016 IEEE computer society annual symposium on VLSI (ISVLSI); 2016. p. 577–82.
2. Benz F, Seffrin A, Huss SA. Bil: a tool-chain for bitstream reverse-engineering. In: 22nd international conference on field programmable logic and applications (FPL); 2012. p. 735–8.
3. Boroumand S, Afshar HP, Brisk P. Approximate quaternary addition with the fast carry chains of FPGAs. In: Proceedings of 2018 design, automation test in europe conference exhibition (DATE); 2018. p. 577–80.
4. Chakraborty RS, Narasimhan S, Bhunia S. Hardware trojan: threats and emerging solutions. In: 2009 IEEE international high level design validation and test workshop; 2009. p. 166–71.
5. Chandra S, Lin Z, Kundu A, Khan L. Towards a systematic study of the covert channel attacks in smartphones. In: Tian J, Jing J, Srivatsa M, editors. International Conference on

Security and Privacy in Communication Networks - 10th International ICST Conference, SecureComm 2014, Beijing, China, September 24–26, 2014, Revised Selected Papers, Part I, volume 152 of Lecture Notes of the Institute for Computer Sciences. Social Informatics and Telecommunications Engineering, Springer; 2014. p. 427–35.

6. Cheremisinov D. Design automation tool to generate edif and VHDL descriptions of circuit by extraction of FPGA configuration. In: East-west design test symposium (EWDTS 2013); 2013. p. 1–4.
7. Crago SP, Yeung D. Reducing data movement with approximate computing techniques. In: 2016 IEEE international conference on rebooting computing (ICRC); 2016. p. 1–4.
8. Dewen S, Wenlan C. Application of HPC technology in the building of a virtual geological visualization system. In: 2010 2nd international conference on future computer and communication, vol. 1; 2010. p. 472–6.
9. Esmaeilzadeh H, Sampson A, Ceze L, Burger D. Neural acceleration for general-purpose approximate programs. IEEE Micro. 2013;33 3:16–27.
10. Frustaci F, Blaauw D, Sylvester D, Alioto M. Better-than-voltage scaling energy reduction in approximate srams via bit dropping and bit reuse. In: Proceedings of 2015 PATMOS; 2015. p. 132–9.
11. Gascón A, Subramanyan P, Dutertre B, Tiwari A, Jovanović D, Malik S. Template-based circuit understanding. In: 2014 formal methods in computer-aided design (FMCAD); 2014. p. 83–90.
12. Grycel J, Schaumont P. Simplifi: hardware simulation of embedded software fault attacks. Cryptography 2021;5(2):15.
13. Gupta V, Mohapatra D, Raghunathan A, Roy K. Low-power digital signal processing using approximate adders. IEEE Trans Comput Aid Des Integr Circuits Syst. 2013;32 1:124–37.
14. Gupta V, Mohapatra D, Park SP, Raghunathan A, Roy K. IMPACT: IMPrecise adders for low-power approximate computing; 2011. p. 409–14.
15. Hasegawa K, Yanagisawa M, Togawa N. A hardware-trojan classification method utilizing boundary net structures. In: 2018 IEEE international conference on consumer electronics (ICCE); 2018. p. 1–4.
16. Hoffmann H, Misailovic S, Sidiroglou S, Agarwal A, Rinard MC. Using code perforation to improve performance, reduce energy consumption, and respond to failures. Tech Rep. MIT-CSAIL-TR-2009-042; 2009.
17. Inoue T, Hasegawa K, Yanagisawa M, Togawa N. Designing hardware trojans and their detection based on a SVM-based approach. In: 2017 IEEE 12th international conference on ASIC (ASICON); 2017. p. 811–4.
18. Kahng AB Kang S. Accuracy-configurable adder for approximate arithmetic designs. In: Proceedings of DAC design automation conference 2012; 2012. p. 820–5.
19. Karaklajić D, Schmidt JM, Verbauwhede I. Hardware designer's guide to fault attacks. IEEE Trans Very Large Scale Integr Syst. 2013;21 12:2295–306.
20. Keshavarz S, Holcomb D. Privacy leakages in approximate adders. In: 2017 IEEE international symposium on circuits and systems (ISCAS); 2017. p. 1–4.
21. Khudia DS, Zamirai B, Samadi M, Mahlke S. Quality control for approximate accelerators by error prediction. IEEE Des Test 2016;33 1:43–50.
22. Liang J, Han J, Lombardi F. New metrics for the reliability of approximate and probabilistic adders. IEEE Trans Comput. 2013;62 9:1760–71.
23. Lin L, Kasper M, Güneysu T, Paar C, Burleson W. Trojan side-channels: lightweight hardware trojans through side-channel engineering. In: Clavier C, Gaj K, editors. Cryptographic hardware and embedded systems - CHES 2009 Berlin: Springer; 2009. p. 382–95.
24. Liu C, Han J, Lombardi F. A low-power, high-performance approximate multiplier with configurable partial error recovery. In: Proceedings of 2014 design, automation test in europe conference exhibition (DATE); 2014. p. 1–4.
25. Liu W, Gu C, O'Neill M, Qu G, Montuschi P, Lombardi F. Security in approximate computing and approximate computing for security: challenges and opportunities. Proc. IEEE 2020;108 12:2214–31.

26. Mengte J, Raghunathan A, Chakradhar S, Byna S. Exploiting the forgiving nature of applications for scalable parallel execution. In: Proceedings of 2010 IPDPS; 2010. p. 1–12.
27. Moein S, Gulliver TA, Gebali F, Alkandari A. A new characterization of hardware trojans. IEEE Access 2016;4:2721–31.
28. Okhravi H, Bak S, King ST. Design, implementation and evaluation of covert channel attacks. In: 2010 IEEE international conference on technologies for homeland security (HST); 2010. p. 481–87.
29. Palomino D, Shafique M, Susin A, Henkel J. Thermal optimization using adaptive approximate computing for video coding. In: Proceedings of 2016 DATE; 2016. p. 1207–12.
30. Raha A, Sutar S, Jayakumar H, Raghunathan V. Quality configurable approximate dram. IEEE Trans Comput. 2017;66 7:1172–87.
31. Rahmati A, Hicks M, Holcomb DE, Fu K. Probable cause: the deanonymizing effects of approximate dram. In: 2015 ACM/IEEE 42nd annual international symposium on computer architecture (ISCA); 2015. p. 604–15.
32. Regazzoni F, Alippi C, Polian I. Security: the dark side of approximate computing? In: 2018 IEEE/ACM international conference on computer-aided design (ICCAD); 2018. p. 1–6.
33. Sampson A, Nelson J, Strauss K, Ceze L. Approximate storage in solid-state memories. In: 2013 46th annual IEEE/ACM international symposium on microarchitecture (MICRO); 2013. p. 25–36.
34. Sekanina L. Introduction to approximate computing: embedded tutorial. In: 2016 IEEE 19th international symposium on design and diagnostics of electronic circuits systems (DDECS); 2016. p. 1–6.
35. Soares LB, Bampi S, Costa E. Approximate adder synthesis for area- and energy-efficient FIR filters in cmos VLSI. In: 2015 IEEE 13th international new circuits and systems conference (NEWCAS); 2015. p. 1–4.
36. Sparsh M. A survey of techniques for approximate computing. ACM Comput Surv. 2016;48(4):1–33.
37. Venkataramani S, Chippa VK, Chakradhar ST, Roy K, Raghunathan A. Quality programmable vector processors for approximate computing. In: 2013 46th annual IEEE/ACM international symposium on microarchitecture (MICRO); 2013. p. 1–12.
38. Venkataramani S, Ranjan A, Roy K, Raghunathan A. AxNN: energy-efficient neuromorphic systems using approximate computing. In: 2014 IEEE/ACM international symposium on low power electronics and design (ISLPED); 2014. p. 27–32.
39. Wallat S, Fyrbiak M, Schlögel M, Paar C. A look at the dark side of hardware reverse engineering - a case study. In: 2017 IEEE 2nd international verification and security workshop (IVSW); 2017. p. 95–100.
40. Wang D, Wu L, Zhang X, Wu X. A novel hardware trojan design based on one-hot code. In: 2018 6th international symposium on digital forensic and security (ISDFS); 2018. p. 1–5.
41. Xiao K, Forte D, Jin Y, Karri R, Bhunia S, Tehranipoor M. Hardware trojans: lessons learned after one decade of research. ACM Trans Des Autom Electron Syst. 2016;22:1–23.
42. Xu S, Schafer BC. Approximate reconfigurable hardware accelerator: adapting the micro-architecture to dynamic workloads. In: Proceedings of IEEE international conference on computer design (ICCD); 2017. p. 113–120.
43. Yellu P, Yu Q. Can we securely use approximate computing? In: 2020 IEEE international symposium on circuits and systems (ISCAS); 2020. p. 1–5.
44. Yellu P, Boskov N, Kinsy MA, Yu Q. Security threats in approximate computing systems. In: Proceedings of the 2019 on great lakes symposium on VLSI, GLSVLSI'19. New York: Association for Computing Machinery; 2019. p. 387–92.
45. Yellu P, Buell L, Xu D, Yu Q. Blurring boundaries: a new way to secure approximate computing systems. New York: Association for Computing Machinery; 2020. p. 327–32.
46. Yellu P, Monjur MR, Kammerer T, Xu D, Yu Q. Security threats and countermeasures for approximate arithmetic computing. In: 2020 25th Asia and South Pacific design automation conference (ASP-DAC); 2020. p. 259–64.

47. Yellu P, Buell L, Mark M, Kinsy MA, Xu D, Yu Q. Security threat analyses and attack models for approximate computing systems: from hardware and micro-architecture perspectives. ACM Trans Des Autom Electron Syst. 2021;26(4):1–31.
48. Zhang T, Wang J, Guo S, Chen Z. A comprehensive FPGA reverse engineering tool-chain: from bitstream to RTL code. IEEE Access 2019;7:38379–89.
49. Zhu N, Goh WL, Yeo KS. An enhanced low-power high-speed adder for error-tolerant application. In: Proceedings of 2009 international symposium on integrated circuits; 2009. p. 69–72.
50. Ziener D, Assmus S, Teich J. Identifying FPGA IP-cores based on lookup table content analysis. In: 2006 international conference on field programmable logic and applications; 2006. p. 1–6.

Part IV
Introduction: Neural Networks and Machine Learning

Weiqiang Liu and Fabrizio Lombardi

This part consists of five chapters; it addresses the application of approximate computing to the emerging fields of neural networks and machine learning. These chapters provide an in-depth overview and innovative research directions of these important topics. The first chapter is "Approximate Computing for Machine Learning Workloads: A Circuits and Systems Perspective" by Sourav Sanyal, Shubham Negi, Anand Raghunathan, and Kaushik Roy. To reap maximum energy benefits as well as ensure the high quality of an outcome for many applications, innovations are needed across the entire computing stack (from circuits and architectures all the way up to algorithms). This chapter discusses different AC techniques in the context of machine learning applications by considering different approximate hardware primitives, such as multipliers, adders, memories, and matrix vector multiplication units, which are indispensable to build a machine learning accelerator. Several algorithm-based techniques that utilize hardware-level approximations to accelerate machine learning workloads are also discussed.

The next chapter, "Approximate Computing for Efficient Neural Network Computation: A Survey," (by Hao Zhang, Mohammadreza Asadikouhanjani, Jie Han, and Seok-Bum Ko) expands the presentation of neural network computation. The implementation of neural network models is hardware expensive; so, to reduce the hardware overhead, many optimization techniques have been explored in the literature, and AC is one of these techniques. With careful design, neural networks with approximate arithmetic units can have similar or even better accuracy than those with conventional exact arithmetic units, while achieving significant improvements in energy efficiency. In this chapter, a comprehensive survey of approximate arithmetic units applied to efficient neural network computation is presented. As a multiplier is more complex than an adder, the focus is on approximate multipliers designed for neural network computation. Design methodologies of approximate multipliers and their performance for neural network computation are discussed; then, a general discussion summarizes the current findings, it also presents the design challenges, and finally, it proposes potential future research directions.

The application of approximate computing to a specific class of neural networks is treated in the next chapter "Enabling Efficient Inference of Convolutional Neural Networks via Approximation" by Georgios Zervakis, Iraklis Anagnostopoulos, Hussam Amrouch, and Jörg Henkel. With the rapid advancement of artificial intelligence, neural networks (NNs) have become the driving force for both general purpose and embedded computing domains. Resource constrained embedded systems progressively rely on multiple NNs to provide on the spot sophisticated services. Nevertheless, supporting NN-based workloads is challenging due to the enormous computational and energy requirements. By exploiting the inherent error resiliency of NNs, this chapter focuses on designing approximate convolutional NN (CNN) inference accelerators, demonstrating that, for negligible accuracy loss, they satisfy tight latency, power, and temperature constraints. This chapter also provides a comprehensive discussion of different aspects of approximate CNN implementations.

The next chapter ("Approximate Computing for Energy-Constrained DNN-Based Speech Recognition" by Bo Liu, Hao Cai, Zhen Wang, and Jun Yang) deals with speech recognition as an innovative application to deep neural networks (DNNs) using AC. In a DNN-based speech recognition system, it has become significant to realize ultra-low-power consumption and real-time keyword spotting (KWS) by dynamically adjusting the computational accuracy. Using approximation approaches, in this chapter DNN-based speech recognition is hierarchically investigated at device, circuit, algorithm, and architectural levels. Energy-performance-area constraints of speech recognition conflicting with the DNN topology optimization are studied and evaluated using advanced CMOS designs; emphasis is also placed on the research perspective of approximate audio signal processing. Therefore, this chapter presents the application of approximate computing in KWS systems at different levels and discusses the hardware and software co-optimization framework to achieve high energy efficiency.

A different applicative area of AC using DNN is addressed in the last chapter "Efficient Approximate DNN Accelerators for Edge Devices: An Experimental Study," by Mohammadreza Asadikouhanjani, Hao Zhang, and Seok Bum Ko. While many approximate multipliers have been proposed, only a few of these approximate designs have been explored for performing inference of DNNs; furthermore, the application of various approximation techniques to different layers of DNNs has not been fully addressed. In this chapter, a step-wise approach for designing a re-configurable approximate Booth multiplier using commonly available approximate techniques is presented. Then, it is shown that for the best accuracy among the available approximation techniques, it is necessary to have a re-configurable multiplier to apply various approximation techniques to the layers of a DNN. Moreover, the multiplier proposed in this chapter is evaluated in an accelerator and compared to other designs.

Chapter 15
Approximate Computing for Machine Learning Workloads: A Circuits and Systems Perspective

Sourav Sanyal, Shubham Negi, Anand Raghunathan, and Kaushik Roy

1 Introduction

Machine Learning workloads have become ubiquitous in today's computing world, ranging from exascale cloud servers in data centers all the way down to tiny edge platforms such as mobile devices and wearables. To service the growing demand in image/text/audio/video recognition problems, deep neural networks (DNN) have become the *de-facto* standard for data-driven computing tasks, owing to their ability to learn from available data with high accuracy as well as scalability. On the other hand, rapid technology scaling have fueled an enormous growth in available on-chip compute resources, resulting in efficient realization of massive neural networks (exhibiting human-like capabilities) but with millions of parameters using billions of transistors. As a consequence, the number of computations has skyrocketed, resulting in unprecedented increase in the power consumption, making machine learning one of the most energy-expensive workloads.

Exploiting the inherent error resiliency of DNNs in order to introduce approximations at different levels of abstractions with the goal of reducing energy consumption is a challenging task. As the degree of approximation plays an important role in the overall accuracy of the application, investigating the overall energy-accuracy trade-offs for different modes of approximation is of prime importance. This chapter focuses on various approximate computing techniques for DNN workloads, considering different approximate hardware and algorithms (that take advantage of the approximate hardware and vice versa) to achieve energy efficiency with little or no degradation in accuracy. Section 2 introduces the basic operating principle behind approximate computing, while Sect. 3 provides the necessary background

S. Sanyal · S. Negi · A. Raghunathan · K. Roy (✉)
Purdue University, West Lafayette, IN, USA
e-mail: kaushik@purdue.edu

W. Liu, F. Lombardi (eds.), *Approximate Computing*,
https://doi.org/10.1007/978-3-030-98347-5_15

on DNNs to the readers. Section 4 discusses the hardware approximation techniques such as designing low complexity approximate adders, multipliers, matrix vector multiplication units, along with approximate low-power memory units (that can have occasional errors). Section 5 elaborates on the approximate hardware-algorithm co-design techniques which include quantization and pruning considering the underlying approximate hardware. Section 6 presents techniques that consider approximate hardware-in-the-loop training to maintain accuracy while achieving energy efficiency. Section 7 summarizes the system-level implications for different combinations of the promising approximation solutions (discussed in this chapter) for inference.

2 What Is Approximate Computing?

Continuous shrinking of the complementary metal-oxide semiconductor (CMOS) device feature size has provided significant performance gains for hardware at different scales from smartphones to supercomputers [1]. However, power dissipation has become a fundamental barrier to scale computing performance across all the hardware [2]. To that effect, an alternate design paradigm, namely *approximate computing*, has been explored to facilitate reduction in on-chip power dissipation by exploiting the intrinsic algorithmic error resiliency [3]. It generates results that are good enough rather than being fully accurate, but with considerably reduced (or approximated) computations to save power.

Applications that are tolerant to error would be a perfect fit for approximate computing. Figure 15.1 illustrates the application resilience due to several factors.

- A unique, golden result does not exist and a range of answers are equally acceptable (common examples include search and recommendation systems).

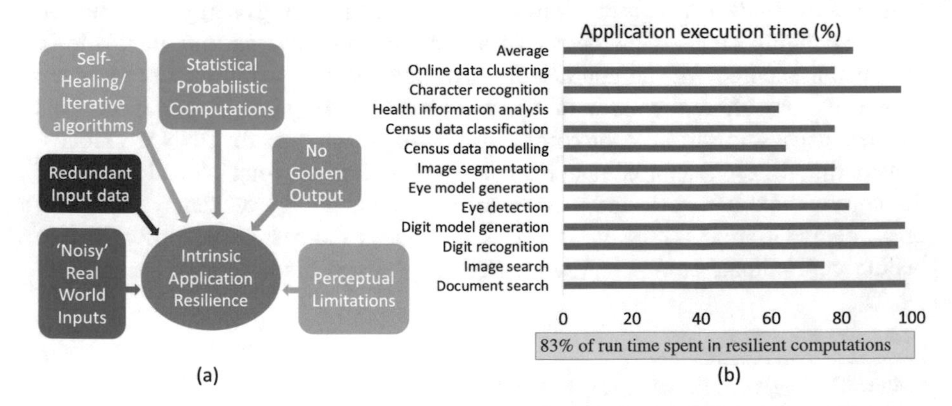

Fig. 15.1 Intrinsic application resilience. (**a**) Resilience Factors (**b**) High Degree of Error Resilient

- Even when a golden answer exists, the best known algorithms may not be guaranteed to find it, and hence, users are conditioned to accept imperfect, but good enough results (most machine learning applications fall into this category).
- Applications are often designed to deal with noisy input data. Noise at the inputs naturally propagates to the intermediate results; qualitatively, approximations have the same effect. In other words, robustness to noisy inputs also endows applications with robustness to approximations in computations.
- Applications frequently use computation patterns such as aggregation or iterative-refinement, which have the property of attenuating or healing the effects of approximations.

Recent studies have quantitatively established the high degree of intrinsic resilience in many applications. For example, one analysis of a benchmark suite of 12 applications shows that on average, 83% of the runtime is spent in computations that can tolerate at least some degree of approximation [4].

Approximate computing platforms should consider the following principles:

- *Measurable notion of quality:* Since both intrinsic resilience and approximate computing arise from the notion of acceptable quality of results, it is important to have a clear, measurable definition of what constitutes acceptable quality. In addition, it is critical to develop methods to ensure that acceptable quality is maintained when approximate computing techniques are used. Broadly speaking, quality specification and verification remains an open challenge. It is important to note that quality metrics do vary across applications (recognition or classification accuracy, relevance of search results, visual quality of images or video, etc.). However, the abstractions and methodology used to specify and validate quality should still be general, and to some extent, re-use tools and concepts from functional verification.
- *Significance-driven:* Not all computations in an application—even the most forgiving ones—may be subject to approximations. Computations that involve pointer arithmetic or affect control flow may lead to catastrophic effects when approximated. Even among the computations that may be approximated, the extent to which they impact application quality when approximated varies greatly. Therefore, it is important to adopt a "significance-driven" approach, i.e., separate resilient and sensitive computations, and approximate resilient computations based on how significantly they impact quality.
- *Disproportionate benefits:* Approximate computing should result in disproportionate benefits, i.e., large improvements in efficiency for little to no impact on quality. To achieve this, it is often beneficial to target bottleneck operations such as global synchronization and communication in software, or critical paths in hardware, for approximation. These sources of disproportionate benefit are inherently spread across various layers of the computing stack. Therefore, a cross-layer approach to approximate computing is more likely to yield a superior overall trade-off.
- *Quality configurable:* Resilience to approximations is not a static property of an application and may depend on both the input data processed and the context

in which the outputs are used. For example, a machine learning algorithm when used in a health-critical medical diagnosis application may have much more stringent quality constraints than when used in a product recommendation system. Similarly, resilience often manifests at scale, i.e., when the input data set is large and more likely to contain redundancy. Since hardware or software components are often re-used across applications, approximate computing techniques need to be quality configurable so that they can be modulated according to the available opportunity. Therefore, the objective should be to design approximate computing platforms that provide the best quality vs. efficiency trade-off across a range of output qualities, rather than optimizing them for a fixed quality.

Most of the above described applications employ machine learning kernels which uses Deep Neural Networks (DNNs or NNs) for different tasks like image recognition, object detection, natural language processing, etc. The basic operation of NNs consists of two phases, training and testing/inference. The training process is usually carried out off-line or in the cloud. The trained NN is then used to process unseen data inputs. For large networks with millions of neurons, the testing process, although less compute-intensive than training, nevertheless requires significant computation. This chapter educates the reader about approximations at different levels of abstraction for achieving energy efficiency. Since different approximations provide different energy-accuracy trade-offs, there is a need to compare these approximations and further explore whether better energy-quality trade-offs can be achieved through cross-layer approximations, from circuits to systems.

3 Neural Networks: Background

The fundamental elements of an artificial neural network are neurons and synapses. The output of an artificial neuron is a weighted sum of its inputs passed through an activation function. The activation function can be hard-limiting (e.g., step function) or soft-limiting (e.g., logistic sigmoid function, tanh function). Soft-limiting functions (Fig. 15.2) are preferred as they allow much more information to be communicated across neurons and greatly improve the neural network modeling capability while reducing network complexity. An artificial neuron can be trained to produce desired outputs for specific inputs by adjusting the weights of the corresponding synapses. This chapter considers fully connected feedforward networks and convolutional neural networks for studying multilevel approximations in DNNs.

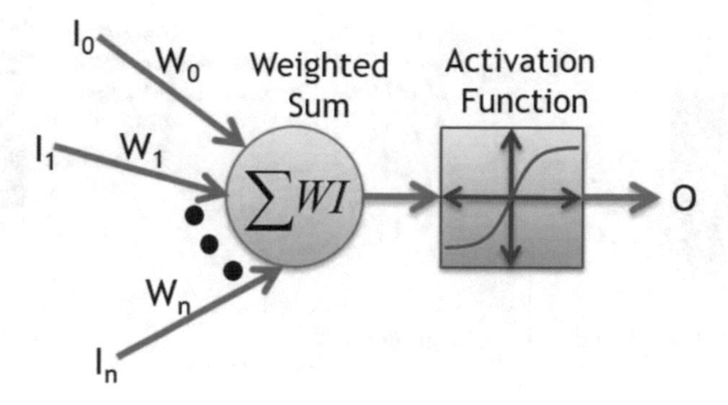

Fig. 15.2 Output of an artificial neuron is the weighted summation of its inputs passed through an activation function

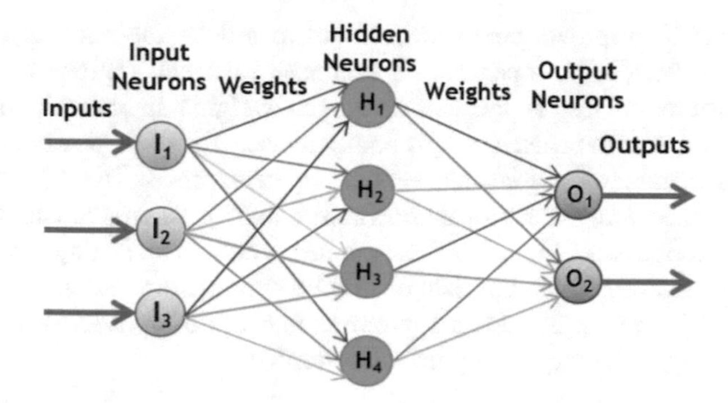

Fig. 15.3 A feedforward FCN, where each neuron in a layer is connected to all the neurons in the following layer as shown by arrows. The different colored arrows indicate that each input is multiplied by different weights

3.1 Fully Connected Networks (FCN)

In FCNs, the neurons are connected in an acyclic (feedforward) manner as illustrated in Fig. 15.3. In such an NN, every neuron in a layer is connected to all the neurons in the following layer via synapses with unique individual connection weights.

3.2 Convolutional Neural Networks (CNN)

CNNs consist of a hierarchical arrangement of alternating convolutional and spatial-pooling layers followed by a fully connected layer, with non-linearity applied at the end of each layer. A typical architecture of a deep CNN is shown in Fig. 15.4. Convolutional layers extract complex high-level features, while spatial-pooling layers are used for dimensionality reduction and fully connected layers are used

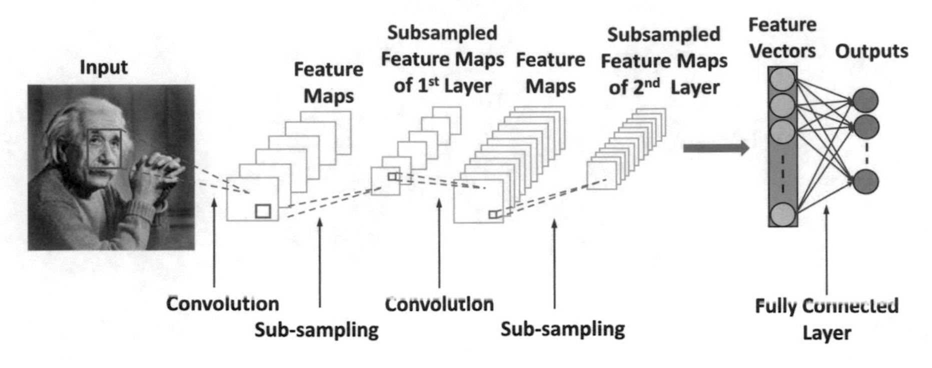

Fig. 15.4 Architecture of a deep CNN

for inference. To improve generalization and to reduce the number of trainable parameters, a convolution operation is exercised on small regions of input. One salient benefit of CNNs is the use of shared weights in convolutional layers, implying that the same filter (weight bank) is used for each pixel of the image; this reduces memory footprint and enhances performance. The NNs (FCNs and CNNs) are trained using backpropagation algorithm [5]. The trained network is then used to test random data inputs, which is done on-chip. The testing phase includes forward propagation, which consists of multiplication, summation, and activation operations. This chapter considers approximations at different levels of DNNs for achieving energy efficiency during inference/testing.

4 Approximate Hardware Design

4.1 Approximate Multiplication for Neuronal Computation

A neuron is a fundamental computational unit of an NN. Typically, a neuron performs a Multiply and Accumulate (MAC) operation (i.e., integrates the product of the incoming inputs and synaptic weights) to obtain a weighted sum, followed by a non-linear activation on the weighted sum, to produce the resultant output. The most power consuming operation among the neuronal computations is multiplication, which by far outweighs the summation and activation operations. To address this issue, an approximate Alphabet Set Multiplier (ASM) was proposed in [6] that achieves significant reduction in neuronal computation energy. In [7], computation sharing is used in conjunction with ASM to design energy-efficient hardware. In ASM, conventional multiplication is substituted by simplified shift and add operations. An ASM consists of a pre-computer bank, an adder, and one or more *"select"* and *"shift"* units. The pre-computer bank computes the product of the input and some smaller bit sequences (e.g., $0001_2, 0011_2, 0101_2, 0111_2$, etc.), which are referred to as alphabets. These alphabets are collectively termed the alphabet set (denoted by $\{1, 3, 5, \ldots\}$). The multiplication of an input and an alphabet is realized by shift and add operations. For example, if X is a multiplier input, then $1X$, $3X$ (equivalent to $X \times 2^1$(1 bit shift)+X) are products of the pre-computer bank which uses the alphabet set $\{1, 3\}$. These products are shared among the "*select*" units of an ASM, thereby utilizing computation sharing for the multiplication operation. Based on the multiplicand (synaptic weight), a proper combination of *select*, *shift* and *add* operations are carried out to perform the complete multiplication. For instance, to realize the multiplication of $Y = 100$ (01100100_2) and X, $0100_2X(4X)$ and $0110_2X(6X) \times 2^4$ (shifted by 4 corresponding to the relative bit position) needs to be generated, and summed up. Note that $4X$ and $6X$ can be generated by selecting $1X$ and $3X$ from the pre-computer bank, and shifting them, respectively, by 2 bits and 1 bit. The multiplication decomposition is demonstrated by the following equation:

$$01100100_2 \times X = (3X \times 2^1) \times 2^4 + (1X \times 2^2) \times 2^0$$

Figure 15.5 illustrates the operation of an 8-bit 4-alphabet ASM. In this example, the alphabet set used is $\{1, 3, 5, 7\}$. Multiplier "I" is supplied to the pre-computer bank, which generates the products 1I, 3I, 5I, and 7I corresponding to the alphabet set. Multiplicand "W" is divided into two parts and fed to the respective control circuits, each of which generates the suitable control logic for the *select* and *shift* units. The *select* unit chooses the appropriate product from the pre-computer bank and feeds it to the *shift* unit, which shifts the product by the required bits. The *adder* unit integrates the output of the individual *shift* units to compute the final result.

It has been shown that 8 alphabets $\{1, 3, 5, 7, 9, 11, 13, 15\}$ are required for a bit sequence size of 4 bits [6] to perform exact computation using the ASM. The number of alphabets used in an ASM directly correlates with the amount of energy expended for the multiplication operation. To achieve higher energy savings, the number of alphabets used in the proposed ASM is fewer than the quantity required for ideal (accurate) operation. As a result, it may not support all the multiplication combinations. For example, a 4 alphabet $\{1, 3, 5, 7\}$ ASM cannot support the operations, 9I, 11I, 13I, and 15I, with (9, 11, 13, 15) being the synaptic

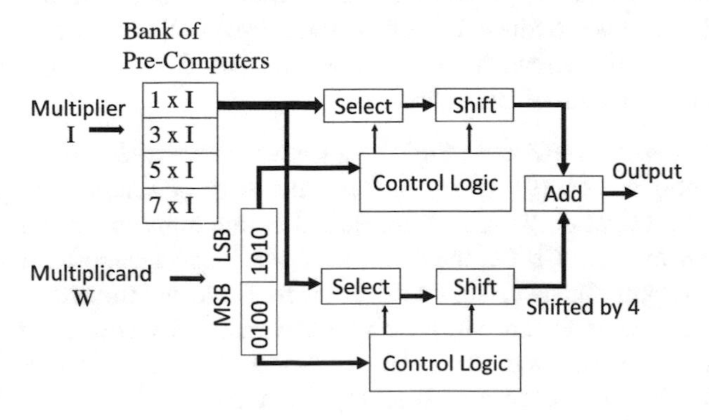

Fig. 15.5 8-bit 4 alphabet ASM [6]

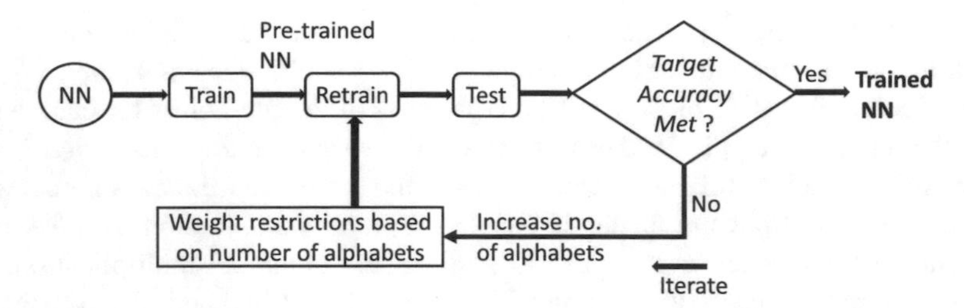

Fig. 15.6 Overview of the NN training methodology for employing ASM based neurons

weights. However, to guarantee proper functioning of the neural network, it must be ensured that the unsupported multiplication combinations do not lead to significant computational errors. For this purpose, the synaptic weights (9, 11, 13, 15) are restricted to the nearest supported values (8, 10, 12, 14). This is similar in effect to quantization, which drops some amount of information resulting in accuracy degradation.

The impact of three different ASMs, namely 4 alphabet $\{1, 3, 5, 7\}$ ASM, 2 alphabet $\{1, 3\}$ ASM, and 1 $\{1\}$ alphabet ASM (Multiplier-less Artificial Neuron [7]), is investigated, on the computational energy consumption of NNs. Each of the ASMs is individually analyzed with NNs of 12-bit, 8-bit, and 4-bit synaptic bit precision. The training methodology for using approximate ASMs is depicted in Fig. 15.6. First, the network is trained and then optimized for the desired bit precision. Then, the network is retrained with the weight restrictions, starting with the minimum number of alphabets (1 alphabet $\{1\}$) and increasing it if the desired accuracy is not met. This process is continued until the target accuracy is met.

4.2 *Approximate Adders*

In this section, different methodologies are presented for designing approximate Full Adder (FA) cells [8]. Since the Mirror Adder (MA) [9] is one of the widely used economical implementations of the FA, it has been used as the basis for proposing different approximations of an FA cell.

(1) *Conventional Mirror Adder*: Fig. 15.7a shows the transistor level schematic of a conventional MA [9], which is a popular way of implementing an FA. It consists of a total of 24 transistors. Note that this implementation is not based on complementary CMOS logic and thus provides an opportunity to cleverly design an approximate version with removal of selected transistors.
(2) *Approximation 1*: In order to get an approximate MA with lesser transistors, transistors are removed from the conventional schematic one by one. In doing so, it must be ensured that any input combination of A, B, and C_{in} does not result in short circuits or open circuits in the simplified schematic.

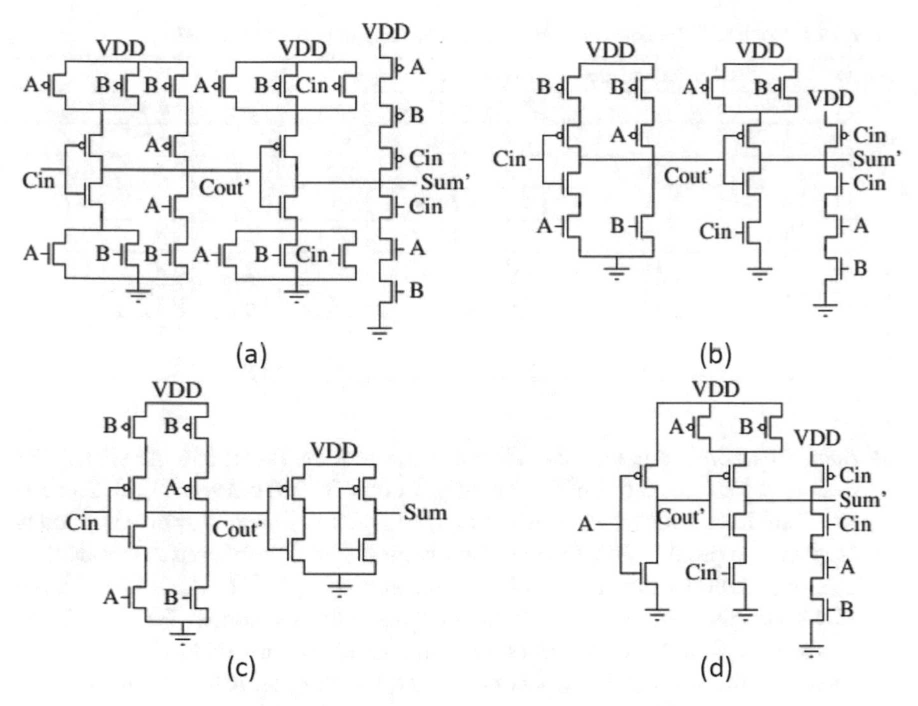

Fig. 15.7 Adder schematic. (**a**) Conventional MA. (**b**) Simplified MA. (**c**) MA approximation 1. (**d**) MA approximation 2

Another criterion that should be imposed is that the resulting simplification should introduce minimal errors in the FA truth table. A judicious selection of transistors to be removed (ensuring no open or short circuits) results in a schematic shown in Fig. 15.7b. Clearly, this schematic has 8 less transistors compared to the conventional MA schematic. A close observation of the truth table of an FA shows that $Sum = C_{out}$ for 6 cases out of 8, except for the input combinations A = 0, B = 0, C_{in} = 0 and A = 1, B = 1, C_{in} = 1. Now, in the conventional MA, C_{out} is computed in the first stage. Thus an elegant way of simplifying the MA further is to discard the Sum circuit completely. Although one can directly set Sum = C_{out} as shown in Fig. 15.1, a buffer stage is introduced after C_{out} (see Fig. 15.7c) to implement the same functionality. The reason for this can be explained as follows. If one sets Sum = C_{out} as it is in the conventional MA, the total capacitance at the *Sum* node would be a combination of 4 source-drain diffusion and 2 gate capacitances. This is an appreciable increase compared to the conventional case. Such a design would lead to a delay penalty in cases where two or more multi-bit approximate adders are connected in a chained fashion. Thus the simplified circuit is combined for C_{out} in Fig. 15.7b with the idea that $Sum = C_{out}$ for 6 cases out of 8. Figure 15.7c shows the simplified MA obtained using this technique. This introduces 1 error in C_{out} and 3 errors in Sum, as shown in Table 15.1.

Table 15.1 Truth table for conventional full adder and approximations 1, 2 and 3

Inputs			Accurate outputs		Approximate outputs					
A	B	C_{in}	Sum	C_{out}	Sum_1	C_{out1}	Sum_2	C_{out2}	Sum_3	C_{out3}
0	0	0	0	0	1 ✗	0✓	0✓	0✓	0✓	0 ✓
0	0	1	1	0	1 ✓	0 ✓	1 ✓	0 ✓	0 ✗	0 ✓
0	1	0	1	0	0 ✗	1 ✗	0 ✗	0 ✓	1 ✓	0 ✓
0	1	1	0	1	0 ✓	1 ✓	1 ✗	0 ✗	1 ✗	0 ✗
1	0	0	1	0	1 ✓	0 ✓	0 ✗	1 ✗	0 ✗	1 ✗
1	0	1	0	1	0 ✓	1 ✓	0 ✓	1 ✓	0 ✓	1 ✓
1	1	0	0	1	0 ✓	1 ✓	0 ✓	1 ✓	1 ✗	1 ✓
1	1	1	1	1	0 ✗	1 ✓	1 ✓	1 ✓	1 ✓	1 ✓

(3) *Approximation 2*: Again, a careful observation of the FA truth table shows that $C_{out} = A$ for 6 cases out of 8. Similarly, Cout = B for 6 cases out of 8. Since A and B are interchangeable, consider $C_{out} = A$. Thus, a second approximation is proposed (approximation 2) where an inverter is used with input A to calculate C_{out} and Sum is calculated similar to the simplified MA in Fig. 15.7b. Figure 15.4 shows the simplified circuit obtained using this technique. This introduces 2 errors in Count and 3 errors in Sum, as shown in Table 15.1. In both approximations 1 and 2, C_{out} is calculated by using an inverter with C_{out} as input.

(4) *Approximation 3*: In approximation 2, there are 3 errors in Sum. This approximation can be taken a step further by allowing 1 more error, i.e., 4 errors in Sum. Another aim is to reduce the dependency of Sum on C_{in} (to save area). This leaves one with 2 choices, $Sum = A$ and $Sum = B$. Also, the approximation $C_{out} = A$ can be used, as in approximation 2. Thus, there can be 2 choices for approximation 3, viz. $Sum = A$, $C_{out} = A$ and $Sum = B$, $C_{out} = A$. If choice 1 is observed, both Sum and C_{out} match with accurate outputs in only 2 out of the 8 cases. In choice 2, Sum and C_{out} match with accurate outputs in 4 out of the 8 cases. Therefore, to minimize errors both in Sum and C_{out}, one must go for choice 2 as approximation 3. The main thrust here is to ensure that for a particular input combination (A, B, and C_{in}), ensuring correctness in Sum also makes C_{out} correct. Now consider the addition of two 20 bit integers a[19:0] and b[19:0] using a Ripple Carry Adder (RCA). Suppose approximate FAs are used for 7 LSBs. Then $C_{in}[7] = C_{out}[6]$. Note that $C_{out}[6]$ is approximate. Applying this approximation to the present example, the carry propagation from bit 0 to bit 6 is entirely eliminated. In addition, the circuitry needed to calculate $C_{out}[0]$ to $C_{out}[5]$ is also saved. To limit the output capacitance at Sum and C_{out} nodes, the approximation 3 is implemented , i.e., $Sum = B$, $Cout = A$ using buffers.

4.3 Voltage-Scaled Approximate Memory for On-chip Storage

In this section, the approximations applied to weighted synapses interconnecting various layers of a deep neural network are discussed. Note that the number of synapses is typically two to three orders of magnitude greater than the number of neurons. The on-chip memory, conventionally designed using 6T SRAM, consumes significant amount of access and leakage energy. The supply voltage of 6T Static Random Access Memory (SRAM) can potentially be scaled to lower the energy consumption. However, 6T bitcells are susceptible to read-access and write failures at scaled voltages. The failures are aggravated in scaled technology nodes due to random process parameter variations [10–12]. The random variations effectively change the relative strength of the transistors constituting the individual bitcells. This negatively impacts the ability to read from (write into) a 6T bitcell within the stipulated time duration, resulting in read-access (write) failures. DNNs, being inherently resilient to small perturbations in the synaptic weights, enable the supply voltage of 6T SRAM based synaptic storage to be scaled moderately. This reduces the energy consumption for a negligible loss in the classification accuracy. The 6T bitcell failures increase exponentially as the supply voltage is scaled. Therefore, aggressive voltage scaling of 6T SRAM could potentially result in unacceptable degradation in accuracy due to corruption of a large fraction of the DNN weights. A solution to this is storing few Most Significant Bits (MSBs) of the weights in reliable 8T bitcells while the relatively tolerant Least Significant Bits (LSBs) are stored in 6T bitcells as shown in Fig. 15.8 [13]. The enhanced stability of an 8T

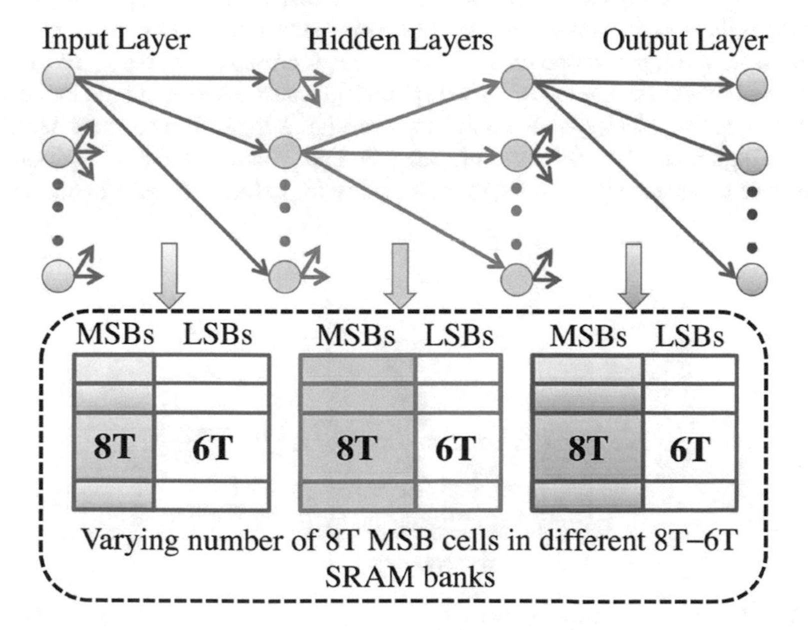

Fig. 15.8 Weight sensitivity driven hybrid 8T-6T memory architecture [13]

bitcell at scaled voltages can principally be attributed to decoupled read and write paths, leading to independent optimization for the respective operations (read and write).

4.4 Analog In-Memory Computing

The fundamental computation of both DNN and FCN is matrix multiplication operation [14]. To accelerate this operation efficiently, different hardwares have been explored in both industry and academia [15–17]. These hardwares use adders and multipliers to perform computation in digital domain. Hence, the techniques from Sects. 4.1, 4.2, 4.3 can be used in these hardwares to accelerate ML workloads efficiently. Further, training a DNN on a particular task requires training with the enormous amount of data. Hence the weights in the DNN can be trained with the approximations in the hardware using the hardware-aware training which is explained further in Sect. 5. These accelerators try to move computation closer to memory but still involve some accesses from the memories. The cost to access one word of data (64 b) from different sized memories in a 45-nm technology to the energy of multiplication operations is shown in Fig. 15.9. Hence, the data movement contributes majorly to the energy cost in these accelerators.

To mitigate the cost of data movement further, In-Memory Computing (IMC) hardware [18] has been explored which performs the matrix multiplication inside the memory array itself as shown in Fig. 15.10a. At the cross-points of the crossbars, Non-volatile Memory (NVM) or SRAMs can be used [18]. Upon applying an input voltage at the crossbar's rows, the MVM result can be obtained as output current at the crossbar's columns based on Kirchhoff's law. A crossbar thus performs MVM in one computational step—including $O(n^2)$ multiplications and additions for an $n \times n$ matrix– which typically takes many steps in digital logic. Hence these hardware achieve higher energy efficiency compared to digital hardware due to high on-chip density of crossbars [19]. Owing to their ability to perform efficient in-situ Matrix

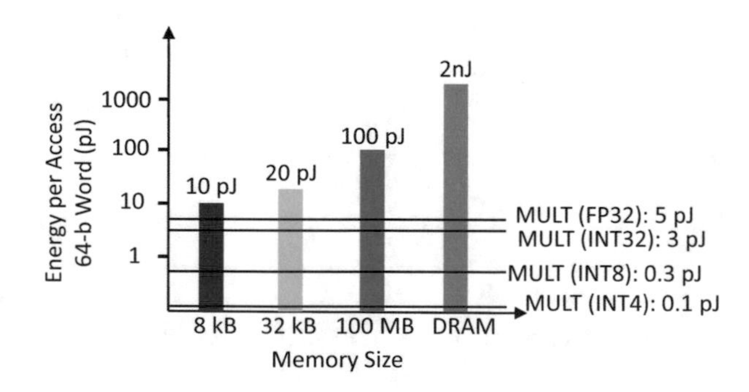

Fig. 15.9 Energy cost of accessing memory [14]

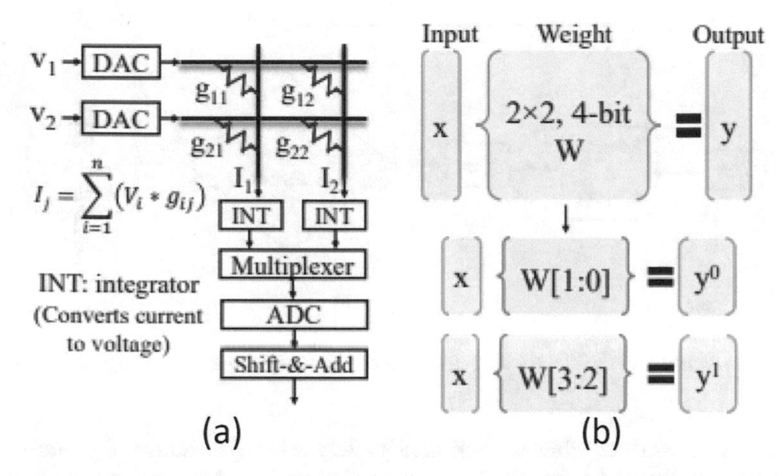

Fig. 15.10 Matrix vector multiplication with Crossbars. (**a**) Analog MVM. (**b**) Bit slicing [18]

Vector Multiplication (MVM) operations [20], NVM based IMC architectures have been adopted in energy-efficient target recognition applications [21]. However, crossbars also suffer from computational errors originating from device and circuit non-idealities such as: parasitic resistance, non-linearity from access transistors and I–V characteristics of NVM device [22]. The degree of error in computation also depends on crossbar size [22]. In large-scale DNNs, these computation errors accumulate and result in severe degradation in classification accuracy. Again the training of the DNNs come to rescue which helps to train the DNN weights with non-idealities and recover the accuracy drop due to the approximations in these memory.

5 Approximate Hardware-Algorithm Co-design

Two algorithm level approximations, viz. lower complexity networks and pruning, are considered in this section. However, these algorithmic techniques are designed taking the underlying hardware into consideration. The algorithmic approximations can potentially lead to a relaxed hardware design, which eventually enables energy-efficient acceleration of DNNs.

5.1 Lower Complexity Networks

For a given baseline DNN, a low complexity network can be trained (containing lesser number of layers and/or neurons per layer) in order to achieve a significant reduction in energy for a small reduction in accuracy. There are two basic

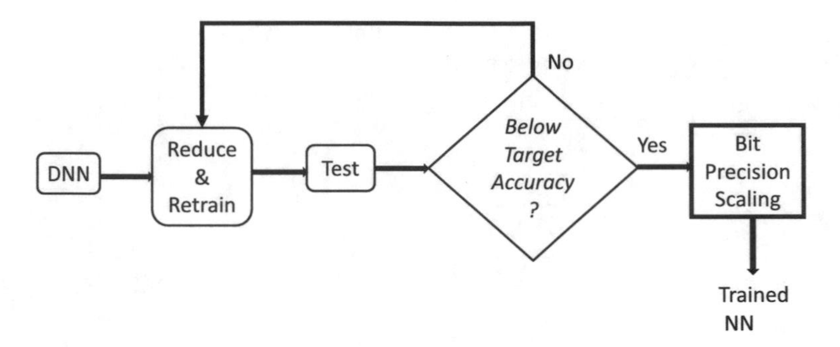

Fig. 15.11 Overview of the NN training methodology for employing network approximations

approaches to deriving lower complexity networks: (i) reduce the number of hidden/convolutional layers, or, (ii) reduce the number of neurons/convolutional kernels in the hidden/convolutional layers. If the achieved accuracy is above the target accuracy, then the network is further reduced and retrained. If the network is slightly less than or equal to the target accuracy, then retraining is terminated. Finally, the network is optimized by bit precision scaling as explained below. During bit precision scaling, the input and synaptic bit-width is reduced until the point where accuracy starts to degrade so as to determine the optimal bit precision for the network. This is also referred to as quantization. Although the discussed techniques can be applied independent of the hardware, a consequence of lower complexity network training places relaxations on the underlying hardware (which can now be approximate). Using such approximate hardware, the low complexity NNs (which is a result of algorithmic approximations) can be executed. The training methodology is depicted in Fig. 15.11.

5.2 Pruning

Pruning has been shown to reduce the complexity of an NN tremendously [23, 24]. Essentially, the insignificant synaptic connections are eliminated (or pruned) to achieve an order of magnitude reduction in the NN computations with minimal impact on accuracy. Figure 15.12 shows the distribution of the synaptic weights for a deep FCN (with 784 input neurons, 2 hidden layers consisting of 1200, 600 neurons and 10 output neurons denoted as [784 1200 600 10]) trained on the MNIST dataset using a precision of 12 bits. It is observed that a substantial portion of the weights carry very small values (experimentally determined to be below 0.04 for this application). Based on this analysis, it is found that these weights are insignificant and can be removed (pruned) without having a significant impact on accuracy. It is further observed that 8-bit synapses follow a similar weight distribution as 12 bit synapses. Hence, pruning the 8 bit synapses (below 0.04) also had negligible impact on accuracy. For an NN with 4 bit synapses, the lower bit precision naturally

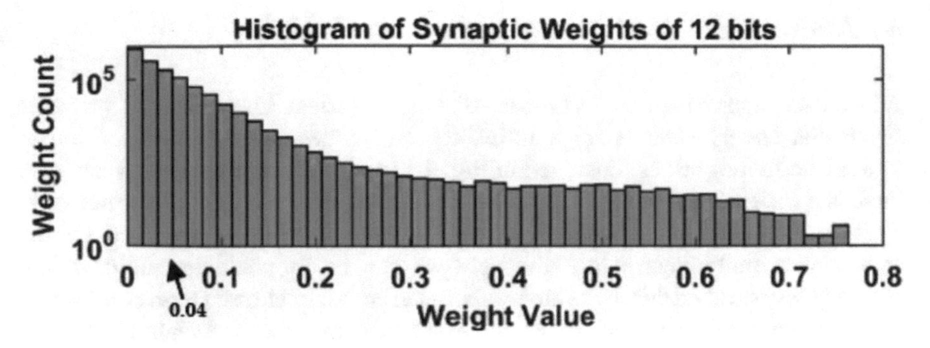

Fig. 15.12 Synaptic weight distribution of a Deep FCN trained on MNIST:[784 1200 600 10] with 12 bit synaptic weights

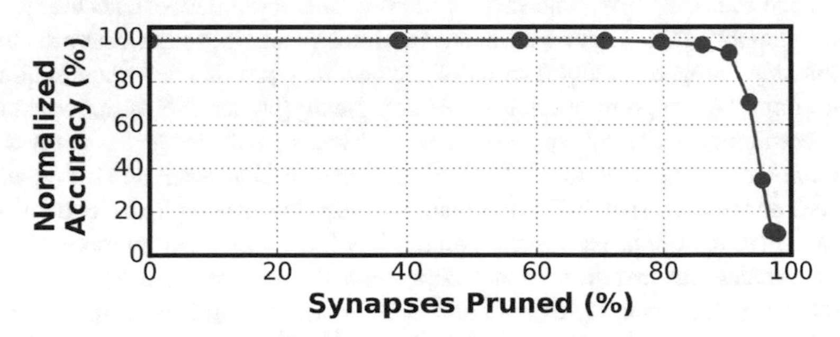

Fig. 15.13 Effect of pruning on accuracy for a Deep FCN trained on MNIST:[784 1200 600 10] with 12 bit synaptic weights

truncates a majority of synaptic weight values to 0. However, a synaptic precision of 4 bits, although more computationally efficient than 8 or 12 bits, is not preferable, since it drastically degrades the accuracy. The energy/accuracy trade-offs offered by NNs with different synaptic bit precisions will be discussed in Sect. 6

A three-step process is used to prune the synaptic weights. First, the NN is trained to learn all the synaptic connections. Then, the unimportant connections are removed (pruned) based on a pruning threshold that is determined from the distribution of the learned synaptic weights. Finally, the NN is retrained to adjust the weights of the remaining connections to reclaim a significant portion of the accuracy lost due to pruning. The percentage of the synaptic weights that can be pruned is estimated by analyzing its impact on the network accuracy. One analysis (Fig. 15.13) indicates that almost 80% of the synaptic connections of a trained deep FCN (for MNIST) can be pruned for negligible accuracy degradation.

6 Approximate Hardware-in-the-Loop Training

While each approximation technique, discussed earlier, independently provides substantial energy benefits for a DNN, the techniques can be combined into a synergistic framework as illustrated in Fig. 15.14 to maximize the energy savings. First, the insignificant synaptic weights can be pruned from the trained network. Next, appropriate weight restrictions can be introduced as necessitated by the approximate multipliers. Then, the network can be retrained to minimize the accuracy loss suffered due to the aforementioned approximations. During retraining, only the non-zero weights are updated while accounting for the weight constraints imposed by the approximate multiplier topology. Finally, bit-flips can be introduced in the resultant synaptic weights to incorporate the read-access and write failures of the voltage-scaled approximate memory and estimate the network accuracy. Note that the memory failures are distributed randomly. Retraining the network further will not help in regaining any fraction of the accuracy lost due to voltage scaling.

It is important to point out that lower complexity/pruned NN is not considered while combining different approximate techniques. This is because, once this approximation is applied to a DNN, it becomes a shallower network with its "degrees of freedom"/flexibility diminished. These lower complexity networks are very sensitive to weight perturbation since they have very small number of learning parameters and therefore cannot cope with further hardware approximations without significant accuracy degradation, unless the training is done in an iterative manner, keeping the hardware in consideration. Therefore, lower complexity/pruned networks (algorithmic approximations) should be considered separately from the hardware approximations, for retraining.

6.1 *Retraining to Mitigate Accuracy Loss*

It was mentioned in previous sections that retraining is used to mitigate the accuracy degradation incurred due to certain approximations including pruning and approximate multiplication. In this section, the importance of adapting the learning

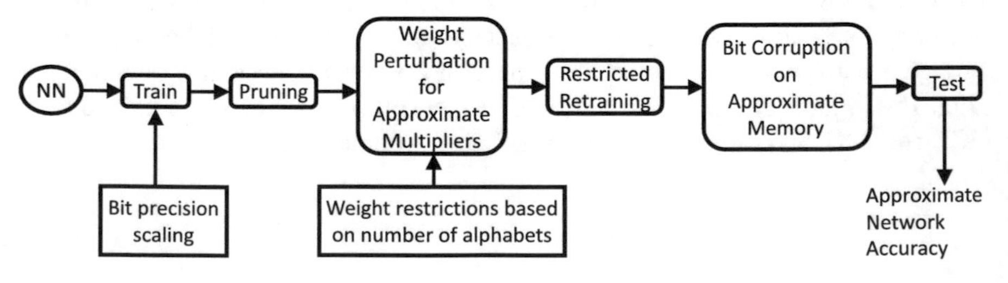

Fig. 15.14 Flow diagram of the proposed combined approximation process of a NN

rate for efficiently retraining the network with these approximations in place is highlighted. The learning rate, which essentially is a multiplication factor in the weight update, influences the speed and quality of learning. It needs to be precisely regulated while using the approximate multiplier due to the non-uniformity in the distance between the allowed weight levels. For instance, considering a 2 alphabet $\{1, 3\}$ ASM, where the permitted weight levels are $0\times$, $1\times$, $2\times$, $3\times$, $4\times$, $6\times$, $8\times$, and $12\times$. It is evident that the distance between the levels $8\times$ and $12\times$ is $4\times$, while that between $6\times$ and $8\times$ is $2\times$. In this case, if the learning rate is too low, the updates might not be substantial enough for the weights to overcome the distance barrier between certain allowed levels. This could potentially cause the weights to get stuck at a specific level, which leads to non-convergence during the learning process. Too high a learning rate might cause the weights to widely oscillate between different levels, which leads to a deterioration in the accuracy. Hence, it is necessary to determine the optimal learning rate for retraining an approximate DNN. Figure 15.15 illustrates the flow diagram of the retraining process, which begins with the highest learning rate that was used to originally train the DNN without approximation. If the accuracy improves, retraining is carried on with the same learning rate for few more iterations, until there is no significant change in the accuracy. On the other hand, if the accuracy does not improve, the learning rate is regulated (reduced by a factor) and the approximate DNN is further retrained. This process of regulating the learning rate is continued until the accuracy improvement saturates. The initial high learning rate during retraining allows the synapses to compensate for the weight perturbations despite the weight constraints still being applied. However, high learning rate incurs oscillations and may not allow the retraining to converge to the optimal solution. Therefore, learning rate is reduced based on the retrained network performance. The low learning rate helps to dampen the oscillations and allow the retraining to converge. It is to be noted that the retraining overhead for an approximate DNN is negligible compared to the number of iterations required to train the original DNN (without approximations). Also, retraining does not affect the energy consumption during inference/testing.

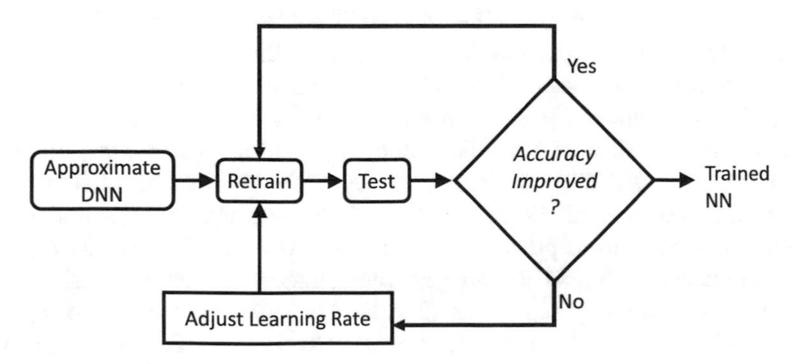

Fig. 15.15 Flow diagram of the retraining process

6.2 Simulation Framework: Circuits to Systems

A circuit to system-level simulation framework was developed [25] to analyze the effectiveness of the approximations on NNs. For the approximate multiplication, the multiplier and adder units were implemented at the Register-Transfer Level (RTL) in Verilog, and mapped the designs to the 45nm technology library using the Synopsys Design Compiler. A neuronal energy computation model was developed based on the number of MAC (multiply-accumulate) operations in the forward propagation of the NN algorithms (FCN and CNN). The power and delay numbers of the individual adder and multiplier units obtained from the Design Compiler were subsequently fed to the energy computation model to estimate the total neuronal energy consumption. For the voltage-scaled approximate memory, the constituent 6T and 8T bitcells were designed and subjected to Monte Carlo SPICE simulations to determine the read-access and write failure probabilities at reduced voltages. The energy consumption of the hybrid 8T-6T SRAM bank including the peripherals is obtained using CACTI [26] for the 45nm process technology. At the system-level, the deep learning toolbox [27] and MatConvNet [28], which are MATLAB based open source neural network simulators, were used to model the approximations and evaluate the performance (classification accuracy) of the DNNs under consideration. The fully connected and convolutional NNs were implemented without data augmentation, batch normalization, and dropout features to primarily single out the effects of different approximations.

6.3 Optimized Baseline Deep Neural Networks

Optimized DNNs are bit precision scaled DNNs without any approximations. First, the DNN architecture that provides the best classification accuracy for a given synaptic bit precision is identified. The DNN configuration thus determined is chosen as the optimized network, which is then subjected to different approximations presented in this chapter. The analysis is performed on deep FCNs trained on the MNIST digit recognition dataset, and deep CNNs trained on the CIFAR-10 image recognition dataset. Figure 15.16 demonstrates the effect of bit precision scaling. The experiments indicate that a deep FCN (with 784 input neurons, 2 hidden layers consisting of 1200, 600 neurons and 10 output neurons, denoted as [784 1200 600 10]) offers the best accuracy of 98.87% on the MNIST dataset for synaptic precision of 12 bits. The accuracy degradation was found to be minimal ($< 0.5\%$) up to precision of 4 bits. For the CIFAR-10 dataset, a CNN was implemented with 3 MLPConv [29] (combination of convolutional and fully connected layers) blocks: [1024×3 $(5 \times 5)192c$ $160fc$ $96fc$ $(3 \times 3)mp$ $(5 \times 5)192c$ $192fc$ $192fc$ $(3 \times 3)mp$ $(3 \times 3)192c$ $192fc$ $10o$]. The input layer is $32 \times 32 \times 3$. The convolutional layers use 5×5 and 3×3 kernel size with different number of feature maps. A 3×3 max pooling window is used between two

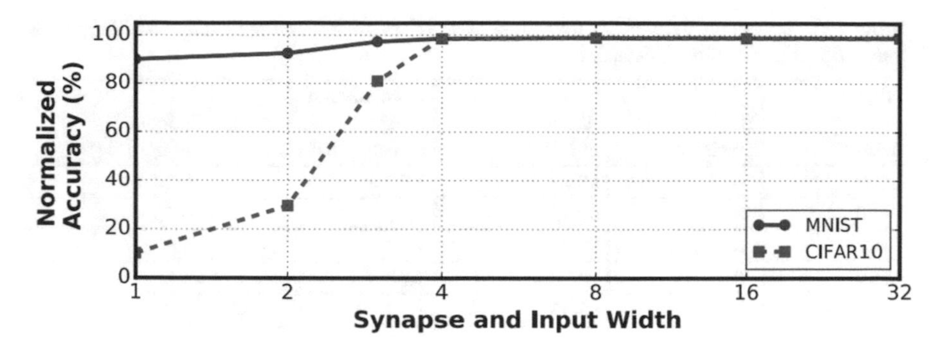

Fig. 15.16 Effect of bit precision scaling (with retraining) on benchmark applications

MLPConv blocks. The final features from the last block are then fully connected to a 10-neuron output layer. This Deep CNN provided the maximum accuracy of 87.3%. There was reasonable deterioration in the accuracy ($< 1.5\%$) for a precision of 4 bits.

6.4 Approximations at Algorithm Level

6.4.1 Lower Complexity Networks

The lower complexity networks that were additionally precision-scaled to reduce energy for a given accuracy were evaluated. In this subsection, the results are presented for lower complexity NNs for the datasets under investigation. First, low complexity FCNs are investigated, trained on the MNIST dataset, consisting of a single hidden layer. The initial configuration had twice the number of neurons in the only hidden layer of the low complexity network compared to the number of neurons in the first hidden layer of the deep FCN. The number of neurons in the hidden layer of the low complexity network is decreased until the desired accuracy is achieved. Bit precision scaling is subsequently applied on the low complexity FCN to determine the optimal synapse bit-width. For a baseline deep FCN [784 1200 600 10] trained on MNIST, a few low complexity FCNs are listed in Table 15.2.

Next, low complexity CNNs on the CIFAR-10 dataset was explored, where the initial setup was begun with a single convolutional layer, and then the number of weight kernels was decreased to attain the desired accuracy. It was observed that such a shallow (only one MLPConv block) CNN failed to match the accuracy of a deep CNN (87.3%) even with greatly increased number of kernels, with 78% being the maximum achievable accuracy. Therefore, two MLPConv blocks were utilized with equal or lesser number of kernels in each block than the Deep CNN, to realize the low complexity CNNs. This enabled the low complexity CNN to

Table 15.2 Low complexity FCNs trained on MNIST. Savings are computed by considering the conventional DNN (12bit) as standard

Hidden layer neurons	Bit precision	Accuracy of NN (%)	Computation energy consumption (J)	Energy savings (%)
2400	9	98.88	1.82	5.35
2400	8	98.85	1.91	0.73
2100	8	98.85	1.40	27.07
1800	9	98.77	1.36	29.01
1550	9	98.73	1.17	38.87
800	9	98.64	0.61	68.45
650	12	98.52	0.60	69.07
425	7	98.36	0.25	86.80
400	10	98.22	0.32	83.23

Table 15.3 Low complexity CNNs trained on CIFAR-10. Savings are computed by considering the conventional DNN (12bit) as standard

Neurons	Synapses	Accuracy of NN (%)	Computation energy consumption (J)	Energy savings (%)
542762	538719	86.0	2.25	65.49
405546	379281	85.6	1.60	75.46
359466	324768	85.3	1.36	79.14
268330	223374	83.6	0.94	85.58
221226	171522	80.5	0.70	89.26

attain an accuracy of 86% for a precision of 8 bits and above, while the deep CNNs consistently delivered an accuracy of greater than 87% for equivalent synaptic bit-widths. For a baseline deep CNN [1024×3 $(5\times5)192c$ $160fc$ $96fc$ $(3\times3)mp$ $(5\times5)192c$ $192fc$ $192fc$ $(3\times3)mp$ $(3\times3)192c$ $192fc$ $10o$] trained on CIFAR-10, a few low complexity CNNs are listed in Table 15.3.

6.4.2 Synaptic Pruning

The synaptic weights were systematically pruned using the procedure described in Sect. 5.2. The energy-accuracy comparison between conventional DNN and pruned DNNs is shown in Fig. 15.17. One analysis on a deep FCN trained on the MNIST dataset using 8-bit and 12-bit synapses showed that $\sim$ 80% of the synaptic connections could be removed with minimal accuracy degradation, which leads to roughly 80% savings in energy compared to the conventional (un-approximated) DNNs. It should be further noted that FCNs with 4-bit synapses allowed up to 84% of the connections to be pruned, resulting in larger energy benefits. Figure 15.17a shows that the accuracy degradation is less than 0.20% for pruned DNNs across different bit-width configurations.

A similar analysis on a deep CNN trained on the CIFAR-10 dataset indicated that 41.24%, 39.51%, and 25.75% of the synaptic connections could be eliminated for 12-bit, 8-bit, and 4-bit synapses, respectively, with negligible accuracy degradation. It can be seen from Fig. 15.17b that pruned networks consistently provide better energy consumption than conventional DNNs.

6.5 Approximations at Hardware Level

6.5.1 Approximate Multiplier Based Networks

Alphabet set multipliers (ASM) [6] based approximate multipliers were deployed in the neural processing units to reduce the computational energy. The computational energy-accuracy trade-offs offered by the ASM based deep FCNs and CNNs are presented along with a comparison with the conventional (un-approximated) DNNs. It is observed from Fig. 15.18 that, ASM based FCNs (trained on MNIST) using 4 alphabets $\{1, 3, 5, 7\}$ do not provide much energy savings over conventional (un-approximated) DNNs of 12, 8, and 4 bit neurons. On the other hand, 18–27% reduction in energy consumption is achieved using only 2 $\{1, 3\}$ alphabets ASM based FCNs. In the case of multiplier-less neurons (using only 1 alphabet $\{1\}$), 26–32% reduction in energy consumption is achieved. The accuracy degradation is less than 0.40% for ASM based DNNs compared to 12 bit conventional DNN baseline.

Figure 15.18b shows that, 5–18% and 31–40% energy savings are achieved using 4 alphabet and 2 alphabet ASM based CNNs (trained on CIFAR-10), respectively. The analysis further indicates that up to 44% energy savings is achieved using multiplier-less neurons (using only 1 alphabet $\{1\}$). It should be noted that although the 4-bit ASM based CNNs offer the lowest energy consumption relative to other ASM configurations, they lead to higher degradation in accuracy (up to 2.3%), which is undesirable. The higher accuracy loss can be attributed mostly to lower bit precision. From the effect of bit precision scaling, it was found that a Deep CNN $[1024 \times 3(5 \times 5)192c\ 160fc\ 96fc\ (3 \times 3)mp\ (5 \times 5)192c\ 192fc\ 192fc\ (3 \times 3)mp\ (3 \times 3)192c\ 192fc\ 10o]$ of 4 bit precision trained on CIFAR-10 can achieve only an accuracy of 86%. If hardware approximations are applied on this network, further accuracy degradation is unavoidable.

6.5.2 Voltage-Scaled Approximate Synaptic Memory

The supply voltage of 6T SRAM based synaptic memory was scaled down to achieve energy efficiency by exploiting the intrinsic error resiliency of neural networks. Simulations on a deep FCN ([784 1200 600 10]) trained for MNIST digit recognition showed that the supply voltage could be lowered up to 0.85 V from the nominal voltage of 0.90 V for negligible accuracy loss. Further reduction in voltage

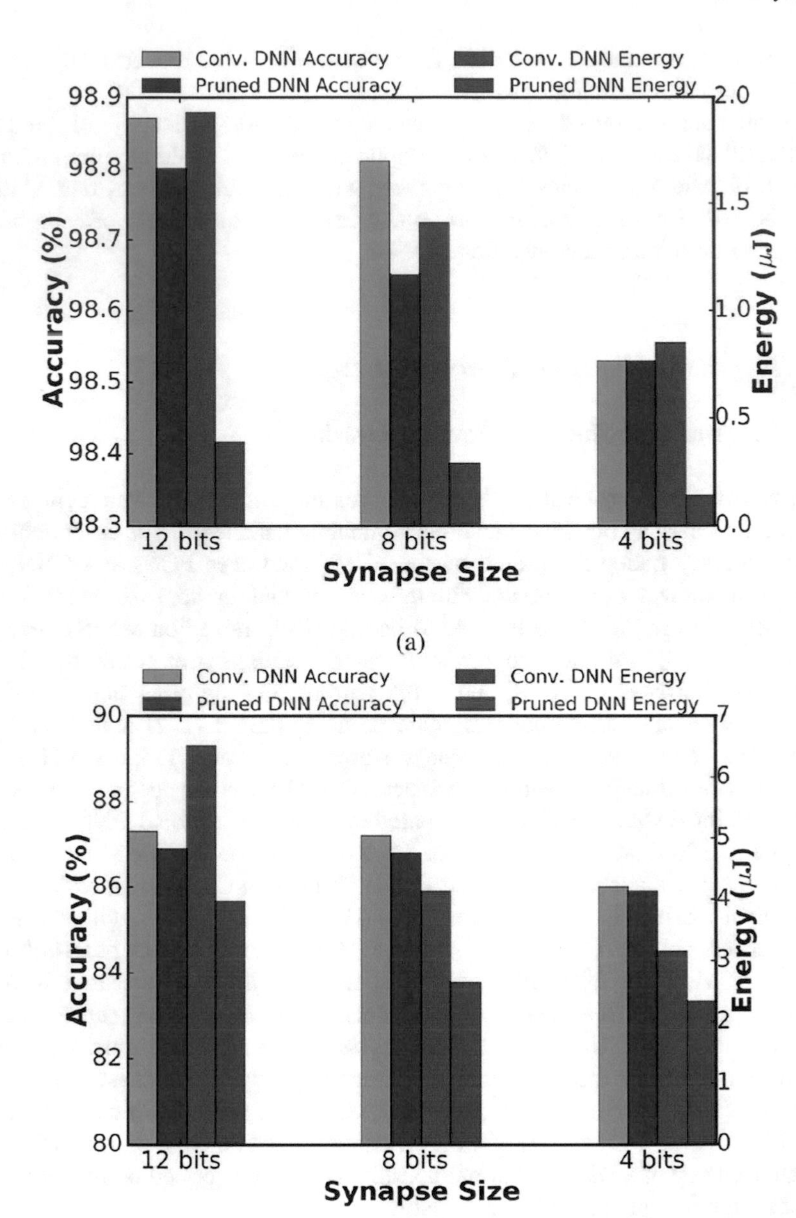

Fig. 15.17 Energy/accuracy trade-off comparison, between pruned DNNs and conventional DNNs, is shown for (**a**) MNIST trained on Deep FCN and (**b**) CIFAR-10 trained on Deep CNN, for different synapse sizes

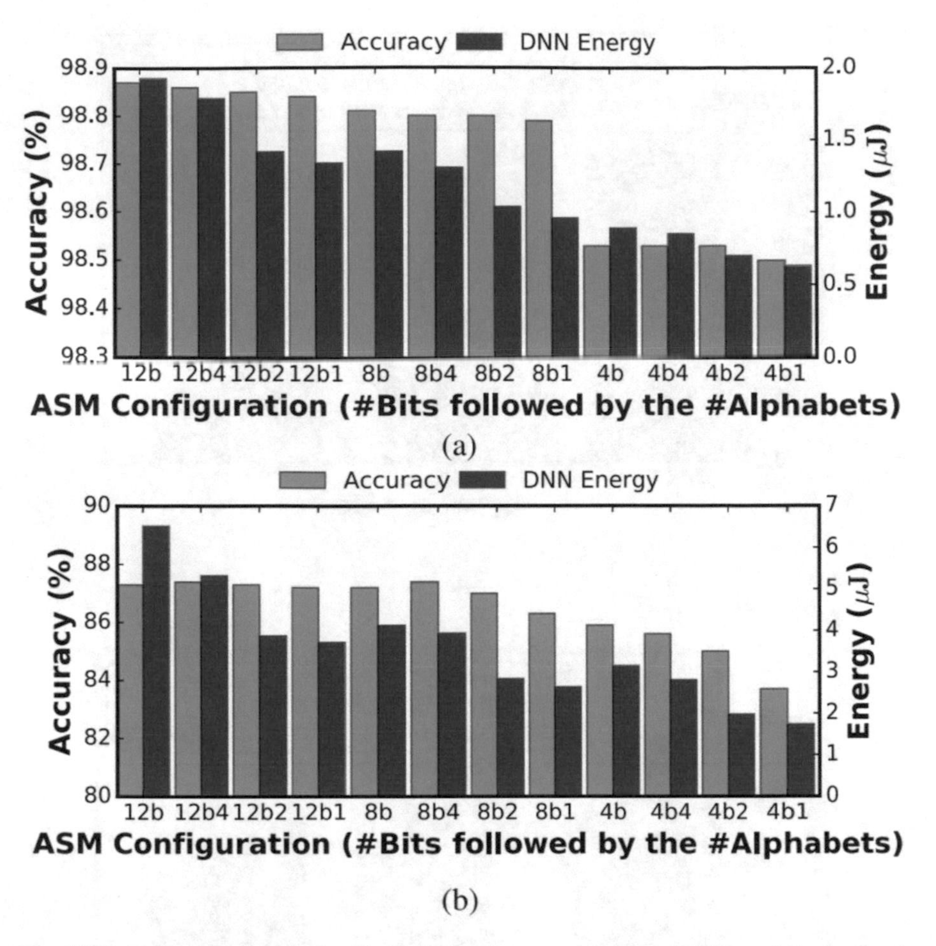

Fig. 15.18 Energy/accuracy trade-off comparison between ASM based DNNs and Conventional DNNs, for (**a**) MNIST dataset on FCN and (**b**) CIFAR 10 dataset on CNN. 12b: 12 bit synapse NN, 12b4: 12 bit synapse NN with 4 alphabet ASM, 12b2: 12 bit synapse NN with 2 alphabet ASM, 12b1: 12 bit synapse NN with 1 alphabet ASM. Similar notations for 8 bit and 4 bit NNs

resulted in substantial accuracy degradation. Therefore, the synaptic sensitivity driven hybrid memory architecture was used for aggressive voltage scaling.

Three different hybrid memory configurations are discussed, corresponding to a synaptic precision of 12 bits, 8 bits, and 4 bits, respectively. Each memory configuration consists of three 8T-6T SRAM banks to store the weight of synapses interconnecting every pair of layers of the deep FCN under investigation. The number of 8T MSB cells in each bank (shown in Table 15.4) is determined based on synaptic sensitivity as described in Sect. 4.3.

One simulation indicated that the hybrid 8T-6T SRAM enable the voltage to be scaled down to 0.80 V, which offers improved memory access energy efficiency compared to an all-6T SRAM that needs to be operated at 0.85 V for negligible

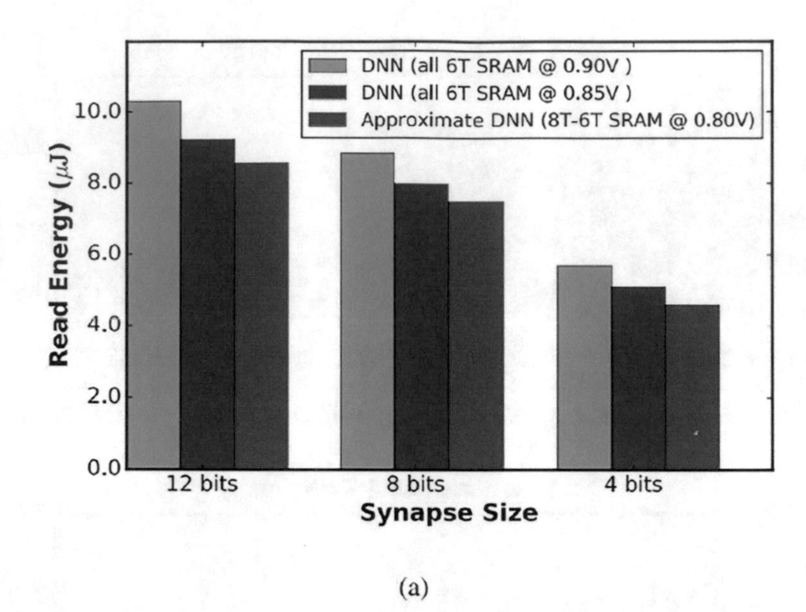

(a)

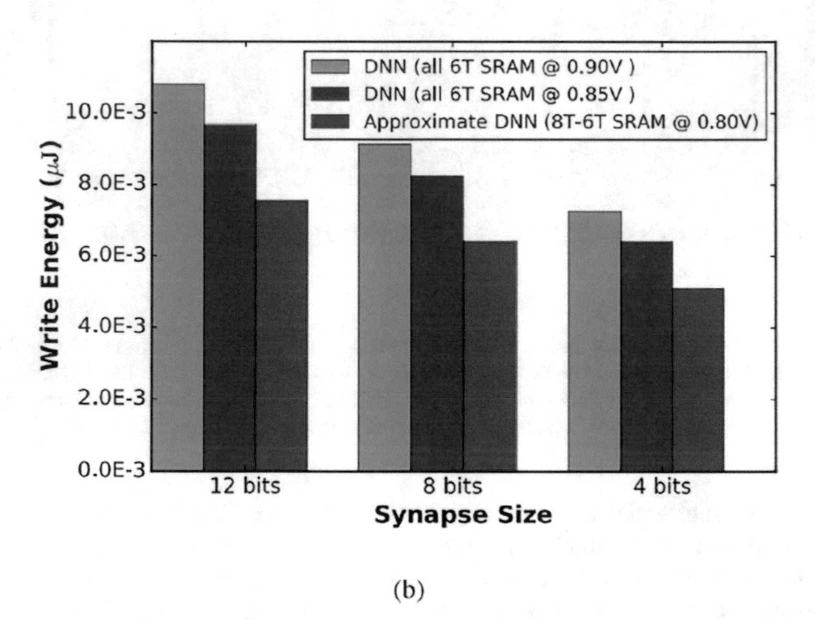

(b)

Fig. 15.19 Comparison of total memory (**a**) read energy and (**b**) write energy, for classification of one image under iso-accuracy condition. Accuracy for 12, 8, and 4 bit networks are $\sim$ 98.9%, $\sim$ 98.8%, and $\sim$ 98.5%, respectively, with degradation up to 0.5% for 12 bit and 8 bit synapse, and $\sim$ 1.25 for 4 bit synapse

Table 15.4 Synaptic sensitivity driven hybrid memory configuration

Synaptic precision	8T MSBs in Bank-1	8T MSBs in Bank-2	8T MSBs in Bank-3
12 bits	4	3	5
8 bits	3	2	3
4 bts	1	1	2

accuracy degradation. This is corroborated by Fig. 15.19, which shows that the deep FCN under consideration using the hybrid 8T-6T SRAM (operating at 0.80 V) consumes lower memory access energy relative to an all-6T SRAM (operating at 0.85 V). However, the improvement in energy consumption decreases as the synaptic precision is lowered from 12 bits to 4 bits. This can be attributed to a reduction in the complexity of the memory peripheral (decoding and sensing) circuitry. The reduced complexity minimizes the improvement in memory access energy achieved by operating at 0.80 V relative to that expended at 0.85 V. It should be noted that deep CNNs (trained on CIFAR-10) also demonstrate a similar trend.

It should be noted that this approximation is non-deterministic, as a result of which retraining cannot be used to retrieve portion of the lost accuracy. Therefore, the voltage was scaled down only till that point where the accuracy degradation is marginal ($< 0.5\%$ for 12 bit and 8 bit synapses, and $\sim 1.25\%$ for 4 bit synapse), for a fair iso-accuracy comparison with the conventional DNN. It was found that the accuracy could occasionally improve ($< 1\%$) even after applying memory approximation due to the non-deterministic nature of the approximation.

6.6 Combined Approximate Networks

The results thus far indicate that approximate DNNs demonstrate improved energy efficiency compared to conventional DNNs both in terms of computation and memory access energy with minimal accuracy degradation. In this subsection, the combination of all the approximation techniques together is discussed. It was found that, the all-inclusive approximate DNNs incurred greater accuracy loss. The memory approximation is the dominant factor for the accuracy degradation since the accuracy lost due to voltage scaling could not be regained by retraining the network. Approximate memory also gave rise to a good amount of non-zero weight values which were previously pruned. This causes reduction in the energy savings, which was achieved in computation by pruning. Even with these setbacks, the combined network provides good amount of energy savings both in computation and memory access. This is illustrated in Fig. 15.20a, which shows that approximate DNNs (FCN trained on MNIST) achieve significant amount of energy savings with synapse pruning, approximate multiplication and approximate memory being applied simultaneously. They provide up to 7.85$\times$ improvement in the computation energy over conventional DNNs of equivalent bit precision.

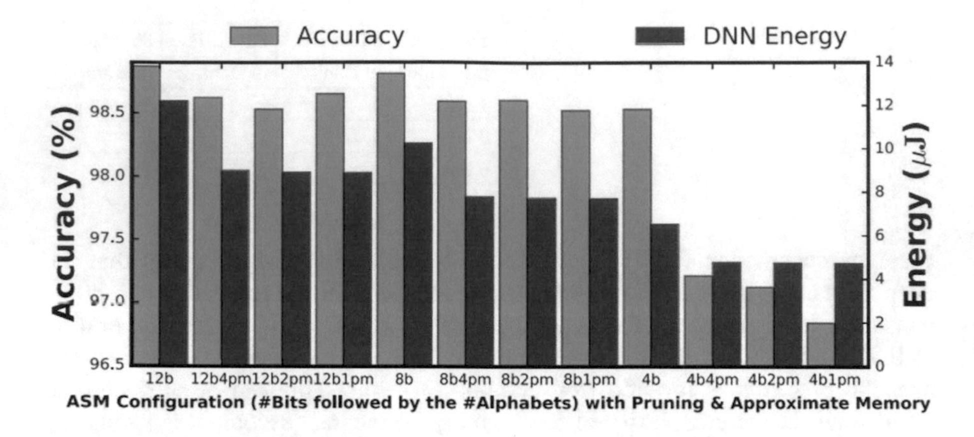

(a)

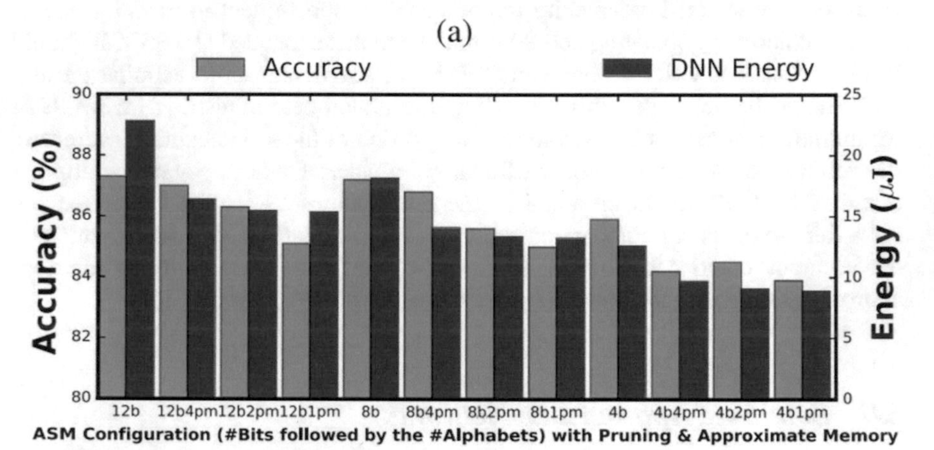

Fig. 15.20 Energy/accuracy trade-off comparison between approx. DNNs and conventional DNNs, for (**a**) MNIST dataset on FCN and (**b**) CIFAR 10 dataset on CNN, where pruning of synapses, approx. multipliers and approx. memory are used simultaneously. 12b:12 bit synapse NN, 12b4pm: 12 bit synapse NN with 4 alphabet ASM, pruning and approximate memory. Similar notations for 8 bit and 4 bit NNs

Approximate Deep CNNs (trained on CIFAR-10) also indicate a similar trend (Fig. 15.20b) and offer up to 2.76× reduction in the computation energy consumption over conventional DNNs of equivalent bit precision. However, in Fig. 15.20a, all the energy consumption bars for the approximate networks look similar as all of them have same memory access energy (for their respective bit precision), which dominates the total energy consumption. On the other hand, in Fig. 15.20b, the energy consumption bars for the approximate networks are distinct since the memory access energy is less dominant in CNNs. Both Approximate DNNs (FCN and CNN) provide up to 19% read energy and 30% write energy savings compared to DNNs operated at nominal voltage (Fig. 15.19).

It is not necessary to combine all the approximation techniques. Based on the energy-accuracy requirements, one or more approximation techniques can be applied.

7 Comparing Approximate Networks

In this section, a comparison is presented between the conventional (un-approximated) DNNs and approximated DNNs in terms of the neuronal computation energy, and synaptic memory accesses. A comparative analysis between the respective NNs is performed in order to validate the effectiveness of the discussed approximations in context of neuromorphic applications.

Approximations cause accuracy degradations and it has an inverse relationship with energy savings. The extent of accuracy degradation is mostly dependent on the intensity of approximations. At the same time, bit precision of the neurons and synapses therefore plays an important part in controlling the energy-accuracy trade-off. Different approximations provide varied energy-accuracy trade-offs. Depending on the energy-accuracy specifications for a given application, different approximations need to be employed. In this subsection, comparison of the discussed approximations is presented in an attempt to help the readers find out maximal beneficial approximation method among different combination schemes.

Figure 15.21 shows energy-accuracy trade-off comparison between Conventional DNNs and different approximate DNNs listed below:

- Conventional DNN (un-approximated baseline)
- Pruned DNN (algorithm level approximation)
- Approximate Memory based DNN (hardware level approximation)
- Approximate Multiplier based DNN (hardware level approximation)
- Low Complexity Network (algorithm level approximation)
- DNN with Approximate Multiplier and Memory
- DNN with Approximate Multiplier and Pruning
- NN with Approximate Memory and Pruning
- Combined Approximated DNN (pruning, approx. multiplication and memory)

From Fig. 15.21a, it is observed that the low complexity networks cannot match the higher baseline accuracy. For high accuracy requirements, pruning and approximate multiplier based DNNs provide better energy efficiency compared to low complexity networks. For low accuracy requirements, low complexity networks are better candidates, if only computation energy consumption is considered. Such substantial deterioration in the accuracy of low complexity networks is not acceptable for practical purposes. Moreover, in Fig. 15.21a, it is observed that memory approximated networks provide better overall energy efficiency compared to low complexity networks. The high accuracy regime of the energy-accuracy graph is mostly considered over the low complexity networks.

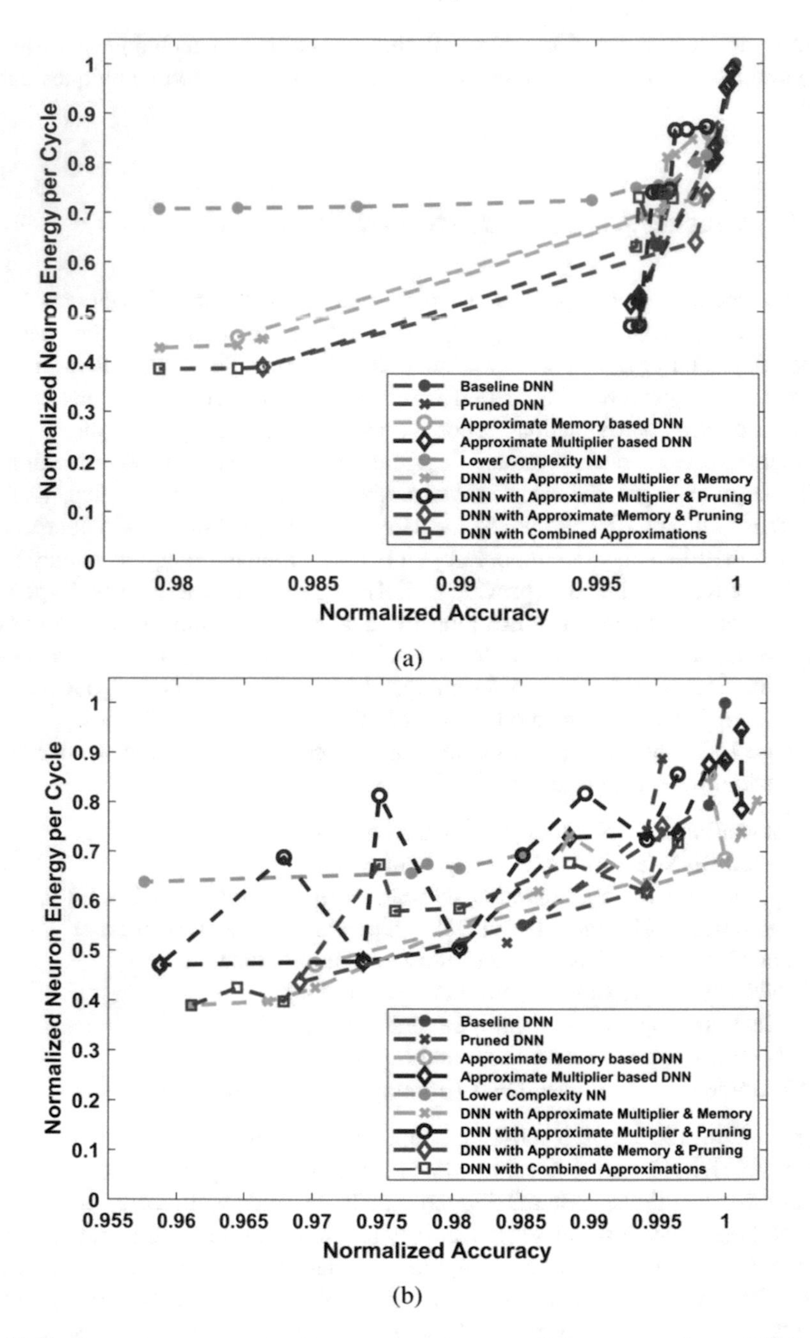

Fig. 15.21 Energy/accuracy trade-off comparison between different approximate DNNs and Conventional DNNs, for (**a**) MNIST dataset on FCN and (**b**) CIFAR-10 dataset on CNN. Here baseline for normalization is a 12 bit un-approximated DNN

Approximate multiplier based DNNs provide good amount of computation energy savings with minimal accuracy loss. Note, pruned DNNs provide better computation energy efficiency than all other approximation techniques. Pruning of synapses along with approximate multipliers can be used to achieve greater savings in computational energy for less than 0.15% drop in accuracy. Note that the order of the approximations (pruning after or before applying weight restrictions for approximate multiplier) does not matter, since retraining is applied after employing both approximations. On the other hand, Approximate memory based DNNs consume similar computation energy as conventional DNNs with slightly lower classification accuracy. However, they provide good amount of memory energy savings.

The combination of three approximation techniques (pruning, approximate multiplier and memory) provides slightly lower computational energy efficiency than the combination of pruned and approximate multiplier based network. This can be attributed to the fact that the memory approximation introduces bit corruption, which converts some of the pruned weights to non-zero values, and thus reduces the computation energy savings. However, memory approximation individually offers large amount of savings in memory access energy as total energy consumption for a deep FCN is greatly dominated by memory accesses.

In Fig. 15.21b, a similar trend can be observed in the case of deep CNN trained on CIFAR-10 dataset. One major difference is the comparatively lesser domination of memory access energy on the total energy consumption due to weight sharing in the convolutional kernels. Another distinctive feature is the less effectiveness of pruning as most of the kernel values are non-zero.

Conclusion The ever-growing complexity of the state-of-the-art deep neural networks (DNNs) together with the explosion in the amount of data to be processed, place significant energy demands on the computing platforms. Approximations at the algorithmic and hardware level can provide energy savings while incurring tolerable accuracy degradation. Retraining the approximate networks, with the approximations in place, helps in mitigating the accuracy loss. This chapter explored algorithmic and hardware level approximations to determine their effectiveness in achieving energy improvements while maintaining the output quality. In particular, synapse pruning, approximate neuronal multiplication and addition, voltage-scaled memory for synaptic storage, along with a brief discussion on analog in-memory computing were presented. Lower complexity networks were also discussed to explore network approximations. The efficacy of the approximations were evaluated by comparing the energy benefits with that of optimized DNNs (without approximations). Algorithm (Pruning) and Hardware (Approximate Multiplication, Approximate Memory) level approximations are energy-efficient for high accuracy requirements, while lower complexity networks are beneficial for low accuracy requirements. Algorithm (Pruning) and Hardware (Approx. Multiplication and Approx. Memory) level approximations can be combined to get higher energy savings while maintaining reasonable quality. On the contrary, low complexity networks, even though energy-efficient, incur severe accuracy loss. The results presented clearly indicate that employing properly selected approximations on DNNs leads to improved energy efficiency with competitive classification accuracy.

References

1. Liu W, Lombardi F, Schulte M. Approximate computing: from circuits to applications. Proc IEEE. 2020;108(12):2103–7
2. Esmaeilzadeh H, Blem E, Amant RS, Sankaralingam K, Burger D. Dark silicon and the end of multicore scaling. In: 2011 38th annual international symposium on computer architecture (ISCA), 2011 Jun 4. New York: IEEE; 2011, pp 365–376
3. Venkataramani S, Chakradhar ST, Roy K, Raghunathan A. Approximate computing and the quest for computing efficiency. In: Proceedings of the 52nd Annu. Des. Automat. Conf., San Francisco, CA, Jun 2015; pp. 120-1–120-6
4. Chippa VK et al. Analysis and characterization of inherent application resilience for approximate computing. In Proceedings of the DAC; 2013
5. Krizhevsky A, Sutskever I, Hinton GE. ImageNet classification with deep convolutional neural networks. Adv Neur Inf Process Syst 2012;25:1097–105
6. Park J, Choo H, Muhammad K, Choi S, Im Y, Roy K. Non-adaptive and adaptive filter implementation based on sharing multiplication. In IEEE international conference on acoustics, speech, and signal processing (ICASSP); 2000, pp 460–463
7. Sarwar SS, Venkataramani S, Raghunathan A, Roy K. Multiplier-less artificial neurons exploiting error resiliency for energy-efficient neural computing. In 2016 design, automation & test in Europe conference & exhibition (DATE); 2016, pp 145–150
8. Gupta V, Mohapatra D, Park SP, Raghunathan A, Roy K. IMPACT: IMPrecise adders for low-power approximate computing. In IEEE/ACM international symposium on low power electronics and Design 2011 Aug 1. New York: IEEE; 2011, pp 409–414
9. Rabaey JM. Digital integrated circuits: a design perspective. Upper Saddle River, NJ: Prentice-Hall, Inc.; 1996
10. Nassif SR. Modeling and analysis of manufacturing variations. In Proc. IEEE conf. custom integr. circuits, May 2001, pp 223–228
11. Visweswariah C. Death, taxes and failing chips. In Proceedings of the 40th annual design automation conference (DAC); 2003, pp 343–347
12. Borkar S, Karnik T, Narendra S, Tschanz J, Keshavarzi A, De V. Parameter variations and impact on circuits and microarchitecture. In Proceedings of the 40th annu. des. automat. conf.; 2003, pp 338–342
13. Srinivasan G, Wijesinghe P, Sarwar SS, Jaiswal A, Roy K. Significance driven hybrid 8T-6T SRAM for energy-efficient synaptic storage in artificial neural networks. In Design, automation & test in Europe conference & exhibition (DATE), 2016 Mar 14. New York: IEEE; 2016, pp 151–156
14. Verma N, Jia H, Valavi H, Tang Y, Ozatay M, Chen LY, Zhang B, Deaville P. In-memory computing: advances and prospects. In IEEE solid-state circuits magazine, 2019 Aug 23; 2019, pp 11(3):43–55
15. Jouppi NP, Young C, Patil N, Patterson D, Agrawal G, Bajwa R, Bates S, Bhatia S, Boden N, Borchers A, Boyle R. In-datacenter performance analysis of a tensor processing unit. In Proceedings of the 44th annual international symposium on computer architecture (ISCA) 2017 Jun 24; 2017, pp 1–12
16. Sharma H, Park J, Suda N, Lai L, Chau B, Chandra V, Esmaeilzadeh H. Bit fusion: bit-level dynamically composable architecture for accelerating deep neural network. In ACM/IEEE 45th annual international symposium on computer architecture (ISCA), 2018 Jun 1; 2018, pp 764–775

17. Ryu S, Kim H, Yi W, Kim JJ. Bitblade: area and energy-efficient precision-scalable neural network accelerator with bitwise summation. In Proceedings of the 56th annual design automation conference (DAC), 2019 Jun 2; 2019, pp 1–6
18. Ankit A, Hajj IE, Chalamalasetti SR, Ndu G, Foltin M, Williams RS, Faraboschi P, Hwu WM, Strachan JP, Roy K, Milojicic DS. PUMA: a programmable ultra-efficient memristor-based accelerator for machine learning inference. In Proceedings of the twenty-fourth international conference on architectural support for programming languages and operating systems, 2019 Apr 4; 2019, pp. 715–731.
19. Hu M, Graves CE, Li C, Li Y, Ge N, Montgomery E, Davila N, Jiang H, Williams RS, Yang JJ, Xia Q. Memristor-based analog computation and neural network classification with a dot product engine. Adv Mater. 2018;30(9):1705914
20. Hu M, Strachan JP, Li Z, Grafals EM, Davila N, Graves C, Lam S, Ge N, Yang JJ, Williams RS. Dot-product engine for neuromorphic computing: programming 1T1M crossbar to accelerate matrix-vector multiplication. In 53nd ACM/EDAC/IEEE design automation conference (DAC), 2016 Jun 5. New York: IEEE; 2016, pp 1–6
21. Sanyal S, Ankit A, Vineyard CM, Roy K. Energy-efficient target recognition using ReRAM crossbars for enabling on-device intelligence. In IEEE workshop on signal processing systems (SiPS); 2020, pp 1–6
22. Chakraborty I, Ali MF, Kim DE, Ankit A, Roy K. Geniex: a generalized approach to emulating non-ideality in memristive xbars using neural networks. In 57th ACM/IEEE design automation conference (DAC), 2020 Jul 20; 2020, pp. 1–6
23. Reagen B, et al. Minerva: enabling low-power, highly-accurate deep neural network accelerators. In Proceedings of 43rd international symposium of computer architecture (ISCA); 2016, pp 267–278
24. Han S, Mao H, Dally WJ. Deep compression: compressing deep neural networks with pruning, trained quantization and Huffman coding. [Online]; 2015. Available: https://arxiv.org/abs/1510.00149
25. Sarwar SS, Srinivasan G, Han B, Wijesinghe P, Jaiswal A, Panda P, Raghunathan A, Roy K. Energy efficient neural computing: a study of cross-layer approximations. IEEE J Emerg Select Top Circ Syst. 2018;8(4):796–809
26. Muralimanohar N, Balasubramonian R, Jouppi NP. CACTI 6.0: a tool to model large caches. HP Lab., Palo Alto, CA; 2009, pp 22–31
27. Palm RB. Prediction as a candidate for learning deep hierarchical models of data. M.S. thesis, Tech. Univ. Denmark, Denmark, Lyngby; 2012, vol 5.
28. Vedaldi A, Lenc K, MatConvNet: convolutional neural networks for MATLAB. In Proc. 23rd ACM int. conf. multimedia; 2015, pp 689–692
29. Lin M, Chen Q, Yan S. Network in network; 2013 [Online]. Available: https://arxiv.org/abs/1312.4400

Chapter 16
Approximate Computing for Efficient Neural Network Computation: A Survey

Hao Zhang, Mohammadreza Asadikouhanjani, Jie Han, Deivalakshmi Subbian, and Seok-Bum Ko

1 Introduction

Neural network (NN) is one of the most popularly used artificial intelligence (AI) technologies. From the early stage artificial neural network (ANN) [1] to the current deep neural network (DNN) [2, 3] and the next generation spiking neural network (SNN) [4], NN has achieved great performance in many fields of applications. To pursue better accuracy, the NN models become deeper and the NN algorithm becomes more complex which make the NN computation intensive and memory intensive. In addition, the recent emerging edge intelligence [5] computing paradigm brings AI to the edge and embedded devices which have limited computing power and tight power budget. All these factors raise a high efficiency requirement for NN processing hardware. With Moore's law [6] and the Dennard scaling [7] approaching to the end, improving performance and energy efficiency through

H. Zhang
Faculty of Information Science and Engineering, Ocean University of China, Qingdao, China
e-mail: hao.zhang@ouc.edu.cn

M. Asadikouhanjani · S.-B. Ko (✉)
Department of Electrical and Computer Engineering, University of Saskatchewan, Saskatoon, SK, Canada
e-mail: m.asadi@usask.ca; seokbum.ko@usask.ca

J. Han
Department of Electrical and Computer Engineering, University of Alberta, Edmonton, AB, Canada
e-mail: jhan8@ualberta.ca

D. Subbian
Department of Electronics and Communication Engineering, National Institute of Technology, Tiruchirappalli, TN, India
e-mail: deiva@nitt.edu

W. Liu, F. Lombardi (eds.), *Approximate Computing*,
https://doi.org/10.1007/978-3-030-98347-5_16

advancing semiconductor technology becomes difficult. Under this situation, the design of novel computer architecture or application-specific architecture becomes imperative to meet the requirements of those emerging applications [8]. In the past few years, a lot of research efforts have been put to the design of architectures that are specific for NN processing [9]. Due to the inherent error tolerance of NN models, approximate computing [10] has become an effective methodology in NN processing unit design to achieve improved performance and energy efficiency.

Approximate computing method usually applies simplified logic circuits or simplified algorithms to trade off computing results accuracy for better computing performance and energy efficiency. This method is quite different from traditional computing methods where a fully accurate result is pursued. As the computing results are inaccurate, approximate computing may not be desired in every applications. However, in some applications, such as image and audio processing, as human perception is not sensitive to small variance in results, those errors derived from approximate computing can be tolerated. In AI and NN applications, the model prediction is generated from the comparison of the final layer's outputs. As a result, as long as the error introduced by approximate computing does not change the order of those outputs, a correct prediction can still be made by the model. In some application scenarios where accuracy requirement is not that strict, more aggressive approximate computing strategy can be applied to achieve more efficiency improvement.

The design of an NN processing system usually contains three design layers: the algorithm, the hardware, and the software. The NN model is developed in the algorithm layer. The corresponding processing hardware then belongs to the hardware layer. Between these two, in the software layer, compiler or optimization tool is used to map the NN model to the processing hardware. In the literature, approximate computing has been applied in all three layers of NN processing systems [11, 12]. In the algorithm layer, precision scaling [13], which uses reduced precision numeric formats, and model compression [14], which removes unnecessary synaptic connections, are widely used. In the hardware layer, architectures that support algorithm-level approximation [15, 16] are designed. In addition, many approximate arithmetic unit architectures, such as [17], and approximate computing in memory architecture [18] are proposed for the hardware design layer. Finally, in the software layer, NN compilers and optimization tools which incorporate approximate computing methods [19, 20] are proposed in the literature. In addition to these design techniques, voltage scaling [21] and approximate logic synthesis [22] are applied in general approximate computing but are not yet used in NN processing.

Several research papers in the literature have reviewed the status of approximate computing techniques in NN processing [12, 18, 23–25]. In [23] and [24], approximate methods in the algorithm layer are reviewed. The use of approximate memory in NN processing is presented in [18]. In [25] and [12], approximate computing techniques applied across multiple design layers are presented. However, to the best of our knowledge, there is no paper reviewing the status of using approximate arithmetic units in NN processing. Designing efficient arithmetic unit is an effective way to improve the efficiency of NN processing [26–28, 30]. It can also be used in conjunction with other approximate computing techniques to achieve further efficiency improvement. To facilitate the design of approximate arithmetic units for NN processing, in this chapter, the use of approximate arithmetic units in NN processing is discussed and summarized as a complementary review to the current literature. As the multiplier is more hardware expensive than adder and accurate accumulation is required in NN computation [29, 31], the design of approximate multipliers, both in the linear domain and in the logarithm domain, for multi-layer perceptron models and deep neural networks will be discussed. Then, the current findings of effective characteristics of approximate multiplier for NN computation will be summarized, and the design challenges and some potential future research directions are presented.

The rest of this chapter is organized as follows: Sect. 2 presents the background information including the computation in NN and the feasibility of applying approximate computing in NN. The approximate multipliers for NN processing are presented in Sect. 3. The design challenges and future research directions are presented in Sect. 4. Finally, Sect. 5 concludes the whole chapter.

2 Background

As the approximate arithmetic units reviewed in this chapter are designed for neural network computation, it is necessary to understand the computations performed in neural networks and their characteristics. In this section, the computations in neural networks and their characteristics will be introduced which will guide the design of approximate arithmetic units to make the computations more efficient.

In addition, to help evaluate the approximate multipliers and their performance in neural network computation, some popularly used evaluation metrics will be discussed as well.

2.1 Neural Network Computation

Neural network is composed of multiple layers of neurons. It is widely used in computer vision applications. In the literature, feedforward neural networks with no more than three layers are commonly referred to as artificial neural network (ANN) or multi-layer perceptron (MLP). Figure 16.1 presents an example of a 3-layer ANN. We will use this example to explain the computation of neural network. The ANN shown in Fig. 16.1 is trained to classify handwritten digit images in the MNIST dataset [32]. The images from the MNIST dataset has a spatial resolution of 28×28, and thus, there are 784 neurons in the input layer. Each of them corresponds to one image pixel. The hidden layer usually contains more neurons than the input layer to process image features. As the neural network is to classify handwritten digits, there are 10 neurons in the output layer. Each of them corresponds to the probability of the input being the corresponding index number, as shown in Fig. 16.1. The ten probability values will be compared and the index corresponding to the largest

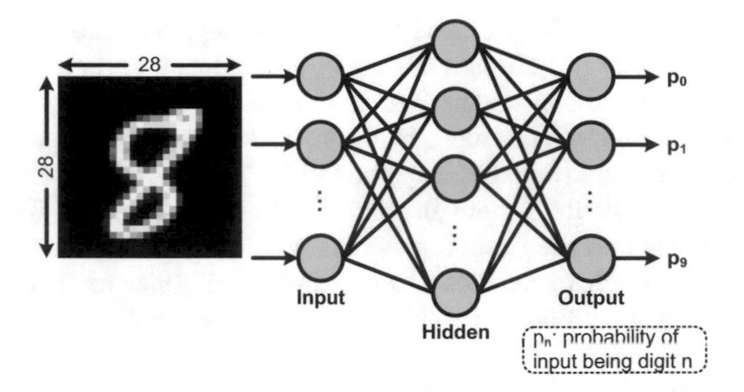

Fig. 16.1 Example neural network in handwritten digit recognition

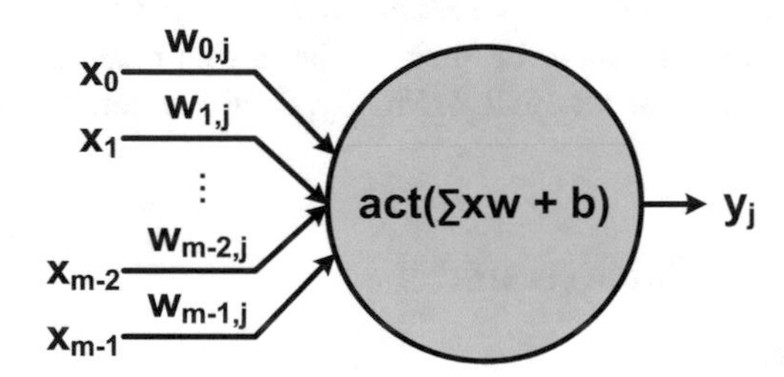

Fig. 16.2 Model of neurons in neural networks

probability is considered as the prediction for the input image. Neurons in these three layers are connected in a fully connected manner, where a neuron in one layer is connected to all neurons in the adjacent layer.

The model of the neurons in neural networks is shown in Fig. 16.2. In Fig. 16.2, x represents the value of a precedent neuron which is also the input to the current neuron, w is the synaptic weight, and y is the output of the current neuron. Suppose the current neuron j is connected to m precedent neurons, its output can be computed with Eq. 16.1:

$$y_j = act(\sum_{i=0}^{m-1} x_i \cdot w_{i,j} + b) \tag{16.1}$$

where $w_{i,j}$ represents the weight between precedent neuron i and current neuron j, b is the bias value, and $act()$ is a nonlinear activation function. The commonly used activation functions are sigmoid function and rectified linear unit (ReLU). The weight w and bias b are obtained during the training process. To calculate the hidden layer neurons and output layer neurons shown in Fig. 16.1, Eq. (16.1) is repeatedly used with corresponding weights and inputs.

According to the computation process, the main operation in ANN or MLP is the dot product between the inputs and the weights. The number of terms of the dot product is determined by the amount of neurons in the input layer, while the amount of dot products is related to the amount of neurons in the current layer. The

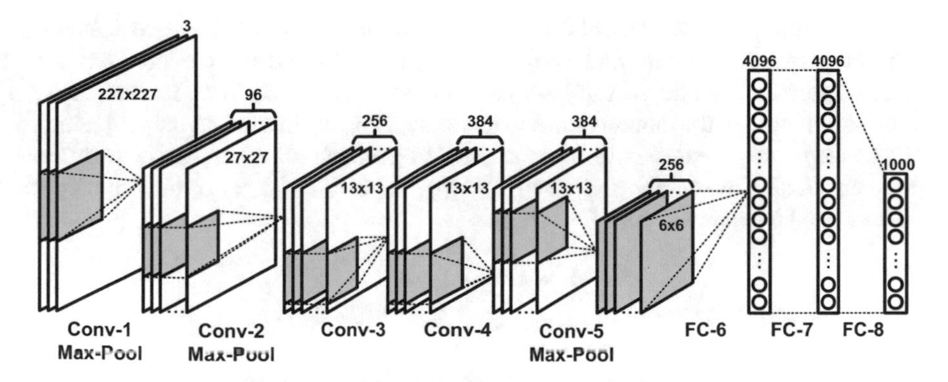

Fig. 16.3 Architecture of the AlexNet model

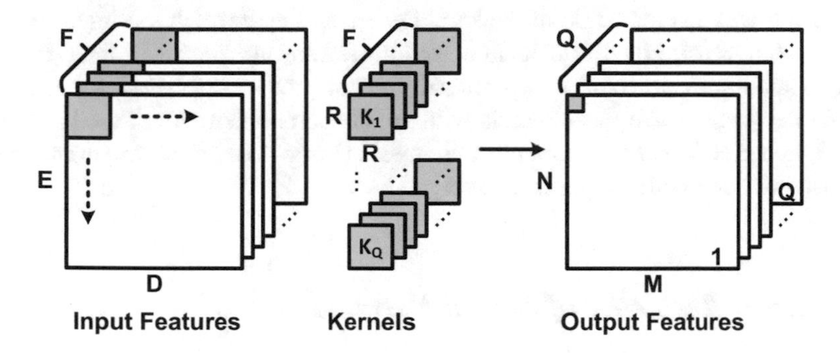

Fig. 16.4 Convolution operations in deep neural networks

dot product can be further decomposed into multiplication operation and addition operation.

When an NN has more than 3 layers, it can be classified as deep neural network (DNN). In recent years, DNN can achieve better performance than ANN and MLP in many applications and gradually becomes the state-of-the-art AI technology. The architecture of AlexNet [33], which is a popular DNN model, is shown in Fig. 16.3. As shown in Fig. 16.3, DNN uses convolution layers to extract features and then uses fully connected layers to perform classification. The classification results are generated in the same way as ANN and MLP where the outputs of the final layer are compared.

The convolution operation is essentially a dot-product operation, but unlike the fully connected operation, only part of the inputs are used to multiply with kernels (weights) to generate output. The same kernel will slide through the whole inputs to complete the computation of all outputs. The convolution operation is depicted in Fig. 16.4.

In Fig. 16.4, Q kernels of size $R \times R \times F$ are applied to input features of size $E \times D \times F$ to generate output features of size $M \times N \times Q$. Here, F and Q usually refer to the amount of input channels and output channels, respectively.

During computation, the kernel K_1 is applied to the top left of the input features. The corresponding inputs and weights are multiplied and all the products are accumulated to generate the top-left pixel in the first output feature. The kernel K_1 then slides through the input features to generate all pixels in the first output feature. These steps are repeated with other kernels to generate all other output features. Mathematically, the output pixel $OF[m][n][q]$ in position (m, n) of the qth output feature can be computed with Eq. (16.2):

$$OF[m][n][q] = act(\sum_{ch=0}^{F-1} \sum_{kh=0}^{R-1} \sum_{kw=0}^{R-1} K_q[kw][kh][ch] \cdot IF[m \cdot S + kw][n \cdot S + kh][ch] + B[q]) \tag{16.2}$$

where S represents the stride of the kernel when sliding through the input features, B is the bias shared by all pixels in an output feature. As shown in Eq. (16.2), the main operation in convolution layer is still the dot-product operation. Modern deep neural networks usually use kernels with small spatial resolutions, such as 1×1, 3×3, and 5×5, but the amount of channel is large. Therefore, the computation intensity of convolution layer is quite high.

2.2 *Error Resilience of Neural Networks*

Error resilience is the main factor that can determine whether approximate computing can be applied. Neural network is error resilient and the resilience comes from many factors [12, 34].

As mentioned above, neural networks generate final results by comparing the output of the final layer neurons. In other words, the result is not determined by a specific (fixed) value. As a result, even though approximate computing is introduced, as long as the order comparison among output neurons is not changed, the neural network can still generate correct results. This characteristic is analyzed in [34] where all results are affected by noises after applying their approximate multiplier, but the order of the outputs does not change for most of the cases, and thus the accuracy of the neural network is not affected.

In another case, there exist approximate arithmetic units that can introduce unbiased errors to the results. That means the approximate units will make some of the results become larger, while the others become smaller. If the amount of multiplication cases is large enough, after accumulating all products, the errors can be canceled out. This is the case for some modern DNN where the amount of accumulation to generate a single output is large.

In addition, neural networks have nonlinear activation functions, such as ReLU and sigmoid. In ReLU, the negative computation results are directly set to 0. In sigmoid, the output values are restricted in a certain range. With a successful training process, these changes of computation results do not affect the accuracy of neural network inference. Therefore, there is a large space to apply approximate computing by utilizing the application of nonlinear activation functions [12].

Moreover, in some applications, such as image captioning or image super-resolution, the quality of the result is subjective and more than one result can be treated as correct. In some other applications, accuracy requirements are not that strict and the top-5 accuracy can be used to take the place of the top-1 accuracy. In these cases, inexact computation will not lead to quality degradation, and thus approximate computing can be applied.

2.3 Evaluation Metrics

Approximate computing may introduce errors to the computation results. These errors are usually measured by some metrics [35], such as error rate (ER), error distance (ED), mean error distance (MED), absolute error distance (AED), and relative error distance (RED).

ER is the probability of the occurrence of the erroneous results. ED is the difference between the approximated result and the exact result. MED is the mean value of EDs of all possible output values in the design. AED is the absolute value of ED. RED is the ED relative to the value of the exact result. The formulas of these metrics are defined in Eq. (16.3):

$$\begin{aligned} ED &= E - A \\ AED &= |E - A| \\ MED &= \frac{1}{N}\sum_{i=1}^{N}(E_i - A_i) \\ RED &= 1 - \frac{A}{E} \end{aligned} \tag{16.3}$$

where E represents the exact result, A is the approximated result, and N is the total number of possible results. In NN computation, the root mean square of the ED (RMS-ED) and the variance of ED (Var-ED) are also important [36] and they are defined by Eq. (16.4):

$$\begin{aligned} RMS\_ED &= \sqrt{\frac{1}{N}\sum_{i=1}^{N}(E_i - A_i)^2} \\ Var\_ED &= \frac{1}{N}\sum_{i=1}^{N}(ED_i - \frac{1}{N}\sum_{i=1}^{N}ED_i)^2 \end{aligned} \tag{16.4}$$

These metrics are used to evaluate the single approximate multiplier unit. For the whole NN model, the accuracy after applying approximate multipliers or the normalized accuracy will be used to measure the performance of using the approximate multiplier in the NN model. The normalized accuracy is defined by the ratio of the accuracy of the NN model after applying approximate arithmetic units over the accuracy of the NN model when using exact units.

As the main purpose of applying approximate arithmetic units is to reduce energy consumption, therefore, in terms of hardware metrics, the power delay product (PDP) is the most popularly used one. An ideal approximate arithmetic unit can significantly reduce the PDP while maintaining a similar or even better accuracy for the NN model.

3 Approximate Multiplier for Neural Network Computation

According to the discussion in Sect. 2, the main computation in NN processing is the dot-product operation which can be further divided into multiplication and addition operations. Multiplier is hardware expensive in terms of area and power consumption. As a result, many research works focused on the design of approximate multiplier to reduce the hardware cost. Compared to the multiplier, adder is less expensive in hardware resources. In addition, many research works found that an accurate accumulation was necessary to maintain the quality of the results for NNs [30, 31]. Therefore, for approximate NN computation, usually approximate multipliers are used together with an exact addition unit. In this section, approximate multipliers used in NN computation are reviewed.

For NN computation, many binary approximate multipliers designed in linear domain are applied. The use of those multipliers in NN is evaluated and characteristics that are suitable for NN computing are proposed. In addition to linear domain multiplication, as multiplication in linear domain can be converted to a simple addition in logarithm domain, many works focus on the approximate logarithm multiplier designs for NN computations.

3.1 Linear Domain Approximate Multiplier

Many approximate multipliers have been proposed for a wide variety of applications [10]. Some of them have been successfully applied in NN applications. Approximate multipliers that are designed specific for NN computation are also available in the literature. In this section, the design methodologies of the approximate multipliers that have been applied for NN computation are presented. After that, the method to select an optimal approximate multiplier among all available options is discussed.

3.1.1 Design Methodology

The binary multiplication in the linear domain contains three steps: partial product generation where each bit of the multiplier will be ANDed with the multiplicand, partial product accumulation where partial products are condensed columnwise to two vectors, and final addition by adding the two vectors to generate the final product. The process to generate $P = A \times B$ is shown in Fig. 16.5a.

In the linear domain, truncated multiplier is one of the popular approximate design choices. In the truncated multiplier, several least significant bits of the partial products are truncated to reduce energy consumption. The truncated multipliers proposed in [37] and [38] are shown in Fig. 16.5b. In this design, only the most significant $(n+k)$-bit result will be reserved, while the least significant $(n-k)$-bit of the partial products (gray dots) is truncated. To compensate for the truncation error, an error correction vector C is generated and added to the final result. The correction vector is a fixed vector and is generated from number occurrence probability in [37], while in [38], a variable vector that is generated according to the amount of one bits appeared in truncated partial products is applied to further reduce error.

The 16-bit version of the truncated multiplier in [38] is applied in [39] together with techniques to exploit the locality properties of deep neural network layers. The whole design achieved a speedup of more than 100 times and an energy reduction of more than 20 times compared to a single instruction multiple data (SIMD) processor. In [40], the same truncated multiplier is applied together with other design techniques such as memory access skipping and precision scaling. Compared to an accurate implementation, the proposed design achieved on average 34–51% energy reduction in various NN applications with minor quality loss. Although the effect of the truncated multiplier is not evaluated separately in these two works, based on the implementation results, approximate multiplier is an effective contributor in reducing the energy consumption of the computation cores.

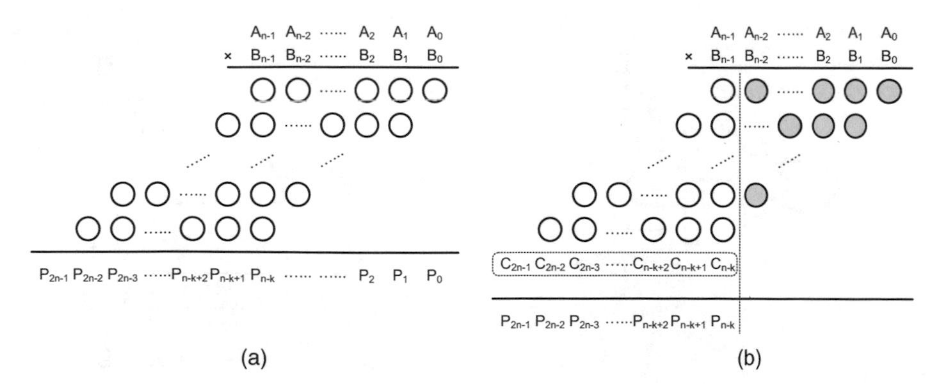

Fig. 16.5 Partial products of normal multiplier and truncated multiplier. (**a**) Normal multiplier. (**b**) Truncated multiplier with correction vector

In some other designs, further logic minimization is applied in addition to partial product truncation to seek for more energy saving. In [30] and [42], a 16-bit version of the truncated multiplier in [38] is implemented with k set to 0 such that both the input and output are 16-bit numbers. It is used as the baseline design. The probabilistic logic minimization (PLM) [41] is then applied to the baseline design. An example of PLM applied to a full adder circuit is shown in Fig. 16.6. PLM looks for the opportunities to perform intentional bit flips for logic minimization. Compared to the standard full adder logic shown in Fig. 16.6a, the simplified full adder cell shown in Fig. 16.6b can achieve a 10 times energy-delay-area product (EDAP) improvement while maintaining a relatively low error rate for the sum S and output carry C_{out}. The design in [30] first applies PLM searching for approximate multiplier with the least error. Then, the resulting multiplier is applied to replace the exact multiplier in NN computation. The neural network model is then retained to compensate for accuracy loss.

A total of eight approximate multipliers, including 1 baseline multiplier and 7 truncated multipliers with logic minimization, are obtained after the searching process. These multipliers are then applied to MLPs for performance evaluation. Without retraining, a large accuracy degradation occurs in the benchmark MLPs. However, accuracy can be recovered by retraining. Implemented using Synopsys Design Compiler and IC Compiler with TSMC-65nm technology, the proposed approximate NN accelerator can achieve up to 2.67 times energy saving while having a delay and area improvement of 1.23 times and 1.46 times, respectively.

In addition to logic minimization, the design of approximate compressor is another method to design approximate multiplier in the literature. In [43], approximate compressors are proposed and used in the approximate multiplier designed for NN applications. An approximate (5, 2) compressor is proposed first. A standard (5, 2) compressor has 6 inputs, including the 5 operand bits and the input carry C_{in} from the compressor in the former bit position, and generates 3 output bits which are the sum S in the current bit position, the carry C in the next bit position, and an output

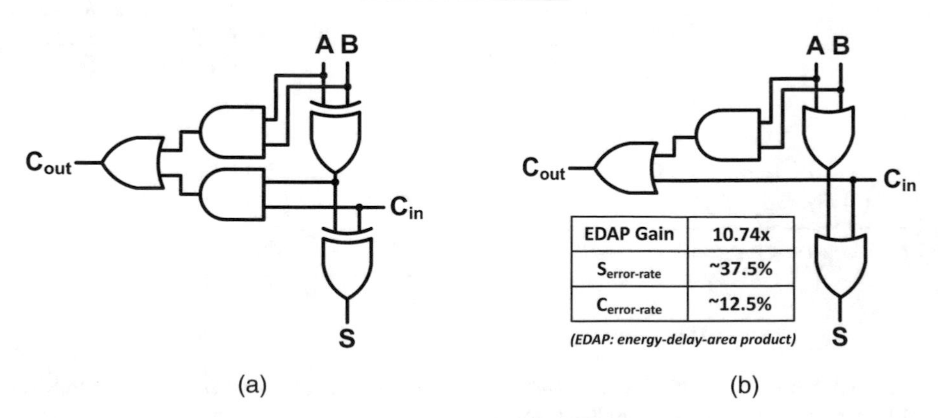

Fig. 16.6 Probabilistic logic minimization in minimizing the logic of a full adder (reproduced from [41]). (**a**) Standard full adder. (**b**) Simplified full adder

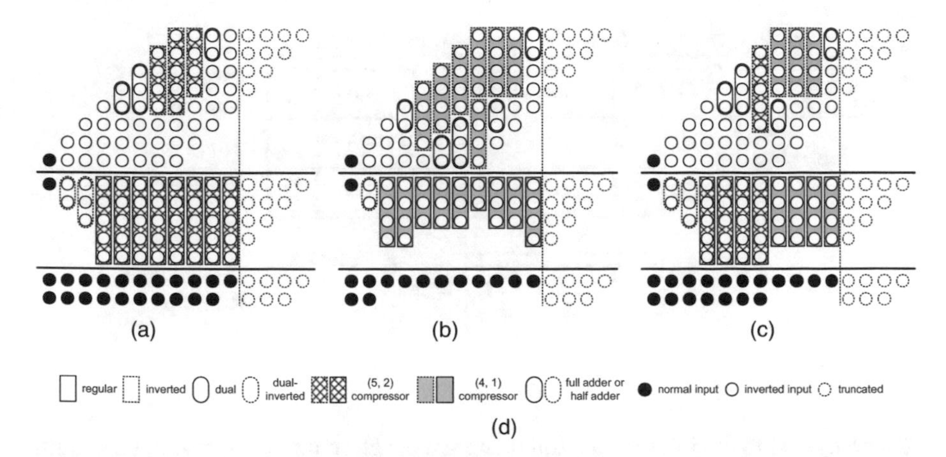

Fig. 16.7 Approximate multiplier using approximate (5, 2) and approximate (4, 1) compressors (reproduced from [43]). (**a**) Use (5,2) compressor only. (**b**) Use (4,1) compressor only. (**c**) Use both compressors. (**d**) Figure legend

carry C_{out} propagating to the compressor in the next bit position. In the proposed design, both C_{in} and C_{out} are ignored, and thus the logic of the (5, 2) compressor is simplified. Similarly, a standard 4-input compressor, the (4, 2) compressor, will generate two outputs, S and C, in addition to C_{out}. In the proposed design, S, C_{in}, and C_{out} are ignored, and thus the resulting compressor becomes a (4, 1) compressor where only the carry C is considered.

In [43], three 8×8 approximate multipliers are proposed which are shown in Fig. 16.7. As shown in Fig. 16.7, the least significant four columns of the partial products are truncated. For the remaining partial product bits, the proposed approximate compressors are applied in the partial product accumulation stage. When generating partial products, the NAND gate is used instead of the AND gate, and therefore, most of the partial product bits are inverted which are shown as the white dots in Fig. 16.7. To deal with these inverted bits, the inverted versions of the proposed approximate compressors and the dual of half adder and full adder are also used in the partial product accumulation. By using these inverted logic, the finally generated two vectors that will be provided to a carry-propagate adder will be normal bits. The first design, shown in Fig. 16.7a, uses the proposed (5, 2) compressor only, while the second design, presented in Fig. 16.7b, uses the proposed (4, 1) compressor only. In the third design, as shown in Fig. 16.7c, the proposed (4, 1) compressors are used in the least significant 8-bit positions (including those truncated positions), while the proposed (5, 2) compressors are used in the most significant 7-bit positions.

The three proposed approximate multipliers are used in an MLP for MNIST [32] classification, which has 784 input neurons, 50 hidden layer neurons, and 10 output neurons, and LeNet [32] for SVHN [44] classification. The accuracy degradation of these two models when using the first and the third approximate multipliers is within

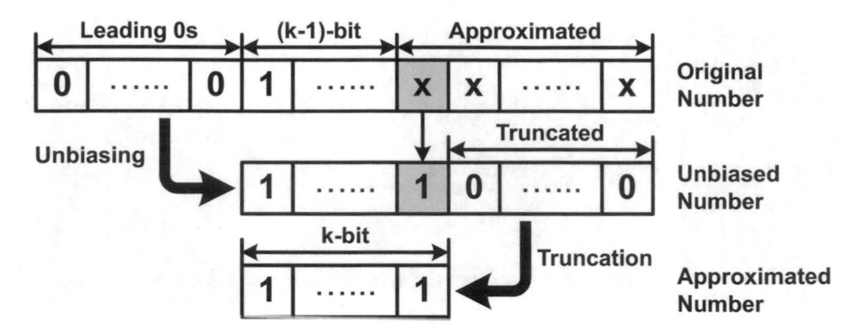

Fig. 16.8 The approximation process of the DRUM design (reproduced from [45])

2% compared to the implementation using the exact multiplier. While the accuracy degradation for the second multiplier is 15% for MLP and 25% for LeNet, it has much better hardware efficiency and is still useful in some application scenarios.

The abovementioned approximate multipliers use simplified logic to reduce energy consumption. In another set of designs, the datapath of the multiplier is not changed. Instead, operands are manipulated at runtime where only the most important part of the operands are extracted and participate in the multiplication. By doing this, the bit-width of the multiplier can be reduced, and thus the energy efficiency can be improved.

A dynamic range unbiased multiplier (DRUM) is proposed in [45]. The approximation process of an operand in the DRUM design is shown in Fig. 16.8. The position of the first nonzero bit is detected and all the leading zeros are then removed because the leading zero bits will not have any contribution to the product. The next $(k-1)$-bit, starting from the first nonzero bit, will then be extracted and used for the actual multiplication. Here, k is a user-defined precision. The remaining least significant bits will be approximated. In DRUM, these bits are not directly truncated as this will lead to a biased error (the approximated value is always no larger than the actual value). To make unbiased error, the expect value which is $100\cdots00$ is used assuming a uniformed distribution for these least significant bits. The one bit is combined with the $(k-1)$-bit to form the operand for the k-bit multiplier. When the leading nonzero bit is within the least significant k-bit of the original, the least significant k-bit will be used instead.

The architecture to perform the abovementioned input process is shown in Fig. 16.9. Leading one detector (LOD) is used to perform logic OR operation on all bits of greater significance for each operand bit to generate a one-hot string where the one bit appears in the position of the leading bit. The one-hot string is then used by an encoder to generate the counting of the leading zero bits. This counting bit is used to select the correct k-bit segment from the original operand. The two count numbers of the two operands are added to generate the final shifting amount for the product. The two generated k-bit segments are multiplied by a multiplier. The $2k$-bit product is then right shifted by the amount of the sum of the two leading zero counts to its correct position to generate the final $2n$-bit product P. Although there are extra

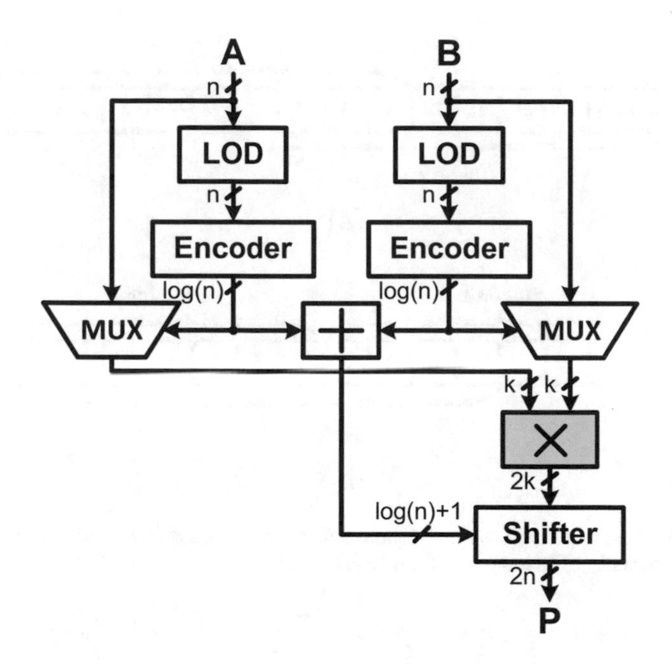

Fig. 16.9 Datapath of the DRUM design (reproduced from [45])

LOD, encoder, multiplexers, adder, and shifter used in addition to the multiplier, as the bit-width of the multiplier is reduced from n-bit to k-bit, the DRUM multiplier is still more efficient than the standard n-bit multiplier. The original DRUM design is not evaluated for NN computation in the original paper [45], and however, the DRUM architecture is used by [46] and [47] where four approximate multipliers with four different k values are applied for NN computation. Implementation results show significant improvement in power consumption with minor accuracy loss compared to exact designs.

The DRUM design [45] can reduce the bit-width of the multiplication from n-bit to smaller k-bit. However, as the leading nonzero bit can be in any position, when the bit-width n becomes large, the extra cost from the LOD and the final dynamic shifter will become more expensive. To solve this problem, a static segment method is proposed in [48]. The segmentation method of [48] is shown in Fig. 16.10. As shown in Fig. 16.10a, to extract $k = 10$ bits from $n = 16$ bits operand, only two options are considered. If the bit positions from 15 to 10 are not all zeros, then bits within the range of option 1 will be extracted. Otherwise, those bits in option 2 will be used. The cost of this operation is much smaller than the cost of a 16-bit LOD, encoder, and multiplexer. In addition, after multiplication, the dynamic shifter used in Fig. 16.9 can be replaced with a constant shifter. When both operands are extracted from option 1 positions, then no shift is needed. If one of the operand is extracted from option 2 positions, then a 6-bit right shift is needed. Finally, if both operands are extracted from option 2 positions, a 12-bit right shift

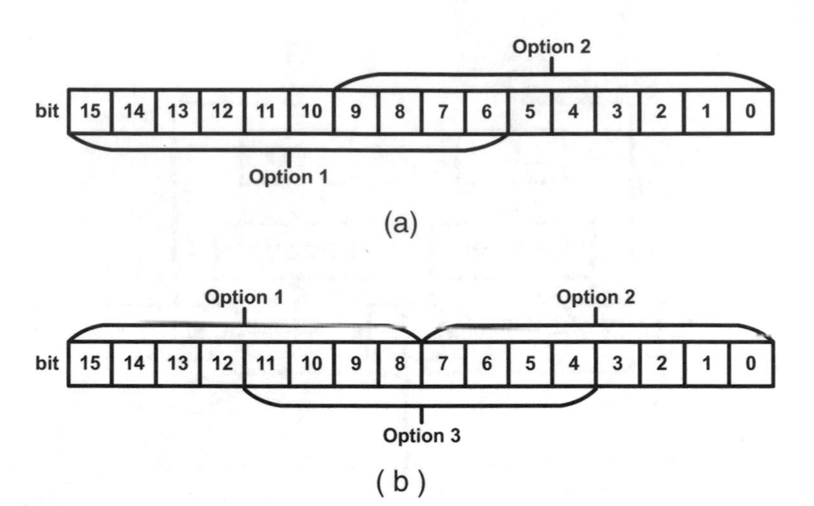

Fig. 16.10 Process of the static segmentation method (reproduced from [48]). (**a**) Extract 10-bit from 16-bit operand. (**b**) Extract 8-bit from 16-bit operand

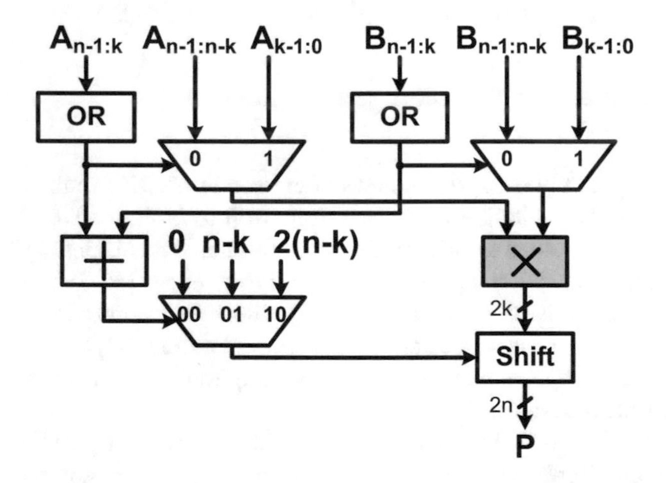

Fig. 16.11 Datapath of the approximate multiplier with static segment method (reproduced from [48])

should be performed. The constant shifter has lower cost than the dynamic shifter. The hardware datapath of the approximate multiplier designed with static segment method is shown in Fig. 16.11.

When $k = n/2$, as shown in Fig. 16.10b, we need to detect whether there is a nonzero bit within bit position 15 and bit position 8. However, in some cases, the leading one may appear in position 9 or position 8. In these cases, many possible nonzero bits within least significant 8-bit positions are discarded and the error of the results may be large. To solve this problem, a third option 3 is added when $k = n/2$

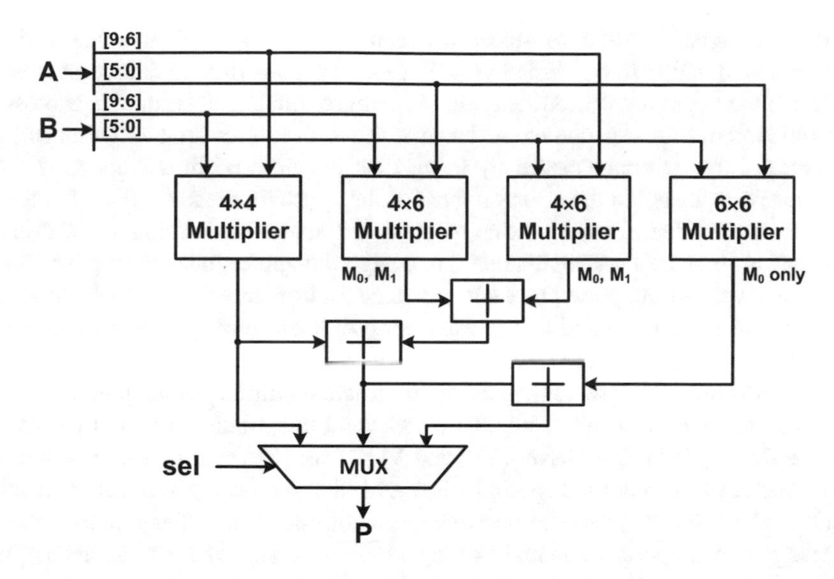

Fig. 16.12 Datapath of the variable-latency approximate multiplier (reproduced from [47])

which is overlapped with option 1 and option 2. In the revised design, if the leading bit is in bit 15:12, then option 1 will be used. Otherwise, if the leading bit is in position 11:8, then option 3 will be selected. For all other cases, option 2 will be extracted. This method ensures as many nonzero bits as possible are included in the final extracted k-bit vector. The hardware architecture, presented in Fig. 16.11, will need to be modified accordingly. Two sets of OR gates are needed for each operand. The multiplexer to extract operand will be changed to a 3-to-1 multiplexer. The multiplexer to determine the shift amount will become a 5-to-1 multiplexer.

In [47], in addition to the 4 variants of the DRUM multiplier [45], a variable-latency approximate multiplier is also proposed and used in neural network computation. The datapath of the variable-latency approximate multiplier for 10-bit operands is shown in Fig. 16.12. Three precision modes, M_0, M_1, M_2, are supported. In the cxact mode M_0, all four multipliers shown in Fig. 16.12 will be used to compute the product. In mode M_1, the 6×6 multiplier is not used, and therefore, the multiplication between the least significant 6-bit of A and B is not performed, while in the lowest precision mode M_2, only the most significant 4-bits of both operands are multiplied. In mode M_2, as the adders shown in Fig. 16.12 are not needed, the latency of the multiplier is the smallest among the three operational modes. In mode M_0, as all three sets of adders are needed, the latency becomes larger. The proposed design is applied in several NN-based image classification applications. Used jointly with the AxTrain method proposed in [47], the proposed variable-latency approximate multiplier achieves minor accuracy loss while significantly improving the energy efficiency.

Wire-by-switch replacement [49] is also a commonly used approximate arithmetic unit design methodology. This technique replaces a wire with a switch.

A control signal is used to select between the wire's original value and an approximated value. In the design of [50], this wire-by-switch replacement is used to design the approximate multiplier. In [50], approximated NN output is expressed as a function of the mean and the variance of the error introduced by multiplication. The mean error is compensated by using bias update, and thus the error of the output is only related to the error variance of the approximate multiplier. To reduce the variance of the error, an exhaustive design space exploration is performed when using the wire-by-switch method to design the approximate multiplier. Three accuracy levels are supported by each generated multiplier: exact multiplication (all switches are turned off) and two levels of approximate multiplication (part of the switches are turned on). At runtime, the accuracy level to be used is determined by the weight values. After searching, the approximate multiplier with minimum power consumption is used in NN computation. Several convolutional neural networks, such as ResNet [51], MobileNet [52], and VGG Net [53], are used to evaluate the performance of the proposed approximate multiplier. According to the experimental results, NNs with the proposed approximate multiplier can achieve improvements on energy consumption while maintaining a high accuracy. The low variance of the approximation error is also proved to be helpful in NN computation in [34].

Another approximate multiplier architecture designed for NN computation is the alphabet set multiplier (ASM) [54–56]. An example of the ASM architecture used in [55] is shown in Fig. 16.13. In ASM method, the multiplier W is divided into segments of a certain bit-width (4 in this example). The multiplicand IF and those segments are not multiplied directly at runtime. Instead, the multiples of the multiplicand IF are precomputed and stored in memory. At runtime, those multiples will be selected according to the value of the multiplier segment and then accumulated with other selected multiples to obtain the final product. This multiplication is feasible for NN computation since the weights or inputs can be reused multiple times, and thus the precomputed multiples can be shared by multiple multiplication operations. When $\{1\times, 3\times, 5\times, 7\times, 9\times, 11\times, 13\times, 15\times\}$ of the multiplicand IF are available, an accurate multiplication can be performed since other multiples can be easily obtained by shifting one of the available multiples [55]. To reduce the hardware cost, only $\{1\times, 3\times, 5\times, 7\times\}$ of the IF are used as shown in Fig. 16.13. By shifting these available multiples, $\{2\times, 4\times, 6\times, 8\times, 10\times, 12\times, 14\times\}$ of the IF can be obtained, and however, $\{9\times, 11\times, 13\times, 15\times\}$ cannot be generated. To compensate errors introduced by those missing multiples, some constraints are applied to the weight values during training. Specifically, the segment values of $\{9, 11, 13, 15\}$ are restricted to their nearest available values $\{8, 10, 12, 14\}$.

During NN inference, as shown in Fig. 16.13, the weight value W is divided into two 4-bit numbers. Each 4-bit number is used by the control module to generate a control signal for the multiplexer to select the correct IF multiple. If the required IF multiple is not directly available, the control module will also generate a control signal for the shifter to shift the selected multiple to generate the required multiple. The two shifted multiples are then added together. Since the higher order 4-bit of the multiplier W has an offset of 4-bit to the least significant bit (LSB) of the multiplier, its corresponding selected multiple also needs to be left shifted by 4-

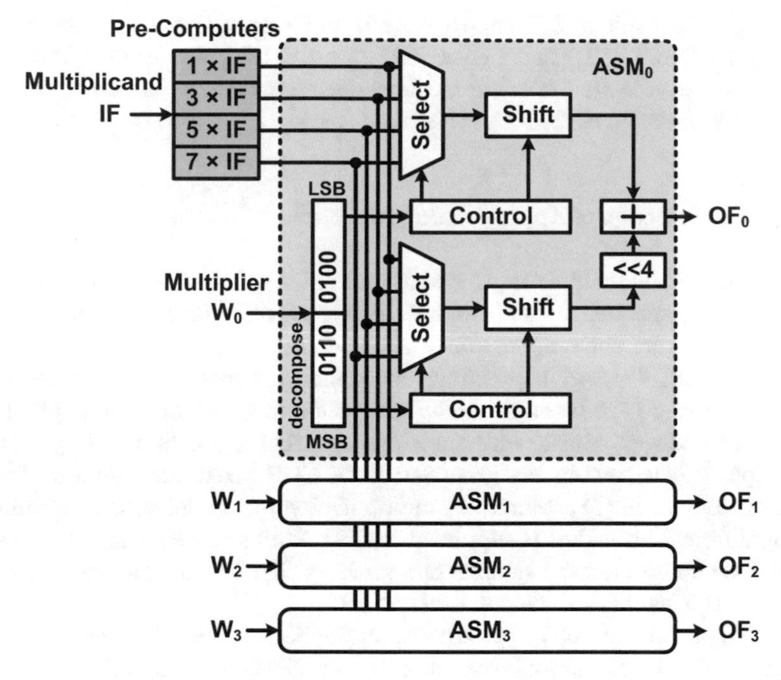

Fig. 16.13 Architecture of alphabet set multiplier (reproduced from [55])

bit before addition. In this example, the weight value is applied to the multiplier port. Weight values are adjusted during training, so that the segment values of $\{9, 11, 13, 15\}$ can be avoided. If input images IF are applied to the multiplier port, as the IF can have any values, the unavailable multiples may be requested very often which leads to a large accuracy degradation. Since IF can be reused in convolution layer and fully connected layer computation, they can be applied to the multiplicand port to generated reused multiples. Moreover, to further exploit the reuse opportunity, multiple weights (W_0 to W_3 in Fig. 16.13) share the same set of precomputed multiples to generate multiple outputs at the same time.

In addition to those manually designed approximate multipliers, some researchers use the Cartesian Genetic Programming (CGP) method to search for a suitable approximate multiplier architecture [57]. The CGP method uses directed acyclic graph to represent a design. The components of the design are represented as 2D programmable nodes. For arithmetic unit, logic functions, such as AND, OR, and XOR, can be used as the nodes. The CGP search process starts with architecture of accurate multipliers. The mutation method is used to randomly change the nodes or connections of up to a certain amount. The generated circuits are then evaluated using the fitness function. The one with the highest fitness score will be used as the starting point for the next round of search. In [57], a good approximate multiplier is considered to have an acceptable error rate and the result is accurate when multiplied by zero. The fitness function is defined based on these two conditions.

By using CGP method, 852 approximate multipliers are generated and used for MLP (for MNIST [32]) and LeNet [32] (for SVHN [44]). With retraining, up to 91% power reduction of multiplication operations can be achieved, while the accuracy degradation of the NNs is less than 2.8%.

3.1.2 The Selection of Approximate Multiplier

We have discussed multiple approximate multiplier architectures that are designed manually or generated by CGP-based search process for NN computation. In addition, there are some approximate arithmetic unit libraries available recently [58, 59]. In [58], 471 8-bit approximate multipliers are created from 6 conventional multipliers using CGP-based methods. In addition to approximate multipliers, 430 CGP generated approximate adders are also available in this library. In [59], many more approximate multipliers generated from CGP-based methods are added in addition to those in [58]. Moreover, multiplication of multiplicand and multiplier having different bit-widths is supported in [59]. With so many available units, the method to choose the most suitable one for a specific NN application needs to be explored. This has been studied in [36] and [60].

In [36], a total of 600 approximate multipliers are used to investigate the critical features of the approximate multiplier for NN computation. Among the 600 approximate multipliers, 100 of them are either manually designed approximate multiplier available in the literature or their variants with different configuration parameters. The other 500 multipliers are designed with CGP-based method. These multipliers are applied to an MLP (for MNIST [32] digit classification) and LeNet [32] (for SVHN [44] classification). After applying approximate multiplier, the model is retrained to recover some accuracy. The accuracy of the models after applying approximate multipliers is recorded for further investigating the critical features of the approximate multipliers. An interesting finding is that the NNs have even better accuracy after applying some of the approximate multipliers.

When analyzing the features of approximate multipliers, nine error characteristics of the approximate multiplier are considered: the ER, the Var-ED, the MED, the RMS-ED, the variance of the RED (Var-RED), the mean value of the RED values (Mean-RED), the root mean square of the RED values (RMS-RED), the variance of the AED (Var-AED), and the mean value of the AED values (Mean-AED). First, feature selection method is performed with Scikit-learn to determine which features are more important than others. With this process, the Var-ED and the RMS-ED are determined to be the two most important features. Then, by using a different number of features as inputs and the quality of the NN (0 or 1) as the outputs, a simple classifier is trained to see whether the selected features can determine whether the approximate multiplier is suitable for NN computation or not. The quality of the NN is defined as follows: the NN accuracy when using exact arithmetic unit is set as the threshold value, and if the NN with approximate multiplier can have higher accuracy, then the approximate multiplier is treated as suitable design and the classifier output (quality of the network) is 1. Otherwise, if the accuracy is degraded, the classifier output is 0. The results confirm that with two features (Var-ED and RMS-ED), the classifier can achieve the highest classification accuracy, and thus Var-ED and RMS-ED are determined to be the most important features. Another evidence to prove this statement is that the Var-ED and RMS-ED are significantly smaller for class 1 than class 0. Then hardware metrics of these multipliers are evaluated. Finally, by jointly considering the accuracy and power efficiency (represented as PDP), 5 best approximate multipliers for these two models are obtained. By comparison, the two models using these 5 multipliers can achieve even better accuracy while having much better power efficiency due to the use of approximate multipliers.

The work in [60] uses similar analysis procedure, but mean relative error (MRE) is used to find the approximate multiplier with good power efficiency. The obtained approximate multipliers are then applied in NN and retraining is performed. The ones with better NN accuracy are selected. Although the NNs using these obtained approximate multipliers can achieve better accuracy than exact computation after retraining, accuracy loss always exists without retraining.

3.2 Logarithm Domain Approximate Multipliers

3.2.1 Design Methodology

Performing multiplication operations in the logarithm domain is an effective way to reduce the hardware cost of the multiplication since the multiplication in the linear domain can be converted to simple addition in the logarithm domain: $log(A \times B) = log(A) + log(B)$. The transform between the linear domain and the logarithm domain may introduce some resource overhead. To reduce this cost, Mitchell logarithm multiplier [61] is proposed, where the transform operation is simplified to obtain better performance and energy efficiency.

In Mitchell multiplier [61], the logarithm transform is performed as follows. An n-bit binary number X can be represented as

$$X = \sum_{i=0}^{n-1} 2^i \cdot x_i = 2^k (1 + \sum_{i=0}^{k-1} 2^{i-k} \cdot x_i) \tag{16.5}$$

where x_i is the binary value at bit position i and k is the position of the first nonzero bit starting from the MSB side which is termed $characteristic$. Then, the base-2 logarithm of X can be expressed in Eq. (16.6):

$$\begin{aligned} log_2 X &= log_2[2^k(1 + \sum_{i=0}^{k-1} 2^{i-k} \cdot x_i)] \\ &= k + log_2(1 + \sum_{i=0}^{k-1} 2^{i-k} \cdot x_i) \end{aligned} \tag{16.6}$$

In Eq. (16.6), as $i < k$, the value of $\sum_{i=0}^{k-1} 2^{i-k} x_i$ is within the range of [0, 1). Therefore, $log_2(1 + \sum_{i=0}^{k-1} 2^{i-k} \cdot x_i)$ can be approximated with $\sum_{i=0}^{k-1} 2^{i-k} x_i$. In the following steps, the two characteristics and the two approximated terms of the two operands are added together, and the result is converted back to the linear domain with an antilogarithm operation.

A low-power hardware architecture to perform Mitchell's multiplication is proposed in [62]. The proposed design strictly follows Eqs. (16.5) and (16.6) to design each module. A revised architecture of [62] is presented in [63]. In the revised design, parameter w is introduced to limit the bit-width of the operand to be added (in logarithm domain). In Mitchell's original architecture, all the lower order k-bits of the operand are involved in the addition. By using parameter w, only w-bits instead of the whole k-bit are used. This is similar to the DRUM multiplier [45] that is discussed in the previous section.

The process of the multiplier proposed in [63] is shown in Fig. 16.14. As Mitchell's algorithm did, the first step is to locate the leading nonzero bit in the operands, and then a left shifting is performed to remove all leading zero bits. The

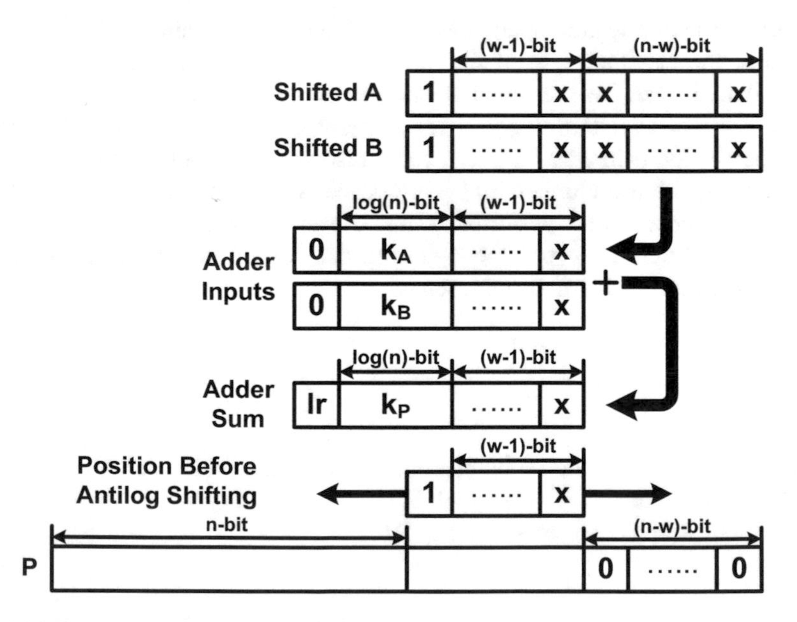

Fig. 16.14 The process of the revised Mitchell logarithm multiplier in [63]

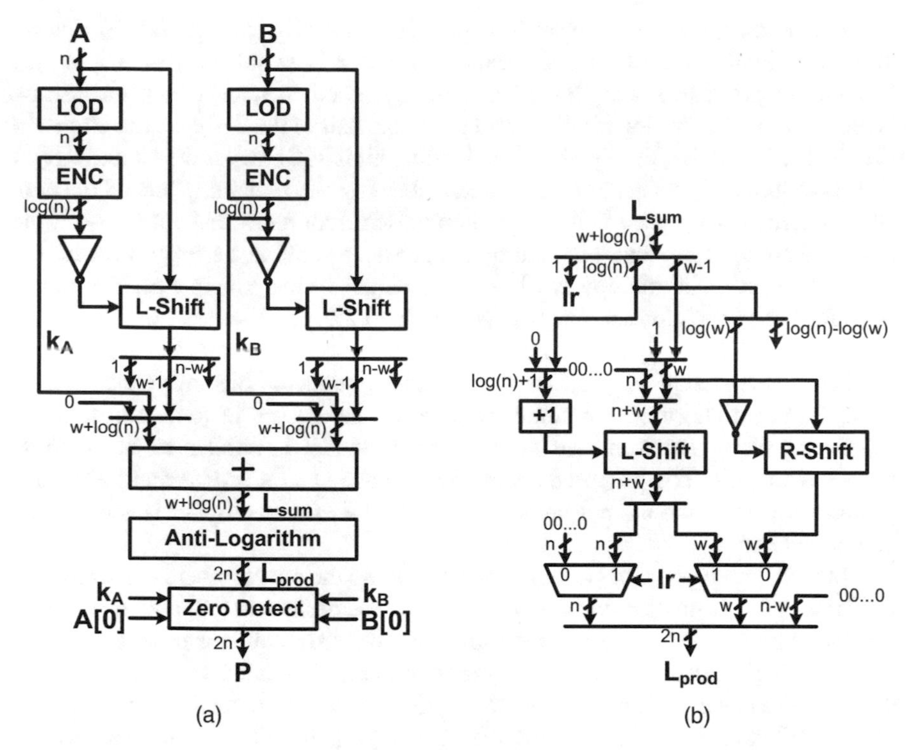

Fig. 16.15 Datapath of the approximate logarithm multiplier in [63] and its antilogarithm module (reproduced from [63]). (**a**) Datapath of the logarithm multiplier. (**b**) Antilogarithm module

shifted operands A and B are shown in the top row of Fig. 16.14. Then, the leading nonzero bit is removed, and its bit position in the original operand, which is called the characteristic, is encoded into $log(n)$-bit k_A and k_B and is prefixed to the bit vectors. Different from Mitchell's algorithm, only the next $(w - 1)$-bit vectors are reserved for further computation. Then, these two $log(n) + w$-bit vectors are added to generate the sum in the logarithm domain. To perform the antilogarithm operation to map the sum to a $2n$-bit vector in the linear domain, the characteristics of the sum, lr and k_P, are replaced with a single one bit. Then, based on the value of lr and k_P, the sum vector is shifted to its correct position in the $2n$-bit vector to complete the multiplication operation.

The datapath of the approximate logarithm multiplier is presented in Fig. 16.15a. A LOD is used to detect the position of the leading nonzero bit, and the position is then encoded by the encoder module. The proposed design in [63] is operated with one's complement encoding. Assume there are k-bit zeros in front of the leading bit; then, the weight of the leading bit should be $n - k - 1$. In one's complement encoding, this can be simply computed by the inverse of k which is implemented by the inverter in Fig. 16.15a. Then, the operands are shifted, combined with their characteristics, and then added in the logarithm domain.

When performing the antilogarithm operation, the design in Fig. 16.15b) is used. The most significant bit of the sum characteristic lr is examined first. If it is one, then the weight of the leading bit of the result is larger than n, and then a left shifting is required to move the logarithm sum to the upper half of the $2n$-bit vector shown in Fig. 16.14. Otherwise, the weight of the leading bit is still smaller than n and a right shift is required. The shift amount is determined by the remaining bits of the sum characteristic (except the lr bit). Implementation results show that with $w = 8$, the proposed approximate logarithm multiplier consumes 88% less energy compared to the 32-bit fixed-point multiplier, while the accuracy of the NN on classification of ILSVRC2012 validation set is only degraded by 0.2%.

In [64], an efficient dynamic range approximate logarithm multiplier is proposed. Similar to the design in [63], only a certain amount of bits from the leading bit is extracted and used. To compensate error, the design in [64] sets the least significant bit of the truncated operand to 1, which is similar to the method proposed in [45]. The proposed multiplier is applied in an MLP and LeNet for performance evaluation. An accuracy similar to the exact multiplier is achieved in both applications.

Mitchell's multiplier [61] always underestimates the results where the computed results are always smaller than the exact results. The designs in [63] and [64] further truncate the operand to a certain bit-width, and the errors of their proposed design are also single-sided although error compensation is applied in [64]. Although these approximate multipliers can be applied in NN computation, an unbiased error distribution is desired for better accuracy [45]. In [17], and an approximate logarithm multiplier with double-sided error distribution is proposed to solve this problem.

By using y to represent the summation part in Eq. (16.5), the n-bit number can be represented by $X = 2^k(1 + y)$. This equation can also be rewritten to

$$X = 2^{k+1}(1 - z) \tag{16.7}$$

if a larger exponent is applied to the radix. In the method proposed in [17], as the actual value of X is in between 2^k and 2^{k+1}, the differences between X and the two bound values are compared. If $X - 2^k < 2^{k+1} - X$, which means X is more close to the lower bound, $log(X)$ is underestimated to $k + y$ according to Eq. (16.6). Otherwise, $log(X)$ is overestimated to $k+1-z$. By doing this conditional operation, double-sided error distribution is generated. When a large amount of computations is performed, the errors are expected to cancel out with each other, and thus the accuracy can be improved.

When performing multiplication, one more approximation is applied. Suppose the operands can be represented as

$$\begin{aligned} A &= 2^{k_1} + q_1 \\ B &= 2^{k_2} + q_2 \end{aligned} \tag{16.8}$$

where q_1 and q_2 are, respectively, the differences between A and B and their selected bounds. Then, the multiplication can be performed with

$$A \times B = 2^{k_1+k_2} + q_1 2^{k_2} + q_2 2^{k_1} + q_1 q_2 \quad (16.9)$$

The first three terms can be calculated efficiently since they are power of two terms. The last term required a multiplier. The design in [17] chooses to ignore the last term in order to reduce cost. In addition to the hardware approximation, a nearest-one detector (NOD) is proposed in [17] to take the place of the LOD to perform the selection between lower bound and upper bound.

The proposed approximate logarithm multiplier is evaluated by using MLP (for MNIST [32] classification) and AlexNet [33] (for CIFAR-10 [65] classification). The performance of using the proposed approximate logarithm multiplier in CIFAR-10 [65] classification can even be higher than the implementation with exact multiplier.

The logarithm approximate multiplier is also applied for floating-point computation in [66]. According to [67], the value of a normal floating-point number can be represented as

$$X = (-1)^s \cdot 2^{exp}(1 + frac) \quad (16.10)$$

where s is the sign, exp is the biased exponent, and $frac$ is the fraction of the number. The sign bit can be processed separately. The remaining part can be processed according to Eq. (16.6):

$$log_2 X = log_2[2^{exp}(1 + frac)] = exp + log_2(1 + frac) \quad (16.11)$$

As $frac$ is in the range [0, 1), the second term can be approximated as $frac$. Then the multiplication of floating-point numbers can be approximated using the approximate logarithm multiplier. The proposed floating-point approximate logarithm multiplier in [66] is applied in NN training, and it shows improved speed performance and power efficiency.

There are some other approximate logarithm multipliers available in the literature but are not yet evaluated for NN computation [68, 69]. In [68], a set-one adder is applied in the addition in the logarithm domain. The set-one adder sets all the approximated bits to 1s. As already discussed, Mitchell's multiplier always underestimates the results. By setting all approximated bits to 1s, the adder actually overestimates its results. The negative error generated by Mitchell's multiplier can be compensated by the positive error introduced by the set-one adder. This feature is beneficial to NN computation. In [69], logarithm multiplier for floating-point operations is proposed. The proposed logarithm multiplier uses integer units to perform multiplication and thus can significantly improve the efficiency. This architecture is expected to be helpful in NN hardware designs with training support.

3.2.2 Comparison of Logarithm and Linear Domain Multipliers

Approximate multipliers in both the linear domain and the logarithm domain have been applied in NN computations. Although both of them achieve acceptable results for some NN models, they have different characteristics that may affect their usage in other NN models.

The comparison of the characteristics of the linear domain approximate multipliers and the logarithm domain multipliers is performed in [68], where approximate Booth multiplier and logarithm multiplier proposed in [68] are compared. Both 8-bit and 16-bit versions of the approximate multipliers are compared in terms of error metrics, such as ED, and hardware metrics, such as PDP. In both cases, the logarithm multipliers are less accurate than the linear multiplier. This can be explained by the fact that the logarithm approximate multiplier has approximation in both the logarithm transform process and the logarithm domain addition process. According to Eq. (16.6), the second logarithm term is approximated with a simple expression. In addition, in the designs, such as [63], only a specific bit-width of the operand is used in the logarithm domain addition. Both of them introduce errors to the computation results. However, in the linear domain, there is no such errors that come from domain transformation. As a result, the linear domain approximate multiplier is generally more accurate than the logarithm domain multiplier.

In terms of PDP, the logarithm domain approximate multipliers perform better than the linear domain multipliers. As the logarithm approximate multiplier is actually accomplished by a simple adder, the cost of the multiplier is much lower, especially when the bit-width of the operands is restricted in some designs. Although the transformation between two domains may introduce some resource overhead, the transformation after approximation becomes hardware efficient. As a result, the logarithm domain approximate multiplier is generally more efficient than linear domain approximate multiplier in terms of hardware metrics.

In addition to those approximate logarithm multipliers introduced in this section, iterative approximate logarithm multiplier, such as [70], is also available in the literature. In these multipliers, the error, which is the difference between the exact result and the approximate result, is once again computed by another round of logarithm multiplication. This process is repeated until the preset amount of iterations is reached. This kind of multipliers can improve the error metrics (compared to the non-iterative logarithm approximate multiplier), and however, the error is usually still larger than the linear domain multiplier as the amount of iterations is usually limited. The hardware metrics, such as PDP, become larger than non-iterative multiplier due to the iterative operations.

In conclusion, although both kinds of multipliers can be applied for NN computation [17, 36] for better energy efficiency, for those models where accuracy is more important, linear domain multiplier is more appropriate. For those application scenarios where relatively larger accuracy degradation is acceptable, logarithm domain approximate multiplier can be applied [68].

4 Discussion and Perspectives

We have so far discussed the approximate multipliers available in the literature that are designed specific for NN computation. Their design methodologies and the summarized implementation results are included in the discussion. In this section, we are going to present the evaluation process, to summarize the current findings, and to propose potential future works.

4.1 Evaluation Process

To evaluate the error metrics of approximate multipliers, a software model that emulates the datapath of the hardware approximate multiplier is needed. The software model could be built with MATLAB, C language, or Python, depending on the framework that the NN is built on. Testing data is then fed to the emulation model to obtain approximated results. Then, error metrics are calculated using the approximated results and the exact results.

To evaluate the performance of the NN, the multiplication in the NN model is replaced with the emulation model of the approximate multiplier. The accuracy of the NN model is tested with test dataset. If accuracy is not acceptable, then retraining is performed on the dataset until the accuracy is recovered. However, as the emulation model is usually not optimized well compared to the framework inherent operations, this experiment process is time consuming, especially for large-scale NN models. This is the reason that many researchers still use MLPs or small-scale DNN to evaluate their proposed approximate arithmetic units.

4.2 Challenges

Basically, the current design and evaluation process treats the NN model as a black box. The approximate arithmetic unit is designed first, and then experiments are performed on NN models to see whether the designed unit is suitable for NN computation or not. A complete in-depth analysis of the internal mechanism is still missing from the literature. Therefore, although some approximate multiplier designs can be applied for their specific NN models, as the systematic analysis is missing, they may not be effective in other NN models or the migration to other NN models needs to go through the whole process again, including a full testing, multiple epochs of retraining, and so on, which is hardware expensive and time consuming.

Without a systematic analysis, when selecting an approximate multiplier for a specific NN model, those experiments need to be performed multiple times to select the one with the best results. If the in-depth analysis is available, then the selection process can be guided by the analysis which will be more efficient and productive.

According to the experiment results in [36] and [17], the accuracy of the NN after incorporating approximate multipliers may be even higher than the implementation with exact multipliers. These accuracy-related phenomenon are not completely explained yet. In [36] and [34], the increase in accuracy is explained by the reduction of overfitting. With appropriate errors introduced, intermediate results become more varied, and thus overfitting is somehow improved by introducing these errors. This explanation can be supported by some other experiments; however, a more rigorous proof is needed. The degree of approximate levels that can lead to a beneficial noise to NN models also needs to be explored. In addition, whether there are other factors leading to the accuracy increase needs to be investigated as well.

For the hardware arithmetic design, a question needs to be answered is that what kind of characteristics of the arithmetic unit will be beneficial to NN computation. In [60], mean relative error is used to find the optimal PDP design point. In [36], two characteristics, Var-ED and RMS-ED, are found to be more relevant to the performance in NN computing. As discussed in [34], the variance of the error should be small enough so that it will not affect the NN accuracy. These three works provide good references for the design, and however, as different characteristics are used, the ones that are most suitable in general or the ones that are most suitable for a specific NN model need to be explored. So that the appropriate approximate multiplier can be selected based on these characteristics.

With just a few approximate multipliers and NN models, this analysis may not be feasible. However, on the one hand, approximate arithmetic libraries, such as the EvoApprox8b [58] and the extended version [59], are available in the literature which eases the design of many different kinds of approximate arithmetic units. In addition, many pretrained NN models are open-sourced and available online. This will also reduce the workload of the experiment process.

Based on the explorations performed in the literature [17, 30, 34, 36], the following approximate arithmetic unit design guidelines can be summarized:

- Approximate multipliers with unbiased error distribution are preferred.
- The variance of the error should be small to not affect the accuracy of NNs.
- Accumulation should be exact computation.
- A large amount of accumulations is needed so that accumulated noise to outputs is at similar level, and thus the order of the outputs is not changed.
- Batch normalization can be helpful in improving the performance of approximate multipliers on NN models.

4.3 Future Research Directions

The current approximate arithmetic units, both linear domain design and logarithm domain design, are all designed for NN inference stage. Approximate computing for the training stage is also necessary. Although for most of the NNs, training is only performed once in the server, while the inference is performed many times in the end devices, the one-time training is quite hardware expensive: the computations involved are more complex and the training time is usually long. For a server that is running training workloads, the cost is also expensive. Approximate computing is expected to be useful in reducing the cost of the training process, and thus it is needed to be investigated.

For most of the current approximate NN implementation, a single approximate arithmetic is used for the whole model process. However, as the data patterns among different layers may be quite different, a single category of approximate arithmetic unit may not the optimal solution for NN computation. Based on the characteristics of different layers [34], different approximate computing strategies may be applied for different layers. This once again needs an in-depth analysis of the neural computing process. In addition, in hardware design, the efficient approximate multipliers that support many different modes are required to support many different approximate computing in a single unit.

5 Conclusion

In this chapter, approximate arithmetic units for efficient neural network computation are reviewed. We start with the discussion of the necessities of approximate computing in neural networks. Then, the fundamentals of neural network computation are discussed to help understanding the designs in the literature. The approximate multipliers that are designed in the linear domain for neural network computations are then presented. A general guide on the selection of approximate multiplier is discussed as well. After that, the approximate logarithm multipliers that are designed for neural network computation are presented, and the comparison between logarithm multiplier and linear multiplier is discussed. Finally, some current design challenges and corresponding future research directions are discussed. This chapter is expected to help the readers to understand the current designs available in the literature and motivate further research projects to solve the proposed questions.

Acknowledgments The authors would like to thank Ocean University of China, University of Saskatchewan, and the Natural Sciences and Engineering Research Council of Canada (NSERC) for their financial support for the related projects and the writing of this chapter.

References

1. Jain A, Mao J, Mohiuddin K. Artificial neural networks: a tutorial. Computer. 1996;29(3):31–44.
2. LeCun Y, Bengio Y, Hinton G. Deep learning. Nature. 2015;521:436–44.
3. Bengio Y, Lecun Y, Hinton G. Deep learning for AI. Commun. ACM. 2021;64(7):58–65.

4. Maass W. Networks of spiking neurons: the third generation of neural network models. Neural Netw. 1997;10(9):1659–71.
5. Zhou Z, Chen X, Li E, Zeng L, Luo K, and Zhang J. Edge intelligence: paving the last mile of artificial intelligence with edge computing. Proc IEEE. 2019;107(8):1738–62.
6. Moore GE. Cramming more components onto integrated circuits, Reprinted from Electronics, volume 38, number 8, April 19, 1965, pp.114 ff. IEEE Solid-State Circ Soc Newsl. 2006;11(3):33–5.
7. Dennard R, Gaensslen F, Yu HN, Rideout V, Bassous E, LeBlanc A. Design of ion-implanted MOSFET's with very small physical dimensions. IEEE J Solid-State Circ. 1974;9(5):256–68.
8. Hennessy JL, Patterson DA. A new golden age for computer architecture. Commun ACM. 2019;62(2):48–60.
9. Sze V, Chen YH, Yang TJ, Emer JS. Efficient processing of deep neural networks: a tutorial and survey. Proc IEEE. 2017;105(12):2295–329.
10. Jiang H, Santiago FJH, Mo H, Liu L, Han J. Approximate arithmetic circuits: a survey, characterization, and recent applications. Proc IEEE. 2020;108(12):2108–35.
11. Liu W, Lombardi F, Shulte M. A retrospective and prospective view of approximate computing [Point of View]. Proc IEEE. 2020;108(3):394–9.
12. Venkataramani S, Sun X, Wang N, Chen CY, Choi J, Kang M, Agarwal A, Oh J, Jain S, Babinsky T, Cao N, Fox T, Fleischer B, Gristede G, Guillorn M, Haynie H, Inoue H, Ishizaki K, Klaiber M, Lo SH, Maier G, Mueller S, Scheuermann M, Ogawa E, Schaal M, Serrano M, Silberman J, Vezyrtzis C, Wang W, Yee F, Zhang J, Ziegler M, Zhou C, Ohara M, Lu PF, Curran B, Shukla S, Srinivasan V, Chang L, Gopalakrishnan K. Efficient AI system design with cross-layer approximate computing. Proc IEEE. 2020;108(12):2232–50.
13. Gupta S, Agrawal A, Gopalakrishnan K, Narayanan P. Deep learning with limited numerical precision. In: Proceedings of the 32nd international conference on international conference on machine learning - volume 37. 2015. pp. 1737–46.
14. Han S, Mao H, Dally WJ. Deep compression: compressing deep neural networks with pruning, trained quantization and Huffman coding. CoRR, vol. abs/1510.00149. 2016. pp. 1–14.
15. Fleischer B, Shukla S, Ziegler M, Silberman J, Oh J, Srinivasan V, Choi J, Mueller S, Agrawal A, Babinsky T, Cao N, Chen CY, Chuang P, Fox T, Gristede G, Guillorn M, Haynie H, Klaiber M, Lee D, Lo SH, Maier G, Scheuermann M, Venkataramani S, Vezyrtzis C, Wang N, Yee F, Zhou C, Lu PF, Curran B, Chang L, Gopalakrishnan K. A scalable multi-TeraOPS deep learning processor core for AI training and inference. In: 2018 IEEE symposium on VLSI circuits. 2018. pp. 35–6.
16. Han S, Liu X, Mao H, Pu J, Pedram A, Horowitz MA, Dally WJ. EIE: efficient inference engine on compressed deep neural network. SIGARCH Comput Archit News. 2016;44(3):243–54.
17. Ansari MS, Cockburn BF, Han J. An improved logarithmic multiplier for energy-efficient neural computing. IEEE Trans Comput. 2021;70(4):614–25.
18. Chakraborty I, Ali M, Ankit A, Jain S, Roy S, Sridharan S, Agrawal A, Raghunathan A, Roy K. Resistive crossbars as approximate hardware building blocks for machine learning: opportunities and challenges. Proc IEEE. 2020;108(12):2276–310.
19. Venkataramani S, Choi J, Srinivasan V, Wang W, Zhang J, Schaal M, Serrano MJ, Ishizaki K, Inoue H, Ogawa E, Ohara M, Chang L, Gopalakrishnan K. DeepTools: compiler and execution runtime extensions for RaPiD AI accelerator. IEEE Micro. 2019;39(5):102–11.
20. NVIDIA. NVIDIA TensorRT Developer Guide. NVIDIA Docs. 2021.
21. Chen J, Hu J. Energy-efficient digital signal processing via voltage-overscaling-based residue number system. IEEE Trans Very Large Scale Integr (VLSI) Systems. 2013;21(7):1322–32.
22. Venkataramani S, Kozhikkottu VJ, Sabne A, Roy K, Raghunathan A. Logic synthesis of approximate circuits. IEEE Trans Comput Aided Des Integr Circuits Syst. 2020;39(10):2503–15.
23. Chen CY, Choi J, Gopalakrishnan K, Srinivasan V, Venkataramani S. Exploiting approximate computing for deep learning acceleration. In: 2018 Design, automation test in Europe Conference Exhibition (DATE). 2018. pp. 821–6.

24. Wang E, Davis , Zhao R, Ng HC, Niu X, Luk W, Cheung PYK, Constantinides GA. Deep neural network approximation for custom hardware: where we've been, where we're going. ACM Comput Surv. 2019;52(2):1–39.
25. Panda P, Sengupta A, SS Sarwar, Srinivasan G, Venkataramani S, Raghunathan A, Roy K. Invited — cross-layer approximations for neuromorphic computing: from devices to circuits and systems. In: 2016 53nd ACM/EDAC/IEEE design automation conference (DAC). 2016. pp. 1–6.
26. Zhang H, Chen D, Ko SB. New flexible multiple-precision multiply-accumulate unit for deep neural network training and inference. IEEE Trans Comput. 2020;69(1):26–38.
27. Zhang H, He J, Ko SB. Efficient posit multiply-accumulate unit generator for deep learning applications. In: 2019 IEEE international symposium on circuits and systems (ISCAS). 2019. pp. 1–5.
28. Zhang H, Lee HJ, Ko SB. Efficient fixed/floating-point merged mixed-precision multiply accumulate unit for deep learning processors. In: 2018 IEEE international symposium on circuits and systems (ISCAS). 2018. pp. 1–5.
29. Venkatachalam S, Adams E, Lee HJ, Ko SB. Design and analysis of area and power efficient approximate booth multipliers. IEEE Trans Comput. 2019;68(11):1697–703.
30. Du Z, Lingamneni A, Chen Y, Palem KV, Temam O, Wu C. Leveraging the error resilience of neural networks for designing highly energy efficient accelerators. IEEE Trans Comput Aided Des Integr Circ Syst. 2015;34(8):1223–35.
31. Mahdiani HR, Haji Seyed Javadi M, Fakhraie SM. Efficient utilization of imprecise computational blocks for hardware implementation of imprecision tolerant applications. Microelectron J. 2017;61(C):57–66.
32. Lecun Y, Bottou L, Bengio Y, Haffner P. Gradient-based learning applied to document recognition. Proc IEEE. 1998;86(11):2278–324.
33. Krizhevsky A, Sutskever I, Hinton GE. ImageNet classification with deep convolutional neural networks. In: Proceedings of the 25th international conference on neural information processing systems - volume 1. 2012, pp. 1097–105.
34. Kim MS, Del Barrio Garcia AA, Kim H, Bagherzadeh N. The effects of approximate multiplication on convolutional neural networks. IEEE Trans Emerg Top Comput. 2021. p. 1.
35. Liang J, Han J, Lombardi F. New metrics for the reliability of approximate and probabilistic adders. IEEE Trans Comput. 2013;62(9):1760–71.
36. MS Ansari, Mrazek V, Cockburn BF, Sekanina L, Vasicek Z, Han J. Improving the accuracy and hardware efficiency of neural networks using approximate multipliers. IEEE Trans Very Large Scale Integr (VLSI) Syst. 2020;28(2):317–28.
37. Schulte M, Swartzlander E. Truncated multiplication with correction constant. In: Proceedings of IEEE workshop on VLSI signal processing. 1993. pp. 388–96.
38. King E, Swartzlander E. Data-dependent truncation scheme for parallel multipliers. In: Conference record of the thirty-first Asilomar conference on signals, systems and computers (Cat. No.97CB36136), vol. 2. 1997. pp. 1178–82.
39. Chen T, Du Z, Sun N, Wang J, Wu C, Chen Y, Temam O. DianNao: A small-footprint high-throughput accelerator for ubiquitous machine-learning. SIGPLAN Not. 2014;49(4):269–84.
40. Zhang Q, Wang T, Tian Y, Yuan F, Xu Q. ApproxANN: an approximate computing framework for artificial neural network. In: 2015 Design, automation test in Europe conference exhibition (DATE). 2015. pp. 701–6.
41. Lingamneni A, Enz C, Palem K, Piguet C. Synthesizing parsimonious inexact circuits through probabilistic design techniques. ACM Trans Embed Comput Syst. 2013;12(2s):1–26.
42. Du Z, Palem K, Lingamneni A, Temam O, Chen Y, Wu C. Leveraging the error resilience of machine-learning applications for designing highly energy efficient accelerators. In: 2014 19th Asia and South Pacific design automation conference (ASP-DAC). 2014. pp. 201–6.
43. Ahmadinejad M, Moaiyeri MH. Energy- and quality-efficient approximate multipliers for neural network and image processing applications. IEEE Trans Emerg Top Comput2021:1. https://ieeexplore.ieee.org/document/9403977

44. Netzer Y, Wang T, Coates A, Bissacco A, Wu B, AY Ng. Reading digits in natural images with unsupervised feature learning. In: NIPS workshop on deep learning and unsupervised feature learning 2011. 2011. pp. 1–9.
45. Hashemi S, Bahar RI, Reda S. DRUM: a dynamic range unbiased multiplier for approximate applications. In: 2015 IEEE/ACM international conference on computer-aided design (ICCAD). 2015. pp. 418–25.
46. He X, Ke L, Lu W, Yan G, and Zhang X. AxTrain: hardware-oriented neural network training for approximate inference. In: Proceedings of the international symposium on low power electronics and design, ser. ISLPED '18. New York: Association for Computing Machinery; 2018.
47. He X, Lu W, Yan G, Zhang X. Joint design of training and hardware towards efficient and accuracy-scalable neural network inference. IEEE J Emerg Sel Top Circuits Syst. 2018;8(4):810–21.
48. Narayanamoorthy S, Moghaddam HA, Liu Z, Park T, Kim NS. Energy-efficient approximate multiplication for digital signal processing and classification applications. IEEE Trans Very Large Scale Integr (VLSI) Syst. 2015;23(6):1180–4.
49. Zervakis G, Amrouch H, Henkel J. Design automation of approximate circuits with runtime reconfigurable accuracy. IEEE Access. 2020;8:53522–38.
50. Tasoulas ZG, Zervakis G, Anagnostopoulos I, Amrouch H, Henkel J. Weight-oriented approximation for energy-efficient neural network inference accelerators. IEEE Trans Circuits Syst I Reg Pap. 2020;67(12):4670–83.
51. He K, Zhang X, Ren S, Sun J. Deep residual learning for image recognition. In: 2016 IEEE conference on computer vision and pattern recognition (CVPR). 2016. pp. 770–8.
52. Sandler M, Howard A, Zhu M, Zhmoginov A, Chen LC. MobileNetV2: inverted residuals and linear bottlenecks. In: 2018 IEEE/CVF conference on computer vision and pattern recognition. 2018. pp. 4510–20.
53. Simonyan K, Zisserman A. Very deep convolutional networks for large-scale image recognition. In: 3rd International conference on learning representations (ICLR 2015). 2015. pp. 1–14.
54. Sarwar SS, Venkataramani S, Raghunathan A, Roy K. Multiplier-less artificial neurons exploiting error resiliency for energy-efficient neural computing. In: 2016 Design, automation test in Europe conference exhibition (DATE). 2016. pp. 145–50.
55. Sarwar SS, Venkataramani S, Ankit A, Raghunathan A, Roy K. Energy-efficient neural computing with approximate multipliers. J Emerg Technol Comput Syst. 2018;14(2):16:1–16:23.
56. Sarwar SS, Srinivasan G, Han B, Wijesinghe P, Jaiswal A, Panda P, Raghunathan A, Roy K. Energy efficient neural computing: a study of cross-layer approximations. IEEE J Emerg Sel Top Circuits Syst. 2018;8(4):796–809.
57. Mrazek V, Sarwar SS, Sekanina L, Vasicek Z, Roy K. Design of power-efficient approximate multipliers for approximate artificial neural networks. In: 2016 IEEE/ACM international conference on computer-aided design (ICCAD). 2016. pp. 1–7.
58. Mrazek V, Hrbacek R, Vasicek Z, Sekanina L. EvoApprox8b: library of approximate adders and multipliers for circuit design and benchmarking of approximation methods. In: Design, automation test in Europe conference exhibition (DATE), 2017. 2017. pp. 258–61.
59. Mrazek V, Sekanina L, Vasicek Z. Libraries of approximate circuits: automated design and application in CNN accelerators. IEEE J Emerg Sel Top Circuits Syst. 2020;10(4):406–18.
60. De la Parra C, Guntoro A, Kumar A. Full approximation of deep neural networks through efficient optimization. In: 2020 IEEE international symposium on circuits and systems (ISCAS). 2020. pp. 1–5.
61. Mitchell JN. Computer multiplication and division using binary logarithms. IRE Trans Electron Comput. 1962;EC-11(4):512–7.
62. Kim MS, Del Barrio AA, Hermida R, Bagherzadeh N. Low-power implementation of Mitchell's approximate logarithmic multiplication for convolutional neural networks. In: 2018 23rd Asia and South Pacific design automation conference (ASP-DAC). 2018. pp. 617–22.

63. Kim MS, Barrio AAD, Oliveira LT, Hermida R, Bagherzadeh N. Efficient Mitchell's approximate log multipliers for convolutional neural networks. IEEE Trans Comput. 2019;68(5):660–75.
64. Yin P, Wang C, Waris H, Liu W, Han Y, Lombardi F. Design and analysis of energy-efficient dynamic range approximate logarithmic multipliers for machine learning. IEEE Trans Sustain Comput. Oct.-Dec. 2021;6(4):612–25.
65. Krizhevsky A. Learning multiple layers of features from tiny images. Tech. Rep. 2009.
66. Cheng T, Yu J, Hashimoto M. Minimizing power for neural network training with logarithm-approximate floating-point multiplier. In: 2019 29th international symposium on power and timing modeling, optimization and simulation (PATMOS). 2019. pp. 91–6.
67. Society IC. IEEE standard for floating-point arithmetic. IEEE Std 754-2019 (Revision of IEEE 754-2008). 2019. pp. 1–84.
68. Liu W, Xu J, Wang D, Wang C, Montuschi P, Lombardi F. Design and evaluation of approximate logarithmic multipliers for low power error-tolerant applications. IEEE Trans Circuits Syst I Reg Pap. 2018;65(9):2856–68.
69. Gustafsson O, Hellman N. Approximate floating-point operations with integer units by processing in the logarithmic domain. In: 2021 28th IEEE symposium on computer arithmetic (ARITH 2021). 2021. pp. 45–52.
70. Kim H, Kim MS, Del Barrio AA, Bagherzadeh N. A cost-efficient iterative truncated logarithmic multiplication for convolutional neural networks. In: 2019 IEEE 26th symposium on computer arithmetic (ARITH). 2019. pp. 108–11.

Chapter 17
Enabling Efficient Inference of Convolutional Neural Networks via Approximation

Georgios Zervakis, Iraklis Anagnostopoulos, Hussam Amrouch, and Jörg Henkel

1 Introduction

In the past years we have experienced a phenomenal boom in Artificial Intelligence (AI) [8], especially in deep learning and Deep Neural Networks (DNNs) that brought superhuman levels of accuracy in many tasks such as natural language processing, object detection, speech recognition, and more [28]. As a result, more and more embedded devices employ DNNs to deliver sophisticated services and we are witnessing an unprecedented rise in Edge AI. Nevertheless, the accuracy improvements of DNNs came at the cost of an immense increase in computational demands. It is noteworthy that state-of-the-art Convolution Neural Networks (CNNs) require billions of multiply and accumulate (MAC) operations for only one inference. Therefore, to enable Edge AI and execute complex AI models, neural network acceleration became mandatory and ASIC DNN accelerators, namely Neural Processing Units (NPUs), are nowadays an integral part of modern systems-on-chip. In order to cope with the computational demands and deliver low latency/high throughput, NPUs integrate thousands of MAC units. For example, Samsung's mobile NPU comprises 6K MAC units [20]. However, such a high number of MAC units operating in parallel and delivering Tera of operations per second results in elevated energy requirements and power consumption [35],

G. Zervakis · J. Henkel (✉)
Karlsruhe Institute of Technology, Karlsruhe, Germany
e-mail: georgios.zervakis@kit.edu; henkel@kit.edu

I. Anagnostopoulos
Southern Illinois University, Carbondale, IL, USA
e-mail: iraklis.anagno@siu.edu

H. Amrouch
University of Stuttgart, Stuttgart, Germany
e-mail: amrouch@iti.uni-stuttgart.de

W. Liu, F. Lombardi (eds.), *Approximate Computing*,
https://doi.org/10.1007/978-3-030-98347-5_17

exacerbating also the power density and forming localized hot-spots and thermal bottlenecks [4].

Approximate computing emerged in recent years as an energy efficient design paradigm. Exploiting the inherent resilience of a large number of applications domains (e.g., machine learning and image processing [32]) approximate computing intelligently trades some quality loss for gains in performance, energy, etc. As a result, the design of approximate adders and multipliers [17, 24, 31] as well as quite larger circuits such as image processing kernels [32] has attracted significant research interest. In this chapter we examine the application of approximate computing on the very complex CNN accelerators. First, in Sect. 1, we discuss the benefits and highlight the limitations of fixed approximation (through approximate multiplications) in CNN inference. To this end, we discuss error compensation approaches that are mandatory to enable the exploitation of fixed approximate multipliers in CNN inference and we highlight the state-of-the-art control variate approximation technique that does not require retraining and enables using highly approximate multipliers while satisfying tight accuracy loss constraints. Next, Sect. 2 presents reconfigurable approximation techniques for CNN inference and discusses how they address the limitations of fixed approximation, at the cost however of reduced savings. Finally, Sect. 3 goes one step beyond energy-efficiency and presents how approximate computing can be employed to address physical constraints in NPUs. In this section we analyze approximate computing as a cooling mechanism and demonstrate its significant contribution toward mitigating the excessive on-chip temperature that is inevitably associated with NPUs. Moreover, we discuss how systematic approximation can be employed to suppress circuit aging effects in NPUs and completely eliminate the aging-induced timing guardbands, boosting thus the NPU performance.

2 CNN Inference with Approximate Multipliers

A significant fraction of CNN inference computations (about 90%) is spent on convolution and matrix multiplication operations [28]. The efficiency of such operations heavily relies on the efficiency of the MAC units. Considering that CNN accelerators integrate thousands of MAC units, designing approximate MAC circuits has gained a lot of attraction in order to boost the efficiency of the overall accelerator. Since the multiplier is the most complex component of the MAC unit, state of the art mainly focuses on replacing the exact multiplier with approximate ones [5, 16, 23].

Modern deep architectures are becoming very complex and error propagation within the network, due to approximate arithmetic, is cumbersome to capture and multiplication errors magnify throughout the CNN layers. As a result, although neural networks feature an inherent tolerance to less accurate computations, modern CNNs are very sensitive to even slight approximation [34]. For example, the authors in [19] evaluated the accuracy of several approximate multipliers across a variety

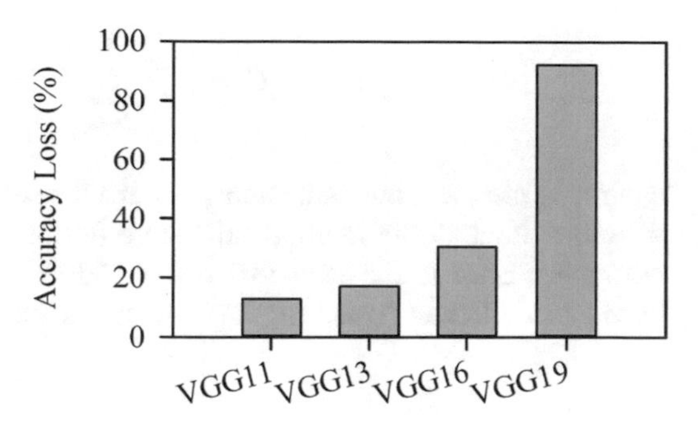

Fig. 17.1 Accuracy loss across several VGG networks that feature incremental size. The approximate multiplier 2P7 of the EvoApprox Library is used to run the inference. All the VGGs are trained on CIFAR100, 8-bit inference is considered, and the MRE of 2P7 is only 0.052%

of CNNs trained on CIFAR10 obtaining considerable power gains for limited accuracy loss. Nevertheless, these energy gains were not maintained in the more challenging CIFAR100 dataset since the accuracy dropped significantly. In addition, Tasoulas et al. [26] demonstrated that, as the network depth increases, the impact of approximate multiplications increases and may lead to devastating accuracy results. In Fig. 17.1, we plot the accuracy degradation with respect to several variants of the VGG network that feature incremental depth (VGG11 to VGG19). The VGGs in Fig. 17.1 are trained on CIFAR100. All the exact multiplications are replaced by the approximate multiplier 2P7 [17] and the tool of [27] is used to obtain the inference accuracy. Note that 2P7 features a tiny mean relative error (MRE) of only 0.052%. From Fig. 17.1 it is obvious that even the slightest approximation can lead to very high accuracy loss at inference level. In addition, as the size of VGG increases, the accuracy drops significantly. The accuracy loss for VGG11 is 13% and jumps to 30% for VGG16 while the accuracy achieved for VGG19 is almost negligible (about 10%). Overall, approximate multipliers appear as a promising solution to improve the energy profile of CNN accelerators. However, the obtained accuracy and consequently the attained gains will highly depend on several parameters such as the network architecture and its complexity, the examined dataset, etc. Despite the significant research activities in the design of approximate arithmetic circuits and especially of approximate multipliers, designing approximate multipliers optimized for CNN inference is still an active field of research. Importantly, complementary error compensation techniques are mandatory in order to achieve high accuracy at inference level.

2.1 *Optimizing Approximate Multipliers*

The most complex computation in CNNs is the convolution operation that is expressed as

$$C = B + \sum_{j=1}^{k} W_j \cdot A_j, \tag{17.1}$$

where A_j are the input activations, W_j are the weights, and B is the neuron's bias. Assuming that all the multiplications are performed by an approximate multiplier with mean error μ and error variance σ and that ϵ_j is the error of multiplying W_j by A_j, the following stand for the convolution error ϵ_C:

$$\epsilon_C = \sum_{j=1}^{k} \epsilon_j, \quad \mathrm{E}[\epsilon_C] = k\mu, \quad \text{and} \quad \mathrm{Var}(\epsilon_C) = k\sigma^2. \tag{17.2}$$

Hence, even if the approximate multiplier features a small error, the error of the convolution can grow very fast since it is proportional to the size of the CNN's filters. It is evident then that achieving high inference accuracy while using approximate multipliers needs special considerations.

Considering (17.1), Tasoulas et al. [26] deduced that the mean error of the convolution can be canceled out with zero cost by just adding a constant value at the bias B of each filter (see Sect. 2). As a result, to maximize the obtained accuracy, when designing approximate multipliers for CNN inference, the main optimization target should be to minimize the error variance of the approximate multiplier [26]. After extensive experimentation and analyzing the impact of several error metrics such as the error rate (ER), the error distance (ED), the absolute ED, and the relative ED, Ansari et al. [5] reached the same conclusion since the authors deduced that the most important features that render an approximate multiplier superior to others for CNN inference are Var(ED) and RMS(ED). In addition, Mrazek et al. [16, 19] observed that in CNN inference multiplication by 0 is very important, mainly due to the high number of such multiplications, and thus should always be performed accurately.

2.2 Error Compensation

Retraining has been repeatedly proven to significantly improve the accuracy of CNN inference accelerators that use approximate multipliers [5, 16, 23]. Retraining is able to adapt the weights of the CNNs to the performed approximate multiplications operations and mitigate the induced inaccuracy. For example, Mrazek et al. [16] showed that after a few retraining epochs most of the accuracy loss is recovered for the examined MNIST and SVHN. Similar results are obtained in [23] for more complex datasets and networks, e.g., ResNet164 on CIFAR100. Interestingly, Ansari et al. [5] showed that using approximate multipliers and applying retraining may even improve the attained accuracy compared with the exact multiplier. CNN retraining is easily applied by just using the approximate multiplier in the forward pass and then typically perform the backpropagation using exact computations. Although retraining is a very promising solution in enabling high inference accuracy and low-power operation through approximate multiplications, it exhibits several limitations that might constraint its exploitation. First, considering the size of the network and of the complexity of the dataset, retraining can become a very time-consuming operation. In addition, if the approximate hardware is not available, we need to emulate the approximate multiplications during the forward pass exacerbating further the time complexity. Moreover, training parameters, e.g., proper learning rate, have to be adequately set. The latter is a complex task and many approximate multipliers might require careful regulation [23]. Finally, in many cases, retraining might be even infeasible due to proprietary datasets and/or models [18, 35].

To avoid retraining, Mrazek et al. [18] proposed an offline, post-training weight adaptation technique. Each weight w is replaced by w' as follows:

$$\underset{\forall w'}{\operatorname{argmin}} \sum_{\forall \alpha} |\mathtt{AM}(\alpha, w') - \alpha \cdot w|, \tag{17.3}$$

where $\mathtt{AM}(a, w')$ is the approximate multiplication a by w'. Mrazek et al. [18] exploits that the weights are known after training and given an approximate multiplier, Mrazek et al. [18] uses (17.3) to replace each weight w with a value w' so that the approximate multiplication w' delivers more accurate products, on average, than the approximate multiplication by w. As a result, (17.3) ensures that after the weight update, the error of the approximate multiplications will be less or equal to the error of the initial approximate multiplication by w. Nevertheless, it does not provide any error guarantees and the efficiency of the error compensation performed by (17.3) depends heavily on the weight values and the characteristics of the approximate multipliers.

All the aforementioned techniques are static and are applied at design time. However, errors due to approximation highly depend on the respective inputs [33]. Hence, since all the aforementioned techniques are based on offline statistics, they fail to efficiently capture the runtime behavior and provide tailored approximation. As a step toward addressing this issue, a control variate approximation technique is proposed in [35]. The control variate technique is used in Monte Carlo methods to achieve variance reduction. The main principle behind the control variate approximation is to estimate the overall convolution error at runtime and compensate it. Therefore, the more precise the estimation is, the more accurate the convolution operation is and thus higher inference accuracy is obtained. For this reason, the control variate approximation employs the approximate perforated multipliers [30]. The perforated multipliers omit the generation and accumulation of the m successive least significant partial products. As a result, high error but also high power reduction is achieved. Though, one of the most important features of partial product perforation is that the induced error can be modeled in a rigorous mathematical manner. Specifically, the error of an m-perforated approximate multiplier is given by

$$\epsilon = W \cdot x, \quad x = A \bmod 2^m, \tag{17.4}$$

where W and A are the two multiplicands. Then, Zervakis et al. [35] proposed to compute C^* instead of C in (17.1), where C^* is

$$C^* = B + \sum_{j=1}^{k} \texttt{PERF}(W_j, A_j) + V, \tag{17.5}$$

where, PERF is the perforated multiplier and V is the control variate. Since a control variate should be easily computed (in order to retain the gains of the approximation), V is expressed as a linear regression of the perforated multiplication error. It can be easily demonstrated that expressing V as

$$V = \overline{W} \cdot \sum_{j=1}^{k} x_j, \quad \overline{W} = \frac{1}{k} \sum_{j=1}^{k} W_j \text{ and } x_j = \bmod 2^m \tag{17.6}$$

results in

$$\mathrm{E}[\epsilon_{C^*}] = \sum_{j=1}^{k} \mathrm{E}\Big[W_j \cdot x_j\Big] - \mathrm{E}[V] = \sum_{j=1}^{k} \mathrm{E}[x_j] \cdot W_j - \overline{W} \cdot \sum_{j=1}^{k} \mathrm{E}[x_j] = 0 \tag{17.7}$$

and

$$\begin{aligned} \mathrm{Var}(\epsilon_{C^*}) &= \sum_{j=1}^{k} \mathrm{Var}\Big(W_j \cdot x_j\Big) + \mathrm{Var}\Big(V\Big) = \sum_{j=1}^{k} \Big((W_j - \overline{W})^2 \cdot \mathrm{Var}(x_j)\Big) \\ &= \underbrace{\frac{(2^m - 1)(2^m + 1)}{12}}_{\mathrm{Var}(x_j)} \sum_{j=1}^{k} (W_j - \overline{W})^2. \end{aligned} \tag{17.8}$$

Hence, the mean convolution error is nullified and considering that the distribution of the weights mainly follows a distribution with low dispersion (i.e., weights are concentrated close to their average value), then the convolution error variance in (17.8) is minimized. As a result, control variate satisfies all the aforementioned conditions that are required to enable high inference accuracy: (1) multiplications by 0 are always performed correctly [30], (2) control variate ensures that the mean convolution error is zero, (3) the convolution error variance is minimized, and (4) the control variate V is built at runtime with respect to the values of the input activations. Finally, note that the hardware implementation of the control variate approach is fairly simple. As (17.6) shows, only k additions and one multiplication are required to compute V. Each MAC unit has to be extended with a small adder that accumulates the m least significant bits of the activations. Then, an additional MAC operation is required to add V with the partial convolution sum and compute C^* as defined in (17.5).

2.3 Inference Accuracy

As described in Sect. 2.2, the control variate method uses a runtime error correction scheme and enables the exploitation of highly approximated multipliers in CNN inference, neglecting also the requirement for retraining. Figure 17.2 presents an accuracy evaluation of the control variate method across six widely used CNNs trained on the CIFAR100 dataset. An 8-bit perforated multiplier with $m = 2$ is considered in Fig. 17.2, and the accuracy is reported as a normalized value with respect to the accuracy achieved by the exact 8-bit multiplier. As shown in Fig. 17.2, the control variate approximation achieved an average normalized accuracy of 0.98. In the meantime, considering a microarchitecture similar to the Google Edge TPU, which comprises a 64×64 systolic MAC array to run the convolution and matrix multiply operations, a power reduction of 35% is achieved [35]. On the other hand, using the same approximate perforated multiplier but without the error compensation, due to the control variate method, results in an average normalized accuracy of only 0.84. Using the control variate improves the accuracy by 1.16× on average and up to 1.28×. Note that control variate can be employed with any approximate multiplier as long as its error can be estimated/expressed by a simple linear relation.

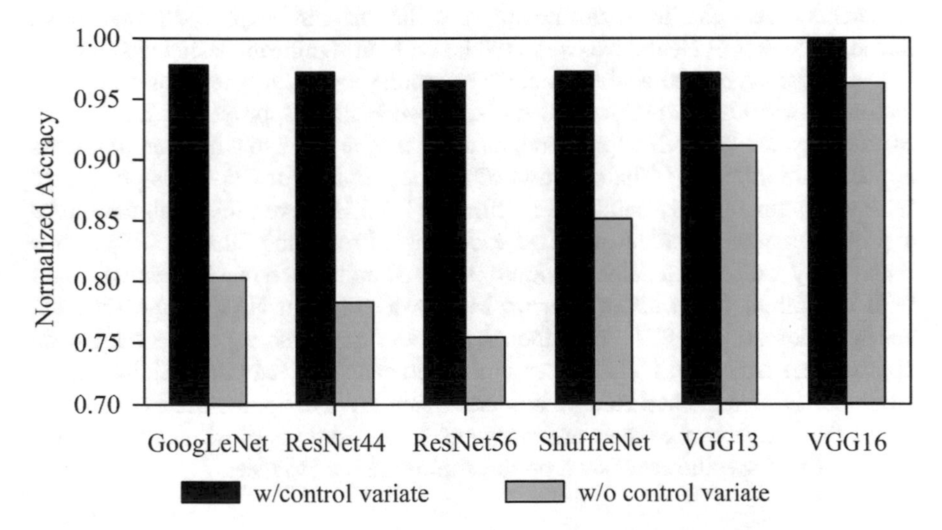

Fig. 17.2 Accuracy evaluation over six CNNs trained on CIFAR100 when using the partial product perforated multipliers w/ and w/o control variate approximation. Two partial products are perforated ($m = 2$) and INT8 inference is used. Accuracy is normalized with respect to the accuracy achieved by the 8-bit exact multiplier

3 CNN-Oriented Reconfigurable Approximation

In Sect. 2, we analyzed impact of fixed approximate multipliers on the inference accuracy and presented the current state-of-the-art error compensation techniques. However, all the aforementioned methods have been designed with static approximation and, once deployed, the generated circuits cannot adapt to input changes. If a change is needed in the error management, it has to be done offline, via a new design of the approximate circuit. On top of that, modern artificial intelligence-oriented applications have different error tolerance and they often require a minimum accuracy threshold, which fixed error approximate circuits cannot guarantee. Thus, since the design of customized approximate circuits per application is challenging, time-consuming, and not flexible, reconfigurability prevails as a solution to this problem. In this section, we present the current state of the art regarding reconfigurable approximate multipliers for CNN inference and we discuss their trade-offs.

3.1 Design of Reconfigurable Multipliers

By bridging the concept of reconfigurability and approximate computing, researchers were able to design flexible circuits that can adapt to different input and architectures of Neural Networks (NNs) without significant design changes.

As a first attempt to achieve accuracy reconfiguration, a heterogeneous architecture is presented in [18], which utilizes several fixed approximate multipliers. Specifically, after a CNN exploration, each layer is mapped to a different but fixed approximate multiplier. The selection of the approximate multiplier is performed by investigating already published multipliers [17]. However, this design requires significant hardware modifications, as each single Processing Element (PE) is now replaced by a tile which combines multiple MAC units, each one introducing error with a different rate a different error. Moreover, different NNs require different configuration for the PEs. Even though this design is more flexible than fixed approximate solutions, it introduces area overheads and requires weight tuning. Additionally, this method follows a coarse-grain layer-to-approximation mapping as all the weights of a layer are assigned to the same approximate multiplier. Consequently, specific accuracy drop thresholds are hard to meet.

3.1.1 Error Variance Based Reconfigurable Approximate Multiplier

In order to address these issues, a method for designing general approximate multipliers that support multiple approximation levels at the same time is presented in [33]. By adopting this design principle, along with the observation that in order to reduce the introduced error the main optimization target should be the minimization of the error variance, Tasoulas et al. [26] present a convolution-specific approximate reconfigurable multiplier. The multiplier, named LVRM (Low-Variance Reconfigurable Multiplier) supports three operating modes: (1) LVRM0, which corresponds to operation at the exact mode, (2) LVRM1, which has low error variance, but it is more energy efficient than LVRM0, and (3) LVRM2, which has larger error variation than LVRM1 (more aggressive approximation), but also higher energy gains at the same time. Additionally, the mean error $\mu(\epsilon_W)$ of each approximate multiplication by a weight W is used to correct the generated convolution error. These derive from the following observation regarding the approximate convolution output C' given by

$$C' = B + \sum_{j=1}^{k} (W_j \cdot A_j - \epsilon(W_j, A_j)) = B - \sum_{j=1}^{k} \epsilon(W_j, A_j) + \sum_{j=1}^{k} W_j \cdot A_j. \tag{17.9}$$

Therefore, Tasoulas et al. [26] used the bias value to compensate the error induced by the employed approximate multiplication. Specifically, the bias B is replaced by $B' = B + \sum_{j=1}^{k} \mu(\epsilon_{W_j})$. So, using the bias update, the approximate convolution output is expressed as

$$\begin{aligned} C' &= B' - \sum_{j=1}^{k} \epsilon(W_j, A_j) + \sum_{j=1}^{k} W_j \cdot A_j \\ &= B + \sum_{j=1}^{k} \mu(\epsilon_{W_j}) - \sum_{j=1}^{k} \epsilon(W_j, A_j) + \sum_{j=1}^{k} W_j \cdot A_j \end{aligned} \tag{17.10}$$

and the error ϵ_C of the output C is given by

$$\epsilon_C = C - C' = \sum_{j=1}^{k} \epsilon(W_j, A_j) - \sum_{j=1}^{k} \mu(\epsilon_{W_j}). \tag{17.11}$$

Therefore, the mean value $\mu(\epsilon_C)$ of ϵ_C, $\forall A_j$, is 0:

$$\mu(\epsilon_C) = \mu\Big(\sum_{j=1}^{k} \epsilon(W_j, A_j) - \sum_{j=1}^{k} \mu(\epsilon_{W_j})\Big) = \sum_{j=1}^{k} \mu(\epsilon_{W_j}) - \sum_{j=1}^{k} \mu(\epsilon_{W_j}) = 0 \tag{17.12}$$

and the variance $Var(\epsilon_C)$ of ϵ_C, $\forall A_j$, is given by

$$Var(\epsilon_Y) = Var\Big(\sum_{j=1}^{k} \epsilon(W_j, A_j) - \sum_{j=1}^{k} \mu(\epsilon_{W_j})\Big) = \sum_{j=1}^{k} Var(\epsilon_{W_j}). \tag{17.13}$$

Hence, the mean error of the convolution can be canceled out with zero cost by just adding a constant value at the bias B of each filter. Considering that the error value ϵ can be viewed as a random variable defined by its mean value $\mu(\epsilon)$ and its variance $Var(\epsilon)$ [15], the convolution error depends mainly on the approximate multiplication variance. Therefore, by controlling the variance $\sum_{j=1}^{k} Var(\epsilon_{W_j})$ using reconfigurable multipliers (e.g., LVRM), we can minimize the convolution error.

3.1.2 Positive/Negative Reconfigurable Approximate Multiplier

In [25], a reconfigurable multiplier is presented that follows a different approach. Instead of trying to support different approximation levels, it proposes the concept of synergistic approximation with positive and negative error as a mean to balance the introduced error. Particularly, based on (17.4), the approximate multipliers in [30, 35] eliminate the generation of m partial products, thus making the approximate product smaller than the exact one and the introduced error always positive.

To that end, and by extending the method in [14] for generating multipliers with multiple approximate modes, the reconfigurable approximate multiplier in [25] supports three different modes: *Zero Error (ZE)*, *Positive Error (PE)*, and *Negative Error (NE)*. The ZE mode corresponds to the exact operation (no error). In the PE mode, the m least partial products are perforated and thus positive error is obtained. In the NE mode, Spantidi et al. [25] forces the generation and accumulation of the m least partial products and thus negative error is obtained. Additionally, in both NE and PE modes result in reduced power consumption due to lower switching activity obtained by forcing (canceling) the generation of the partial products. The average error and error variance of the proposed multiplication by W is given by

$$\mathrm{E}[\epsilon] = s\frac{2^m - 1}{2}W \quad \text{and} \quad \mathrm{Var}(\epsilon) = \frac{2^{2m} - 1}{12}W, \tag{17.14}$$

where $s = 1$ in the PE mode, $s = -1$ in the NE mode, and $m = 0$ in the ZE mode. Therefore, considering (17.14), the average convolution error $\mathrm{E}[\epsilon_C]$ and the convolution error variance $\mathrm{Var}(\epsilon_C)$ are given by

$$\mathrm{E}[\epsilon_C] = \sum_{j=1}^{k} s_j\frac{2^{m_j} - 1}{2}W_j \quad \text{and} \quad \mathrm{Var}(\epsilon_C) = \sum_{j=1}^{k} \frac{2^{2m_j} - 1}{12}W_j. \tag{17.15}$$

At this point, it is important to mention that the selection of the approximate mode for both reconfigurable multipliers [25] and [26] (described in Sect. 3.1.1) is performed based on a control signal that drives instantly the corresponding switches. The control signal is stored along with the respective weight increasing the overall model size. However, setting the accuracy level at runtime does not require any additional cycles nor induces any additional latency (as for example in power gating).

3.2 Weight-Based Mapping

Now that reconfigurable approximate multipliers can be designed without significant area overhead, the next step it so decide which approximation mode will be enabled for the different weights of a CNN optimizing the utilization of approximation in a more fine-grain way compared to coarse-grain methods [18]. Particularly, given an accuracy drop threshold, the next optimization step is to decide which approximation mode will be used for each weight value for each layer of the NN, such that the final accuracy of the NN during inference satisfies the error threshold and the energy consumption is minimized. The complexity of this problem is very big as modern CNNs are composed of tens to hundreds convolutional layers with thousands to millions different weight values, thus making an exhaustive exploration for each weight infeasible. To that end, both approaches presented in [25, 26] propose a heuristic method. Their analysis is based on the observation that the weights' values per layer follow a distribution that features low dispersion. This was observed after examining many CNNs from different families trained on several datasets.

The method in [26] is based on the concepts of layer significance and weight magnitude [7, 21]. The first step is to identify the significance of each layer of the CNN, by mapping all the weights for each one of them separately on the highest approximation (LVRM2) and record the accuracy drop. Layers with low significance value are not considered important as they do not affect the accuracy of the CNN. Then, starting from the least significant layer, all weights are mapped to LVRM2 until the accuracy drop threshold is satisfied. For the remaining layers, a range of weights around zero is defined and mapped on LVRM2 until the accuracy threshold is satisfied. Finally, the same step is performed for LVRM1.

The method in [25] starts with identifying the layer resilience. For each layer of the NN separately, starting from the first one, the occurrences of each weight value are recorded. Then, by taking advantage of the positive/negative architecture of the multiplier, it follows a filter-oriented error balancing method in which half of the occurrences of each weight per filter are mapped on PE mode and the other half on NE. Then, the next step is to find how many layers can be mapped to high approximation (high m) simultaneously, using the aforementioned filter-oriented error balancing method. For the remaining layers, the same concept is followed for lower values of m. Note that high m results in high multiplication error but also high energy reduction. Finally, the weights that appear only once are grouped together per filter, and partitioned into two balanced summation sets using the Largest Differencing Method [11]. Then, all weight values in the first set are mapped on the PE mode, and the weight values in the second set on the NE mode.

3.3 Energy Gains Under Accuracy Drop Thresholds

This section presents a comparison regarding the energy gains of four different approximate multipliers under specific accuracy drop thresholds. The following approximate multipliers are considered: (1) the ALWANN method [18], which utilizes approximate multipliers from the library in [17]; (2) the LVRM method [26], described in Sect. 3.1.1; (3) the ConVar method [35], described in Sect. 2.2; and (4) the Positive/Negative method [25], described in Sect. 3.1.2. Additionally, five state-of-the-art CNNs are considered, GoogleNet, Mobilenetv2, ResNet20, ResNet44, and Shufflenet on four different datasets, CIFAR10, CFIAR100, GTSRB, and LISA. Finally, 0.5% and 1% are used as accuracy drop thresholds w.r.t. the accuracy achieved by the 8-bit quantized model with exact multiplications.

Figures 17.3, 17.4, 17.5, and 17.6 depict the energy gains for all the approximate multipliers at the MAC level for the four selected datasets. Interestingly, the Positive/Negative reconfigurable multiplier achieved on average the highest energy gains, while ALWANN achieved the lowest on average. This can be explained by the fact that ALWANN follows a more coarse-grain mapping (one approximate multiplier per layer), thus restricting the actual benefits. Another important observation is the behavior of ConVar. In many cases it achieves the highest energy savings (e.g., MobileNetv2 for CIFAR-10); however, there are many cases in which it cannot

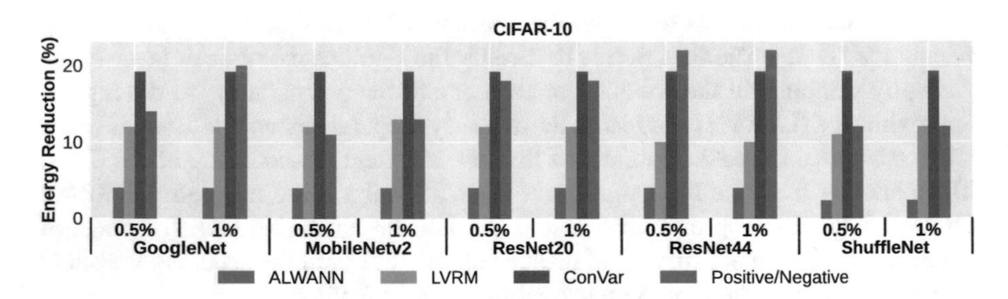

Fig. 17.3 Energy savings at the MAC level for CIFAR-10 dataset

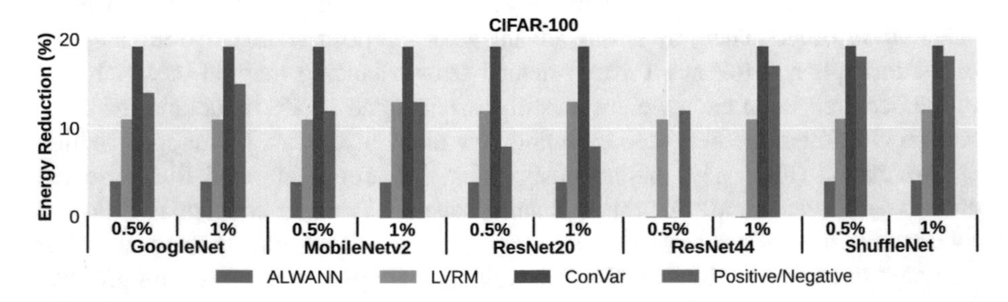

Fig. 17.4 Energy savings at the MAC level for CIFAR-100 dataset. For the 0.5% threshold for the ResNet44 case, ConVar could not produce acceptable solutions, while ALWANN resulted in nearly 0% gains in energy for ResNet44

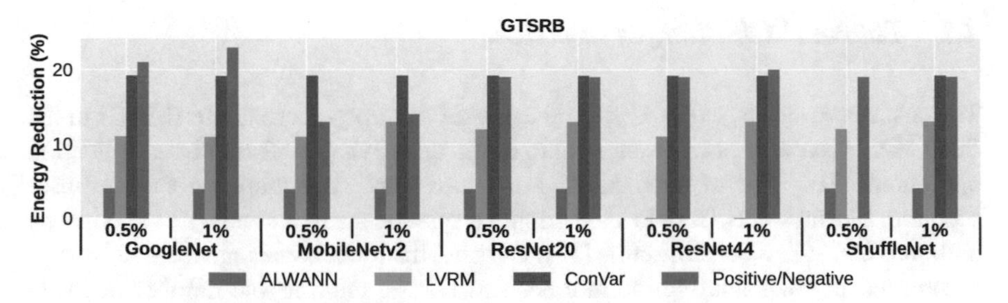

Fig. 17.5 Energy savings at the MAC level for GTSRB dataset. For the 0.5% threshold for the ShuffleNet case, ConVar could not produce acceptable solutions, while ALWANN resulted in nearly 0% gains in energy for ResNet44

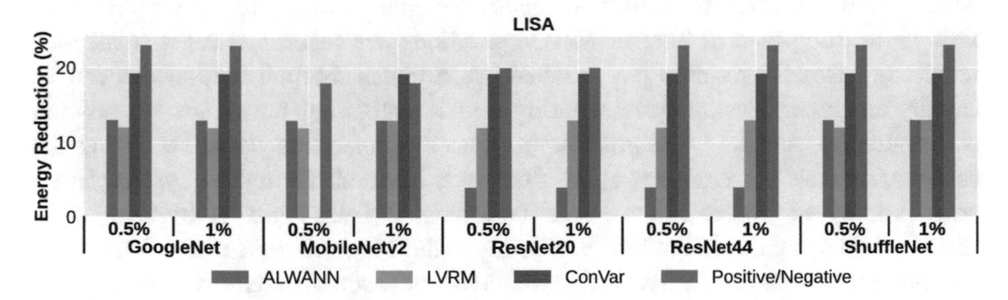

Fig. 17.6 Energy savings at the MAC level for LISA dataset. For the 0.5% threshold for the MobileNetv2 case, ConVar could not produce acceptable solutions, while ALWANN resulted in nearly 0% gains in energy for ResNet44

satisfy the accuracy drop thresholds, such as the 0.5% accuracy drop threshold for ResNet44/CIFAR-100 dataset, ShuffleNet/GTSRB dataset, and MobileNetv2/LISA dataset. This can be explained by the fact that even though the employed error correction is very efficient, by not having the ability to control at runtime the applied approximation results in accuracy violation when considering such low thresholds.

4 Mitigating Temperature and Aging Effects in NPUs Through Approximation

In this section, we discuss the challenges that temperature effects as well as aging effects bring to NPUs. Then, we demonstrate how principles from approximate computing can play a major role in mitigating the deleterious effects of temperature and aging as well as reducing the efficiency losses incurred by the required timing guardbands.

4.1 *Thermal Challenges in NPUs*

The main core of any NPU is a large array of multiply-accumulate (MAC) units. Such MACs are very power hungry akin to the underlying performed multiplication operations. The size of such MAC array can vary depending on the end-user scenario. For instance, Google TPU adopts a systolic array that consists of 64×64 and even 256×256 MAC circuits [10]. Despite, the great advantage of MAC arrays in significantly accelerating the inference of DNNs, the massive number of MAC operations executed in parallel within a very small confined area quickly leads to a profound challenge when it comes to on-chip temperatures. This is inevitable because of the inherent structure of MAC arrays, where an enormous number of MAC circuits are tightly connected together inside a small area. As a matter of fact, performing tens of Tera of MAC operations per second inside a small area results in excessive on-chip power densities in which thermal bottlenecks can be rapidly emerged. Toward demonstrating and investigating how high the on-chip power density and the corresponding generated temperature could be during the runtime, we have presented in [4], the first study in which a complete systolic MAC array is analyzed. To this end, we have implemented (from logic synthesis to layout and physical design) a 128×128 MAC array at the 14 nm FinFET technology node. The power and area results have been extracted from accurate sign-off tools and then employed to calculate the on-chip power density at the maximum clock frequency. To cover different scenarios, we have considered both nominal voltage of 0.7 V and Turbo-boost voltage of 0.8 V. Our analysis revealed that the on-chip power density at the systolic MAC array can reach 187 and 312 $\mathrm{W/cm^2}$, at 0.7 and 0.8 V, respectively (Fig. 17.7). It is noteworthy that in this analysis, we have employed an advanced FinFET technology that is carefully calibrated with measurement data from Intel 14nm technology node [3, 4].

To accurately translate the corresponding on-chip temperatures that could be generated from such high on-chip power densities, we have performed accurate multi-physics simulations that employ finite element methods. The commercial ANSYS software has been used for this purpose. To take the impact of air cooling into account, we simulate a Heat Transfer Coefficient (HTC) of 100 $\mathrm{W/m^2K}$, which represents the maximum forced-convection of air and hence it allows us to estimate the resulting temperature under the maximum capability that conventional cooling can ever deliver. Our temperature analysis revealed that the above-mentioned power densities (i.e., 187 and 312 $\mathrm{W/cm^2}$) can result in unsustainable on-chip temperatures that go beyond the critical temperature of 105 °C reaching around 125 and 175 °C, respectively. In Fig. 17.7, we demonstrate the induced on-chip power densities at 0.7 and 0.8 V and the corresponding on-chip temperatures.

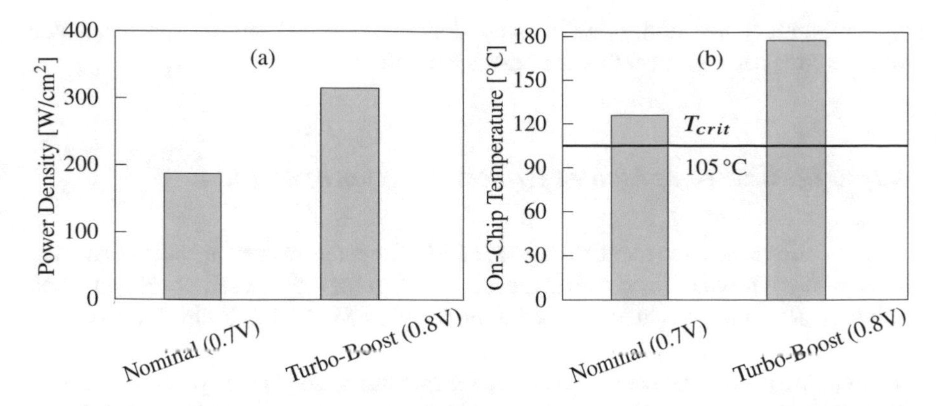

Fig. 17.7 On-chip power density and on-chip temperature analysis of a systolic array that consists of 128 × 128 MACs [4]. The 14 nm FinFET technology node is considered in this study. The MAC circuits operate at the maximum clock frequency. Standard commercial sign-off tool flows for power, delay, and area estimations have been employed for accurate analysis. In addition, commercial multi-physics tool flows heat simulations have been used to accurately analyze the corresponding resulting temperatures. As noticed, MAC arrays can impose a serious thermal bottleneck to NPUs

4.2 Aging Effects in Transistors

During the operation of transistors, applied vertical and horizontal electrical fields induce over time various types of defects. Those defects are unwanted charges that are caused by interface traps and oxide traps [13]. Over time, the generated defects accumulate inside the transistor and lead to a considerable degradation in its switching speed and hence in its performance. Positive/Negative Bias Temperature Instability (P/NBTI) as well as Hot-Carrier Injection (HCI) are the most important aging phenomena that are responsible for stimulating the underlying defect generation mechanisms. In practice, BTI and HCI shift the electrical characteristics of transistors such as threshold voltage (V_T) and carrier mobility (μ) [2, 13]. This reduces the ON current of transistors. As a result, aging over time increases the delay of the circuits' paths and, consequently, timing errors due to violations start to appear because of the unsustainable clock frequencies. In order to overcome aging-induced degradations and sustain proper functionality for the circuit during its entire projected lifetime, the so-called timing guardbands need to be carefully estimated and then included on top of the maximum clock period. However, such timing guardbands result directly in efficiency losses because circuits will be operated at lower frequency clocks than its maximum potential.

The Relation Between Temperature and Aging Effects: Elevated on-chip temperature provides a higher activation energy for the underlying defect generation mechanisms [2]. Therefore, transistors that are located in the chip where excessive temperatures may be generated are subject to much higher aging-induced degradations (i.e., larger ΔV_{th}). In [2, 13], we illustrated the relation between higher temperatures and aging-induced degradations and showed how elevated temperatures can significantly accelerate aging effects.

4.3 Fighting Temperature Effects via Approximation

As earlier discussed, the massive number of MAC operations that are being executed in parallel within a confined area is the main reason for the excessive heat generated in NPUs. Reducing the precision of the underlying MAC circuit directly results in savings in the dynamic power consumption. This leads to reductions in the on-chip power density and hence in the generated temperatures. In general, precision scaling (PS) is a very famous technique that is adopted in many systems that can tolerate errors. It aims at trading-off energy with some errors. The main concept is reducing the precision of the input through clock-gating some input bits. This allows the reduction of the circuit's switching activity and thus the dynamic power.

When it comes to CNNs, precision scaling can be done via *post-training quantization* [29]. In quantization, weights and activations of the CNN model can be translated into a lower-precision numerical representation (e.g., INT8) than what has been originally used during training (i.e., 32-bit FP). Several works demonstrated that INT8 delivers almost the same accuracy as FP32. Note that reducing the precision of the MAC circuit from 32-bit down to 8-bit leads to a significant power reduction. Nowadays, most of the CNN models employ 8-bit in their inference. This is why in our temperature and power density analysis (presented in Fig. 17.7), we have assumed a systolic array in which the underlying MACs are 8-bit, similar to Google TPU [10] and Samsung NPU [20].

To explore the impact of quantization and hence reduce the precision further (i.e., lower than the 8-bit baseline) on the accuracy of DNN models, we employ the open-source machine learning Pytorch framework to train the NN using the default FP32 representation and then we apply n-bit quantization. We use an asymmetric min/max post-training quantization method by using a zero-point (ZP) together with a scale factor (S) [9].

Impact of Precision Scaling on Temperature: In Fig. 17.8, we summarize the impact of PS on reducing the temperature of the systolic array. The latter in this scenario operates at the maximum frequency of 1887 MHz at the nominal voltage of 0.7 V. On the one hand, scaling the MAC precision reduces the dynamic power and consequently the temperature due to the lesser power density. On the other hand, scaling the precision impacts the inference accuracy due to the lower numerical representation used. In our work [4], we have analyzed 20 NN models trained for the ImageNet dataset [6] and in Fig. 17.8 we present the average accuracy of them at different precision levels. As can be observed, decreasing the inference precision from 8-bit to 7-bit, 6-bit, and 5-bit reduces the temperature from the baseline 120 °C, to 116, 111 and 105 °C, respectively, which leads to a drop in the average accuracy from the baseline 89% to 85%, 69%, and 28%, respectively.

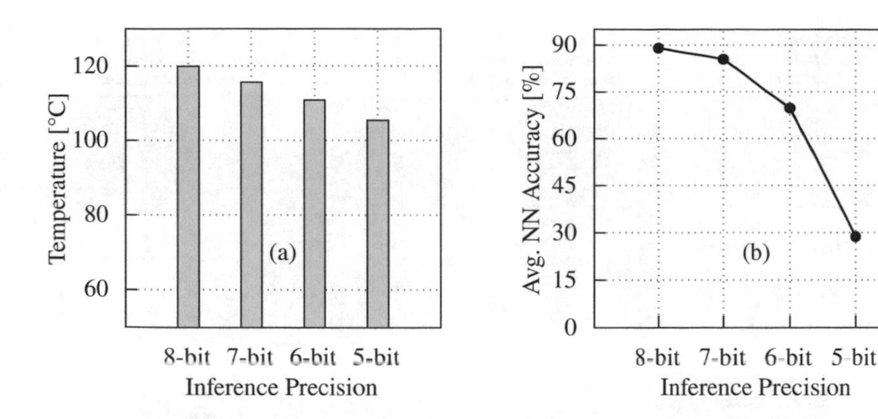

Fig. 17.8 Impact of precision scaling on (**a**) the temperature and (**b**) the inference accuracy [4]

The Need for Additional Cooling Means: As can be noticed in Fig. 17.8, the capability of PS standalone in mitigating the temperature is significant but limited. It can also lead to an unacceptable loss in the accuracy of NN inference. Therefore, in our work [4], we have additionally investigated the capability of frequency scaling and advanced on-chip cooling techniques to mitigate further the on-chip temperature.

1. *Frequency Scaling (FS):* Similar to many existing power and temperature management schemes in processors, FS aims at limiting the clock frequency in which the dynamic power is reduced at the cost of speed/performance that is being scarified. In the context of NPUs, reducing the frequency of MAC array directly (and often linearly) reduces the throughput of the inference that the NPU provides.
2. *Advanced On-Chip Cooling:* Existing traditional cooling techniques are inherently inefficient because the large heat-sink together with fan aim at dissipating the heat from the entire chip's die despite the hot spot is localized at a certain location across the chip. To overcome such a limitation, advanced on-chip cooling aims at providing localized cooling only at the location where it is really needed. This can be realized through a novel material called *superlattice thermoelectric*, which employs Peltier effects to generate controlled cooling. Details on the effectiveness and functionality of on-chip superlattice thermoelectric cooling can be found in [4, 12]. In practice, we envision that the MAC array is covered by such thermoelectric devices, which allows the system to dissipate heat beyond the maximum capability of forced-convection air. Note that thermoelectric devices are active cooling in which power is needed to create cooling effects. The larger the current that is provided to the superlattice TE, the higher the cooling and the larger the consumed power.

The Available Design Space for NPU Thermal Management: In the presence of three different cooling means, which are frequency scaling, precision scaling,

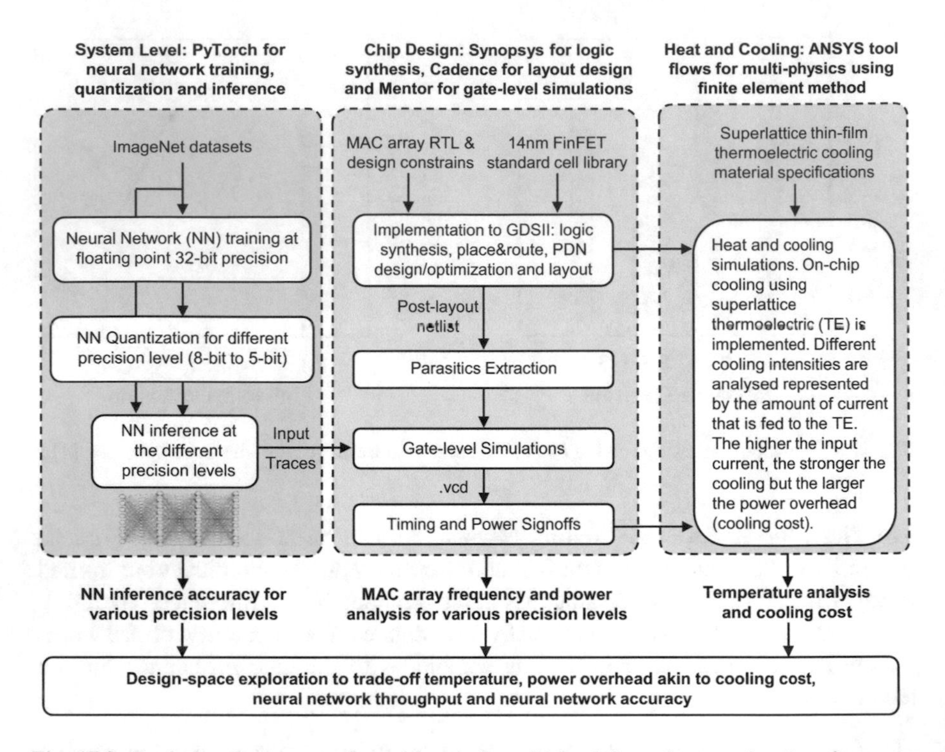

Fig. 17.9 Exploring the impact of precision scaling obtained through approximation, frequency scaling, and on-chip cooling. Figure is obtained and modified from [4]

and advanced on-chip cooling, a large design space becomes available to manage the temperature of NPUs. In this design space, different kinds of trade-offs can be explored such as energy, throughput, temperature, inference accuracy, etc. Under tight temperature constraints, standalone individual cooling techniques lead to sub-optimal results in which either low inference accuracy is delivered, or a high power overhead is incurred, or a very low throughput is achieved. However, we have demonstrated in [4] that a joint hybrid solution to manage the temperature of NPUs is a key. In Fig. 17.9, we illustrate our implemented framework that allows the above-mentioned design-space exploration. It consists of three major parts: (1) NN post-training quantization to explore the impact of precision scaling on the NN inference accuracy, (2), RTL implementation and evaluation that provides us with accurate hardware results like power, area, energy, delay, etc., extracted from standard sign-off tools, and (3) cooling and temperature analysis using multi-physics simulations. Based on a certain given temperature constraint that needs to be fulfilled, the framework provides the corresponding Pareto frontier.

We demonstrate in Fig. 17.10 an example on the obtained efficiency improvement using our hybrid cooling solution for a requested temperature constraint of 105 °C. The MAC array in this scenario operates at the 0.8 V (i.e., Turbo-Boost mode). As shown, a speedup between 30% and 80% can be achieved based on the scaled

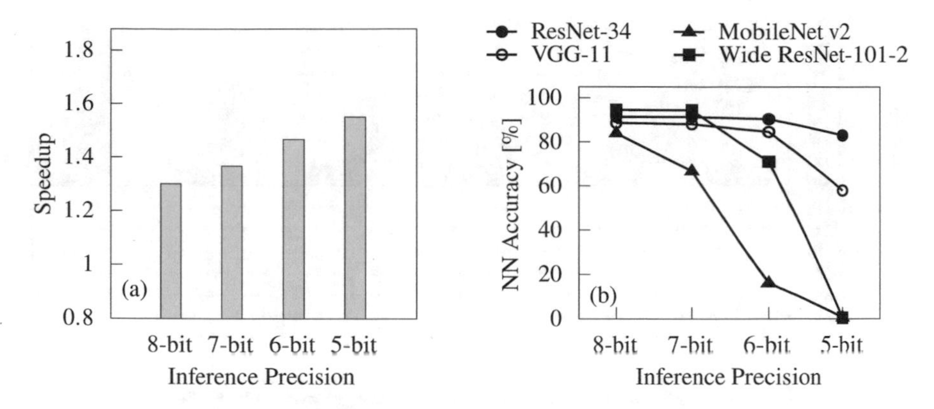

Fig. 17.10 Obtained speedup (**a**) along with the resulting accuracy loss (**b**) in NN inference using our hybrid cooling solution for a temperature constraint of 105 °C [4]

precision level. Some NN models (e.g., ResNet) can tolerate precision scaling more than others (e.g., MobileNet). Therefore, a hybrid cooling solution is a key because it offers different combination of cooling techniques toward maximizing the efficiency while still fulfilling temperature constraints.

4.4 Fighting Aging Effects via Approximation

With technology scaling, wider and wider timing guardbands are needed to overcome aging-induced degradations. These guardbands directly result in efficiency losses. Removing or narrowing guardbands leads to errors due to timing violations. Such errors are unpredictable and lead to unacceptable error rates as demonstrated in [1, 22]. To overcome this challenge, we have introduced in [1, 22], the concept of *aging-induced approximation* in which we aim at completely removing the guardbands through exploring approximate computing principles. In the context of NPUs, we showed in [22] how compressing the inputs of the MAC to achieve faster operation enable us to compensate aging. For a certain degradation that can be caused by aging at the end of the projected lifetime (e.g., 40, 50 mV), a certain level of approximation can be introduced to compensate the induced delay increase. This enables us to tradeoff the aging-induced timing errors (in the absence of timing guardbands) with approximation through quantization that can cause an acceptable and smooth loss in the CNN accuracy. As Fig. 17.11a shows, our aging-aware quantization technique adaptively compresses the inputs over time to gracefully compensate the aging-induced delay increase. As seen, the normalized delay is less than or equal to 1. Hence, our aging-aware quantization technique increases the resiliency against aging effects. Furthermore, aging-aware quantization exhibits almost a constant delay from "fresh" until the end of the

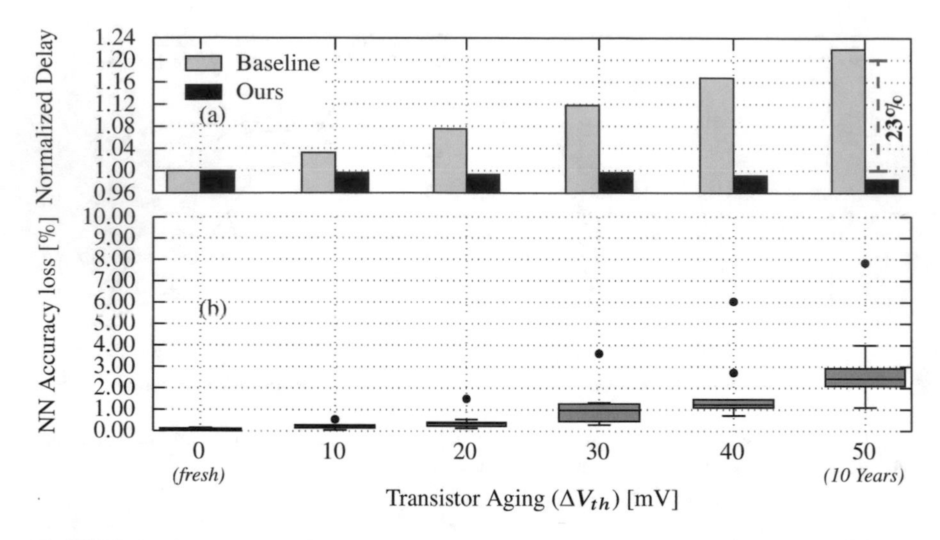

Fig. 17.11 (**a**) Normalized delay from the absence of aging (i.e., fresh) until the end of lifetime (e.g., 10 years). (**b**) Our aging-induced approximation provides a graceful accuracy degradation, from the beginning of the lifetime until the end of the lifetime, presented by box plots w.r.t. several CNN models. Figure is obtained and modified from [22]

projected lifetime (10 years). As a result, timing guardbands can be fully removed, which, in turn, provides a 23% delay gain compared to baseline design. Finally, this progressive input compression through aging-aware quantization results in a graceful accuracy loss over time (Fig. 17.11b).

Acknowledgments This work is partially supported by the German Research Foundation (DFG) through the project "ACCROSS: Approximate Computing aCROss the System Stack."

References

1. Amrouch H, Khaleghi B, Gerstlauer A, Henkel J. Towards aging-induced approximations. In: Design automation conference. 2017.
2. Amrouch H, Ehsani SB, Gerstlauer A, Henkel J. On the efficiency of voltage overscaling under temperature and aging effects. IEEE Trans Comput. 2019;68(11):1647–62.
3. Amrouch H, Pahwa G, Gaidhane AD, Dabhi CK, Klemme F, Prakash O, Chauhan YS. Impact of variability on processor performance in negative capacitance finfet technology. IEEE Trans Circuits Syst I Reg Pap. 2020;67(9):3127–37.
4. Amrouch H, Zervakis G, Salamin S, Kattan H, Anagnostopoulos I, Henkel J. Npu thermal management. IEEE Trans Comput Aided Des Integr Circuits Syst. 2020;39(11):3842–55. https://doi.org/10.1109/TCAD.2020.3012753.
5. Ansari MS, Mrazek V, Cockburn BF, Sekanina L, Vasicek Z, Han J. Improving the accuracy and hardware efficiency of neural networks using approximate multipliers. IEEE Trans Very Large Scale Integr (VLSI) Syst 2019;28(2):317–28.

6. Deng J, Dong W, Socher R, Li L, Kai Li, Li Fei-Fei. Imagenet: A large-scale hierarchical image database. In: Conf. on Comp. Vis. and Pat. Recogn. (CVPR). 2009. pp. 248–55.
7. Han S, Pool J, Tran J, Dally WJ. Learning both weights and connections for efficient neural networks. Preprint. arXiv:150602626. 2015.
8. Hao C, Dotzel J, Xiong J, Benini L, Zhang Z, Chen D (2021) Enabling design methodologies and future trends for edge ai: Specialization and codesign. IEEE Design Test 38(4):7–26. https://doi.org/10.1109/MDAT.2021.3069952.
9. Jacob B, Kligys S, Chen B, Zhu M, Tang M, Howard A, Adam H, Kalenichenko D. Quantization and training of neural networks for efficient integer-arithmetic-only inference. In: Proceedings of the IEEE conference on computer vision and pattern recognition. 2018. pp. 2704–2713.
10. Jouppi NP, et al. In-datacenter performance analysis of a tensor processing unit. In: Int. Symp. on computer architecture. 2017. pp 1–12.
11. Karmarkar N, Karp RM. An efficient approximation scheme for the one-dimensional bin-packing problem. In: 23rd Annual symposium on foundations of computer science (SFCS 1982). IEEE; 1982. pp. 312–20.
12. Kattan H, Chung SW, Henkel J, Amrouch H. On-demand mobile cpu cooling with thin-film thermoelectric array. IEEE Micro. 2021;41(4):67–73.
13. Klemme F, Amrouch H. Machine learning for on-the-fly reliability-aware cell library characterization. IEEE Trans Circ Syst I Reg Pap. 2021;68(6):2569–79.
14. Leon V, Zervakis G, Xydis S, Soudris D, Pekmestzi K (2018) Walking through the energy-error pareto frontier of approximate multipliers. IEEE Micro. 38(4):40–9.
15. Li C, Luo W, Sapatnekar SS, Hu J. Joint precision optimization and high level synthesis for approximate computing. In: Proceedings of the 52nd annual design automation conference. 2015. pp. 1–6.
16. Mrazek V, Sarwar SS, Sekanina L, Vasicek Z, Roy K. Design of power-efficient approximate multipliers for approximate artificial neural networks. In: Int. Conf. computer-aided design. 2016. pp. 1–7.
17. Mrazek V, Hrbacek R, Vasicek Z, Sekanina L. Evoapproxsb: Library of approximate adders and multipliers for circuit design and benchmarking of approximation methods. In: Design, automation & test in Europe conference & exhibition. 2017. pp. 258–261.
18. Mrazek V, Vasicek Z, Sekanina L, Hanif MA, Shafique M. ALWANN: automatic layer-wise approximation of deep neural network accelerators without retraining. In: Int. Conf. computer-aided design. 2019. pp. 1–8.
19. Mrazek V, Sekanina L, Vasicek Z. Libraries of approximate circuits: Automated design and application in CNN accelerators. IEEE J Emerg Sel Top Circuits Syst. 2020;10(4):406–18.
20. Park JS, Jang JW, Lee H, Lee D, Lee S, Jung H, Lee S, Kwon S, Jeong K, Song JH, et al. 9.5 a 6k-mac feature-map-sparsity-aware neural processing unit in 5nm flagship mobile soc. In: IEEE international solid-state circuits conference (ISSCC), vol. 64. 2021. pp. 152–4.
21. Renda A, Frankle J, Carbin M. Comparing rewinding and fine-tuning in neural network pruning. Preprint. arXiv:200302389. 2020.
22. Salamin S, Zervakis G, Spantidi O, Anagnostopoulos I, Henkel J, Amrouch H. Reliability-aware quantization for anti-aging npus. In: Design, automation & test in Europe conference & exhibition. 2021. pp. 1460–65. https://doi.org/10.23919/DATE51398.2021.9474094.
23. Sarwar SS, Venkataramani S, Ankit A, Raghunathan A, Roy K. Energy-efficient neural computing with approximate multipliers. ACM J Emerg Technol Comput Syst (JETC). 2018;14(2):1–23.
24. Shafique M, Ahmad W, Hafiz R, Henkel J. A low latency generic accuracy configurable adder. In: 2015 52nd ACM/EDAC/IEEE design automation conference (DAC). 2015. pp 1–6. https://doi.org/10.1145/2744769.2744778.
25. Spantidi O, Zervakis G, Anagnostopoulos I, Amrouch H, Henkel J. Positive/negative approximate multipliers for DNN accelerators. In: IEEE/ACM International Conference On Computer Aided Design (ICCAD). 2021. pp. 1–9. https://doi.org/10.1109/ICCAD51958.2021.9643491.

26. Tasoulas ZG, Zervakis G, Anagnostopoulos I, Amrouch H, Henkel J (2020) Weight-oriented approximation for energy-efficient neural network inference accelerators. IEEE Trans Circuits Syst I Reg Pap. 67(12):4670–83. https://doi.org/10.1109/TCSI.2020.3019460.
27. Vaverka F, Mrazek V, Vasicek Z, Sekanina L, Hanif MA, Shafique M. Tfapprox: Towards a fast emulation of DNN approximate hardware accelerators on GPU. In: Design, automation and test in Europe conference (DATE). 2020. p. 4.
28. Venkataramani S, et al. Efficient ai system design with cross-layer approximate computing. Proc IEEE. 2020;108(12):2232–50. https://doi.org/10.1109/JPROC.2020.3029453.
29. Yang J, Shen X, Xing J, Tian X, Li H, Deng B, Huang J, Hua X. Quantization networks. In: Conf. on Comp. Vis. and Pat. Recogn. (CVPR). 2019. pp. 7300–8.
30. Zervakis G, Tsoumanis K, Xydis S, Soudris D, Pekmestzi K. Design-efficient approximate multiplication circuits through partial product perforation. IEEE Trans Very Large Scale Integr (VLSI) Syst. 2016;24(10):3105–17.
31. Zervakis G, Koliogeorgi K, Anagnostos D, Zompakis N, Siozios K. Vader: Voltage-driven netlist pruning for cross-layer approximate arithmetic circuits. IEEE Trans Very Large Scale Integr (VLSI) Syst. 2019;27(6):1460–64. https://doi.org/10.1109/TVLSI.2019.2900160.
32. Zervakis G, Xydis S, Soudris D, Pekmestzi K. Multi-level approximate accelerator synthesis under voltage island constraints. IEEE Trans Circ Syst II Express Briefs. 2019;66(4):607–11. https://doi.org/10.1109/TCSII.2018.2869025.
33. Zervakis G, Amrouch H, Henkel J. Design automation of approximate circuits with runtime reconfigurable accuracy. IEEE Access. 2020;8:53522–38.
34. Zervakis G, Saadat H, Amrouch H, Gerstlauer A, Parameswaran S, Henkel J. Approximate computing for ML: state-of-the-art, challenges and visions. In: Asia and South Pacific design automation conference. 2021. pp. 189–96.
35. Zervakis G, Spantidi O, Anagnostopoulos I, Amrouch H, Henkel J. Control variate approximation for dnn accelerators. In: 58th ACM/IEEE Design Automation Conference (DAC). 2021. pp. 481–486. https://doi.org/10.1109/DAC18074.2021.9586092.

Chapter 18
Approximate Computing for Energy-Constrained DNN-Based Speech Recognition

Bo Liu, Hao Cai, Zhen Wang, and Jun Yang

The KWS system is a widely used speech-triggered interface for human–machine interaction and plays an increasingly important role in nowadays practical products, such as the wearable devices and IoT. In these battery-powered devices, the KWS system is usually required to be always-on, and therefore the ultra-low-power and real-time processing with high recognition accuracy are the critical requirements. In addition, DNNs are now demonstrating much more advantages in speech recognition than traditional models (i.e., hidden Markov models (HMMs) and Gaussian mixture models (GMMs)); however, the push towards more complex DNN models conflicts with the power–performance–area (PPA) constraints of KWS. To overcome this challenge, approximate computing is emerging as a design paradigm that can substantially improve the energy efficiency of DNN-based KWS systems. Due to the intrinsic error tolerance characteristics of neural networks, applying approximation in DNN-based recognition applications does not cause too much accuracy loss and can reduce the high computation and implementation costs. By modifying, removing, or adding certain elements to reduce DNN model size or simplify specific circuits, approximate computing can significantly decrease the hardware cost of the system [1]. Currently, approximate computing has been widely used in all levels of the KWS system to increase the system performance. This chapter presents the application of approximate computing in KWS systems at different levels and discusses the hardware and software co-optimization framework to achieve high energy efficiency. The rest part of this chapter is organized as follows: Sect. 1 is devoted to the approximation in algorithm level and describes the approximate schemes in feature extraction, DNN models, quantization, and compression.

B. Liu · H. Cai (✉) · J. Yang
National ASIC System Engineering Research Center, Southeast University, Nanjing, China
e-mail: liubo_cnasic@seu.edu.cn; hao.cai@seu.edu.cn

Z. Wang
Nanjing Prochip Electronic Technology Co. Ltd, Nanjing, China

W. Liu, F. Lombardi (eds.), *Approximate Computing*,
https://doi.org/10.1007/978-3-030-98347-5_18

Section 2 presents the approximation in circuit level from analog / analog–digital mixed computing, digital computing, and new devices for approximate computing. Then, Sect. 3 discusses the accuracy-reconfigurable approximate architecture and co-design framework of software and hardware approximate computing in KWS systems. Finally, this chapter is concluded in Sect. 4.

1 Algorithm-Level Approximate Computing for KWS

The algorithm of KWS mainly includes two parts: speech feature extraction and keywords classification. The feature extraction module of KWS is used to extract the features of the input speech signal, and then the keywords classification module classifies the extracted features to determine which keyword (or unknown word) it is. For feature extraction, it mainly includes: Mel-scale frequency cepstral coefficient (MFCC) [2, 3], perceptual linear production (PLP) [4], relative spectral–perceptual linear predictive (RASTA-PLP) [5], and linear prediction coding coefficient (LPCC) [6]. These feature extraction algorithms are evaluated in work [7] through experiments. The experimental results show that when the signal-to-noise ratio (SNR) is low or the background noise changes greatly, MFCC has the best performance among the evaluated algorithms due to its low computational complexity and high robustness. For keywords classification, DNN has been shown to outperform traditional models in the past decades. Since KWS is usually embedded in IoT devices with limited energy consumption and computing resources, high energy efficiency and extremely low power consumption are critical to the deployment of MFCC and DNN in KWS. Therefore, approximation design technologies in the algorithm (MFCC and DNN) are applied to reduce the data access and computational effort of the KWS. Common approximation schemes include approximation of MFCC, quantization, and compression of DNN models.

1.1 Feature Extraction and DNN-Based Keywords Classification

MFCC is the most commonly used feature extraction algorithm in KWS systems. Optimizing the complex algorithm of MCFC, which leads to a large amount of resource consumption, can significantly improve the energy efficiency of the system. In addition, the output data of feature extraction is the input of the neural network classifier. Since neural networks are error-tolerant, the approximation of MFCC can achieve energy consumption reduction and efficiency improvement with less impact on accuracy. Work [8] used rectangular bandpass Mel filters to replace the conventional triangular band-pass Mel filters and used fast Fourier transform (FFT) with no frame overlapping to simplify the computing. The optimized MFCC can

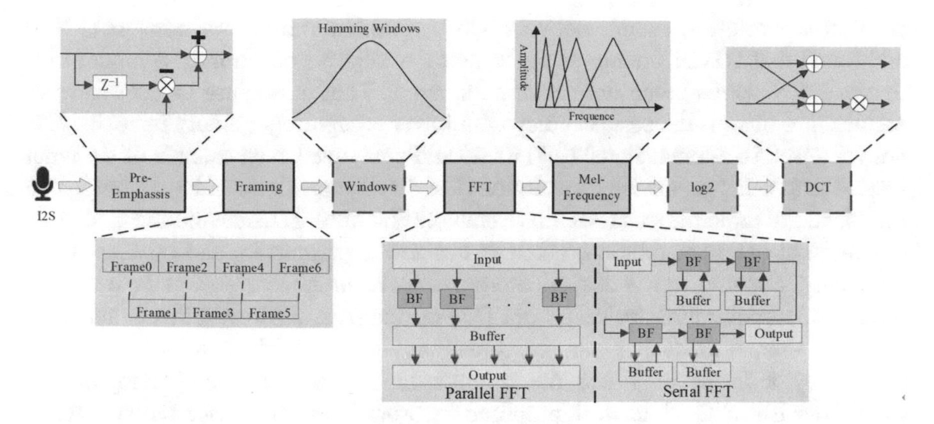

Fig. 18.1 MFCC algorithm framework

effectively reduce the required computation by less than 40% of the standard MFCC with little impact on the recognition accuracy of the KWS system. The complete MFCC algorithm block diagram is shown in Fig. 18.1. The operations in the dashed box can be removed to reduce computational complexity. Work [9] proposed that ultra-low-power speech recognition can be achieved without windowing in MFCC. Work [10] removed the windowing, log2, and discrete Cosine transform (DCT), which have little impact on the recognition accuracy of KWS, and used a two-stage serial pipelined FFT circuit to replace the conventional N single-port serial pipelined FFT circuit for 2^n points FFT operations, which can compress the memory by a factor of 4 and the power consumption by a factor of 2. Since many operations used in the MFCC algorithm rely on complex functions (e.g., square and logarithmic functions), their outputs are associated with a large dynamic range. A floating-point representation can cover a larger dynamic range with fewer bits than a fixed-point representation. In work [11], a floating-point logarithmic unit has been proposed to cover a larger dynamic range with a smaller bit-width and minimize noise, of which the unit and memory are shared with many processes, thus significantly reducing hardware complexity and energy consumption.

DNNs have recently achieved great success in many speech recognition tasks. However, existing DNN models are computationally expensive and memory intensive, hindering their deployment in low memory resource devices or applications with stringent latency requirements. Therefore, a natural idea is to perform approximation and acceleration in DNN models without significantly degrading the performance. The approximation and acceleration methods in speech recognition algorithms and neural networks are described below.

The introduction of approximation operations in MFCC can effectively reduce the complexity of feature extraction. In order to minimize the complexity and power consumption of speech recognition, the approximation and acceleration of neural network classifiers are required. Work [12] proposed a deep separable convolutional neural network (DSCNN) and compared it with neural network architectures such as DNN, convolutional neural network (CNN), and recurrent neural network (RNN) and found that DSCNN achieves an accuracy of 95.4%, which is 10% higher than that of DNN with similar number of parameters. That is because its deeper depth architecture obtains the best accuracy with lower computing/memory intensity. The network used two-dimensional (2D) filters and convolved each channel of the input feature map using pointwise convolution (i.e., 1 × 1 convolution). By decomposing the standard three-dimensional (3D) convolution into 2D convolution and one-dimensional (1D) convolution, DSCNN has fewer parameters and operates more efficiently, allowing for a deeper, more extensive architecture even in resource-limited microcontroller devices. Work [13] normalized and symbolically activated the data after each deep separable convolutional layer, which reduced the data storage by a factor of 7 and the computation by a factor of 7 compared to CNN. The BinaryCmd network proposed by work [14] can reduce the number of parameters by 60% and the computational effort by 50% with only 3.4% decrease in accuracy compared to DSCNN, in which the weights are not learned directly but are dynamically generated by using a linear combination of deterministic binary weighting coefficients. The neural network architecture mentioned above is shown in Fig. 18.2.

Many approximation paradigms have also been proposed for the operation within neural network models. Work [15] made a simplification for multiplier CNNs by setting a preclassifier at each convolutional layer, which is connected to a confidence logic control (CLC) unit to determine whether the classification junction has sufficient confidence. If not, proceed to the next convolutional layer; if the confidence requirement is satisfied, stop the convolutional operation to reduce the convolutional operations. Precise computing of the activation function in neural networks often requires significant resource consumption. Works [8, 16] proposed the use of segmented linear functions to approximate the activation function. Work

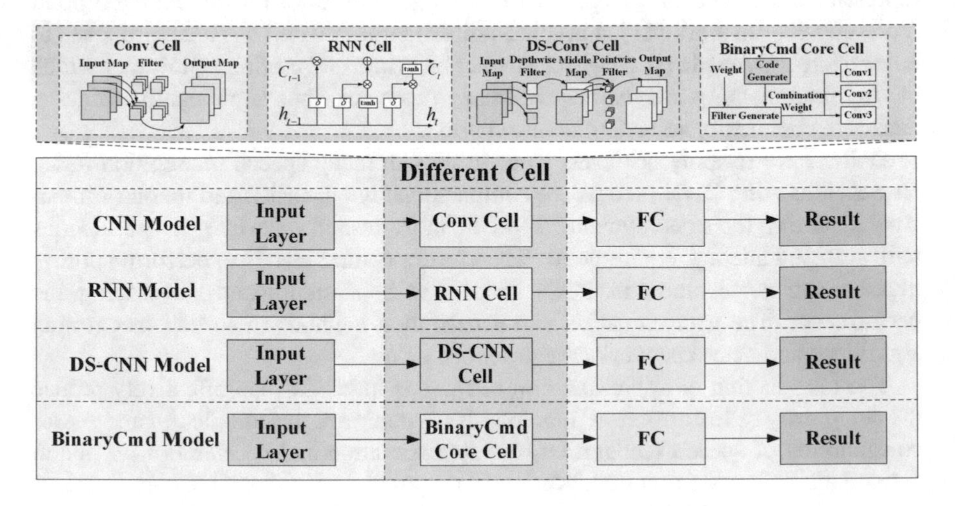

Fig. 18.2 Comparison of CNN, RNN, DSCNN, and BinaryCmd topologies

[16] approximated the tanh and sigmoid activation functions with eight linear functions stored in the LUT and obtained good accuracy. Work [8] adopted different numbers of segmented linear functions to approximate the activation functions to match different accuracies in order to minimize the resource consumption while meeting the system accuracy requirements. Work [17] proposed the concept of similarity score for long short-term memory (LSTM) networks, which measures the degree of similarity and disables highly similar LSTM operations for two adjacent LSTM units and directly outputs the previous results to reduce the cost of speech recognition.

1.2 Quantization and Compression for DNN

As larger neural networks with more layers and nodes are considered, reducing their storage and computational cost becomes critical, especially for network quantization compresses the original network by reducing the bit-width of each weight. Work [18] applied k-means scalar quantization to the parameter values. Work [19] proposed the three-stage compression methods (pruning, quantization, and Huffman encoding) to compress the network. As is shown in Fig. 18.3, the network size can obtain $35\times -49\times$ reduction after three-stage compression.

Work [20] showed that 8-bit quantization of the parameters can result in significant speed-up with minimal loss of accuracy. Work [21] used 16-bit fixed-point representation in stochastic rounding-based CNN training, which significantly reduced memory usage and float-point operations with little loss in classification accuracy. The ResNet-50 [22] with 50 convolutional layers needs over 95MB memory for storage and over 3.8 billion floating number multiplications when processing an image. Discarding some redundant weights, the network still works as usual but saves more than 75% of parameters and 50% computational time.

Due to the limitations in human speech perception and the redundancy of neural networks, the use of quantized fixed-point numbers in neural networks instead of the original 32-bit floating-point numbers does not bring a significant performance

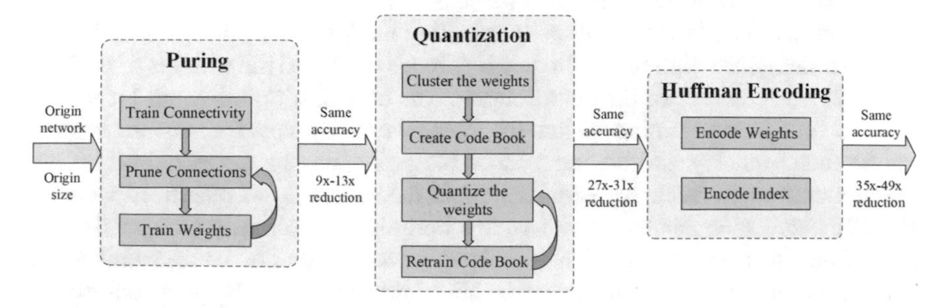

Fig. 18.3 The three-stage compression methods proposed in work [19]

degradation in KWS systems [12]. On the contrary, it results in a smaller model size, which is more favorable for deployment in edge computing. In work [23], a binarized neural network (BNN) for KWS systems with 1-bit-width for both activation and weight has been proposed, which can significantly reduce the storage space and power consumption of activation and weight compared to conventional DNN with 16-bit data and weight bit-width. But this BNN only supports the recognition of one keyword under low background noise, while the recognition accuracy for complex speech environment is greatly reduced. To overcome the shortcomings of this BNN, works [8, 24] proposed a binarized weight neural network (BWN) for KWS with 1-bit quantization for weights only, and work [24] pointed out that reducing the weight bit-width is more effective than reducing the activation bit-width because weights can be pre-quantized and thus the storage can be reduced very effectively. Therefore, increasing the activation bit-width can effectively improve the system accuracy. Work [8] proposed an exact control method for SNR prediction and post-detection of network output confidence based on background noise, using which the data path of the proposed DNN accelerator can be dynamically and adaptively reconfigured to 4, 8, or 16 bits to suit various situations with different noise types and SNRs. Works [12], [16], and [25] quantized all the weights and activations of the network to 8 bits. Due to the regularization caused by the quantization, the accuracies of the quantized networks are the same or slightly improved compared to the full accuracy network. To determine the optimal quantization bit-width for each layer of the neural network, work [15] proposed that for CNNs, the quantization bit-width be started from the first convolutional layer until the network accuracy is lower than the set relative error rate, followed by quantizing the subsequent layers layer by layer using this approach. This method provided better results compared to the traditional way of quantizing all layers in the network to the same bit-width.

The quantization of activation values, especially the quantization of weights, is very important for hardware implementation. Appropriate quantization bit-width can reduce the power consumption of the KWS system and achieve a compromise between power consumption and accuracy. Work [26] proposed a bit-by-bit, layer-by-layer quantization method to quantize the weights during the training process to avoid the loss of recognition accuracy. In order to make the quantized weights closer to the ideal values during the training process, a bit-by-bit quantization method has been proposed, i.e., instead of directly selecting the low bit-width quantization for the first training, the high bit-width quantization is used, and the weights of the high bit-width are reserved for the next training. The quantization bit-width is reduced bit-by-bit in the next training, so that the most favorable points for network training can be found quickly and the most favorable points can be quantized bit by bit, which improves the accuracy of training and the reliability of quantized weights. The CNN activation and weight bit-widths optimized in this way are quantized to 8 and 7 bits, respectively, which can achieve the recognition of 10 keywords under low background noise. The quantization bit-width can be slowly quantized from 8 bits to 4 bits, then 2 bits to 1 bit eventually.

2 Circuit-Level Approximate Computing for KWS

There are a large number of multiplication and addition operations in DNN-based speech recognition. Therefore, approximate computing is introduced in KWS to reduce power consumption or computational cost. Approximate computing used in KWS at the circuit level can be generally classified into analog and digital approximate computing. The two main forms of approximation in analog computational circuit are: functional inaccuracies caused by inherent device and circuit non-idealities, and approximation of peripheral circuits intended to minimize circuit cost. In addition, emerging devices have great potential in designing approximate computing logic at the circuit level due to their non-volatility, zero standby power consumption, unlimited durability, and high density.

2.1 *Analog/Analog–Digital Mixed Approximate Computing*

Analog approximate computing are mainly implemented through processing in memory, specific-applicable analog neural network processors, and analog multipliers, which can achieve low-power performance in specific applications. In work [27], an analog-based voltage-domain multiplier was proposed, which has a low power consumption when compared with a standard multiplier. However, the common approximate multipliers are mainly digital based due to the low robustness of analog computing under process variations and noise. The KWS of work [28] consists of an 85 nW acoustic feature extractor based on spike-domain splitting energy normalized background noise and process variation tolerance together with an analog spiking neural network (SNN) classification chip, which achieves accurate identification of keywords in different noise environments with very low power consumption. The typical computing process of CNN involves a large number of sliding convolution operations. In this regard, computing units supporting parallel multiplicative accumulation computations are highly needed. This need has led to the redesign of conventional computing systems to run CNNs with higher performance and lower power consumption, ranging from general application platforms, such as graphics processing units [29], to application-specific accelerators [30, 31]. However, further improvements in computational efficiency will ultimately be limited by the von Neumann architecture of these systems, where the physical separation of memory and processing units leads to significant energy consumption and large latency for data shuffling between units. In contrast, memristive-supported neuromorphic computing offers a promising non-von Neumann computing paradigm, where both data storage and computing are done within the memory, thereby eliminating the cost of data transfer. Works [32–34] proposed a construction of CNN based on a five-layer crossed memristor array for MNIST10 image recognition, where the weight data matrix is stored on the memristor as a conductor and the input data matrix is converted to an input voltage acting on the memristor to produce an output current through a digital to analog converter (DAC). The currents on the array are superimposed to operate the dot product of the two matrices. The crossed memristor achieves a high accuracy rate over 96%. The energy efficiency of the memristor-based CNN neuromorphic system is more than two orders of magnitude higher than that of the most advanced graphics processing units and is able to scale to larger networks such as residual neural networks, which provides a viable memristor-based non-Von Neumann hardware solution for DNNs and edge computing.

Memristor-crossbar-based computing systems address the disadvantages of traditional von Neumann architecture storage walls. Due to the non-ideal nature of analog circuits, the approximation allows for broader trade-off between accuracy and power consumption between different computing cases. However, analog signals are less tolerant to noise, and the results of analog computations usually need to be digitized before being sent to other modules for processing. Analog computing units need peripheral circuits, such as the DAC on the input side and the analog to digital converter (ADC) on the output side. The peripheral circuits of the analog computing unit, especially the ADC, have been shown to account for a significant portion (up to 80%) of the total energy consumption of the analog computing unit. In addition, the ADC consumes close to 70% of the matrix-vector multiplication (MVM) unit area [35]. Besides, device variations and process deviations are major challenges. The cumulative effect of the non-idealities of device can lead to functional errors at the circuit cell level. In the case of large-scale systems, these errors can accumulate and significantly degrade the application accuracy [36, 37]. Therefore, digital computing is more suitable for DNN-based power-constrained KWS design and optimization. For example, the state-of-the-art KWS [23–28] in recent years has adopted digital computing for design and optimization.

2.2 Digital Approximate Computing

Due to the fault-tolerant nature of neural networks, approximate adders and multipliers are commonly integrated in DNN accelerator to replace traditional accurate addition and multiplication. These operations reduce latency and lower power consumption while meeting system requirements [38]. An approximate computing circuit can be obtained in the following three aspects:

1. **Simplification of algorithm**
   Complex computings can be converted into some simpler operations to improve circuit performance and energy efficiency. Mitchells binary-logarithm-based algorithm is able to implement multipliers and dividers using adders and subtractors, respectively [39], which is the origin of most current simplified algorithms for approximate multiplier and divider designs [40, 41]. However, the accuracy of this kind of design is relatively low and may require many

peripheral circuits to achieve high accuracy, which may limit the hardware efficiency. In fact, several approximation techniques are usually used simultaneously in hybrid approximation circuits [42].

2. **Modification of logic circuits**
   A more common approach to approximate computing circuit design is to derive an approximate design from an accurate circuit by modifying, removing, or adding some elements. For example, some transistors in the mirror adder are removed to implement low power and high speed full adders [43]. Alternatively, an approximate circuit can be obtained by simplifying a truthtable or a Karnaugh map (K-map) [44, 45]. These approaches generate circuits with deterministic error characteristics.
3. **Optimization of transistor parameters**
   Take voltage for example, voltage scaling can exploit the energy-saving potential of the system as much as possible by providing a lower voltage supply to effectively reduce the power consumption of the circuit without changing the circuit structure [46]. However, reducing the voltage increases the critical path delay, which may lead to timing errors [47]. As a result, the output may be erroneous due to the violation of timing constraints. In addition, the error characteristics of this approximation operation are uncertain and subject to parameter variations [48]. When the most significant bit (MSB) is affected, the output error must not be overlooked [49].

2.2.1 Design Metrics of Approximate Computing Units

For approximate circuits, error characteristics and circuit measurement metrics are the most important characteristics. Works [50–59] introduce error characteristics—the basic error metrics are error rate (ER) and error distance (ED). ER indicates the probability of generating an erroneous result, and ED shows the arithmetic difference between the approximate and the exact result. Denoting the exact and approximate calculation results by M and M', respectively, the ED can be expressed as $ED = |M - M'|$; furthermore, the relative error (RED) represents the relative difference with respect to the exact result and is given by $RED = |ED/M|$. In addition, mean error deviation (MED) and mean relative error deviation (MRED) are usually used to assess the accuracy of the approximate design and they are expressed as

$$MED = \sum_{i=1}^{N} ED_i \cdot P(ED_i) \tag{18.1}$$

and

$$MRED = \sum_{i=1}^{N} RED_i \cdot P(RED_i) \tag{18.2}$$

where N is the number of all possible combinations of inputs, ED_i and RED_i are the ED and RED of the i-th input, and P is the probability of occurrence of ED_i and RED_i and also the probability of occurrence of the i-th input. The mean squared error (MSE) and root-mean-square error (RMSE) are also widely used to measure the arithmetic error magnitude.

Other basic circuit metrics include critical path delay, power consumption, and area. Some composite metrics include power-delay product (PDP), area-delay product (ADA), and energy-delay product (EDP). However, the PPA constraints of approximate circuits conflict with the ED or RED for different applications and scenarios, so the goal of designing approximate circuits is usually to optimize the circuit measurement metrics within a tolerable error range.

2.2.2 Approximate Adder

Adders are used to perform the addition of two binary numbers. Approximate adders make a trade-off between hardware efficiency and accuracy, while conventional adders typically bring a large cost in circuit area and power consumption. Commonly used approximate adder schemes include: segmented approximate adder [60–63], approximate carry-select adder [64, 65], and approximate full adder [10, 13, 31].

Segmented Approximate Adder
Currently known parallel adders such as carry lookahead adder, Brent–Kung adder, Kogge–Stone adder, etc. have considerable asymptotic properties. The critical path delay of an N-bit parallel adder is asymptotically proportional to $\log(N)$. The computing must consider all input bits in order to obtain the correct final result, but the significant carry chain is much shorter in most cases. Therefore, a faster adder can be constructed to approximate the result by considering the first k inputs only instead of all the previous input bits of the current carry. Based on the above idea, the adder can be divided into several parallel sub-adders that have their own independent carry inputs. In addition, for the addition of two n-bit numbers, the lowest 1 bit of the approximate full adder is approximated and the higher $n - 1$ bits are computed exactly to shorten the critical path and reduce power consumption, since the lower bits of the result have less impact on the accuracy. The schematic of the segmented approximate adder is shown in Fig. 18.4.

Experiments in work [66] showed that the accuracy of the approximate adder is much higher than the probability derived using random data inputs. For random data, the expected prediction accuracy is about 65% for a 32-bit addition with a 4-bit carry chain. However, simulation results showed that nearly 90% of the additions were correct using an approximate adder with 4-bit carry and inputs from real

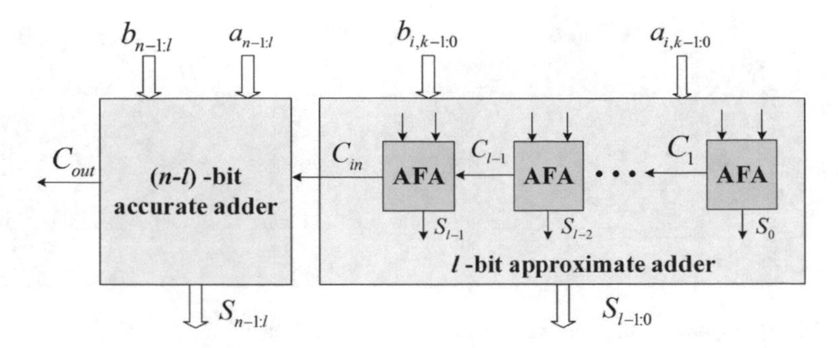

Fig. 18.4 Basic structure of segmented approximate adder

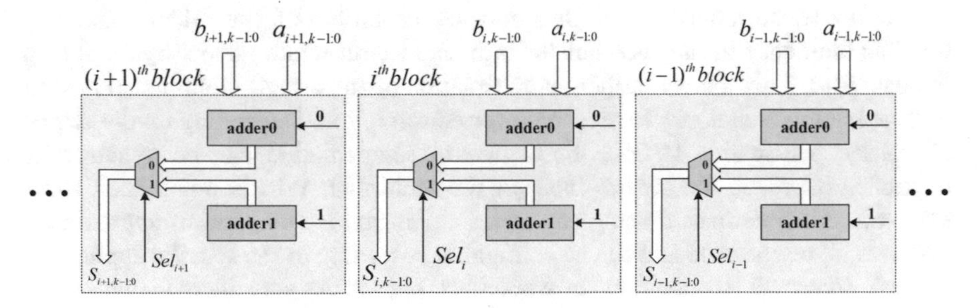

Fig. 18.5 Approximate carry-select adder with sum selection

applications. There is no carry input between the individual subadders in work [60], so there is a large loss of accuracy. This approximate adder can be applied to a low-power neural network accelerator. There is also no carry input between subadders in work [64], but an overlap term is added between adjacent submodules to improve the accuracy. Work [62] used a carry generator to predict the carry of the next adder submodule, which showed significant improvement in accuracy but higher circuit complexity. Work [63] approximated the 16-bit addition in a CNN network, which was a combination of one 3b-RCA, two 4b-RCAs, and one approximate 5b-RCA. The approximate 5b-RCAs carry chain was to propagate the carry correctly, thus ensuring the accuracy of the incremental addition. The accuracy of the approximate adders mentioned in works [61–63] has significant improvement and they can be applied to speech recognition in high noise environments to improve recognition accuracy.

Approximate Carry-Select Adder

Similar to the segmented approximate adder, the approximate carry-select adder [67, 68] also segments the addition and introduces an approximation between the carry and the sum of each individual subadders. Figure 18.5 shows the structure of approximate carry-select adder with sum selection, and Fig. 18.6 shows the structure of approximate carry-select adder with carry-in selection.

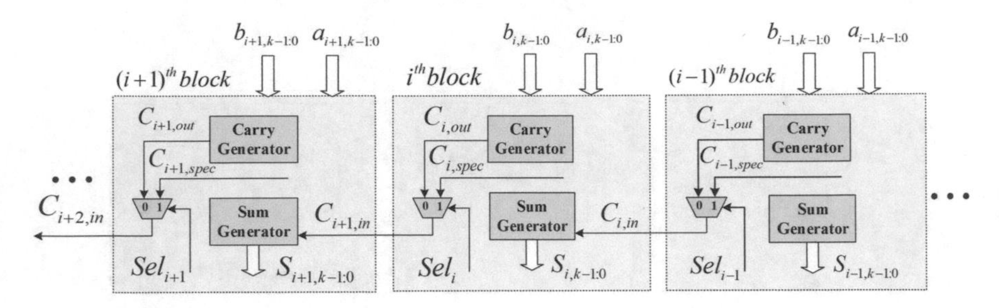

Fig. 18.6 Approximate carry-select adder with carry-in selection

The above structure allows dynamic segmentation of the adder, including dividing the adder in the accumulator into smaller bit-width adders by bit slicing the data path. This allows further approximation of the circuit in conjunction with voltage scaling techniques to reduce power consumption. Depending on the degree of voltage overscaling (VOS), the degree of segmentation can be dynamically controlled based on the control input of the multiplier. What's more, each adder stage can be implemented using any adder architecture. This kind of approximate adders is highly flexible and can be configured according to the specific application scenario of speech recognition. In work [64], one of the subadders 0 (with carry "0") and the subadders 1 (with carry "1") was selected as a partial sum and the carry output of the former submodule was used as the selecting signal for the latter module. Work [65] selected whether to skip the carry of the partial submodule by judging the carry-propagation signal.

Approximate Full Adder

The OR-gate approximate adder is usually used to improve the energy efficiency of hardware accelerator [38, 69, 70], which uses OR-gates instead of adders to implement the addition of two bits. It may go wrong only when both operands have the same bit "1", with an error probability of only 25%. In work [69], a low-part OR adder (LOA) was proposed, which uses OR-gates in the low part and uses AND-gates for the carry input of the exact computing part. In work [71], the low part OR-gates were replaced with approximate mirror adders; compared to LOA, more transistors were used in exchange for the gain in accuracy. Work [10] proposed an accuracy adaptive approximate adder unit for KWS systems, where the system uses LOA adders with different approximate bit-widths depending on the configuration signal when the SNR is large and uses standard full adder units when the SNR is small, dynamically adjusting for greater gains.

The detailed structure of the 1-bit full adder in [10] is shown in Fig. 18.7c, and this approximate adder can be adaptively reconfigured to dynamically adapt to approximate computing requirements under various input workloads. The structure of the 4-bit OR adder is shown in Fig. 18.7d, and this approximate adder can reduce the transistor count, area, and power consumption significantly to about 20%. In the BWN proposed in work [10], an accuracy adaptive adder wad used to offset the

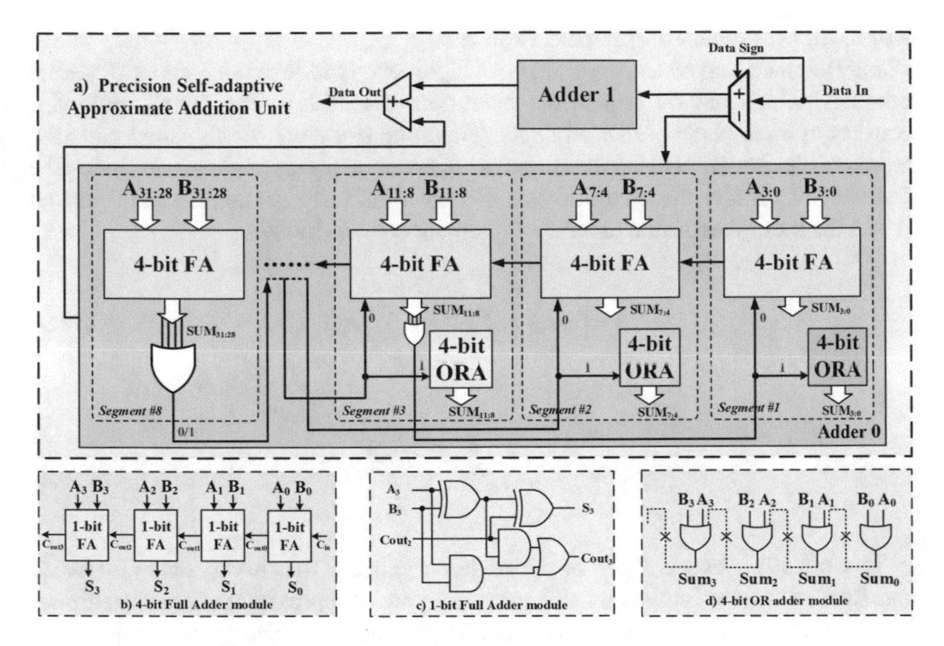

Fig. 18.7 Precision self-adaptive OR-gate-based approximate addition

mismatch caused by positive and negative data / weights to improve the computing accuracy. The accuracy adaptive approximate adder unit is shown in Fig. 18.7a. Compared with the reference design using a standard full adder, the loss of KWS recognition accuracy of the proposed approximate computing is within 0.5% under different background noise.

Other designs of approximate full adders include mirror adders [56], approximate XNOR / XNOR-based adders [72], inexact adder units proposed in work [73], and approximate backward carry-propagation adders [74]. Additionally, emerging technologies such as magnetic tunnel junctions have been considered for the design of approximate full adders (AFAs) for a shorter delay, a smaller area, and a lower power consumption [75, 76]. Finally, a simply truncated adder (TruA) that works with a lower precision is considered as a baseline design.

2.2.3 Approximate Multiplier

Traditional standard multipliers with high latency and power consumption cannot meet the requirements for low power consumption and high energy efficiency in neural network accelerator. Therefore, many approximate multipliers with low power and low latency have been proposed to meet this challenge. These approximate multipliers can be generally classified into approximate generating partial products [77–79], approximate computing partial products [23, 69, 80], and approximate logarithmic multipliers [39, 81–89].

Approximate Generating Partial Products

Using Booth coding on the multiplier can significantly reduce the number of partial products, which greatly reduce the operation of addition. The Booth multiplier consists of three parts: partial products generating using the Booth encoder, partial products accumulation using the compressor, and final products generation using the fast adder. Consider the multiplication of two N-bit integers, i.e., the multiplicand A and the multiplier B in a binary complement code as follows:

$$A = -a_{N-1}2^{N-1} + \sum_{i=0}^{N-2} a_i 2^i \tag{18.3}$$

$$B = -b_{N-1}2^{N-1} + \sum_{i=0}^{N-2} b_i 2^i \tag{18.4}$$

In work [77], two designs of approximate radix-4 Booth encoders (R4ABE1 and R4ABE2) were introduced and analyzed, and an approximate Booth multiplier has been proposed using the approximate Booth encoders. The approximate Booth encoder method of R4ABE1 is to introduce a small error to make the truth table as symmetrical as possible. Thus, the advantage of the R4ABE1 is that a very small error occurs, as only four entries are modified; however, all modifications change a "1" to a "0", so the absolute value of approximate product is always smaller than its exact counterpart. Another approximate radix-4 Booth encoding (R4ABE2) has eight entries in the K-map whose value are modified to simplify the logic of the Booth encoding. Compared with the first design, R4ABE2 introduces nearly double the error, not only by changing "1" to "0" but also by changing "0" to "1" to achieve the modification. Thus, the deviation between the approximate product and the exact result can be positive or negative, and the errors can complement each other in the process of summing partial products. Therefore, when using R4ABE2 in a Booth multiplier, the error may not be larger than for a Booth multiplier with R4ABE1. Figure 18.8 shows the comparison of metrics between exact and approximate Booth encoders to generate one partial product bit, and the power consumption and area of the proposed two types approximate Booth encoders have great reduction, which makes them very suitable for the power and area-constrained KWS.

Similarly, in work [78], three different simplification schemes have been proposed by simplifying the logic truth table and its logic circuit. Compared with the standard multiplier, the proposed three approximate Booth multipliers are significantly reduced in terms of power consumption, delay, and area. In work [78], the radix-8 Booth coding was approximately optimized by splitting +3X into the form of X+2X and using the LOA circuit in the addition. A Booth multiplier structure based on approximate hybrid encoding has been proposed in work [79]. In its approximate encoding scheme, the approximation is done by approximating the +3X case as +2X or +4X for radix-8 Booth encoding.

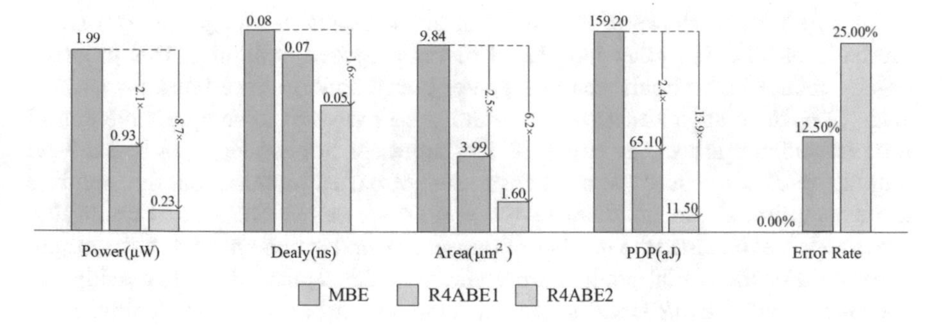

Fig. 18.8 Performance comparison between exact and approximate booth encoders to generate one partial product bit

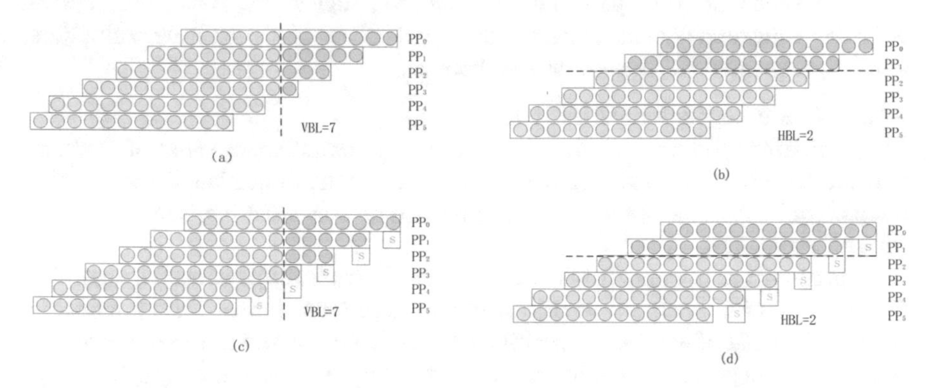

Fig. 18.9 The Broken-Booth Multiplier type 0 (**a**) and type 1 (**b**) for VBL = 7 adopted from work[80], and the Broken-Booth Multiplier type 0 (**c**) and type 1 (**d**) for HBL = 2 adopted from work [69]

Approximate Computing Partial Products

Approximate computing partial products means performing rounding, approximate summation, approximate compression, etc. on partial products.

A low-power multiplier has been proposed in work [80], where vertical breaking level (VBL) has been introduced in the summation array of partial products (as shown in Fig. 18.9a, b). In this approximation method, all the dot products positioned at the right-hand side of the VBL are replaced by zero. This multiplier uses the Broken–Booth multiplier approximation method based on the conventional modified Booth multiplier, which reduces the total power consumption of the multiplier by 58%, but the output accuracy is slightly degraded. In addition, to better evaluate the efficiency of the proposed multiplier, it has been used to design a 30-tap order low-pass FIR filter. Comparing it with a filter using a traditional Booth multiplier, the simulation results have shown that the implementation using Broken-Booth Multiplier reduces power consumption by 17.1% at the expense of only 0.4 dB SNR.

In work [69], as shown in Fig. 18.9c, d, the horizontal breaking level (HBL) on the basis of VBL is further introduced in the proposed multiplier. This proposal greatly reduced the circuit area and power consumption of the breaking part. In work [23], three approximate compressors were proposed to compress the partial product addition array. The precise 4-2 compressor has a 4-bit input and a 3-bit output, which can reduce the number of rows of partial products, but the circuit is more complicated. An approximate 4-2 compressor can be obtained by modifying the truth table to reduce the number of output bits and the complexity of the output equation. For the partial product array addition, the segmentation processing has been performed, the MSB used an accurate compressor, and the Least Significant Bit (LSB) used an approximate 4-2 compressor. The image sharpening algorithm was then investigated as an application of the proposed multiplier design. The simulation results showed that the proposed design achieved a high signal-to-noise ratio (SNR > 35 dB) compared with the precise counterpart and other approximate multipliers, while significantly reducing the area and power.

Approximate Logarithmic Multiplier
The approximate logarithmic multiplier uses the approximate value of a binary numbers logarithm and anti-logarithm to implements multiplication. By performing a logarithmic approximation, multiplication can be converted into summation and shift operation [39].

Logarithmic multipliers (LMs) were applied in work [85]. Mitchell's algorithm utilizes the logarithmic and anti-logarithmic approximation of binary numbers, which is the basis of the logarithmic multiplier [80]. To improve the accuracy of LM, approximate logarithmic multipliers (ALMs) using a truncated binary logarithm converter and a set-one adder (SOA) for addition has been proposed in work [82]. The SOA simply sets the LSB to a constant "1" and uses an AND-gate to generate the carry for the MSB. In addition, an improved algorithm using exact and approximate adders (ILM-EA and ILM-AA, respectively) have been proposed in work [87]. In work [88], the input operands between two consecutive powers of 2 have been split into several segments, and then an error reduction factor was determined for each segment analysis and compensated to the result of the underlying linear model.

Dynamic range approximate LMs (DR-ALMs) for machine learning applications have been proposed in work [89]. Based on the Mitchell approximation, dynamic range operand truncation scheme and the worst-case (absolute and relative) error analysis, the best approximation scheme based on different metrics according to the accuracy of the provided design and hardware overhead was selected. The proposed DR-ALMs were compared with the conventional LM with exact operands and previous approximate multipliers; the results showed that PDP of the best proposed design has been decreased by up to 54.07% with the MRED decreasing by 21.30% compared with 16-bit conventional design. Case studies for three applications showed the feasibility of the proposed DR-ALMs. Compared with the exact multiplier and its conventional counterpart, the back-propagation classifier with DR-ALMs in the truncation length larger than 4 had a similar classification result; the K-means clustering with all DR-ALMs had a similar clustering result; and the handwritten number recognition with DR-ALM-5 or DR-ALM-6 for LeNet-5 remained similar or even slightly higher recognition rate.

In work [90], a multi-objective Cartesian genetic programming (CGP) algorithm was used to generate 471 high-performance 8×8 approximate multipliers. Since CGP can provide much better implementations of circuits than common circuit design and optimization tools, it has been adopted as the design method. Approximate circuits were generated by randomly removing some connections from several exact designs. Genetic algorithms were then applied to design space exploration to obtain the best approximate circuit with respect to the MRED. In work [91], the generated 8×8 approximate multiplier was used to synthesize an approximate multiplier with a larger bit-width, which has excellent performance in power consumption and area. The approximate multiplier generated by CGP is superior to the approximate multiplier designed by the traditional method in area, power, and performance. In the future, this is a hot topic of approximate multiplier.

2.3 *Approximate Computing with Emerging Microelectronic Device*

Approximate computing has shown great potential in the next generation of computing systems, and more and more work has begun to study how to use the emerging microelectronic device with new material to design approximate computing logic to acquire higher energy efficiency.

In work [92], two approximate full adders based on non-volatile logic-in-memory structure have been proposed, called non-volatile AFA. The two adders used spin torque transfer magnetic tunnel junction (STT-MTJ) as the storage element to form a magnetic full adder. Figure 18.10 (a) shows the circuit implementation of the two approximate MFAs: AX-MFA1 (without blue box) and AX-MFA2. Figure 18.10 (b), (c) show the conventional CMOS approximate adders: AXA1, AXA2. The first approximate AX-MFA1 used MTJ to replace the logic circuit under the traditional CMOS technology and simplified the logic to eliminate the carry input C_i. The second approximate AX-MFA2 realized approximate computing by controlling the MTJ write current, and MTJ's unique switch function allowed the full adder to switch between the precise mode and the approximate mode. As shown in Fig. 18.11, this dual mode approximate AX-MFA2 can save 78% of dynamic power in the approximate mode compared with AXA1.

Figure 18.11 summarizes the simulation results of conventional AFA (AXA1, AXA2) and proposed NV-AFA (AX-MFA1, AX-MFA2). Two types of AX-MFA can reduce leakage power by $3.4\times - 17.5\times$ compared with AXA2, which are very suitable for power-constrained IoT applications including KWS. Another advantage of this technology is that it can switch between approximate mode and precise mode by changing the supply voltage of the device, which greatly increases its flexibility and versatility.

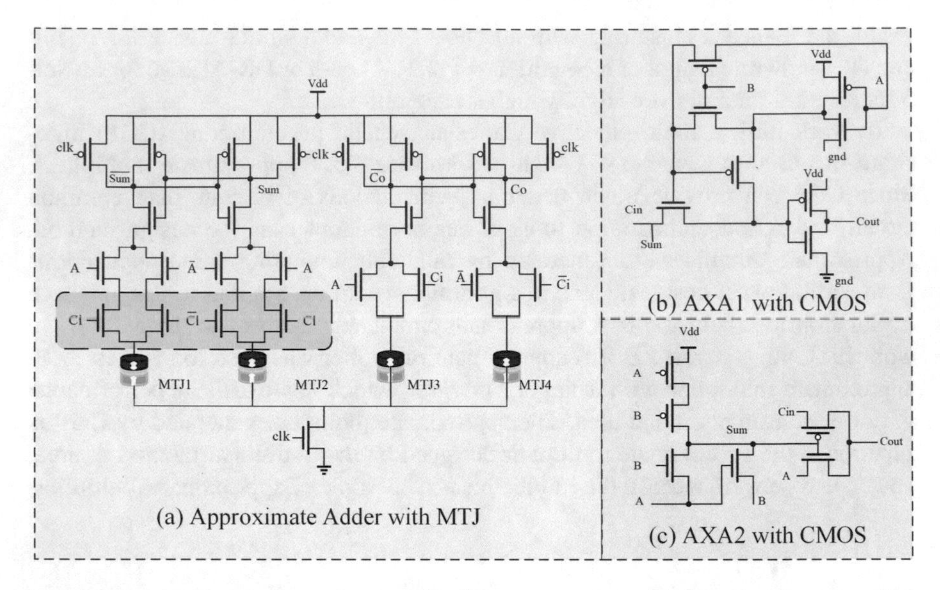

Fig. 18.10 (**a**) Circuit implementation of two approximate MFAs: AX-MFA1 (without blue box) and AX-MFA2. AX-MFA1 is implemented with simplified logic: input C_i in dashed rectangle is eliminated. The second dual-mode approximate AX-MFA2 is implemented with the whole schematic (**b**), (**c**): conventional CMOS approximate adders: AXA1, AXA2

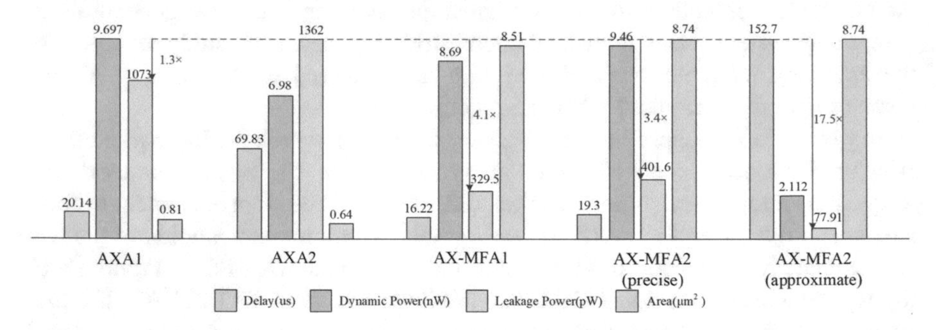

Fig. 18.11 Performance comparison of conventional AFA (AXA1, AXA2) and proposed NV-AFA (AX-MFA1, AX-MFA2)

In work [93], based on the write operation of four types of magnetic random access memory, four types of NV-AFA have been proposed. The approximate computing circuit is built using a storage unit, so no peripheral circuit is needed, and the area overhead of the circuit is small. At the same time, the approximate computing structure can be converted into a memory by simply changing the wire connection, and the structure is more flexible. In work [94], a random number generator circuit (SNM) was proposed using the random characteristics of STT-MTJ. Due to the infinite durability and non-volatility of STT-MTJ, the SNM circuit can be used to build reliable circuits and systems. In work [95], toggle spin torques magnetic tunnel junction (TST-MTJ) switching mechanism was used in MRAN to implement voting logic. Voting logic and write-only logic realize toggle spin torque magnetoresistive random access memory (TST-MRAM) approximate full adder and TST-MRAM multi-bit approximation multiplier. Compared with traditional adders and multipliers, these two designs can achieve low latency and low power with less accuracy loss. The traditional approximate computing circuit implemented by voltage scaling and simplified logic is also suitable for the emerging microelectronic device with new material. In work [96], two schemes for implementing approximate circuits using emerging microelectronic devices of new materials were discussed. One is to use imprecise MTJ write operations to generate approximate operands, and the other is to allow the device to work near the threshold voltage to achieve approximate computing. In order to achieve a better trade-off between accuracy and power consumption, the MTJ precision adder can be used in the important bits, and the MTJ approximate adder can be used in the unimportant bits. The accurate bits and the approximate bits can be dynamically selected according to the accuracy and power consumption. In work [97], a device-circuit-algorithm joint optimization scheme was proposed, and the weight sparseness of CNN was realized in MRAM to alleviate the problems of MRAM access failure and insufficient voltage supply. In work [98], it was proposed to implement classic arithmetic and logic circuits in resistive random access memory to solve the problem of frequent data movement caused by the separation of computing and storage in the traditional Von Neumann architecture, which provides a new scheme for computing-in-memory (CIM). In work [99], a hybrid scheme of STT-MTJ and CMOS was proposed to reduce dynamic and static power, combining the advantages of STT-MTJ's extremely low switching power consumption, zero static leakage power consumption, high density, and unlimited durability. Based on this scheme, a non-volatile flip-flop was designed, which can reduce dynamic power by 76% and static power by 79%.

3 Architecture-Level Approximate Computing for KWS

In order to meet the design requirements of high performance and low power consumption of the speech recognition system, it is usually ncccssary to design a special neural network accelerator architecture for the speech recognition system. The architecture encompasses the hardware organization of the processing node as well as the dynamics between its components. The objective at this level is to dynamically adapt the combination of algorithms and circuits in different application scenarios, or find the optimal combination of algorithms and circuits in a specific application scenario. One example of such architectural optimization is the use of low-precision arithmetic hardware that has been used extensively for feature extraction [89] and machine learning classification [22] reducing the computational and memory footprint compared to floating-point hardware. In addition, to improve the flexibility and versatility of KWS, precision reconfigurable architectures have been proposed in work [10, 15, 100]. KWS in work [10] can dynamically select high-precision and low-power computing modes according to the SNR ratio in different application scenarios.

3.1 Reconfigurable Architecture Using Approximate Computing

The speech recognition system using traditional DSP / CPU has poor performance and cannot meet the requirements of real time, and the hardware architecture is solidified so that the adaptive ability is poor. At the same time, the energy consumption of DSP / CPU scheme is much higher than that of traditional customized circuit scheme, which cannot meet the application requirements of low power consumption. In addition, the software and hardware architecture of the traditional customized circuit scheme is solidified. This traditional scheme lacks the ability of scene adaptation and cannot be applied on a large scale. Therefore, more and more designs begin to consider the reconfigurability, low power consumption, high energy efficiency, and other characteristics of speech recognition system. And then, we will take two typical designs as examples to introduce reconfigurable architecture using approximate computing.

Work [15] proposes an approximate and network configurable computing architecture called Accuracy-Reconfigurable Architecture (ARA). The proposed system integrates the system controller implemented by ARM7TDMI, convolution neuron processing unit (CNPU), a scratch-pad memory (SPM), and several auxiliary modules for system scheduling. Among them, CNPU is composed of two types of neural network processing units, the first type is multiplication and accumulation neural processing element array (MA-NPEA), which is a two-dimensional pulsating array composed of approximate multiplier and approximate adder with error compensation. The approximate computing unit of error compensation and the control logic module jointly realize the reconfigurable characteristics and precision controllable of processing element (PE) array. At the same time, the two-dimensional array also optimizes the data path, which can reuse the feature map and convolution core and improve the energy efficiency of neural network processor. The approximate computing unit of error compensation and the control logic module jointly realize the reconfigurable characteristics and precision controllable of PE array. The second type is look-up table (LUT) multiplier-based neural processing element array (LUT-NPEA) as shown in the right half of Fig. 18.12. LUT-NPEA is composed of multiplier, arithmetic unit, result register, and control logic composed of LUT.

Based on the CNPU composed of the above computing units, a reconfigurable data path is designed, which can be configured for different neural networks. In work [15], ARA was used to process four most typical CNN networks (image recognition, face recognition, target detection, and image segmentation). The experimental results showed that ARA achieved a good compromise between performance and power consumption, and ARA can improve the energy efficiency ratio by $1.51\times$ $-4.36\times$ compared with other architectures.

In work [100], in order to achieve high energy efficiency, a precision adaptive energy-efficient reconfigurable architecture (E-ERA) based on approximate computing unit and control logic was proposed. The top-level architecture of E-

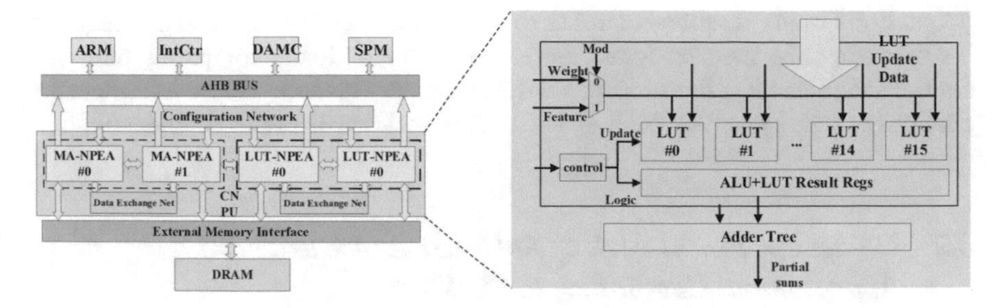

Fig. 18.12 The approximate and network configurable computing architecture of ARA for CNNs

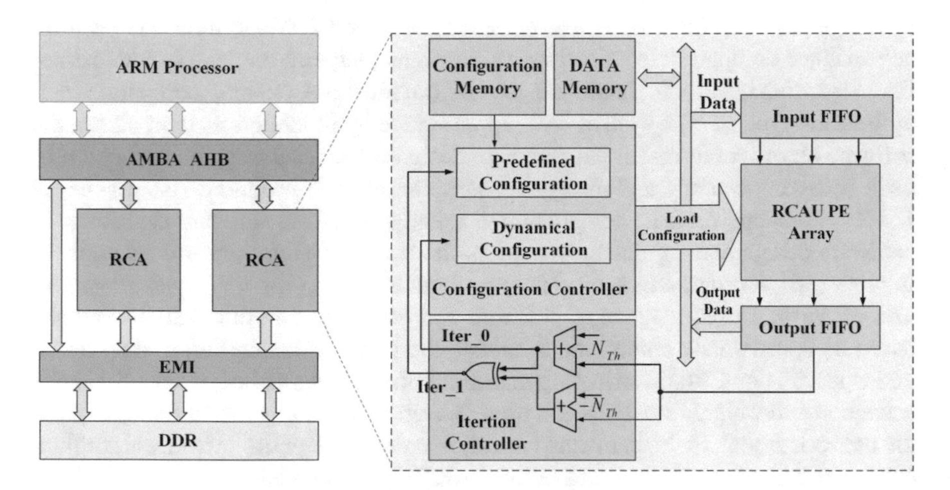

Fig. 18.13 Prototype system of precision adaptive E-ERA using approximate computing

ERA is shown in the left half of Fig. 18.13. It is composed of a system controller implemented by arm, reconfigurable computing array, and multiple auxiliary modules. The structure of RCA is shown in the right half of Fig. 18.13, in which the configuration controller is composed of two parts: the predefined configurator and the dynamic configuration module. The predefined configurator is responsible for loading the data stored in the configuration memory. The dynamic configuration module can determine the number of iterations required for the current computing according to the previous output of RCA and dynamically configure the number of iterations. The specific computing process is that when one frame computing is completed, the iterative controller will generate the iteration number used in the next frame computing by evaluating the current network output, write this information into the dynamic configuration module, and transfer the configuration information to the RCAU PE array through the dynamic configuration module to complete the configuration of the approximate multiplier. In addition, the computing result of the approximate unit array is temporarily stored in the output first in first out (FIFO), and then the computing result of the current frame is transmitted from the output FIFO to the iteration controller to generate the number of iterations required for the computing the next frame. Based on the approximate computing unit with controllable accuracy, the proposed E-ERA can significantly improve the energy efficiency of the system on the premise that the accuracy loss can be tolerated.

3.2 Co-design Framework of Software and Hardware Approximate Computing for KWS

The above two cases introduce the application of approximate computing in reconfigurable architecture. However, a well-designed KWS system can perform approximate computing from software algorithms (network models, feature extraction algorithms, etc.) to hardware design (architecture design, computing unit optimization, etc.). This system-level approximate optimization method allows the software algorithm to match the hardware design while achieving the best compromise between accuracy and energy consumption. In the proposed state-of-the-art KWS system, approximate computation is widely used from software algorithms to hardware design and significantly reduces the overall system power consumption. In work [16], a programmable LSTM accelerator (Laika) for KWS was proposed. The accelerator used a series of software and hardware co-optimization methods based on approximate computing to reduce the power consumption to only 5 μW under the 65 nm CMOS technology. In the software algorithm, it quantified the activation and weight into 8 bits and used clustering technology to further compress the network model. In the hardware design, in order to adapt the software algorithm, it approximated the nonlinear activation through the piecewise approximation function based on the LUT function to reduce the large amount of overhead in calculating accurate nonlinear activation functions. In work [23], a binarized convolutional neural network (BCNN)-based speech recognition processor was proposed, which quantized the data except for the activation of the first convolutional (Conv) layer and the weight and output of the final fully connected (FC) layer to 1 bit to reduce the storage and power of the model. In addition, according to the characteristics of the data in BCNN, the system applied a customized approximate addition unit to support accurate 1-bit incremental addition and 16-bit approximate addition. Based on this co-design framework of software and hardware approximate computing, the power consumption of this system is only 141 μW under the 28 nm CMOS process.

Different from BCNN, a BWN model that only quantized the weight bit-width to 1 bit was proposed in work [10]. While reducing power consumption, the number of recognizable keywords increases to 10. The system used approximate computing technology to design a compact MFCC module, removed redundant operations such as windowing, log2, and DCT, and designed a two-stage serial pipeline FFT circuits to reduce data storage and operation. In the hardware, the system adopted an adaptive dual computing mode structure based on the SNR prediction module, which further compressed the data to 12 bits in the low-precision

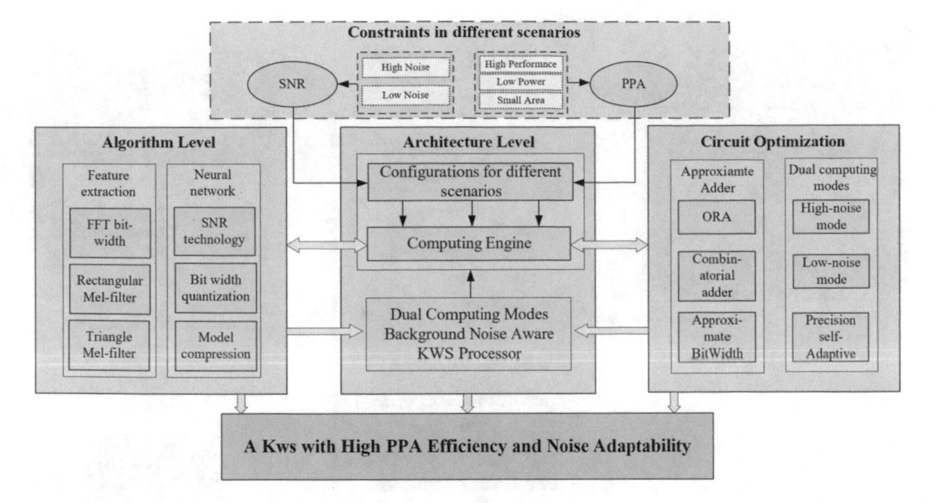

Fig. 18.14 Co-design framework of software and hardware approximate computing for KWS

mode (16 bits in the high-precision mode), and used two precisions self-adaptive approximate computing units to process the positive and negative data separately, significantly reducing system power consumption while ensuring accuracy. As shown in Fig. 18.14, a hardware–software co-design framework was used in work [10] to achieve energy-efficient KWS for various background environments with the trade-off between high accuracy and low power consumption. The bit-width quantization of the input data and the settings of the approximate computing circuits could be dynamically configured according to the PPA design requirements in different application scenarios and different SNR levels. In work [13], a binarized DSCNN model has been used to decompose the standard 3D convolution into 2D convolution with a sevenfold reduction in data storage and computation, while an approximate MFCC module based on serial FFT and log2 budget was utilized. At the hardware level, the system adopted a hardware architecture that eliminates redundant memory and computing and customized a register file-based storage block for near-threshold design of the entire chip. Based on the approximate design of this software and hardware collaboration, the system finally achieved extremely low power of 510 nW. In order to solve the power consumption and latency problems caused by data access of the traditional Von Neumann architecture, a CIM architecture has been proposed. CIM-based neural network accelerators commonly require on-chip training to achieve the integration of algorithms and new storage devices. In work [101], a CIM-based RNN training process was proposed to adjust the weight value to reduce CIM power consumption. This method of software and hardware collaborative optimization has achieved an average energy consumption reduction of 9.9%, while the speech recognition accuracy rate exceeded 90.2%.

The above case of KWS system-level approximate computing optimization proves that approximate computing has great potential in the next generation

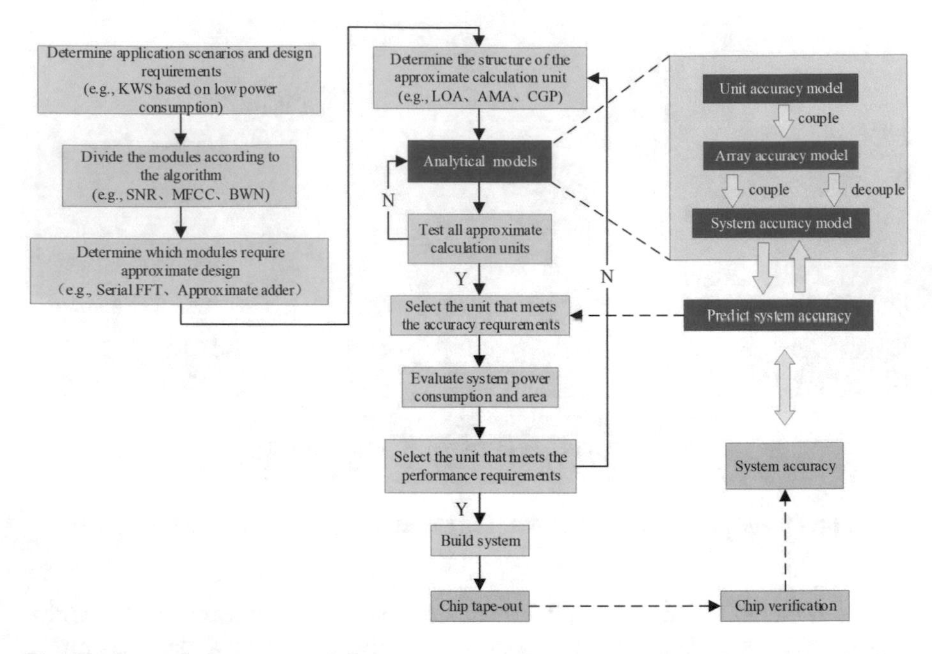

Fig. 18.15 The design process of KWS using analytical models to evaluate different combinations of approximate computing technologies

computing system, so it is necessary to establish a complete set of approximate computing design process. We propose to use the system-level error loss and accuracy control evaluation model in the KWS design to optimize the energy efficiency of the system. The error analysis model of the approximate computing unit of the intelligent speech recognition neural network proposed by us for specific application scenarios is shown in Fig. 18.15.

Firstly, the corresponding design indicators need to be formulated according to the specific application scenarios of KWS, such as low-power design for scenes with low accuracy requirements, and noise-adaptive multi-mode computing design for scenes with more complex environmental noise. Secondly, after selecting the application scenarios and design requirements, a specific feature extraction algorithm and keywords recognition neural network need to be carefully designed, and the corresponding hardware modules that need to be approximately designed should be divided according to the software algorithm. Finally, an accuracy evaluation model is built to predict the accuracy of the system using the alternative approximate computing structures, and the chip area and power consumption are evaluated with EDA tools to find an approximate computing unit structure that meets the system requirements. Based on the proposed scheme, work [10] can achieve 21.7% reduction in power consumption within the accuracy loss of 1% under low background noise.

4 Conclusion

Approximate computing is one of the most promising and energy-efficient emerging computing paradigms used to design DNN-based speech recognition systems, and its importance has become more and more obvious in recent years. Due to the certain fault tolerance brought by the self-learning ability of neural network, the DNN-based speech recognition system using approximate computing can achieve high energy efficiency with acceptable accuracy loss for different speech recognition application scenarios. The optimization methods using approximate computing at algorithm level, circuit level, and architecture level of the DNN-based speech recognition system have been extensively studied, and it has been proved that the application of approximate computing can significantly reduce the energy consumption with acceptable loss of accuracy. In addition, emerging circuits with non-volatile memory device have been proved to have the potential in further improving the energy efficiency of power-constrained KWS; this is because they have storage data retention characteristics, infinite durability, zero standby power consumption, and high density. In the future, the hardware and software co-design framework with approximate computing can be used to further improve and achieve optimal design of KWS for different application scenarios. In this framework, the approximate computing technologies at different levels should be considered and evaluated together to achieve the best combinations of these technologies. For this framework, the analytical models for accuracy loss and hardware benefits analysis at the approximate computing unit level, the approximate computing array level, and the system level should be built to perform early exploration and evaluation of different combinations of approximate computing technologies and settings.

References

1. Liu B, Wang Z, Zhu W et al. An ultra-low power always-on keyword spotting accelerator using quantized convolutional neural network and voltage-domain analog switching network-based approximate computing. IEEE Access 2019;7:186456–186469.
2. Hidayat R, Bejo A, Sumaryono S et al. Denoising speech for MFCC feature extraction using wavelet transformation in speech recognition system. In: 2018 10th international conference on information technology and electrical engineering (ICITEE). Piscataway: IEEE; 2018. p. 280–284.
3. Li Q, Zhu H, Qiao F et al. Energy-efficient MFCC extraction architecture in mixed-signal domain for automatic speech recognition. In: 2018 IEEE/ACM international symposium on nanoscale architectures (NANOARCH). Piscataway: IEEE; 2018. p. 1–3.
4. Hermansky H. Perceptual linear predictive (PLP) analysis of speech. J Acoust Soc Am. 1990;87(4):1738–1752.
5. Hermansky H, Morgan N, Bayya A et al. The challenge of inverse-E: the RASTA-PLP method. In: Conference record of the twenty-fifth Asilomar conference on signals, systems & computers Piscataway: IEEE Computer Society; 1991. p. 800–801.
6. Gupta H, Gupta D. LPC and LPCC method of feature extraction in speech recognition system. In: 2016 6th international conference-cloud system and big data engineering (confluence) Piscataway: IEEE; 2016. p. 498–502.

7. Këpuska VZ, Elharati HA. Robust speech recognition system using conventional and hybrid features of MFCC, LPCC, PLP, RASTA-PLP and hidden Markov model classifier in noisy conditions. J Comput Commun 2015;3(06):1.
8. Liu B, Wang Z, Fan H et al. EERA-KWS: A 163 TOPS/W always-on keyword spotting accelerator in 28nm CMOS using binary weight network and precision self-adaptive approximate computing. IEEE Access 2019;7:82453–82465.
9. Giraldo JP, Lauwereins S, Badami K et al. 18μW SoC for near-microphone keyword spotting and speaker verification. In: 2019 symposium on VLSI circuits Piscataway: IEEE; 2019. p. C52–C53.
10. Liu B, Cai H, Wang Z et al. A 22nm, 10.8 μW/15.1 μW dual computing modes high power-performance-area efficiency domained background noise aware keyword-spotting processor. IEEE Trans Circuits Syst I Regu Pap. 2020;67(12):4733–4746.
11. Jo J, Yoo H, Park IC. Energy-efficient floating-point MFCC extraction architecture for speech recognition systems. IEEE Trans Very Large Scale Integr Syst 2015;24(2):754–758.
12. Zhang Y, Suda N, Lai L et al. Hello edge: keyword spotting on microcontrollers 2017. Preprint arXiv: 1711.07128.
13. Shan W, Yang M, Xu J et al. 14.1 A 510*n*W 0.41 V low-memory low-computation keyword-spotting chip using serial FFT-based MFCC and binarized depthwise separable convolutional neural network in 28*n*m CMOS. In: 2020 IEEE international solid-state circuits conference-(ISSCC) Piscataway: IEEE; 2020. p. 230–232.
14. Fernandez-Marques J, Tseng VWS, Bhattachara S et al. On-the-fly deterministic binary filters for memory efficient keyword spotting applications on embedded devices. In: Proceedings of the 2nd international workshop on embedded and mobile deep learning 2018. p. 13–18.
15. Gong Y, Liu B, Ge W et al. ARA: cross-layer approximate computing framework based reconfigurable architecture for CNNs. Microelectron J. 2019;87:33–44.
16. Giraldo JSP, Verhelst M. Laika: a 5μW programmable LSTM accelerator for always-on keyword spotting in 65*n*m CMOS. In: ESSCIRC 2018-IEEE 44th European solid state circuits conference (ESSCIRC). Piscataway: IEEE; 2018. p. 166–169.
17. Jo J, Kung J, Lee Y. Approximate LSTM computing for energy-efficient speech recognition. Electronics 2020;9(12):2004.
18. Lin M, Chen Q, Yan S. Network in network. 2013. Preprint arXiv: 1312.4400.
19. Han S, Mao H, Dally WJ. Deep compression: compressing deep neural networks with pruning, trained quantization and Huffman coding. 2015. Preprint arXiv: 1510.00149.
20. He K, Zhang X, Ren S et al. Deep residual learning for image recognition. In: Proceedings of the IEEE conference on computer vision and pattern recognition 2016. p. 770–778.
21. Zoph B, Le QV. Neural architecture search with reinforcement learning. 2016. Preprint arXiv: 1611.01578.
22. He K, Zhang X, Ren S et al. Deep residual learning for image recognition. In: Proceedings of the IEEE conference on computer vision and pattern recognition. 2016. p. 770–778.
23. Yin S, Ouyang P, Zheng S et al. A 141 μw, 2.46 *p*J/neuron binarized convolutional neural network based self-learning speech recognition processor in 28*n*m CMOS. In: 2018 IEEE symposium on VLSI circuits. Piscataway: IEEE; 2018. p. 139–140.
24. Ko JH, Fromm J, Philipose M et al. Limiting numerical precision of neural networks to achieve real-time voice activity detection. In: 2018 IEEE international conference on acoustics, speech and signal processing (ICASSP) Piscataway: IEEE; 2018. p. 2236–2240.
25. Lu Y, Shan W, Xu J. A depthwise separable convolution neural network for small-footprint keyword spotting using approximate MAC unit and streaming convolution reuse. In: 2019 IEEE Asia pacific conference on circuits and systems (APCCAS) Piscataway: IEEE; 2019. p. 309–312.
26. Liu B, Sun Y, Cai H et al. An ultra-low power keyword-spotting accelerator using circuit-architecture-system co-design and self-adaptive approximate computing based BWN. In: Proceedings of the 2020 on great lakes symposium on VLSI 2020. p. 193–198.
27. Liu B, Wang Z, Zhu W et al. An ultra-low power always-on keyword spotting accelerator using quantized convolutional neural network and voltage-domain analog switching network-based approximate computing. IEEE Access 2019;7:186456–186469.

28. Wang D, Kim SJ, Yang M et al. A background-noise and process-variation-tolerant 109nW acoustic feature extractor based on spike-domain divisive-energy normalization for an always-on keyword spotting device. In: 2021 IEEE international solid-state circuits conference (ISSCC). vol. 64. Piscataway: IEEE; 2021. p. 160–162.
29. Coates A, Huval B, Wang T et al. Deep learning with COTS HPC systems. In: International conference on machine learning New York: PMLR; 2013. p. 1337–1345.
30. Jouppi NP, Young C, Patil N et al. In-datacenter performance analysis of a tensor processing unit. In: Proceedings of the 44th annual international symposium on computer architecture 2017. p. 1–12.
31. Chen YH, Krishna T, Emer JS et al. Eyeriss: an energy-efficient reconfigurable accelerator for deep convolutional neural networks. IEEE J Solid-State Circuits 2016;52(1):127–138.
32. Ielmini D, Wong HSP. In-memory computing with resistive switching devices. Nat Electron 2018;1(6):333–343.
33. Wong HSP, Salahuddin S. Memory leads the way to better computing. Nat Nanotechnol. 2015;10(3):191–194.
34. Yao P, Wu H, Gao B et al. Fully hardware-implemented memristor convolutional neural network. Nature 2020;577(7792):641–646.
35. Ni L, Wang Y, Yu H et al. An energy-efficient matrix multiplication accelerator by distributed in-memory computing on binary RRAM crossbar. In: 2016 21st Asia and south pacific design automation conference (ASP-DAC). Piscataway: IEEE; 2016. p. 280–285.
36. Jain S, Raghunathan A. CxDNN: hardware-software compensation methods for deep neural networks on resistive crossbar systems. ACM Trans Embed Comput Syst. 2019;18(6):1–23.
37. Chakraborty I, Roy D, Roy K. Technology aware training in memristive neuromorphic systems for nonideal synaptic crossbars. IEEE Trans Emerg Topics Comput Intell. 2018;2(5):335–344.
38. Han J, Orshansky M. Approximate computing: an emerging paradigm for energy-efficient design. In: 2013 18th IEEE European test symposium (ETS) Piscataway: IEEE; 2013. p. 1–6.
39. Mitchell JN. Computer multiplication and division using binary logarithms. IRE Trans Electron Comput. 1962;EC-11(4):512–517.
40. Low JYL, Jong CC. Non-iterative high speed division computation based on Mitchell logarithmic method. In: 2013 IEEE international symposium on circuits and systems (ISCAS) Piscataway: IEEE; 2013. p. 2219–2222.
41. Ansari MS, Cockburn BF, Han J. A hardware-efficient logarithmic multiplier with improved accuracy. In: 2019 design, automation & test in europe conference & exhibition (DATE). Piscataway: IEEE; 2019. p. 928–931.
42. Zervakis G, Xydis S, Tsoumanis K et al. Hybrid approximate multiplier architectures for improved power-accuracy trade-offs. In: 2015 IEEE/ACM international symposium on low power electronics and design (ISLPED). Piscataway: IEEE; 2015. p. 79–84.
43. Gupta V, Mohapatra D, Raghunathan A et al. Low-power digital signal processing using approximate adders. IEEE Trans Comput-Aided Des Integr Circuits Syst. 2012;32(1):124–137.
44. Kulkarni P, Gupta P, Ercegovac M. Trading accuracy for power with an underdesigned multiplier architecture. In: 2011 24th international conference on VLSI design. Piscataway: IEEE; 2011. p. 346–351.
45. Lin CH, Lin C. High accuracy approximate multiplier with error correction. In: 2013 IEEE 31st international conference on computer design (ICCD) Piscataway: IEEE; 2013. p. 33–38.
46. Liu Y, Zhang T, Parhi KK. Computation error analysis in digital signal processing systems with overscaled supply voltage. IEEE Trans Very Large Scale Integr Syst. 2009;18(4):517–526.
47. Liu Y, Zhang T, Parhi KK. Analysis of voltage overscaled computer arithmetics in low power signal processing systems. In: 2008 42nd Asilomar conference on signals, systems and computers. Piscataway: IEEE; 2008. p. 2093–2097.
48. Ghosh S, Roy K. Parameter variation tolerance and error resiliency: new design paradigm for the nanoscale era. Proc IEEE 2010;98(10):1718–1751.

49. Chippa VK, Mohapatra D, Roy K et al. Scalable effort hardware design. IEEE Trans Very Large Scale Integr Syst 2014;22(9):2004–2016.
50. Liu C, Han J, Lombardi F. An analytical framework for evaluating the error characteristics of approximate adders. IEEE Trans Comput. 2014;64(5):1268–1281.
51. Liang J, Han J, Lombardi F. New metrics for the reliability of approximate and probabilistic adders. IEEE Trans Comput. 2012;62(9):1760–1771.
52. Huang J, Lach J, Robins G. A methodology for energy-quality tradeoff using imprecise hardware. In: DAC design automation conference 2012. Piscataway: IEEE; 2012. p. 504–509.
53. Miao J, He K, Gerstlauer A et al. Modeling and synthesis of quality-energy optimal approximate adders. In: Proceedings of the international conference on computer aided design. 2012. p. 728–735.
54. Venkatesan R, Agarwal A, Roy K et al. MACACO: modeling and analysis of circuits for approximate computing. In: 2011 IEEE/ACM international conference on computer-aided design (ICCAD). Piscataway: IEEE; 2011. p. 667–673.
55. Mazahir S, Hasan O, Hafiz R et al. Probabilistic error modeling for approximate adders. IEEE Trans Comput. 2016;66(3):515–530.
56. Ayub MK, Hasan O, Shafique M. Statistical error analysis for low power approximate adders. In: Proceedings of the 54th annual design automation conference 2017. 2017. p. 1–6.
57. Qureshi A, Hasan O. Formal probabilistic analysis of low latency approximate adders. IEEE Trans Comput-Aided Des Integr Circuits Syst 2018;38(1):177–189.
58. Liu W, Zhang T, McLarnon E et al. Design and analysis of majority logic based approximate adders and multipliers. IEEE Trans Emerg Topics Comput. 2019;9:1609–1624.
59. Liang J, Han J, Lombardi F. New metrics for the reliability of approximate and probabilistic adders. IEEE Trans Comput. 2013;62(9):1760–1771. https://doi.org/10.1109/TC.2012.146
60. Liang J, Han J, Lombardi F. New metrics for the reliability of approximate and probabilistic adders. IEEE Trans Comput. 2012;62(9):1760–1771.
61. Kahng AB, Kang S. Accuracy-configurable adder for approximate arithmetic designs. In: Proceedings of the 49th annual design automation conference. 2012. p. 820–825.
62. Zhu N, Goh WL, Wang G et al. Enhanced low-power high-speed adder for error-tolerant application. In: 2010 international SoC design conference. Piscataway: IEEE; 2010. p. 323–327.
63. Yang Z, Han J, Lombardi F. Approximate compressors for error-resilient multiplier design. In: 2015 IEEE international symposium on defect and fault tolerance in VLSI and nanotechnology systems (DFTS). Piscataway: IEEE; 2015. p. 183–186.
64. Du K, Varman P, Mohanram K. High performance reliable variable latency carry select addition. In: 2012 design, automation & test in Europe conference & exhibition (DATE). Piscataway: IEEE; 2012. p. 1257–1262.
65. Kim Y, Zhang Y, Li P. An energy efficient approximate adder with carry skip for error resilient neuromorphic VLSI systems. In: 2013 IEEE/ACM international conference on computer-aided design (ICCAD). Piscataway: IEEE; 2013. p. 130–137.
66. Lu SL. Speeding up processing with approximation circuits. Computer 2004;37(3):67–73.
67. Du K, Varman P, Mohanram K. High performance reliable variable latency carry select addition. In: 2012 design, automation & test in Europe conference & exhibition (DATE). Piscataway: IEEE; 2012. p. 1257–1262.
68. Camus V, Schlachter J, Enz C. A low-power carry cut-back approximate adder with fixed-point implementation and floating-point precision. In: 2016 53nd ACM/EDAC/IEEE design automation conference (DAC). Piscataway: IEEE; 2016. p. 1–6.
69. Mahdiani HR, Ahmadi A, Fakhraie SM et al. Bio-inspired imprecise computational blocks for efficient VLSI implementation of soft-computing applications. IEEE Trans Circuits Syst I Regul Pap. 2009;57(4):850–862.
70. John V, Sam S, Radha S et al. Design of a power-efficient Kogge–Stone adder by exploring new OR gate in 45nm CMOS process. Circuit World 2020;46:257–269.

71. Gupta V, Mohapatra D, Raghunathan A et al. Low-power digital signal processing using approximate adders. IEEE Trans Comput-Aided Des Integr Circuits Syst. 2012;32(1):124–137.
72. Yang Z, Jain A, Liang J et al. Approximate XOR/XNOR-based adders for inexact computing. In: 2013 13th IEEE international conference on nanotechnology (IEEE-NANO 2013). Piscataway: IEEE; 2013. p. 690–693.
73. Almurib HA, Kumar TN, Lombardi F. Inexact designs for approximate low power addition by cell replacement. In: 2016 design, automation & test in Europe conference & exhibition (DATE). Piscataway: IEEE; 2016. p. 660–665.
74. Pashaeifar M, Kamal M, Afzali-Kusha A et al. Approximate reverse carry propagate adder for energy-efficient DSP applications. IEEE Trans Very Large Scale Integr Syst. 2018;26(11):2530–2541.
75. Cai H, Wang Y, Naviner LA et al. Approximate computing in MOS/spintronic non-volatile full-adder. In: 2016 IEEE/ACM international symposium on nanoscale architectures (NANOARCH). Piscataway: IEEE; 2016. p. 203–208.
76. Angizi S, Jiang H, DeMara RF et al. Majority-based spin-CMOS primitives for approximate computing. IEEE Trans Nanotechnol. 2018;17(4):795–806.
77. Liu W, Qian L, Wang C et al. Design of approximate radix-4 booth multipliers for error-tolerant computing. IEEE Trans Comput. 2017;66(8):1435–1441.
78. Boro B, Reddy KM, Kumar YN et al. Approximate radix-8 Booth multiplier for low power and high speed applications. Microelectron J. 2020;101:104816.
79. Waris H, Wang C, Liu W. Hybrid low radix encoding-based approximate booth multipliers. IEEE Trans Circuits Syst II Express Briefs 2020;67(12):3367–3371.
80. Zhao Y, Li T, Dong F et al. A new approximate multiplier design for digital signal processing. In: 2019 IEEE 13th international conference on ASIC (ASICON). Piscataway: IEEE; 2019. p. 1–4.
81. Yin P, Wang C, Waris H et al. Design and analysis of energy-efficient dynamic range approximate logarithmic multipliers for machine learning. IEEE Trans Sustain Comput. 2020;6:612–625.
82. Liu W, Xu J, Wang D et al. Design and evaluation of approximate logarithmic multipliers for low power error-tolerant applications. IEEE Trans Circuits Syst I Regul Pap. 2018;65(9):2856–2868.
83. Liu W, Xu J, Wang D et al. Design of approximate logarithmic multipliers. In: Proceedings of the on great lakes symposium on VLSI 2017. 2017. p. 47–52.
84. Yin P, Wang C, Liu W et al. Design of dynamic range approximate logarithmic multipliers. In: Proceedings of the 2018 on great lakes symposium on VLSI. 2018. p. 423–426.
85. Lotri U, Pilipovi R, Buli P. A hybrid radix-4 and approximate logarithmic multiplier for energy efficient image processing. Electronics 2021;10(10):1175.
86. Alla N, Ahmed SE. An area and delay efficient logarithmic multiplier. In: 2020 international conference on contemporary computing and applications (IC3A). Piscataway: IEEE; 2020. p. 169–174.
87. Ansari MS, Cockburn BF, Han J. An improved logarithmic multiplier for energy-efficient neural computing. IEEE Trans Comput. 2020;70(4):614–625.
88. Saadat H, Javaid H, Ignjatovic A et al. Realm: reduced-error approximate log-based integer multiplier. In: 2020 design, automation & test in Europe conference & exhibition (DATE). Piscataway: IEEE; 2020. p. 1366–1371.
89. Yin P, Wang C, Waris H et al. Design and analysis of energy-efficient dynamic range approximate logarithmic multipliers for machine learning. IEEE Trans Sustain Comput. 2020;6:612–625.
90. Mrazek V, Hrbacek R, Vasicek Z et al. Evoapprox8b: library of approximate adders and multipliers for circuit design and benchmarking of approximation methods. In: Design, automation & test in Europe conference & exhibition (DATE). Piscataway: IEEE; 2017. p. 258–261.

91. Mrazek V, Vasicek Z, Sekanina L et al. Scalable construction of approximate multipliers with formally guaranteed worst case error. IEEE Trans Very Large Scale Integr Syst. 2018;26(11):2572–2576.
92. Cai H, Wang Y, Naviner LA et al. Approximate computing in MOS/spintronic non-volatile full-adder. In: 2016 IEEE/ACM international symposium on nanoscale architectures (NANOARCH). Piscataway: IEEE; 2016. p. 203–208.
93. Cai H, Jiang H, Han M et al. Pj-AxMTJ: process-in-memory with joint magnetization switching for approximate computing in magnetic tunnel junction. In: 2019 IEEE computer society annual symposium on VLSI (ISVLSI). Piscataway: IEEE; 2019. p. 111–115.
94. de Barros Naviner LA, Cai H, Wang Y et al. Stochastic computation with spin torque transfer magnetic tunnel junction. In: 2015 IEEE 13th international new circuits and systems conference (NEWCAS). Piscataway: IEEE; 2015. p. 1–4.
95. Xiong K X, Cai H. A novel In-MRAM multiplier using toggle spin torques switching. In: 2020 IEEE 15th international conference on solid-state & integrated circuit technology (ICSICT). Piscataway: IEEE; 2020. p. 1–3.
96. Cai H, Wang Y, Naviner LADB et al. Robust ultra-low power non-volatile logic-in-memory circuits in FD-SOI technology. IEEE Trans Circuits Syst I Regul Pap. 2016;64(4):847–857.
97. Cai H, Chen J, Zhou Y et al. Sparse realization in unreliable spin-transfer-torque RAM for convolutional neural network. IEEE Trans Magn. 2020;57(2):1–5.
98. Xie L, Cai H, Yang J. REAL: logic and arithmetic operations embedded in RRAM for general-purpose computing. In: 2019 IEEE/ACM international symposium on nanoscale architectures (NANOARCH). Piscataway: IEEE; 2019. p. 1–4.
99. Cai H, Wang Y, de Barros Naviner LA et al. Exploring hybrid STT-MTJ/CMOS energy solution in near-/sub-threshold regime for IoT applications. IEEE Trans Magn. 2017;54(2):1–9.
100. Liu B, Dong W, Xu T et al. E-ERA: an energy-efficient reconfigurable architecture for RNNs using dynamically adaptive approximate computing. IEICE Electron Express 2017;14;20170637.
101. Bang S, Wang J, Li Z et al. 14.7 a 288μw programmable deep-learning processor with 270kb on-chip weight storage using non-uniform memory hierarchy for mobile intelligence. In: 2017 IEEE international solid-state circuits conference (ISSCC). Piscataway: IEEE; 2017. p. 250–251.

Chapter 19
Efficient Approximate DNN Accelerators for Edge Devices: An Experimental Study

Mohammadreza Asadikouhanjani, Hao Zhang, Kyunghwan Cho, Young-Jin Park, and Seok Bum Ko

1 Introduction

Although the training phase of DNNs comes with billions of multiply and accumulate (MAC) operations when compared to inference. But, it is the inference that is usually subject to more strict design constraints, for example, equipping edge devices with DNN applications that usually come with resources and energy constraints. Accordingly, efficient implementation of DNNs is vital before integrating them into edge devices [1].

Many proposed DNN accelerators until now have shown significant improvements while performing the computations of DNN inferences [2, 3]. In [4], the authors have shown that performing the computations in memory results in remarkable energy efficiency. These types of accelerators try to avoid data movement between memory and processing engine. However, their non-generic layouts are expensive to build since it is usually a mix of CMOS and other technologies. The authors in [5] and Eyeriss-V2 [6] have shown that efficient Processing Element (PE) and Network on Chip (NoC) designs are both critically important to keep the performance of the whole design high in the NoC-based DNN accelerators. There are also other works that have focused on efficient data movement inside the computation engine [7].

M. Asadikouhanjani · S. B. Ko (✉)
Department of Electrical and Computer Engineering, University of Saskatchewan, Saskatoon, SK, Canada
e-mail: m.asadi@usask.ca; seokbum.ko@usask.ca

H. Zhang
Department of Electrical and Computer Engineering, Ocean University of China, Qingdao, China
e-mail: hao.zhang@ouc.edu.cn

K. Cho · Y.-J. Park
Korea Electrotechnology Research Institute (KERI), Uiwang, South Korea
e-mail: chokh@keri.re.kr; yjpark@keri.re.kr

W. Liu, F. Lombardi (eds.), *Approximate Computing*,
https://doi.org/10.1007/978-3-030-98347-5_19

The authors in [8, 9] have proposed an algorithm to reduce the number of multiplication operations. Eyeriss [10] is an efficient accelerator equipped with the row stationary dataflow which increases the chance of local reuse of data inside the computation engine.

In [11], the authors have presented a method for transparently identifying ineffectual computations during inference with deep learning models. Specifically, by decomposing multiplications down to the bit level. The proposed architecture in [11] targets bit sparsity, that is zero bits since processing zero bits in a MAC operation does not affect the outcome. However, in these types of accelerators, the focus is only on skipping the ineffectual computations [12, 13]. In [14], the authors have proposed a real-time architecture for skipping non-effectual and possible effectual computations. It has been shown that when the MAC operations are decomposed down to bit level, there is still room to prune identical effectual computations.

One of the promising solutions to perform inference of DNNs efficiently is utilizing approximate multipliers inside the computation engine of DNN accelerators[15]. The previous researches have reported that DNNs are resilient against small arithmetic errors [16]. The authors in [17] have shown that the costs of applying DNNs to edge devices can be mitigated via approximate computing. Exploiting the resilience of DNNs to numerical errors from approximate computing, they have demonstrated a reduction in communication overhead of distributed deep learning training via Adaptive residual gradient Compression (AdaComp) and a reduction in computation cost for deep learning inference via Parameterized clipping Activation (PAct)-based network quantization.

In [18], the authors have proposed a novel automated framework for High-Level Synthesizing (HLS) of approximate accelerators by providing a library of approximate functional units. The authors in [19] have proposed a novel approximate multiplier that can be used for reducing the energy consumption of MAC-oriented signal processing algorithms. To give more detail, two approximate 4 to 2 compressors with opposite error directions have been developed to minimize the hardware costs, and then the corresponding approximate multipliers have been analyzed to predict the amount of error in a probabilistic way. Compared to previous works, suffering from the accumulated errors, the proposed interleaving method in [19] generates a narrow and balanced error distribution.

In [20], the authors have introduced approximate multipliers based on alphabet-set multiplication. The weights are divided into parts, having 4 bits. Multiplication by each 4-bit part of the weight is implemented by shifting a precomputed input value and followed by summation. This architecture uses an algorithm to approximate multiplication in the specified PEs of the computation engine when training DNNs. Similarly in [21], the authors have proposed an algorithm that determines bit precision for all PEs to minimize power consumption for given target accuracy. While all of the proposed algorithms in [20, 21] are very progressive and interesting research topics that propose different paradigms of efficient DNN processing, it is not easy to integrate them into the edge devices equipped with DNN applications.

In [22], the authors have evaluated how the error caused by an approximate logarithm-based multiplier affects deep DNN inferences. In more detail, the authors have explained that how convolution and fully connected layers maintain their intended functionalities despite applying a logarithm-based approximation into the computation engine of DNN accelerators. Also, they have explained that the total computed error will not converge properly while performing the MAC operations of different layers using approximate adders. In this chapter, a detailed stepwise approach for designing a re-configurable approximate Booth multiplier suitable for performing inference of DNNs. Furthermore, applying various approximation techniques into different layers of DNNs is investigated.

The rest of this chapter is as follows: Sect. 2 provides some background information on radix-4 Booth multiplication. Then, it presents the considered stepwise approach for designing a re-configurable approximate Booth multiplier. Section 3 explains the proposed Booth multiplier architecture, and the general architecture of the accelerator is built upon the explored approximate multipliers in this chapter. Section 4 compares the performance of the developed Booth multiplier to the most competitive designs in terms of area and energy consumption. Finally, Sect. 5 concludes the chapter.

2 Designing a Re-configurable Approximate Booth Multiplier for Inference of DNNs

DNNs can work with smaller data types with less precision, such as 8-bit integers, which makes them suitable for integration into edge devices [23]. In other words, it is possible to achieve an acceptable accuracy by utilizing 8-bit integer multipliers instead of regular 32-bit floating-point multipliers. Radix-4 Booth multiplication is a popular multiplication algorithm that reduces the size of the partial product array by half. Figure 19.1a shows the dot-diagram of a signed 8×8 radix-4 Booth multiplier. As the considered bit-width of the multiplier operand is 8, the radix-4 Booth encoded of each weight will have 4 elements. In Fig. 19.1a, the partial product bits are organized into 4 rows and 16 columns. In this figure, the numbers given to the rows and columns of partial products are based on their significance.

The weights of each layer of DNNs have a Gaussian distribution around zero which makes them suitable to be encoded as the multiplier operand in the radix-4 Booth multiplication algorithm. To add more detail, by choosing weights as the multiplier operands, the produced error caused by applying different approximation techniques into partial products of the multiplier has opportunities to be both positive and negative which accordingly causes an unbiased distribution of error around zero. Unbiased distribution of error has a direct effect on inference accuracy, especially, large DNNs. The authors in [24, 25] provided a detailed comparison.

Figure 19.1b shows the partial product generator circuit in radix-4 Booth algorithm which tends to be more logically complex than a single AND gate [26].

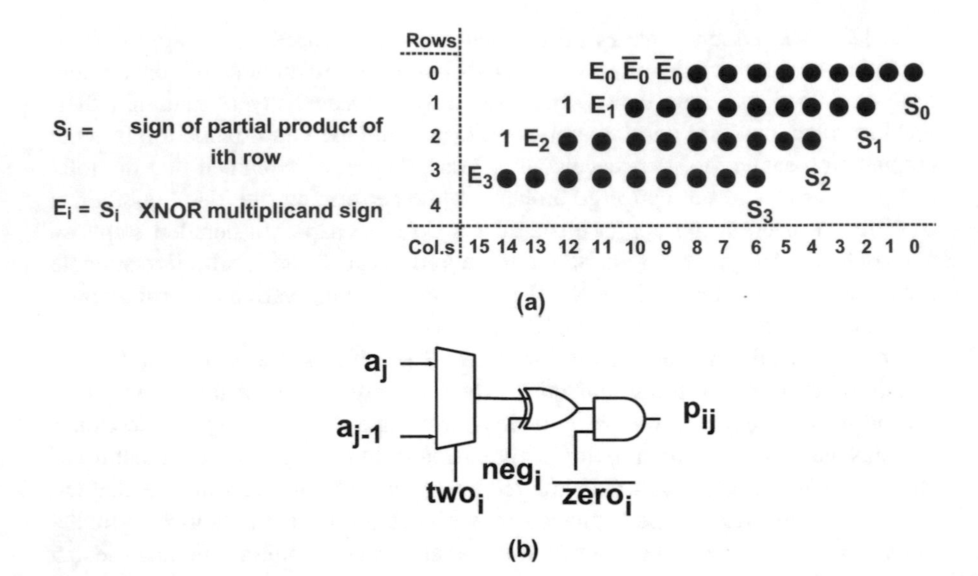

Fig. 19.1 The (**a**) dot-diagram of a signed 8×8 radix-4 Booth multiplier and the (**b**) internal architecture of the partial product generator in a radix-4 Booth multiplier. Filled circle: exact partial products

Table 19.1 Radix-4 booth terms corresponding to various inputs and encoded signal values

Inputs			Encoded signals			
b_{2i+1}	b_{2i}	b_{2i-1}	neg_i	two_i	$zero_i$	Value
0	0	0	0	0	1	0
0	0	1	0	0	0	+1
0	1	0	0	0	0	+1
0	1	1	0	1	0	+2
1	0	0	1	1	0	−2
1	0	1	1	0	0	−1
1	1	0	1	0	0	−1
1	1	1	0	0	1	0

However, the Booth multiplication algorithm makes up for this in that it reduces the actual number of partial products that need to be generated and therefore accumulated [26]. In Fig. 19.1b, $\{a_{i-1}, a_i\}$ are two following bits of each input feature map (ifmap) and $\{p_{ij}\}$ is jth bit of the ith row of partial products. As mentioned earlier in this section, the multiplier operand is grouped in sets of bits $\{b_{2i-1}, b_{2i}, b_{2i+1}\}$, which corresponds to one of the values from 0, ±1, ±2. The radix-4 Booth encoder encodes these bit groupings into three signals neg_i, two_i, and $zero_i$. Table 19.1 illustrates how the encoding terms and multiplier bits are associated with radix-4 Booth terms.

To evaluate the function of the multipliers explored in the following sections, an accelerator has been implemented which is discussed in detail in Sect. 3. This

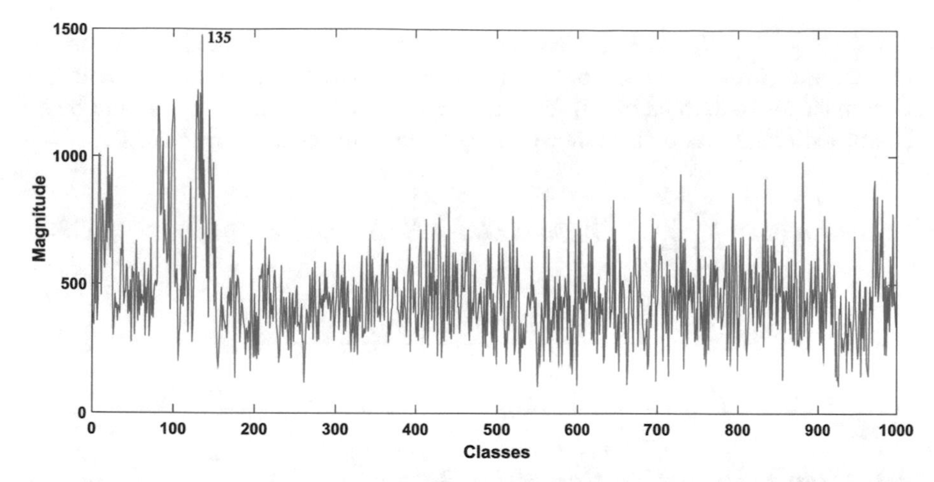

Fig. 19.2 Output of the proposed accelerator in Sect. 3 when a picture of a Crane is fed into this accelerator that runs SqueezeNet in exact mode

accelerator can perform the computations of different layers of DNNs. The output of this accelerator is an input for a Softmax calculator. As an example, a picture of a Crane, which is shown in Fig. 19.3a, is given to this accelerator when it runs SqueezeNet [27]. SqueezeNet is an object image classifier that classifies 1000 object images. Crane is class number 135 of the SqueezeNet. Figure 19.2a shows the computed output of the proposed accelerator when exact Booth multipliers are used. In this figure, index 135, which represents the Crane class, has the highest value among other indexes. In addition, the value of most of the non-relevant classes is pretty low compared to the peak index.

There are many different types of approximate multipliers and MAC units out there, with various costs and error characteristics [26, 28–32]. Without studying the error characteristics of an approximation method, there is a low chance to get the optimum accuracy after applying that specific approximation into DNN accelerators. Furthermore, it is necessary to study the error characteristics of approximate design using real benchmark DNNs, since applying approximation into the computation engine of DNN accelerators is somehow different. In more detail, DNNs come with a high number of layers that each layer has a various number of channels. Also, the error could propagate across layers. In what follows, the effect of the applied some common approximate techniques is discussed in more detail by calculating and analyzing the error distribution, normalized mean error distance (NMED), and variance of the error.

To analyze the error characteristics of different approximation methods, various approximate multipliers are simulated based on the convolution operation using $f_w \times f_h$ filter kernels, where the internal MAC operations can be formulated as

$\sum_{l=1}^{c}\sum_{i=1}^{w}\sum_{j=1}^{h} I_{lij} \times K_{lij}$. Note that the number of accumulated multiplication results varies with the number of channels in the convolution operations, denoted as C. Similar to the authors in[19] for each filter, the accumulated error would be as (1). In addition, the equations shown in (2) are used to calculate the NMED.

$$E_{MAC} = \sum_{l=1}^{c}\sum_{i=1}^{w}\sum_{j=1}^{h} Approx.(I_{lij} \times K_{lij}) - (I_{lij} \times K_{lij}) \tag{19.1}$$

$$NMED = \frac{Mean(|E_{MAC}|)}{MAX(\sum_{l=1}^{c}\sum_{i=1}^{w}\sum_{j=1}^{h} I_{lij} \times K_{lij})} \tag{19.2}$$

2.1 Step 1: Removing Correction Terms

Each signal S_i shown in Fig. 19.1a is a correction term that is identical to signal neg_i shown in Fig. 19.1b. Removing these correction terms from partial product rows comes with decreasing the length of the critical path by $0.18ns$ inside the design shown in Fig. 19.1a. However, it also comes with increasing the magnitude of each negative partial product row slightly. It is like performing 1's complement instead of 2's complement. Figure 19.5 is the output of the proposed accelerator in Sect. 3, which shows the computed value of all the object classes when this accelerator is fed by a Crane picture shown in Fig. 19.3a, and all the correction terms are removed from the Booth multipliers. In this figure, index 101 is the peak index. This index represents the Black Swan class which is shown in Fig. 19.3b. Index 135 is still among the 5 indexes with the highest values though.

Figure 19.4 shows the NMED of 4 hidden layers and the final average-pooling layer of SqueezeNet for different explored approximation techniques in this section. The input test benches are created using ImageNet validation dataset [33]. To present all the values in Fig. 19.5 in a more readable manner, the NMED values of the output layers for all the approximate methods in this figure are divided by 32. As expected, when all the correction terms are removed, the normalized error distance is high, especially for the layers with a high number of channels.

As this approximation only increases the magnitude of negative partial products, it seems that the error caused by applying this method should only be distributed on negative values. However, this is only true for the first layer of DNNs. Figure 19.6a shows the error distribution of layer Fire 4 of SqueezeNet when all the correction terms are removed and the DNN is fed by the mentioned Crane picture above. Based on this figure, the error is mostly distributed on negative values, and however, the error is still distributed on positive values too. This is caused by the propagated error from former layers. To add more detail, since the number of output features that are set to zero after applying nonlinearity increases in former layers, the input image of

(a) (b) (c)

Fig. 19.3 Pictures of (**a**) a Crane, (**b**) a Black Swan, and (**c**) a Feather Boa

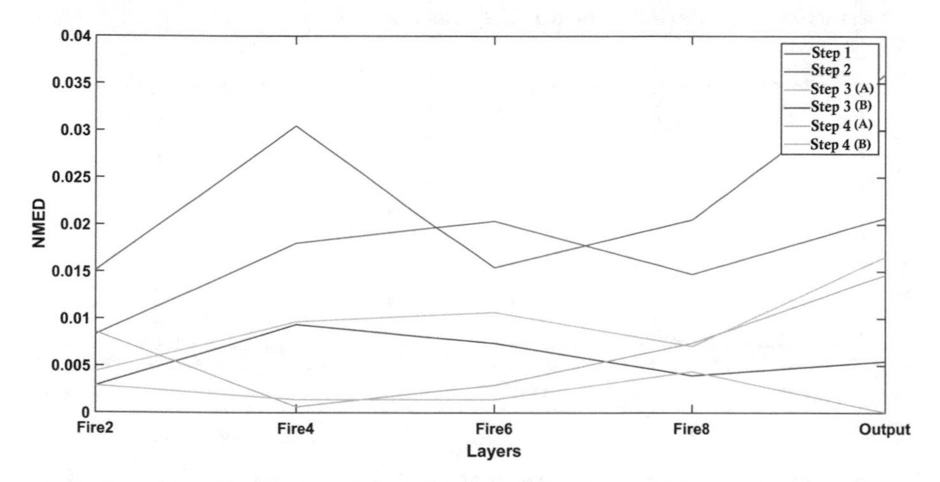

Fig. 19.4 The computed NMED of a few layers of SqueezeNet [27] for the various approximation techniques explored in Sect. 2 while running ImageNet dataset [33]

hidden layers would be different compared to the exact mode. The variance of error distribution for this layer is $1.13e + 03$.

This technique can be used in the layers where its effect is negligible on DNN inference accuracy such as layers that contain a few number of channels, or when the Booth-encoded element of the weights is mostly positive, especially those regarding the least significant partial product rows. Since in such layers, the sum of the accumulated error would not be that large enough to affect the inference accuracy. However, as shown in the following steps, that it is possible to apply this technique into large DNNs if the effect of this approximation is somehow will be compensated by other applied approximation techniques.

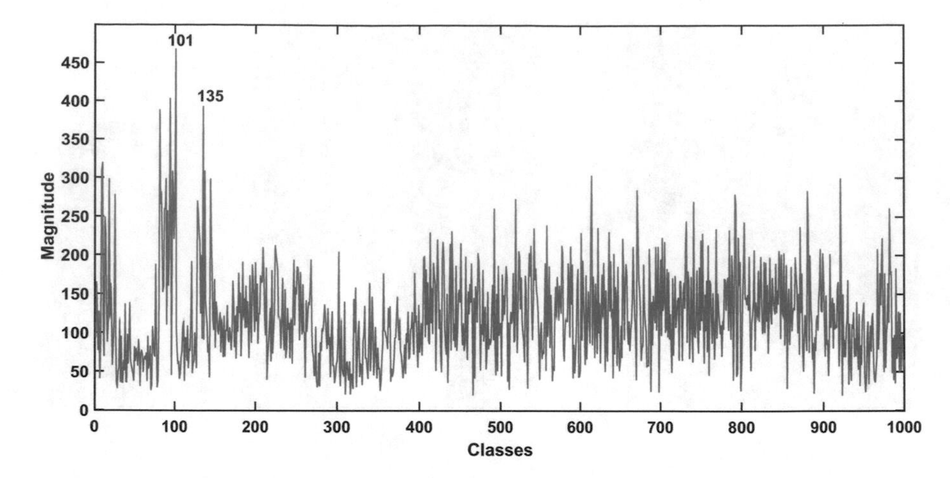

Fig. 19.5 Output of the proposed accelerator when a picture of a Crane is fed into this accelerator, and all the correction terms are removed in all the Booth multipliers

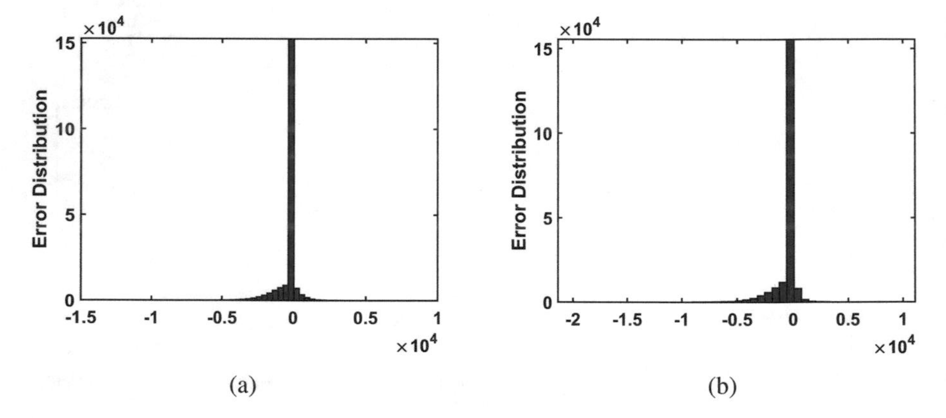

Fig. 19.6 The error distribution of layer Fire 4 (**a**) when all the correction terms are removed and (**b**) when the 5 least significant partial product columns are truncated in all the Booth multipliers

2.2 Step 2: Truncation of Partial Products

Truncation is one of the most efficient ways of applying approximation into multipliers in terms of reducing power consumption, area, and complexity. Replacing a few partial products in the least significant columns of the multiplier with zeros comes with removing the circuit that generates and the circuit that is considered for the accumulation of those partial products. Furthermore, truncation of the least significant columns in a Booth multiplier also means reducing the magnitude of positive partial products and increasing the magnitude of negative partial products. Therefore, the error direction for many of the output features, especially those in the high-depth layers, would be negative. Combining this approximation with the

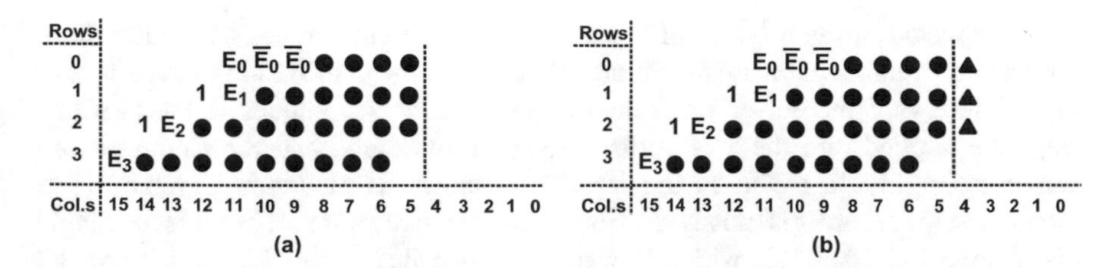

Fig. 19.7 The dot-diagram of an approximate signed radix-4 Booth multiplier explored in step 1, when (**a**) the 5 least significant partial product columns are truncated and (**b**) the partial products in the 5 least significant columns are replaced with a single inexact partial product [26]. Filled triangle: inexact compressed partial products using an OR reduction operation. Filled circle: exact partial products

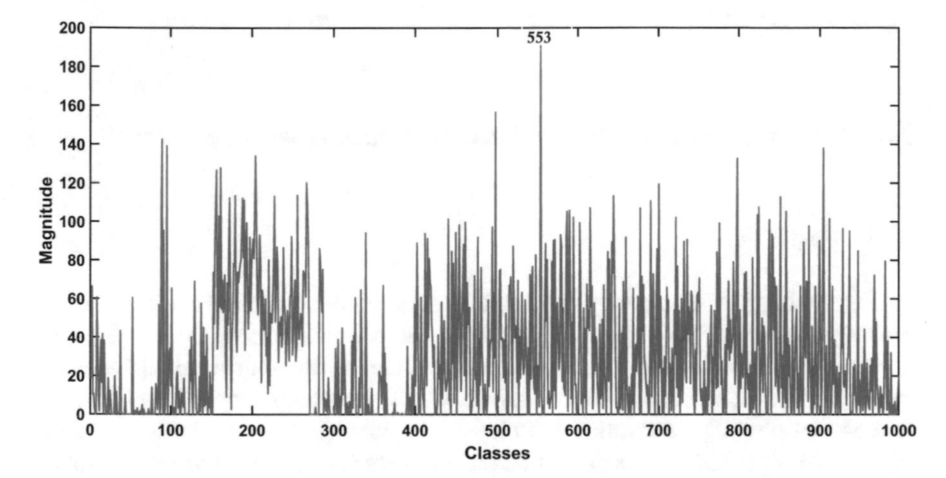

Fig. 19.8 Output of the proposed accelerator when a picture of a Crane is fed into this accelerator, and the introduced approximation in steps 1 and 2 is applied into the Booth multipliers

approximation explored in the previous step, the total accumulated error direction would still be toward negative.

Figure 19.7a shows the dot-diagram of a signed 8×8 radix-4 Booth multiplier when 5 least significant columns of partial products are truncated in excess of the introduced approximation technique explored in the previous step. Figure 19.8 shows the output of layer Fire 4 when the picture of a Crane shown in Fig. 19.3 is fed into the proposed accelerator, and the introduced approximate techniques both in steps 1 and 2 are applied into all the Booth multipliers of our proposed accelerator. In Fig. 19.8, not only there is no obvious peak at index 135, but also this output is noisy. In more detail, the computed value of so many object classes that are not even close to the given input image is much higher than that of the previous design. In this figure, the peak index is 553, and this index represents the Feather Boa class which is shown in Fig. 19.3c.

As expected in Fig. 19.4, NMED increases for so many layers, especially those with a large number of channels. Figure 19.6b shows the distribution of the error for layer Fire 4 of SqueezeNet when both approximation techniques in the previous steps are applied into the 8×8 radix-4 Booth multipliers, and the Crane picture shown in Fig. 19.3a is fed to the DNN accelerator as an input. Similar to the previous step, the error is mostly distributed on negative values. The variance of this distribution is $1.43e + 03$, which is even more than that of the first step. Although the error direction of the explored approximation techniques in steps 1 and 2 is toward negative values, it is still possible to use these approximation techniques in the layers that their number of channels is low. Furthermore, as explained in the following steps, it is applicable to use these types of approximations with other approximate techniques that come with a positive error direction, since there is a chance that the total accumulated error would be balanced around zero.

2.3 Step 3: Replacing Exact Partial Products with Inexact Ones

2.3.1 Step 3 (A)

By decreasing the number of truncated partial product columns, the inference accuracy increases for the explored design in the previous step. However, this step shows that by studying and considering the behavior of the total accumulated error, there is a high chance to end up to a design which is not only very efficient in terms of area and power consumption but also at the same time accurate. The authors in[26] have proposed an approximation technique for the partial product generation stage. However, they have not evaluated their method by running any DNN as their benchmark. In more detail, a group of exact bits of the multiplicands is replaced with a single inexact partial product using an OR reduction operation.

Figure 19.9 shows the gate-level architecture of this approximation technique. The partial product pp_{ij} is computed by performing an OR reduction on the $j + 1$ least significant bits of multiplicand and finally checking the $zero_i$ signal in case the partial product ought to be zeroed out. Figure 19.7b shows the dot-diagram of an 8×8 radix-4 Booth multiplier when, in each of the least 3 significant rows, the partial product bits that are in the 5 least significant columns of the dot-diagram are replaced with a single inexact partial product. As this figure shows, this approximation is applied in excess of the applied approximation in step 1.

Applying the proposed approximation technique in [26] comes with reducing the magnitude of positive partial product rows that the logical value of their jth bit and also the logical value of at least one of the bits $0, \ldots, (j - 1)$ is "1." However, this approximation technique increases the magnitude of many positive partial product rows that the logical value of their jth bit is "0" and at least the logical value of one of the bits $0, \ldots, (j - 1)$th is "1." In addition, this approximation technique would increase or decrease the magnitude of negative

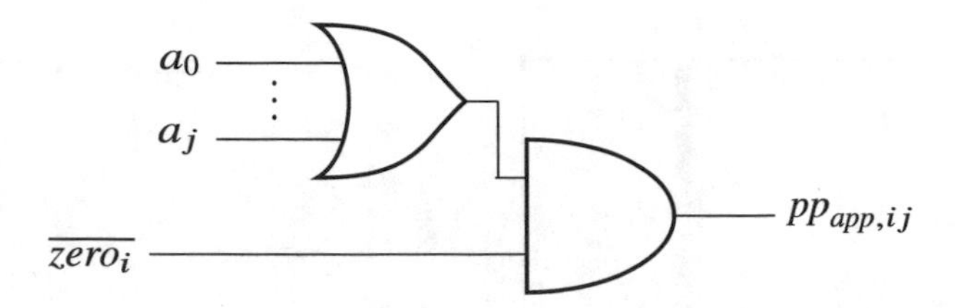

Fig. 19.9 The gate-level architecture of the approximate partial product generator which compresses a few partial products into a single inexact partial product using an OR reduction operation [26]

partial product rows in the same way as well. Furthermore, as shown in Fig. 19.9, the multiplicand operand is approximated before applying neg_i signals, and since the weights that are chosen as the multiplier operands in the multiplication operations have a Gaussian distribution around zero, the error direction would be balanced around zero as well. Moreover, although the error direction caused by the discussed approximate techniques in the previous step is toward negative values for many of the output features. But, combining the explored approximations in both steps, the total accumulated error distribution is more balanced around zero compared to just applying the approximation methods explored in the previous steps.

Figure 19.11 shows the output of the proposed accelerator when the same Crane picture used for benchmarking former approximate architectures is used as an input of the accelerator, and the approximation technique shown in Fig. 19.7b is applied to all the multipliers of this accelerator. In this figure, there is an obvious peak at index 135 and the value of other object classes is pretty low than that of the explored design in the previous step. Figure 19.10a shows the error distribution of layer Fire 4 of the SqueezeNet, while a group of exact bits in the 5 least significant columns of the dot-diagram is replaced with a single partial product in excess of applied approximation techniques in step 1. The variance of this distribution is $0.99e + 03$. This distribution is more balanced compared to previous methods but not significantly. However, for some other layers like Fire 2, the error distribution is more balanced around zero compared to explored approximation in the previous steps (Fig. 19.7a).

Figure 19.4 shows the NMED of a few layers of SqueezeNet when the approximation techniques shown in Fig. 19.7b are applied to the multipliers of our proposed accelerator. As expected, by replacing a group of exact bits of partial product rows with a single inexact partial product and removing all the correction terms, the computed NMED of most of the layers is smaller compared to only applying the discussed approximation techniques in the previous steps. Furthermore, the propagated error from low depth layers to high-depth layers is less as well.

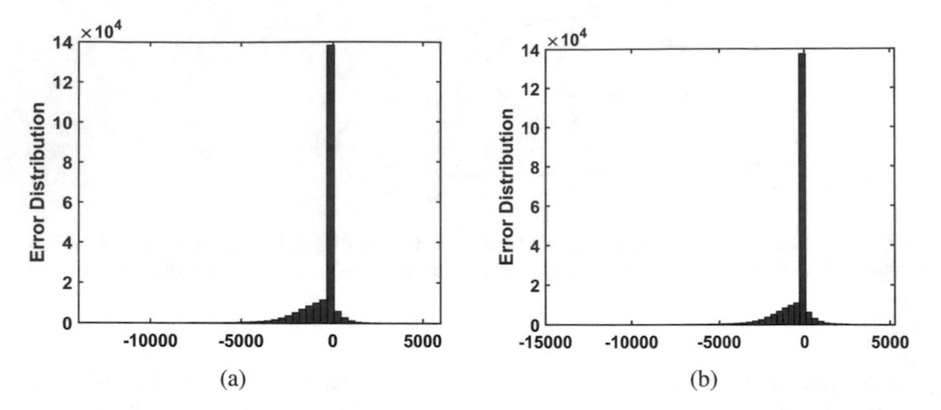

Fig. 19.10 The error distribution of layer Fire 4 (**a**) when the partial products in the 5 least columns are compressed to a single inexact partial product using an OR reduction operation and (**b**) when the 5 least significant partial products are replaced with a constant logical value "1" in all the Booth multipliers. The mentioned approximation is applied in excess of the explored approximation in step 1

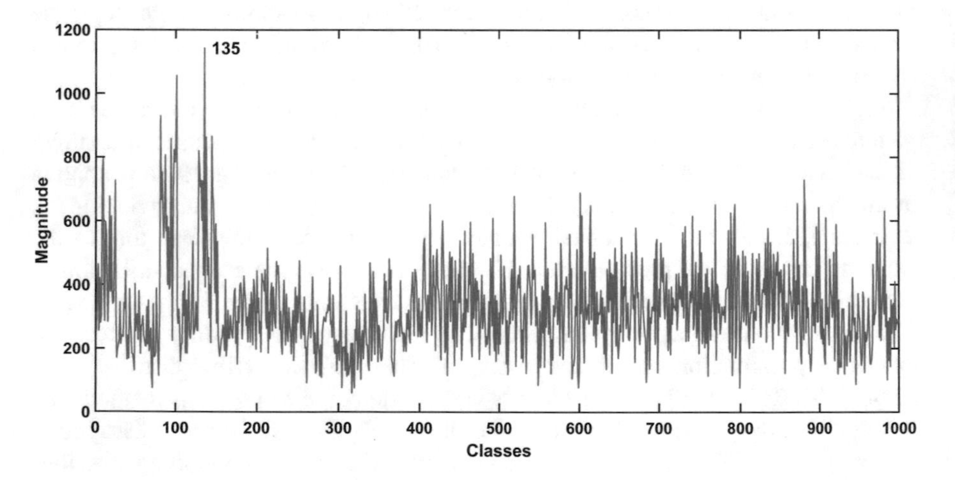

Fig. 19.11 The output of the proposed accelerator when a picture of a Crane is fed into this accelerator, and the partial products in the 5 least columns are compressed to a single inexact partial product using an OR reduction operation in all the Booth multipliers. The mentioned approximation is applied in excess of the explored approximation in step 1

2.3.2 Step 3 (B)

It is possible to implement a more efficient design by rethinking the mentioned approximate technique shown in Fig. 19.7b. This approximation technique increases the magnitude of so many positive and negative partial product rows. However, some partial product rows still remain unchanged. As an example, the magnitude of the partial product rows that their 5 least significant columns are zero is

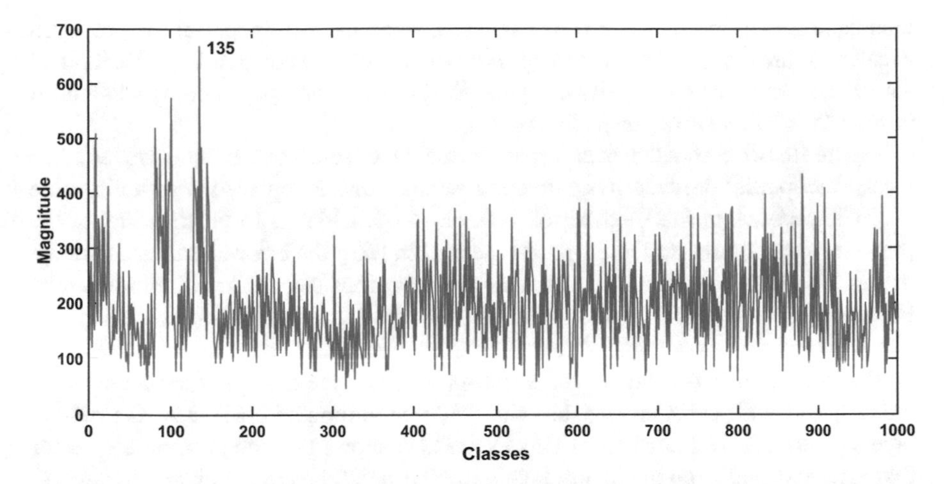

Fig. 19.12 The output of our proposed accelerator when a picture of a Crane is given to this accelerator, and the partial products in the 5 least columns are replaced with a single constant logical value "1." The mentioned approximation is applied in excess of the explored approximation in step 1

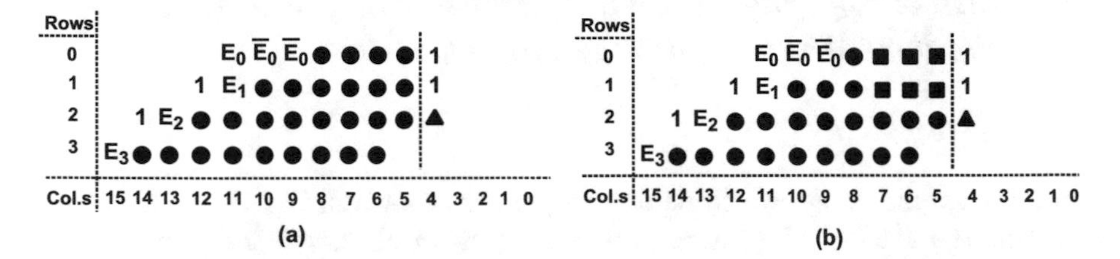

Fig. 19.13 The dot-diagram of an approximate signed radix-4 Booth multiplier explored in step 1, when (**a**) the 5 least significant partial product columns are replaced with a constant logical value "1" and when (**b**) approximate partial product generators are used in the columns 5 to 7 of the 2 least significant rows. Filled triangle: inexact compressed partial products using an OR reduction operation. Filled square: inexact partial product generator by removing the multiplexers considered for shifting two adjacent partial product bits. Filled circle: exact partial products

identical before and after applying approximation. Based on our evaluation, the magnitude of unchanged partial products must change as well to have better accuracy. Replacing the inexact partial product in the first and second rows of the approximate dot-diagram shown in Fig. 19.7b with a constant logical value "1" increases the magnitude of unchanged partial products as well. This change is also very efficient in terms of implementation. Since, for all the multipliers, the OR reduction operation is replaced by logical constant "1." Figure 19.13a shows the dot-diagram of this approximation technique.

Figure 19.12 shows the output of the proposed accelerator when the same Crane picture that is used for benchmarking the former approximate multipliers is fed to this accelerator, and all its multipliers are equipped with the approximation technique shown in Fig. 19.13a. As expected, when the multipliers of the proposed accelerator are equipped with the approximate technique shown in Figs. 19.13a and 19.7b, the computed probability using the Softmax function provided by Matlab for index 135 is 94 and 85%, respectively.

Figure 19.10b shows the error distribution of layer Fire 4 of the SqueezeNet when the inexact partial product in the first and second rows is replaced with logic value "1." The variance of this distribution is $0.94e + 03$, which is lower than that of the previous step. Similar to the proposed method in [26], the error distribution shown in Fig. 19.10b is more balanced compared to previous steps, but not significantly. However, for many other layers such as Fire 8, the error distribution is much more balanced around zero compared to explored approximation in the previous steps.

Based on Fig. 19.4, when exact partial products in the first and the second rows are replaced with constant logical value "1," the computed NMED of the hidden layers of SqueezeNet decreases. This figure also shows that the accumulated error that is propagated from low depth layers to high-depth layers is lower compared to previous explored designs, especially for layers with a high number of channels.

2.4 Step 4: Applying Dynamic Approximation into Partial Products While Performing Shift Operation

2.4.1 Step 4 (A)

In this step, the design which its dot-diagram has shown in Fig. 19.13a is pushed further. The authors in[26] have explored the effect of removing the multiplexer in the exact partial product generator shown in Fig. 19.1b. However, they have not evaluated their method by running any DNN as their benchmark. In more detail, applying this approximation into j least significant partial products of the ith row means if the value of the ith Booth-encoded element of the multiplier operand is $+2$ or -2, the shift operation will not be applied on the j least significant partial products of the ith row of the Booth multiplier.

Similar to the former step, this approximation is applied before applying the neg_i signals, meaning the error distribution caused by this method, depending on the weights, has the potential to have a balanced distribution around zero. However, this method is somehow different since the value of only a few percentages of the whole Booth-encoded weights of the DNNs is $+2$ or -2. This issue has not been considered in [26]. For example, the value of only 2% of the fourth Booth-encoded element of the weights of the SqueezeNet is $+2$ or -2. Accordingly, applying this approximation into the fourth row of the partial products only affects a low percentage of the output features in SqueezeNet. However, the fourth partial product row is the most significant row of the design which its dot-diagram is shown in Fig. 19.13a. Based on our evaluation, applying this approximation to the fourth row of the partial product of the Booth multipliers leads to a sharp inference accuracy loss while running SqueezeNet on the proposed accelerator. A high percentage of the Booth-encoded elements of the weights that their value is +2 or −2 is among the 2 least significant elements of the weights of the benchmark networks.

By applying the approximation technique explained above, depending on the logical value of $j-1$th bit, the magnitude of the partial product rows could increase or decrease after a shift operation. For example, if the ith partial product row is positive and the logical value of $j-1$th bit is "1," the magnitude of the partial product will increase after applying a shift operation. Since the logical value of $j-1$th bit will be repeated as jth and $j-1$th bits of that partial product row through a shift operation. The same thing is true when the partial product row is negative and the logical value of $j-1$th bit is "0." When the logical value of $j-1$th bit is "0" and the partial product is positive, the magnitude of the partial product will decrease after applying shift operation. The same thing is true when the partial product is negative and the logical value of the $j-1$th bit is "1."

Figure 19.13b shows the dot-diagram of the Booth multipliers when the multiplexers inside columns 5 to 7 of the 2 least significant rows are removed in excess of the approximation techniques evaluated in the previous step. Figure 19.14a shows the error distribution of layer Fire 4 of the SqueezeNet when the approximation technique shown in Fig. 19.13b is applied to all the multipliers, and the Crane picture shown in Fig. 19.3a is fed to the proposed accelerator. This distribution is more balanced around zero compared to previous methods. But the variance of this distribution is $1e+03$, which is slightly larger than that of the designs explored in the previous step. However, for the high-depth layers with a large number of channels, the error distribution is not balanced around zero.

Figure 19.4a shows the NMED for a few layers of SqueezeNet when the approximation techniques shown in Fig. 19.13b are applied to the multipliers of the proposed accelerator. In more detail, the NMED of the high-depth layers of SqueezeNet, when the introduced approximation in this step is applied to all the

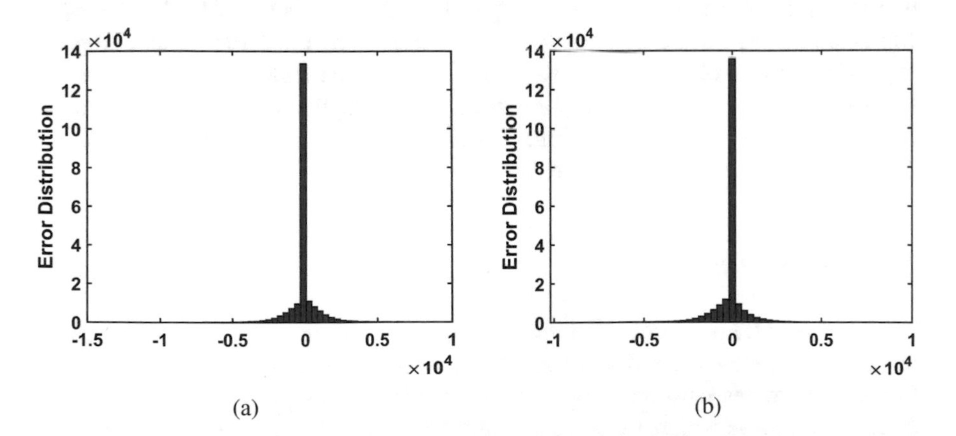

Fig. 19.14 The error distribution of layer Fire 4 when (**a**) the fixed version of the approximation introduced in step 4 (A) and (**b**) the dynamic version of the approximation introduced in step 4 (B) is applied to all the Booth multipliers of the proposed accelerator

multipliers, is higher than that of the approximation technique in the previous step. However, the NMED of mid-depth layers is lower when the introduced approximation in this step is applied to all the multipliers compared to that of the explored approximation technique in the previous step. Based on our evaluation, the approximation method in this step compensates the error caused by the former approximation methods in the mid-depth layers.

The reason why the approximation method shown in Fig. 19.13b is not effective for the high-depth layers is mostly shift operation itself. As described in the previous step, a few least significant bits of the partial products are replaced with an inexact bit which its logical value is "1." Shifting this constant logical value to columns with a greater significance increases the amount of error for some output features mostly in the high-depth layers. For the mid-depth layers shown in Fig. 19.4, the introduced approximation in this step is effective because the magnitude of the ifmaps is such a way that after performing a shift operation, the magnitude of the 2 least partial product rows decreases as explained at the beginning of this section. In more detail, by selecting $j - 1$ as 7, the magnitude of those partial product rows decreases after shift operation for the 2 least partial product rows. In other words, applying approximation shown in Fig. 19.13b compensates the mentioned error above caused by shift operation itself in mid-depth layers like Fire 4 and Fire 6.

However, for the high-depth layers where the magnitude of the ifmaps decreases, applying this approximation while performing shift operation would not compensate the error caused by shift operation itself. In fact, the shift operation still should be applied in an exact way for some ifmaps and should not be performed for some others while performing the computations of high-depth layers. For example, for the ifmaps that their magnitude is very small, shifting the constant logical value "1" from column 4 to 5 increases the computed NMED of these layers.

As shown in Fig. 19.4, when the approximation which its dot-diagram is shown in Fig. 19.10b is applied to the multipliers of the proposed accelerator, the NMED of the layers of SqueezeNet up to layer Fire 2 is the lowest. Furthermore, as explained above, for the mid-depth layers such as Fire 4 and Fire 6, applying the approximation introduced in this step while performing shift operations decreases the NMED of these layers. In addition, for the high-depth layers, the shift operation should be performed for some ifmaps and should not be performed for others, depending on the magnitude of the ifmaps.

2.4.2 Step 4 (B)

To apply a various degree of the approximation introduced in step 4 (A), it is not possible to use a fixed architecture. A fixed version of the design shown in Fig. 19.13b comes with reducing area and power consumption [26]. However, as discussed, to have a higher chance to compensate the error of previous steps while running different DNNs, a dynamic version of this approximation technique should be implemented.

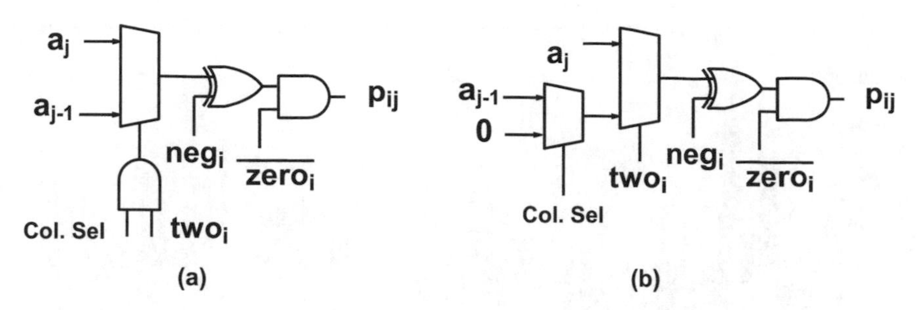

Fig. 19.15 The gate-level architecture of the approximate partial product generator (**a**) which can avoid shift operation dynamically and (**b**) which can coordinate to shift a specific logical value dynamically

To perform the shift operation in an exact mode for the layers of SqueezeNet up to Fire 2 and to decrease the computed NMED of the mid-depth layers, the approximate partial product generators shown in Fig. 19.15a are proposed. As shown in Fig. 19.15a, the signal two_i in the partial product generator is gated for the partial products in columns 5 and 7 through some control signals. To add more detail, using this control signal, it is possible to apply the approximation introduced in this step whenever it is needed. For the high-depth layers, replacing the constant logical value "1" with "0" in column 4 avoids shifting the logical value "1" from column 4 to 5 for the small feature maps. In other words, it is like applying the truncation method evaluated in step 2 of this section for some ifmaps. It is also possible to use the approximate partial product generator shown in Fig. 19.15b for the partial product generators in column 5 to compensate the error in high-depth layers in a more accurate way. To give more detail, the approximate partial product generator in Fig. 19.15b sets the logical value of the partial products in column 5 of the 2 least significant partial product rows to a specific value.

The ifmap distributor unit of the proposed accelerator which is explained in Sect. 3 coordinates applying a various degree of this approximation in columns 5 and 7 for different layers. Furthermore, this unit coordinates the value of constant logical value in the fourth column while performing a shift operation. To give more detail, this unit coordinates the select control signal of the fifth column automatically based on the logical value of the bits a_3 to a_0. For example, in the lest significant row, when the partial product is positive and the logical values of these bits are "11xx," then the constant logical value "1" will be shifted to column 5 like the exact mode but, when the logical values of these bits are "0001," the constant logical value "0" will be shifted to column 5.

Figure 19.16 shows the output of the proposed accelerator when the same Crane picture used for benchmarking former approximate architectures is used as an input of the accelerator, and the dynamic version of the approximation technique shown in Fig. 19.13b is applied to all the multipliers of this accelerator. The computed probability using the Softmax function provided by Matlab for index 135 is 99%, which is higher than that of the previous designs. Figure 19.14b shows the

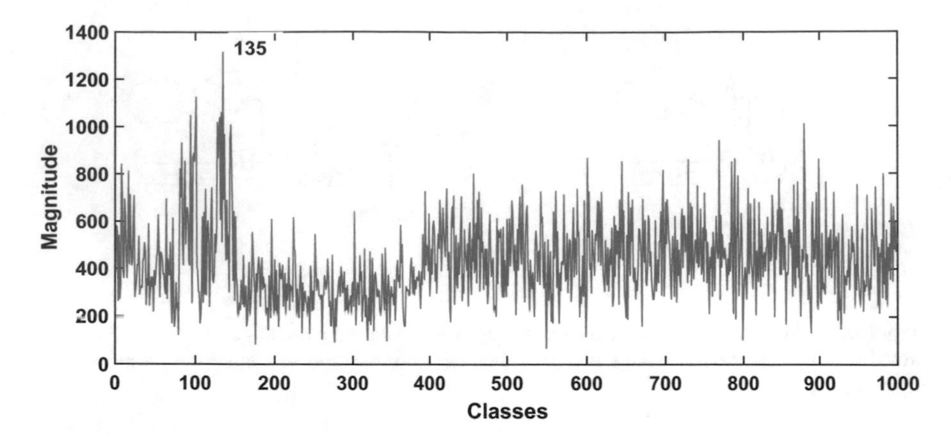

Fig. 19.16 The output of the proposed accelerator, when a picture of a Crane is fed to this accelerator, and the dynamic version of the approximation introduced in step 4 (B), is applied to all the Booth multipliers in excess of the applied approximation in the previous steps

error distribution of layer Fire 4 of the SqueezeNet when the introduced dynamic approximation in this step is applied to all the Booth multipliers, and the Crane picture shown in Fig. 19.3a is fed to the proposed accelerator. This distribution is balanced around zero compared to that of previous methods. The variance of this distribution is $0.83e + 03$, which is the lowest compared to that of other explored designs. Furthermore, for almost all the layers with a different number of channels, the error distribution is balanced. Figure 19.4 shows the NMED of a few layers of SqueezeNet when the dynamic version of the approximation techniques shown in Fig. 19.13b is applied to the multipliers of the proposed accelerator. In more detail, the average NMED of all the layers is the lowest compared to that of the previous steps. Moreover, the propagated error remains low through the DNN compared to previous techniques.

In this chapter, the approximation techniques targeting the accumulation stages of the multipliers are not evaluated. There are many kinds of approximate half-adders, full-adders, 4 to 2 compressors. However, the function of a few of these designs was evaluated while running DNNs [19, 34]. In [19], the authors proposed two 4 to 2 compressors, each of them can compensate the error caused by the other type. They also built up two types of multiplier upon these compressors. In more detail, the authors in [19] perform the computations of a specific number of the channels of each layer with multiplier type 1 and they perform the rest of the computations of that layer with multiplier type 2. The proposed designs in [19] are simulated and discussed in Sect. 4 of this chapter.

3 Proposed Architecture

The general architecture of the proposed accelerator is shown in Fig. 19.17. The proposed computation engine contains 16 PEs, 4 column global buffers (GBs), 4 row GBs, an ifmap, and a filter distributor units. Each regular PE contains 3 MAC units. Each column of the PEs computes an output feature of 4 different filters.

3.1 MAC Unit and PE Specifications

A MAC unit is built upon the re-configurable proposed Booth multiplier introduced in part B of step 4 of Sect. 2. Figure 19.18a shows the internal architecture of this MAC unit. In each MAC unit, there is a re-configurable Booth multiplier, an adder, an accumulator, and an internal control unit. The bit-width of both the input ifmap and the filter weight is 8-bit. In the Booth multiplier, the approximation method explored in step 1 is already applied in a fixed format, since we believe the error regarding this approximation can be compensated using other approximate techniques. The proposed MAC unit supports all the approximation techniques explored in Sect. 2. Signal Trunc./Const1 is considered for replacing the constant logical value "1" with "0" in column 4 of the partial products. In short, using this signal, it is possible to apply the truncation method, which is explained in step 2, as well as the evaluated approximation technique in step 3. Signal Column Sel is considered for applying the dynamic format of the approximation technique explored in part B of step 4, while performing the MAC operations of different layers. The Shared PSum port in each MAC unit is considered for accumulating the partial sums computed by other MAC units within each PE. Each MAC unit has a control unit that coordinates the computation process within that MAC unit.

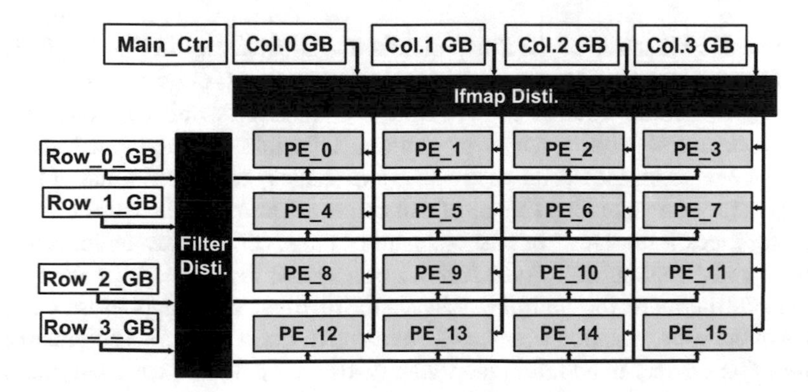

Fig. 19.17 The general architecture of the proposed accelerator

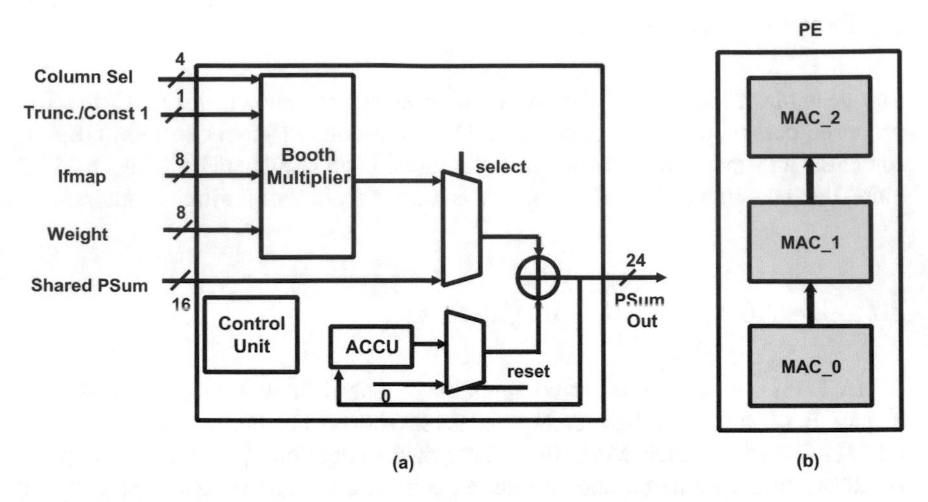

Fig. 19.18 The internal architecture of (**a**) the proposed re-configurable MAC unit that supports all the approximation methods explored in Sect. 2 dynamically and (**b**) each PE inside the computation engine

Figure 19.18b shows the internal architecture of each PE in the proposed accelerator. To give more detail, each PE includes 3 MAC units, and the computed partial sum by each MAC unit could be an input for another MAC unit within each PE. Accordingly, the ifmap and filter rows are divided into rows containing 3 elements before mapping into the computation engine. Zero padding is applied wherever it is needed.

3.2 Distributor Units

The internal architecture of the ifmap distributor unit is shown in Fig. 19.19. There are 3 main sub-units inside each distributor unit. These sub-units are a comparator, a scratchpad, and an internal control unit. The distributor unit can load up to 4 ifmaps concurrently from each of the column GBs into the scratchpad unit. The comparator unit considered for each column evaluates the logical value of the bits a_3 to a_0 of the corresponding ifmaps of that column. The internal control unit, based on the feedback it receives from the comparator units, coordinates the logical value of the signals Column Sel, Trunc./Const1 considered for the MAC units in each column. Furthermore, the control unit inside the distributor unit applies the dynamic format of the approximation techniques evaluated in step 4 to various columns of the computation engine. In addition, the ifmap distributor unit, in each cycle, can read four ifmaps from column 0 to column 3 GBs concurrently. This process continues

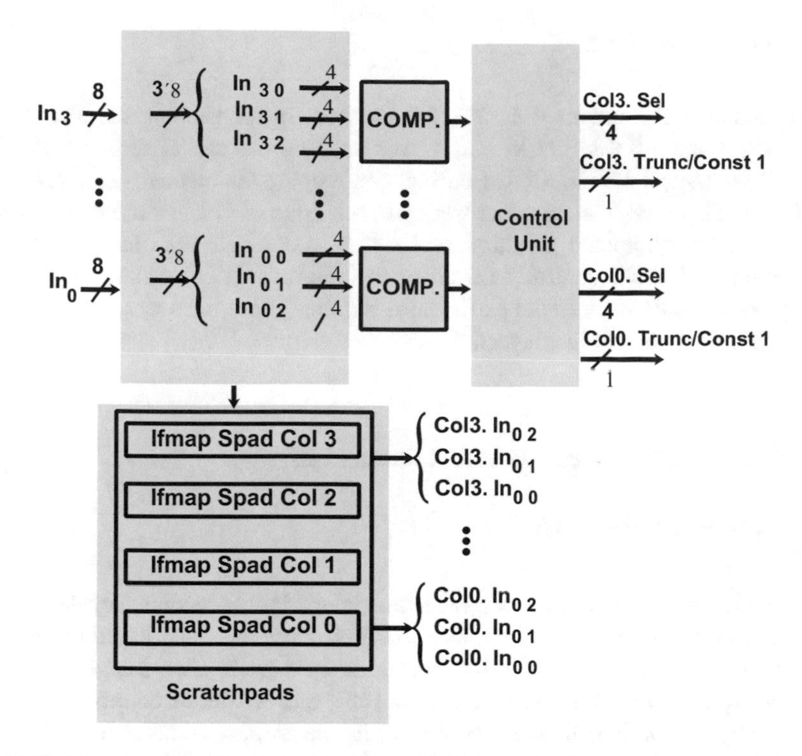

Fig. 19.19 The internal architecture of the ifmap distributor unit of the proposed accelerator

based on the considered dataflow for performing the computations of each layer. Then, this unit feeds the PEs inside the computation engine with the Booth-encoded format of the loaded ifmaps. Moreover, the scratchpad unit in the ifmap distributor unit contains 4 ifmap sub-units. Each ifmap scratchpad sub-unit contains 16 8-bit registers to store the elements of the encoded ifmaps.

The internal architecture of the filter distributor units is somehow similar to ifmap distributor unit. However, there are no comparator units in the filter distributor unit. Accordingly, there are no Trunc./ Cosnt1 and Column Sel signals. In addition, the control unit inside each distributor unit coordinates the write and read data process into and from the mentioned scratchpads above. Furthermore, the main control unit starts each processing cycle by notifying the control unit inside the ifmap and filter distributor units to load their first stored ifmaps or encoded weights. Then, the main control unit notifies the internal control unit of the ifmap distributor unit to load all the remained ifmaps into the PEs in each cycle. The main control unit repeats this process for the remained encoded weights inside the filter distributor units.

3.3 Global Buffers

Each column GB contains 4 1.5 KB SRAM buffers for keeping the elements of ifmap rows and 1.5 KB SRAM buffers for keeping the partial sum values. Each row GB contains 3 1.5 KB SRAM buffers for keeping the elements of each row of the filters. There are 4 small adder trees in each column GB that are considered to accumulate the computed partial sums by PEs to compute the final partial sums. Furthermore, 4 small control units in each column GB coordinate the write or read processes into or from the partial sum buffers. These units also coordinate the computations of the pooling layers.

4 Experimental Results and Discussion

4.1 Accuracy Analysis

To evaluate the accuracy of the explored multipliers in this section, a cycle-accurate simulator is implemented in C#. This simulator is implemented based on the RTL design of the proposed accelerator explained in Sect. 3. Two image classifiers, namely SqueezeNet [27] and GoogleNet [35], and a speech command detector provided by Matlab [36] are used for evaluating the proposed architecture. The main reason for choosing these classifiers is to evaluate almost all different layer types of DNNs and DNNs both with low and high number of layers. As well as, exploring the effectiveness of the proposed approximation method on different applications. ImageNet ILSVRC2012 validation dataset [33] is used for evaluating the image classifiers, and Google speech commands dataset [37] is used for evaluating the speech command detector. As the goal of this research is evaluating the effect of approximation while doing inference of DNNs, pretrained DNN coefficients provided by Mat lab Deep Learning Tool are quantized into 8 bits and also sorted into rows containing 3 elements.

As the authors showed in [31, 32], logarithmic multiplication is a promising approach for applying approximation into neural network computation. These types of multipliers convert the multiplication operations into additions by converting the inputs to the logarithm domain. These types of multipliers are efficient in terms of area and power consumption. Accordingly, two competitive logarithm-based designs, which already have shown promising accuracy while performing inference of DNNs, are chosen for comparison.

Figures 19.20, 19.21, and 19.22 demonstrate the NMED of different layers of the benchmark DNNs when the exact multipliers in the proposed accelerator are replaced with various approximate multipliers. Due to the large number of layers of SqueezeNet and GoogleNet, only the computed NMED of a few layers of these benchmark networks is shown in these figures. The given name to the layers of these DNNs is the same as the given name in Matlab Deep Learning Toolbox. As shown

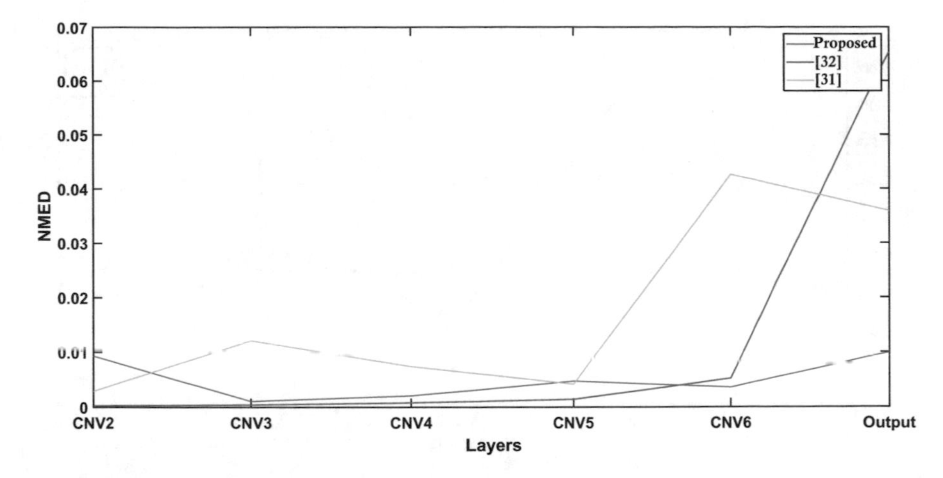

Fig. 19.20 The computed NMED of a few layers of the speech command detector network for different approximate multipliers

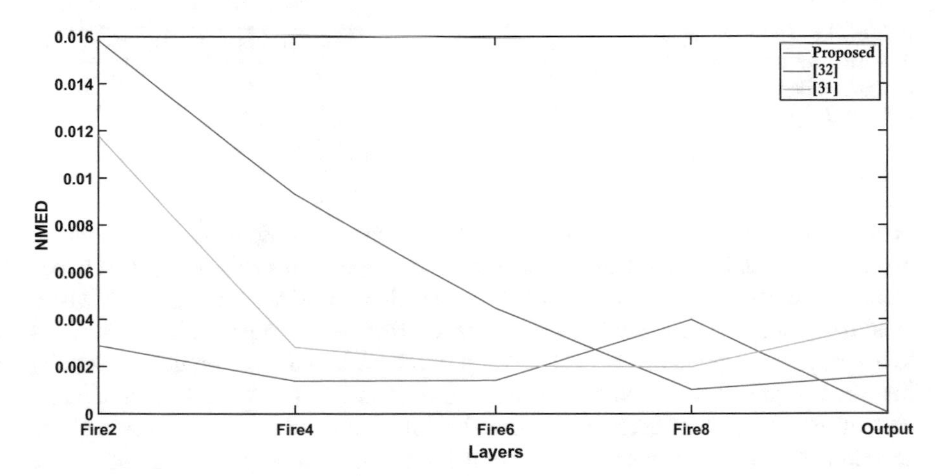

Fig. 19.21 The computed NMED of a few layers of SqueezeNet for different approximate multipliers

in these figures, when the exact multipliers are replaced with the proposed Booth multiplier, the NMED of all the output layers of the benchmark DNNs is lower than that of other approximate designs. Accordingly, when the exact multipliers are replaced with the proposed Booth multiplier, the inference accuracy is higher compared to the computed inference accuracy of other approximate designs, while running benchmark DNNs.

Table 19.2 shows the accuracy of the benchmark DNNs when exact multipliers in the proposed accelerator are replaced with approximate multipliers. In general, the function of the proposed approximate multiplier is better when the number

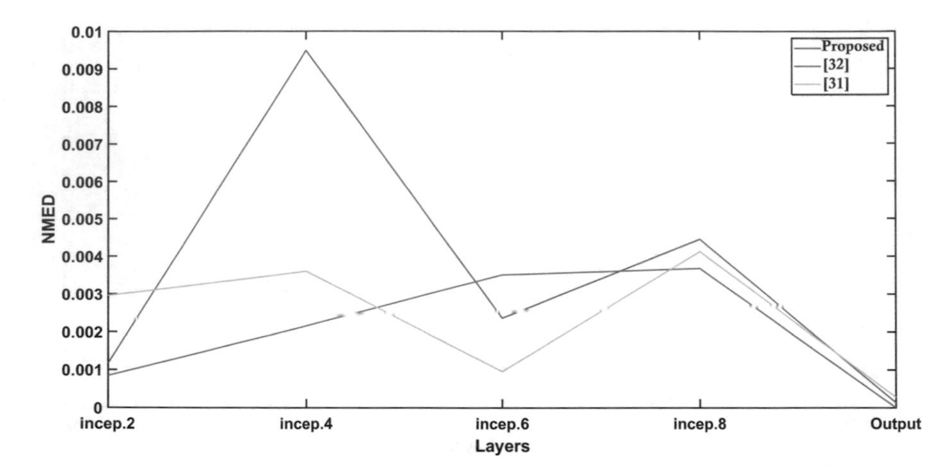

Fig. 19.22 The computed NMED of a few layers of GoogleNet for different approximate multipliers

Table 19.2 Accuracy report of the explored approximate designs while running benchmark DNNs

DNN	Exact	[31]	[32]	Proposed
Speech command detector	79.3	77.1	77.7	78.4
SqueezeNet	57.5	54.9	55.8	56.5
GoogleNet	68.7	66.7	67.1	67.6

of MAC operations that have to be done regarding an output feature increases. To give more detail, when the number of MAC operations increases, the chance of having error in both positive and negative directions increases as well, which eventually ends up in a smaller accumulated error. We also have simulated the 4 to 2 compressors proposed in [19] and applied them into the accumulation stage of the 8 least significant columns of the exact Booth multiplier. We performed the first half of the channels of each layer using the first proposed multiplier type and the second half of the channels of each layer using the second proposed multiplier type in [19]. However, the inference accuracy drops sharply while running SqueezeNet and GoogleNet. The main reason is that the Booth algorithm reduces the number of partial product rows. Accordingly, there is only room to apply this approximation into the 4 to 2 compressors of columns 5–7 of the exact Booth multiplier. To add more detail, the chance of applying this approximation into low significant columns of an 8×8 radix-4 Booth multiplier is low.

4.2 Hardware Implementation

All the evaluated multipliers are implemented in Verilog and are synthesized using SYNOPSYS Design Compiler and TSMC 65 nm technology library. ARM Artisan 65 nm SRAM and register file generator are used for creating the global buffer

Table 19.3 Synthesis report of the explored multipliers

Multiplier	Area (μm^2)	Power (mw)	Critical path (ns)
Exact	844.5	0.310	1.29
[31]	812.4	0.240	1.13
[32]	781	0.253	1.51
Proposed	670	0.212	1.11

Table 19.4 Synthesis report of the explored DNN accelerators

Hardware spec.	Accelerator 1	Accelerator 2	Accelerator 3	Accelerator 4
Mult. type	Exact	[31]	[32]	Proposed
Frequency	200	200	200	200
Core voltage/tech.	1.1 v/65 nm	1.1 v/65 nm	1.1 v/65 nm	1.1 v/65 nm
Reg. file	0.125 KB	0.125 KB	0.125 KB	0.125 KB
Comp. engine area	178K (μm^2)	173K (μm^2)	168K (μm^2)	150K (μm^2)
Comp. engine power	32.3 (mw)	28.8 (mw)	27.5 (mw)	25.2 (mw)

and scratchpads. To measure the power consumption precisely, the activity of the signals inside the computation engine is stored in a SAIF file during the functional simulation of the design. Then, we synthesized the design using Design Compiler and used the mentioned SAIF file to report the power of each unit and its sub-units.

Table 19.3 shows the synthesis report of the evaluated multipliers in this section. The power consumption of all the multipliers is measured when the working frequency is set to 200 MHz for all the multipliers. When compared to the exact multiplier, the proposed design is $\times$ 1.26 smaller and $\times$ 1.46 more energy efficient. The same authors in [32] implemented an efficient version of this multiplier in [15], which is very optimized in terms of both area and energy consumption. The area of their proposed approximate multiplier in [15] is slightly smaller than our proposed design, and its power consumption is almost the same as that of our design. However, since they apply approximation in both the logarithm transform process and the logarithm domain addition process, their design is less accurate compared to the proposed design in [32].

A few accelerators, which their general architecture is shown in Fig. 19.17, are built upon the explored multipliers, and their specifications are reported in Table 19.3. In accelerator 4, which is built upon the proposed Booth multiplier, regardless of the multipliers, the ifmap distributor unit is also somehow different compared to other designs. To give more detail, the ifmap distributor unit in other designs does not include any comparator units. The size of the used Global Buffer is equal in all the developed accelerators. Therefore, to have a better understanding of the cost of different approximate multipliers, the power consumption and area of the Global Buffers are not reported in Table 19.4.

Although the area and power consumption of the ifmap distributor unit in accelerator 4 are higher, however, as shown in Table 19.4, the total area and power consumption of this accelerator are smaller than those of other designs. To give more detail, the proposed design achieves an area efficiency of $\times$ 1.19 and an energy efficiency of × 1.28 compared to the exact design while running benchmark DNNs. The inference processing delay for all the developed accelerators while running different benchmark networks is the same. The inference processing delay while running the speech command detector network, SqueezeNet, and GoogleNet is 2.3, 25.1, and 86 ms, respectively.

It is also possible to improve the area and energy efficiency of the proposed multiplier by sharing the Booth encoder part of the multipliers over all the multipliers in the same row of the computation engine. Therefore, instead of sharing weights, the filter distributor unit shares the signals neg_i, two_i, and $zero_i$ regarding each Booth encoded element of each weight. Sharing part of the multiplier is not possible in every design, and it is challenging for some others. For example, in the evaluated logarithm-based multiplier in [32], the critical path is long, and sharing the encoder part comes with increasing the delay of the critical path more.

5 Conclusion

In this chapter, applying various approximate techniques into the multipliers of DNN accelerators is studied experimentally. We start by explaining why applying approximation into DNNs is somehow different from other applications. Then, a stepwise approach for designing a re-configurable Booth multiplier is presented. And then, the effect of applying the introduced approximation in each step is experimentally studied. Furthermore, the effect of applying a combination of all the explored approximation techniques while running different layers of DNNs is studied. After that, a re-configurable multiplier architecture is proposed which supports all the approximation techniques explored in the proposed stepwise approach. Finally, a DNN accelerator is built upon the proposed multiplier and a few other competitive multipliers that are suitable for performing inference of DNNs. The proposed design achieves an area efficiency of ×1.19 and an energy efficiency of ×1.28 compared to the exact design while running benchmark DNNs. This chapter is expected to help the readers understand the necessities of studying the error characteristics of approximation methods, and the behavior of total accumulated error when performing the computations of different layers of DNNs, before applying these methods into DNN accelerators.

Acknowledgments This work was supported by NSERC of Canada, the R&D program of MOTIE/KEIT (No. 10077609, Developing Processor Memory Storage Integrated Architecture for Low Power, High Performance Big Data Servers) and Korea Electrotechnology Research Institute (An Energy-Efficient DNN-Based Environmental Sound Classifier).

References

1. Zhang H, Ko SB. Design of power efficient posit multiplier. IEEE Trans Circuits Syst II Exp Briefs 2020;67 5:861–5.
2. Chen Y, Luo T, Liu S, Zhang S, He L, Wang J, Li L, Chen T, Xu Z, Sun N. et al. DaDianNao: a machine-learning supercomputer. In: 2014 47th annual IEEE/ACM international symposium on microarchitecture. Piscataway: IEEE; 2014. p. 609–22.
3. Choi Y, Bae D, Sim J, Choi S, Kim M, Kim LS. Energy-efficient design of processing element for convolutional neural network. IEEE Trans Circuits Syst II Exp Briefs 2017;64 11:1332–1336.
4. Tang Y, Zhang J, Verma N. Scaling up in-memory-computing classifiers via boosted feature subsets in banked architectures. IEEE Trans Circuits Syst II Exp Briefs 2018;66 3:477–81.
5. Chen KC, Ebrahimi M, Wang TY, Yang YC. NoC-based DNN accelerator: a future design paradigm. In: Proceedings of the 13th IEEE/ACM international symposium on networks-on-chip; 2019. p. 1–8.
6. Chen YH, Yang TJ, Emer J, Sze V. Eyeriss v2: a flexible accelerator for emerging deep neural networks on mobile devices. IEEE J Emerg Sel Top Circuits Syst. 2019;9 2:292–308.
7. Chen YH, Emer J, Sze V. Using dataflow to optimize energy efficiency of deep neural network accelerators. IEEE Micro 2017;37 3:12–21.
8. Cheng C, Parhi KK. Fast 2d convolution algorithms for convolutional neural networks. IEEE Trans Circuits Syst I Regul Pap. 2020;67 5:1678–91.
9. Lavin A, Gray S. Fast algorithms for convolutional neural networks. In: Proceedings of the IEEE conference on computer vision and pattern recognition; 2016. p. 4013–21.
10. Chen YH, Krishna T, Emer JS, Sze V. Eyeriss: an energy-efficient reconfigurable accelerator for deep convolutional neural networks. IEEE J Solid-State Circuits 2016;52 1:127–38.
11. Sharify S, Lascorz AD, Mahmoud M, Nikolic M, Siu K, Stuart DM, Poulos Z, Moshovos A. Laconic deep learning inference acceleration. In: 2019 ACM/IEEE 46th annual international symposium on computer architecture (ISCA). Piscataway: IEEE; 2019. p. 304–17.
12. Albericio J, Delmás A, Judd P, Sharify S, O'Leary G, Genov R, Moshovos A. Bit-pragmatic deep neural network computing. In: Proceedings of the 50th annual IEEE/ACM international symposium on microarchitecture; 2017. p. 382–94.
13. Delmas Lascorz A, Judd P, Stuart DM, Poulos Z, Mahmoud M, Sharify S, Nikolic M, Siu K, MoshovosA. Bit-tactical: a software/hardware approach to exploiting value and bit sparsity in neural networks. In: Proceedings of the twenty-fourth international conference on architectural support for programming languages and operating systems;2019. p. 749–63.
14. Asadikouhanjani M, Zhang H, Gopalakrishnan L, Lee HJ, Ko SB. A real-time architecture for pruning the effectual computations in deep neural networks. IEEE Trans Circuits Syst I Regul Pap. 2021;68 5:2030–41.
15. Kim MS, Barrio AAD, Oliveira LT, Hermida R, Bagherzadeh N. Efficient Mitchell's approximate log multipliers for convolutional neural networks. IEEE Trans Comput. 2019;68 5:660–75.
16. Du Z, Lingamneni A, Chen Y, Palem KV, Temam O, Wu C. Leveraging the error resilience of neural networks for designing highly energy efficient accelerators. IEEE Trans Comput Aid Des Integr Circuits Syst. 2015;34 8:1223–35.
17. Chen CY, Choi J, Brand D, Agrawal A, Zhang W, Gopalakrishnan K. AdaComp: adaptive residual gradient compression for data-parallel distributed training. In: Proceedings of the AAAI conference on artificial intelligence vol. 32, no. 1; 2018.
18. Vaverka F, Hrbacek R, Sekanina L. Evolving component library for approximate high level synthesis. In: 2016 IEEE symposium series on computational intelligence (SSCI); 2016. p. 1–8.
19. Park G, Kung J, Lee Y. Design and analysis of approximate compressors for balanced error accumulation in MAC operator. IEEE Trans Circuits Syst I Regul Pap. 2021;68 7:2950–61.

20. Sarwar SS, Venkataramani S, Ankit A, Raghunathan A, Roy K. Energy-efficient neural computing with approximate multipliers. ACM J Emerg Technol Comput Syst. 2018;14 2:1–23.
21. Venkataramani S, Ranjan A, Roy K, Raghunathan A. AxNN: energy-efficient neuromorphic systems using approximate computing. In: 2014 IEEE/ACM international symposium on low power electronics and design (ISLPED). Piscataway: IEEE; 2014. p. 27–32.
22. Kim MS, Del Barrio Garcia AA, Kim H, Bagherzadeh N. The effects of approximate multiplication on convolutional neural networks. IEEE Trans Emerg Top Comput. 2021:1–1. https://doi.org/10.1109/TETC.2021.3050989.
23. Chen Y, Xie Y, Song L, Chen F, Tang T. A survey of accelerator architectures for deep neural networks. Engineering 2020;6 3:264–274.
24. He X, Ke L, Lu W, Yan G, Zhang X. AxTrain: hardware-oriented neural network training for approximate inference. In: Proceedings of the international symposium on low power electronics and design; 2018. p. 1–6.
25. He X, Lu W, Yan G, Zhang X. Joint design of training and hardware towards efficient and accuracy-scalable neural network inference. IEEE J Emerg Sel Top Circuits Syst. 2018;8 4:810–21.
26. Venkatachalam S, Adams E, Lee HJ, Ko SB. Design and analysis of area and power efficient approximate booth multipliers. IEEE Trans Comput 2019;68 11:1697–703.
27. Iandola FN, Han S, Moskewicz MW, Ashraf K, Dally WJ, Keutzer K. SqueezeNet: alexNet-level accuracy with 50x fewer parameters and< 0.5 mb model size; 2016. arXiv:1602.07360.
28. Venkatachalam S, Ko SB. Design of power and area efficient approximate multipliers. IEEE Trans Very Large Scale Integr Syst. 2017;25 5:1782–86.
29. Esposito D, Strollo AG, Alioto M. Low-power approximate mac unit. In: 2017 13th conference on Ph. D. research in microelectronics and electronics (PRIME). Piscataway: IEEE; 2017. p. 81–4.
30. Yang T, Sato T, Ukezono T. A low-power approximate multiply-add unit. In: 2019 2nd international symposium on devices, circuits and systems (ISDCS). Piscataway: IEEE; 2019. p. 1–4.
31. Lu Y, Shan W, Xu J. A depthwise separable convolution neural network for small-footprint keyword spotting using approximate mac unit and streaming convolution reuse. In: 2019 IEEE Asia Pacific conference on circuits and systems (APCCAS). Piscataway: IEEE;2019. p. 309–12.
32. Kim MS, Del Barrio AA, Hermida R, Bagherzadeh N. Low-power implementation of Mitchell's approximate logarithmic multiplication for convolutional neural networks. In: 2018 23rd Asia and South Pacific design automation conference (ASP-DAC); 2018. p. 617–22.
33. https://image-net.org/challenges/LSVRC/2012/index.php (2012). Accessed 3 Aug 2021.
34. Ahmadinejad M, Moaiyeri MH. Energy-and quality-efficient approximate multipliers for neural network and image processing applications. IEEE Trans Emerg Top Comput. 2021:1–1. https://doi.org/10.1109/TETC.2021.3072666.
35. Szegedy C, Liu W, Jia Y, Sermanet P, Reed S, Anguelov D, Erhan D, Vanhoucke V, Rabinovich A. Going deeper with convolutions. In: Proceedings of the IEEE conference on computer vision and pattern recognition; 2015. p. 1–9.
36. https://www.mathworks.com/help/deeplearning/ug/deep-learning-speech-recognition.html (2017). Accessed 3 Aug 2021.
37. http://download.tensorflow.org/data/speech_commands_v0.02.tar.gz (2016). Accessed 3 Aug 2021.

Part V
Introduction: Applications

Weiqiang Liu and Fabrizio Lombardi

The last part of this book consists of three chapters; they deal with several innovative applications to show that AC has a great potential to expand its domain to computational environments that can tolerate errors in operation.

The first chapter is "Cross-Level Design of Approximate Computing for Continuous Perception System" by Zheyu Liu, Qin Li, Xinghua Yang, and Fei Qiao. In this chapter, the authors introduce the technology of AC (analog and digital) to a continuous perception system, including arithmetic circuit design, storage units (with an appropriate scheduling strategy), and cross-layer integrated design methods. Initially, a general framework for introducing AC at various levels is proposed. The relationship among the cross-layer design is explained. Then, approximate arithmetic circuits for analog and digital processing are introduced. An approximate memory (SRAM and DRAM) with a corresponding scheduling strategy is also presented. Finally, all these cross-layer designs are integrated into a general framework, and training methods in approximate neutral networks are introduced as an application.

The second chapter is "Approximate Computing in Image Compression and Denoising" by Huang Junqi, Haider Abbas F Almrib, Nandha Kumar Thulasiraman, and Fabrizio Lombardi. This chapter presents two energy-efficient approximate techniques: an image compression technique and an image denoising technique. The image compression algorithm modifies the conventional DCT compression scheme to achieve a nearly 70% reduction in required adders (and therefore energy), while maintaining the quality of compressed images. The image denoising technique introduces an additional step length parameter to the conventional unconstrained total variation-based inexact Newton scheme; the proposed scheme retains quality and requires fewer iterations, making it a faster and more energy-efficient alternative.

The last chapter of this part is "Approximate Computation for Baseband Processing" by Chuan Zhang and Huizheng Wang. With the ever-increasing requirements of wireless communication, emerging techniques are expected to provide better performance, higher throughput, and lower cost. However, the resulting baseband systems are rather massive requiring more modules, complicated algorithms, and higher complexity. To address these issue and balance performance and cost, new paradigms such as AC are considered. Though in the existing literature possible implementation approaches of AC have been reported for baseband processing modules, an overview of this topical area and design methodologies have not been provided; in this chapter, the authors give a state-of-the-art research progress of this area from a very focused perspective.

Chapter 20
Cross-Level Design of Approximate Computing for Continuous Perception System

Zheyu Liu, Qin Li, Xinghua Yang, and Fei Qiao

In the intelligent Internet-of-things (iIoT) systems, a large number of smart sensors will be involved in the whole system to generate useful or user-specific information. For the common sensors without intelligent processing ability, they will produce massive data to be transferred to the cloud system for further intelligent processing, which puts forward extremely high requirements on the processing capacity of cloud systems. As a promising solution, smart sensors can transfer the processing ability from the cloud to the sensor end, which means that large power consumption for the massive data transmission could be saved. However, due to the limitations of the hardware resources and "always-on" operating mode for the smart sensors, conventional architectures are difficult to meet the extremely high requirements for energy efficiency in continuous perception applications. Approximate computing would be a promising technology to balance the computing accuracy with energy efficiency and hardware cost for such continuous perception system. In fact, both in analog and digital domain, approximate computing could achieve large performance improvements or reduce the power consumption of the system in different levels.

In this chapter, we introduce the technology of approximate computing in analog and digital domain to continuous perception system, including arithmetic circuit design, storage units with its scheduling strategy, and cross-layer integrated design methods. Section 1 presents the general framework of approximate computing in various levels and explains the relationship among the cross-layer design. Section 2 introduces the approximate arithmetic circuits in analog and digital domain

Z. Liu · Q. Li · F. Qiao (✉)
Department of Electronic Engineering and BNRist (Beijing National Research Center for Information Science and Technology), Tsinghua University, Beijing, China
e-mail: qiaofei@tsinghua.edu.cn

X. Yang (✉)
College of Science in Beijing Forestry University, Beijing, China
e-mail: yangxh@bjfu.edu.cn

W. Liu, F. Lombardi (eds.), *Approximate Computing*,
https://doi.org/10.1007/978-3-030-98347-5_20

processing. Section 3 presents approximate memory in SRAM and DRAM design with the corresponding scheduling strategy. Section 4 puts all the cross-layer design into a general framework and introduces the training methods in approximate neutral networks.

1 General Framework of Approximate Computing in Continuous Perception System

A large part of the researches in electronic and information technology is about how the electronic circuit system simulates or replaces the five senses of humans. This is no different in smart sensor design, in which the camera or microphone will collect the image or voice signals and try to extract the inner useful information contained in the original signals. Figure 20.1 shows the data processing in different levels for visual continuous perception system. In the low-level processing, the original signal is pre-processed by simple filtering or binarization, and then some features like scale-invariant feature transform (SIFT) or histogram of oriented gradients (HoG) will be extracted in the middle-level processing. At last, these features will be sent to a high-level processing like support vector machine (SVM) or convolutional neural network (CNN) algorithms to obtain the final useful information. In fact, most power consumption will be consumed in low and middle processing levels since massive data will be processed. In smart sensor design, all the processing from these three levels are expected to in involved in the sensor circuit system in order to avoid large amounts of data transmission to the cloud processing.

In fact, three characteristics of smart sensor applications should be noticed. First, the circuit for these applications should be "always-on," which means that the whole system should always keep working to capture the signals of the environment.

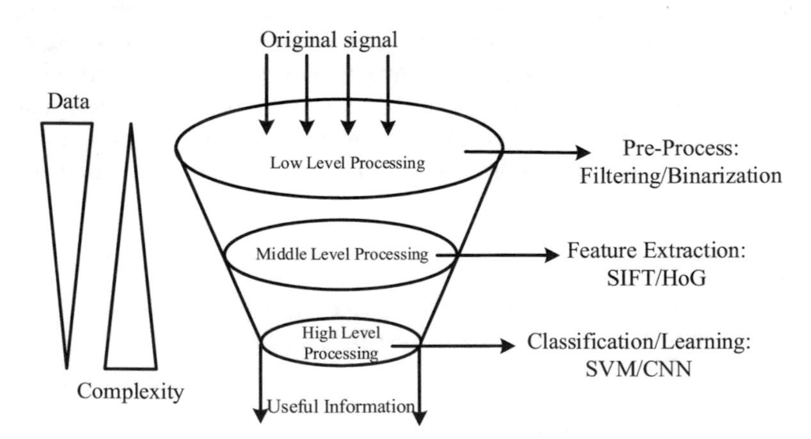

Fig. 20.1 Data processing in different levels for visual continuous perception system

Second, since most of the smart sensors are powered by the battery and the hardware resources are limited, the energy consumption of the circuit system should be low enough. However, the performance of the whole system should keep a relatively high level as the applications of smart sensors always need "real-time" processing ability. Thus, it can be seen that the first and the second characteristics are contradictory in a certain degree. But it is lucky that the third characteristic will make the approximate computing technology to satisfy the design challenge. It should be noted that the applications of image or voice processing have no "golden standard" to the final output and present error tolerance in nature. That means we can introduce certain errors or approximation to the whole processing. In a conventional design, the researchers concentrated on the trade-off between the energy and performance in the circuit system design, while little efforts have been paid for the output quality effect since most of the works are based on the premise of accurate calculation. However, as the continuous perception systems like image or voice processing are widespread, approximate computing in these areas can also obtain closed output quality compared with accurate computing [1, 2].

In general, the digital circuit system mainly consists of arithmetic units like adder, multiplier, and memory devices like SRAM and DRAM. The whole algorithms for image or voice processing will be put on these devices in real scenarios. In fact, approximate computing could be applied to the all these circuit designs. It should be noted that in digital domain processing, approximate computing means that the corresponding approximate adders, multipliers, and memory will produce some outputs with errors or approximation due to the specific approximate circuit structure. This definition of approximate computing can also be extended to analog domain processing since the analog circuit always produces approximate results compared with its accurate digital counterparts [3]. In fact, analog circuits present far more improvements in performance and energy efficiency since they usually need no clock to complete the processing compared with digital ones. Nevertheless, the last and the most import question for approximate computing is that how to control the error or approximation to ensure the algorithms operating on these approximate circuits are running correctly, or in other words, the output of the approximate circuit system can satisfy the user's specific requirements. Thus, a unified cross-layer optimization method, which involves algorithm and approximate circuits, should be adopted. In a conventional design, the error or approximation caused by approximate computing will directly degrade the output performance, such as peak signal-to-noise ratio (PSNR) in image processing. Although the error tolerance in these applications can be used, the results may still be unacceptable. We hope there is an optimization method to repair the performance losses, and it seems that the machine learning algorithms will complete this difficult task. The key technique is re-training process in machine learning, and we will discuss the details in Sect. 4.

2 Approximate Digital and Analog Arithmetic Units

No matter in accurate or approximate circuit system design, basic arithmetic units like adder and multiplier are essential because most of the whole complex processing can be divided into addition and multiplication. Figure 20.2 shows a processing diagram of gated recurrent unit (GRU), which is a famous neutral network used in speech recognition tasks. The GRU network consists of neurons that can be described by input feature vector x_t, current hidden vector h_{t-1}, reset gate r_t, update gate z_t, and candidate hidden vector h_t, in which the key modules include multiply-accumulation (MAC), nonlinear operations, and element-wise multiplication. As the MAC unit dominates the computing energy in a whole GRU computing process [4, 5], it is important to design fast and energy-efficient adder and multiplier.

2.1 *Analog Circuit Design for Approximate Adder and Multiplier*

In analog circuit design, the multiply-accumulation (MAC) could be implemented with several symmetrical NMOS and PMOS transistors as shown in Fig. 20.3. The overview of the current-mode MAC circuit structure is shown in Fig. 20.3a, where the input current vectors ($I1$, $I2$, ..., Im) represent the extracted features from the original input signals and the digital 5-bit input wires (W_{m4}, W_{m3}, W_{m2}, W_{m1}, W_{m0}) represent the trained weights, which are obtained from offline training processing for the GRU network. The MAC unit multiplies the feature vectors and the weights and then accumulates the output current directly by shorting the output node. Weights are stored in the on-chip memory and loaded before the MAC

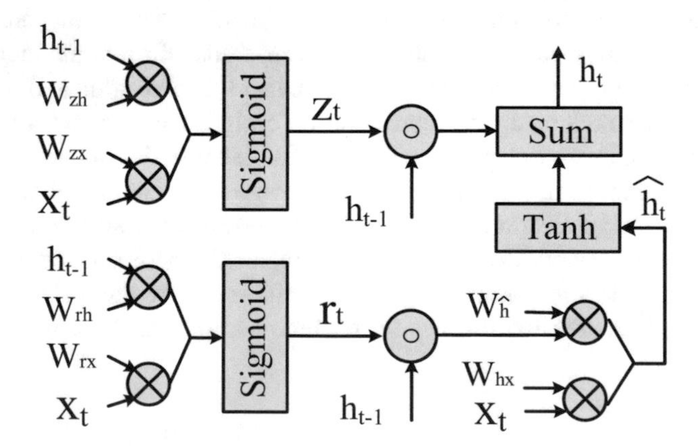

Fig. 20.2 Diagram of GRU network

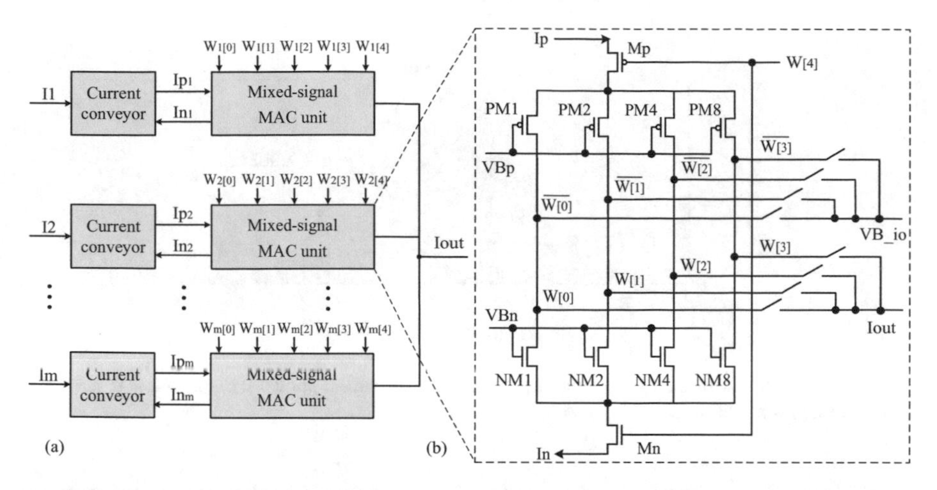

Fig. 20.3 (**a**) Overview circuit diagram of MAC, (**b**) circuit structure of analog MAC unit

operation. As an example, we realize the MAC operations with the number of m in parallel, and all of the MAC unit outputs are connected together at Iout for current-mode addition. The output current Iout is then sent to the next module for following non-linearity operation.

The detailed circuit structure of the MAC unit is shown in Fig. 20.3b. There are two directions of input currents (Ip and In), which are controlled by the switch transistors Mp and Mn and represent the positive and negative of input feature value. Besides, the input digital line W4 represents the positive and negative of the input weight. If the weight value is positive, then transistor Mp would be turned on, transistor Mn would be turned off, and vice versa. Taking the positive input current Ip as an example, the width-to-length ratio of the transistors (W/L_{PM1}):(W/L_{PM2}):(W/L_{PM4}):(W/L_{PM8}) is 1:2:4:8 so that the input current Ip is divided into three groups with the ratio of 1:2:4:8. The current is then controlled by the transistor switches with the digital control line W0/W1/W2/W3 to determine which currents are added to the output Iout. With this circuit structure, the multiplication of the input current and the weight and the summation of multiplication results will be achieved. Although we set the computing precision as 5 bits in this example, the precision of MAC unit can be changed according to the actual application requirements.

It should be noted that this analog MAC circuit could directly process the input feature currents without extra current–voltage converter. For systems with current-mode feature input, the proposed MAC circuit avoids additional conversion error, improves the processing precision, and can be implemented in small areas. The current mirror structure of this MAC circuit further ensures the high precision and temperature independence of the multiplication operation by proportional transistors. Moreover, the working current of the circuit is equal to the input current, which does not introduce extra power consumption for multiplication.

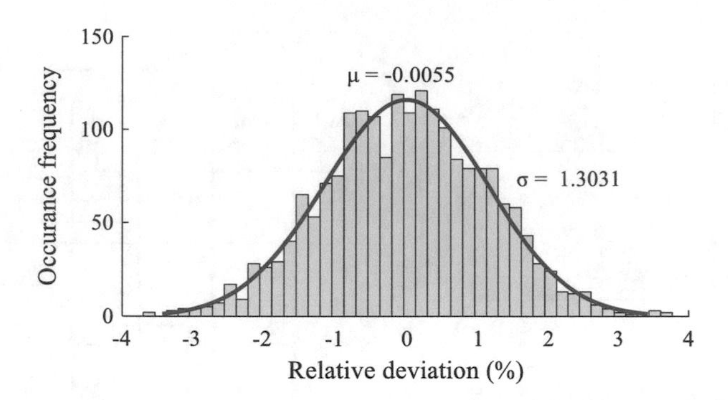

Fig. 20.4 Monte Carlo analysis results of MAC circuit, which shows the statistical result of a 2000-point simulation

However, due to the interference of non-ideal factors such as process deviation, temperature drift, voltage fluctuation, etc. after the design and manufacture of the analog processing circuit, a completely accurate calculation cannot be realized. Therefore, the analog processing circuit can only achieve limited computing precision. Worse still, computing precision constraint caused by analog imperfection is a key indicator to influence the recognition performance of GRU network. Limited computing precision could cause a serious decline in the recognition performance of the GRU network. In order to evaluate the computing precision of the proposed MAC circuit, a 2000-point Monte Carlo simulation for this analog MAC circuit with the current conveyor is performed, and the analysis results are shown in Fig. 20.4. The mismatch of proportional transistors of the current conveyor and MAC circuit introduces relative deviation, which is obtained by dividing the maximum deviation by the output dynamic range. The majority of relative deviation is within the 2σ, which corresponds to the equivalent computing precision more than 5 bits. In order to reduce or even eliminate the impact of low computing precision caused by analog processing deviation on recognition performance, it is necessary to retrain the GRU network with low-precision computing units.

Computation in memory (CIM) is a method of achieving energy-efficient MAC operations [6, 7]. CIM is essentially a non–von Neumann calculation paradigm, because its calculations occur directly in the memory block instead of reading data from the memory before the calculation. Since MAC operations and the required memory access operations dominate neural network algorithms, the CIM architecture is expected to increase energy efficiency by reducing data movement. However, there are two main shortcomings that hinder the widespread application of CIM technology in hardware design. First of all, a large number of analog-to-digital converters (ADCs) and digital-to-analog converters (DACs) as the interface between sensors, external memory, and processing units have become one of the bottlenecks. These interface overheads usually dominate the energy consumption of CIM systems, especially when multi-bit input/output accuracy is required. Second, the physical defects of analog circuits, such as mismatch, process changes, and nonlinearity, fundamentally limit the computational SNR and scalability of these CIM architectures. Therefore, how to remove the interface overhead brought by ADC/DAC and directly input the analog signal collected by the active pixel sensor (APS) to the CIM-based analog MAC circuit is a design challenge.

Figure 20.5a shows the structure of a current-mode SRAM with nine transistors. The structure is composed of a 6T SRAM cell plus a 3T switch current circuit, where the switch current circuit is composed of two NMOS switches and a mirror transistor. The switch is controlled by Q and $\overline{Q}$ of the 6T SRAM cell so that the current mirror transistor is connected to ether BL or BLB. As shown in the simplified circuit model in Fig. 20.5b, the mirror transistor in each unit acts as a current source, and the current generated by it is controlled by the drive line voltage VDL. This VDL is the output from the current conveyor (CC). The function of CC is to transfer the current of the APS to the CIM array and make the VDL proportional to the light intensity. CC converts the pixel output current into a voltage signal, thereby providing a copy of the pixel current to BL or BLB. One end of the two switches is connected to the drain of the mirror current transistor, and the other end is connected to BL and BLB, respectively. The mirror current transistor actually copies the input current to the storage computing unit. Since Q and $\overline{Q}$ are always in opposite states, there is only one current path in each unit structure. In fact, the 1-bit weight stored in 6T SRAM determines the direction of the drain current of the mirror current transistor: when the stored value is 1, the current switch at the Q terminal is turned on, and the current flows in through the BL line; on the contrary, when the stored value is 0, the current switch at the $\overline{Q}$ terminal is turned on, and the current flows in through the BLB line. In this way, we get the calculation function of the entire 6T+3T unit: copy the current of a pixel circuit, and drain the current from BL or BLB according to the stored 1-bit weight. This operation is actually equivalent to a 1-bit signed weighting operation on the input current, that is, $y = w \cdot x, w \in \{-1,+1\}$.

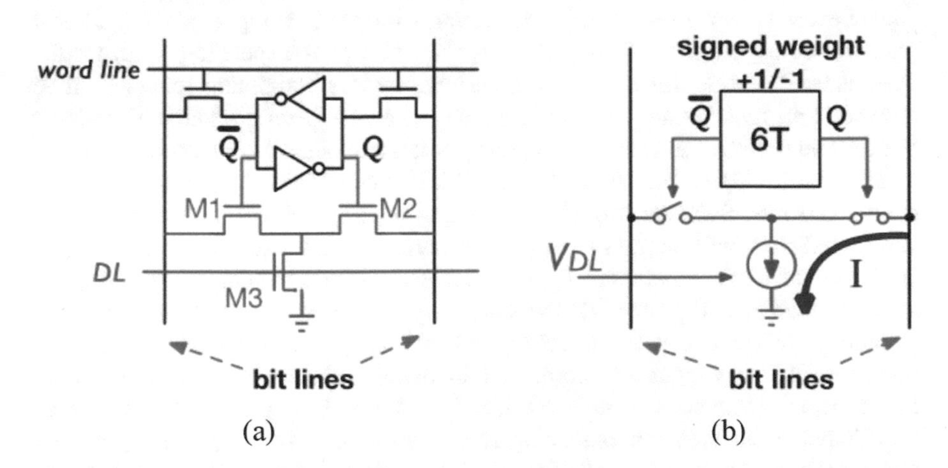

Fig. 20.5 6T+3T SRAM cell for MAC operation: (**a**) the structure of 1-bit current-mode SRAM, (**b**) the simplified circuit model of 1-bit current-mode SRAM

It is worth noting that the logic "1" in the storage unit represents the +1 weight, the logic "0" represents the −1 weight, and the input signal x represents the current signal. Expanding this unit in the vertical and horizontal directions can intuitively realize the multiplication between the input vector and the weight matrix.

Analog computing presents high performance and energy efficiency. Moreover, the ADCs/DACs in conventional digital design could be eliminated as the analog MAC or other analog processing circuits could directly receive the original analog signals from the smart sensors. However, analog computing often suffers from circuit imperfections, such as channel length modulation effect, transistor mismatch, and process variation. These approximations could be solved by feedback circuits or re-training the weights to effectively mitigate the accuracy degradation caused by analog imperfections.

2.2 Digital Circuit Design for Approximate Adder

In digital design, approximate computing technique could also be used to adders and multipliers. In fact, from transistor level to logic level, approximate adder and multipliers have been proposed extensively [8, 9]. In this part, we want to introduce variable latency adder, which employs approximate computing [10]. In conventional implementation, the arithmetic unit design is based on fixed latency and accurate computing methodology. Ripple carry adder (RCA) and carry look-ahead adder (CLA) are two typical cases with this design methodology. In fact, the RCA achieves low power consumption, but it has a long processing latency. On the contrary, CLA presents short processing latency but has a large area and power consumption. For both of RCA and CLA, the worst-case delay of the whole circuit will be defined as the critical path, from which the operating frequency cannot exceed this worse-case limit. However, with practical input data, this critical path will be triggered with low probability since it needs a specific pattern for the input data. Based on this observation, variable latency computing scheme can be applied to arithmetic unit design. Figure 20.6 shows a general scheme of variable latency adder with approximate computing technique. Take 32-bit variable latency adder as an example: The whole 32-bit RCA is divided into eight stages and each stage has 4 bits. Between two adjacent stages, a predictor will be inserted. Thus, seven predictors will be used to the whole variable latency adder. In Fig. 20.6, it can be seen that the input data ($A_{32...1}$ and $B_{32...1}$) and the output result $S_{32...1}$ are controlled by D flip-flops (DFFs). For every new input data, each predictor will generate a speculative value as a carry signal ($spec\_c_i$) to the upper stage. This prediction of carry signals is computed in parallel during the first clock cycle. Since the carry signals are predicted, it means some of the prediction values may be different from their corresponding real carry signals ($real\_c_i$). Thus, another clock cycle has been utilized to the variable latency adder to correct the wrong predictions. At this time, the predictor with wrong speculation in the first clock cycle will produce error signals (err_i) and make it high, and then the clr signal will

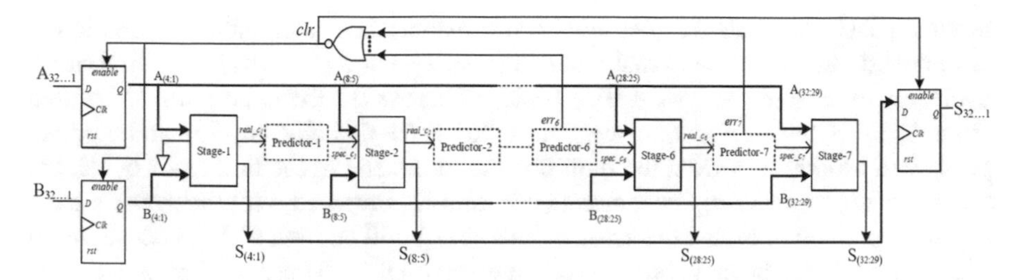

Fig. 20.6 Scheme of approximate variable latency adder

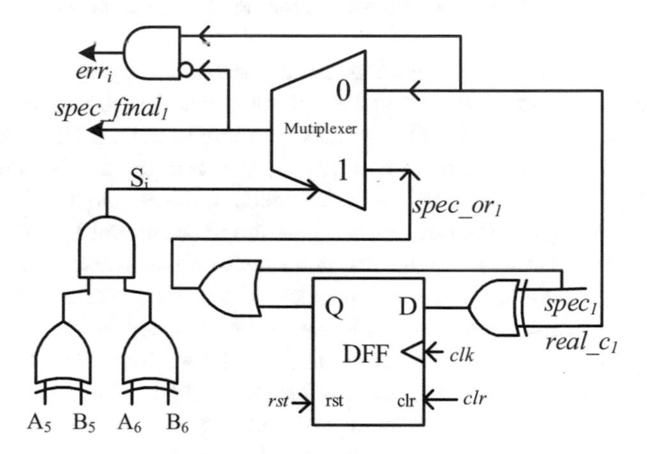

Fig. 20.7 Scheme of predictor in approximate adder

become high to lock the DFFs. With this correction method, the second clock cycle will be used and the real carry signals will replace the wrong prediction. This whole process will not stop until there are no wrong predictions any more, which means that there may be multiple cycles for the whole computation.

The predictors could be used as shown in Fig. 20.7. Take the predictor-1 between stage-1 and stage-2 as an example. The stage-1 makes the addition of the input data ($A_{4...1}$ and $B_{4...1}$) and the stage-2 makes the addition the input data ($A_{8...5}$ and $B_{8...5}$). There are many kinds of ways to generate the speculative carry signal; here, we take $spec_1 = A_4B_4 + (A_4 + B_4)\ A_3B_3$. This $spec_1$ will pass the OR-gate with output of the DFF to be one of the signals $spec\_or_1$ in the multiplexer. The other input signal for the multiplexer is the real carry signal ($real\_c_1$) from lower stage-1. The select-signal (S_1) of the multiplexer is generated by another structure of circuit and determined by logic ($A_5 \bigoplus B_5$) or ($A_5 \bigoplus B_5) \cap (A_6 \bigoplus B_6$).

The basic idea of the predictor is that when $A_5 = B_5$ or $A_6 = B_6$, the select-signal (S_1) will be low and the real carry signal ($real\_c_1$) will be selected. The $spec\_or_1$ is ignored because $A_5 = B_5$ or $A_6 = B_6$ has killed the propagation of carry signal from the lower stage, which means that the stage-1 and stage-2 can be computed in parallel naturally without any speculation. As the real carry signal ($real\_c_1$) is used, the error signals (err_i) will not be effective. With this detection circuit, massive redundant cycles will be eliminated since potential wrong prediction is removed. Another case is that when $A_5! = B_5$ and $A_6! = B_6$, in the first clock cycle, $spec_1 = A_4B_4 + (A_4 + B_4)\ A_3B_3$ will pass the OR-gate with "0" output from the DFF (the DFF will always output "0" at the first clock cycle), and the final $pred_1$ will be $spec\_or_1$ from the multiplexer. At the end of the first clock cycle, the real carry signal ($real\_c_1$) becomes steady and is compared with the $pred_1$ signal. If wrong prediction exists, the error signals (err_i) will be high and the *clr* signal in Fig. 20.6 will also become high to lock the input DFFs. Thus, another clock cycle will be allocated to the adder to correct the wrong the prediction.

With this variable latency computing scheme, approximate computing can be easily introduced to make trade-off between output quality and performance. The general idea of approximate computing to the variable latency adder is that we keep all the predictors in the adder so that each stage could be computed in parallel. However, since the upper bits of the results are more important than the lower bits, the correction for the wrong prediction for the lower stages will be removed. This can be realized just simply by no longer connecting the err_i signals in Fig. 20.7 to the NOR-gate in Fig. 20.6, which means that the upper bits of the addition output will always be correct with the prediction–correction mechanism, while the lower bits of the output may be wrong without any extra clock to make correction. With this approximate computing scheme, less clock cycles will be consumed, and the whole corresponding performance of the adder will be improved with certain error. It should be noted that this approximate computing scheme is also quite convenient for the designer to make different settings of output quality and performance, which is useful in some application processing as different levels of approximation are needed. The scheme can even be modified to achieve adaptive ability as the settings for the circuit are simple.

This approximate circuit structure of variable latency computing could also be applied to other arithmetic units like multipliers or MAC operation. It can be seen that both in analog and digital processing domain, approximate computing technique could be used to improve the performance of the circuit for computation. The main difference is that ADCs/DACs are no longer needed in analog arithmetic units, which will cut off large amount of power consumption. The drawback for analog domain processing is also clear, and that the whole design process is much more difficult and the error or approximation in analog circuit is more complicated.

3 Approximate Memory Design

In embedded systems where the power resources are limited, the memory, including off-chip DRAM and on-chip SRAM, is a key problem in image/voice processing, which accounted for more than 90% workload in these applications. For example, the image data is collected by the sensors and then sent to the off-chip DRAM. Part of the DRAM data is cached to the on-chip SRAM for calculation and processing by the on-chip computing units, and some of the output results may be written back

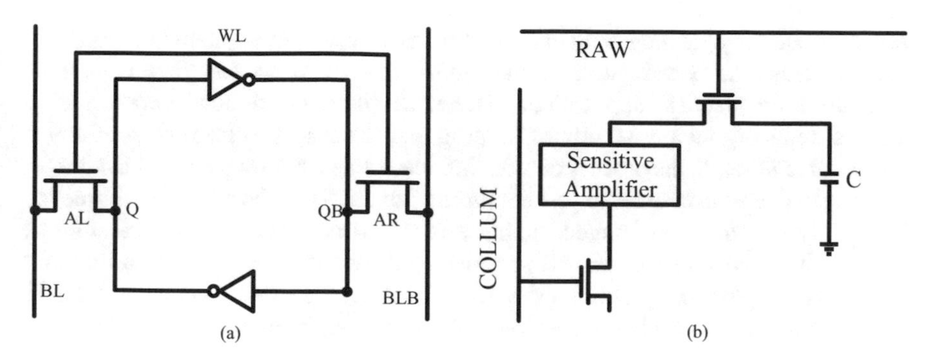

Fig. 20.8 (**a**) One-bit cell in SRAM, (**b**) one-bit cell in DRAM

to the off-chip DRAM [11]. The traditional 1-bit SRAM storage structure is shown in Fig. 20.8a. It can be seen that this structure consists of six CMOS transistors, referred to as 6T structure. When writing new data, the word-line signal WL is at a high level, the NMOS transistor where it is located is turned on, and the input new data is written to the inverter ring through the bit-line signals BL and BLB. Since the original inverter loop is a positive feedback structure, during the write operation, if the data is different from the previous state, the latch structure formed by the inverter will be forcibly reversed. In the flip process, the capacitor will be charged and discharged, which will consume energy and constitute the power consumption of the SRAM write operation. In the reading process, the bit-line signals BL and BLB are charged first, and then WL is set to a high level, so the load capacitance of BL and BLB is charged and discharged according to the stored data. The power consumption of SRAM is composed of three parts, namely, the power consumption of write and read operation and the leakage power consumption of the device itself. At the same time, it should be noted that the entire latch needs to be flipped during the write operation, but there is no such problem in the read operation. Therefore, the power consumption of the write operation is much larger than the read operation. The power consumption of a single write operation is more than three times that of the read operation, and the leakage power consumption is much less than that of the read operation. Therefore, reducing the power consumption of SRAM write operations is of great significance, which is proportional to several parameters as

$$P_{\text{switching}} \propto \alpha C V_{\text{DD}}^2 f \tag{20.1}$$

where α is the activity factor (switches between bit-“1” and bit-“0”), C is the effective capacitance, V_{DD} is the supply voltage, and f is the operating frequency.

From the perspective of reducing power consumption, it is more difficult to reduce the circuit capacitance, which is mainly related to the circuit structure. Once the circuit structure is determined, the space for the optimization of the capacitance is smaller. Reducing the operating frequency can certainly achieve the purpose of low-power design, but for applications such as image and voice processing, the reduction of the operating frequency will greatly reduce the system performance, and the real-time performance of the circuit system to process data cannot be guaranteed in practical applications. Therefore, in terms of low-power on-chip storage, reducing the probability of flipping and the supply voltage is an effective method. However, in the process of reducing the voltage, the frequency of the circuit also has to decrease in order to prevent timing violations. If the operating frequency is forcibly to remain unchanged, timing errors will occur during the operation of the circuit. This timing error will seriously pollute the original data and greatly reduce the output quality of the application itself. At the same time, as the CMOS process continues to evolve, under low voltage, the logic output of the circuit will experience errors under the influence of noise and process deviations. Unlike the timing error, this error cannot be compensated even if the frequency is reduced. In other words, the designer has to reserve a part of the voltage margin to ensure the correct operation of the memory cell. Since the power consumption of the SRAM write operation is proportional to the activity probability at the same time, it is also possible to reduce the power consumption of the write operation by reducing the data activity probability. Using approximate computing, a most straightforward method is to set the low-order bits of the original pixels to "0," that is, the data truncation method. However, simple data truncation will make the output quality of image processing have a great decline and error compensation is required.

The storage structure of off-chip DRAM is shown in Fig. 20.8b. When writing new data, the row address line and column address line become high, and the data will charge and discharge the capacitor through the bit-line with the sensitive amplifier. Due to the charge leakage problem from the capacitors, the entire DRAM needs to be refreshed at intervals to ensure the correctness of the original stored data. Essentially, the dynamic refresh process is the process of data rewriting. Therefore, when the stored data is a high-level logic "1," the refresh circuit needs to recharge the capacitor after a period. If the original stored data is a low-level logic "0," there is no need to recharge during the refresh process, that is, no further dynamic energy consumption will be generated. Therefore, the operating power consumption of off-chip DRAM is linearly related to the refresh frequency and the number of high-potential logic values.

With all the analysis above, a strategy with approximate computing could be applied to the DRAM and SRAM simultaneously; take image processing as an example: One pixel from the image, which contains 8 bits named bit_7, bit_6, ..., bit_0 from high order to low order, is divided into accurate and approximate parts. The higher part will be accurate, which contains the bits from bit_7 to bit_(k), where k is an integer ranging from 0 to 8 ($k = 7$ means no bits are pushed into the accurate part), while the lower part will be approximate, which contains the bits from bit_($k - 1$) to bit_0 ($k = 0$ means no bits are pushed into the approximate part). With the data structure, the lower approximate parts will be processed as follows: The first appeared bit-"1" is reserved and the remaining bit-"1" will be enforced to be changed into bit-"0." As illustrated in Table 20.1 where $k = 4$, as the time interval increases, the original pixel data listed in the second column will be processed with approximation, where in the lower part from bit_3 to bit_0, the first appeared bit-"1"

Table 20.1 Storage data processing with approximation

Time	Original data	Approximate data	Error
Reset	0000_0000	0000_0000	0
Time-1	0100_0001	0100_0001	0
Time-2	0110_0010	0110_0010	0
Time-3	0111_0011	0111_0010	1
Time-4	1110_0100	1110_0100	0
Time-5	1000_0101	1000_0100	1
Time-6	1010_0110	1010_0100	2
Time-7	1011_0111	0110_0100	3
Time-8	0110_1000	0110_1000	0
Time-9	1110_1001	1110_1000	1
Time-10	1111_1010	1111_1000	2
Time-11	1101_1011	1101_1000	3
Time-12	1001_1100	1001_1000	4
Time-13	1000_1101	1000_1000	5
Time-14	1100_1110	1100_1000	6
Time-15	1101_1111	1101_1000	7

is reserved and the remaining bit-“1” will be changed into bit-“0.” The approximate results are listed in the third column with the corresponding errors listed in the last column.

With this approximate processing to the original data, the overall number of bit-“1” in LSBs of one pixel will be reduced. After the strategy is applied to all the image pixel data, these approximate data will be pushed into DRAM and SRAM. The refresh power for DRAM will get lower as the number of bit-“1” is reduced. Meanwhile, the switch probability for write operation in SRAM also decreases as there are more bit-“0.” With more lower bits involved into the approximate processing, more power savings for DRAM and SRAM will be achieved. However, the corresponding error will also become large with strong approximation. Thus, with various k values, a trade-off between output quality and power savings can be obtained as shown in Fig. 20.9.

It can be seen that this strategy is efficient for trade-off between output quality and power savings. No changes to the scheme of DRAM and SRAM are needed, and the overall control flow is simple. Furthermore, the logic implementation for the approximate processing is also simple as shown in Fig. 20.10. Take $k = 4$ as an illustration, and suppose the original image data is $P_t[7:0]$ and the data after approximate computing is $P_{t,\text{approximate}}[7:0]$. Only several AND-gates and inverters are used for circuit implementation.

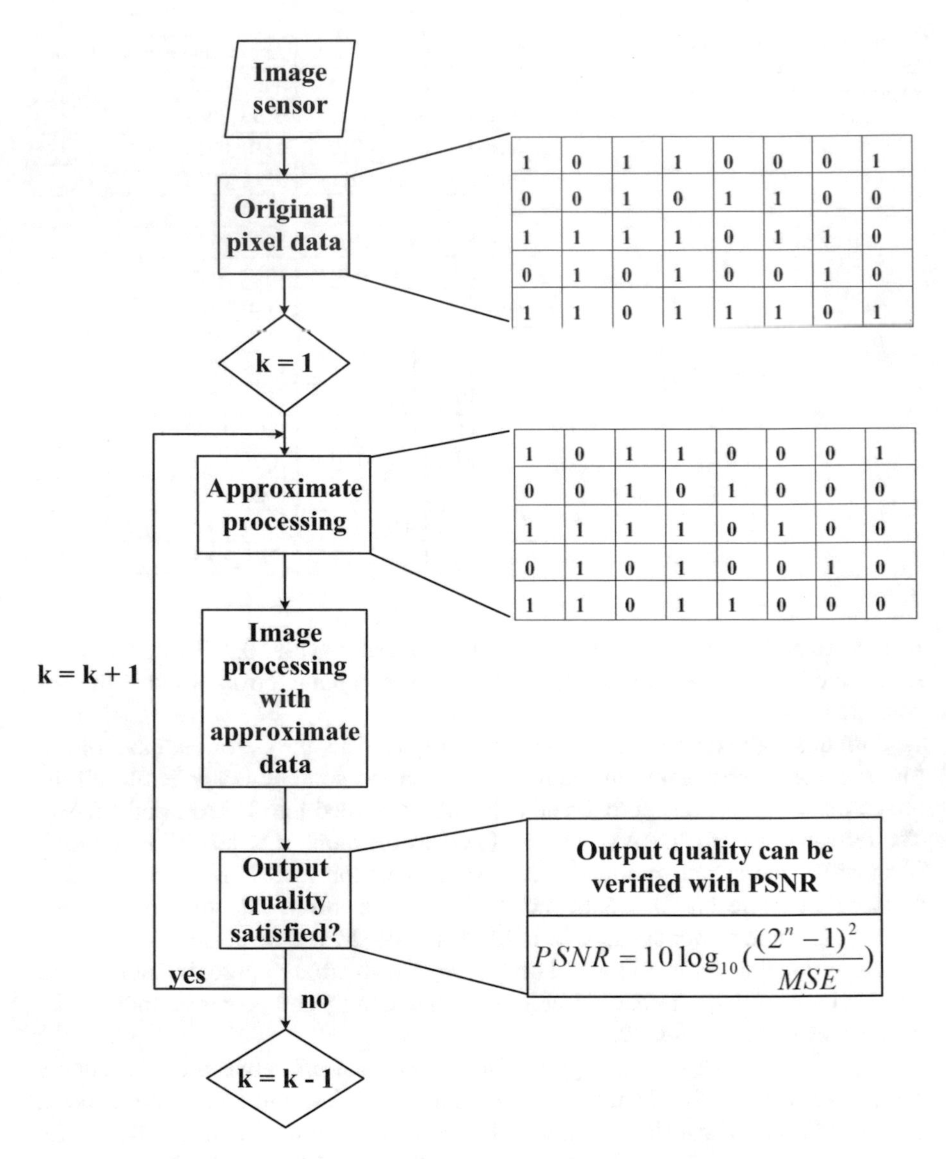

Fig. 20.9 Trade-off design for the approximate storage strategy

4 Cross-Layer Optimization of Approximate Computing

With approximate arithmetic units and memory, the algorithms should be co-designed with these approximate computing units as the output quality of the algorithm may decrease sharply and become unacceptable. If the approximate arithmetic units and memory scheme described above are applied to conventional image/voice processing applications, like discrete cosine transform (DCT) process-

Circuit implementation for approximate processing
Input: $P_t[7:0]$; $k=4$;
Output: $P_{t.approximate}[7:0]$;
$P_{t.approximate}[7:3] = P_t[7:3]$;
$P_{t.approximate}[2] = \overline{P_t[3]} \cap P_t[2]$;
$P_{t.approximate}[1] = \overline{P_t[3]} \cap \overline{P_t[2]} \cap P_t[1]$;
$P_{t.approximate}[0] = \overline{P_t[3]} \cap \overline{P_t[2]} \cap \overline{P_t[1]} \cap P_t[0]$;
Return $P_{t.approximate}[7:0]$

Fig. 20.10 Logic implementation for the approximate storage

ing, which means that the addition, multiplication, and memory in DCT processing will be implemented with approximate computing technique, the algorithm of DCT should be modified and designed carefully [12, 13]. Since there are many different approximate structures for arithmetic units and memory, the most straightforward method for the designer to construct an approximate computing scheme for the algorithm is to traverse all the possible combinations based on the optional approximate arithmetic units and memory schemes. In the process of traversal, the metrics including performance, power consumption, energy consumption, and output quality will be used to find out the optimal combination. For example, in DCT processing, the metric of PSNR will be used to measure the output quality. All the combinations with approximate computing for the DCT will be verified with the PSNR and other metrics like performance or power consumption. The optimal combination will satisfy the basic output quality and present the most improvements of performance or power savings. In order to measure the output quality, the behavior model of each optional approximate arithmetic unit and memory scheme should be built with C/C++, Matlab, or Python platforms. The algorithm should be realized with these behavior models so that the metric of output quality can be achieved. As for the performance and power consumption evaluation, the corresponding approximate circuit of the computing units or memory schemes should be implemented in a real hardware platform, from which the real performance or power consumption will be obtained, or at last, the corresponding simulation results based on HSPICE or DesignCompiler should be presented. With all the above description, a table contained all the metrics for each combination of approximate circuit will be clearly generated, from which the optimal design will be found out.

It should be noted that there are many kinds of approximate arithmetic units and memory schemes, which will make a huge design space when we traverse all the possible combinations [14, 15]. Thus, some strategies, like heuristic searching, could be used to accelerate the whole searching process. In heuristic searching, the error model of the approximate computing units or memory and their corresponding

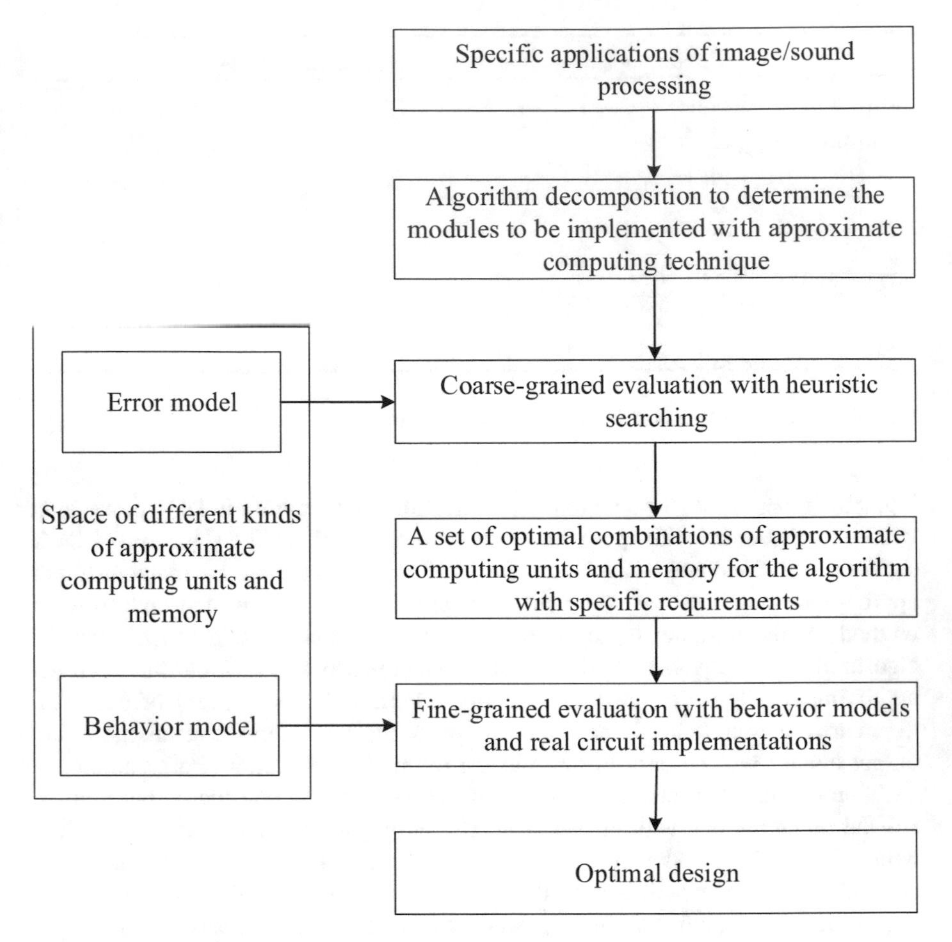

Fig. 20.11 Process of optimization for the approximate circuit system design

performance or power consumption in circuits should be built in advance. Then, the error model will be applied to the processing of the algorithms to fast achieve the output quality with corresponding performance and power consumption. All these evaluations will be inserted into the heuristic searching, and we will quickly obtain a set of optimal combinations of approximate computing units and memory for the algorithm with specific requirements. This is called coarse-grained cross-layer optimization, which will eliminate large amount of the optional combinations. At last, fine-grained cross-layer optimization will be used, in which the behavior models and real circuit implementations for the remaining combinations will be utilized to get the accurate evaluation for the output quality, performance, and power consumption. The whole process is shown in Fig. 20.11.

It can be seen that when approximate computing units and memory are applied to conventional algorithm processing, the error or approximation will not disappear

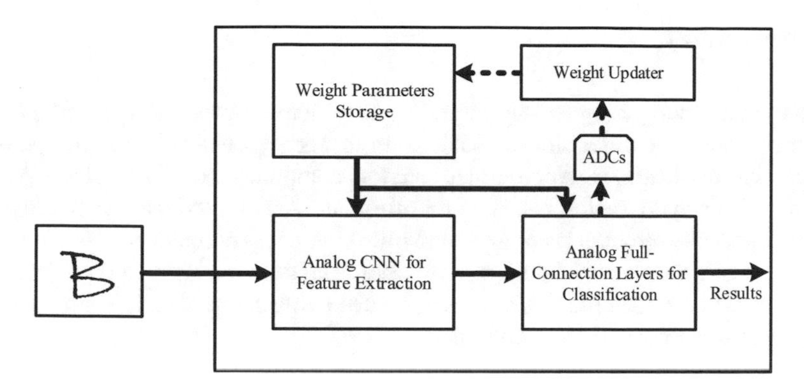

Fig. 20.12 Process of optimization for the analog circuit system design

or be alleviated. However, as the machine learning algorithm for image/voice processing is developing, we find that the re-training method in machine learning algorithm can offset the error or approximation. In some extreme cases, the final output quality with approximate computing after the re-training could have the same level to the accurate computing. The re-training method is as follows. First, the image/voice processing algorithms, like image/voice classification or recognition, are trained with floating-point data with specific neutral networks such as convolutional neural network (CNN) and recurrent neural network (RNN). Second, the trained parameters will be quantized to fixed-point data and the first round of re-training will be carried out to make up for the output quality loss caused by quantification. Third, the behavior model of the approximate units or memory will be inserted into the inference processing of the machine learning algorithm, which means the real error or approximation will be enrolled in the inference processing. Then, the second round of re-training will be carried out. With these three steps, the error or approximation will disappear or be alleviated, and the output quality for the algorithm with approximate computing will become acceptable.

When the machine algorithm is implemented with analog circuits, the last problem is how to combine the design in algorithm and circuit to complete the cross-layer optimization. The solution is a little more complex and illustrated in Fig. 20.12, in which an analog circuit of LeNet-5 is implemented to classify the digitals. The re-training method for this analog circuit system is as follows: First, the fixed-point parameters of LeNet-5 are achieved by offline training and quantization using Caffe platform. Second, the analog circuit of LeNet-5 is realized in HSPICE with transistor-level simulation, in which the trained parameters from the first step will be turned into analog value by DAC and the analog input signal with these turned parameters will be sent to the analog circuit to complete the inference processing. Third, the output of the neutral network will be sampled by ADC and turned into digital data. These digital data will be put back to the back-propagation to update the weight in Caffe. Repeat these three steps until the convergence is finished.

5 Summary

Approximate computing in digital and analog domain processing could provide another important dimension to achieve high energy efficiency for image/voice processing applications. Approximate analog computing could complete most of the algorithm with analog circuits and eliminate ADCs in signal processing, but the whole design process is complex and inflexible. Approximate digital computing could provide various levels of trade-offs between output quality and performance or power savings. Compared with approximate analog computing, it is more flexible and the design process is less difficult.

References

1. Liu W, Lombardi F, Shulte M. A retrospective and prospective view of approximate computing [point of view}. Proc IEEE. 2020;108(3):394–399.
2. Han J, Orshansky M. Approximate computing: an emerging paradigm for energy-efficient design. In: 2013 18th IEEE European Test Symposium (ETS). IEEE; 2013.
3. Jia K, et al. Calibrating process variation at system level with in-situ low-precision transfer learning for analog neural network processors. In: Proceedings of the 55th Annual Design Automation Conference. 2018.
4. Li Q, et al. NS-FDN: near-sensor processing architecture of feature-configurable distributed network for beyond-real-time always-on keyword spotting. IEEE Trans Circuits Syst I Regul Pap. 2021;68(5):1892–905.
5. Liu B, et al. A 22nm, 10.8 μW/15.1 μW dual computing modes high power-performance-area efficiency domained background noise aware keyword-spotting processor. IEEE Trans Circuits Syst I Regul Pap. 2020;67(12):4733–46.
6. Liu Z, et al. NS-CIM: a current-mode computation-in-memory architecture enabling near-sensor processing for intelligent IoT vision nodes. IEEE Trans Circuits Syst I Regul Pap. 2020;67(9):2909–22.
7. Si X, et al. 24.5 A twin-8T SRAM computation-in-memory macro for multiple-bit CNN-based machine learning. In: 2019 IEEE International Solid-State Circuits Conference-(ISSCC). IEEE; 2019.
8. Liu W, et al. Design and evaluation of approximate logarithmic multipliers for low power error-tolerant applications. IEEE Trans Circuits Syst I Regul Pap. 2018;65(9):2856–68.
9. Liu W, et al. Design of approximate radix-4 booth multipliers for error-tolerant computing. IEEE Trans Comput. 2017;66(8):1435–41.
10. Yang X, et al. Multistage latency adders architecture employing approximate computing. J Circuits Syst Comput. 2017;26(3):1750039.
11. Yang X, et al. A priority-based selective bit dropping strategy to reduce DRAM and SRAM power in image processing. IEICE Electron Express. 2016;13(23):20160990.
12. Liu Z, et al. INA: incremental network approximation algorithm for limited precision deep neural networks. In: 2019 IEEE/ACM International Conference on Computer-Aided Design (ICCAD). IEEE; 2019.
13. Yuan T, et al. High performance CNN accelerators based on hardware and algorithm co-optimization. IEEE Trans Circuits Syst I Regul Pap. 2020;68(1):250–63.

14. Chan W-TJ, et al. Statistical analysis and modeling for error composition in approximate computation circuits. In: 2013 IEEE 31st International Conference on Computer Design (ICCD). IEEE; 2013.
15. Pashaeifar M, et al. A theoretical framework for quality estimation and optimization of DSP applications using low-power approximate adders. IEEE Trans Circuits Syst I Regul Pap. 2018;66(1):327–40.

Chapter 21
Approximate Computing in Image Compression and Denoising

Junqi Huang, Haider Abbas F. Almrib, Thulasiraman Nandha Kumar, and Fabrizio Lombardi

1 Introduction

In image processing applications, computation is highly complicated; when these high-power consumption applications are performed on mobile devices, a reduced battery life is often encountered [1]. Thus, integrated circuit designers and manufacturers have exploited to decrease the feature size to address this problem, so also improving the performance of these processors. Improvements in energy efficiency and performance of CMOS integrated circuits at nanometer scales are often difficult to simultaneously achieve [2], because the increase in power dissipation inevitably limits the improvement of these devices. Image processing systems are considered error tolerant due to the limitations in human vision; human eyes are only sensitive to significant changes. In many cases, it is hard to establish the differences between accurate and slightly flawed results; moreover, due to the probabilistic nature of image processing algorithms, it is difficult for inaccurate results to appear frequently, and a noticeable degradation in performance is less likely to be caused by small errors. That is the reason image processors often use Lossy algorithms.

In addition to using Lossy algorithms, image processing can benefit from the use of energy-efficient approximate computing that decreases circuit complexity and power dissipation. Approximate computing can be applied at different abstractions

J. Huang
School of Opto-Electronic and Communication Engineering, Xiamen University of Technology, Xiamen, China

H. A. F. Almrib (✉) · T. Nandha Kumar
Department of Electrical & Electronic Engineering, University of Nottingham Malaysia Campus, Semenyih, Malaysia
e-mail: haider.abbas@nottingham.edu.my

F. Lombardi
Department of Electrical and Computer Engineering, Northeastern University, Boston, MA, USA

W. Liu, F. Lombardi (eds.), *Approximate Computing*,
https://doi.org/10.1007/978-3-030-98347-5_21

in image processing: circuit, logic, and/or algorithm. At the circuit level, researchers mainly focus on the simplification of circuit designs for arithmetic units, such as full adder cells [3]; therefore, the number of transistors in a single adder cell is reduced. This produces faster adders that dissipate less energy and occupy less circuit area [1, 4]. Another form of circuit level for approximate computing is voltage overscaling (VOS) that operates by scaling the supply voltage of CMOS circuits to save energy [5, 6]. Logic-level approximate computing concentrates on simplifying multi-bit adders and multiplier [7–9] and improving parallelism [10, 11]. Authors in [9], for example, have proposed a design technique to prune portions of multi-bit adder circuits that have lower probability of being active during operation. Another example is the work in [11] in which the authors proposed an approximate adder that reconfigures itself based on the quality requirements during operation. When quality is not of primary concern, the design reconfigures the adder to eliminate unnecessary computations.

Different from circuit and logic levels, approximate computing techniques at the algorithmic level have been developed by focusing on the image processing algorithm itself [3]. Current research mainly focuses on approximate discrete cosine transform (ADCT) designs [12–16], approximate Newton method for unconstrained optimization [17–19], algorithmic noise tolerance (ANT) [20, 21], and significance-driven computation (SDC) [22]. Two ADCT matrices have been introduced in [13, 15], and both require the lowest number of additions (14 additions) to perform the 8×8 DCT using a fast butterfly algorithm (in comparison, the conventional 8×8 DCT requires 56 additions and 64 multiplications). When used, the output images of both ADCTs have resulted in acceptable peak signal-to-noise ratio (PSNR) values, with study in [13] producing better results. Authors in [16] have proposed a low-complexity algorithm to avoid multiplications in the process of an adaptive multiple transform, which involves a DCT matrix of sizes from 8×8 to 128×128 for post-HEVC technology.

The image compression algorithm modifies the conventional DCT compression scheme to achieve approximately a 70% reduction in the number of required adders (and therefore energy) while maintaining the quality of the compressed images. Also, the image denoising technique introduces an additional step length parameter to the conventional unconstrained total variation (TV)-based inexact Newton scheme; the proposed scheme retains quality and requires fewer iterations making it a faster and more energy-efficient alternative.

In this chapter, two novel algorithmic-level approximate computing techniques for image processing applications are proposed: the zigzag low-complexity approximate DCT (ZLCADCT) image compression technique [3] and the approximate Newton method for unconstrained total variation–based image denoising technique [23]. Both schemes are developed to improve existing approximate computing techniques such that computation, required resources, energy dissipation, and processing time can be reduced. These two methods are presented in detail in the following sections.

2 Zigzag Low-Complexity Approximate DCT

Different from ADCT [14], the proposed ZLCADCT is a deterministic approach to configure the T matrix by establishing the relationship between the number of retained coefficients in T and the number of rows of the same T matrix, such that computations for unused coefficients can be totally avoided. Therefore, ZLCADCT eliminates the zigzag scanning process and significantly reduces the number of required hardware (adders). In addition, it reduces figures of merit such as delay and energy consumption, while retaining a nearly similar image quality compared with ADCT [13] and ADCT [14] (as measured by the PSNR) at the same number of retained coefficients. Also, the output image quality of ZLCADCT is controllable and adjustable by changing the number of coefficients to be retained from the input according to the specified requirements.

2.1 Preliminaries of DCT in Image Compression

Due to its energy properties, the 8×8 integer DCT has become one of the most used transforms in a video compression system. Equations (21.1) and (21.2) show the basic steps for the DCT and inverse DCT (IDCT), respectively, where X is the 8×8 matrix from the input image. Y is the 8×8 output matrix of the DCT and A denotes the conventional floating point 8×8 DCT matrix. For reducing the computational complexity, A is divided into D and T, such that the DCT can be calculated in the integer domain (D is the 8×8 diagonal matrix and T is the 8×8 integer matrix). Then, the matrix D is extracted to generate $E_f = DE_1D^T$. E_f is computed as part of a quantization process ($\bigotimes$ denotes the dot product between two matrices, while E_1 denotes the 8×8 matrix whose elements are all "1"). Therefore, $T \cdot X \cdot T^T$ is the main computation of the integer DCT, and by simplifying the matrix T, it is then possible to decrease the computational complexity of DCT.

$$Y = AXA^T = (DT)X\left(T^TD^T\right) = \left(TXT^T\right) \otimes \left(DE_1D^T\right) = \left(TXT^T\right) \otimes E_f \tag{21.1}$$

$$X = A^TYA = \left(T^TD^T\right)Y(DT) = T^T\left(Y \otimes \left(DE_1D^T\right)\right)T = T^T\left(Y \otimes E_f\right)T \tag{21.2}$$

The computational steps of the entire 8×8 integer DCT can be described as follows: The input image is initially divided into 8×8 sub-blocks X. Then, every sub-block is transformed by the integer ADCT ($T \cdot X \cdot T^T$). Next, quantization ($\bigotimes E_f$) is pursued for every sub-block. After quantization is completed, every

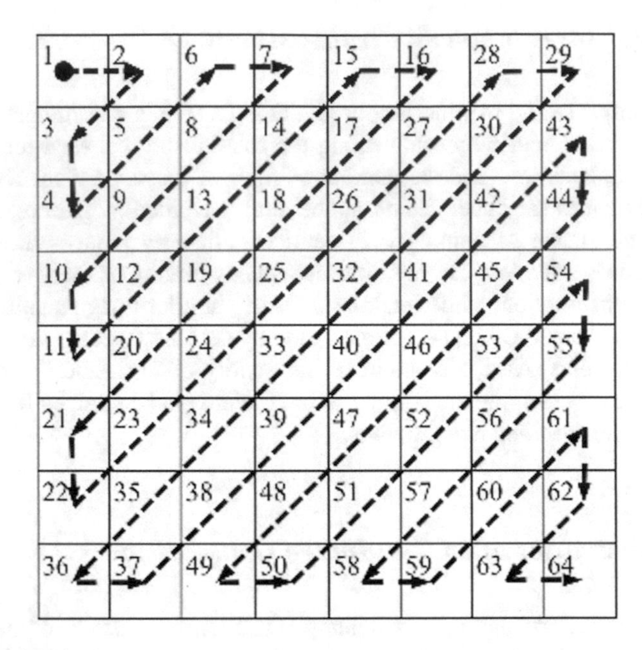

Fig. 21.1 Fundamental zigzag scanning rule for image compression algorithm (such as JPEG)

compressed sub-block must undergo the so-called zigzag scanning (shown in Fig. 21.1) for image coding (such as JPEG). The zigzag scanning finally determines the number of coefficients that must be retained as compressed data; for example, if 10 coefficients are retained, all coefficient data numbered from 11 to 64 are removed and set to zero.

The 8×8 low-complexity ADCT matrix T proposed in [13] requires the least number of computational resources; it only requires 14 additions when using the butterfly algorithm. In this section, the matrix T of ADCT [13] is applied for analysis and making comparisons with proposed ZLCADCT method. The applied D matrix can be also found in [13].

2.2 Principle of ZLCADCT Technique

The proposed ZLCADCT utilizes a deterministic method to achieve a low-complexity approximate DCT by preprocessing the transform matrix T such that a new T_p matrix is found as a function of only the number of retained coefficients (achieved by zigzag scanning) and by reducing the number of additions to be performed. In zigzag scanning (with an ordering from the upper left to the bottom right), only some coefficients are retained; therefore, processing the coefficients that are not retained is redundant. In ZLCADCT, the implementation of the required

Table 21.1 TCR range and output vector for different m values

m	TCR	Output vector $Z_{0\ldots m-1,j}(1\ldots m)$
1	TCR = 1	$Z_{0,j}(1)$
2	TCR ∈ [2, 3]	$Z_{0,j}(1)$ to $Z_{1,j}(2)$
3	TCR ∈ [4, 6]	$Z_{0,j}(1)$ to $Z_{2,j}(3)$
4	TCR ∈ [7, 10]	$Z_{0,j}(1)$ to $Z_{3,j}(4)$
5	TCR ∈ [11, 15]	$Z_{0,j}(1)$ to $Z_{4,j}(5)$
6	TCR ∈ [16, 21]	$Z_{0,j}(1)$ to $Z_{5,j}(6)$
7	TCR ∈ [22, 28]	$Z_{0,j}(1)$ to $Z_{6,j}(7)$
8	TCR ∈ [29, 36]	$Z_{0,j}(1)$ to $Z_{7,j}(8)$

zigzag scanning is performed in the earlier stage of DCT, that is, in the T matrix. So, it is possible to reduce the number of calculations by adjusting the matrix T, while computing the integer DCT ($T \cdot X \cdot T^T$) for only those coefficients that are retained. The resulting matrix is now denoted as T_p. The following definitions are initially introduced:

1. The *actual coefficient retained* (*ACR*) is defined as the number of coefficients generated when calculating the new output matrix $Y = T_p \cdot X \cdot {T_p}^T$.
2. The *targeted coefficient retained* (*TCR*) is defined as the number of coefficients of the new output matrix Y that are planned to be retained, that is, TCR is contained in ACR, and obviously, TCR < ACR.

T_p is generated based on the targeted coefficients to be retained. Table 21.1 shows a few TCR values and their corresponding number of rows (m) to be retained in the T_p matrix, as well as the output vectors generated by T_p in $X1 = T_p \cdot X$. During the operation $X1 = T_p \cdot X$, for example, if the number of targeted coefficients to be retained (TCR) in the process of zigzag scanning is only 6, then only the first three rows ($m = 3$) of the T matrix (ADCT [13]) are retained (Table 21.1) by using truncation from [14], while the coefficients for the other rows are set to zero, thus systematically generating the new T_p matrix (as in Eq. (21.3)). Its butterfly flow diagram (Fig. 21.2a) shows that only 9 additions ($X1$ requires 72 additions) are required (the orange dashed lines identify the redundant operations not performed by ZLCADCT). Upon applying the new T_p matrix on the input image X, the resulting matrix $X1$ has coefficients only for three rows (the same as the size of the T_p matrix). Therefore, the coefficients in the other rows are not computed; in the first step, T_p substantially reduces 40 addition operations compared with ADCT [13]. As a further example, consider when 21 coefficients are retained in the zigzag scanning; then, only the first six rows are calculated. The new T_p matrix is given in Eq. (21.4), and its butterfly flow diagram is shown in Fig. 21.2b.

$$T_{p\_m=3} = \begin{pmatrix} 1 & 1 & 1 & 1 & 1 & 1 & 1 & 1 \\ 0 & 1 & 0 & 0 & 0 & 0 & -1 & 0 \\ 1 & 0 & 0 & -1 & -1 & 0 & 0 & 1 \end{pmatrix} \tag{21.3}$$

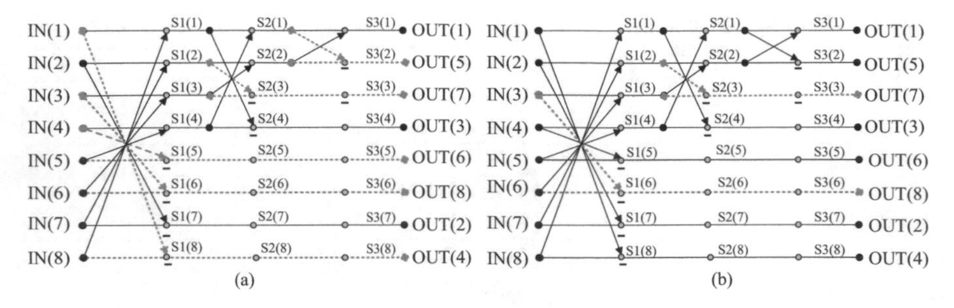

Fig. 21.2 Butterfly flow diagram of T_p using the truncation measure from [14] for the proposed ZLCADCT when 6 coefficients (**a**) or 21 coefficients (**b**) are retained

$$T_{p\_m=6} = \begin{pmatrix} 1 & 1 & 1 & 1 & 1 & 1 & 1 & 1 \\ 0 & 1 & 0 & 0 & 0 & 0 & -1 & 0 \\ 1 & 0 & 0 & -1 & -1 & 0 & 0 & 1 \\ 1 & 0 & 0 & 0 & 0 & 0 & 0 & -1 \\ 1 & -1 & -1 & 1 & 1 & -1 & -1 & 1 \\ 0 & 0 & 0 & 1 & -1 & 0 & 0 & 0 \end{pmatrix} \tag{21.4}$$

Next, for the operation $Y = X1 \cdot T_p^T$, the resulting size of the Y matrix is $m \times m$. For example, TCR = 6 results in a Y matrix of size 3×3 with the number of actual coefficients retained (ACR) of 9; therefore, the resulting three coefficients are not necessary. So, a look-up table (LUT) is needed (Table 21.2) for the number of coefficients retained in each row (k) as function of TCR and m. This LUT is used to check the number of coefficients to be retained in each row of Y, such that T_p can be adjusted to be of $k \times 8$ size, so avoiding the generation of additional and unnecessary coefficients in each row. Thus, for TCR = 6, as gray highlighted but with the right slash too in Table 21.2, the numbers of retained coefficients k are 3 in the first row (because TCR $\in$ [6, 7), 2 in the second row (TCR $\in$ [5, 8), and 1 for the third row (TCR $\in$ [4, 9)). Furthermore, Eqs. (21.5), (21.6), and (21.7) illustrate the difference for the operation of Y in the presence or absence of an LUT. With an LUT, the last coefficient ($\sum_{m=0}^{7} T_{2m} \sum_{n=0}^{7} T_{1n} X_{nm}$) in the second row and the last two coefficients ($\sum_{m=0}^{7} T_{1m} \sum_{n=0}^{7} T_{2n} X_{nm}$ and $\sum_{m=0}^{7} T_{2m} \sum_{n=0}^{7} T_{2n} X_{nm}$) in the third row of Y are zero. The final 8×8 output matrix Y is given in Eq. (21.8); the 6 retained coefficients are the same for both ZLCADCT and when employing ADCT [13] by applying the zigzag scanning to retain these coefficients. Hence, ZLCADCT retains the targeted coefficients with no zigzag scanning, while substantially reducing the number of addition operations (as shown in detail in later sessions). The Algorithm 21.1 of ZLCADCT is shown below. Figure 21.3 shows the flow diagram in terms of the entire ZLCADCT.

Based on Algorithm 21.1, the step of processing an image by ZLCADCT is given as follows: Initially, read the image and divide it into 8×8 sub-blocks (input matrix X). For a given TCR value provided by the user, calculate the value of m from Table

Table 21.2 LUT for the TCR range

TCR ∈		Number of retained coefficients for each row (*k*)							
		1	2	3	4	5	6	7	8
Number of rows (*m*)	1	[1,2)	[2,6)	[6,7)	[7,15)	[15,16)	[16,28)	[28,29)	[29,64]
	2	[3,5)	[5,8)	[8,14)	[14,17)	[17,27)	[27,30)	[30,43)	[43,64]
	3	[4,9)	[9,13)	[13,18)	[18,26)	[26,31)	[31,42)	[42,44)	[44,64]
	4	[10,12)	[12,19)	[19,25)	[25,32)	[32,41)	[41,45)	[45,54)	[54,64]
	5	[11,20)	[20,24)	[24,33)	[33,40)	[40,46)	[46,53)	[53,55)	[55,64]
	6	[21,23)	[23,34)	[34,39)	[39,47)	[47,52)	[52,56)	[56,61)	[61,64]
	7	[22,35)	[35,38)	[38,48)	[48,51)	[51,57)	[57,60)	[60,62)	[62,64]
	8	[36,37)	[37,49)	[49,50)	[50,58)	[58,59)	[59,63)	[63,64)	64

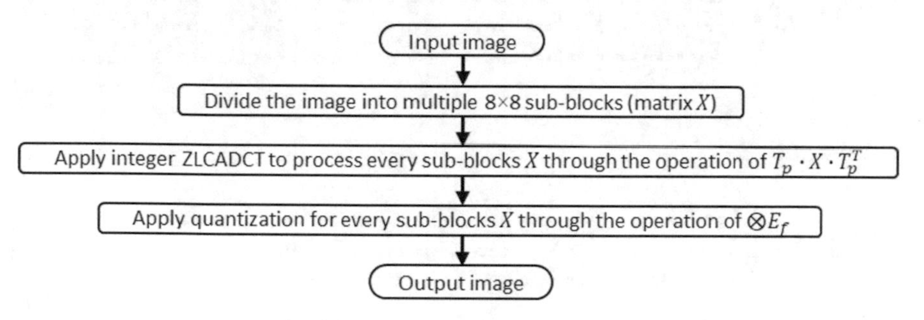

Fig. 21.3 Computational flow diagram for 8 × 8 ZLCADCT

21.1 to determine T_p ($m \times 8$ size). Next, for every sub-block matrix X, compute every column of X by T_p (using the fast butterfly algorithm) in turn and assign a "0" to the not required coefficients; then, save the results to matrix $X1$ and repeat the processing until the calculation for each sub-block X is completed. After matrix $X1$ is found, as for every m rows of $X1$, determine the range that covers the value of TCR by using the LUT in Table 21.2. As per the range, find the corresponding k value as the number of retained coefficients; adjust T_p to $k \times 8$ size and set "0" to the coefficients that are not required to be calculated. Then, complete processing for each m rows of $X1$ until $Y = X1 \cdot T_p^T$ is completed. Finally, quantize every sub-block matrix Y to find the final output results.

Algorithm 21.1 ZLCADCT Algorithm

```
1: procedure ZLCADCT (X, TCR)
2:        m ← TCR range
3:          Tp←Tp m × 8
4:   for j←1, 8 do
5:        X1(:, j)← Tp · X(:, j)
6:   end for
7:   for i←1, m do
8:        k ← TCR∨i ∈ range in LUT
```

```
9:          Tp ← Tp k × 8
10:         Y(i, :) ← X1(i, :) · Tp^T
11:    end for
12:         return Y
13: end procedure
```

$$T_p \cdot X = \begin{pmatrix} \sum_{n=0}^{7} T_{0n} X_{n0} & \sum_{n=0}^{7} T_{0n} X_{n1} & \cdots & \sum_{n=0}^{7} T_{0n} X_{n7} \\ \sum_{n=0}^{7} T_{1n} X_{n0} & \sum_{n=0}^{7} T_{1n} X_{n1} & \cdots & \sum_{n=0}^{7} T_{1n} X_{n7} \\ \sum_{n=0}^{7} T_{2n} X_{n0} & \sum_{n=0}^{7} T_{2n} X_{n1} & \cdots & \sum_{n=0}^{7} T_{2n} X_{n7} \end{pmatrix} \tag{21.5}$$

Without using LUT:

$$X1 \cdot T_p^T = \begin{pmatrix} \sum_{m=0}^{7} T_{0m} \sum_{n=0}^{7} T_{0n} X_{nm} & \sum_{m=0}^{7} T_{1m} \sum_{n=0}^{7} T_{0n} X_{nm} & \sum_{m=0}^{7} T_{2m} \sum_{n=0}^{7} T_{0n} X_{nm} \\ \sum_{m=0}^{7} T_{0m} \sum_{n=0}^{7} T_{1n} X_{nm} & \sum_{m=0}^{7} T_{1m} \sum_{n=0}^{7} T_{1n} X_{nm} & \boxed{\boldsymbol{\sum_{m=0}^{7} T_{2m} \sum_{n=0}^{7} T_{1n} X_{nm}}} \\ \sum_{m=0}^{7} T_{0m} \sum_{n=0}^{7} T_{2n} X_{nm} & \boxed{\boldsymbol{\sum_{m=0}^{7} T_{1m} \sum_{n=0}^{7} T_{2n} X_{nm}}} & \boxed{\boldsymbol{\sum_{m=0}^{7} T_{2m} \sum_{n=0}^{7} T_{2n} X_{nm}}} \end{pmatrix} \tag{21.6}$$

Using LUT:

$$X1 \cdot T_p^T = \begin{pmatrix} \sum_{m=0}^{7} T_{0m} \sum_{n=0}^{7} T_{0n} X_{nm} & \sum_{m=0}^{7} T_{1m} \sum_{n=0}^{7} T_{0n} X_{nm} & \sum_{m=0}^{7} T_{2m} \sum_{n=0}^{7} T_{0n} X_{nm} \\ \sum_{m=0}^{7} T_{0m} \sum_{n=0}^{7} T_{1n} X_{nm} & \sum_{m=0}^{7} T_{1m} \sum_{n=0}^{7} T_{1n} X_{nm} & 0 \\ \sum_{m=0}^{7} T_{0m} \sum_{n=0}^{7} T_{2n} X_{nm} & 0 & 0 \end{pmatrix} \tag{21.7}$$

$$\begin{aligned} Y &= T_p \cdot X1 \cdot T_p^T \\ &= \begin{pmatrix} \sum_{m=0}^{7} T_{0m} \sum_{n=0}^{7} T_{0n} X_{nm} & \sum_{m=0}^{7} T_{1m} \sum_{n=0}^{7} T_{0n} X_{nm} & \sum_{m=0}^{7} T_{2m} \sum_{n=0}^{7} T_{0n} X_{nm} & 0 & \cdots & 0 \\ \sum_{m=0}^{7} T_{0m} \sum_{n=0}^{7} T_{1n} X_{nm} & \sum_{m=0}^{7} T_{1m} \sum_{n=0}^{7} T_{1n} X_{nm} & 0 & \cdots & \cdots & 0 \\ \sum_{m=0}^{7} T_{0m} \sum_{n=0}^{7} T_{2n} X_{nm} & 0 & 0 & \cdots & \cdots & 0 \\ 0 & \vdots & \vdots & \ddots & \cdots & 0 \\ \vdots & \vdots & \vdots & \vdots & \ddots & \vdots \\ 0 & 0 & 0 & 0 & 0 & 0 \end{pmatrix} \end{aligned} \tag{21.8}$$

2.3 Modeling of ZLCADCT

Consider the traditional DCT when computed on 8 × 8 image blocks. The DCT matrix T ($T_{0,0}, \cdots, T_{i,j}$) and the input block matrix X ($X_{0,0}, \cdots, X_{i,j}$) are given by Eq. (21.9); then, the DCT results in a matrix $Y(Y_{0,0}, \cdots, Y_{i,j}) = T \cdot X \cdot T^T$ with coefficients are given in Eq. (21.10). The zigzag scanning (Fig. 21.1) is then performed to select the coefficients from Eq. (21.10) for data storage or transmission.

$$T = \begin{pmatrix} T_{00} & T_{01} & \cdots & T_{07} \\ T_{10} & T_{11} & \cdots & T_{17} \\ \vdots & \vdots & \ddots & \vdots \\ T_{70} & T_{71} & \cdots & T_{77} \end{pmatrix} \qquad X = \begin{pmatrix} X_{00} & X_{01} & \cdots & X_{07} \\ X_{10} & X_{11} & \cdots & X_{17} \\ \vdots & \vdots & \ddots & \vdots \\ X_{70} & X_{71} & \cdots & X_{77} \end{pmatrix} \tag{21.9}$$

$$Y = T \cdot X \cdot T^T$$
$$= \begin{pmatrix} \sum_{m=0}^{7} T_{0m} \sum_{n=0}^{7} T_{0n} X_{nm} & \sum_{m=0}^{7} T_{1m} \sum_{n=0}^{7} T_{0n} X_{nm} & \cdots & \sum_{m=0}^{7} T_{7m} \sum_{n=0}^{7} T_{0n} X_{nm} \\ \sum_{m=0}^{7} T_{0m} \sum_{n=0}^{7} T_{1n} X_{nm} & \sum_{m=0}^{7} T_{1m} \sum_{n=0}^{7} T_{1n} X_{nm} & \cdots & \sum_{m=0}^{7} T_{7m} \sum_{n=0}^{7} T_{1n} X_{nm} \\ \vdots & \vdots & \ddots & \vdots \\ \sum_{m=0}^{7} T_{0m} \sum_{n=0}^{7} T_{7n} X_{nm} & \sum_{m=0}^{7} T_{1m} \sum_{n=0}^{7} T_{7n} X_{nm} & \cdots & \sum_{m=0}^{7} T_{7m} \sum_{n=0}^{7} T_{7n} X_{nm} \end{pmatrix} \tag{21.10}$$

where

$$Y_{ij} = \sum_{m=0}^{7} T_{jm} \sum_{n=0}^{7} T_{in} X_{nm} \quad \text{for } i, j = 0 \ldots 7 \tag{21.11}$$

In ZLCADCT, instead of T, a new matrix T_p is first generated based on the number of retained coefficients. This new matrix is designed to eliminate those calculations performed on the unused (not retained) coefficients. The *non-zero partition* (*NZP*) is defined as the partition of the new output matrix Y in which a non-zero multiplication of $Y = T_p \cdot X \cdot {T_p}^T$ takes place, that is, *NZP* does not contain those zero multiplications that always results in zero vectors, rows, or columns.

Remark 1: The partition matrix NZP is a square matrix, that is, of $m \times m$ size, where $1 \leq m \leq 8$. Hence, ACR $= m^2$. For example, for a TCR of value 10, NZP is of size 4×4, or $m = 4$, and ACR is of value 16.

Next, consider the first multiplication operation to calculate Y, that is, $X1 = T \cdot X$. The matrix T is replaced by the matrix T_p of $m \times 8$ size. Therefore, the new $X1 = T_p \cdot X$ has 8 columns (denoted by $X1_{i,0}, X1_{i,1}, \cdots, X1_{i,7}$) that contain the necessary first m elements, that is, $i = 0 \ldots m - 1$. The remaining elements $i = m \ldots 7$ are all zero by definition. Similarly, the number of final elements in each row $(Y_{0,j}, Y_{1,j}, \cdots, Y_{7,j})$ of Y is determined by the processing of $Y = X1 \cdot {T_p}^T$. The use of T_p instead of T allows each row of Y to retain only the first m elements; as only m rows are considered in $X1$, then only these m rows must be used when calculating $Y = X1 \cdot {T_p}^T$.

Remark 2: Since m is limited in the range $1 \leq m \leq 8$, then TCR can have eight ranges. Mathematically, TCR $\in \left[1 + \sum_{t=0}^{m-1} t,\ \sum_{t=0}^{m} t\right]$.

Table 21.1 shows the ranges that TCR takes depending on the value of m. For example, if $m = 4$, TCR can take one of the values in the range from 7 to 10. The reciprocal relationship is also true, that is, if $7 \leq \text{TCR} \leq 10$, then m is 4 and the size of NZP is 4×4. So, $Y = T_p \cdot X \cdot T_p{}^T$ contains at least $8 \times 8 - m \times m = 48$ coefficients of zero value. These 48 coefficients are not stored/transmitted, and therefore, they are not calculated in ZLCADCT. Table 21.1 is then used to obtain the value of m for determining the size of T_p, that is, $m \times 8$ once TCR is chosen.

Consider Table 21.1, for example, when $m = 1$, the resulting T_p has one row, that is, $T_p = T_{p\,i,0\ldots7} = [T_{p0,0}, T_{p0,1}, \cdots, T_{p0,7}] = [1,1,1,1,1,1,1,1]$ where $i = 0$. Then for a given input matrix X, the element of the output vector $Z_{i,j}$ for each column (j-th) of the output matrix $X1$ is given by $Z_{m-1,j}(m) = Z_{0,j}(1) - T_{p0,0} \cdot X_{0,j} + T_{p0,1} \cdot X_{1,j} + \cdots + T_{p0,7} \cdot X_{7,j} - X_{0,j} + X_{1,j} + \cdots + X_{7,j}$ where $j = 0\ldots7$. Similarly, when $m = 5$, T_p has five rows, and for each column (j-th) of $X1$, five output vector elements $Z_{0\ldots m-1,j}(1\ldots m)$ ($Z_{0,j}(1)$ to $Z_{4,j}(5)$) are generated. For example, consider the fifth row of T_p, ($T_{p\,4,0\ldots7} = [T_{p\,4,0}, T_{p\,4,1}, \cdots, T_{p\,4,7}] = [1,-1,-1,1,1,-1,-1,1]$). The fifth output element is given by $Z_{m-1,j}(m) = Z_{4,j}(5) = T_{p4,0} \cdot X_{0,j} + T_{p4,1} \cdot X_{1,j} + \cdots + T_{p4,7} \cdot X_{7,j} = X_{0,j} - X_{1,j} - X_{2,j} + X_{3,j} + X_{4,j} - X_{5,j} - X_{6,j} + X_{7,j}$. By generalizing the above two examples, the first term $\left(\sum_{t=0}^{3}\left[\left(X_{t,j} + X_{7-t,j}\right) \cdot (-1)^{\left\lfloor \log_2^{t+1} \right\rfloor \cdot \left(\frac{m-1}{4}\right)}\right]\right)$ of Eq. (21.13) is obtained. $\sum_{t=0}^{3}\left[\left(X_{t,j} + X_{7-t,j}\right)\right]$ is used to determine the input elements, while $(-1)^{\left\lfloor \log_2^{t+1} \right\rfloor \cdot \left(\frac{m-1}{4}\right)}$ is used to determine the sign of the input elements according to $T_{p\,m-1,0\ldots7}$ (the m-th row of T_p). In general, the calculation of the output vector $Z_{i,j}$ ($i = 0\ldots m-1$ and $j = 0\ldots7$) for each column (j-th) of $X1$ (given by Eq. (21.12)) requires the general expression given by Eq. (21.13). In Eq. (21.13), $\left\lfloor \log_2^{t+1} \right\rfloor$ denotes the ceiling function for the highest integer value that is not larger than $\log_2^{t+1}$.

$$X1_{m\times8} = T_{p\,m\times8} \cdot X_{8\times8} = \begin{pmatrix} Z_{0,0}(1) & Z_{0,1}(1) & \cdots & Z_{0,7}(1) \\ Z_{1,0}(2) & Z_{1,1}(2) & \cdots & Z_{1,7}(2) \\ \vdots & \vdots & \ddots & \vdots \\ Z_{m-1,0}(m) & Z_{m-1,1}(m) & \cdots & Z_{m-1,7}(m) \end{pmatrix} \tag{21.12}$$

$$Z_{m-1,j}(m) = \begin{cases} \sum_{t=0}^{3}\left[\left(X_{t,j} + X_{7-t,j}\right) \cdot (-1)^{\left\lfloor \log_2^{t+1} \right\rfloor \cdot \left(\frac{m-1}{4}\right)}\right], & m \in \{1,5\} \\ \begin{pmatrix} X_{\left\lfloor \log_6^{m-1} \right\rfloor,j} - X_{3-\left\lfloor \log_6^{m-1} \right\rfloor,j} \\ - X_{4+\left\lfloor \log_6^{m-1} \right\rfloor,j} + X_{7-\left\lfloor \log_6^{m-1} \right\rfloor,j} \end{pmatrix} \cdot (-1)^{\left\lfloor \log_6^{m-1} \right\rfloor}, & m \in \{3,7\} \\ X_{\log_2^{6-m}-1,j} - X_{8-\log_2^{6-m},j}, & m \in \{2,4\} \\ X_{\log_2^{10-m}+1,j} - X_{6-\log_2^{10-m},j}, & m \in \{6,8\} \end{cases} \tag{21.13}$$

Equation (21.13) calculates the j-th column vector of $X1 = T_p \cdot X = [X1_{0,j}, X1_{1,j}, \cdots, X1_{m-1,j}]$ by using a $m \times 8$ matrix T_p to process j-th column vector of input X, that is, $[X_{0,j}, X_{1,j}, \cdots, X_{7,j}]$, and $X1 = [Z_{0,j}(1), Z_{1,j}(2), \ldots, Z_{m-1,j}(m)]$ is finally produced. All eight column vectors of X are processed using Eq. (21.13) and the final $m \times 8$ matrix $X1$ is shown in Eq. (21.12).

The resulting $m \times m$ NZP matrix of $Y = T_p \cdot X \cdot {T_p}^T$ is given in Table 21.2 (highlighted in gray) for TCR = 6, and then, m is 3 and ACR is 9. The remaining 55 coefficients are not calculated. However, a conventional DCT calculation (using the matrix T in Eqs. (21.9) and (21.10)) requires the calculation of all 64 coefficients for the entire 8×8 matrix and then retaining the 6 coefficients (after performing zigzag scanning). Therefore, a considerable improvement is accomplished by generating only the NZP matrix; so, the scanning process for transmitting/storing the selected coefficients can be removed.

By definition, the number of coefficients (9 coefficients) highlighted in gray in Table 21.2 is not always the same as those generated using ADCT [13] with a zigzag scanning (highlighted in gray with right slashes). The NZP matrix generates additional terms, except for the case when the number of retained coefficients is one ($m = 1$). Therefore, the use of the NZP matrix may incur in an additional overhead.

Remark 3: For a size m block, the TCR range is constant, and therefore, Table 21.2 can be used as an LUT, so the unnecessary coefficients found for $Y = X1 \cdot {T_p}^T$ can be entirely avoided.

To remove the additional terms generated by the NZP matrix during the process of calculating $Y = X1 \cdot {T_p}^T$, an LUT (Table 21.2) is introduced by ZLCADCT. The size of T_p is determined by the number of retained coefficients $k \in \{k_1, k_2, \cdots, k_m\}$ for the different m row vectors. For example, when $k = 1$, for the i-th row vector of input $X1$, T_p has only one row, that is, $T_p = T_{p\,0,0\ldots7} = [T_{p\,0,0}, T_{p\,0,1}, \ldots, T_{p\,0,7}] = [1, 1, 1, 1, 1, 1, 1, 1]$. Therefore, the output vector $Y_{i,j}$ for the i-th row is given by only one vector element $Y_{i,k-1}(k) = Y_{i,0}(1) = T_{p0,0} \cdot Z_{i,0} + T_{p0,1} \cdot Z_{i,1} + \cdots + T_{p0,7} \cdot Z_{i,7} = Z_{i,0} + Z_{i,1} + \cdots + Z_{i,7}$, where $0 \le i \le m - 1$ and $j = 0$. Similarly, when $k = 5$, T_p has five rows, and five output vector elements $Y_{i,k-1}(k)$ ($Y_{i,0}(1)$ to $Y_{i,4}(5)$) are generated for the i-th row. Therefore, for the fifth row of T_p, $T_{p\,4,0\ldots7} = [T_{p\,4,0}, T_{p\,4,1}, \cdots, T_{p\,4,7}] = [1, -1, -1, 1, 1, -1, -1, 1]$, and the fifth output element $Y_{i,k-1}(k) = Y_{i,4}(5) = T_{p4,0} \cdot Z_{i,0} + T_{p4,1} \cdot Z_{i,1} + \cdots + T_{p4,7} \cdot Z_{i,7} = Z_{i,0} - Z_{i,1} - Z_{i,2} + Z_{i,3} + Z_{i,4} - Z_{i,5} - Z_{i,6} + Z_{i,7}$ can be found. Generalizing the above expressions leads to the first term $\left(\sum_{t=0}^{3}\left[\left(Z_{i,t} + Z_{i,7-t}\right) \cdot (-1)^{\left\lfloor \log_2^{t+1} \right\rfloor \cdot \left(\frac{k_m - 1}{4}\right)}\right]\right)$ of Eq. (21.15). $\sum_{t=0}^{3}\left[\left(Z_{i,t} + Z_{i,7-t}\right)\right]$ is used to determine the input elements, while $(-1)^{\left\lfloor \log_2^{t+1} \right\rfloor \cdot \left(\frac{k_m - 1}{4}\right)}$ is used to determine the sign of the input elements according to $T_{p\,k-1,0\ldots7}$ (the k-th row of T_p). In general, the calculation of the output vector $Y_{i,j}$ ($i = 0 \ldots m - 1$ and $j = 0 \ldots k - 1$) for each (i-th) row of Y (given by Eq. (21.14)) requires the general expression given by Eq. (21.15).

$$Y = X1_{m\times 8}\cdot T^{T}_{p\ k_m\times 8} = \begin{pmatrix} Y_{0,0}(1) & Y_{0,1}(2) & \cdots & Y_{0,k_1-1}(k_1) \\ Y_{1,0}(1) & Y_{1,1}(2) & \cdots & Y_{1,k_2-1}(k_2) \\ \vdots & \vdots & \ddots & \vdots \\ Y_{m-1,0}(1) & Y_{m-1,1}(2) & \cdots & Y_{m-1,k_m-1}(k_m) \end{pmatrix} \tag{21.14}$$

$$Y_{i,k_m-1}(k_m) = \begin{cases} \sum_{t=0}^{3}\left[\left(Z_{i,t}+Z_{i,7-t}\right)\cdot(-1)^{\left\lfloor \log_2^{t+1}\right\rfloor\cdot\left(\frac{k_m-1}{4}\right)}\right], & k_m\in\{1,5\} \\ \begin{pmatrix} Z_{i,\left\lfloor \log_6^{k_m-1}\right\rfloor} - Z_{i,3-\left\lfloor \log_6^{k_m-1}\right\rfloor} \\ -Z_{i,4+\left\lfloor \log_6^{k_m-1}\right\rfloor} + Z_{i,7-\left\lfloor \log_6^{k_m-1}\right\rfloor} \end{pmatrix}\cdot(-1)^{\left\lfloor \log_6^{k_m-1}\right\rfloor}, & k_m\in\{3,7\} \\ Z_{i,\log_2^{6-k_m}-1} - Z_{i,8-\log_2^{6-k_m}}, & k_m\in\{2,4\} \\ Z_{i,\log_2^{10-k_m}+1} - Z_{i,6-\log_2^{10-k_m}}, & k_m\in\{6,8\} \end{cases} \tag{21.15}$$

where

$$k\in\{k_1,k_2,\cdots k_m\} \text{ and } k\le m \tag{21.16}$$

Equation (21.15) calculates the i-th row vector of $Y = X1\cdot {T_p}^T = [Y_{0,0}, Y_{1,1},\cdots, Y_{m-1,k-1}]$ by using the $k\times 8$ matrix T_p to process the i-th vector of the input $X1$, that is, $[Z_{i,0}, Z_{i,1},\cdots, Z_{i,7}]$, and then, the output matrix$Y$ is finally generated by $Y = \left[Y_{0,0\ldots k_1-1}(1\ldots k_1), Y_{1,0\ldots k_2-1}(1\ldots k_2), \cdots, Y_{m-1,0\ldots k_m-1}(1\ldots k_m)\right]$. All m row vectors of $X1$ are processed using Eq. (21.15) and the final matrix Y is shown in Eq. (21.14).

2.4 *Evaluation by Simulation for ZLCADCT*

Five input images (Lena, Cameraman, baboon, eagle, and moon) are processed by ADCT [13], ADCT [14], and ZLCADCT for different numbers of retained coefficients. The required number of additions, computational complexity, energy dissipation, delay, and output images are then analyzed. For evaluation purposes, the selection of the rows to be pruned for the random technique of [14] is executed using the proposed technique.

2.4.1 Resources and Complexity Analysis

Performance metrics such as the number of additions, energy consumption, and delay of the output image for the entire computation of $T\cdot X\cdot T^T$ are determined initially; the results are presented in Table 21.3. The number of coefficients for transmission/storage increases by using [14], so ZLCADCT is also evaluated to keep the number of transmitted coefficients the same as [13]. In Table 21.3, row 1 shows the number of TCRs (i.e., TCR); row 2 reports the number of ACRs for [14]. Row 3 shows the number of retained rows for DCT matrix. Rows 4–6 show the number of additions required by ADCT [13], ADCT [14], and ZLCADCT. The number of additions required by ZLCADCT is significantly smaller (28% for a single retained coefficient) at a low number of retained coefficients compared with ADCT [13]; and it can be even lower than ADCT [14]. By increasing the number of retained coefficients, the difference between the numbers of additions by ZLCADCT and ADCT [13] decreases.

Also, the results for the number of additions of ADCT [13], ADCT [14], and ZLCADCT can be verified by analyzing the computational complexity. ZLCADCT removes the unnecessary additions when calculating both $X1 = T_p \cdot X$ and $Y = X1 \cdot {T_p}^T$ so making their executions faster than ADCT [13, 14]. In ZLCADCT, for $X1 = T_p \cdot X$, each column of $X1$ is calculated using $14 - (8 - m) = m + 6$ additions ($1 \leq m \leq 8$); so, the total number of additions required by $X1$ is $8(m + 6) = 8m + 48$. Next, for computing $Y = X1 \cdot {T_p}^T$, each non-zero row of $X1$ must be checked from the LUT. If one row is required to retain k ($1 \leq k < m$) rather than m coefficients, then the number of additions for processing this row can be reduced from $m + 6$ to $k + 6$ by removing the unnecessary (redundant) additions. For example, if TCR is 6 (as explained previously), the number of additions required by using the LUT for calculating the second row of $X1$ decreases from 9 to 8 ($k = 2$) and from 9 to 7 ($k = 1$) for the third row. Thus, the total number of additions required for computing $T_p \cdot X \cdot T_p^T$ using ZLCADCT is 96, a further saving of 3 additions when compared with [14]. The generic equation to determine the number of additions required by ZLCADCT is given in Eq. (21.17); note that NA is dependent on both m and TCR.

$$\mathrm{NA}_{\mathrm{ZLCADCT}}(m, \mathrm{TCR}) = \begin{cases} m^2 + 13m - 2\sum_{t=0}^{m-1} t + \mathrm{TCR} + 48, & m \in \{1, 3, 5, 7\} \\ m^2 + 13m - 2\sum_{t=0}^{m-1} t + \mathrm{TCR} + 48 - 6\left\lceil 1 - \frac{\mathrm{TCR}}{\sum_{t=0}^{m} t} \right\rceil, & m \in \{2, 4, 6, 8\} \end{cases} \tag{21.17}$$

where the value of m can be obtained according to the TCR value. The range covering a TCR can be found in Table 21.1. $\left\lceil 1 - \mathrm{TCR}/\sum_{t=0}^{m} t \right\rceil$ denotes the ceiling function for the least integer that is not smaller than $1 - \mathrm{TCR}/\sum_{t=0}^{m} t$.

The total number of additions required by ADCT [14] is $m^2 + 14m + 48$ in [14]. As per *Remark 2* in a previous section, the largest TCR value for a specific m is $\sum_{t=0}^{m} t$. Thus, it can be found that $-2\sum_{t=0}^{m-1} t + \mathrm{TCR} = -2\sum_{t=0}^{m-1} t + \sum_{t=0}^{m} t = m - \sum_{t=0}^{m-1} t = m - m(m-1)/2 = \left(m - m^2\right)/2 \leq 0$ in Eq. (21.17); then, the number of additions for ZLCADCT will be smaller than that for ADCT [14]. For

Table 21.3 Comparison among ADCT [13], ADCT [14], and ZLCADCT when computing $T \cdot X \cdot T^T$ for different TCR for 512×512 Lena as an input image

TCR		1	3	6	10	15	21	28
ACR		1	4	9	16	25	36	49
m		1	2	3	4	5	6	7
Number of additions	ADCT [13]	224	224	224	224	224	224	224
	ADCT [14]	63	80	99	120	143	168	195
	ZLCADCT	63	79	96	114	133	153	174
Estimated energy consumption (J)	ADCT [13]	2.569E-12	2.569E-12	2.569E-12	2.569E-12	2.569E-12	2.569E-12	2.569E-12
	ADCT [14]	7.227E-13	9.177E-13	1.135E-12	1.376E-12	1.640E-12	1.927E-12	2.237E-12
	ZLCADCT	7.227E-13	9.062E-13	1.101E-12	1.307E-12	1.525E-12	1.755E-12	1.996E-12
Matlab implementation delay(s)	ADCT [13]	0.89174	0.89174	0.89174	0.89174	0.89174	0.89174	0.89174
	ADCT [14]	0.66657	0.70945	0.75237	0.76163	0.79125	0.82011	0.87752
	ZLCADCT	0.64775	0.69138	0.71599	0.74088	0.75129	0.79121	0.84789

ADCT [13], $T \cdot X \cdot T^T$ for an 8×8 input block requires executing 16 times the full fast butterfly algorithm that, in turn, requires 14 additions each time, so resulting in a total of 224 additions for any value of TCR. As explained previously, compared with [13], if TCR = 6, ADCT [14] requires 99 additions and ZLCADCT requires the least (96 additions).

In Table 21.3, as a hardware, a 32-bit ripple carry adder (*RCA*) (made of mirror full adder cells) is used to perform the addition. The eight input cases (i.e., the inputs A, B, and C in: from 000 to 111) of an exhaustive simulation are applied to a 32-nm single full adder cell. LTSPICE is used to establish the average energy consumption; the average energy consumption of a single full adder is found to be 3.58492E-16 J. The energy consumption of DCT using the matrix T_p is given by using the total number of additions in the butterfly algorithm; rows 7–9 of Table 21.3 show the energy consumption of ADCT [13], ADCT [14], and ZLCADCT. When the number of coefficients increases, the energy consumption gradually increases; significant savings in energy consumption are achieved by ZLCADCT, especially at a lower number of retained coefficients compared to ADCT.

As a measure of performance, the delay must also be assessed. The delay of the butterfly algorithm is determined by its critical path full adder at each level (so not depending on the number of adders in each level of addition); for the proposed matrix T_p, the critical path of the butterfly algorithm is always along the 3 additions. Therefore, when the number of retained coefficients increases, the delay of DCT using the proposed matrix T_p depends only on the 3 level additions. Rows 10, 11, and 12 of Table 21.3 show the software-based delay evaluation by Matlab for processing the Lena image. For any TCR value, ZLCADCT has the lowest delay, because its required calculation has the least complexity compared with ADCT [13, 14]. For both ZLCADCT and ADCT [14], the delay increases as TCR increases, while the delay of ADCT [13] is unchanged.

2.4.2 Output Image Evaluation

Next, ADCT [13], ADCT [14], and ZLCADCT are considered and evaluated by employing the peak signal-to-noise ratio (PSNR) as metric [1] for output images; the PSNR is used to assess the quality of processed images using Eqs. (21.18) and (21.19).The *mean squared error* (*MSE*) (defined in Eq. (21.19)) is first calculated; it evaluates the variation of pixel value between two images; $p_{i,j}$ refers to the pixel value of the original image of size $a \times b$, while $\hat{p}_{i,j}$ is the pixel value for the final processed image.

$$\text{PSNR} = 10 \log \frac{(2^n - 1)^2}{\text{MSE}} \tag{21.18}$$

where

$$\text{MSE} = \frac{1}{a \times b} \sum_{i=1}^{a} \sum_{j=1}^{b} \left(p_{i,j} - \hat{p}_{i,j}\right)^2 \tag{21.19}$$

Based on Eqs. (21.18) and (21.19), both ADCT [13] and ZLCADCT generate the same output matrix (i.e., Eq. (21.8)). So, the values of $\hat{p}_{i,j}$ for both these two methods are identical, and therefore, both the MSE and the PSNR have the same values.

However, for ADCT [14], the additional non-zero coefficients $\hat{p}_{i,j}$ (e.g., those coefficients highlighted in bold font in Eq. (21.6)) cause a difference $p_{i,j} - \hat{p}_{i,j}$ for these pixels to be smaller than $p_{i,j} - 0$, because the values of these coefficients for both ADCT [13] and ZLCADCT are all zero. Therefore, the MSE of ADCT [14] is smaller than ADCT [13] and ZLCADCT, and the PSNR of [14], in turn, is higher for these two methods.

Table 21.4 compares ADCT [13], ADCT [14], and ZLCADCT in terms of output images for Cameraman when the numbers of retained coefficients are 1, 3, 6, 10, 15, 21, 28, and 36. Table 21.5 shows the values of the PSNR obtained for different TCRs when applied on five input images (Lena, Cameraman, Baboon, Eagle, and Moon) and processed using ADCT [13], ADCT [14], and ZLCADCT. As observed from the results, when only one coefficient is retained, the outputs of ZLCADCT and ADCT [14] are the same; therefore, their PSNR values are equal only in this case. However, when the number of retained coefficients increases, the PSNR of ADCT [14] can be higher than ADCT [13] and ZLCADCT, because they retain more coefficients than ADCT [13] using zigzag scanning and ZLCADCT. For example, when 6 coefficients are required for compressing images, only 6 coefficients are retained using ADCT [13] (after zigzag scanning) and ZLCADCT, but ADCT [14] retains three more coefficients (highlighted by a bold font in Eq. (21.6)). ADCT [14] can have a marginally higher PSNR due to the additional elements in the resulting matrix. However, by using ZLCADCT, the additional elements are avoided, and ZLCADCT has a PSNR whose value is the same as ADCT [13]. Moreover, when the number of retained coefficients is larger than 1, this scheme reduces the number of additions and energy consumption. Next, the average PSNR value (rows 6 and 13) and its percentage (rows 7 and 14) for different TCR shown in Table 21.5 are used to determine the desired value of the TCR depending on the percentage of the required PSNR. The percentage values are calculated by using the average PSNR values divided by the PSNR value of TCR = 63. For example, using the ZLCADCT approach, if the output image quality is required to be at 60% of the PSNR value, then the TCR value of 15 must be selected.

Table 21.4 Output images of ADCT [13], ADCT [14], and ZLCADCT for different number of retained coefficients

TCR	ADCT [14]	ADCT [13] or ZLCADCT	TCR	ADCT [14]	ADCT [13] or ZLCADCT
1			3		
6			10		
15			21		
28			36		

Table 21.5 PSNR of four different input images computed by ADCT [13], ADCT [14], and ZLCADCT for different TCR

Applied measure	PSNR / Input images	TCR						
		1	3	6	10	15	21	28
ADCT [14]	Lena	23.665	24.6831	25.5064	29.0783	29.5925	30.2545	31.1538
	Cameraman	19.4785	20.319	21.1865	23.5678	24.4284	25.284	26.6829
	Baboon	19.7051	20.1917	20.9795	22.4631	23.5461	24.6208	26.2839
	Eagle	19.6944	20.6155	21.4448	24.3211	24.9546	25.7347	26.5863
	Moon	25.4545	26.5577	27.3203	31.3366	31.8525	32.6273	33.3232
	Average value	21.5995	22.4734	23.2875	26.15338	26.87482	27.70426	28.80602
	Average (%)	50.30%	52.34%	54.23%	60.91%	62.59%	64.52%	67.09%
ADCT [13] or ZLCADCT	Lena	23.665	24.6146	25.2641	27.9908	28.7141	29.4039	30.2238
	Cameraman	19.4785	20.2603	20.9831	22.792	23.5013	24.2461	25.2993
	Baboon	19.7051	20.1185	20.6814	21.708	22.4431	23.2003	24.1371
	Eagle	19.6944	20.5464	21.1677	23.3922	24.1524	24.8455	25.5907
	Moon	25.4545	26.5051	27.1862	30.4515	31.1244	31.9182	32.657
	Average value	21.5995	22.40898	23.0565	25.2669	25.98706	26.7228	27.58158
	Average (%)	50.30%	52.19%	53.70%	58.84%	60.52%	62.24%	64.24%

3 Approximate Newton Method for Unconstrained Total Variation–Based Image Denoising

This section presents the second proposed algorithm for approximate image denoising. The inexact Newton method is useful because it can be applied to total variation (TV)-based image denoising algorithm [24]. The total processing time of an approximate Newton method is determined by the number of iterations (NOI) and processing time of each iteration. If the number of required iterations can be reduced, the number of calculations required for solving the linear equation matrices can be also reduced, hence saving considerable processing time and floating-point operations. Therefore, for reducing the total processing time, hardware, and energy dissipation for a TV-based image denoising technique, this work introduces an additional step length parameter into the original algorithm to reduce the computation complexity.

3.1 Preliminaries of the Approximate Newton Method

The total variation of a noisy image is significantly higher than the original image [19]. A technique that decreases the total variation of a noisy image has been proposed in [19], and the total variation is defined as the integration of the gradient magnitude in Eq. (21.20):

$$J(x) = \int_{D_x} \sqrt{x_i^2 + x_j^2} didj \tag{21.20}$$

where $x_i = \partial x/\partial i$, $x_j = \partial x/\partial j$, D_x is the image support region, and x is the input image. Since digital images are two-dimensional discrete signals, the total variation of a discrete signal can be written as in Eq. (21.21) [19]:

$$J(x) = \sum_{i,j} \sqrt{\left|x_{i+1,j} - x_{i,j}\right|^2 + \left|x_{i,j+1} - x_{i,j}\right|^2} = \sum_{i,j} \left\| \text{grad}\,(x)_{i,j} \right\| \tag{21.21}$$

where $x_{i,j}$ is the pixel value of an image, i and j are the row and column number, and grad$(x)_{i,j}$ is the gradient value of an input image x for the i direction (horizontal) and the j direction (vertical). For any input image x, $J(x)$ is a scalar and the gradient of x is a three-dimensional matrix.

In the presence of noise, differences between two neighboring pixels ($|x_{i+1,j} - x_{i,j}|$ and $|x_{i,j+1} - x_{i,j}|$) increase in the noisy region of the image; therefore, the total variation term $J(x)$ increases. However, [19] has shown that denoising an image is achieved by minimizing the total variation $J(x)$ of a noisy image to smooth the image signal. Moreover, the difference $E(x, y)$ (given in Eq. (21.22)) between the recovered image (x) and the original noisy image (y) is introduced in a TV denoising technique to minimize the image quality degradation [18, 25, 26]. $E(x, y)$ is also a scalar.

$$E\,(x, y) = \frac{1}{2} \sum_{i,j} \left(y_{i,j} - x_{i,j}\right)^2 = \frac{1}{2} \|y - x\|^2 \tag{21.22}$$

Therefore, a TV-based image denoising technique can be viewed as the solution of Eq. (21.23) [25]:

$$\min_x \left\{ f(x) = E\,(x, y) + \lambda \cdot J(x) = \frac{1}{2} \|y - x\|^2 + \lambda \cdot J(x) \right\} \tag{21.23}$$

where λ is a positive constant number and $f(x)$ is a scalar. The parameter λ is used to adjust the influence between the two constrained terms ($E(x, y)$ and $J(x)$). When $E(x, y)$ and $J(x)$ are both convex functions, the minimization of Eq. (21.23) for finding the minimum value of $f(x)$ corresponds to solving an unconstrained optimization problem. When an unconstrained optimization is applied for the minimization of Eq. (21.23), the gradient of $f(x)$ (given in Eq. (21.24)) must be found [25], because the problem is equivalent to finding the solution for $\nabla f(x) = 0$. The gradient of $J(x)$ from Eq. (21.21) is now given in Eq. (21.25):

$$\nabla f(x) = x - y + \lambda \nabla J(x) \tag{21.24}$$

$$\nabla J(x) = \text{div}\left(\text{grad}\,(x)_{i,j} / \left\|\text{grad}\,(x)_{i,j}\right\|\right) \tag{21.25}$$

where div(grad$(x)_{i,j}$/‖grad$(x)_{i,j}$‖) is the divergence of grad$(x)_{i,j}$/‖grad$(x)_{i,j}$‖, and both $\nabla f(x)$ and $\nabla J(x)$ have the same size of the matrix x. $J(x)$ should be regularized and smoothed to avoid grad$(x)_{i,j} = 0$. The smoothed $J(x)$ is given now in Eq. (21.26) [25, 26]; then, the gradient of the smoothed $J(x)$ can be written as Eq. (21.27):

$$J(x) = \sum_{i,j} \sqrt{\left|x_{i+1,j} - x_{i,j}\right|^2 + \left|x_{i,j+1} - x_{i,j}\right|^2 + \varepsilon^2} = \sum_{i,j} \left\|\text{grad}\,(x)_{i,j}\right\|_{\varepsilon} \tag{21.26}$$

$$\nabla J(x) = \text{div}\left(\text{grad}\,(x)_{i,j} / \left\|\text{grad}\,(x)_{i,j}\right\|_{\varepsilon}\right) \tag{21.27}$$

where ε is the so-called regularization parameter.

For finding the solution of $\nabla f(x) = 0$ in unconstrained optimization, the Newton method (whose equations are given in Eqs. (21.28) and (21.29)) has been widely used:

$$x^{(l+1)} = x^{(l)} - d \tag{21.28}$$

$$d = H_l^{-1} \nabla f\left(x^{(l)}\right) \tag{21.29}$$

where l is the iteration, $x^{(l)}$ is the resulting image matrix generated at the l iteration, and d is the correction matrix. The Newton method is viewed as a quasi-Newton scheme when H_l is an approximation of the Hessian function for $f(x)$.

For solving Eq. (21.29) in a large-scale unconstrained optimization, an approximate Newton method (also known as the Newton-conjugate gradient [CG] method) has been proposed in [17]; this solves (Eq. 21.29) by using the CG method. The search directions of the matrix d are computed by applying the CG method [17]; the CG method guarantees that the d matrix can keep reducing the $f(x)$ value toward the minimum value in each iteration using Eq. (21.28).

Reference [18] has presented the first algorithm of the approximate Newton method using CG (Newton-CG method) for solving the unconstrained optimization problem for TV-based image denoising. In this paper, the Newton-CG method [18] is utilized for removing Gaussian noise for TV-based image denoising in Matlab. The process of the Newton-CG method [18] for TV-based image denoising is shown below as Algorithm 21.2; φ is the precision control parameter for determining the termination of the iterations, $f(x)_{\text{previous}}$ is the $f(x)$ value of the previous iteration, and num is the number of iterations. The criteria "$\varphi > 0.1$ and num < 20" are added in this paper as conditions to terminate the loop. ε in Eq. (21.27) and λ are selected as 0.001 and 0.06, respectively, as per [18].

Algorithm 21.2 Newton-CG Method for TV-Based Denoising

1. Noisy image y, make $x = y$
2. Calculate the TV function $J(x)$ and the image data error function $E(x, y)$
3. Calculate $f(x) = E(x, y) + \lambda J(x)$
4. Calculate $\nabla f(x)$ (the derivative of $f(x)$) as $\nabla f(x) = x - y + \lambda \nabla J(x)$
5. Calculate the approximate Hessian matrix $\nabla^2 f(x)$
6. Calculate $d = \nabla f(x)/\nabla^2 f(x)$ by solving equation $\nabla^2 f(x) \cdot d - \nabla f(x)$ using the conjugate gradient
7. Calculate $x = x - d$
8. Calculate $f(x) = E(x, y) + \lambda J(x)$
9. Calculate $\varphi = f(x) - f(x)_{previous}$
10. If $\varphi > 0.1$ and $num < 20$, go back to step 4
11. Output x

Prior to performing the numerical calculation, the pixels of an image are normalized, and their values are converted into a range from 0 to 1. The solution of Eq. (21.29) involves many floating-point calculations; therefore, a floating-point adder is required for image denoising. Throughout this manuscript, a 32-bit single precision floating adder (in accordance with the IEEE 754 standard) is utilized in the operation of the Newton method. A 32-bit single precision floating point adder receives operands that are defined by the IEEE 754 standard, that is, 1 sign bit, 8 bits for the exponent, and 23 bits for the fraction (so, 24 bits for the mantissa with the hidden bit). Noisy images with different Gaussian noise levels are generated and simulated using Matlab; the pixel values of the original image (so with no noise) are normalized in the range from 0 to 1. Then, the normalized input image matrix is added with the Gaussian noise matrix; the element values of the Gaussian noise matrix are subject to a zero-mean Gaussian distribution. The range of element values for the Gaussian noise matrix is given by [−0.25 0.25] when the noise level is 25% (so [−0.5 0.5] for 50% noise, [−0.75 0.75] for 75% noise, and [−1 1] for 100% noise).

3.2 *Approximate Newton Method Using Additional Step Length Parameter*

In each iteration of the approximate Newton method using CG, the image matrix x is subtracted from the correction matrix d of Eq. (21.28). An element of the matrix d is the searching direction calculated by Eq. (21.29) using the CG method for finding the minimum value of $f(x)$. $f(x)$ decreases by subtraction (as in Eq. (21.28)), so as the number of iterations increases, $f(x)$ gradually approaches the desired minimum value, and therefore, the element value of the correct matrix d should also gradually decrease to zero. Figure 21.4 shows the one-dimensional Newton method; the d

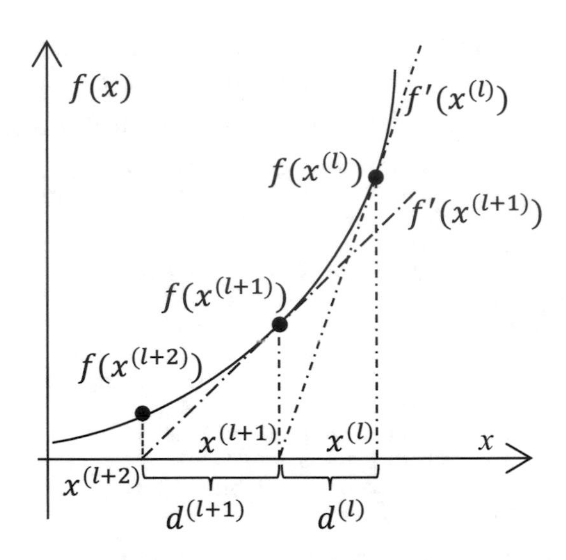

Fig. 21.4 Traditional one-dimensional Newton method

value in each iteration gradually decreases toward zero. This implies that after each iteration, $x^{(l+1)} = x^{(l)} - d = x^{(l)} - d^{(l)}$ approaches the desired final value. So, if d can be increased by multiplying it with a positive number $\alpha > 1$, then it is possible to further decrease the x value in an iteration. For example, for the first iteration in Fig. 21.4, if d is increased as αd, $x^{(l+1)} = (x^{(l)} - \alpha d) < x^{(l+1)} = (x^{(l)} - d)$, that is, $x^{(l+1)}$ will be closer to the desired minimum value compared with only applying $x^{(l+1)} = x^{(l)} - d$. This means that $f(x^{(l+1)} = (x^{(l)} - \alpha d)) < f(x^{(l+1)} = (x^{(l)} - d))$; $f(x)$ can be reduced to a lower value. By choosing a suitable α value larger than 1, d in each iteration is increased, and the final minimum $f(x)$ value is likely obtained with a smaller number of iterations.

Therefore, for reducing the number of iterations, the additional step length parameter α is utilized in Eq. (21.28) now given by Eq. (21.30) in each iteration. By changing the value of α in each iteration, the total number of iterations changes too.

$$x^{(l+1)} = x^{(l)} - \alpha d \tag{21.30}$$

Figure 21.5 shows the change of element values for four different random elements ((a), (b), (c), and (d)) of the d matrix when the number of iterations rises for different α (1, 1.2, 1.4, 1.6, 1.8, and 2) to process the Cameraman image with 50% noise. For all elements, their values gradually approach 0 as the number of iterations rises at $\alpha \leq 1.8$. When $1.2 \leq \alpha \leq 1.6$, the element value requires a smaller number of iterations compared with the case of $\alpha = 1$. Figure 21.6 shows the norm

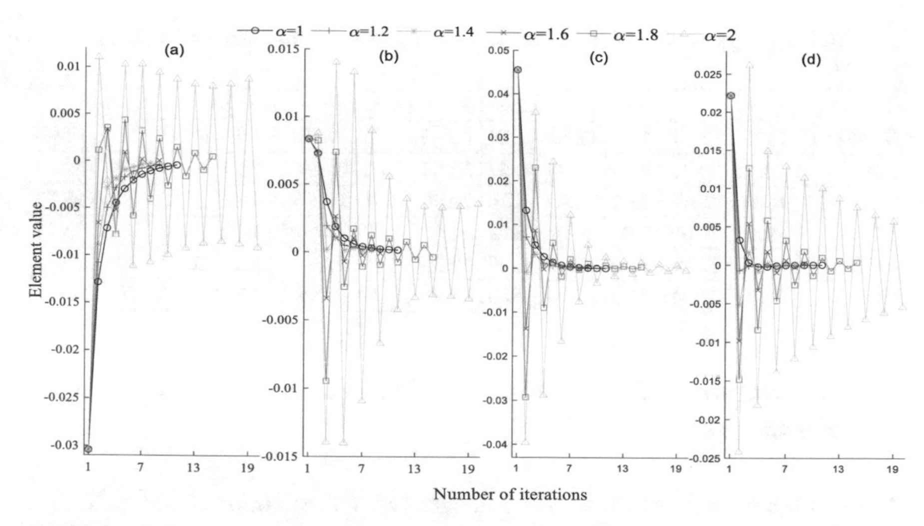

Fig. 21.5 The change of element value for four different elements ((**a**) for NO.1 element, (**b**) for NO.78 element, (**c**) for NO.278 element, and (**d**) for NO.400 element) of d matrix when the number of iterations increases for different α values by processing Cameraman image with 50% noise

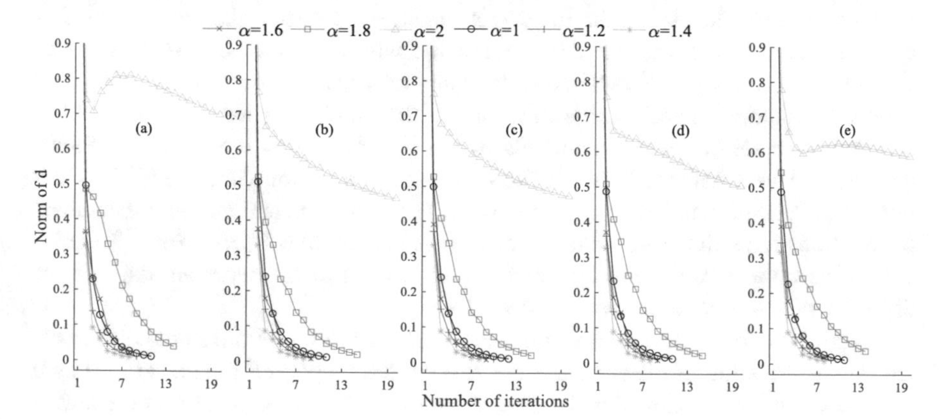

Fig. 21.6 The norm value of the d matrix for five different images with 50% noise ((**a**) Lena, (**b**) Cameraman, (**c**) eagle, (**d**) house, and (**e**) rice) when the number of iterations changes at different α values

value of the d matrix when the number of iterations increases for different input images and different α values. The norm value of the d matrix gradually decreases to zero when the number of iterations increases for $\alpha \leq 1.8$.

Table 21.6 Average number of iterations (NOI) and PSNR for the image set [27] when α is changed

	α = 1		α = 1.2		α = 1.4		α = 1.6		α = 1.8		α = 2	
Noise level	NOI	PSNR	NOI	PSNR	NOI	PSNR	NOI	PSNR	NOI	PSNR	NOI	PSNR
25%	9.22	30.91	7.44	30.90	6.57	30.89	8.62	30.87	15.09	30.87	20	30.27
50%	10.96	28.49	9.03	28.49	7.76	28.50	8.51	28.52	14.37	28.52	20	28.29
75%	9.94	24.06	8.07	24.07	6.99	24.08	8.00	24.09	13.10	24.08	20	24.12
100%	8.99	20.29	7.16	20.30	6.00	20.30	7.06	20.31	12.24	20.31	20	20.38
Mean	9.78	25.94	7.93	25.94	6.83	25.94	8.05	25.95	13.70	25.95	20	25.76

3.3 *Evaluation by Simulation for Additional Step Length Parameter*

To assess the impact of α, a set of 68 images (with a resolution of 256 × 256) is considered; these images are from the database of the computer vision group (CVG) [27]. This image set is diverse to include different types, subjects, and backgrounds. Table 21.6 shows the average number of iterations and PSNR when α is changed at different noise levels for the image set; when α increases, the average number of iterations at α = 1.4 is the lowest compared with other cases. Also, there is no significant difference in PSNR when the value of α varies from 1 to 2. Table 21.7 shows the maximum and minimum values in the number of iterations and PSNR for the image set; the results show that when the noise level increases, the PSNR decreases. The difference between the minimum and the maximum PSNR values reflects the variation of image characteristics. The minimum and maximum numbers of iterations have the lowest values for α = 1.4, while the PSNRs for different α values are quite close. The noisy and output images of Cameraman and rice for different values of α are shown in Table 21.8.

Figure 21.7 shows the average number of iterations (NOI) at different noise levels and the mean value of different noise levels for the employed image set [27] as α increases; the results show that the average value of iterations (6~8) is at the lowest when α = 1.4. This means that the number of iterations for α = 1.4 is reduced by 27.3–33.3% when compared with the value (9~11) for α = 1. Then, the point values in Fig. 21.7 are used to estimate and calculate the equations of the fitted curves (shown in Table 21.9) by using linear regression; the minimum values of the estimated fitted curves are found in Table 21.9.

Table 21.9 shows that the minimum values are found when the α value is close to 1.4 (varying from 1.39 to 1.45). Figure 21.8 shows the $f(x)$ of different α values for Cameraman with 50% noise when the NOI increases; the initial $f(x)$ value (at NOI = 0) is 799 for all α values. Also, for any α value, the $f(x)$ value decreases when the NOI increases. When α = 1.4, the $f(x)$ value is always lower than for other α values in each iteration.

Figure 21.9 shows the number of iterations for denoising the set of 68 images with 25%, 50%, 75%, and 100% noise levels by applying α between 1 and 1.8. Fifty

Table 21.7 Maximum and minimum values in number of iterations (NOI) and PSNR of the image set [27] when α is changed

α	25% Noise				50% Noise				75% Noise				100% Noise			
	PSNR		NOI		PSNR		NOI		PSNR		NOI		PSNR		NOI	
	Min	Max	Min	Max	Min	Max	Min	Max	Min	Max	Min	Max	Min	Max	Min	Max
1	23.31	47.33	8	12	22.18	38.65	10	12	20.02	31.61	9	10	17.54	27.19	8	9
1.2	23.31	47.36	7	10	22.18	38.67	8	10	20.02	31.63	7	9	17.54	27.20	7	8
1.4	23.30	47.39	6	8	22.18	38.68	7	8	20.02	31.63	6	7	17.55	27.20	6	6
1.6	23.27	47.40	8	9	22.17	38.74	8	9	20.02	31.66	8	8	17.55	27.21	7	8
1.8	23.26	47.38	14	17	22.17	38.76	13	15	20.02	31.67	13	14	17.55	27.20	12	13
2	23.18	44.56	20	20	22.09	38.41	20	20	20.01	31.74	20	20	17.57	27.31	20	20

Table 21.8 Noisy and output images by using different α values for Cameraman and rice with 50% noise

α		1	1.2	1.4	1.6	1.8	2
Original image	Noisy image	Output image					
Cameraman	PSNR	27.40	27.41	27.42	27.43	27.43	27.27
	NOI	11	9	8	9	15	20
Rice	PSNR	30.13	30.14	30.14	30.15	30.15	29.92
	NOI	11	9	8	8	14	20

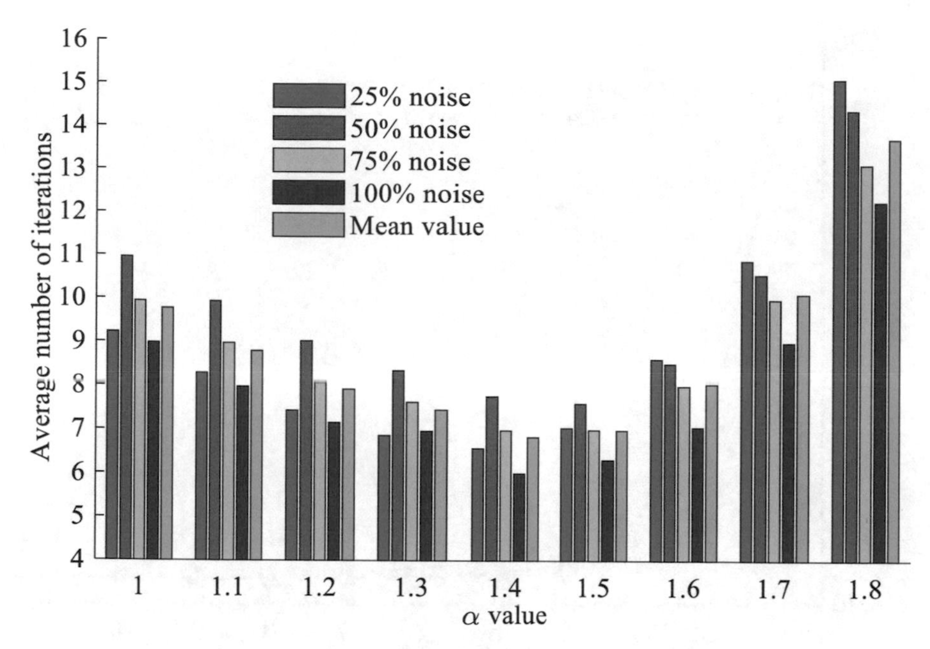

Fig. 21.7 Average number of iterations of 68 images and mean value of different noise values versus α

Table 21.9 Minimized values and equations of fitted curves for the average number of iterations for the image set at different noise levels

Noise level (%)	Equation of fitted curves between number of iterations (NOI) and α	Minimum value	
		α	NOI
25	$NOI = 39.686\alpha^3 - 132.22\alpha^2 + 137.8\alpha - 36.078$	1.39	6.58
50	$NOI = 47.881\alpha^3 - 170.18\alpha^2 + 191.61\alpha - 58.466$	1.45	7.54
75	$NOI = 36.963\alpha^3 - 127\alpha^2 + 136.37\alpha - 36.455$	1.43	6.94
100	$NOI = 38.461\alpha^3 - 134.09\alpha^2 + 147.3\alpha - 42.8$	1.43	6.11
Mean	$NOI = 40.748\alpha^3 - 140.87\alpha^2 + 153.27\alpha - 43.45$	1.42	6.82

(50) different α values were selected randomly with a normal distribution between 1 and 1.8. Then, these α values are applied to the process of denoising; Fig. 21.9 shows that when α varies between 1.39 and 1.45, the number of iterations is the lowest for all noise levels. In Fig. 21.9, the best α value (between 1.39 and 1.45) is depicted by the enclosed range between the two vertical red lines in each sub-plot. This result agrees with the minimization value results presented previously in Table 21.9.

Finally, images with three different types of noises (salt and pepper noise, multiplicative noise, and Poisson noise [28]) are processed by the Newton-CG method with the proposed step length parameter α. Table 21.10 shows the output images of Cameraman with different noises for cases of $\alpha = 1$ and $\alpha = 1.4$.

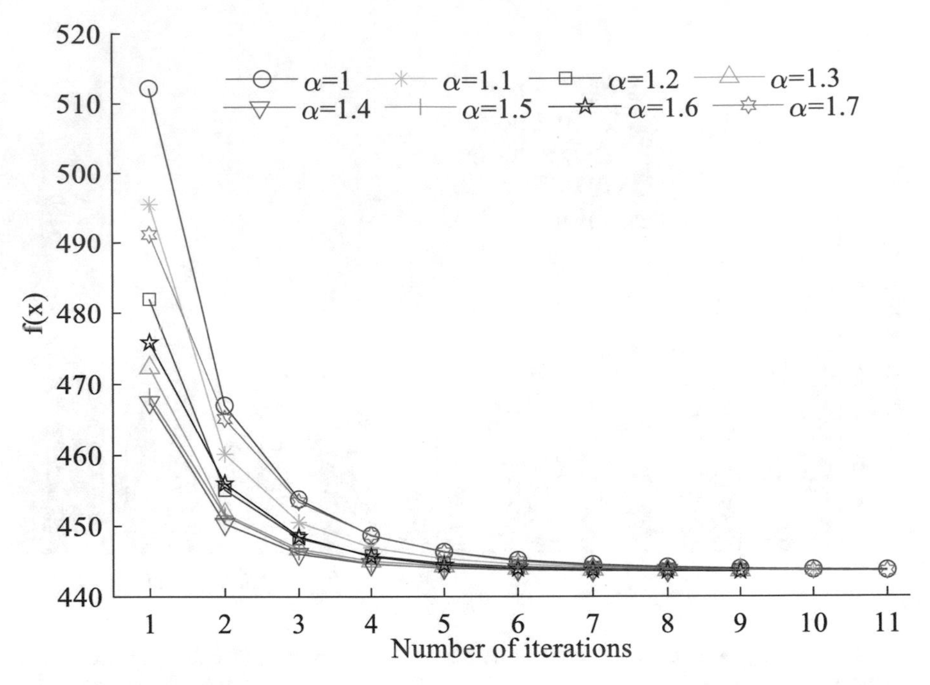

Fig. 21.8 $f(x)$ by processing Cameraman at 50% noise when α is changed

Table 21.11 shows the average values of the used image set in terms of PSNR, the number of iterations, and final $f(x)$ values for different types of noises when the α value varies from 1 to 2 by an increment of 0.1. The quality of the output images for the salt and pepper noise is not as good as the Gaussian noise of Table 21.8; this occurs because the Gaussian noise is an additive noise, while the salt and pepper noise (impulse noise [29]) is not additive. As reported in [19], the TV-based image denoising technique is designed to process a noise model with an additive component; hence, these results confirm this finding. The output performance for images with other types of noise may not be as good as additive noise when using the TV-based image denoising technique. However, as shown in Table 21.11, irrespective of the type of noise, the average number of iterations for $\alpha = 1.4$ is the lowest compared with other α values, and there is no significant change in output image quality, PSNR, and final $f(x)$ value.

4 Summary

In this chapter, two novel energy-efficient and high-performance algorithmic-level approximate computing techniques (ZLCADCT for image compression and approximate Newton method for TV-based image denoising) were presented. For image

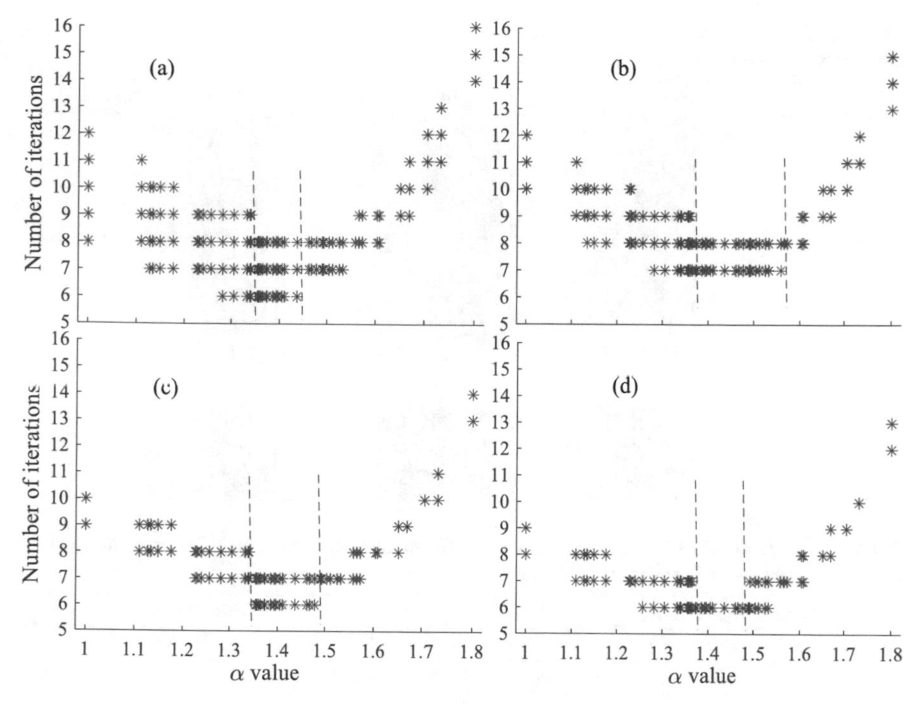

Fig. 21.9 Number of iterations when applying random α values between 1 and 1.8 for a set of images with different noise levels ((**a**) 25% noise, (**b**) 50% noise, (**c**) 75% noise, and (**d**) 100% noise)

compression, ZLCADCT configures the T matrix by establishing the relationship between the number of retained coefficients and the number of rows of the T matrix. It significantly reduces the computational complexity of an approximate DCT scheme (such as ADCT [13]) and avoids the additional unnecessary coefficients generated by [14]. A detailed mathematical analysis of ZLCADCT has been presented to show the properties of T_p and a novel method for removing the additions required to process the unused coefficients of a compressed image. An LUT is employed in ZLCADCT to enhance performance; ZLCADCT does not require coefficient scanning and its subsequent processing, while keeping the number of coefficients the same as ADCT [13]. The performance of ZLCADCT has been extensively evaluated by the simulation. The simulation results have shown that in the best case for the number of targeted coefficients to be retained (i.e., TCR = 1), ZLCADCT requires only 28.12% of the adders compared with ADCT [13], and the energy utilized by ADCT [13] is 3.5 times larger than that for ZLCADCT. Larger values of TCR modestly increase both the percentages of the required number of adders and the energy consumption by ZLCADCT; moreover, for the entire range of TCR values, ZLCADCT retains the same image quality of the compressed image as [13].

Table 21.10 Noisy images and output images under salt and pepper noise, multiplicative noise, and Poisson noise when $\alpha = 1$ and $\alpha = 1.4$

	Noisy input image	Output image for $\alpha = 1$	Output image for $\alpha = 1.4$
Salt and pepper noise			
Multiplicative noise			
Poisson noise			

Table 21.11 Average values of image set for PSNR, number of iterations (NOI), and final $f(x)$ for salt and pepper, multiplicative, and Poisson noises

	Salt and pepper noise			Multiplicative noise			Poisson noise		
α	PSNR	NOI	$f(x)$	PSNR	NOI	$f(x)$	PSNR	NOI	$f(x)$
1	26.69	8.62	207.65	26.53	8.78	203.94	26.77	8.75	202.78
1.1	26.69	7.79	207.63	26.53	7.97	203.92	26.77	7.93	202.76
1.2	26.69	7.18	207.62	26.53	7.32	203.91	26.76	7.26	202.74
1.3	26.68	6.56	207.61	26.52	6.69	203.90	26.76	6.66	202.74
1.4	26.68	6.34	207.59	26.52	6.46	203.89	26.76	6.41	202.72
1.5	26.67	6.84	207.57	26.51	7.00	203.87	26.75	6.94	202.70
1.6	26.66	8.28	207.57	26.50	8.50	203.86	26.74	8.43	202.69
1.7	26.66	10.51	207.59	26.50	10.76	203.88	26.74	10.68	202.71
1.8	26.66	14.51	207.65	26.50	14.87	203.95	26.74	14.74	202.79
1.9	26.64	19.41	208.55	26.48	19.90	204.87	26.72	19.71	203.71
2	26.41	19.44	233.88	26.25	20.00	230.68	26.48	19.72	229.68

Next, for image denoising, the approximate Newton method using the conjugate gradient (also known as the Newton-CG method) is used for solving unconstrained optimization problems, such as the total variation–based image denoising. The new technique introduces an additional step length parameter α (of value greater than 1) to decrease the number of iterations (reduced by 27.3~33.3% to the maximum) for the entire algorithm, such that the total processing time is reduced. In each iteration of the approximate Newton method using CG, the image matrix x is subtracted by the correction matrix d; the element of matrix d is the searching direction calculated by using the CG method for finding the minimum value by subtraction. Iteratively, the desired minimum value is gradually reached, and therefore, the element value of the correct matrix d should also gradually decrease to zero. After each iteration, the desired final value is reached faster because α is greater than 1, that is, by choosing a suitable value for α larger than 1, d in each iteration is increased and the final x value is likely obtained with a smaller number of iterations. The proposed technique has been tested on a set of images (68 images) taken from a public domain library [27] and found that when α varies in a range of 1.39–1.45, the iteration is at the lowest number.

References

1. Almurib HA, Kumar TN, Lombardi F. Inexact designs for approximate low power addition by cell replacement. In: 2016 Design, Automation & Test in Europe Conference & Exhibition (DATE), Dresden, Germany. IEEE; 2016. p. 660–5.
2. Jiang H, Han J, Lombardi F. A comparative review and evaluation of approximate adders. In: Proceedings of the 25th edition on Great Lakes Symposium on VLSI. ACM; 2015. p. 343–8.
3. Junqi H, Kumar TN, Abbas H, Lombardi F. A deterministic low-complexity approximate (multiplier-less) technique for DCT computation. IEEE Trans Circuits Syst I Regul Pap. 2019;66(8):3001–14.
4. Gupta V, Mohapatra D, Raghunathan A, Roy K. Low-power digital signal processing using approximate adders. IEEE Trans Comput Aided Des Integr Circuits Syst. 2013;32(1):124–37.
5. Hegde R, Shanbhag NR. Soft digital signal processing. IEEE Trans Very Large Scale Integr VLSI Syst. 2001;9(6):813–23.
6. Chakrapani LN, Muntimadugu KK, Lingamneni A, George J, Palem KV. Highly energy and performance efficient embedded computing through approximately correct arithmetic: a mathematical foundation and preliminary experimental validation. In: 2008 International Conference on Compilers, Architectures and Synthesis for Embedded Systems, Atlanta, GA, USA. ACM; 2008. p. 187–96.
7. Verma AK, Brisk P, Ienne P. Variable latency speculative addition: a new paradigm for arithmetic circuit design. In: Proceedings of the Conference on Design, Automation and Test in Europe. IEEE; 2008. p. 1250–5.
8. Kulkarni P, Gupta P, Ercegovac MD. Trading accuracy for power in a multiplier architecture. J Low Power Electron. 2011;7(4):490–501.
9. Lingamneni A, Enz C, Nagel J-L, Palem K, Piguet C. Energy parsimonious circuit design through probabilistic pruning. In: 2011 Design, Automation & Test in Europe, Grenoble, France. IEEE; 2011. p. 1–6.
10. Kahng AB, Kang S. Accuracy-configurable adder for approximate arithmetic designs. In: Proceedings of the 49th Annual Design Automation Conference, San Francisco, California. IEEE; 2012. p. 820–5.
11. Ye R, Wang T, Yuan F, Kumar R, Xu Q. On reconfiguration-oriented approximate adder design and its application. In: 2013 IEEE/ACM International Conference on Computer-Aided Design (ICCAD), San Jose, CA, USA. IEEE; 2013. p. 48–54.
12. Bouguezel S, Ahmad MO, Swamy M. A low-complexity parametric transform for image compression. In: 2011 IEEE International Symposium of Circuits and Systems (ISCAS), Rio de Janeiro, Brazil. IEEE; 2011. p. 2145–8.
13. Sadhvi Potluri U, Madanayake A, Cintra RJ, Bayer FM, Kulasekera S, Edirisuriya A. Improved 8-point approximate DCT for image and video compression requiring only 14 additions. IEEE Trans Circuits Syst I Regul Pap. 2014;61(6):1727–40.
14. Cintra RJ, Bayer FM, Coutinho VA, Kulasekera S, Madanayake A, Leite A. Energy-efficient 8-point DCT approximations: theory and hardware architectures. Circuits Syst Signal Process. 2016;35(11):4009–29.
15. Bayer FM, Cintra RJ. DCT-like transform for image compression requires 14 additions only. Electron Lett. 2012;48(15):919–21.

16. Jdidia SB, Jridi M, Belghith F, Masmoudi N. Low-complexity algorithm using DCT approximation for POST-HEVC standard. In: Pattern Recognition and Tracking XXIX, Orlando, Florida, United States, vol. 10649. SPIE; 2018. p. 106490Y.
17. Nocedal J, Wright S. Numerical optimization. New York: Springer Science & Business Media; 2006. p. 664.
18. Peyré G. The numerical tours of signal processing – advanced computational signal and image processing. IEEE Comput Sci Eng. 2011;13(4):94–7.
19. Rudin LI, Osher S, Fatemi E. Nonlinear total variation based noise removal algorithms. Physica D. 1992;60(1–4):259–68.
20. Hegde R, Shanbhag NR. Energy-efficient signal processing via algorithmic noise-tolerance. In: Proceedings of the 1999 International Symposium on Low Power Electronics and Design, San Diego, CA, USA. IEEE; 1999. p. 30–5
21. Hegde R, Shanbhag NR. A voltage overscaled low-power digital filter IC. IEEE J Solid-State Circuits. 2004;39(2):388–91.
22. Mohapatra D, Karakonstantis G, Roy K. Significance driven computation: a voltage-scalable, variation-aware, quality-tuning motion estimator. In: Proceedings of the 2009 ACM/IEEE International Symposium on Low Power Electronics and Design, San Francisco, CA, USA. ACM; 2009. p. 195–200.
23. Junqi H, Abbas H, Kumar TN, Fabrizio L. An inexact Newton method for unconstrained total variation-based image denoising by approximate addition. In: IEEE Transactions on Emerging Topics in Computing (Early Access). IEEE; 2021. p. 1.
24. Peyré G. The numerical tours of signal processing. Comput Sci Eng. 2011;13(4):94–7.
25. Vogel C, Oman M. Iterative methods for total variation denoising. SIAM J Sci Comput. 1996;17(1):227–38.
26. Chambolle A, Levine SE, Lucier BJ. Some variations on total variation-based image smoothing. Institute for Mathematics and Its Applications, University of Minnesota; 2009.
27. C. V. Group. CVG-UGR image database. 2019. Available: http://decsai.ugr.es/cvg/index2.php.
28. Gonzalez RC, Woods RE, Eddins SL. Digital image processing using MATLAB. 2nd ed. New Delhi: McGraw Hill Education; 2010. p. 738.
29. Gonzalez RC, Woods RE. Digital image processing. 4th ed. New York: Pearson; 2018. p. 1019.

Chapter 22
Approximate Computation for Baseband Processing

Chuan Zhang and Huizheng Wang

1 Stochastic Computing

As the computing systems become increasingly mobile and embedded, approximate computing (AC) has emerged as a promising paradigm to the low-complexity design of some applications, which are inherently error-tolerant or error-resilient, such as machine learning and digital signal processing [1]. By relaxing the command for fully precise operations [2, 3], the AC usually exhibits substantially improved energy efficiency.

As a special and important branch of AC, stochastic computing, which represents and processes data in the format of pseudo-random bit-streams denoting probabilities, was first proposed in the 1960s and regarded as an alternative to traditional deterministic computing [4]. Compared with the binary counterpart, stochastic-computing-based implementations can provide lower power consumption, lower hardware area overhead, and higher tolerance for soft errors. In stochastic computing, each data x that has been converted into the probability $P(x)$ in range of [0, 1] is represented by an unweighted binary bit-stream where the occurrence probability of bit "1" is equal to $P(x)$. For instance, a binary number $x = 0.011$ that is interpreted as $P(x) = 3/8$ can be represented by a bit-stream 10011000 in which the number of 1s appeared in the bit-stream and the overall length of the bit-stream are three and eight, respectively.

One attractive feature of stochastic computing lies in its strong ability of error tolerance. A casual bit flip in binary representation would result in a severe error, while a bit flip in an unweighted bit-stream can impose only a minute influence on the final value. For instance, the flip of an arbitrary bit in the output stream of

C. Zhang (✉) · H. Wang
Southeast University and the Purple Mountain Laboratories, Jiangsu, China
e-mail: chzhang@seu.edu.cn; huizhwang@seu.edu.cn

W. Liu, F. Lombardi (eds.), *Approximate Computing*,
https://doi.org/10.1007/978-3-030-98347-5_22

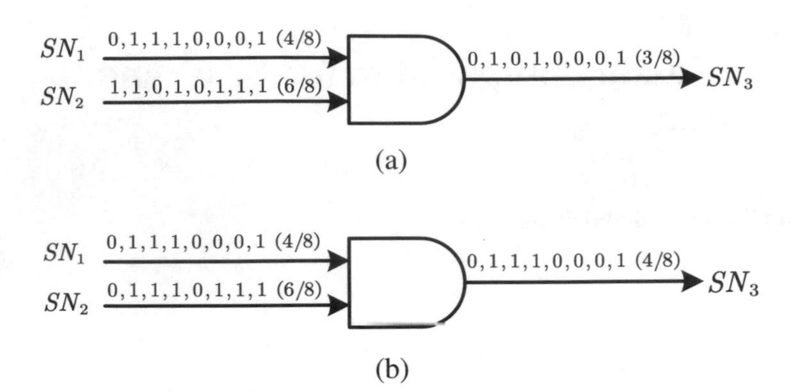

Fig. 22.1 Stochastic multiplier using AND gate: (**a**) exact result and (**b**) approximate result

the AND gate of Fig. 22.1 changes its value from $3/8$ to $2/8$ or $4/8$. By contrast, considering the same number $3/8$ with traditional binary form 0.011, a single bit flip would lead to an evident error if it influences a high-order bit. Specifically, a change from 0.011 to 0.111 alters the corresponding value from $3/8$ to $7/8$. Stochastic numbers (SNs) own no high-order bits as such, and the unweighted bit-stream representation makes the stochastic-computing-based numbers more robust to the soft errors, compared with the traditional binary representation.

Simplicity of hardware implementation is another desirable advantage of stochastic computing. Most arithmetic operations can be implemented by low-cost logic in stochastic computing [5, 6]. For example, the multiplication can be performed by a simple AND gate for unipolar stochastic computing. Supposing there are two uncorrelated bit-streams, whose probabilities of observing 1s are p_1 and p_2, respectively, the probability of 1s at the output of the AND gate is $p_1 \times p_2$. Specifically, as illustrated in Fig. 22.1, the inputs of the AND gate stand for probabilities $4/8$ and $6/8$, respectively. As shown in Fig. 22.1a, we can obtain an output bit-stream representing $4/8 \times 6/8 = 3/8$. Additionally, Fig. 22.1b depicts other two possible stochastic-computing representations of the same inputs $4/8$ and $6/8$. As can be seen, the output bit-stream denotes $4/8$, which introduces negligible fluctuated error but still can be interpreted as an approximated result to the exact product $3/8$.

Based on the advantages aforementioned, stochastic computing has been successfully applied to those fields that do not require exact computation, including artificial neural networks (ANNs) [7–9], the decoding of modern error-correcting codes (ECCs) [10–13], and some computation-intensive scenarios, such as multiple-input multiple-output (MIMO) detection [14, 15], convolution computation [16, 17], and image processing [18]. Some of these applications are characterized by the requirement for a large number of arithmetic operations, which can take advantage of the simple circuits provided by stochastic computing. Additionally, the other applications exhibit low-accuracy requirements for the final results, which can avoid utilizing excessively long bit-streams to represent the data values. In this chapter, we will review three successes of stochastic computing in the baseband processing: (1) polar decoding; (2) low-density parity check (LDPC) decoding; and (3) MIMO detection. Admittedly, the existing works are not limited to the above three parts. However, due to the space limit, not all works are discussed in this chapter. If readers are interested in more details, they can check the corresponding papers for reference.

2 Stochastic-Computing-Based Decoders

In this section, we will introduce two stochastic-computing-based ECCs, namely polar decoding and LDPC decoding.

2.1 *Stochastic Polar Code Decoder*

Polar codes have attracted recent attention because they are the first provable capacity-achieving channel codes with an explicit construction. In general, polar codes can be decoded exerting either successive-cancellation (SC) algorithm or belief propagation (BP) algorithm. The SC decoding was proposed initially [19], while successive-cancellation list (SCL) decoding was proposed to further enhance the performance of SC decoding [20, 21]. Although the deterministic-computing-based SC decoder has achieved spectacular success in recent years, it is facing serious challenges in the current nanoscale CMOS era that has stringent constraints on area overhead and power and energy consumption. To this end, stochastic-computing-based SC decoder is a promising alternative to deterministic SC decoder.

2.1.1 Stochastic SC (SSC) Decoder

Polar Codes

Exerting the channel polarization, polar codes were proposed and developed in [19]. Typically, the decoding for (n, k) polar codes is composed of two major steps, where n, k denote the code length and the number of information bits, respectively. First, construct the intermediate vector $\mathbf{u} = (u_1, u_2, \ldots, u_n)$ from a length-k source message. Considering the reliability of bits is polarized according to their positions in the polar codeword, a great polar code encoder usually allocates the bits of source message to the most reliable k positions and forces the other $(n - k)$ positions as bit “0”. Next, $\mathbf{u}$

(continued)

is multiplied by a generator matrix **G** to attain the transmitted polar codeword as $\mathbf{x} = \mathbf{uG}$.

Deterministic SC Decoding

At the receiver, the received codeword $\mathbf{y} = (y_1, y_2, \ldots, y_n)$ is usually a damaged version of the transmitted polar codeword $\mathbf{x}$. Exerting the likelihood ratio (LR) of y_i, conventional deterministic SC decoders execute decoding procedure to recover **u**. Figure 22.2 exemplifies an SC decoding procedure for $n = 4$ polar codes. It can be observed that the SC decoder is composed of two fundamental nodes, namely **f** node and **g** node, respectively. It is noteworthy that the operations of the two nodes are built on the deterministic calculations that are represented by Eqs. (22.1) and (22.2), respectively. Additionally, in Fig. 22.2, the number in each node is the times index that denotes when the node is executed. Further, at the 2nd, 3rd, 5th, and 6th clock cycles, either the **f** node or **g** node in the stage 2 sends the LR values to the hard-decision units to calculate the final outputs.

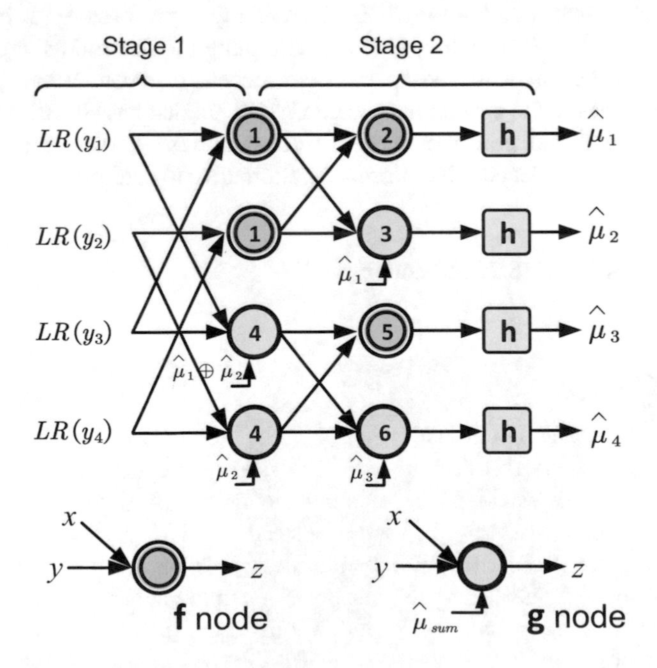

Fig. 22.2 The illustration for LR-based SC decoding for $n = 4$

$$f(x, y) = \frac{1 + xy}{x + y}. \tag{22.1}$$

$$g(x, y, \hat{u}_{\text{sum}}) = x^{1-2\hat{u}_{\text{sum}}} y. \tag{22.2}$$

Channel Message Conversion

Since stochastic computing processes values in probability domain, to design a stochastic SC decoder, it is inevitable to convert the original channel information into probability form. Considering the channel messages are always obtained in log-likelihood ratio (LLR) format, the likelihood information can be deduced by exerting the LLR as Eq. (22.3):

$$Pr(y_i = 1) = \frac{1}{1 + \mathrm{e}^{-LLR(y_i)}}. \tag{22.3}$$

Since the $Pr(y_i = 1)$, i.e., probability is always within range [0, 1], it can be legitimately represented by a stochastic bit-stream. Therefore, in the stochastic SC decoder, the bit-stream representing $Pr(y_i = 1)$ is set as input instead of the original $LLR(y_i)$.

The Reconstruction for **f** Node in Likelihood Domain

For the compatibility with stochastic bit-streams, the primitive deterministic **f** and **g** nodes need to be transformed into the corresponding stochastic forms accordingly.

According to the function of **f** node that owns two LR-based inputs x and y, as described in Eq. (22.1), and the LR definition, we can rewrite Eq. (22.1) as [10]

$$\begin{aligned} z = \frac{Pr(z=0)}{Pr(z=1)} = f(x, y) = \frac{1 + xy}{x + y} = \frac{1 + \frac{Pr(x=0)}{Pr(x=1)} \frac{Pr(y=0)}{Pr(y=1)}}{\frac{Pr(x=0)}{Pr(x=1)} + \frac{Pr(y=0)}{Pr(y=1)}}, \\ = \frac{Pr(x=0)Pr(y=0) + Pr(x=1)Pr(y=1)}{Pr(x=0)Pr(y=1) + Pr(x=1)Pr(y=0)}. \end{aligned} \tag{22.4}$$

It can be observed from Eq. (22.4) that the output of **f** node is the ratio of numerator to denominator, whose sum exactly equals 1. Therefore, we can further obtain

$$\begin{aligned} P_z \triangleq Pr(z = 1) &= Pr(x = 0)Pr(y = 1) + Pr(x = 1)Pr(y = 0) \\ &= P_y(1 - P_x) + P_x(1 - P_y), \end{aligned} \tag{22.5}$$

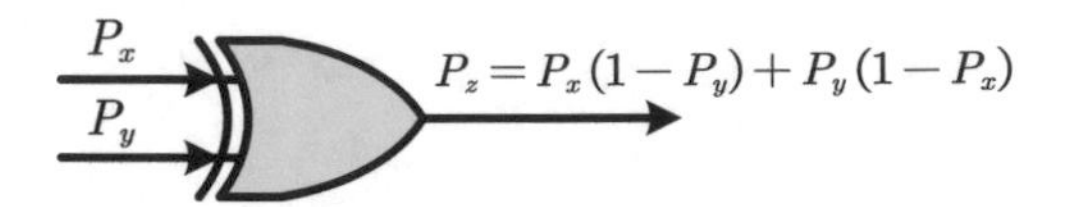

Fig. 22.3 The architecture for stochastic **f** node

where $P_x \triangleq Pr(x = 1)$ and $P_y \triangleq Pr(y = 1)$. Therefore, the mathematical expression of stochastic-computing-based **f** node is shown in Eq. (22.5), where we utilize $P_z \triangleq Pr(z = 1)$ as the output. Further, according to Eq. (22.5), we can observe that it can be implemented by a simple XOR gate, as illustrated in Fig. 22.3.

The Reconstruction for g Node in Likelihood Domain

Likewise, the function of the deterministic **g** node, which is shown in Eq. (22.2), can also be reformulated from LR-based form to the likelihood format. First, considering the case that $\hat{u}_{\text{sum}} = 0$, it can be deduced that:

$$z = \frac{Pr(z=0)}{Pr(z=1)} = g(x, y, 0) = xy = \frac{Pr(x=0)Pr(y=0)}{Pr(x=1)Pr(y=1)}. \tag{22.6}$$

However, it can be observed from Eq. (22.6) that the sum of denominator and numerator is not equal to 1. To this end, Eq. (22.6) is scaled, and the $Pr(c = 1)$ is obtained as follows:

$$\begin{aligned} P_z = Pr(z=1) &= \frac{Pr(x=1)Pr(y=1)}{Pr(x=1)Pr(y=1) + Pr(x=0)Pr(y=0)} \\ &= \frac{P_x P_y}{P_x P_y + (1-P_x)(1-P_y)}. \end{aligned} \tag{22.7}$$

Similarly, for the case that $\hat{u}_{\text{sum}} = 1$, it can be obtained:

$$z = \frac{Pr(z=0)}{Pr(z=1)} = g(x, y, 1) = \frac{y}{x} = \frac{Pr(x=1)Pr(y=0)}{Pr(x=0)Pr(y=1)}. \tag{22.8}$$

Based on this, $Pr(z = 1)$ can be acquired as follows:

$$\begin{aligned} P_z = Pr(z=1) &= \frac{Pr(x=0)Pr(y=1)}{Pr(x=0)Pr(y=1) + Pr(x=1)Pr(y=0)} \\ &= \frac{(1-P_x)P_y}{(1-P_x)P_y + P_x(1-P_y)}. \end{aligned} \tag{22.9}$$

Therefore, according to Eqs. (22.7) and (22.9), the corresponding hardware architecture for **g** node is designed in Fig. 22.4, which consists of one J-K flip-flop (JK-FF), AND, NAND, inverter, and one multiplexer (MUX).

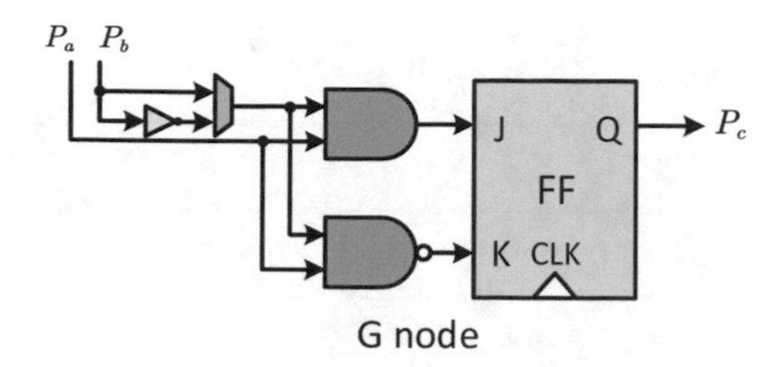

Fig. 22.4 The architecture for stochastic **g** node

However, due to the approximation characteristic of stochastic computing, the straightforward applications of stochastic **f** and **g** nodes would result in severe performance degradation. To this end, the authors of [10] further introduce several optimization schemes to enhance the decoding performance of stochastic SC decoder, including:

1. *Channel message scaling*. In this scheme, the original LLR information is scaled by $LLR(y_i)' = \alpha N_0 LLR(y_i)$, where the α is usually 0.5. Then, the scaled LLR information is utilized to generate input bit-streams.
2. *Increasing the length of bit-streams*. The precision of stochastic computing can be enhanced with the increase of the length of bit-streams. In general, a length-2^k-bit-stream can offer $1/2^s$ precision.
3. *Re-randomizing bit-streams*. Considering the randomness of bit-streams is gradually lost in each calculation stage of the stochastic SC decoder, the bit-streams are re-randomized after each iteration.

2.1.2 Stochastic SCL (SSCL) Decoder

Deterministic SCL Decoding
The basic idea of SCL decoder is similar to that of K-best MIMO detector [22]. For the classical SC decoder, only the most likely code bit is retained at each decoding stage. By contrast, the SCL decoder always maintains a list of L code bits at each stage and outputs the most likely codeword when the final step is completed. In other words, the SCL decoder enhances the performance at the expense of linearly increasing hardware. An illustration for the decoding steps of SCL decoder with $L = 2$ is shown in Fig. 22.5.

To narrow the performance gap between SC decoder and maximum likelihood (ML) decoder, the successive-cancellation list (SCL) decoder is proposed in [20, 21], which can be regarded as a generalization of the classical SC decoder. Although the SCL can outperform the SC decoder and achieve near-ML performance, the list size L usually grows sharply in order to realize satisfying performance, thus hindering

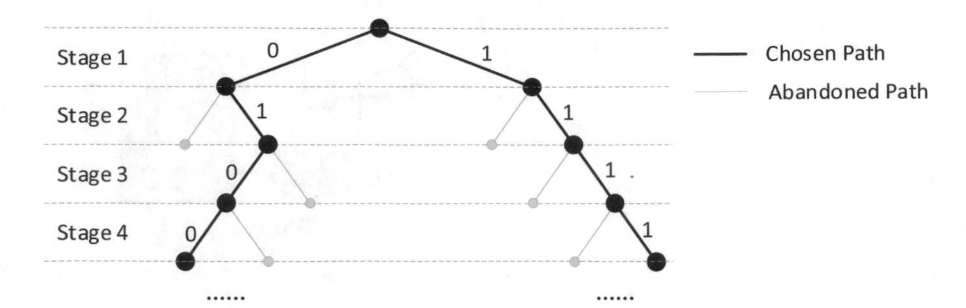

Fig. 22.5 The decoding step of SCL decoder with $L = 2$. Reprinted by permission from Springer Nature Customer Service Centre GmbH: [Springer] [Science China Information Sciences] [11] (Efficient stochastic successive-cancellation list decoder for polar codes, Xiao Liang, Huizheng Wang, Yifei Shen, et al.), © (2021)

its practical application in embedded systems and internet-of-things (IoT). To tackle the issue, the stochastic SCL decoder is proposed in [11, 23].

Since SCL decoder is a generalization of SC decoder, the two basic nodes, i.e., **f** and **g** nodes, are implemented in stochastic domain with the same way mentioned in Sect. 2.1.1. Therefore, the key point lies in how to choose L-best decoding paths at each decoding stage.

To further unify the formulas that describe the functions of **f** node and **g** node, Liang et al. [23] re-conclude the functions of **f** and **g** nodes as Eq. (22.10):

$$\begin{cases} f'(Pr_a, Pr_b) = Pr_f = Pr_a(1 - Pr_b) + Pr_b(1 - Pr_a), \\ g'(Pr_a, Pr_b, \mathbf{b}) = Pr_g = \dfrac{(\mathbf{b}Pr_a + (1-\mathbf{b})(1 - Pr_a))Pr_b}{(1-\mathbf{b})Pr_f + \mathbf{b}(1 - Pr_f)}, \end{cases} \tag{22.10}$$

where **b** is 1 when $\hat{u}_{\text{sum}} = 0$; otherwise $\mathbf{b} = 0$. Further, based on Eq. (22.10), the recursive expression of SC decoding based on bit conditional probability is concluded as Eq. (22.11):

$$\begin{cases} Pr_n^{2i-1}(y_1^n, \hat{u}_1^{2i-2}) = f'\left(Pr_{n/2}^{(i)}(y_1^{n/2}, \hat{u}_{1,o}^{2i-2} \oplus \hat{u}_{1,e}^{2i-2}),\ Pr_{n/2}^{(i)}(y_{n/2+1}^{n}, \hat{u}_{1,e}^{2i-2})\right), \\ Pr_n^{2i}(y_1^n, \hat{u}_1^{2i-1}) = g'\left(Pr_{n/2}^{(i)}(y_1^{n/2}, \hat{u}_{1,o}^{2i-2} \oplus \hat{u}_{1,e}^{2i-2}), Pr_{n/2}^{(i)}(y_{n/2+1}^{n}, \hat{u}_{1,e}^{2i-2}), \mathbf{b}\right). \end{cases} \tag{22.11}$$

Accordingly, the bit conditional probability for each decoding stage can be attained by Eq. (22.12),

$$p(u_i = \hat{u}_i | \hat{u}_1^{i-1}) = \begin{cases} Pr_n^i, & \text{if } i \in \mathcal{A} \text{ and } \hat{u}_i = 1, \\ 1 - Pr_n^i, & \text{if } i \in \mathcal{A} \text{ and } \hat{u}_i = 0, \\ \mathbf{1}_{u_i=0}, & \text{if } i \in \mathcal{A}^c. \end{cases} \tag{22.12}$$

For stochastic SC decoder, as long as the bit conditional probability at each decoding stage is obtained, the hard decision can be performed. However, in stochastic SCL decoder, we have to find L-best paths at each decoding stage. To this end, Liang et al. [23] introduce the path probabilities as the decision metric. Specifically, the path probability in the i-th stage is denoted as $P(\hat{u}_1^i)$, which is calculated by Eq. (22.13):

$$P(\hat{u}_1^i) = P(\hat{u}_1^{i-1})p(u_i = \hat{u}_i|\hat{u}_1^{i-1}) = \prod_{n=1}^{i} p(u_n = \hat{u}_n|\hat{u}_1^{n-1}). \tag{22.13}$$

Specifically, the path probabilities at the i-th decoding stage are obtained by multiplying the path probabilities at the $(i-1)$-th stage with the corresponding bit conditional probability. Additionally, the calculation for the bit conditional probability involves the recursive operations of f' function and g' function. In the successive way, the best L paths are always maintained at each decoding stage, and the best path is picked up at the final stage. An illustration for the calculation of the ordered probabilities for a stochastic SCL decoder with $L = 2$ is depicted in Fig. 22.6. As can be seen, at each decoding stage, the SSCL decoding tree continuously expands paths, updates the corresponding path probabilities, and then picks out the L paths with the largest L path probabilities instead of only maintaining the best path.

Further, it can be observed in Eq. (22.13) that only one AND gate is required to generate the path probability, which is depicted in Fig. 22.7.

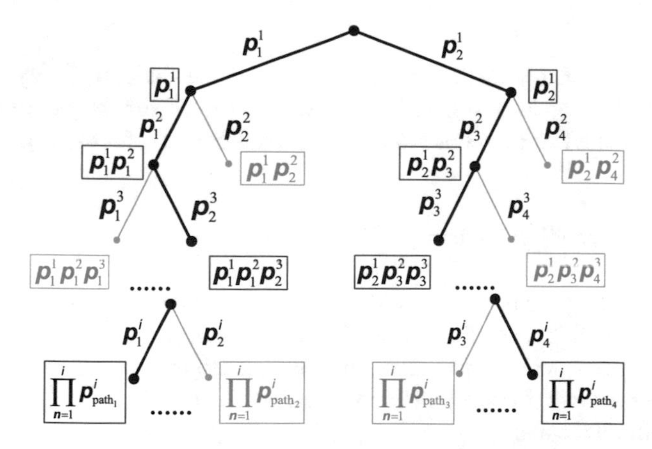

Fig. 22.6 The path probabilities illustration for stochastic SCL decoder with $L = 2$. Reprinted by permission from Springer Nature Customer Service Centre GmbH: [Springer] [Science China Information Sciences] [11] (Efficient stochastic successive-cancellation list decoder for polar codes, Xiao Liang, Huizheng Wang, Yifei Shen, et al.), © (2021)

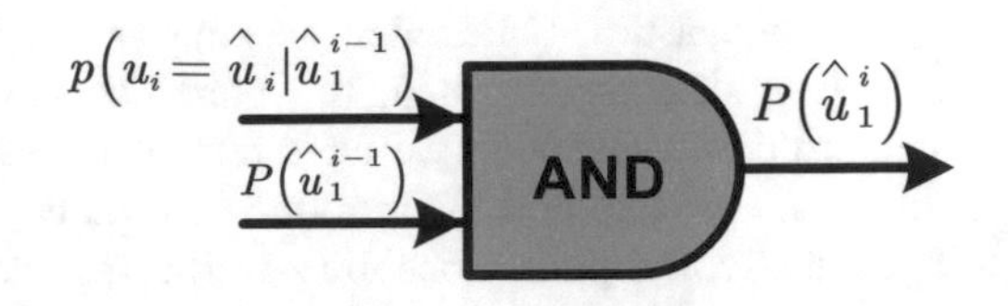

Fig. 22.7 The architecture to calculate the path probability

Doubling Probability Approach

However, since the ordered probability $P(\hat{u}_1^i)$ is the product of i bit conditional probabilities and the value of each bit conditional probability is less than 1, the value of $P(\hat{u}_1^i)$ would shrink evidently with the increase of i. To make matters worse, for stochastic representation, the value of $P(\hat{u}_1^i)$ is represented by the proportion of bit 1 in a fixed-length bit-stream. Therefore, once the value of $P(\hat{u}_1^i)$ becomes extremely small, the bit-stream would lose its randomness and thus resulting in severe performance degradation. To this end, Liang et al. [23] proposed a *doubling probability approach* to alleviate the corresponding degradation.

Specifically, when all ordered probabilities of the chosen paths are less than 0.5, they are doubled once. It is noteworthy that since all the ordered probabilities are only utilized to be compared with each other, doubling them together would not affect the result of path selection. Therefore, by exerting the *doubling probability approach*, the ordered probabilities of the chosen paths are always kept in the region of [0.5, 1].

Distributed Sorting

Distributed sorting is a method to reasonably reduce the complexity of the path sorting in SCL decoding. Figure 22.8 shows a process for the path selection by exerting distributed sorting. For more details, readers can refer to [24].

Double-Level Decoding Method

The long decoding latency is a critical limitation for the stochastic SCL decoder. Direct stochastic SCL decoding makes the decoding for each bit consume considerable clock periods, which equals around to the length of bit-streams. To tackle this issue, a *double-level decoding* method is proposed in [11], in which two bits are estimated simultaneously.

According to Eq. (22.11), only when the values of $Pr_{n/2}^{(i)}\left(y_1^{n/2}, \hat{u}_{1,o}^{2i-2} \oplus \hat{u}_{1,e}^{2i-2}\right)$ and $Pr_{n/2}^{(i)}\left(y_{n/2+1}^{n}, \hat{u}_{1,e}^{2i-2}\right)$ have been obtained, $Pr_n^{(2i-1)}\left(y_1^n, \hat{u}_1^{2i-2}\right)$ can be calculated. Further, only after the estimated value $\hat{u}^{2i-1}$ is acquired, $Pr_n^{(2i)}\left(y_1^n, \hat{u}_1^{2i-1}\right)$ can be further computed. Since there are only two possible

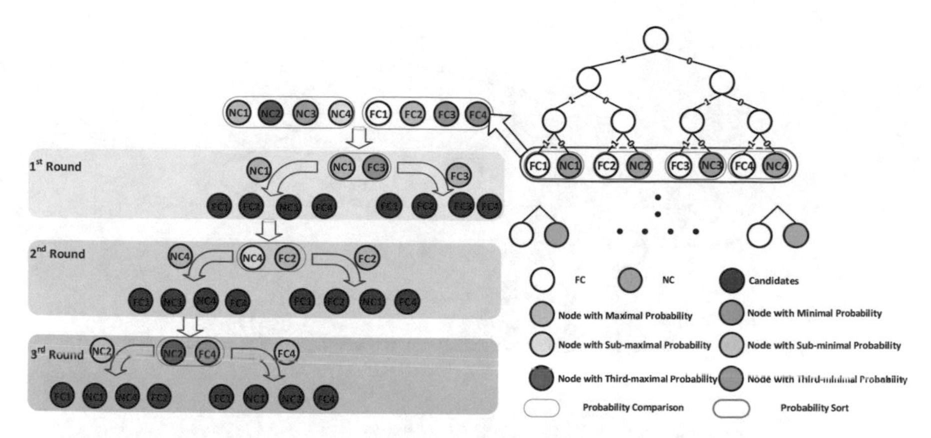

Fig. 22.8 The path selection by distributed sorting ($L = 4$). Reprinted by permission from Springer Nature Customer Service Centre GmbH: [Springer] [Science China Information Sciences] [11] (Efficient stochastic successive-cancellation list decoder for polar codes, Xiao Liang, Huizheng Wang, Yifei Shen, et al.), © (2021)

values for a, i.e., "0" or "1", it is possible to calculate the values of $Pr_n^{(2i)}\left(y_1^n, 0\right)$ and $Pr_n^{(2i)}\left(y_1^n, 1\right)$ in advance based on the two cases. That is to say, the bit-stream-based values of $Pr_n^{(2i-1)}\left(y_1^n, \hat{u}_1^{2i-2}\right)$, $Pr_n^{(2i)}\left(y_1^n, 0\right)$ and $Pr_n^{(2i)}\left(y_1^n, 1\right)$ can be output simultaneously, which also means that two adjacent bit conditional probabilities $p\left(u_{2i-1} = \hat{u}_{2i-1}|\hat{u}_1^{2i-2}\right)$ and $p\left(u_{2i} = \hat{u}_{2i}|\hat{u}_1^{2i-1}\right)$ can be obtained at the same time. Accordingly, the formula as shown in Eq. (22.13) to calculate the path probability is reformulated in the double-level decoding scheme, which is expressed as Eq. (22.14).

$$
\begin{aligned}
P\left(\hat{u}_1^{2i}\right) &= P\left(\hat{u}_1^{2i-2}\right) p\left(u_{2i-1} = \hat{u}_{2i-1}|\hat{u}_1^{2i-2}\right) p\left(u_{2i} = \hat{u}_{2i}|\hat{u}_1^{2i-1}\right) \\
&= \prod_{n=1}^{2i} p\left(u_n = \hat{u}_n|\hat{u}_1^{n-1}\right).
\end{aligned}
\tag{22.14}
$$

An illustration for the double-level decoding with $L = 2$ is shown in Fig. 22.9. In the decoding tree, each father node at the $(2i - 2)$-th level owns two children nodes at the $(2i - 1)$-th level and four sub-children nodes at the $(2i)$-th level. Further, according to distributions of frozen bits and information bits, there are a total of three distribution situations for the double-level decoding scheme. As depicted in Fig. 22.9, the first case indicates that the $(2i - 1)$-th level is a frozen bit, and the $(2i)$-th level is an information bit. By contrast, the second case denotes that the $(2i - 1)$-th level is an information bit, while the $(2i)$-th level is a frozen bit. In addition, the third case means that there are two information bits in both the $(2i - 1)$-

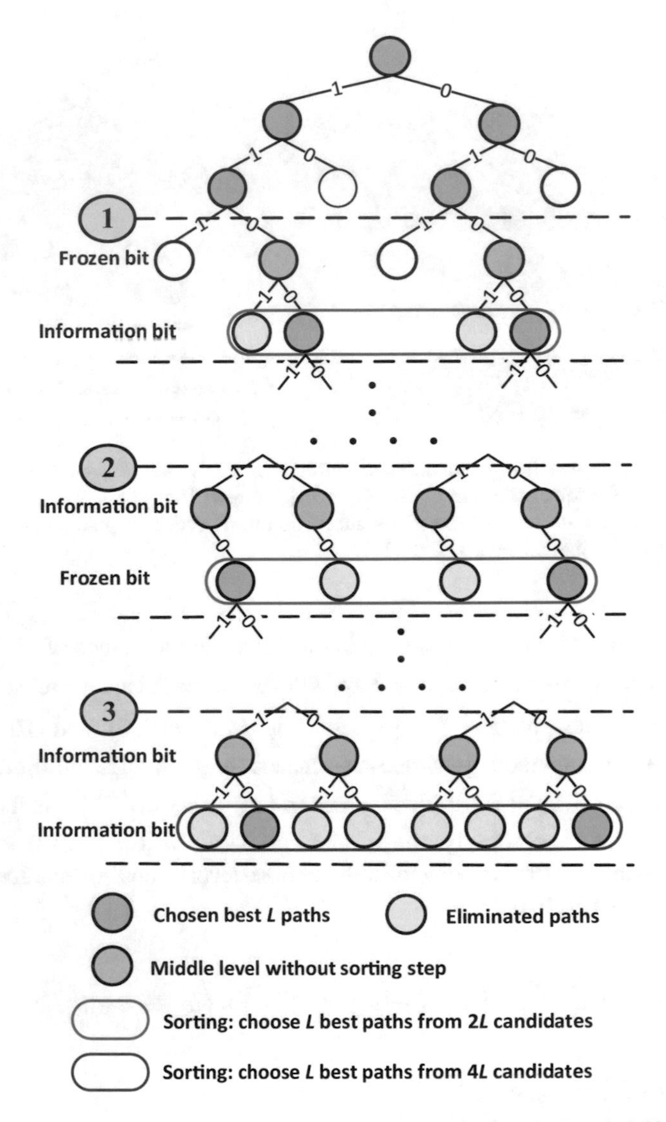

Fig. 22.9 Binary tree of stochastic SCL decoder in double level ($L = 2$). Reprinted by permission from Springer Nature Customer Service Centre GmbH: [Springer] [Science China Information Sciences] [11] (Efficient stochastic successive-cancellation list decoder for polar codes, Xiao Liang, Huizheng Wang, Yifei Shen, et al.), © (2021)

th level and $(2i)$-th level. For the first and second cases, the optimal 2 (L) paths need to be picked up from 4 ($2L$) paths. In contrast, for the third case, the best 2 (L) paths need to be selected from 8 ($4L$) paths. It is noteworthy that conventional distributed sorting [24] is only suitable to pick up L-best paths from $2L$ paths. Therefore, Liang et al. [11] further propose an *adaptive distributed sorting* for the double-level decoding scheme.

Adaptive Distributed Sorting (ADS)

Figure 22.10 provides an illustration for the steps of the *adaptive distributed sorting* algorithm to choose the best L paths from $4L$ candidate paths. Each father node corresponds to four sub-children nodes, which are further classified into one first children (FC) node and three next children (NC) nodes. The FC node denotes the node with the largest ordered probability among other children nodes from one father node:

- First, L FC nodes are extracted, and the remaining $3L$ nodes are NC nodes.
- Next, in the first round of comparison, the NC node (NC1) with the maximal path probability metric (PPM) (requiring $L - 1$ comparisons) and the FC node (FC3) with the minimal PPM (requiring $3L - 1$ comparisons) are first picked

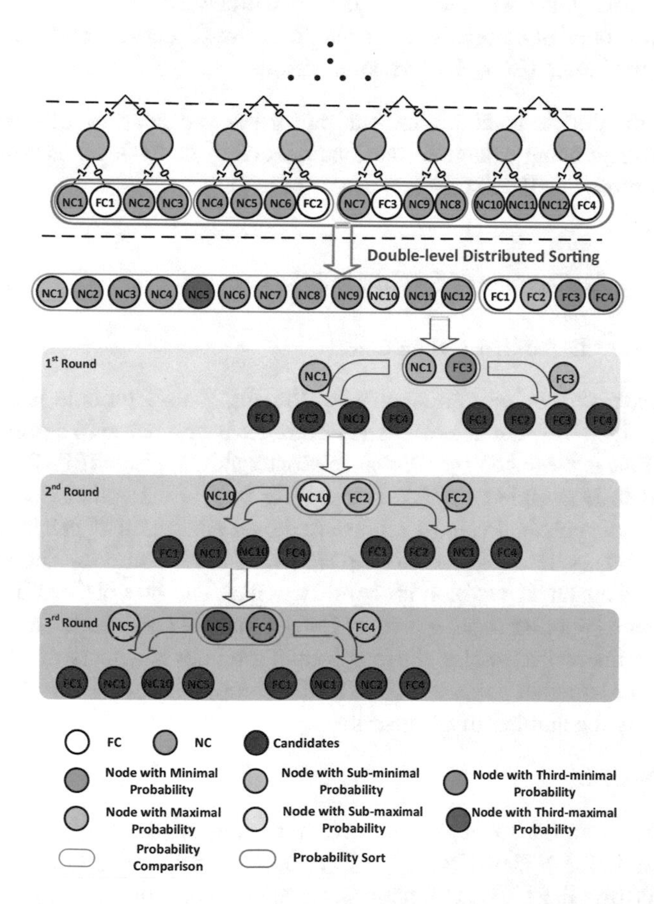

Fig. 22.10 The illustration for paths' selection in SCL decoder with double-level decoding ($L = 4$). Reprinted by permission from Springer Nature Customer Service Centre GmbH: [Springer] [Science China Information Sciences] [11] (Efficient stochastic successive-cancellation list decoder for polar codes, Xiao Liang, Huizheng Wang, Yifei Shen, et al.), © (2021)

out and then compared the NC1 with FC3 (requiring 1 comparison). If the PPM of FC3 is larger than that of NC1, the sorting process will complete with the selected candidates: FC1, FC2, FC3, and FC4. Under this situation, only one round of comparison is required, and the total number of comparisons is $4L-1$. Otherwise, replace the FC3 with NC1, and the two PPMs will not be compared in the next steps.

- The second round of comparison follows a similar pattern. Above all, the NC node (NC10) with sub-maximal PPM (requiring $3L-2$ comparisons) and the FC node (FC2) with sub-maximal PPM (requiring $L-2$ comparisons) are selected out. Next, compare the PPM of FC2 with that of NC10 (requiring 1 comparison). Likewise, if the PPM of FC2 is greater than that of NC10, the sorting procedure will finish with the chosen candidates: FC1, FC2, NC1, and FC4. In this situation, two rounds of comparisons are required with total $8L-4$ comparisons. Otherwise, replace the FC2 with NC10.
- The third round of comparison repeats the above procedure, and the final results are attained after $12L-9$ PPM comparisons.

Further, the double-level scheme can be further expanded to arbitrary 2^p-level, and the corresponding adaptive distributed sorting in 2^p-level is also designed in [11]. For more details, the readers can refer to [11].

The Hardware Architecture of Stochastic SCL Decoder

1. Binary to stochastic (B-to-S) module:

To convert a binary value into stochastic domain, B-to-S module is required. As depicted in Fig. 22.11, the deterministic number x is compared to a pseudo-random number R that is produced by a linear feedback shift register (LFSR). Denote the output of B-to-S module as $x^s(t)$. If $x > R$, $x^s(t) = 1$; otherwise $x^s(t) = 0$. Accordingly, the practical value x represented by a bit-stream x^s can be obtained by $x = \frac{1}{l}\sum_t^l x^s(t)$, where l represents the length of the bit-stream. The input of this module is the channel transition probability, which can be obtained by Eq. (22.3). To decode an n-bit polar code, n parallel B-to-S modules are needed as the interface connected to the main decoder. Since n B-to-S modules can share one LFSR as the pseudo-random number generator (RNG), the complexity of the interface is mainly determined by the number of comparators.

2. Mixed Node I:

To make the overall decoder architecture more concise, the **f** node and **g** node are first merged as the Mixed Node I [11], as illustrated in Fig. 22.12. Although this node can perform the f' and g' functions based on stochastic computing, it is unable to perform the function of double-level decoding. To this end, an improved version called Mixed node II is further proposed in [11].

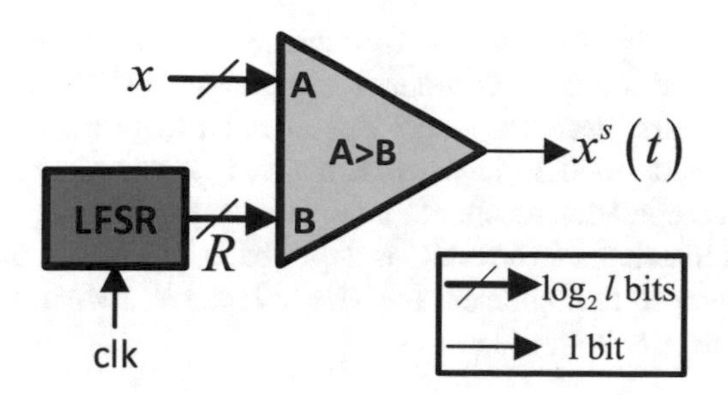

Fig. 22.11 The architecture of B-to-S module. Reprinted by permission from Springer Nature Customer Service Centre GmbH: [Springer] [Science China Information Sciences] [11] (Efficient stochastic successive-cancellation list decoder for polar codes, Xiao Liang, Huizheng Wang, Yifei Shen, et al.), © (2021)

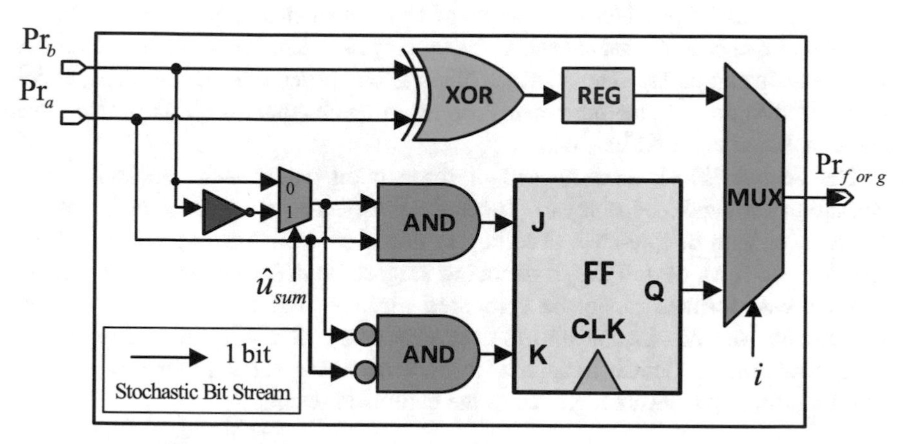

Fig. 22.12 The architecture of the Mixed Node I. Reprinted by permission from Springer Nature Customer Service Centre GmbH: [Springer] [Science China Information Sciences] [11] (Efficient stochastic successive-cancellation list decoder for polar codes, Xiao Liang, Huizheng Wang, Yifei Shen, et al.), © (2021)

3. Mixed Node II

To implement the double-level decoding, the corresponding Mixed Node II is designed as depicted in Fig. 22.13, in which the bit-stream representing the bit conditional probability $p(u_{2i} = \hat{u}_{2i}|\hat{u}_1^{2i-1})$ will be output simultaneously with the bit-stream denoting $p(u_{2i-1} = \hat{u}_{2i-1}|\hat{u}_1^{2i-2})$. Therefore, for the double-level decoding, the Mixed Node I modules at the last stage need to be replaced by the Mixed Node II modules. It is also noteworthy that for 2^p-level decoding, the Mixed Node I modules at last p stages should be replaced by the corresponding Mixed Node II modules accordingly.

4. Mixed Node III

Traditionally, to implement an SCL decoder with list size L, L parallel basic SC decoders need to be configured. However, the direct parallel design would increase the hardware complexity dramatically. By exerting the characteristics of the outputs for the first stage of the SCL decoder, Liang et al. [11] simplify the hardware architecture of the first decoding stage. Specifically, there are only three possible outputs for each node module in the first stage, i.e., one **f** node output and two possible **g** node computation results. In other words, the first stage does not require the L parallel architecture. As a result, as depicted in Fig. 22.13, the architecture in green box is employed to implement the Mixed Node III, which will be applied in the first stage of the SSCL decoder.

5. Architecture of Stochastic SC Decoder with Double-level Decoding

As shown in Fig. 22.14, the architecture of the stochastic SC decoder with double-level decoding scheme consists of two main blocks: interface block and decoder block. Figure 22.14 provides an example of an 8-bit stochastic SC decoder, and the design idea can be extended to design an arbitrary n-bit stochastic SC decoder based on stochastic computing. The interface block is composed of n SG modules, which are utilized to produce the bit-streams representing channel transition probabilities as the inputs of the decoder block.

The decoder block is composed of three main parts: node processing part, estimation part, and feedback part. The node processing part for an n-bit stochastic SC decoder with double-level decoding requires $(n - 2)$ Mixed Node I and one Mixed Node II on overall $\log_2 n$ decoding stages. Considering the values denoted by bit-stream formats cannot be compared directly, four counters are employed to transform the bit-streams into the corresponding deterministic binary values. Next, a maximum module is applied to pick out the greatest probability, and its corresponding two bits are regarded as the estimated results.

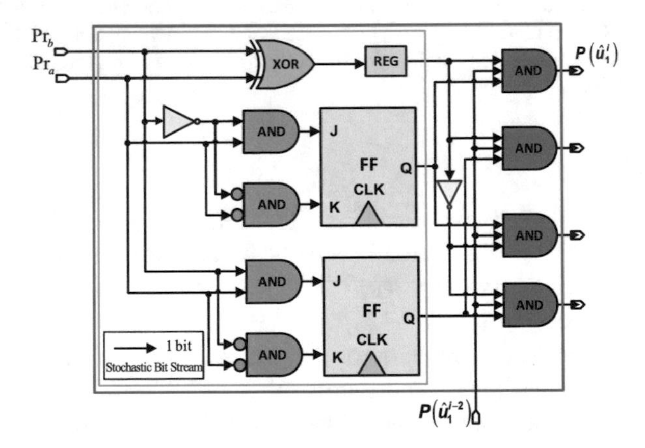

Fig. 22.13 The architecture of the Mixed Node II. Reprinted by permission from Springer Nature Customer Service Centre GmbH: [Springer] [Science China Information Sciences] [11] (Efficient stochastic successive-cancellation list decoder for polar codes, Xiao Liang, Huizheng Wang, Yifei Shen, et al.), © (2021)

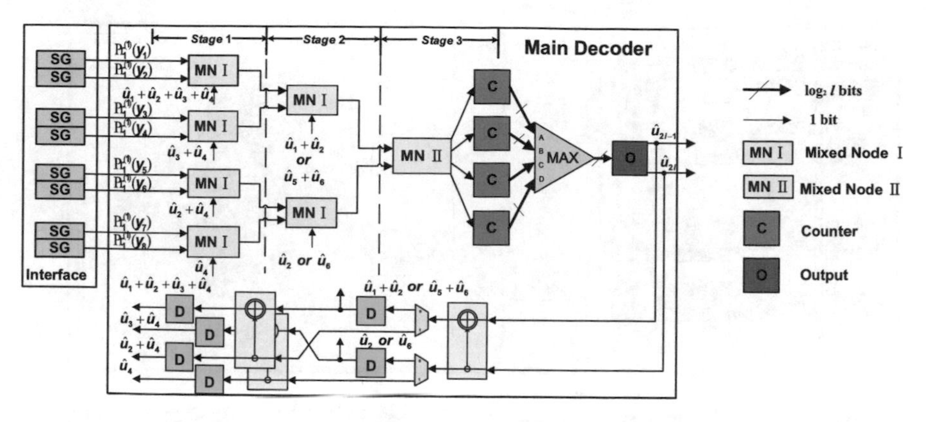

Fig. 22.14 The 8-bit stochastic SC decoder with double-level decoding. Reprinted by permission from Springer Nature Customer Service Centre GmbH: [Springer] [Science China Information Sciences] [11] (Efficient stochastic successive-cancellation list decoder for polar codes, Xiao Liang, Huizheng Wang, Yifei Shen, et al.), © (2021)

The function of feedback part is to generate the partial sum signals $\hat{u}_{\text{sum}}$ for the calculations of **g** nodes. For more details about the feedback architecture, the readers can refer to [25]. Additionally, it is noteworthy that the clock frequency utilized in the interface part, node processing part, and estimation part is l times that of the feedback part.

5. Architecture of SSCL Decoder with Double-level Decoding

The architecture of the SSCL decoder with double-level decoding is depicted in Fig. 22.15, whose basic components are similar to that of stochastic SC decoder. The interface block consists of n SG modules as well, while the decoder block is extended to L paralleling counterparts. Since each **g** node in the first stage can only produce two possible results, it is unnecessary for the first stage of the decoder block to adopt L paralleling design. Therefore, only $n/2$ Mixed Node III modules are required in the first stage, while all nodes in the stage 2 to stage m employ the parallel scheme, where $m = \log_2 n$. Different from the stochastic SC decoder, in the estimation module of stochastic SCL decoder, the list core (LC) module is designed to implement the adaptive distributed sorting method to pick up L-best paths among $4L$ candidates, which will be introduced later.

6. The LC module

The function of LC module is to implement the ADS and enlarging probability schemes. First, as shown in Fig. 22.16, to realize the ADS, the PPMs are divided into two categories: NCs and FCs. Next, the $3L$ NCs are stored in memory block I, while the L FCs are saved in memory block II. As the previous descriptions for the ADS scheme, PPMs in memory block II own higher priorities to be chosen compared to those in memory block I. Then, for each round of comparison, two MUXes read data from the memory block I and memory block II in a certain order. Specifically,

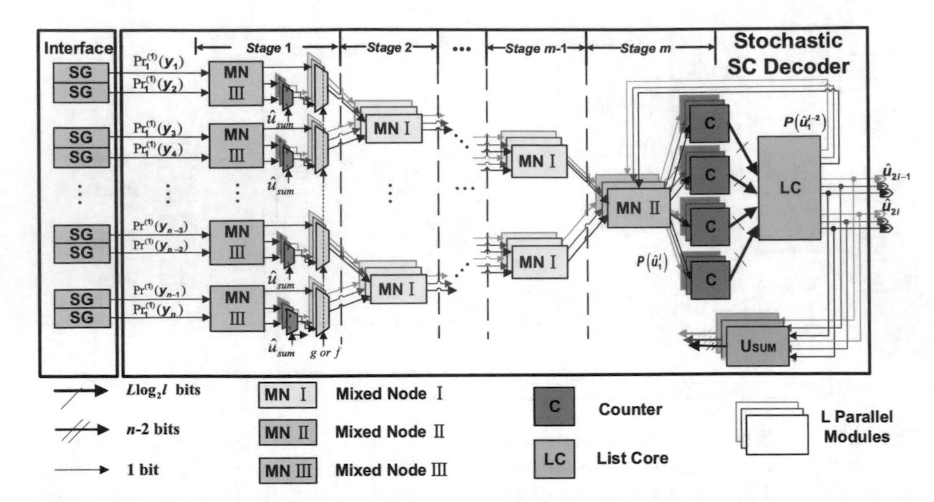

Fig. 22.15 The n-bit SSCL decoder with double-level decoding. Reprinted by permission from Springer Nature Customer Service Centre GmbH: [Springer] [Science China Information Sciences] [11] (Efficient stochastic successive-cancellation list decoder for polar codes, Xiao Liang, Huizheng Wang, Yifei Shen, et al.), © (2021)

select the NC_i with the maximum PPM value from the memory block I and the FC_j with the minimum PPM value from the memory block II. If the NC_i is greater than FC_j, the control signal is activated to perform the instructions ① and ②. Instruction ① is responsible for removing the NC_i from memory block I, while instruction ② is responsible for replacing the value of FC_j with the value of NC_i. Likewise, the second round of comparison is based on the updated memory blocks. If FC_j is greater than NC_i, the sorting process is finished. The values in memory block II are the L optimal PPMs, and execute the instruction ③. Instruction ③ indicates the end of sorting and starts the probability enlarging scheme. Then, L optimal paths corresponding to the L-best PPMs are output.

When instruction ③ is carried out to perform the probability enlarging scheme, first, it is necessary to judge whether the condition for probability enlarging is met. If the best PPM among the selected paths is smaller than $l/2$, which means that the corresponding probability is smaller than 0.5, instruction ④ is performed to execute left shift operations for all PPMs in the memory block II. Next, instruction ⑤ regenerates the bit-streams corresponding to the L new PPMs. It is noteworthy that if the best PPM is still greater than $l/2$, the instruction ④ is skipped and instruction ⑤ is executed directly.

2.1.3 Stochastic BP (SBP) Decoder

Although stochastic SC and SCL decoders deliver a significant advantage in overhead reduction compared to the deterministic counterparts, there is still one

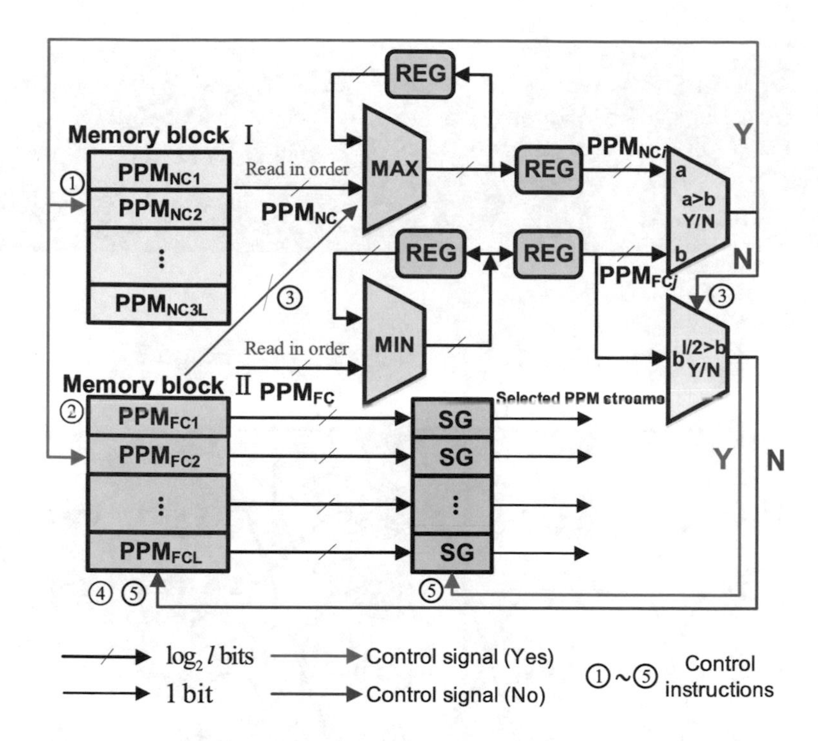

Fig. 22.16 The architecture of LC module. Reprinted by permission from Springer Nature Customer Service Centre GmbH: [Springer] [Science China Information Sciences] [11] (Efficient stochastic successive-cancellation list decoder for polar codes, Xiao Liang, Huizheng Wang, Yifei Shen, et al.), © (2021)

crucial drawback for SC and SCL decoding. That is, both SC and SCL decoding work in a serial way, meaning their throughput is potentially lower than that of parallel schemes, such as BP decoding that deploys on a fully parallel factor graph. Therefore, the parallel computing property of BP decoding makes it more suitable for stochastic-computing-based implementations.

Deterministic BP Decoding

The deterministic BP decoding was proposed in [26] to decode (n, k) polar codes by exerting their factor graph containing $\log_2 n$ stages and $n(\log_2 n + 1)$ nodes, while each stage owns $n/2$ basic computation blocks. For example, Fig. 22.17 exemplifies the factor graph for $n = 8$, in which each node is assigned with a coordinate (i, j) where i denotes the stage number and j indicates the node number in the column. The decoding scheme is the process of passing messages iteratively through the factor graph. In each

(continued)

iteration, the left-to-right and right-to-left soft messages between adjacent nodes are updated by the basic computation blocks, each of which contains an F node and a G node, and the LR-based function of the basic computation block is expressed by Eq. (22.15), where $f_{BP}(x, y) = (1 + xy)/(x + y)$, $g_{BP}(x, y) = x \cdot y$.

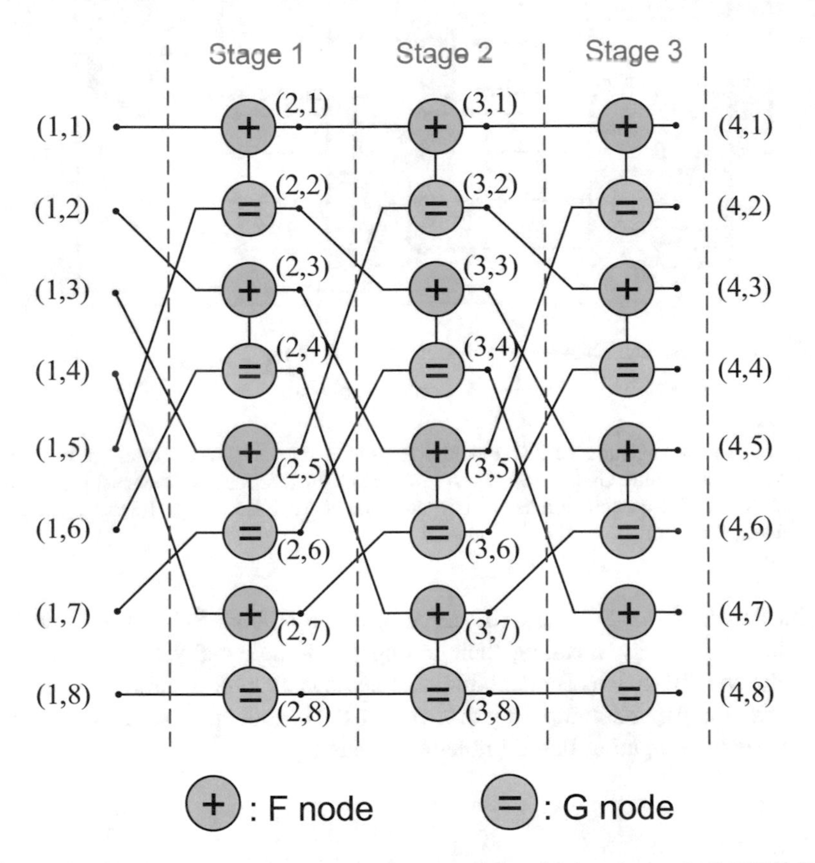

Fig. 22.17 The factor graph of BP decoding for polar codes with length $n = 8$. © [2021] IEEE. Reprinted, with permission, from Ref. [27]

$$
\begin{cases}
L_{i,j} = f_{\mathrm{BP}}\left(L_{i+1,2j-1}, g_{\mathrm{BP}}\left(L_{i+1,2j} R_{i,j+N/2}\right)\right), \\
L_{i,j+N/2} = g_{\mathrm{BP}}\left(f_{\mathrm{BP}}\left(R_{i,j}, L_{i+1,2j-1}\right), L_{i+1,2j}\right), \\
R_{i+1,2j-1} = f_{\mathrm{BP}}\left(R_{i,j}, g_{\mathrm{BP}}(L_{i+1,2j} R_{i,j+N/2})\right), \\
R_{i+1,2j} = g_{\mathrm{BP}}\left(f_{\mathrm{BP}}\left(L_{i,j+1}, R_{i,j}\right), R_{i+n/2^j,j}\right).
\end{cases}
\tag{22.15}
$$

Reformulation for G and F Nodes

The crucial step to design an SBP decoder is to reformulate Eq. (22.15) into stochastic form. Specifically, it can be observed that the F node is the same as the **f** node in SC decoding, which has been introduced in Sect. 2.1.1. Therefore, the F node can also be implemented by an XOR gate. Different from the **g** node in SC decoding, the G node has no feedback signal $\hat{u}_{\mathrm{sum}}$ as input; hence, it can be regarded as a particular case of the **g** node but only needs to calculate $x \cdot y$. Therefore, it can be implemented with the same architecture for the **g** node of stochastic SC decoder, but without the inverter and multiplexer, which is illustrated in Fig. 22.18.

Schemes to Enhance the Performance of SBP Decoder

Further, the authors of [12] also proposed an efficient scheme called *splitting bit-stream-based re-randomization*, to enhance the decoding performance and decrease the decoding latency of the SBP decoder. First, all the bit-streams of the SBP decoder are split into s segments. Accordingly, each basic computation block is also duplicated by s copies to process all the segments in parallel and simultaneously. As a result, the decoding latency is decreased by s times at the expense of s times increase in area cost, which is acceptable as the ultra-small footprint of stochastic implementations provides a sufficient area budget. Additionally, those segments are shuffled after each iteration to assure the bit-streams to re-gain the randomness. As the shuffling only introduces small footprint of wire, the proposed *splitting bit-stream-based re-randomization* can effectively enhance the randomness of SBP decoder at the cost of small overhead.

However, the re-randomization of the strategy that bit-streams are re-randomized between two neighboring iterations is not sufficient and in time. To this end, [27] further proposed a *stage-wise re-randomization* scheme, which keeps re-

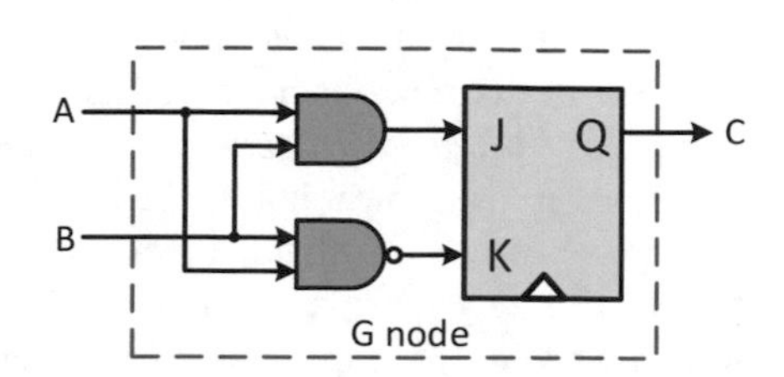

Fig. 22.18 The architecture for stochastic G node

randomizing all bit-streams after each stage during iterations. Table 22.1 shows the specific schedule of the stage-wise re-randomization scheme. The bit-streams in red nodes are updated in the first step; hence, they are re-randomized first. Then, bit-streams in blue nodes are updated in the second step, so they are also re-randomized as well. Likewise, all bit-streams are re-randomized in each stage instead of after each iteration.

Moreover, to accelerate the re-randomization process, [27] further proposed an efficient re-randomization hardware architecture using directive register, which is shown in Fig. 22.19. Specifically, supposing there are two M-bit bit-streams A and B, they are both divided into 4 segments and each segment has $M/4$ bits:

(a) In the first $M/4$ clock cycles, the first segments A.1 and B.1 of the bit-streams A and B are stored in the directive register, respectively.
(b) In the $(M/4+1)$ to $M/2$ clock cycles, A.1 is moved out of the directive register in bit order and gets involved in the next stage of decoding. The vacant positions in the directive register are filled in bit order by the second segment A.2 of the bit-stream A; on the other hand, the first segment B.1 of bit-stream B remains in the directive register, and the second segment B.2 of bit-stream B is directly aligned with A.1 and enters the next stage of decoding.
(c) In the $M/2+1$ to $3M/4$ clock cycles, the second segment A.2 of bit-stream A and the first segment B.1 of bit-stream B are removed from the directive register in bit order and enter the next stage of decoding. Then, the vacant positions in the directive register are filled in bit order by the third bit segments A.3 and B.3 of the bit-streams A and B.
(d) Similar to the step 2, in the $3M/4+1$ to M clock cycles, the third segment A.3 of bit-stream A is moved out of the directive register in bit order and gets involved in the next stage of decoding. The vacant positions in the directive register are filled in bit order by the fourth segment A.4. Then, the third segment B.3 of bit-stream B remains in the directive register, and the fourth segment B.4 of bit-stream B is directly aligned with A.3 and enters the next stage of decoding.
(e) In the $M+1$ to $5M/4$ clock cycles, the fourth segment A.4 of bit-stream A and the third segment B.3 of bit-stream B are moved out of the boot register in bit order for the next stage of decoding.

Partial Stage-Wise Re-randomization

Recently, according to the research in [28], the correlation among bit-streams in stochastic BP decoding is almost resulted by the G nodes, while F nodes can alleviate the correlation to a certain extent. Therefore, to reduce the hardware complexity, a *partial stage-wise re-randomization* scheme was proposed in [28]. Specifically, as shown in Fig. 22.20, the bit-streams with green circles are re-randomized when the channel messages pass from left to right. By contrast, while the channel messages pass from right to left, those bit-streams with blue circles are re-randomized. Since only those bit-streams output by G nodes are re-randomized,

Table 22.1 Schedule of the stage-wise re-randomization. © [2021] IEEE. Reprinted, with permission, from Ref. [27]

(1, 1)	(1, 2)	(1, 3)	(1, 4)	...	$(1, 1+n/2)$	$(1, 2+n/2)$	...	$(1, n)$
Re-randomization								
(2, 1)	(2, 2)	(2, 3)	(2, 4)	...	$(2, 1+n/2)$	$(2, 2+n/2)$	...	$(2, n)$
Re-randomization								
(3, 1)	(3, 2)	(3, 3)	(3, 4)	...	$(3, 1+n/2)$	$(3, 2+n/2)$	...	$(3, n)$
......								
$(m^{①}+1, 1)$	$(m+1, 2)$	$(m+1, 3)$	$(m+1, 4)$	...	$(m+1, 1+n/2)$	$(m+1, 2+n/2)$	...	$(m+1, n)$

① $m = \log n$

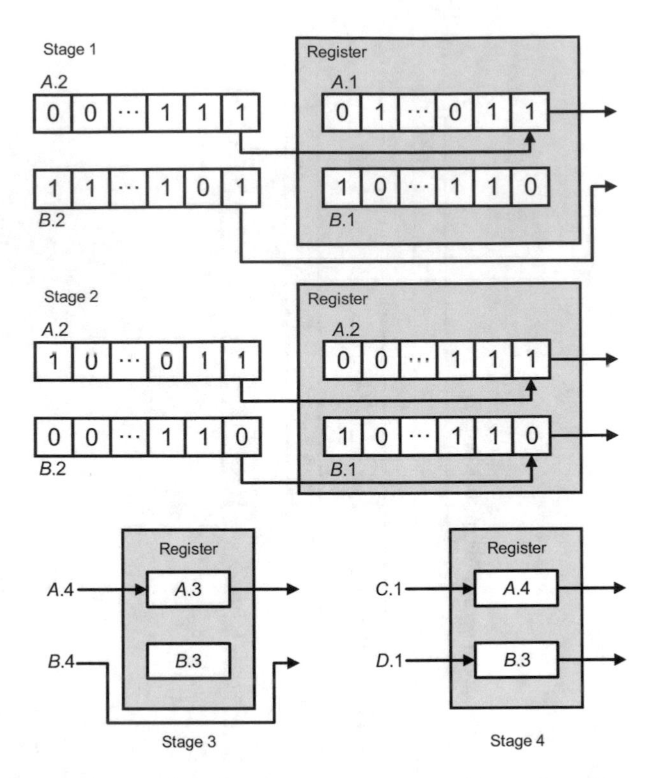

Fig. 22.19 Process of the exchanging in directive register. © [2021] IEEE. Reprinted, with permission, from Ref. [27]

the scheme achieves around 50% overhead reduction compared to the original *stage-wise re-randomization* scheme.

2.2 Stochastic LDPC Decoder

LDPC codes are another attractive solution for the ECCs, not only due to their close to Shannon's limit performances [29] but also the corresponding high-parallelism decoding algorithms [30, 31]. The LDPC decoding exerts a probabilistic scheme that passes messages on the Tanner graph while iteratively performing two basic operations, namely *parity checking* and *equality checking* [32]. Additionally, it has been pointed out in [33] that the two operations can also be efficiently implemented by the same stochastic circuits as that of the BP decoder, which are shown in Figs. 22.3 and 22.18, respectively.

In addition to lowering the overhead of basic processing units, stochastic computing can also decrease the routing area. For traditional binary implementations, the communication for a probability value with k bit width between two nodes requires

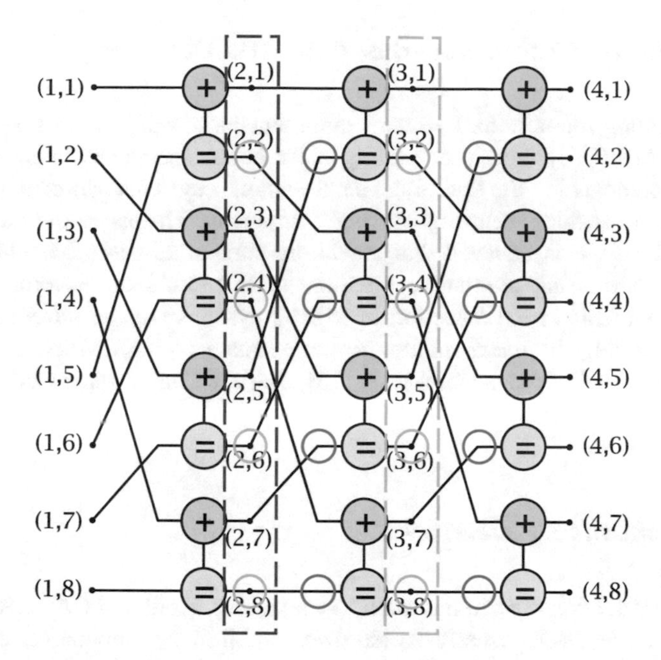

Fig. 22.20 Schedule of the partial stage-wise re-randomization with $N = 8$. © [2021] IEEE. Reprinted, with permission, from Ref. [28]

k wires, while only one wire is needed in stochastic-computing-based implementations due to the bit-serial property of stochastic computation. Furthermore, another advantage of stochastic computing lies in its support of asynchronous pipelines. To be more specific, due to the bit-stream-based representations for values, we only care about the proportion of "1" bits instead of the bit order. Therefore, in stochastic LDPC designs, the inputs of variable nodes and parity nodes do not need to be the output bits of the previous cycles. Based on this, Tehrani et al. [34] proposed an efficient fully parallel stochastic LDPC decoder, which allocates different edges with different numbers of pipeline stages. For decoding an irregular (1056, 528) code, implementation results on FPGA suggest that the stochastic decoder can achieve 1.66 Gb/s throughput and 0.5 dB loss compared to the float-point sum–product algorithm (SPA) at BER of 10^{-8}.

However, to accelerate the convergence, Tehrani et al. [34] utilized a module named edge memory (EM) for each edge in the Tanner graph, thus resulting in huge overhead burden. To alleviate this burden, tracking forecast memories (TFMs) was introduced in [35], which can deliver much less hardware complexity compared with EMs and, hence, considerably decrease the overall complexity. Moreover, Tehrani et al. [36] proposed the majority-based tracking forecast memory (MTFMs) to further reduce the overhead.

3 Stochastic-Computing-Based MIMO Detector

By transmitting multiple data streams simultaneously within the same frequency band, the MIMO systems successfully offer better spectral efficiency and link reliability compared to the traditional single-antenna systems. Unfortunately, these benefits come at the expense of soaring complexity. Thanks to the advantage in low complexity, it is believed that stochastic-computing-based MIMO detection provides a promising alternative solution to this problem. Several successful applications of SC in the detection of MIMO systems will be introduced in this section, including the linear minimum mean square error (MMSE) detector [37] and non-linear belief propagation (BP) [15], Markov Chain Monte Carlo (MCMC) detectors [14].

3.1 Stochastic BP Detector

Traditional detection algorithms, such as maximum likelihood (ML), K-best, and sphere decoding (SD), usually suffer from prohibitive computation complexity, which hinders their practical applications in large-scale MIMO detection. To tackle this bottleneck, some low-complexity algorithms such as reactive tabu search (RTS), likelihood ascent search (LAS), and BP algorithms are proposed. Among these candidates, BP can provide better performance and is more robust, which means it is less vulnerable to the local minimum problem. However, BP detection still has the potential to further reduce its hardware complexity by exerting stochastic computing.

Deterministic BP Detector for MIMO Systems
Consider a MIMO uplink model with M transmitting antennas and N receiving antennas in real domain as the following:

$$\mathbf{r} = \mathbf{H}\mathbf{x} + \mathbf{n},$$

where $\mathbf{r} = [r1, r2, \ldots, r_{2N}]^{\mathrm{T}}$, $\mathbf{x} = [x1, x2, \ldots, x_{2M}]^{\mathrm{T}}$, and $\mathbf{H} = \{h_{j,i}\}_{1\le j\le 2N, 1\le i\le 2M}$. $\mathbf{n}$ denotes the additive white Gaussian noise (AWGN) and $\mathbf{n} \sim \mathcal{N}(0, \sigma^2\mathbf{I}_{2\mathrm{N}})$. Furthermore, the transmitted signals $\{x_1, x_2, \ldots, x_{2M}\}$ and received signals $\{r_1, r_2, \ldots, r_{2N}\}$ are regarded as "symbol nodes" and "observation nodes," respectively. The messages transmitting from x_i to r_j are denoted as $\boldsymbol{\alpha}_{x_i\to r_j}$, and the messages from r_j to x_i are represented by $\boldsymbol{\beta}_{r_j\to x_i}$, respectively. The essence of BP algorithm lies in the message passing between symbol nodes and observation nodes, and a brief message procedure is depicted in Fig. 22.21. For more details about deterministic BP detection, the readers please refer to [38].

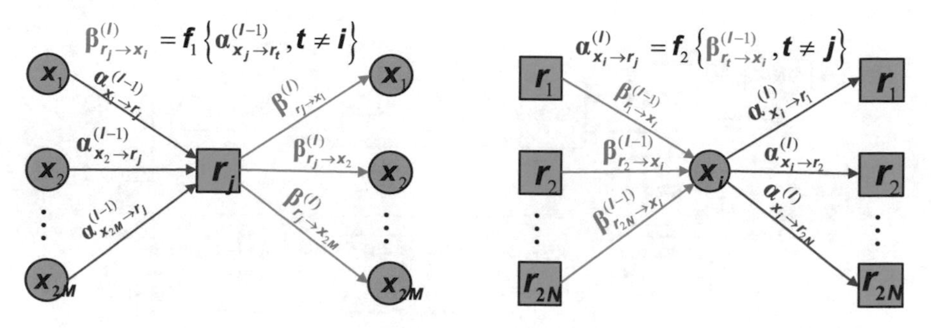

Fig. 22.21 Message passing between variable nodes and observation nodes. © [2021] IEEE. Reprinted, with permission, from Ref. [15]

3.1.1 The Stochastic Architecture for Observation Nodes

According to the conversion from deterministic BP detection [38] into stochastic version, the stochastic message updating function for observation nodes is expressed by Eq. (22.16):

$$\beta_{j,i}^{s} = \frac{\varepsilon^{s} h_{j,i}^{s}\left(r_{j}^{s} - \mu_{z_{j,i}}^{s}\right)}{\left(\sigma_{z_{j,i}}^{2}\right)^{s}} \approx \frac{\varepsilon^{s} h_{j,i}^{s}\left(r_{j}^{s} - \mu_{z_{j,i}}^{s}\right)}{\varepsilon^{s} h_{j,i}^{s}\left(r_{j}^{s} - \mu_{z_{j,i}}^{s}\right) + \left(\sigma_{z_{j,i}}^{2}\right)^{s}}, \tag{22.16}$$

in which the superscript $(\cdot)^{s}$ denotes it is in stochastic bit-stream format. Especially, the $\mu_{z_{j,i}}$ and $\sigma_{z_{j,i}}$ represent the marginal mean and variance of an interference term, which is assumed to obey the Gaussian distribution. Further, to implement the stochastic division by a JK-FF, a scaled factor ε is introduced to ensure $|\varepsilon \times h_{j,i}(r_j - \mu_{z_{j,i}})| \ll \sigma_{z_{j,i}}^{2}$, and the empirical value for ε is $1/32$.

Based on stochastic computing, the overall architecture hardware for the message updating in observation nodes is depicted in Fig. 22.22, in which the RR and -X represent the re-randomization unit and the bit flipping unit, respectively. From the left side to right, first, the stochastic number generator (SNG) converts the scaled factor ε into its bit-stream version, and then the first stochastic real multiplication (SRM) unit calculates the product between ε^{s} and $h_{j,i}^{s}$. Meanwhile, the stochastic real addition (SRA) unit computes the result of $(r_j^{s} - \mu_{z_{j,i}}^{s})$. Next, the second SRM outputs the result of $\varepsilon^{s} h_{j,i}^{s}\left(r_{j}^{s} - \mu_{z_{j,i}}^{s}\right)$, which then is re-randomized through the RR unit. Finally, the JK-FF calculates the approximated result of $\beta_{j,i}^{s}$.

3.1.2 The Stochastic Architecture for Symbol Nodes

The overall architecture for the message updating in symbol nodes is designed as shown in Fig. 22.23, in which the γ_i can be effectively computed by a feedback

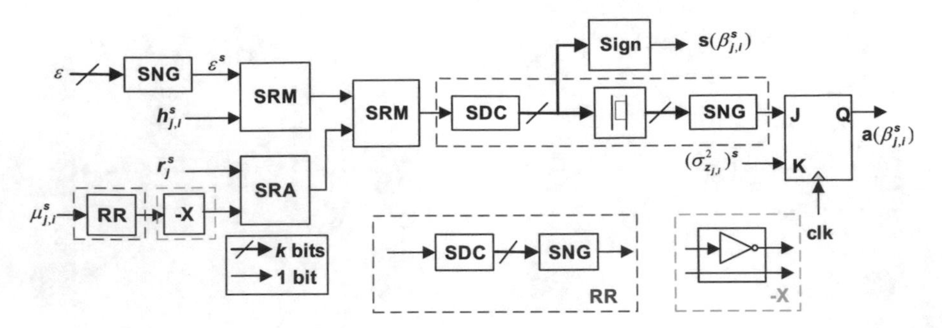

Fig. 22.22 Stochastic architecture for the message updating in observation nodes (PE7). © [2021] IEEE. Reprinted, with permission, from Ref. [15]

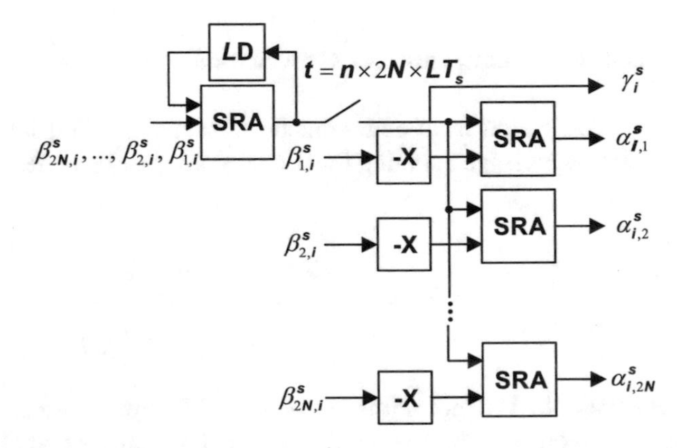

Fig. 22.23 Stochastic architecture for the message updating in symbol nodes (PE8). © [2021] IEEE. Reprinted, with permission, from Ref. [15]

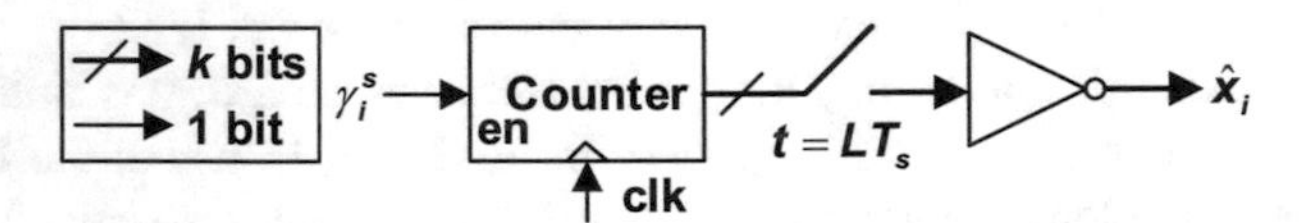

Fig. 22.24 Architecture for stochastic output decision (PE9). © [2021] IEEE. Reprinted, with permission, from Ref. [15]

architecture by exerting only one SRA and T_s stands for the system clock. Accordingly, the $\alpha_{i,j}$ is obtained by subtracting the corresponding $\beta_{j,i}$ from γ_i.

3.1.3 The Stochastic Architecture of Decision Module

The decision module is implemented as shown in Fig. 22.24, which contains a controllable up–down counter to convert the γ_i^s into deterministic value. Then, the hard decision is performed based on the sign of deterministic value γ_i.

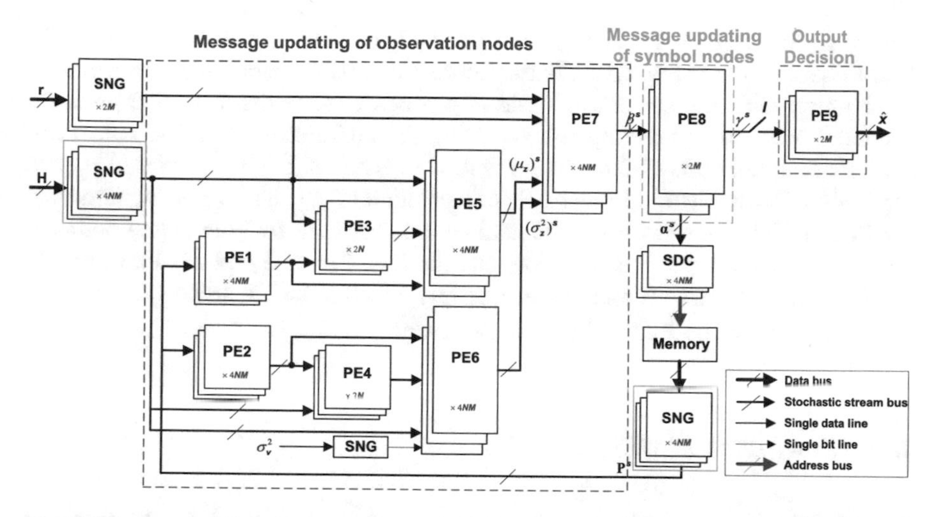

Fig. 22.25 The overall architecture of stochastic BP detector. © [2021] IEEE. Reprinted, with permission, from Ref. [15]

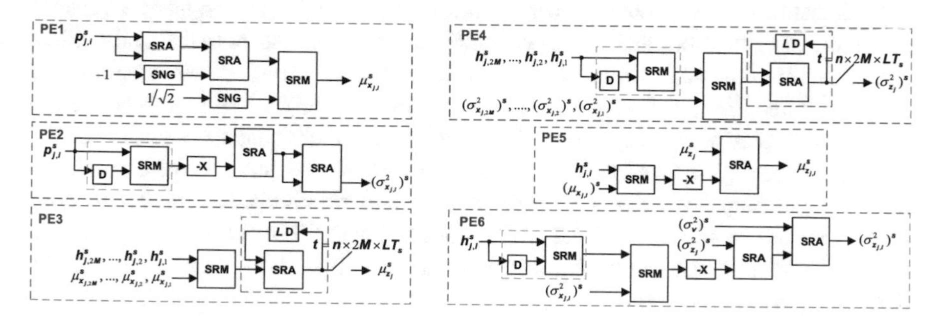

Fig. 22.26 Architectures of PE1 to PE6. © [2021] IEEE. Reprinted, with permission, from Ref. [15]

3.1.4 The Overall Architecture for Stochastic BP Detector

Targeting at 4-QAM modulation MIMO systems, the overall hardware architecture for the SBP detector is illustrated in Fig. 22.25, which consists of three main modules: message updating module for observation nodes, message updating module for symbol nodes, and the decision module. Additionally, the details of PE1 to PE6 are provided in Fig. 22.26, and the PE7, PE8, and PE9 are shown in Figs. 22.22, 22.23, and 22.24, respectively.

Considering the real model of 4-QAM MIMO systems, the energy normalization factor for the transmitted vector is $1/\sqrt{2}$; hence, the mean and variance of x can be expressed by

$$\mu_x = \frac{1}{2}\left(p + p + (-1)\right), \quad \sigma_x^2 = \left(p - p^2\right) + \left(p - p^2\right), \tag{22.17}$$

where $\mu_x, \sigma_x^2 \in [-1, 1]$. Therefore, Eq. (22.17) can be implemented by the PE1 and PE2 as shown in Fig. 22.26. Furthermore, the square operation is performed by the hardware in the green dotted bordered rectangle, where the register D produces another uncorrelated bit-stream representing the probability $p_{j,i}^s$. The number of both PE1 modules and PE2 modules required to generate the bit-streams of $\boldsymbol{\mu}_x$ and $\boldsymbol{\sigma}_x^2$ is $4NM$. Moreover, the stochastic computation of μ_{z_j} and $\sigma_{z_j}^2$ is implemented by PE3 and PE4, respectively. The hardware highlighted by blue dotted bordered rectangles executes the accumulation, and the total $2N$ PE3 and $2N$ PE4 produce the bit-streams for the $\boldsymbol{\mu}_z$ and $\boldsymbol{\sigma}_z^2$, respectively. Finally, the bit-streams for $\mu_{j,i}$ and $\sigma_{z_{i,j}}^2$ are generated by PE5 and PE6, respectively.

3.2 Stochastic MMSE Detector

For MIMO systems, matrix decomposition is the major bottleneck for the hardware implementation of linear MIMO detectors, such as MMSE detector and zero-force (ZF) detector. Further, lower–upper decomposition (LUD) is a significant matrix decomposition method, and it plays an indispensable role in both ZF and MMSE detectors. To this end, Chen et al. [37] proposed an efficient stochastic-based LUD scheme. First, a dual partition computation (DPC) scheme is proposed to reduce computation latency from 2^k to $2^{k/2+1}$. Further, based on the DPC, the corresponding high-accuracy multiplier and divider are designed, which can achieve high computation accuracy with short stream and relatively low hardware cost. Finally, the stochastic LUD (SLUD) is synthesized with CMOS 130-nm technology, which shows that the hardware efficiency of SLUD is 1.5× higher that of the deterministic LUD methods.

3.3 Stochastic Markov Chain Monte Carlo (MCMC) Detector

As a particular type of Monte Carlo sampling technique, the MCMC method has been successfully applied as a MIMO detection tool [39], which can deliver satisfying detection performance. However, the prohibitive computational complexity is the fatal bottleneck for its practical implementation. Specifically, the traditional MCMC iteration is composed of two main steps: first, utilize current data to evaluate the conditional probability. Next, based on the given conditional probability distribution, update the samples. Both the two steps impose huge computational load on classical MCMC detection.

To this end, Chen et al. [14] proposed a low-complexity MCMC MIMO detector based on stochastic computing. Two novel techniques, namely a sliding window generator (SWG) and a log-likelihood ratio-based updating method (LUM), are introduced to enhance the corresponding performance. Specifically, the SWG is employed to decrease the length of the bit-streams so that the overall system latency can be reduced. Further, sample updating is produced by exerting the LUM; hence, the calculation of LLRs and the process of updating samples can share one unit, which can effectively decrease the hardware consumption. The synthesis results targeting the 4×4 16-QAM MIMO system with 130-nm CMOS technology suggest that the stochastic MCMC detector can achieve 1.5 Gbps throughput with merely 0.2 dB performance degradation compared to conventional float-point detection. Additionally, it also improves the ratio of gate count to scaled throughput by around 30% compared to other conventional MIMO detectors.

Acknowledgments The authors would like to thank Wuqiong Zhao and Xiaoran Jiang for their help in editing this chapter. This work was supported in part by the National Key R&D Program of China under Grant 2020YFB2205503, in part by NSFC under Grants 62122020 and 61871115, in part by the Jiangsu Provincial NSF under Grant BK20211512, in part by the Six Talent Peak Program of Jiangsu Province under Grant 2018-DZXX-001, in part by the Distinguished Perfection Professorship of Southeast University, and in part by the Fundamental Research Funds for the Central Universities.

References

1. Liu W, Lombardi F, Shulte M. A retrospective and prospective view of approximate computing point of view. Proc IEEE. 2020;108(3):394–9.
2. Liu W, Qian L, Wang C, et al. Design of approximate radix-4 booth multipliers for error-tolerant computing. IEEE Trans Comput. 2017;66(8):1435–41.
3. Liu W, Xu J, Wang D, et al. Design and evaluation of approximate logarithmic multipliers for low power error-tolerant applications. IEEE Trans Circuits Syst I Regul Pap. 2018;65(9):2856–68.
4. Gaines BR. Stochastic computing systems. Adv Inf Syst Sci. 1969;2:37–172.
5. Qian W, Li X, Riedel MD, et al. An architecture for fault-tolerant computation with stochastic logic. IEEE Trans Comput. 2010;60(1):93–105.
6. Qian W, Wang C, Li P, et al. An efficient implementation of numerical integration using logical computation on stochastic bit streams. In: International conference on computer-aided design; 2012. p. 156–62.
7. Ren A, Li Z, Ding C, et al. SC-DCNN: highly-scalable deep convolutional neural network using stochastic computing. ACM SIGPLAN Notices. 2017;52(4):405–18.
8. Li Z, Li J, Ren A, et al. HEIF: highly efficient stochastic computing-based inference framework for deep neural networks. IEEE Trans Comput Aided Des Integr Circuits Syst. 2018;38(8):1543–56.
9. Ardakani A, Leduc-Primeau F, Onizawa N, et al. VLSI implementation of deep neural network using integral stochastic computing. IEEE Trans Very Large Scale Integr Syst. 2017;25(10):2688–99.
10. Yuan B, Keshab KP. Successive cancellation decoding of polar codes using stochastic computing. In: IEEE international symposium on circuits and systems; 2015. p. 3034–43.
11. Liang X, Wang H, Shen Y, et al. Efficient stochastic successive cancellation list decoder for polar codes. Sci China Inf Sci. 2020;63(10):1–19.
12. Yuan B, Keshab KP. Belief propagation decoding of polar codes using stochastic computing. In: IEEE International symposium on circuits and systems; 2016. p. 157–60.
13. Han K, Wang J, Gross WJ, et al. Stochastic bit-wise iterative decoding of polar codes. IEEE Trans Signal Process. 2018;67(5):1138–51.
14. Chen J, Hu J, Sobelman GE. Stochastic MIMO detector based on the Markov chain Monte Carlo algorithm. IEEE Trans Signal Process. 2014;62(6):1454–63.

15. Yang J, Zhang C, Xu S, et al. Efficient stochastic detector for large-scale MIMO. In: IEEE international conference on acoustics, speech and signal processing; 2016. p. 6550–4.
16. Wang H, Zhang Z, You X, et al. Low-complexity Winograd convolution architecture based on stochastic computing. In: IEEE international conference on digital signal processing; 2018. p. 1–5.
17. Xu R, Yuan B, You X, et al. Efficient fast convolution architecture based on stochastic computing. In: IEEE international conference on wireless communications and signal processing; 2017. p. 1–6.
18. Li P, Lilja DJ, Qian W, et al. Computation on stochastic bit streams digital image processing case studies. IEEE Trans Very Large Scale Integr Syst. 2013;22(3):449–62.
19. Arikan E. Channel polarization: a method for constructing capacity-achieving codes for symmetric binary-input memoryless channels. IEEE Trans Inf Theory. 2009;55(7):3051–73.
20. Tal I, Alexander V. List decoding of polar codes. IEEE Trans Inf Theory. 2015;61(5):2213–26.
21. Chen K, Niu K, Lin JR. List successive cancellation decoding of polar codes. Electron Lett. 2012;48(9):500–1.
22. Bakulin M, Kreyndelin V, Rog A, et al. MMSE based K-best algorithm for efficient MIMO detection. In: IEEE International Congress on Ultra Modern Telecommunications and Control Systems and Workshops; 2017. p. 258–63.
23. Liang X, Zhang C, Xu M, et al. Efficient stochastic list successive cancellation decoder for polar codes. In: IEEE international system-on-chip conference; 2015. p. 421–6.
24. Liang X, Yang J, Zhang C, et al. Hardware efficient and low-latency CA-SCL decoder based on distributed sorting. In: IEEE global communications conference; 2016. p. 1–6.
25. Zhang C, Keshab KP. Low-latency sequential and overlapped architectures for successive cancellation polar decoder. IEEE Trans Signal Process. 2013;61(10):2429–41.
26. Arı́kan E. Polar codes: a pipelined implementation. In: Proc. 4th ISBC; 2010. p. 11–4.
27. Xu M, Liang X, Zhang C, et al. Stochastic BP polar decoding and architecture with efficient re-randomization and directive register. In: IEEE international workshop on signal processing systems; 2016. p. 315–20.
28. Xu M, Liang X, Yuan B, et al. Stochastic belief propagation polar decoding with efficient re-randomization. IEEE Trans Veh Technol. 2020;69(6):6771–6.
29. MacKay DJC, Neal RM. Near Shannon limit performance of low density parity check codes. Electron Lett. 1996;32(18):1645.
30. Chen Y, Zhang Q, Wu D. An efficient multi-rate LDPC-CC decoder with a layered decoding algorithm for the IEEE 1901 standard. IEEE Trans Circuits Syst II Express Briefs. 2014;61(12):992–6.
31. Bao D, Xiang B, Shen R, et al. Programmable architecture for flexi-mode QC-LDPC decoder supporting wireless LAN/MAN applications and beyond. IEEE Trans Circuits Syst I Regul Pap. 2009;57(1):125–38.
32. Gallager R. Low-density parity-check codes. IRE Trans Inf Theory 1962;8(1):21–8.
33. Gaudet VC, Rapley AC. Iterative decoding using stochastic computation. Electron Lett. 2003;39(3):299–301.
34. Tehrani SS, Mannor S, Gross WJ. Fully parallel stochastic LDPC decoders. IEEE Trans Signal Process. 2008;56(11):5692–703.
35. Tehrani SS, Naderi A, Kamendje GA, et al. Tracking forecast memories in stochastic decoders. In: IEEE international conference on acoustics, speech and signal processing; 2009. p. 561–4.
36. Tehrani SS, Naderi A, Kamendje GA, et al. Majority-based tracking forecast memories for stochastic LDPC decoding. IEEE Trans Signal Process. 2010;58(9):4883–96.
37. Chen J, Hu J, Zhou J. Hardware and energy-efficient stochastic LU decomposition scheme for MIMO receivers. IEEE Trans Very Large Scale Integr Syst. 2015;24(4):1391–401.
38. Yang J, Song W, Zhang S, et al. Low-complexity belief propagation detection for correlated large-scale MIMO systems. J Signal Process Syst. 2018;90(4):585–99.
39. Farhang-Boroujeny B, Zhu H, Shi Z. Markov chain Monte Carlo algorithms for CDMA and MIMO communication systems. IEEE Trans Signal Process. 2006;54(5):1896–909.

Index

W. Liu, F. Lombardi (eds.), *Approximate Computing*,
https://doi.org/10.1007/978-3-030-98347-5

F

G

H

Printed in the United States
by Baker & Taylor Publisher Services